D0713011

A NUMERICAL LIBRARY IN JAVA FOR SCIENTISTS & ENGINEERS

HANG T. LAU

CHAPMAN & HALL/CRC

A CRC Press Company

Boca Raton   London   New York   Washington, D.C.

Cover design by Vill Mak

Library of Congress Cataloging-in-Publication Data

Lau, H. T. (Hang Tong), 1952-
A numerical library in Java for scientists and engineers / Hang T. Lau.
p. cm.
Includes bibliographical references and index.
ISBN 1-58488-430-4 (alk. paper)
1. Java (Computer program language) I. Title.

QA76.73.J38L363 2003
005.13'3—dc21   2003055149

Visit the CRC Press Web site at www.crcpress.com

No claim to original U.S. Government works
International Standard Book Number 1-58488-430-4
Library of Congress Card Number 2003055149
Printed in the United States of America 1 2 3 4 5 6 7 8 9 0
Printed on acid-free paper

To my dearest wife, Helen,
and our children Matthew, Lawrence, and Tabia
for their patience, love, and support

The Author

Hang T. Lau is a faculty lecturer in the Department of Career and Management Studies, Centre for Continuing Education, McGill University, Montreal, Canada. Dr. Lau has over twenty years of industrial experience in research and development of telecommunication.

Contents

Introduction

The following collection of procedures is a translation into Java of the library *NUMAL* (*NUM*erical procedures in *AL*gol 60) [He80] developed at the Mathematical Centre, Amsterdam. The scope of the library is indicated by the list of procedure descriptions in Appendix B and refers to linear algebra, the numerical solution of ordinary and partial differential equations, optimisation, parameter estimation, special functions of mathematical physics, and indeed most concerns of an active research group. Each main procedure is assembled from auxiliary submodules and for this reason the library is highly compact. The submodules implement routine numerical processes and are of use in the development of new methods. A number of addenda containing some fast Fourier transform procedures, and procedures for time series analysis from other sources have been added to NUMAL.

The expert is offered both a collection of procedures for implementing numerical methods which are by now standard and with whose theory the user will be perfectly familiar and a powerful research tool. The novice is offered a prodigious source of entertainment and instruction: the user may run the test programs which carry out the worked examples provided with each critical procedure or group of procedures, the user is free to construct other examples and apply the procedures to them, and the user is even at liberty to devise examples which provoke the procedures into emitting one of the failure indications consistently provided.

The use of some Algol procedures in the NUMAL library was explained in detail in a seminar on numerical software. Furthermore, the commercially available program libraries IMSL (IMSL Inc., Houston, Texas, USA) and NAG (Numerical Algorithms Group, Oxford, United Kingdom) were compared with the Algol NUMAL library. The proceedings of that seminar appeared (in Dutch) in [Bus76, Rie77].

NUMAL was originally constructed with the aid of a CDC-Cyber 70 system. However, personal computers, equipped with free Java compilers, are now within the purchasing capability of the professionally active. The Java version of NUMAL, being highly compact, is peculiarly suited to implementation upon a personal computer.

The original order of the NUMAL procedures has been preserved. The documentations are taken partly from the original NUMAL source and partly from extended forms given in a FORTRAN version of NUMAL [Wy81] constructed, with considerable help from the author, by P. Wynn for his own use. To ensure maximal compactness of the main library, the worked examples have been grouped together at the end of the Java version library. The references are collected in Appendix A. Appendix B contains a capsule statement of what each procedure does in the library.

Each procedure in the library is described using the following scheme:

Top line	Each procedure in the library is identified by its name as the top line.
Module description	Following the top line are a description of the procedure and a brief discussion on the algorithm used. References are given for further details of the underlying mathematical method.
Procedure parameters	This is the specification of the procedure, followed by the description of each parameter of the procedure.
Procedures used	List of other procedures in the library used by this procedure.
Remarks	Supplementary remarks on the method and performance of the module.
Program listing	The Java code listing of the procedure.

The programs have been tested using the Java 2 SDK, Standard Edition, version 1.4.1, on a personal computer using the Windows XP operation system.

In conclusion it should perhaps be emphasized that the library is offered as a research tool. The documentation to the library procedures briefly describes the algorithms used; it is in no way intended as a treatise upon numerical methods. For fuller theory the user is directed to the references given and to further research texts — in particular to the relevant Mathematical Centre colloquium publications. Reference [Pr92] also serves as an excellent elementary introduction to part of the library.

1. Elementary Procedures

This chapter contains elementary operations for vectors and matrices such as the assignment of initial values to vectors and matrix slices, duplication and interchange of such slices, rotations, and reflections. The procedures are taken from [Dek68, DekHo68]. Most of them are used in subsequent chapters. The elementary procedures are all quite short; prepared versions may be recoded, making optimal use of machine capabilities in assembly language.

All procedures in this chapter are class methods of the following *Basic* class.

```
package numal;

import numal.*;

public class Basic extends Object {
  // all procedures in this chapter are to be inserted here
}
```

1.1 Real vector and matrix — Initialization

A. inivec

Initializes part of a vector a by setting $a_i = x$ $(i=l,l+1,...,u)$.

Procedure parameters:
$$\text{void inivec } (l,u,a,x)$$
l,u: int; lower and upper index of the vector a, respectively;
a: double $a[l:u]$; the array to be initialized;
x: double; initialization constant.

```
public static void inivec(int l, int u, double a[], double x)
{
  for (; l<=u; l++) a[l]=x;
}
```

B. inimat

Initializes a rectangular submatrix a by setting
$$a_{i,j}=x \ (i=lr,lr+1,...,ur; \ j=lc,lc+1,...,uc).$$

Procedure parameters:

void inimat (*lr,ur,lc,uc,a,x*)

lr,ur,lc,uc: int; lower and upper row index, and lower and upper column index of the matrix *a*, respectively;

a: double *a[lr:ur,lc:uc]*; the matrix to be initialized;

x: double; initialization constant.

```
public static void inimat(int lr, int ur, int lc, int uc, double a[][], double x)
{
  int j;

  for (; lr<=ur; lr++)
    for (j=lc; j<=uc; j++) a[lr][j]=x;
}
```

C. inimatd

Initializes a forward diagonal sequence of elements of a rectangular matrix *a* by setting

$$a_{i,i+shift}=x \ (i=lr,lr+1,...,ur).$$

Procedure parameters:

void inimatd (*lr,ur,shift,a,x*)

lr,ur: int; lower and upper row index of the codiagonal to be initialized;

shift: int; distance between diagonal and codiagonal;

a: double *a[lr:ur,lr+shift:ur+shift]*; the array to be initialized;

x: double; initialization constant.

```
public static void inimatd(int lr, int ur, int shift, double a[][], double x)
{
  for (; lr<=ur; lr++) a[lr][lr+shift]=x;
}
```

D. inisymd

Initializes a forward diagonal sequence of elements of a symmetric square matrix *b*. The upper triangle of *b* is stored columnwise in a linear array *a* such that $b_{i,j}$ is $a_{(j-1)j/2+i}$.

Procedure parameters:

void inisymd (*lr,ur,shift,a,x*)

lr,ur: int; lower and upper row index of a codiagonal of a symmetric matrix of order n to be initialized, *lr* and *ur* should satisfy: $lr \geq 1$, $ur \leq n$;

shift: distance between diagonal and codiagonal, -n < *shift* < n;

a: double $a[1:n(n+1)/2]$; the linear array containing the columnwise stored
        upper triangle of a symmetric matrix;

x: double; initialization constant.

```
public static void inisymd(int lr, int ur, int shift, double a[], double x)
{
  shift=Math.abs(shift);
  ur += shift+1;
  shift += lr;
  lr += ((shift-3)*shift)/2;
  lr += shift;
  while (shift < ur) {
    a[lr]=x;
    shift++;
    lr += shift;
  }
}
```

E. inisymrow

Initializes part of a row of a symmetric square matrix b. The upper triangle of b is
stored columnwise in a linear array a such that $b_{i,j}$ is $a_{(j-1)j/2+i}$.

Procedure parameters:

$$\text{void inisymrow } (l,u,i,a,x)$$

l,u: int; lower and upper index of row-element of a codiagonal of a symmetric
        matrix of order n to be initialized, l and u should satisfy: $1 \le l \le n$, $1 \le u \le n$;

i: row index, $1 \le i \le n$;

a: double $a[1:n(n+1)/2]$; the linear array containing the columnwise stored
        upper triangle of a symmetric matrix;

x: double; initialization constant.

```
public static void inisymrow(int l, int u, int i, double a[], double x)
{
  int k;

  if (l <= i) {
    k=((i-1)*i)/2;
    l += k;
    k += (u<i) ? u : i;
    for (; l<=k; l++) a[l]=x;
    l=i+1;
  }
  if (u > i) {
    k=((l-1)*l)/2+i;
    do {
      a[k]=x;
```

```
      l++;
      k += l-1;
    } while (l <= u);
  }
}
```

1.2 Real vector and matrix — Duplication

A. dupvec

Duplicates part of a vector b by setting
$$a_i = b_{i,i+shift} \quad (i=l,l+1,...,u).$$

Procedure parameters:
$$\text{void dupvec } (l,u,shift,a,b)$$
l,u: int; lower and upper vector-index, respectively;
shift: int; index-shifting parameter;
a,b: double $a[l:u]$, $b[l+shift:u+shift]$; the array b is copied into a.

```
public static void dupvec(int l, int u, int shift, double a[], double b[])
{
    for (; l<=u; l++) a[l]=b[l+shift];
}
```

B. dupvecrow

Duplicates part of a row vector of a rectangular matrix b by setting
$$a_j = b_{i,j} \quad (j=l,l+1,...,u).$$

Procedure parameters:
$$\text{void dupvecrow } (l,u,i,a,b)$$
l,u: int; lower and upper vector-index, respectively;
i: int; row index of the matrix b;
a,b: double $a[l:u]$, $b[i:i,l:u]$; b is copied into a.

```
public static void dupvecrow(int l, int u, int i, double a[], double b[][])
{
    for (; l<=u; l++) a[l]=b[i][l];
}
```

C. duprowvec

Replaces part of a row vector of a rectangular matrix a by a corresponding part of a vector b by setting

$$a_{i,j}=b_j \ (j=l,l+1,...,u).$$

Procedure parameters:

$$\text{void duprowvec } (l,u,i,a,b)$$

l,u: int; lower and upper vector-index, respectively;
i: int; row index of the matrix a;
a,b: double $a[i:i,l:u]$, $b[l:u]$; b is copied into a.

```
public static void duprowvec(int l, int u, int i, double a[][], double b[])
{
    for (; l<=u; l++) a[i][l]=b[l];
}
```

D. dupveccol

Duplicates part of a column vector of a rectangular matrix b by setting

$$a_i=b_{i,j} \ (i=l,l+1,...,u).$$

Procedure parameters:

$$\text{void dupveccol } (l,u,j,a,b)$$

l,u: int; lower and upper vector-index, respectively;
j: int; column index of the matrix b;
a,b: double $a[l:u]$, $b[l:u,j:j]$; b is copied into a.

```
public static void dupveccol(int l, int u, int j, double a[], double b[][])
{
    for (; l<=u; l++) a[l]=b[l][j];
}
```

E. dupcolvec

Replaces part of a column vector of a rectangular matrix a by a corresponding part of a vector b by setting

$$a_{i,j}=b_i \ (i=l,l+1,...,u).$$

Procedure parameters:

$$\text{void dupcolvec } (l,u,j,a,b)$$

l,u: int; lower and upper vector-index, respectively;
j: int; column index of the matrix a;
a,b: double $a[l:u,i:i]$, $b[l:u]$; b is copied into a.

```
public static void dupcolvec(int l, int u, int j, double a[][], double b[])
{
  for (; l<=u; l++) a[l][j]=b[l];
}
```

F. dupmat

Replaces a submatrix of the rectangular matrix a by a corresponding submatrix of the rectangular matrix b by setting

$$a_{m,n}=b_{m,n} \ (m=l,l+1,...,u; \ n=i,i+1,...,j).$$

Procedure parameters:
$$\text{void dupmat } (l,u,i,j,a,b)$$
l,u: int; lower and upper row index, respectively;
i,j: int; lower and upper column index, respectively;
a,b: double $a[l:u,i:j]$, $b[l:u,i:j]$; b is copied into a.

```
public static void dupmat(int l, int u, int i, int j, double a[][], double b[][])
{
  int k;

  for (; l<=u; l++)
    for (k=i; k<=j; k++) a[l][k]=b[l][k];
}
```

1.3 Real vector and matrix — Multiplication

A. mulvec

Forms a constant multiple of part of a vector by setting
$$a_i=xb_{i+shift} \ (i=l,l+1,...,u).$$

Procedure parameters:
$$\text{void mulvec } (l,u,shift,a,b,x)$$
l,u: int; lower and upper vector-index, respectively;
$shift$: int; subscript-shifting parameter;
a,b: double $a[l:u]$, $b[l+shift:u+shift]$; the product of the contents of b are stored in
         a;
x: multiplication factor.

```
public static void mulvec(int l, int u, int shift, double a[], double b[], double x)
{
   for (; l<=u; l++) a[l]=b[l+shift]*x;
}
```

B. mulrow

Replaces a row sequence of elements of a rectangular matrix a by a constant multiple of a row sequence of elements of a rectangular matrix b by setting
$$a_{i,k}=xb_{j,k}\ (k=l,l+1,...,u).$$

Procedure parameters:

void mulrow (l,u,i,j,a,b,x)

l,u: int; lower and upper column index, respectively;
i,j: int; row indices of a and b;
a,b: double $a[i:i,l:u]$, $b[j:j,l:u]$; the contents of b multiplied by x are stored into a;
x: multiplication factor.

```
public static void mulrow(int l, int u, int i, int j,
                          double a[][], double b[][], double x)
{
   for (; l<=u; l++) a[i][l]=b[j][l]*x;
}
```

C. mulcol

Replaces a column sequence of elements of a rectangular matrix a by a constant multiple of a column sequence of elements of a rectangular matrix b by setting
$$a_{k,i}=xb_{k,j}\ (k=l,l+1,...,u).$$

Procedure parameters:

void mulcol (l,u,i,j,a,b,x)

l,u: int; lower and upper row index, respectively;
i,j: int; column indices of a and b;
a,b: double $a[l:u,i:i]$, $b[l:u,j:j]$; the contents of b multiplied by x are stored into a;
x: multiplication factor.

```
public static void mulcol(int l, int u, int i, int j, double a[][], double b[][],
                          double x)
{
   for (; l<=u; l++) a[l][i]=b[l][j]*x;
}
```

D. colcst

Replaces the elements of a column sequence taken from a rectangular matrix a by constant multiples of the same elements, by setting
$$a_{ij}=xa_{ij}\ (i=l,l+1,...,u).$$

Procedure parameters:
$$\text{void colcst } (l,u,j,a,x)$$
l,u: int; lower and upper row index, respectively;
j: int; column index;
a: double $a[l:u,j:j]$; the rectangular matrix;
x: multiplication factor.

```
public static void colcst(int l, int u, int j, double a[][], double x)
{
    for (; l<=u; l++) a[l][j] *= x;
}
```

E. rowcst

Replaces the elements of a row sequence taken from a rectangular matrix a by constant multiples of the same elements, by setting
$$a_{ij}=xa_{ij}\ (j=l,l+1,...,u).$$

Procedure parameters:
$$\text{void rowcst } (l,u,i,a,x)$$
l,u: int; lower and upper column index, respectively;
i: int; row index;
a: double $a[i:i,l:u]$; the rectangular matrix;
x: multiplication factor.

```
public static void rowcst(int l, int u, int i, double a[][], double x)
{
    for (; l<=u; l++) a[i][l] *= x;
}
```

1.4 Real vector vector products

A. vecvec

Forms the inner product s of part of a vector a and part of a vector b by setting

$$s = \sum_{i=l}^{u} a_i * b_{i+shift} \; .$$

Procedure parameters:

double vecvec $(l,u,shift,a,b)$

vecvec: given the value of s in the above;
l,u: int; lower and upper bound of the running subscript;
shift: int; index-shifting parameter of the vector b;
a,b: double $a[l:u]$, $b[l+shift:u+shift]$.

```
public static double vecvec(int l, int u, int shift, double a[], double b[])
{
   int k;
   double s;

   s=0.0;
   for (k=l; k<=u; k++) s += a[k]*b[k+shift];
   return (s);
}
```

B. matvec

Forms the inner product s of part of a row of a rectangular matrix a and the corresponding part of a vector b by setting

$$s = \sum_{k=l}^{u} a_{i,k} * b_k \; .$$

Procedure parameters:

double matvec (l,u,i,a,b)

matvec: given the value of s in the above;
l,u: int; lower and upper bound of the running subscript;
i: int; row index of a;
a,b: double $a[i:i,l:u]$, $b[l:u]$.

```
public static double matvec(int l, int u, int i, double a[][], double b[])
{
   int k;
   double s;

   s=0.0;
   for (k=l; k<=u; k++) s += a[i][k]*b[k];
   return (s);
}
```

C. tamvec

Forms the inner product s of part of a column of a rectangular matrix a and the corresponding part of a vector b by setting

$$s = \sum_{k=l}^{u} a_{k,i} * b_k \ .$$

Procedure parameters:

$$\text{double tamvec } (l,u,i,a,b)$$

tamvec: given the value of s in the above;
l,u: int; lower and upper bound of the running subscript;
i: int; column index of a;
a,b: double $a[l{:}u,i{:}i]$, $b[l{:}u]$.

```
public static double tamvec(int l, int u, int i, double a[][], double b[])
{
    int k;
    double s;

    s=0.0;
    for (k=l; k<=u; k++) s += a[k][i]*b[k];
    return (s);
}
```

D. matmat

Forms the inner product s of part of a row of a rectangular matrix a and the corresponding part of a column of a rectangular matrix b by setting

$$s = \sum_{k=l}^{u} a_{i,k} * b_{k,j} \ .$$

Procedure parameters:

$$\text{double matmat } (l,u,i,j,a,b)$$

matmat: given the value of s in the above;
l,u: int; lower and upper bound of the running subscript;
i,j: int; row index of a and column index of b;
a,b: double $a[i{:}i,l{:}u]$, $b[l{:}u,j{:}j]$.

```
public static double matmat(int l, int u, int i, int j, double a[][], double b[][])
{
    int k;
    double s;

    s=0.0;
    for (k=l; k<=u; k++) s += a[i][k]*b[k][j];
```

```
   return (s);
}
```

E. tammat

Forms the inner product s of part of a column of a rectangular matrix a and the corresponding part of a column of a rectangular matrix b by setting

$$s = \sum_{k=l}^{u} a_{k,i} * b_{k,j} \; .$$

Procedure parameters:
$$\text{double tammat } (l,u,i,j,a,b)$$

tammat:	given the value of s in the above;
l,u:	int; lower and upper bound of the running subscript;
i,j:	int; column indices of a and b, respectively;
a,b:	double $a[l:u,i:i]$, $b[l:u,j:j]$.

```
public static double tammat(int l, int u, int i, int j, double a[][], double b[][])
{
   int k;
   double s;

   s=0.0;
   for (k=l; k<=u; k++) s += a[k][i]*b[k][j];
   return (s);
}
```

F. mattam

Forms the inner product s of part of a row of a rectangular matrix a and the corresponding part of a row of a rectangular matrix b by setting

$$s = \sum_{k=l}^{u} a_{i,k} * b_{j,k} \; .$$

Procedure parameters:
$$\text{double mattam } (l,u,i,j,a,b)$$

mattam:	given the value of s in the above;
l,u:	int; lower and upper bound of the running subscript;
i,j:	int; row indices of a and b, respectively;
a,b:	double $a[i:i,l:u]$, $b[j:j,l:u]$.

```
public static double mattam(int l, int u, int i, int j, double a[][], double b[][])
{
```

```
    int k;
    double s;

    s=0.0;
    for (k=1; k<=u; k++) s += a[i][k]*b[j][k];
    return (s);
}
```

G. seqvec

Forms the sum

$$s = \sum_{j=l}^{u} a_{il+(j+l-1)*(j-l)/2} * b_{j+shift} .$$

Procedure parameters:

> double seqvec (*l,u,il,shift,a,b*)

seqvec: given the value of *s* in the above;
l,u: int; lower and upper bound of the running subscript;
il: int; lower bound of the vector *a*;
shift: int; index-shifting parameter of the vector *b*;
a,b: double *a[p:q]*, *b[l+shift:u+shift]*; where $p \le il$ and $q \ge il+(u+l-1)(u-l)/2$.

```
public static double seqvec(int l, int u, int il, int shift, double a[], double b[])
{
    double s;

    s=0.0;
    for (; l<=u; l++) {
        s += a[il]*b[l+shift];
        il += l;
    }
    return (s);
}
```
                                                                                    •

H. scaprd1

Forms the stepped inner product

$$s = \sum_{j=1}^{n} a_{la+(j-1)*sa} * b_{lb+(j-1)*sb}$$

from the elements of two vectors *a* and *b*.

Procedure parameters:

> double scaprd1 (*la,sa,lb,sb,n,a,b*)

scaprd1:   given the value of *s* in the above;
la,lb: int; lower bounds of the vectors *a* and *b*, respectively;
sa,sb: int; index-shifting parameters of the vectors *a* and *b*, respectively;
n: int; upper bound of the running subscript;
a,b: double *a[p:q], b[r:s]*; the subscripts above and the values of *la+(j-1)*sa*
                and *lb+(j-1)*sb, j=1,...,n* should not contradict each other.

```
public static double scaprd1(int la, int sa, int lb, int sb, int n, double a[],
                             double b[])
{
  int k;
  double s;

  s=0.0;
  for (k=1; k<=n; k++) {
    s += a[la]*b[lb];
    la += sa;
    lb += sb;
  }
  return (s);
}
```

I. symmatvec

Forms the inner product of part of a row of a symmetric square matrix *c*, whose
elements are stored columnwise in the linear array *a*, and the corresponding part
of a vector *b*, by setting

$$s = \sum_{k=l}^{u} c_{i,k} * b_k \ .$$

Procedure parameters:
                    double symmatvec (*l,u,i,a,b*)
symmatvec: given the value of *s* in the above;
l,u: int; lower and upper bound of the vector *b*, $l \geq 1$;
i: int; row index of *a*, $i \geq 1$;
a: double *a[p:q]*; array with:
                if $i > l$ then *p=(i-1)i/2+l* else *p=(l-1)l/2+i* and
                if $i > u$ then *q=(i-1)i/2+u* else *q=(u-1)u/2+i*;
b: double *b[l:u]*.

Procedures used: vecvec, seqvec.

```
public static double symmatvec(int l, int u, int i, double a[], double b[])
{
  int k, m;
```

```
   m=(l>i) ? l : i;
   k=(m*(m-1))/2;
   return (vecvec(1, (i<=u) ? i-1 : u, k,b,a) + seqvec(m,u,k+i,0,a,b));
}
```

1.5 Real matrix vector products

A. fulmatvec

Premultiplies part of a vector by a submatrix by setting

$$c_i = \sum_{j=lc}^{uc} a_{i,j} * b_j, \qquad i = lr,...,ur.$$

Procedure parameters:
                          void fulmatvec (lr,ur,lc,uc,a,b,c)
lr,ur: int; lower and upper bound of the row index;
lc,uc: int; lower and upper bound of the column index;
a: double a[lr:ur,lc:uc]; the matrix;
b: double b[lc:uc]; the vector;
c: double c[lr:ur]; the result a*b is delivered in c.

```
public static void fulmatvec(int lr, int ur, int lc, int uc, double a[][],
                        double b[], double c[])
{
   for (; lr<=ur; lr++) c[lr]=matvec(lc,uc,lr,a,b);
}
```

B. fultamvec

Premultiplies part of a vector by a transposed submatrix by setting

$$c_j = \sum_{i=lr}^{ur} a_{i,j} * b_i, \qquad j = lc,...,uc.$$

Procedure parameters:
                          void fultamvec (lr,ur,lc,uc,a,b,c)
lr,ur: int; lower and upper bound of the row index;
lc,uc: int; lower and upper bound of the column index;
a: double a[lr:ur,lc:uc]; the matrix;
b: double b[lr:ur]; the vector;
c: double c[lc:uc]; the result a'*b is delivered in c, a' denotes the transposed of
         the matrix a.

```
public static void fultamvec(int lr, int ur, int lc, int uc, double a[][],
                             double b[], double c[])
{
  for (; lc<=uc; lc++) c[lc]=tamvec(lr,ur,lc,a,b);
}
```

C. fulsymmatvec

Premultiplies part of a vector by a submatrix of a linearly stored symmetric matrix
by setting

$$c_i = \sum_{j=lc}^{uc} a_{i,j} * b_j, \qquad i = lr,...,ur.$$

Procedure parameters:

$$\text{void fulsymmatvec } (lr,ur,lc,uc,a,b,c)$$

lr,ur: int; lower and upper bound of the row index, $lr \geq 1$;
lc,uc: int; lower and upper bound of the column index, $lc \geq 1$;
a: double $a[l:u]$; where
             $l = min\ (lr(lr-1)/2+lc,\ lc(lc-1)/2+lr),$
             $u = max\ (ur(ur-1)/2+uc,\ uc(uc-1)/2+ur)$
             and the (i,j)-th element of the symmetric matrix should be given in
             $a_{j(j-1)/2+i}$;
b: double $b[lc:uc]$; the vector;
c: double $c[lr:ur]$; the result $a*b$ is delivered in c.

```
public static void fulsymmatvec(int lr, int ur, int lc, int uc, double a[],
                                double b[], double c[])
{
  for (; lr<=ur; lr++) c[lr]=symmatvec(lc,uc,lr,a,b);
}
```

D. resvec

Determines a residual vector by setting

$$c_i = \sum_{j=lc}^{uc} a_{i,j} * b_j + xc_i, \qquad i = lr,...,ur.$$

Procedure parameters:

$$\text{void resvec } (lr,ur,lc,uc,a,b,c,x)$$

lr,ur: int; lower and upper bound of the row index;
lc,uc: int; lower and upper bound of the column index;
a: double $a[lr:ur,lc:uc]$; the matrix;

b: double *b[lc:uc]*; the vector;
c: double *c[lr:ur]*; the result $a*b+x*c$ is overwritten on *c*;
x: double; the value of the multiplying scalar.

```
public static void resvec(int lr, int ur, int lc, int uc, double a[][], double b[],
                          double c[], double x)
{
   for (; lr<=ur; lr++)  c[lr]=matvec(lc,uc,lr,a,b)+c[lr]*x;
}
```

E. symresvec

Determines a residual vector from a submatrix of a linearly stored symmetric matrix by setting

$$c_i = \sum_{j=lc}^{uc} a_{i,j} * b_j + xc_i, \qquad i = lr,...,ur.$$

Procedure parameters:
                      void symresvec (*lr,ur,lc,uc,a,b,c,x*)
lr,ur: int; lower and upper bound of the row index, $lr \geq 1$;
lc,uc: int; lower and upper bound of the column index, $lc \geq 1$;
a: double *a[l:u]*; where
            $l = min\ (lr(lr-1)/2+lc,\ lc(lc-1)/2+lr)$,
            $u = max\ (ur(ur-1)/2+uc,\ uc(uc-1)/2+ur)$,
            and the (i,j)-th element of the symmetric matrix should be given in
            $a_{j(j-1)/2+i}$;
b: double *b[lc:uc]*; the vector;
c: double *c[lr:ur]*; the result $a*b+x*c$ is overwritten on *c*;
x: double; the value of the multiplying scalar.

```
public static void symresvec(int lr, int ur, int lc, int uc, double a[], double b[],
                             double c[], double x)
{
   for (; lr<=ur; lr++)  c[lr]=symmatvec(lc,uc,lr,a,b)+c[lr]*x;
}
```

1.6 *Real matrix matrix products*

A. hshvecmat

Premultiplies a submatrix by an elementary reflector:
$a=(I-xu^Tu)a$ where $a_{i-lr+1,j-lc+1}=a_{i,j}$ (*i=lr,...,ur; j=lc,...,uc*), $u_{i-lr+1}=u_i$ (*i=lr,...,ur*).

Procedure parameters:
$$\text{void hshvecmat } (lr,ur,lc,uc,x,u,a)$$
lr,ur: int; lower and upper row-indices;
lc,uc: int; lower and upper column-indices;
x: double; the Householder constant;
u: double *u[lr:ur]*; the Householder vector;
a: double *a[lr:ur,lc:uc]*; the matrix to be premultiplied by the Householder matrix.

Procedures used: tamvec, elmcolvec.

```
public static void hshvecmat(int lr, int ur, int lc, int uc, double x, double u[],
                       double a[][])
{
  for (; lc<=uc; lc++)
    elmcolvec(lr,ur,lc,a,u,tamvec(lr,ur,lc,a,u)*x);
}
```

B. hshcolmat

Premultiplies one submatrix by an elementary reflector formed from a column of another:
$$a=(I-xu^{T}u)a \text{ where } a_{k-lr+1,j-lc+1}=a_{kj} \ (k=lr,...,ur; \ j=lc,...,uc), \ u_{k-lr+1}=U_{k,i} \ (k=lr,...,ur).$$

Procedure parameters:
$$\text{void hshcolmat } (lr,ur,lc,uc,i,x,u,a)$$
lr,ur: int; lower and upper row-indices;
lc,uc: int; lower and upper column-indices;
i: int; the column index of the Householder vector;
x: double; the Householder constant;
u: double *u[lr:ur,i:i]*; the Householder vector;
a: double *a[lr:ur,lc:uc]*; the matrix to be premultiplied by the Householder matrix.

Procedures used: tammat, elmcol.

```
public static void hshcolmat(int lr, int ur, int lc, int uc, int i, double x,
                       double u[][], double a[][])
{
  for (; lc<=uc; lc++)
    elmcol(lr,ur,lc,i,a,u,tammat(lr,ur,lc,i,a,u)*x);
}
```

C. hshrowmat

Premultiplies one submatrix by an elementary reflector formed from a row of another:
$$a=(I-xu^Tu)a \text{ where } a_{k-lr+1,j-lc+1}=a_{kj} \ (k=lr,...,ur; \ j=lc,...,uc), \ u_{k-lr+1}=U_{i,k} \ (k=lr,...,ur).$$

Procedure parameters:
$$\text{void hshrowmat } (lr,ur,lc,uc,i,x,u,a)$$
lr,ur: int; lower and upper row-indices;
lc,uc: int; lower and upper column-indices;
i: int; the row index of the Householder vector;
x: double; the Householder constant;
u: double *u[i:i,lr:ur]*; the Householder vector;
a: double *a[lr:ur,lc:uc]*; the matrix to be premultiplied by the Householder matrix.

Procedures used: matmat, elmcolrow.

```
public static void hshrowmat(int lr, int ur, int lc, int uc, int i, double x,
                     double u[][], double a[][])
{
  for (; lc<=uc; lc++)
    elmcolrow(lr,ur,lc,i,a,u,matmat(lr,ur,i,lc,u,a)*x);
}
```

D. hshvectam

Postmultiplies a submatrix by an elementary reflector:
$$a=(I-xu^Tu)a \text{ where } a_{i-lr+1,j-lc+1}=a_{i,j} \ (i=lr,...,ur; \ j=lc,...,uc), \ u_{i-lc+1}=u_i \ (i=lc,...,uc).$$

Procedure parameters:
$$\text{void hshvectam } (lr,ur,lc,uc,x,u,a)$$
lr,ur: int; lower and upper row-indices;
lc,uc: int; lower and upper column-indices;
x: double; the Householder constant;
u: double *u[lc:uc]*; the Householder vector;
a: double *a[lr:ur,lc:uc]*; the matrix to be postmultiplied by the Householder matrix.

Procedures used: matvec, elmrowvec.

```
public static void hshvectam(int lr, int ur, int lc, int uc, double x, double u[],
                     double a[][])
{
  for (; lr<=ur; lr++)
    elmrowvec(lc,uc,lr,a,u,matvec(lc,uc,lr,a,u)*x);
}
```

E. hshcoltam

Postmultiplies one submatrix by an elementary reflector formed from a column of another:

$a=(I-xu^Tu)a$ where $a_{k-lr+1,j-lc+1}=a_{kj}$ $(k=lr,...,ur;\ j=lc,...,uc),\ u_{k-lc+1}=U_{k,i}$ $(k=lc,...,uc)$.

Procedure parameters:
$$\text{void hshcoltam } (lr,ur,lc,uc,i,x,u,a)$$
lr,ur: int; lower and upper row-indices;
lc,uc: int; lower and upper column-indices;
i: int; the column index of the Householder vector;
x: double; the Householder constant;
u: double *u[lc:uc,i:i]*; the Householder vector;
a: double *a[lr:ur,lc:uc]*; the matrix to be postmultiplied by the Householder matrix.

Procedures used: matmat, elmrowcol.

```
public static void hshcoltam(int lr, int ur, int lc, int uc, int i, double x,
                        double u[][], double a[][])
{
  for (; lr<=ur; lr++)
    elmrowcol(lc,uc,lr,i,a,u,matmat(lc,uc,lr,i,a,u)*x);
}
```

F. hshrowtam

Postmultiplies one submatrix by an elementary reflector formed from a row of another:

$a=(I-xu^Tu)a$ where $a_{k-lr+1,j-lc+1}=a_{kj}$ $(k=lr,...,ur;\ j=lc,...,uc),\ u_{k-lc+1}=U_{i,k}$ $(k=lc,...,uc)$.

Procedure parameters:
$$\text{void hshrowtam } (lr,ur,lc,uc,i,x,u,a)$$
lr,ur: int; lower and upper row-indices;
lc,uc: int; lower and upper column-indices;
i: int; the row index of the Householder vector;
x: double; the Householder constant;
u: double *u[i:i,lc:uc]*; the Householder vector;
a: double *a[lr:ur,lc:uc]*; the matrix to be postmultiplied by the Householder matrix.

Procedures used: mattam, elmrow.

```
public static void hshrowtam(int lr, int ur, int lc, int uc, int i, double x,
                        double u[][], double a[][])
{
```

```
for (; lr<=ur; lr++)
    elmrow(lc,uc,lr,i,a,u,mattam(lc,uc,lr,i,a,u)*x);
}
```

1.7 Real vector and matrix — Elimination

A. elmvec

Adds a constant multiple of part of a vector b to part of a vector a by setting
$$a_i = a_i + x b_{i+shift} \quad (i=l,l+1,...,u).$$

Procedure parameters:
$$\text{void elmvec } (l,u,shift,a,b,x)$$
l,u: int; lower and upper bound of the running subscript;
shift: int; index-shifting parameter of the vector b;
a,b: double $a[l:u]$, $b[l+shift:u+shift]$;
x: double; elimination factor.

```
public static void elmvec(int l, int u, int shift, double a[], double b[], double x)
{
    for (; l<=u; l++) a[l] += b[l+shift]*x;
}
```

B. elmcol

Adds a constant multiple of part of a column of a rectangular matrix b to part of a column of a rectangular matrix a by setting
$$a_{k,i} = a_{k,i} + x b_{k,j} \quad (k=l,l+1,...,u).$$

Procedure parameters:
$$\text{void elmcol } (l,u,i,j,a,b,x)$$
l,u: int; lower and upper bound of the running subscript;
i: int; column index of a;
j: int; column index of b;
a,b: double $a[l:u,i:i]$, $b[l:u,j:j]$;
x: double; elimination factor.

```
public static void elmcol(int l, int u, int i, int j, double a[][], double b[][],
                    double x)
{
    for (; l<=u; l++) a[l][i] += b[l][j]*x;
}
```

C. elmrow

Adds a constant multiple of part of a row of a rectangular matrix b to part of a row of a rectangular matrix a by setting
$$a_{i,k} = a_{i,k} + xb_{j,k} \quad (k=l, l+1, \ldots, u).$$

Procedure parameters:
$$\text{void elmrow } (l, u, i, j, a, b, x)$$

l, u:	int; lower and upper bound of the running subscript;
i:	int; row index of a;
j:	int; row index of b;
a, b:	double $a[i:i, l:u], b[j:j, l:u]$;
x:	double; elimination factor.

```
public static void elmrow(int l, int u, int i, int j, double a[][], double b[][],
                double x)
{
    for (; l<=u; l++) a[i][l] += b[j][l]*x;
}
```

D. elmveccol

Adds a constant multiple of part of a column of a rectangular matrix b to part of a vector a by setting
$$a_k = a_k + xb_{k,i} \quad (k=l, l+1, \ldots, u).$$

Procedure parameters:
$$\text{void elmveccol } (l, u, i, a, b, x)$$

l, u:	int; lower and upper bound of the running subscript;
i:	int; column index of b;
a, b:	double $a[l:u], b[l:u, i:i]$;
x:	double; elimination factor.

```
public static void elmveccol(int l, int u, int i, double a[], double b[][],
                double x)
{
    for (; l<=u; l++) a[l] += b[l][i]*x;
}
```

E. elmcolvec

Adds a constant multiple of part of a vector b to part of a column of a rectangular matrix a by setting

$$a_{k,i} = a_{k,i} + xb_k \quad (k=l,l+1,...,u).$$

Procedure parameters:
$$\text{void elmcolvec } (l,u,i,a,b,x)$$

l,u: int; lower and upper bound of the running subscript;
i: int; column index of *a*;
a,b: double $a[l:u,i:i]$, $b[l:u]$;
x: double; elimination factor.

```
public static void elmcolvec(int l, int u, int i, double a[][], double b[],
                        double x)
{
    for (; l<=u; l++) a[l][i] += b[l]*x;
}
```

F. elmvecrow

Adds a constant multiple of part of a row of a rectangular matrix *b* to the corresponding part of a vector *a* by setting
$$a_k = a_k + xb_{i,k} \quad (k=l,l+1,...,u).$$

Procedure parameters:
$$\text{void elmvecrow } (l,u,i,a,b,x)$$

l,u: int; lower and upper bound of the running subscript;
i: int; row index of *b*;
a,b: double $a[l:u]$, $b[i:i,l:u]$;
x: double; elimination factor.

```
public static void elmvecrow(int l, int u, int i, double a[], double b[][],
                        double x)
{
    for (; l<=u; l++) a[l] += b[i][l]*x;
}
```

G. elmrowvec

Adds a constant multiple of part of a vector *b* to the corresponding part of a row of a rectangular matrix *a* by setting
$$a_{i,k} = a_{i,k} + xb_k \quad (k=l,l+1,...,u).$$

Procedure parameters:
$$\text{void elmrowvec } (l,u,i,a,b,x)$$

l,u: int; lower and upper bound of the running subscript;
i: int; row index of *a*;
a,b: double $a[i:i,l:u]$, $b[l:u]$;
x: double; elimination factor.

```
public static void elmrowvec(int l, int u, int i, double a[][], double b[],
                             double x)
{
    for (; l<=u; l++) a[i][l] += b[l]*x;
}
```

H. elmcolrow

Adds a constant multiple of part of a row of a rectangular matrix b to the corresponding part of a column of a rectangular matrix a by setting

$$a_{k,i} = a_{k,i} + xb_{j,k} \quad (k=l,l+1,...,u).$$

Procedure parameters:

void elmcolrow (l,u,i,j,a,b,x)

l,u: int; lower and upper bound of the running subscript;
i: int; column index of a;
j: int; row index of b;
a,b: double $a[l:u,i:i]$, $b[j:j,l:u]$, when $a=b$ then correct elimination is guaranteed only when the row and column are disjunct;
x: double; elimination factor.

```
public static void elmcolrow(int l, int u, int i, int j, double a[][], double b[][],
                             double x)
{
    for (; l<=u; l++) a[l][i] += b[j][l]*x;
}
```

I. elmrowcol

Adds a constant multiple of part of a column of a rectangular matrix b to the corresponding part of a row of a rectangular matrix a by setting

$$a_{i,k} = a_{i,k} + xb_{k,j} \quad (k=l,l+1,...,u).$$

Procedure parameters:

void elmrowcol (l,u,i,j,a,b,x)

l,u: int; lower and upper bound of the running subscript;
i: int; row index of a;
j: int; column index of b;
a,b: double $a[i:i,l:u]$, $b[l:u,j:j]$, when $a=b$ then correct elimination is guaranteed only when the row and column are disjunct;
x: double; elimination factor.

```
public static void elmrowcol(int l, int u, int i, int j, double a[][], double b[][],
                             double x)
```

```
{
  for (; l<=u; l++) a[i][l] += b[l][j]*x;
}
```

J. maxelmrow

Performs the same operations as *elmrow* and, in addition, sets *maxelmrow=l* if $l>u$ and otherwise *maxelmrow=m* where m is that value of k for which $\left| a_{i,k} \right|$ is a maximum over $k=l,l+1,...,u$.

Procedure parameters:
$$\text{int maxelmrow } (l,u,i,j,a,b,x)$$
maxelmrow: delivers the index of the maximal element after elimination step upon a;
l,u: int; lower and upper bound of the running subscript;
i: int; row index of a;
j: int; row index of b;
a,b: double $a[i:i,l:u]$, $b[i:i,l:u]$;
x: double; elimination factor.

```
public static int maxelmrow(int l, int u, int i, int j, double a[][], double b[][],
                    double x)
{
  int k;
  double r, s;

  s=0.0;
  for (k=l; k<=u; k++) {
    r=(a[i][k] += b[j][k]*x);
    if (Math.abs(r) > s) {
      s=Math.abs(r);
      l=k;
    }
  }
  return (l);
}
```

1.8 Real vector and matrix —Interchanging

A. ichvec

Interchanges two parts of the same vector a by setting
$$a_i \leftrightarrow a_{i+shift} \ (i=l,l+1,...,u).$$

Procedure parameters:

$$\text{void ichvec } (l,u,shift,a)$$

l,u: int; lower and upper bound of the running subscript;

shift: int; index-shifting parameter;

a: double $a[p:q]$; p and q should satisfy: $p \leq l$, $q \geq u$, $p \leq l+shift$, and $q \geq u+shift$.

```
public static void ichvec(int l, int u, int shift, double a[])
{
  double r;

  for (; l<=u; l++) {
    r=a[l];
    a[l]=a[l+shift];
    a[l+shift]=r;
  }
}
```

B. ichcol

Interchanges corresponding parts of two columns of the same rectangular matrix a by setting

$$a_{k,i} \leftrightarrow a_{k,j} \ (k=l,l+1,...,u).$$

Procedure parameters:

$$\text{void ichcol } (l,u,i,j,a)$$

l,u: int; lower and upper bound of the running subscript;

i,j: int; column indices of a;

a: double $a[l:u,p:q]$; p and q should satisfy: $p \leq i$, $p \leq j$, $q \geq i$, and $q \geq j$.

```
public static void ichcol(int l, int u, int i, int j, double a[][])
{
  double r;

  for (; l<=u; l++) {
    r=a[l][i];
    a[l][i]=a[l][j];
    a[l][j]=r;
  }
}
```

C. ichrow

Interchanges corresponding parts of two rows of the same rectangular matrix a by setting

$$a_{i,k} \leftrightarrow a_{j,k} \ (k=l,l+1,...,u).$$

Procedure parameters:
$$\text{void ichrow } (l,u,i,j,a)$$

l,u: int; lower and upper bound of the running subscript;
i,j: int; row indices of a;
a: double $a[p{:}q,l{:}u]$; p and q should satisfy: $p{\leq}i$, $p{\leq}j$, $q{\geq}i$, and $q{\geq}j$.

```
public static void ichrow(int l, int u, int i, int j, double a[][])
{
   double r;

   for (; l<=u; l++) {
     r=a[i][l];
     a[i][l]=a[j][l];
     a[j][l]=r;
   }
}
```

D. ichrowcol

Interchanges corresponding parts of a row and a column of the same rectangular matrix a by setting
$$a_{i,k} \leftrightarrow a_{k,j} \ (k=l,l+1,...,u).$$

Procedure parameters:
$$\text{void ichrowcol } (l,u,i,j,a)$$

l,u: int; lower and upper bound of the running subscript;
i: int; row index of a;
j: int; column index of a;
a: double $a[p{:}q,r{:}s]$; p, q, r, and s should satisfy: $p{\leq}i$, $p{\leq}l$, $q{\geq}i$, $q{\geq}u$, $r{\leq}j$, $r{\leq}l$, $s{\geq}j$, and $s{\geq}u$, furthermore the row and column to be interchanged should be disjoint.

```
public static void ichrowcol(int l, int u, int i, int j, double a[][])
{
   double r;

   for (; l<=u; l++) {
     r=a[i][l];
     a[i][l]=a[l][j];
     a[l][j]=r;
   }
}
```

E. ichseqvec

Interchanges two parts of the same vector by setting
$$a_{il+(j+l-1)(j-l)/2} \leftrightarrow a_{j+shift} \; (j=l,l+1,...,u).$$

Procedure parameters:

void ichseqvec (l,u,il,shift,a)

l,u: int; lower and upper bound of the running subscript;
il: int; lower bound of the vector *a*;
shift: int; index-shifting parameter;
a: double *a[p:q]*; it is assumed that the values of *l+shift*, *u+shift*, and
        il+(u+l-1)(u-l)/2 are not out of range.

```
public static void ichseqvec(int l, int u, int il, int shift, double a[])
{
  double r;

  for (; l<=u; l++) {
    r=a[il];
    a[il]=a[l+shift];
    a[l+shift]=r;
    il += l;
  }
}
```

F. ichseq

Interchanges two parts of the same vector by setting
$$a_{il+(j+l-1)(j-l)/2} \leftrightarrow a_{shift+il+(j+l-1)(j-l)/2} \; (j=l,l+1,...,u).$$

Procedure parameters:

void ichseq (l,u,il,shift,a)

l,u: int; lower and upper bound of the running subscript;
il: int; lower bound of the vector *a*;
shift: int; index-shifting parameter;
a: double *a[p:q]*; it is assumed that the values of *il+shift+(j+l-1)(j-l)/2, j=l,...,u*,
        do not go out of range.

```
public static void ichseq(int l, int u, int il, int shift, double a[])
{
  double r;

  for (; l<=u; l++) {
    r=a[il];
```

```
      a[il]=a[il+shift];
      a[il+shift]=r;
      il += 1;
    }
  }
}
```

1.9 Real vector and matrix — Rotation

A. rotcol

Rotates two columns of a rectangular matrix by setting

$$a_{k,i} = ca_{k,i} + sa_{k,j}, \quad a_{k,j} = ca_{k,j} - sa_{k,i} \quad (k=l,l+1,...,u).$$

Procedure parameters:

$$\text{void rotcol } (l,u,i,j,a,c,s)$$

l,u: int; lower and upper bound of the running subscript;
i,j: int; column indices of the array a;
a: double $a[l:u,p:q]$; p and q should satisfy: $p{\le}i$, $p{\le}j$, $q{\ge}i$, and $q{\ge}j$;
c,s: double; rotation factors.

```
public static void rotcol(int l, int u, int i, int j, double a[][], double c,
                          double s)
{
  double x, y;

  for (; l<=u; l++) {
    x=a[l][i];
    y=a[l][j];
    a[l][i]=x*c+y*s;
    a[l][j]=y*c-x*s;
  }
}
```

B. rotrow

Rotates two rows of a rectangular matrix by setting

$$a_{i,k} = ca_{i,k} + sa_{j,k}, \quad a_{j,k} = ca_{j,k} - sa_{i,k} \quad (k=l,l+1,...,u).$$

Procedure parameters:

void rotrow (l,u,i,j,a,c,s)
l,u: int; lower and upper bound of the running subscript;
i,j: int; row indices of the array a;
a: double $a[p:q,l:u]$; p and q should satisfy: $p{\leq}i$, $p{\leq}j$, $q{\geq}i$, and $q{\geq}j$;
c,s: double; rotation factors.

```
public static void rotrow(int l, int u, int i, int j, double a[][], double c,
                          double s)
{
  double x, y;

  for (; l<=u; l++) {
    x=a[i][l];
    y=a[j][l];
    a[i][l]=x*c+y*s;
    a[j][l]=y*c-x*s;
  }
}
```

1.10 Real vector and matrix — Norms

A. infnrmvec

Computes the ∞-norm of part of a vector, and determines the index of the element with largest absolute value (ρ = max$|a_i|$ $(l{\leq}i{\leq}u)$ and the smallest k such that $|a_k|{=}\rho$ $(l{\leq}k{\leq}u)$).

Procedure parameters:
double infnrmvec (l,u,k,a)
infnrmvec: delivers the ∞-norm of the vector a, the value of ρ above;
l,u: int; lower and upper bound of the index of the vector a, respectively;
k: int $k[0:0]$;
               exit: the value of k above;
a: double $a[l:u]$.

```
public static double infnrmvec(int l, int u, int k[], double a[])
{
  double r, max;

  max=0.0;
  k[0]=l;
  for (; l<=u; l++) {
    r=Math.abs(a[l]);
    if (r > max) {
      max=r;
```

```
      k[0]=1;
    }
  }
  return (max);
}
```

B. infnrmrow

Computes the ∞-norm of part of a row of a rectangular matrix, and determines the index of the element with largest absolute value (ρ = max$|a_{i,j}|$ ($l \leq j \leq u$) and the smallest k such that $|a_{i,k}| = \rho$ ($l \leq k \leq u$)).

Procedure parameters:
$$\text{double infnrmrow } (l,u,i,k,a)$$
infnrmrow: delivers the ∞-norm of the row vector of a, the value of ρ above;
l,u: int; lower and upper bound of the column index of a, respectively;
i: int; the row index;
k: int $k[0:0]$;
        exit: the value of k above;
a: double $a[i:i,l:u]$.

```
public static double infnrmrow(int l, int u, int i, int k[], double a[][])
{
  double r, max;

  max=0.0;
  k[0]=l;
  for (; l<=u; l++) {
    r=Math.abs(a[i][l]);
    if (r > max) {
      max=r;
      k[0]=l;
    }
  }
  return (max);
}
```

C. infnrmcol

Computes the ∞-norm of part of a column of a rectangular matrix, and determines the index of the element with largest absolute value (ρ = max$|a_{i,j}|$ ($l \leq i \leq u$) and the smallest k such that $|a_{k,j}| = \rho$ ($l \leq k \leq u$)).

Procedure parameters:

double infnrmcol (l,u,j,k,a)

infnrmcol:	delivers the ∞-norm of the column vector of a, the value of ρ above;
l,u:	int; lower and upper bound of the row index of a, respectively;
j:	int; the column index;
k:	int $k[0:0]$;
	exit: the value of k above;
a:	double $a[l:u,j:j]$.

```
public static double infnrmcol(int l, int u, int j, int k[], double a[][])
{
  double r, max;

  max=0.0;
  k[0]=l;
  for (; l<=u; l++) {
    r=Math.abs(a[l][j]);
    if (r > max) {
      max=r;
      k[0]=l;
    }
  }
  return (max);
}
```

D. infnrmmat

Computes the ∞-norm of a submatrix and determines the index of the row whose 1-norm is equal to the ∞-norm of the submatrix:

$$\rho = \max \sum_{j=lc}^{uc} \left| a_{i,j} \right|, \; lr \le i \le ur,$$

and the smallest kr such that

$$\rho = \sum_{j=lc}^{uc} \left| a_{kr,j} \right|.$$

Procedure parameters:

double infnrmmat (lr,ur,lc,uc,kr,a)

infnrmmat:	delivers the ∞-norm of the matrix a, the value of ρ above;
lr,ur:	int; lower and upper bound of the row index, respectively;
lc,uc:	int; lower and upper bound of the column index, respectively;
kr:	int $kr[0:0]$;
	exit: the value of kr above;
a:	double $a[lr:ur,lc:uc]$.

Procedure used: onenrmrow.

```
public static double infnrmmat(int lr, int ur, int lc, int uc, int kr[],
                               double a[][])
{
  double r, max;

  max=0.0;
  kr[0]=lr;
  for (; lr<=ur; lr++) {
    r=onenrmrow(lc,uc,lr,a);
    if (r > max) {
      max=r;
      kr[0]=lr;
    }
  }
  return (max);
}
```

E. onenrmvec

Computes the 1-norm of part of a vector:

$$\rho = \sum_{i=l}^{u} \left| a_i \right|.$$

Procedure parameters:
$$\text{double onenrmvec } (l,u,a)$$
onenrmvec: delivers the 1-norm of the vector *a*, the value of ρ above;
l,u: int; lower and upper bound of the index of the vector *a*, respectively;
a: double *a[l:u]*.

```
public static double onenrmvec(int l, int u, double a[])
{
  double sum;

  sum=0.0;
  for (; l<=u; l++) sum += Math.abs(a[l]);
  return (sum);
}
```

F. onenrmrow

Computes the 1-norm of part of a row of a rectangular matrix:

$$\rho = \sum_{j=l}^{u} \left| a_{i,j} \right|.$$

Procedure parameters:

double onenrmrow (l,u,i,a)

onenrmrow: delivers the 1-norm of a row vector, the value of ρ above;

l,u: int; lower and upper bound of the column index of a, respectively;

i: int; the row index;

a: double $a[i:i,l:u]$.

```
public static double onenrmrow(int l, int u, int i, double a[][])
{
  double sum;

  sum=0.0;
  for (; l<=u; l++) sum += Math.abs(a[i][l]);
  return (sum);
}
```

G. onenrmcol

Computes the 1-norm of part of a column of a rectangular matrix:

$$\rho = \sum_{i=l}^{u} \left| a_{i,j} \right|.$$

Procedure parameters:

double onenrmcol (l,u,j,a)

onenrmcol: delivers the 1-norm of a column vector, the value of ρ above;

l,u: int; lower and upper bound of the row index of a, respectively;

j: int; the column index;

a: double $a[l:u,j:j]$.

```
public static double onenrmcol(int l, int u, int j, double a[][])
{
  double sum;

  sum=0.0;
  for (; l<=u; l++) sum += Math.abs(a[l][j]);
  return (sum);
}
```

H. onenrmmat

Computes the 1-norm of a submatrix and determines the index of the column whose 1-norm is that of the submatrix:

$$\rho = \max \sum_{i=lr}^{ur} \left| a_{i,j} \right|, \ lc \le j \le uc,$$

and the smallest kc such that

$$\rho = \sum_{i=lr}^{ur} \left| a_{i,kc} \right|.$$

Procedure parameters:

double onenrmmat (lr,ur,lc,uc,kc,a)

onenrmmat: delivers the 1-norm of the matrix a, the value of ρ above;

lr,ur: int; lower and upper bound of the row index, respectively;

lc,uc: int; lower and upper bound of the column index, respectively;

kc: int $kc[0:0]$;

exit: the value of kc above;

a: double $a[lr:ur,lc:uc]$.

Procedure used: onenrmcol.

```
public static double onenrmmat(int lr, int ur, int lc, int uc, int kc[],
                               double a[][])
{
  double r, max;

  max=0.0;
  kc[0]=lc;
  for (; lc<=uc; lc++) {
    r=onenrmcol(lr,ur,lc,a);
    if (r > max) {
      max=r;
      kc[0]=lc;
    }
  }
  return (max);
}
```

I. absmaxmat

Computes the absolute value of an element with largest absolute value belonging to a submatrix, and the row and column indices of the element in question ($\rho = \max \left| a_{k,h} \right|$, $(lr \le k \le ur, \ lc \le h \le uc)$ and the smallest i for the smallest j such that $\rho = \left| a_{i,j} \right|$).

Procedure parameters:
$$\text{double absmaxmat } (lr,ur,lc,uc,i,j,a)$$

absmaxmat: delivers for a given matrix *a* the modulus of an element which is of maximum absolute value, the value of ρ above;

lr,ur: int; lower and upper bound of the row index, respectively;

lc,uc: int; lower and upper bound of the column index, respectively;

i,j: int *i[0:0]*, *j[0:0]*;
    exit: the row and column index of an element for which the modulus is maximal;

a: double *a[lr:ur,lc:uc]*.

Procedure used: infnrmcol.

```
public static double absmaxmat(int lr, int ur, int lc, int uc, int i[], int j[],
                               double a[][])
{
  int ii[] = new int[1];
  double r, max;

  max=0.0;
  i[0]=lr;
  j[0]=lc;
  for (; lc<=uc; lc++) {
    r=infnrmcol(lr,ur,lc,ii,a);
    if (r > max) {
      max=r;
      i[0]=ii[0];
      j[0]=lc;
    }
  }
  return (max);
}
```

1.11 Real vector and matrix — Scaling

reascl

Normalizes the (non-null) columns of a matrix in such a way that, in each column, an element of maximum absolute value equals one. The normalized vectors are delivered in the corresponding columns of the matrix.

Procedure parameters:
$$\text{void reascl } (a,n,n1,n2)$$

a: double *a[1:n,n1:n2]*;
    entry: the *n2-n1+1* column vectors must be given in *a*;

exit: the normalized vectors (i.e., in each vector an element of maximum absolute value equals one) are delivered in the corresponding columns of *a*;

n: int; the number of rows of *a*;

n1,n2: int; the lower and upper bound of the column indices of *a*.

```
public static void reascl(double a[][], int n, int n1, int n2)
{
  int i, j;
  double s;

  for (j=n1; j<=n2; j++) {
    s=0.0;
    for (i=1; i<=n; i++)
      if (Math.abs(a[i][j]) > Math.abs(s)) s=a[i][j];
    if (s != 0.0)
      for (i=1; i<=n; i++) a[i][j] /= s;
  }
}
```

1.12 Complex vector and matrix — Multiplication

A. comcolcst

Replaces part of a column of a matrix with complex elements by a constant complex multiple of that part:

$$a_{i,j}=xa_{i,j}, \ i=l,...,u.$$

Procedure parameters:

$$\text{void comcolcst } (l,u,j,ar,ai,xr,xi)$$

l,u: int; lower and upper bound of the column vector;

j: int; column index of the column vector;

ar,ai: double *ar[l:u,j:j]*, *ai[l:u,j:j]*;

      entry: ar: real part;

                ai: imaginary part of the column vector;

      exit: the transformed complex column;

xr,xi: double;

      entry: xr: real part of the multiplication factor;

                xi: imaginary part of the multiplication factor.

Procedure used: commul.

```
public static void comcolcst(int l, int u, int j, double ar[][], double ai[][],
                        double xr, double xi)
```

```
{
  double br[] = new double[1];
  double bi[] = new double[1];

  for (; l<=u; l++) {
    commul(ar[l][j],ai[l][j],xr,xi,br,bi);
    ar[l][j] = br[0];
    ai[l][j] = bi[0];
  }
}
```

B. comrowcst

Replaces part of a row of a matrix with complex elements by a constant complex multiple of that part:

$$a_{i,j}=xa_{i,j}, \ j=l,...,u.$$

Procedure parameters:

$$\text{void comrowcst } (l,u,i,ar,ai,xr,xi)$$

l,u: int; lower and upper bound of the row vector;
i: int; row index of the row vector;
ar,ai: double *ar[i:i,l:u], ai[i:i,l:u]*;
           entry: ar: real part;
                   ai: imaginary part of the row vector;
           exit: the transformed complex row;
xr,xi: double;
           entry: xr: real part of the multiplication factor;
                   xi: imaginary part of the multiplication factor.

Procedure used: commul.

```
public static void comrowcst(int l, int u, int i, double ar[][], double ai[][],
                      double xr, double xi)
{
  double br[] = new double[1];
  double bi[] = new double[1];

  for (; l<=u; l++) {
    commul(ar[i][l],ai[i][l],xr,xi,br,bi);
    ar[i][l] = br[0];
    ai[i][l] = bi[0];
  }
}
```

1.13 Complex vector and matrix — Scalar products

A. commatvec

Forms the inner product of part of a row of a matrix and part of a vector, both having complex elements:

$$r = \sum_{j=l}^{u} a_{i,j} b_j .$$

Procedure parameters:

$$\text{void commatvec } (l,u,i,ar,ai,br,bi,rr,ri)$$

l,u: int; lower and upper bound of the vectors;
i: int; the row index of *ar* and *ai*;
ar,ai: double *ar[i:i,l:u],ai[i:i,l:u]*;
       entry: ar: real part and
                ai: imaginary part of the matrix;
br,bi: double *br[l:u],bi[l:u]*;
       entry: br: real part and
                bi: imaginary part of the vector;
rr,ri: double *rr[0:0],ri[0:0]*;
       exit: rr: the real part and
              ri: the imaginary part of the scalar product.

Procedure used: matvec.

```
public static void commatvec(int l, int u, int i, double ar[][], double ai[][],
                            double br[], double bi[], double rr[], double ri[])
{
  double mv;

  mv=matvec(l,u,i,ar,br)-matvec(l,u,i,ai,bi);
  ri[0]=matvec(l,u,i,ai,br)+matvec(l,u,i,ar,bi);
  rr[0]=mv;
}
```

B. hshcomcol

Given the components $a_{i,j}$ $(i=l+1,...,n)$ of the j-th column of the complex valued matrix *a*, and a real tolerance *tol*, and if

$$\sum_{i=l+1}^{u} \left| a_{i,j} \right|^2 > tol \tag{1}$$

this procedure determines a complex vector *c* such that, the j-th column of

$$a = \left(I - 2c\bar{c}^T / (\bar{c}^T c) \right) a \qquad \text{(where } \bar{c}^T \text{ denotes conjugating and transposing)}$$

has zero elements in the second,..., last positions, and computes

$$t = (\bar{c}^T c)/2$$

and the first component of c.

Procedure parameters:

boolean hshcomcol $(l,u,j,ar,ai,tol,k,c,s,t)$

$hshcomcol$:   if the condition (1) above is satisfied then a transformation is performed and $hshcomcol$=true, otherwise $hshcomcol$=false;

l,u,j: int; the complex vector to be transformed, must be given in the j-th column from row l until row u of a complex matrix;

ar,ai: double $ar[l:u,j:j],ai[l:u,j:j]$;

entry: the real part and the imaginary part of the vector to be transformed must be given in the arrays ar and ai, respectively;

exit: if the condition (1) above is satisfied then the real part and the imaginary part of the vector c of the Householder matrix are delivered in the arrays ar and ai, respectively, otherwise the arrays ar and ai are unaltered;

tol: double;

entry: a tolerance (for example, the square of the machine precision times a norm of the matrix under consideration);

k,c,s: double $k[0:0],c[0:0],s[0:0]$;

exit: if the condition (1) above is satisfied then the modulus, cosine and sine of the argument of the first element of the transformed vector are delivered in k, c and s, respectively, otherwise the modulus, cosine and sine of the argument of the complex number $ar[l,j]+ai[l,j]*I$ are delivered;

t: double $t[0:0]$;

exit: if the condition (1) above is satisfied then it has the value t in the above, otherwise it has the value -1.

Procedures used: carpol, tammat.

```
public static boolean hshcomcol(int l, int u, int j, double ar[][], double ai[][],
                                double tol, double k[], double c[], double s[],
                                double t[])
{
  double vr, h, arlj, ailj;
  double mod[] = new double[1];

  vr=tammat(l+1,u,j,j,ar,ar)+tammat(l+1,u,j,j,ai,ai);
  arlj=ar[l][j];
  ailj=ai[l][j];
  carpol(arlj,ailj,mod,c,s);
  if (vr > tol) {
    vr += arlj*arlj+ailj*ailj;
    h = k[0] = Math.sqrt(vr);
    t[0]=vr+mod[0]*h;
    if (arlj == 0.0 && ailj == 0.0)
```

```
      ar[l][j]=h;
    else {
      ar[l][j]=arlj + c[0] * k[0];
      ai[l][j]=ailj + s[0] * k[0];
      s[0] = - s[0];
    }
    c[0] = - c[0];
    return (true);
  } else {
    k[0]=mod[0];
    t[0] = -1.0;
    return (false);
  }
}
```

C. hshcomprd

Given the components of the complex vector c, and
$$t = \left(c\bar{c}^T\right)/2 \quad \text{(where } \bar{c}^T \text{ denotes conjugating and transposing)}$$
and the submatrix a composed of the elements $a_{k,h}$ ($k=i,...,ii$; $h=l,...,u$) of the complex matrix a, forms the product
$$a = \left(I - t\bar{c}^T c\right)a,$$
it being known that c is such that the elements in rows $l+1,...,u$ of the j-th column of a are zero.

Procedure parameters:
$$\text{void hshcomprd } (i,ii,l,u,j,ar,ai,br,bi,t)$$

i,ii,l,u: int; the complex matrix to be premultiplied must be given in the l-th to u-th column from row i to row ii of a complex matrix;

j: int; the complex vector c of the Householder matrix must be given in the j-th column from row i to row ii of a complex matrix given in (br,bi);

ar,ai: double $ar[i:ii,l:u],ai[i:ii,l:u]$;
        entry: the real part and the imaginary part of the vector to be premultiplied must be given in the arrays ar and ai, respectively;
        exit: the real part and the imaginary part of the resulting matrix are delivered in the arrays ar and ai, respectively, otherwise the arrays ar and ai are unaltered;

br,bi: double $br[i:ii,j:j], bi[i:ii,j:j]$;
        entry: the real part and the imaginary part of the complex vector c of the Householder matrix must be given in the arrays br and bi, respectively;

t: double;
        entry: the scalar t above of the Householder matrix (for example, as delivered by *hshcomcol*).

Procedures used: tammat, elmcomcol.

```
public static void hshcomprd(int i, int ii, int l, int u, int j, double ar[][],
                             double ai[][], double br[][], double bi[][], double t)
{
  for (; l<=u; l++)
    elmcomcol(i,ii,l,j,ar,ai,br,bi,
              (-tammat(i,ii,j,l,br,ar)-tammat(i,ii,j,l,bi,ai))/t,
              (tammat(i,ii,j,l,bi,ar)-tammat(i,ii,j,l,br,ai))/t);
}
```

1.14 Complex vector and matrix — Elimination

A. elmcomveccol

Adds a constant complex multiple of part of a column of a rectangular matrix to part of a vector, both having complex elements: $a_i = a_i + xb_{i,j}$, $i=l,...,u$.

Procedure parameters:
                    void elmcomveccol (l,u,j,ar,ai,br,bi,xr,xi)

l,u: int; lower and upper bounds of the vectors;
j: int; column index of br and bi;
ar,ai: double $ar[l:u]$, $ai[l:u]$;
         entry: ar: real part and
                 ai: imaginary part of the vector;
         exit: the resulting vector (adds $xr+xi{*}i$ times the complex column vector
                 given in arrays br and bi to the complex vector given in arrays ar
                 and ai);
br,bi: double $br[l:u,j:j]$, $bi[l:u,j:j]$;
         entry: br: real part and
                 bi: imaginary part of the column vector;
xr,xi: double;
         entry: xr: real part and
                 xi: imaginary part of the elimination factor.

Procedure used: elmveccol.

```
public static void elmcomveccol(int l, int u, int j, double ar[], double ai[],
                                double br[][], double bi[][], double xr, double xi)
{
  elmveccol(l,u,j,ar,br,xr);
  elmveccol(l,u,j,ar,bi,-xi);
  elmveccol(l,u,j,ai,br,xi);
  elmveccol(l,u,j,ai,bi,xr);
}
```

B. elmcomcol

Adds a constant complex multiple of part of a column of one complex matrix to part of a column of another: $a_{k,i}=a_{k,i}+(xr+xi*i)b_{k,j}$, $k=l,...,u$.

Procedure parameters:
$$\text{void elmcomcol } (l,u,i,j,ar,ai,br,bi,xr,xi)$$

l,u: int; lower and upper bounds of the vectors;
i,j: int;
          i: column index of *ar* and *ai*;
          j: column index of *br* and *bi*;
ar,ai: double *ar[l:u,i:i]*, *ai[l:u,i:i]*;
          entry: ar: real part and
                  ai: imaginary part of the column vector;
          exit: the resulting vector (adds *xr+xi*i* times the complex column vector given in arrays *br* and *bi* to the complex column vector given in arrays *ar* and *ai*);
br,bi: double *br[l:u,j:j]*, *bi[l:u,j:j]*;
          entry: br: real part and
                  bi: imaginary part of the column vector;
xr,xi: double;
          entry: xr: real part and
                  xi: imaginary part of the elimination factor.

Procedure used: elmcol.

```
public static void elmcomcol(int l, int u, int i, int j, double ar[][],
                            double ai[][], double br[][], double bi[][], double xr,
                            double xi)
{
    elmcol(l,u,i,j,ar,br,xr);
    elmcol(l,u,i,j,ar,bi,-xi);
    elmcol(l,u,i,j,ai,br,xi);
    elmcol(l,u,i,j,ai,bi,xr);
}
```

C. elmcomrowvec

Adds a constant complex multiple of part of a vector to part of a row of a rectangular matrix, both having complex elements: $a_{i,j}=a_{i,j}+xb_j$, $j=l,...,u$.

Procedure parameters:
$$\text{void elmcomrowvec } (l,u,i,ar,ai,br,bi,xr,xi)$$

l,u: int; lower and upper bounds of the vectors;
i: int; row index of *ar* and *ai*;
ar,ai: double *ar[i:i,l:u]*, *ai[i:i,l:u]*;

entry: ar: real part and
              ai: imaginary part of the row vector;

exit: the resulting vector (adds $xr+xi*i$ times the complex vector given in arrays *br* and *bi* to the complex row vector given in arrays *ar* and *ai*);

br,bi: double *br[l:u]*, *bi[l:u]*;

       entry: br: real part and
                   bi: imaginary part of the vector;

xr,xi: double;

       entry: xr: real part and
                   xi: imaginary part of the elimination factor.

Procedure used: elmrowvec.

```
public static void elmcomrowvec(int l, int u, int i, double ar[][], double ai[][],
                                double br[], double bi[], double xr, double xi)
{
  elmrowvec(l,u,i,ar,br,xr);
  elmrowvec(l,u,i,ar,bi,-xi);
  elmrowvec(l,u,i,ai,br,xi);
  elmrowvec(l,u,i,ai,bi,xr);
}
```

1.15 Complex vector and matrix — Rotation

A. rotcomcol

Rotates two columns of a matrix with complex elements:
$$a_{k,i} = (cr+ci*i)a_{k,i} + sa_{k,j}, \quad a_{k,j} = (cr+ci*i)a_{k,i} - sa_{k,j} \quad (k=l,...,u).$$

Procedure parameters:
$$\text{void rotcomcol } (l,u,i,j,ar,ai,cr,ci,s)$$

l,u,i,j: int; the rotation is performed on the column vectors *ar[l:u,i:i]*, *ai[l:u,i:i]* and *ar[l:u,j:j]*, *ai[l:u,j:j]*;

ar,ai: double *ar[l:u,i:j]*, *ai[l:u,i:j]*;

       entry: ar: real part and
                 ai: imaginary part of the column vectors;

       exit: the resulting vectors;

cr,ci,s: double; rotation factors.

```
public static void rotcomcol(int l, int u, int i, int j, double ar[][],
                             double ai[][], double cr, double ci, double s)
{
  double arli,aili,arlj,ailj;
```

```
for (; l<=u; l++) {
  arli=ar[l][i];
  aili=ai[l][i];
  arlj=ar[l][j];
  ailj=ai[l][j];
  ar[l][i]=cr*arli+ci*aili-s*arlj;
  ai[l][i]=cr*aili-ci*arli-s*ailj;
  ar[l][j]=cr*arlj-ci*ailj+s*arli;
  ai[l][j]=cr*ailj+ci*arlj+s*aili;
}
}
```

B. rotcomrow

Rotates two rows of a matrix with complex elements:
$$a_{i,k} = (cr+ci*i)a_{i,k} + sa_{j,k}, \quad a_{j,k} = (cr+ci*i)a_{j,k} - sa_{i,k} \quad (k=l,...,u).$$

Procedure parameters:
$$\text{void rotcomrow } (l,u,i,j,ar,ai,cr,ci,s)$$
l,u,i,j: int; the rotation is performed on the row vectors $ar[i:i,l:u]$, $ai[i:i,l:u]$ and $ar[j:j,l:u]$, $ai[j:j,l:u]$;
ar,ai: double $ar[i:j,l:u]$, $ai[i:j,l:u]$;
  entry: ar: real part and
    ai: imaginary part of the row vectors;
  exit: the resulting vectors;
cr,ci,s: double; rotation factors.

```
public static void rotcomrow(int l, int u, int i, int j, double ar[][],
                             double ai[][], double cr, double ci, double s)
{
  double aril,aiil,arjl,aijl;

  for (; l<=u; l++) {
    aril=ar[i][l];
    aiil=ai[i][l];
    arjl=ar[j][l];
    aijl=ai[j][l];
    ar[i][l]=cr*aril+ci*aiil+s*arjl;
    ai[i][l]=cr*aiil-ci*aril+s*aijl;
    ar[j][l]=cr*arjl-ci*aijl-s*aril;
    ai[j][l]=cr*aijl+ci*arjl-s*aiil;
  }
}
```

C. chsh2

Computes the complex Householder matrix that maps the complex vector (a_1, a_2) into the direction $(1,0)$: given complex a_1 and a_2, it determines real c and complex s such that

$$\begin{pmatrix} -c & s \\ s & c \end{pmatrix}\begin{pmatrix} a_1 \\ a_2 \end{pmatrix} = \begin{pmatrix} z \\ 0 \end{pmatrix}$$

where z is nonzero if $|a_1| + |a_2| \neq 0$.

Procedure parameters:
$$\text{void chsh2 } (a1r, a1i, a2r, a2i, c, sr, si)$$

$a1r$:   double;
        entry: the real part of the first vector component a_1 above;
$a1i$:   double;
        entry: the imaginary part of the first vector component a_1 above;
$a2r$:   double;
        entry: the real part of the second vector component a_2 above;
$a2i$:   double;
        entry: the imaginary part of the second vector component a_2 above;
c: double $c[0:0]$;
        exit: the value of c above;
sr, si: double $sr[0:0]$, $si[0:0]$;
        exit: the real part and imaginary part of s above.

```
public static void chsh2(double a1r, double a1i, double a2r, double a2i,
                         double c[], double sr[], double si[])
{
  double r;

  if (a2r != 0.0 || a2i != 0.0) {
    if (a1r != 0.0 || a1i != 0.0) {
      r=Math.sqrt(a1r*a1r+a1i*a1i);
      c[0]=r;
      sr[0]=(a1r*a2r+a1i*a2i)/r;
      si[0]=(a1r*a2i-a1i*a2r)/r;
      r=Math.sqrt(c[0] * c[0] + sr[0] * sr[0] + si[0] * si[0]);
      c[0] /= r;
      sr[0] /= r;
      si[0] /= r;
    } else {
      si[0] = c[0] = 0.0;
      sr[0]=1.0;
    }
  } else {
    c[0]=1.0;
    sr[0] = si[0] = 0.0;
  }
}
```

1.16 Complex vector and matrix — Norms

comeucnrm

Computes the Euclidean norm

$$\rho = \sqrt{\sum_{i=1}^{n}\sum_{j=1}^{n}\left|a_{i,j}\right|^2}$$

of the $n{\times}n$ complex band matrix a for which $a_{i,j}=0$ when $|i\text{-}j|>lw$.

Procedure parameters:

$$\text{double comeucnrm } (ar,ai,lw,n)$$

comeucnrm: double; delivers the Euclidean norm of matrix a with lw codiagonals, value of ρ above;

ar,ai: double $ar[1{:}n,1{:}n],ai[1{:}n,1{:}n]$; the real part and imaginary part of the complex matrix with lw lower codiagonals;

lw: int; the number of lower codiagonals;

n: int; the order of the matrix.

Procedure used: mattam.

```
public static double comeucnrm(double ar[][], double ai[][], int lw, int n)
{
  int i,l;
  double r;

  r=0.0;
  for (i=1; i<=n; i++) {
    l=(i>lw) ? i-lw : 1;
    r += mattam(l,n,i,i,ar,ar)+mattam(l,n,i,i,ai,ai);
  }
  return (Math.sqrt(r));
}
```

1.17 Complex vector and matrix — Scaling

A. comscl

Scales a sequence of columns, some complex, others pure real of an n-rowed matrix a. The real and imaginary components of the complex columns, and the real elements of the pure real columns are stored in succession in columns $n1$ to $n2$ of the matrix a. The complex and pure real columns of the original matrix may be distinguished by inspection of the values stored in locations $n1$ to $n2$ of the real vector *im*. In succession, starting with $j_1=n1$, if $im(j_k)=0$, column j_k of a contains the elements of a real column of the input a; if $im(j_k)\neq0$ then $im(j_k+1)=-im(j_k)$, and columns j_k and j_k+1 of a contain the real and imaginary parts respectively of a complex column of the input a; in the first case $j_{k+1}=j_k+1$, in the second $j_{k+1}=j_k+2$. The columns are scaled by taking $\mu_j=a_{k,j}$ where $|\mu_j|=\max|a_{i,j}|$ $(1\leq i\leq n)$ and $a_{i,j}=a_{i,j}/\mu_j$ $(i=1,...,n)$ where j runs through the column suffices of the stored columns of a.

Procedure parameters:
$$\text{void comscl } (a,n,n1,n2,im)$$
a: double $a[1:n,n1:n2]$;
            entry: each real eigenvector must be given in a column of array a whose corresponding element of array im equals 0; the real and imaginary part of each complex eigenvector must be given in consecutive columns of array a whose corresponding elements of array im are not equal to 0;
            exit: the normalized eigenvectors (i.e., in each eigenvector an element of maximum modulus equals one) are delivered in the corresponding columns of a;
n: int; the number of rows of array a;
n1,n2: int; the lower and upper bound of the column indices of array a;
im: double $im[n1:n2]$; the imaginary parts of the eigenvalues of which the eigenvectors are given in the corresponding columns of array a must be given in array im.

```
public static void comscl(double a[][], int n, int n1, int n2, double im[])
{
  int i,j,k;
  double s,u,v,w,aij,aij1;

  k = 0;
  for (j=n1; j<=n2; j++) {
    s=0.0;
    if (im[j] != 0.0) {
      for (i=1; i<=n; i++) {
        aij=a[i][j];
        aij1=a[i][j+1];
        u=aij*aij+aij1*aij1;
        if (u > s) {
          s=u;
          k=i;
        }
      }
    }
    if (s != 0.0) {
      v=a[k][j]/s;
      w = -a[k][j+1]/s;
      for (i=1; i<=n; i++) {
        u=a[i][j];
        s=a[i][j+1];
        a[i][j]=u*v-s*w;
        a[i][j+1]=u*w+s*v;
```

```
          }
        }
      j++;
    } else {
      for (i=1; i<=n; i++)
        if (Math.abs(a[i][j]) > Math.abs(s)) s=a[i][j];
      if (s != 0.0)
        for (i=1; i<=n; i++)
          a[i][j] /= s;
    }
  }
}
```

B. sclcom

Scales the columns $n1$ to $n2$ of an n-rowed matrix a with complex elements by dividing the elements in each column by the element of largest modulus in that column (with $\mu_j = a_{k,j}$ where $|\mu_j| = \max|a_{i,j}|$ $(1 \leq i \leq n)$, $a_{i,j} = a_{i,j}/\mu_j$ $(i=1,...,n)$ when $\mu_j \neq 0$; $j = n1,...,n2$)).

Procedure parameters:
$$\text{void sclcom } (ar, ai, n, n1, n2)$$

ar,ai: double $ar[1:n,n1:n2]$, $ai[1:n,n1:n2]$;
       entry: the real part and the imaginary part of the matrix of which the columns are to be scaled must be given in the arrays ar and ai, respectively;
       exit: the real part and imaginary part of the matrix with scaled columns are delivered in the arrays ar and ai;
n: int; the order of the matrix;
n1,n2: int; the $n1$-th to $n2$-th column vectors are to be scaled.

Procedure used: comcolcst.

```
public static void sclcom(double ar[][], double ai[][], int n, int n1, int n2)
{
  int i,j,k;
  double s,r,arij,aiij;

  k = 0;
  for (j=n1; j<=n2; j++) {
    s=0.0;
    for (i=1; i<=n; i++) {
      arij=ar[i][j];
      aiij=ai[i][j];
      r=arij*arij+aiij*aiij;
      if (r > s) {
        s=r;
```

```
      k=i;
    }
  }
  if (s != 0.0) comcolcst(1,n,j,ar,ai,ar[k][j]/s,-ai[k][j]/s);
  }
}
```

1.18 Complex monadic operations

A. comabs

Computes the modulus of a complex number: $\rho = |xr+xi*i|$.

Procedure parameters:
$$\text{double comabs } (xr, xi)$$
comabs: double; delivers the modulus of the complex number, value of ρ above;
xr,xi: double; the real part and imaginary part of the complex number, respectively.

```
public static double comabs(double xr, double xi)
{
  double temp;

  xr=Math.abs(xr);
  xi=Math.abs(xi);
  if (xi > xr) {
    temp=xr/xi;
    return (Math.sqrt(temp*temp+1.0)*xi);
  }
  if (xi == 0.0)
    return (xr);
  else {
    temp=xi/xr;
    return (Math.sqrt(temp*temp+1.0)*xr);
  }
}
```

B. comsqrt

Forms the square root of a complex number.

Procedure parameters:

$$\text{void comsqrt } (ar, ai, pr, pi)$$

ar,ai: double;

       entry: the real part and imaginary part of the complex number, respectively;

pr,pi: double *pr[0:0], pi[0:0]*;

       exit: the real part and imaginary part of the square root, respectively.

```
public static void comsqrt(double ar, double ai, double pr[], double pi[])
{
  double br,bi,h,temp;

  if (ar == 0.0 && ai == 0.0)
    pr[0] = pi[0] = 0.0;
  else {
    br=Math.abs(ar);
    bi=Math.abs(ai);
    if (bi < br) {
      temp=bi/br;
      if (br < 1.0)
        h=Math.sqrt((Math.sqrt(temp*temp+1.0)*0.5+0.5)*br);
      else
        h=Math.sqrt((Math.sqrt(temp*temp+1.0)*0.125+0.125)*br)*2;
    } else {
      if (bi < 1.0) {
        temp=br/bi;
        h=Math.sqrt((Math.sqrt(temp*temp+1.0)*bi+br)*2)*0.5;
      } else {
        if (br+1.0 == 1.0)
          h=Math.sqrt(bi*0.5);
        else {
          temp=br/bi;
          h=Math.sqrt(Math.sqrt(temp*temp+1.0)*bi*0.125+br*0.125)*2;
        }
      }
    }
    if (ar >= 0.0) {
      pr[0]=h;
      pi[0]=ai/h*0.5;
    } else {
      pi[0] = (ai >= 0.0) ? h : -h;
      pr[0] = bi/h*0.5;
    }
  }
}
```

C. carpol

Determines the polar form of a complex number: obtains r, $c=\cos\varphi$, and $s=\sin\varphi$ where

$$ar+ai*\text{i}=re^{\text{i}\varphi}.$$

Procedure parameters:
$$\text{void carpol } (ar,ai,r,c,s)$$
ar,ai: double;

         entry: the real part and imaginary part of the complex number, respectively;

r,c,s: double $r[0:0]$, $c[0:0]$, $s[0:0]$;

         exit: the modulus of the complex number is delivered in r and the cosine and the sine of the argument are delivered in c and s, respectively; when $ar=ai=0$ then $c=1$ and $r=s=0$.

```
public static void carpol(double ar, double ai, double r[], double c[], double s[])
{
  double temp;

  if (ar == 0.0 && ai == 0.0) {
    c[0] = 1.0;
    r[0] = s[0] = 0.0;
  } else {
    if (Math.abs(ar) > Math.abs(ai)) {
      temp=ai/ar;
      r[0] = Math.abs(ar)*Math.sqrt(1.0+temp*temp);
    } else {
      temp=ar/ai;
      r[0] = Math.abs(ai)*Math.sqrt(1.0+temp*temp);
    }
    c[0] = ar / r[0];
    s[0] = ai / r[0];
  }
}
```

1.19 Complex dyadic operations

A. commul

Forms the product of two complex numbers: $rr+ri*\text{i} = (ar+ai*\text{i})(br+bi*\text{i})$.

Procedure parameters:
$$\text{void commul } (ar,ai,br,bi,rr,ri)$$
ar,ai: double;

entry: the real part and imaginary part of the first complex number, respectively;

br,bi: double;

entry: the real part and imaginary part of the second complex number, respectively;

rr,ri: double *rr[0:0],ri[0:0]*;

exit: the real part and imaginary part of the product of the two complex numbers, respectively.

```
public static void commul(double ar, double ai, double br, double bi, double rr[],
                         double ri[])
{
  rr[0]=ar*br-ai*bi;
  ri[0]=ar*bi+ai*br;
}
```

B. comdiv

Forms the quotient of two complex numbers $zr+zi^*i = (xr+xi^*i)/(yr+yi^*i)$. It is assumed that $yr+yi^*i \neq 0$.

Procedure parameters:
$$\text{void comdiv } (xr,xi,yr,yi,zr,zi)$$

xr,xi: double;

entry: the real part and imaginary part of the first complex number, respectively;

yr,yi: double;

entry: the real part and imaginary part of the second complex number, respectively;

zr,zi: double *zr[0:0],zi[0:0]*;

exit: the real part and imaginary part of the quotient of the two complex numbers, respectively.

```
public static void comdiv(double xr, double xi, double yr, double yi, double zr[],
                         double zi[])
{
  double h,d;

  if (Math.abs(yi) < Math.abs(yr)) {
    if (yi == 0.0) {
      zr[0]=xr/yr;
      zi[0]=xi/yr;
    } else {
      h=yi/yr;
      d=h*yi+yr;
      zr[0]=(xr+h*xi)/d;
      zi[0]=(xi-h*xr)/d;
```

```
    }
  } else {
    h=yr/yi;
    d=h*yr+yi;
    zr[0]=(xr*h+xi)/d;
    zi[0]=(xi*h-xr)/d;
  }
}
```

1.20 *Long integer arithmetic*

A. Ingintadd

Forms the sum of two multilength integers, each expressed in the form

$$\sum_{j=1}^{i_0} i_j B^{j-1}$$

where the i_j are single length nonnegative integers, and B is a single length positive integer:

$$\sum_{j=1}^{s_0} s_j B^{j-1} = \sum_{j=1}^{u_0} u_j B^{j-1} + \sum_{j=1}^{v_0} v_j B^{j-1},$$

$s_0 \leq \max\{u_0,v_0\}+1$; B^2+B not greater than the largest integer having a machine representation.

Procedure parameters:

$$\text{void lngintadd } (u,v,sum)$$

u,v: int $u[0{:}u[0]]$, $v[0{:}v[0]]$;
           entry: the long integers to be added, values of u and v above;
sum: int $sum[0{:}\max(u[0],v[0])+1]$;
           exit: the multilength sum of u and v, while u and v remain unchanged.

Remark: See the procedure *lngintpower*.

```
static final int BASE = 100;    /* value of B in the above */

public static void lngintadd(int u[], int v[], int sum[])
{
  int lu,lv,diff,carry,i,t,max;

  lu=u[0];
  lv=v[0];
  if (lu >= lv) {
    max=lu;
    diff=lu-lv+1;
```

```
      carry=0;
      for (i=lu; i>=diff; i--) {
        t=u[i]+v[i-diff+1]+carry;
        carry = (t < BASE) ? 0 : 1;
        sum[i]=t-carry*BASE;
      }
      for (i=diff-1; i>=1; i--) {
        t=u[i]+carry;
        carry = (t < BASE) ? 0 : 1;
        sum[i]=t-carry*BASE;
      }
    } else {
      max=lv;
      diff=lv-lu+1;
      carry=0;
      for (i=lv; i>=diff; i--) {
        t=v[i]+u[i-diff+1]+carry;
        carry = (t < BASE) ? 0 : 1;
        sum[i]=t-carry*BASE;
      }
      for (i=diff-1; i>=1; i--) {
        t=v[i]+carry;
        carry = (t < BASE) ? 0 : 1;
        sum[i]=t-carry*BASE;
      }
    }
    if (carry == 1) {
      for (i=max; i>=1; i--) sum[i+1]=sum[i];
      sum[1]=1;
      max=max+1;
    }
    sum[0]=max;
}
```

B. Ingintsubtract

Forms the difference of two multilength integers, each expressed in the form

$$\sum_{j=1}^{i_0} i_j B^{j-1}$$

where the i_j are single length nonnegative integers, and B is a single length positive integer:

$$\sum_{j=1}^{d_0} d_j B^{j-1} = \sum_{j=1}^{u_0} u_j B^{j-1} - \sum_{j=1}^{v_0} v_j B^{j-1},$$

$d_0 \leq u_0$; B^2+B not greater than the largest integer having a machine representation.

Procedure parameters:

$$\text{void lngintsubtract } (u,v,\textit{difference})$$

u,v: int $u[0:u[0]]$, $v[0:v[0]]$;

                    entry: the long integers to be subtracted, values of u and v above;

$\textit{difference}$:   int $\textit{difference}[0:u[0]]$;

                    exit: the multilength difference $u-v$, if $u<v$ then $\textit{difference}[0]=0$;
                              u and v remain unchanged.

Remark: See the procedure *lngintpower*.

```
static final int BASE = 100;    /* value of B in the above */

public static void lngintsubtract(int u[], int v[], int difference[])
{
  int lu,lv,diff,i,t,j,carry;

  lu=u[0];
  lv=v[0];
  if ((lu < lv) || ((lu == lv) && (u[1] < v[1]))) {
    difference[0]=0;
    return;
  }
  diff=lu-lv+1;
  carry=0;
  for (i=lu; i>=diff; i--) {
    t=u[i]-v[i-diff+1]+carry;
    carry = (t < 0) ? -1 : 0;
    difference[i]=t-carry*BASE;
  }
  for (i=diff-1; i>=1; i--) {
    t=u[i]+carry;
    carry = (t < 0) ? -1 : 0;
    difference[i]=t-carry*BASE;
  }
  if (carry == -1) {
    difference[0]=0;
    return;
  }
  i=1;
  j=lu;
  while ((difference[i] == 0) && (j > 1)) {
    j--;
    i++;
  }
  difference[0]=j;
  if (j < lu)
    for (i=1; i<=j; i++) difference[i]=difference[lu+i-j];
}
```

C. Ingintmult

Forms the product of two multilength integers, each expressed in the form

$$\sum_{j=1}^{i_0} i_j B^{j-1}$$

where the i_j are single length nonnegative integers, and B is a single length positive integer:

$$\sum_{j=1}^{p_0} p_j\, B^{j-1} = \left(\sum_{j=1}^{u_0} u_j\, B^{j-1} \right) \left(\sum_{j=1}^{v_0} v_j\, B^{j-1} \right)$$

$p_0 \le u_0 + v_0$; $B^2 + B$ not greater than the largest integer having a machine representation.

Procedure parameters:
$$\text{void lngintmult } (u,v,product)$$

u,v: int $u[0:u[0]]$, $v[0:v[0]]$;
           entry: the long integers to be multiplied, values of u and v above;

product: int *product[0:u[0]+v[0]]*;
           exit: the multilength product of u and v, while u and v remain unchanged.

Remark: See the procedure *lngintpower*.

```
static final int BASE = 100;   /* value of B in the above */

public static void lngintmult(int u[], int v[], int product[])
{
  int lu,lv,luv,i,j,carry,t;

  lu=u[0];
  lv=v[0];
  luv=lu+lv;
  for (i=lu+1; i<=luv; i++) product[i]=0;
  for (j=lu; j>=1; j--) {
    carry=0;
    for (i=lv; i>=1; i--) {
      t=u[j]*v[i]+product[j+i]+carry;
      carry=t/BASE;
      product[j+i]=t-carry*BASE;
    }
    product[j]=carry;
  }
  if (product[1] == 0) {
    for (i=2; i<=luv; i++) product[i-1]=product[i];
    luv--;
  }
  product[0]=luv;
}
```

D. lngintdivide

Forms the quotient and remainder from two multilength integers, each expressed in the form

$$\sum_{j=1}^{i_0} i_j B^{j-1}$$

where the i_j are single length nonnegative integers, and B is a single length positive integer:

$$\sum_{j=1}^{u_0} u_j B^{j-1} = \left(\sum_{j=1}^{q_0} q_j B^{j-1}\right)\left(\sum_{j=1}^{v_0} v_j B^{j-1}\right) + \sum_{j=1}^{r_0} r_j B^{j-1}$$

$q_0 = u_0 - v_0 + 1$; $r_0 \le v_0$; $B^2 + B$ not greater than the largest integer having a machine representation.

Procedure parameters:

	void lngintdivide (*u,v,quotient,remainder*)	
u,v:	int *u[0:u[0]]*, *v[0:v[0]]*;	
	entry:	*u* contains the dividend, *v* the divisor ($v \ne 0$);
quotient,remainder:	int *quotient[0:u[0]-v[0]+1]*, *remainder[0:v[0]]*;	
	exit:	results of the division, *u* and *v* remain unchanged.

Remark: See the procedure *lngintpower*.

```
static final int BASE = 100;    /* value of B in the above */

public static void lngintdivide(int u[], int v[], int quotient[], int remainder[])
{
  int lu,lv,v1,diff,i,t,scale,d,q1,j,carry;

  lu=u[0];
  lv=v[0];
  v1=v[1];
  diff=lu-lv;

  if (lv == 1) {
    carry=0;
    for (i=1; i<=lu; i++) {
      t=carry*BASE+u[i];
      quotient[i]=t/v1;
      carry=t-quotient[i]*v1;
    }
    remainder[0]=1;
    remainder[1]=carry;
    if (quotient[1] == 0) {
      for (i=2; i<=lu; i++) quotient[i-1]=quotient[i];
      quotient[0]=lu - ((lu == 1) ? 0 : 1);
    } else
      quotient[0]=lu;
```

```
    return;
  }

  if (lu < lv) {
    quotient[0]=1;
    quotient[1]=0;
    for (i=0; i<=lu; i++) remainder[i]=u[i];
    return;
  }

  int uu[] = new int[lu+1];
  int a[] = new int[lv+1];
  for (i=0; i<=lu; i++) uu[i]=u[i];
  scale=BASE/(v1+1);
  if (scale > 1) {
    /* normalize u */
    carry=0;
    for (i=lu; i>=1; i--) {
      t=scale*uu[i]+carry;
      carry=t/BASE;
      uu[i]=t-carry*BASE;
    }
    uu[0]=carry;
    /* normalize v */
    carry=0;
    for (i=lv; i>=1; i--) {
      t=scale*v[i]+carry;
      carry=t/BASE;
      v[i]=t-carry*BASE;
    }
    v1=v[1];
  } else
    uu[0]=0;

  /* compute quotient and remainder */
  for (i=0; i<=diff; i++) {
    d=uu[i]*BASE+uu[i+1];
    q1 = (uu[i] == v1) ? BASE-1 : d/v1;
    if (v[2]*q1 > (d-q1*v1)*BASE+uu[i+2]) {
      q1--;
      if (v[2]*q1 > (d-q1*v1)*BASE+uu[i+2]) q1--;
    }
    /* uu[i:i+lv]=u[i:i+lv]-q1*v[1:lv] */
    carry=0;
    for (j=lv; j>=1; j--) {
      t=q1*v[j]+carry;
      carry=t/BASE;
      a[j]=t-carry*BASE;
    }
    a[0]=carry;
```

```
    carry=0;
    for (j=lv; j>=0; j--) {
      t=uu[i+j]-a[j]+carry;
      carry = (t < 0) ? -1 : 0;
      uu[i+j]=t-carry*BASE;
    }
    /* if carry=-1 then q1 is one too large,
       and v must be added back to uu[i:i+lv] */
    if (carry == -1) {
      q1--;
      carry=0;
      for (j=lv; j>=1; j--) {
        t=uu[i+j]+v[j]+carry;
        carry = (t < BASE) ? 0 :1;
        uu[i+j]=t-carry*BASE;
      }
    }
    quotient[i]=q1;
  }

/* correct storage of quotient */
if (quotient[0] != 0) {
  for (i=diff; i>=0; i--) quotient[i+1]=quotient[i];
  quotient[0]=diff+1;
} else
  if (quotient[1] != 0)
    quotient[0]=diff;
  else {
    for (i=1; i<=diff-1; i++) quotient[i]=quotient[i+1];
    quotient[0]=diff-1;
  }

/* remainder=uu[diff+1:lu]/scale */
if (scale > 1) {
  carry=0;
  for (i=1; i<=lv; i++) {
    t=carry*BASE+uu[diff+i];
    remainder[i]=t/scale;
    carry=t-remainder[i]*scale;
  }
} else
  for (i=1; i<=lv; i++) remainder[i]=uu[diff+i];

/* correct storage of remainder */
i=1;
j=lv;
while (remainder[i] == 0 && j > 1) {
  j--;
  i++;
}
```

```
  remainder[0]=j;
  if (j < lv)
    for (i=1; i<=j; i++) remainder[i]=remainder[lv+i-j];

  /* unnormalize the divisor v */
  if (scale > 1) {
    carry=0;
    for (i=1; i<=lv; i++) {
      t=carry*BASE+v[i];
      v[i]=t/scale;
      carry=t-v[i]*scale;
    }
  }
}
```

E. Ingintpower

Forms the single length positive integer power of a multilength integer expressed in the form

$$\sum_{j=1}^{i_0} i_j B^{j-1}$$

where the i_j are single length nonnegative integers, and B is a single length positive integer:

$$\sum_{j=1}^{r_0} r_j B^{j-1} = \left(\sum_{j=1}^{u_0} u_j B^{j-1} \right)^e$$

$r_0 \leq eu_0$; B^2+B not greater than the largest integer having a machine representation.

Procedure parameters:
$$\text{void lngintpower } (u, exponent, result)$$

u: int *u[0:u[0]]*;
              entry: *u* must contain the long integer which has to be raised to the power exponent;
exponent: int;
              entry: the (positive) power to which the long integer *u* will be raised;
result: int *result[0:u[0]*exponent]*;
              exit: contains the value of *u* raised to the power of *exponent*, *u* remains unchanged.

Procedure used: lngintmult.

Remark: For the method of the procedures *lngintadd, lngintsubtract, lngintmult,* and *lngintdivide*, see [K69]; the procedure *lngintpower* uses the binary method for exponentiation [K69].

```
public static void lngintpower(int u[], int exponent, int result[])
{
  int max,i,n,exp;

  exp=exponent;
  max=u[0]*exp;
  int y[] = new int[max+1];
  int z[] = new int[max+1];
  int h[] = new int[max+1];

  y[0]=y[1]=1;
  for (i=u[0]; i>=0; i--) z[i]=u[i];
  for (;;) {
    n=exp/2;
    if (n+n != exp) {
      lngintmult(y,z,h);
      for (i=h[0]; i>=0; i--) y[i]=h[i];
      if (n == 0) {
        for (i=y[0]; i>=0; i--) result[i]=y[i];
        return;
      }
    }
    lngintmult(z,z,h);
    for (i=h[0]; i>=0; i--) z[i]=h[i];
    exp=n;
  }
}
```

2. Algebraic Evaluations

All procedures in this chapter are class methods of the following *Algebraic_eval* class.

```
package numal;

import numal.*;

public class Algebraic_eval extends Object {
   // all procedures in this chapter are to be inserted here
}
```

2.1 Evaluation of polynomials in Grunert form

A. pol

Computes the sum of the polynomial

$$p(x) = \sum_{i=0}^{n} a_i x^i$$

using Horner's rule. The error growth is given by a linear function of the degree of the polynomial [Wi63].

Procedure parameters:
$$\text{double pol } (n,x,a)$$
pol: given the value of *p(x)* above;
n: int; the degree of the polynomial;
x: double; the argument of the polynomial;
a: double *a[0:n]*;
           entry: the coefficients of the polynomial.

```
public static double pol(int n, double x, double a[])
{
  double r;

  r=0.0;
  for (; n>=0; n--) r=r*x+a[n];
  return (r);
}
```

B. taypol

Computes the values of the terms $x^j D^j p(x)/j!$ $(j=0,...,k \leq n)$ where

$$p(x) = \sum_{i=0}^{n} a_i x^i .$$

Procedure parameters:

$$\text{void taypol } (n,k,x,a)$$

n: int; the degree of the polynomial;

k: int; the first k terms of the above are to be calculated;

x: double; the argument of the polynomial;

a: double $a[0:n]$;

     entry: the coefficients of the polynomial;

     exit: the j-th term $x^j *(j$-th derivative$)/j!$ is delivered in $a[j]$, $j=0,1,...,k \leq n$, the other elements of a are generally altered.

Remark: The method of evaluation is given in [ShT74]. The more sophisticated algorithm based on divisors of n+1 in [ShT74] was not implemented because of the more complex appearance of the implementation and because of the difficulty in choosing the most efficient divisor. In this implementation of the one-parameter family of algorithms, the linear number of multiplications is preserved. See [Wo74] for the k-th normalized derivative.

```
public static void taypol(int n, int k, double x, double a[])
{
  int i,j,nm1;
  double xj,aa,h;

  if (x != 0.0) {
    xj=1;
    for (j=1; j<=n; j++) {
      xj *= x;
      a[j] *= xj;
    }
    aa=a[n];
    nm1=n-1;
    for (j=0; j<=k; j++) {
      h=aa;
      for (i=nm1; i>=j; i--) h = a[i] += h;
    }
  } else {
    for (; k>=1; n--) a[k]=0;
  }
}
```

C. norderpol

Computes the first k normalized derivatives $D^j p(x)/j!$, $(j=0,...,k \leq n)$ of the polynomial

$$p(x) = \sum_{i=0}^{n} a_i x^i .$$

Procedure parameters:
void norderpol (n,k,x,a)

n: int; the degree of the polynomial;
k: int; the first k normalized derivatives (j-th derivative / j factorial) are to be
          calculated;
x: double; the argument of the polynomial;
a: double $a[0:n]$;
          entry: the coefficients of the polynomial;
          exit: the j-th normalized derivative is delivered in $a[j]$, $j=0,1,...,k \leq n$, the
                  other elements of a are generally altered.

Remark: See the procedure *taypol*.

```
public static void norderpol(int n, int k, double x, double a[])
{
  int i,j,nm1;
  double xj,aa,h;

  if (x != 0.0) {
    double xx[] = new double[n+1];
    xj=1;
    for (j=1; j<=n; j++) {
      xx[j] = xj *= x;
      a[j] *= xj;
    }
    h=aa=a[n];
    nm1=n-1;
    for (i=nm1; i>=0; i--) h = a[i] += h;
    for (j=1; j<=k; j++) {
      h=aa;
      for (i=nm1; i>=j; i--) h = a[i] += h;
      a[j]=h/xx[j];
    }
  }
}
```

D. derpol

Computes the first k derivatives $D^j p(x)$ $(j=0,...,k \leq n)$ of the polynomial

$$p(x) = \sum_{i=0}^{n} a_i x^i .$$

Procedure parameters:

$$\text{void derpol } (n,k,x,a)$$

n: int; the degree of the polynomial;

k: int; the first k derivatives are to be calculated;

x: double; the argument of the polynomial;

a: double $a[0:n]$;

      entry: the coefficients of the polynomial;

      exit: the j-th derivative is delivered in $a[j]$, $j=0,1,...,k \leq n$, the other elements of a are generally altered.

Procedure used: norderpol.

Remark: See the procedure *taypol*.

```
public static void derpol(int n, int k, double x, double a[])
{
  int j;
  double fac;

  fac=1.0;
  norderpol(n,k,x,a);
  for (j=2; j<=k; j++) {
    fac *= j;
    a[j] *=fac;
  }
}
```

2.2 *Evaluation of general orthogonal polynomials*

This section contains six procedures for evaluating orthogonal polynomials [AbS65, Lu69]. The procedures ending with *sym* are versions for symmetric polynomials.

A. ortpol

Evaluates the single polynomial $p_n(x)$, where

$$p_0(x) = 1, \quad p_1(x) = x - b_0,$$
$$p_{k+1}(x) = (x - b_k)p_k(x) - c_k p_{k-1}(x), \quad k=1,...,n-1$$

with given b_k and c_k.

Procedure parameters:

                              double ortpol (n,x,b,c)
ortpol: given the value of the orthogonal polynomial in the above;
n: int; the degree of the polynomial;
x: double; the argument of the orthogonal polynomial;
b,c: double *b[0:n-1], c[1:n-1]*;
            entry: the recurrence coefficients of the polynomial.

```
public static double ortpol(int n, double x, double b[], double c[])
{
  int k,l;
  double r,s,h;

  if (n == 0) return (1.0);
  r=x-b[0];
  s=1.0;
  l=n-1;
  for (k=1; k<=l; k++) {
    h=r;
    r=(x-b[k])*r-c[k]*s;
    s=h;
  }
  return (r);
}
```

B. ortpolsym

Given the c_k and x, evaluates the single polynomial $p_n(x)$, where
$$p_0(x) = 1, \quad p_1(x) = x,$$
$$p_{k+1}(x) = xp_k(x) - c_kp_{k-1}(x), \quad k=1,...,n-1.$$

Procedure parameters:
                              double ortpolsym (n,x,c)
ortpolsym: given the value of the orthogonal polynomial in the above;
n: int; the degree of the polynomial;
x: double; the argument of the orthogonal polynomial;
c: double *c[1:n-1]*;
               entry: the recurrence coefficients of the polynomial.

```
public static double ortpolsym(int n, double x, double c[])
{
  int k,l;
  double r,s,h;

  if (n == 0) return (1.0);
  r=x;
  s=1.0;
  l=n-1;
```

```
  for (k=1; k<=1; k++) {
    h=r;
    r=x*r-c[k]*s;
    s=h;
  }
  return (r);
}
```

C. allortpol

Evaluates the sequence of polynomials $p_k(x)$ for $k=0,...,n$, where
$$p_0(x) = 1, \quad p_1(x) = x-b_0,$$
$$p_{k+1}(x) = (x-b_k)p_k(x) - c_k p_{k-1}(x), \quad k=1,...,n-1$$
with given b_k and c_k.

Procedure parameters:
$$\text{void allortpol } (n,x,b,c,p)$$
n: int; the degree of the polynomial;
x: double; the argument of the orthogonal polynomial;
b,c: double $b[0:n-1]$, $c[1:n-1]$;
          entry: the recurrence coefficients of the polynomial;
p: double $p[0:n]$;
          exit: $p[k]$ contains, for the argument, the value of the k-th orthogonal
          polynomial as defined by the recurrence coefficients.

```
public static void allortpol(int n, double x, double b[], double c[], double p[])
{
  int k,k1;
  double r,s,h;

  if (n == 0) {
    p[0]=1.0;
    return;
  }
  r=p[1]=x-b[0];
  s=p[0]=1.0;
  k=1;
  for (k1=2; k1<=n; k1++) {
    h=r;
    p[k1]=r=(x-b[k])*r-c[k]*s;
    s=h;
    k=k1;
  }
}
```

D. allortpolsym

Given the c_k and x, evaluates the sequence of polynomials $p_k(x)$ for $k=0,...,n$, where
$$p_0(x) = 1, \quad p_1(x) = x,$$
$$p_{k+1}(x) = xp_k(x) - c_kp_{k-1}(x), \quad k=1,...,n-1.$$

Procedure parameters:
$$\text{void allortpolsym } (n,x,c,p)$$
n: int; the degree of the polynomial;
x: double; the argument of the orthogonal polynomial;
c: double $c[1:n-1]$;
       entry: the recurrence coefficients of the polynomial;
p: double $p[0:n]$;
       exit: $p[k]$ contains, for the argument, the value of the k-th orthogonal
           polynomial as defined by the recurrence coefficients.

```
public static void allortpolsym(int n, double x, double c[], double p[])
{
  int k;
  double r,s,h;

  if (n == 0) {
    p[0]=1.0;
    return;
  }
  r=p[1]=x;
  s=p[0]=1.0;
  for (k=2; k<=n; k++) {
    h=r;
    p[k]=r=x*r-c[k-1]*s;
    s=h;
  }
}
```

E. sumortpol

Evaluates the sum

$$s = \sum_{k=0}^{n} a_k p_k(x)$$

where $p_0(x) = 1, \quad p_1(x) = x-b_0,$
$$p_{k+1}(x) = (x-b_k)p_k(x) - c_kp_{k-1}(x), \quad k=1,...,n-1,$$
with given a_k, b_k, and c_k.

Procedure parameters:
$$\text{double sumortpol } (n,x,b,c,a)$$
sumortpol: given the value of the sum s in the above;

n:	int; the degree of the polynomial;
x:	double; the argument of the orthogonal polynomial;
b,c:	double *b[0:n-1], c[1:n-1]*;
	entry: the recurrence coefficients of the polynomial;
a:	double *a[0:n]*;
	entry: the value of a_k in the above.

```
public static double sumortpol(int n, double x, double b[], double c[], double a[])
{
  int k;
  double h,r,s;

  if (n == 0) return (a[0]);
  r=a[n];
  s=0.0;
  for (k=n-1; k>=1; k--) {
    h=r;
    r=a[k]+(x-b[k])*r+s;
    s = -c[k]*h;
  }
  return (a[0]+(x-b[0])*r+s);
}
```

F. sumortpolsym

Given the a_k, c_k, and x, evaluates the sum

$$s = \sum_{k=0}^{n} a_k p_k(x)$$

where $p_0(x) = 1,\quad p_1(x) = x,$
$p_{k+1}(x) = xp_k(x) - c_k p_{k-1}(x),\quad k=1,...,n-1.$

Procedure parameters:

	double sumortpolsym (*n,x,c,a*)
sumortpolsym:	given the value of the sum *s* in the above;
n:	int; the degree of the polynomial;
x:	double; the argument of the orthogonal polynomial;
c:	double *c[1:n-1]*;
	entry: the recurrence coefficients of the polynomial;
a:	double *a[0:n]*;
	entry: the values of a_k in the above.

```
public static double sumortpolsym(int n, double x, double c[], double a[])
{
  int k;
  double h,r,s;
```

```
if (n == 0) return (a[0]);
r=a[n];
s=0.0;
for (k=n-1; k>=1; k--) {
  h=r;
  r=a[k]+x*r+s;
  s = -c[k]*h;
}
return (a[0]+x*r+s);
}
```

2.3 Evaluation of Chebyshev polynomials

This section contains four procedures for evaluating Chebyshev polynomials of the first kind [FoP68, Riv74].

A. chepolsum

Uses the Clenshaw or generalized Horner algorithm [Cle62] to evaluate a sum of Chebyshev polynomials

$$S = \sum_{i=0}^{n} a_i T_i(x)$$

where $T_i(x)$ is a Chebyshev polynomial, by use of the backward recursion
$$S_{n+1} = 0, \quad S_n = a_n,$$
$$S_k = a_k + 2xS_{k+1} - S_{k+2}, \quad k=n-2,n-3,...,1;$$
$$S = a_0 + S_1 x - S_2$$
for $n \geq 2$.

Procedure parameters:
                        double chepolsum (n,x,a)
chepolsum: given the value of the Chebyshev sum S in the above;
n: int; the degree of the polynomial represented by the Chebyshev sum, $n \geq 0$;
x: double; the argument of the Chebyshev polynomials, $|x| \leq 1$;
a: double $a[0:n]$;
               entry: $a[k]$ is the coefficient of the Chebyshev polynomial of degree k, $0 \leq k \leq n$.

```
public static double chepolsum(int n, double x, double a[])
{
  int k;
  double h,r,s,tx;

  if (n == 0) return (a[0]);
  if (n == 1) return (a[0]+a[1]*x);
```

```
tx=x+x;
r=a[n];
h=a[n-1]+r*tx;
for (k=n-2; k>=1; k--) {
   s=r;
   r=h;
   h=a[k]+r*tx-s;
}
return (a[0]-r+h*x);
}
```

B. oddchepolsum

Given the a_i and x, uses the Clenshaw or generalized Horner algorithm [Cle62] to evaluate a sum of Chebyshev polynomials of odd degree

$$S = \sum_{i=0}^{n} a_i T_{2i+1}(x) ;$$

the T_i being Chebyshev polynomials.

Procedure parameters:
                        double oddchepolsum (n,x,a)

oddchepolsum: given the value of the Chebyshev sum S in the above;

n: int; the degree of the polynomial represented by the Chebyshev sum is $2n+1$, $n{\geq}0$;

x: double; the argument of the Chebyshev polynomials, $|x|{\leq}1$;

a: double $a[0{:}n]$;
                    entry: $a[k]$ is the coefficient of the Chebyshev polynomial of degree $2k+1$, $0{\leq}k{\leq}n$.

```
public static double oddchepolsum(int n, double x, double a[])
{
   int k;
   double h,r,s,y;

   if (n == 0) return (x*a[0]);
   if (n == 1) return (x*(a[0]+a[1]*(4.0*x*x-3.0)));
   y=4.0*x*x-2.0;
   r=a[n];
   h=a[n-1]+r*y;
   for (k=n-2; k>=0; k--) {
      s=r;
      r=h;
      h=a[k]+r*y-s;
   }
   return (x*(h-r));
}
```

C. chepol

Computes the value of the Chebyshev polynomial $T_n(x)$ by use of the recursion [St72]

$$T_0(x) = 1, \quad T_1(x) = x,$$
$$T_i(x) = 2xT_{i-1}(x) - T_{i-2}(x), \quad i=2,...,n.$$

Procedure parameters:

double chepol (n,x)

chepol: given the value of the Chebyshev polynomial of the first kind, $T_n(x)$ above;

n: int; the degree of the polynomial, $n \geq 0$;

x: double; the argument of the Chebyshev polynomial, $|x| \leq 1$.

```
public static double chepol(int n, double x)
{
  int i;
  double t1,t2,h,x2;

  if (n == 0) return (1.0);
  if (n == 1) return (x);
  t2=x;
  t1=1.0;
  x2=x+x;
  h=0.0;
  for (i=2; i<=n; i++) {
    h=x2*t2-t1;
    t1=t2;
    t2=h;;
  }
  return (h);
}
```

D. allchepol

Computes the values of the Chebyshev polynomials $T_i(x)$, $i=1,...,n$, by use of the recursion [St72]

$$T_0(x) = 1, \quad T_1(x) = x,$$
$$T_i(x) = 2xT_{i-1}(x) - T_{i-2}(x), \quad i=2,...,n.$$

Procedure parameters:

void allchepol (n,x,t)

n: int; the degree of the last polynomial, $n \geq 0$;

x: double; the argument of the Chebyshev polynomials, $|x| \leq 1$;

t: double $t[0:n]$;

         exit: the value of the Chebyshev polynomial of the first kind of degree k, for the argument x, is delivered in $t[k]$, $k=0,...,n$.

```
public static void allchepol(int n, double x, double t[])
{
  int i;
  double t1,t2,h,x2;

  if (n == 0) {
    t[0]=1.0;
    return;
  }
  if (n == 1) {
    t[0]=1.0;
    t[1]=x;
    return;
  }
  t[0]=t1=1.0;
  t[1]=t2=x;
  x2=x+x;
  for (i=2; i<=n; i++) {
    t[i]=h=x2*t2-t1;
    t1=t2;
    t2=h;;
  }
}
```

2.4 Evaluation of Fourier series

A. sinser

Computes the value of a sine series

$$s = \sum_{j=1}^{n} b_j \sin(j\theta)$$

where θ and the b_j are real. When $-\frac{1}{2} \leq \cos(\theta) \leq \frac{1}{2}$, the recursion

$$u_{n+2} = u_{n+1} = 0,$$
$$u_k = 2\cos(\theta)u_{k+1} - u_{k+2} + b_k, \quad k=n,\dots,1$$

is used. When $\cos(\theta) < -\frac{1}{2}$, the equivalent recursion

$$u_{n+1} = d_{n+1} = 0,$$
$$d_k = 2(1 + \cos(\theta))u_{k+1} - d_{k+1} + b_k, \quad k=n,\dots,1$$
$$u_k = d_k - u_{k+1},$$

and when $\cos(\theta) > \frac{1}{2}$, the further recursion

$$u_{n+1} = e_{n+1} = 0,$$
$$e_k = 2(\cos(\theta) - 1)u_{k+1} - e_{k+1} + b_k, \quad k=n,\dots,1$$
$$u_k = e_k - u_{k+1}$$

are used. In each case $s = u_1 \sin(\theta)$.

Procedure parameters:

$$\text{double sinser } (n,theta,b)$$

sinser: given the value of the sine series, the value of *s* above;
n: int;
           entry: the number of terms in the sine series;
theta: double;
           entry: the argument of the sine series;
b: double *b[1:n]*;
           entry: the coefficients of the sine series.

Remark: See the procedure *comfouser2*.

```
public static double sinser(int n, double theta, double b[])
{
  int k;
  double c,cc,lambda,h,dun,un,un1,temp;

  c=Math.cos(theta);
  if (c < -0.5) {
    temp=Math.cos(theta/2.0);
    lambda=4.0*temp*temp;
    un=dun=0.0;
    for (k=n; k>=1; k--) {
      dun=lambda*un-dun+b[k];
      un=dun-un;
    }
  } else {
    if (c > 0.5) {
      temp=Math.sin(theta/2.0);
      lambda = -4.0*temp*temp;
      un=dun=0.0;
      for (k=n; k>=1; k--) {
        dun += lambda*un+b[k];
        un += dun;
      }
    } else {
      cc=c+c;
      un=un1=0.0;
      for (k=n; k>=1; k--) {
        h=cc*un-un1+b[k];
        un1=un;
        un=h;
      }
    }
  }
  return (un*Math.sin(theta));
}
```

B. cosser

Computes the value of a cosine series

$$s = \sum_{j=0}^{n} a_j \cos(j\theta)$$

where θ and the a_j are real. The method used is based upon recursion of the documentation to *sinser*, where now $s = a_0 + 2u_1\cos(\theta) - u_2$, and upon further recursions similar to those given in that documentation.

Procedure parameters:

$$\text{double cosser } (n,theta,a)$$

cosser: given the value of the cosine series, value of s above;

n: int;

            entry: the degree of the trigonometric polynomial;

theta: double;

            entry: the argument of the cosine series;

a: double $a[0:n]$;

            entry: the coefficients of the cosine series.

Remark: See the procedure *comfouser2*.

```
public static double cosser(int n, double theta, double a[])
{
  int k;
  double c,cc,lambda,h,dun,un,un1,temp;

  c=Math.cos(theta);
  if (c < -0.5) {
    temp=Math.cos(theta/2.0);
    lambda=4.0*temp*temp;
    un=dun=0.0;
    for (k=n; k>=0; k--) {
      un=dun-un;
      dun=lambda*un-dun+a[k];
    }
    return (dun-lambda/2.0*un);
  } else {
    if (c > 0.5) {
      temp=Math.sin(theta/2.0);
      lambda = -4.0*temp*temp;
      un=dun=0.0;
      for (k=n; k>=0; k--) {
        un += dun;
        dun += lambda*un+a[k];
      }
      return (dun-lambda/2.0*un);
    } else {
      cc=c+c;
```

```
    un=un1=0.0;
    for (k=n; k>=1; k--) {
      h=cc*un-un1+a[k];
      un1=un;
      un=h;
    }
    return (a[0]+un*c-un1);
  }
 }
}
```

C. fouser

Computes the value of

$$s = \sum_{j=0}^{n} a_j \cos(j\theta) + \sum_{j=1}^{n} a_j \sin(j\theta)$$

where θ and the a_j are real, by methods similar to those described in the documentations to *sinser* and *cosser*.

Procedure parameters:

                         double fouser (*n,theta,a*)

fouser: given the value of the fourier series, value of *s* above;
n: int;
          entry: the degree of the trigonometric polynomial;
theta: double;
          entry: the argument of the fourier series;
a: double *a[0:n]*;
          entry: the coefficients of the (finite) fourier series.

```
public static double fouser(int n, double theta, double a[])
{
  int k;
  double c,cc,lambda,h,dun,un,un1,c2,s2;

  c=Math.cos(theta);
  if (c < -0.5) {
    c2=Math.cos(theta/2.0);
    lambda=4.0*c2*c2;
    un=dun=0.0;
    for (k=n; k>=0; k--) {
      un=dun-un;
      dun=lambda*un-dun+a[k];
    }
    return (dun+2.0*c2*(Math.sin(theta/2.0)-c2)*un);
  } else {
    if (c > 0.5) {
```

```
        s2=Math.sin(theta/2.0);
        lambda = -4.0*s2*s2;
        un=dun=0.0;
        for (k=n; k>=0; k--) {
            un += dun;
            dun += lambda*un+a[k];
        }
        return (dun+2.0*s2*(s2+Math.cos(theta/2.0))*un);
    } else {
        cc=c+c;
        un=un1=0.0;
        for (k=n; k>=1; k--) {
            h=cc*un-un1+a[k];
            un1=un;
            un=h;
        }
        return (a[0]-un1+(c+Math.sin(theta))*un);
    }
  }
}
```

D. fouser1

Computes the value of

$$s = \sum_{j=0}^{n} a_j \cos(j\theta) + \sum_{j=1}^{n} b_j \sin(j\theta)$$

where θ, a_j, and b_j are real, using a Horner scheme.

Procedure parameters:
                        double fouser1 (n,theta,a,b)
fouser1: given the value of the fourier series, value of s above;
n: int;
            entry: the degree of the trigonometric polynomial;
theta: double;
            entry: the argument of the fourier series;
a,b: double a[0:n], b[1:n];
            entry: the coefficients of the (finite) fourier series,
                    with a_k coefficient of cos(k*θ), k=0,...,n,
                    and b_k coefficient of sin(k*θ), k=1,...,n.

Remark: See the procedure *comfouser2*.

```
public static double fouser1(int n, double theta, double a[], double b[])
{
    int i;
    double r,s,h,co,si;
```

```
r=s=0.0;
co=Math.cos(theta);
si=Math.sin(theta);
for (i=n; i>=1; i--) {
   h=co*r+si*s+a[i];
   s=co*s-si*r+b[i];
   r=h;
}
return (co*r+si*s+a[0]);
}
```

E. fouser2

Computes the value of

$$s = \sum_{j=0}^{n} a_j \cos(j\theta) + \sum_{j=1}^{n} b_j \sin(j\theta)$$

where θ, a_j, and b_j are real, by methods similar to those described in the documentations to *sinser* and *cosser*.

Procedure parameters:

                        double fouser2 (n,theta,a,b)
fouser2:   given the value of the fourier series, value of *s* above;
n: int;
             entry: the degree of the trigonometric polynomial;
theta: double;
             entry: the argument of the fourier series;
a,b: double *a[0:n]*, *b[1:n]*;
             entry: the coefficients of the (finite) fourier series,
                      with a_k coefficient of cos(k*θ), k=0,...,n,
                      and b_k coefficient of sin(k*θ), k=1,...,n.

Procedures used: sinser, cosser.

Remark: See the procedure *comfouser2*.

```
public static double fouser2(int n, double theta, double a[], double b[])
{
   return (cosser(n,theta,a)+sinser(n,theta,b));
}
```

F. comfouser

Computes the value of

$$s = \sum_{j=0}^{n} a_j \, e^{ij\theta}$$

where θ and the a_j are real, by methods similar to those described in the documentations to *sinser* and *cosser*.

Procedure parameters:

void comfouser (*n,theta,a,rr,ri*)

n: int;
           entry: the degree of the polynomial in $e^{i\theta}$;
theta: double;
           entry: the argument of the fourier series;
a: double *a[0:n]*;
           entry: the real coefficients in the series;
rr,ri: double *rr[0:0], ri[0:0]*;
           exit: the real part and the imaginary part of *s* above are delivered in *rr[0]*
                   and *ri[0]*, respectively.

```
public static void comfouser(int n, double theta, double a[], double rr[],
                          double ri[])
{
  int k;
  double c,cc,lambda,h,dun,un,un1,temp;

  c=Math.cos(theta);
  if (c < -0.5) {
    temp=Math.cos(theta/2.0);
    lambda=4.0*temp*temp;
    un=dun=0.0;
    for (k=n; k>=0; k--) {
      un=dun-un;
      dun=lambda*un-dun+a[k];
    }
    rr[0]=dun-lambda/2.0*un;
  } else {
    if (c > 0.5) {
      temp=Math.sin(theta/2.0);
      lambda = -4.0*temp*temp;
      un=dun=0.0;
      for (k=n; k>=0; k--) {
        un += dun;
        dun += lambda*un+a[k];
      }
      rr[0]=dun-lambda/2.0*un;
    } else {
      cc=c+c;
      un=un1=0.0;
      for (k=n; k>=1; k--) {
        h=cc*un-un1+a[k];
        un1=un;
```

```
        un=h;
      }
      rr[0]=a[0]+un*c-un1;
    }
  }
  ri[0]=un*Math.sin(theta);
}
```

G. comfouser1

Computes the value of

$$s = \sum_{j=0}^{n} a_j \, e^{ij\theta}$$

where θ is real and the a_j are complex, using a Horner scheme.

Procedure parameters:
                    void comfouser1 (n,theta,ar,ai,rr,ri)
n: int;
          entry: the degree of the polynomial in $e^{i\theta}$;
theta: double;
          entry: the argument of the fourier series;
ar,ai: double ar[0:n], ai[0:n];
          entry: the real part and the imaginary part of the complex coefficients in
                  the series must be given in arrays ar and ai, respectively;
rr,ri: double rr[0:0], ri[0:0];
          exit: the real part and the imaginary part of s above are delivered in rr[0]
                  and ri[0], respectively.

Remark: See the procedure comfouser2.

```
public static void comfouser1(int n, double theta, double ar[], double ai[],
                              double rr[], double ri[])
{
  int k;
  double h,hr,hi,co,si;

  hr=hi=0.0;
  co=Math.cos(theta);
  si=Math.sin(theta);
  for (k=n; k>=1; k--) {
    h=co*hr-si*hi+ar[k];
    hi=co*hi+si*hr+ai[k];
    hr=h;
  }
  rr[0]=co*hr-si*hi+ar[0];
  ri[0]=co*hi+si*hr+ai[0];
}
```

H. comfouser2

Computes the value of

$$s = \sum_{j=0}^{n} a_j e^{ij\theta}$$

where θ is real and the a_j are complex, by methods similar to those described in the documentations to *sinser* and *cosser*.

Procedure parameters:
$$\text{void comfouser2 } (n, theta, ar, ai, rr, ri)$$

n: int;
       entry: the degree of the polynomial in $e^{i\theta}$;
theta: double;
       entry: the argument of the fourier series;
ar,ai: double *ar[0:n]*, *ai[0:n]*;
       entry: the real part and the imaginary part of the complex coefficients in the series must be given in arrays *ar* and *ai*, respectively;
rr,ri: double *rr[0:0]*, *ri[0:0]*;
       exit: the real part and the imaginary part of *s* above are delivered in *rr[0]* and *ri[0]*, respectively.

Procedure used: comfouser.

Remark: For the evaluation of a finite fourier series two algorithms are used: (a) the Horner scheme in the procedure *fouser1*, and (b) a combination of the Clenshaw algorithm [Ge69, Lu69, St72] and the modification of Reinsch [Re67, St72] in the procedures *sinser*, *cosser*, *fouser*, and *fouser2*. A modification of the idea of Newbery is not implemented because of the introduction of sine (cosine) terms in a cosine (sine) series and the inefficiency of the algorithm [N73]. For the evaluation of a finite complex fourier series, two algorithms, in real arithmetic, are used: (a) the Horner scheme in the procedure *comfouser1*, and (b) a combination of the Clenshaw algorithm and the modification of Reinsch in procedures *comfouser* and *comfouser2*. The Horner scheme is implemented because of the simplicity of the algorithm (although this algorithm is less efficient than the Goertzel-Watt-Clenshaw-Reinsch algorithm) and the stable nature of orthogonal transformations. A combination of the algorithm of Goertzel-Watt-Clensaw and the modification of Reinsch is implemented because of the efficiency of the Goertzel-Watt-Clensaw algorithm and the stability of the modification of Reinsch, especially for small values of the argument. An upper bound for the error growth is given by a linear function of the degree for both implemented algorithms.

```
public static void comfouser2(int n, double theta, double ar[], double ai[],
                         double rr[], double ri[])
{
   double car[] = new double[1];
```

```
double cai[] = new double[1];
double sar[] = new double[1];
double sai[] = new double[1];

comfouser(n,theta,ar,car,sar);
comfouser(n,theta,ai,cai,sai);
rr[0]=car[0]-sai[0];
ri[0]=cai[0]+sar[0];
}
```

2.5 Evaluation of continued fractions

jfrac

Computes the value of the convergent

$$C_n = b_0 + \frac{a_1}{b_1 +} \frac{a_2}{b_2 +} \dots \frac{a_n}{b_n}$$

by use of the recursion

$$D_1 = b_n,$$
$$D_{i+1} = b_{n-i} + a_{n-i+1}/D_i, \quad i=1,\dots,n,$$

when $C_n = D_{n+1}$.

Procedure parameters:
$$\text{double jfrac } (n,a,b)$$

jfrac: delivers the value of the terminating continued fraction, the value of C_n above;

n: int; the upper index of the arrays a and b;

a,b: double $a[1:n],b[0:n]$; the elements of the continued fraction, a_i and b_i above.

```
public static double jfrac(int n, double a[], double b[])
{
  int i;
  double d;

  d=0.0;
  for (i=n; i>=1; i--) d=a[i]/(b[i]+d);
  return (d+b[0]);
}
```

2.6 Transformation of polynomial representation

A. polchs

Given the a_k, derives the b_k occurring in the relationship

$$\sum_{k=0}^{n} a_k x^k = \sum_{k=0}^{n} b_k T_k(x) ;$$

the $T_k(x)$ being Chebyshev polynomials.

Procedure parameters:
$$\text{void polchs } (n,a)$$
n: int; the degree of the polynomial;
a: double $a[0:n]$;
         entry: the coefficients of the power sum;
         exit: the coefficients of the Chebyshev sum.

Remark: Although the transformation of representations of polynomials could
         have been obtained by fast evaluation and fast interpolation, the
         algorithm of Hamming [H73] was implemented here because of its simple
         appearance.

```
public static void polchs(int n, double a[])
{
  int k,l,twopow;

  if (n > 1) {
    twopow=2;
    for (k=1; k<=n-2; k++) {
      a[k] /= twopow;
      twopow *= 2;
    }
    a[n-1]=2.0*a[n-1]/twopow;
    a[n] /= twopow;
    a[n-2] += a[n];
    for (k=n-2; k>=1; k--) {
      a[k-1] += a[k+1];
      a[k]=2.0*a[k]+a[k+2];
      for (l=k+1; l<=n-2; l++) a[l] += a[l+2];
    }
  }
}
```

B. chspol

Given the b_k, derives the a_k occurring in the relationship

$$\sum_{k=0}^{n} b_k T_k(x) = \sum_{k=0}^{n} a_k x^k \; ;$$

the $T_k(x)$ being Chebyshev polynomials.

Procedure parameters:

$$\text{void chspol } (n, a)$$

n: int; the degree of the polynomial;
a: double $a[0{:}n]$;
          entry: the coefficients of the Chebyshev sum;
          exit: the coefficients of the power sum.

Remark: See the procedure *polchs*.

```
public static void chspol(int n, double a[])
{
  int k,l,twopow;

  if (n > 1) {
    for (k=0; k<=n-2; k++) {
      for (l=n-2; l>=k; l--) a[l]  -= a[l+2];
      a[k+1] /= 2.0;
    }
    twopow=2;
    for (k=1; k<=n-2; k++) {
      a[k]  *= twopow;
      twopow *= 2;
    }
    a[n-1] *= twopow;
    a[n]  *= twopow;
  }
}
```

C. polshtchs

Given the a_k, derives the b_k occurring in the relationship

$$\sum_{k=0}^{n} a_k x^k = \sum_{k=0}^{n} b_k S_k(x) \; ;$$

the $S_k(x)$ being shifted Chebyshev polynomials defined by $S_k(x)=T_k(2x-1)$, $T_k(x)$ being a Chebyshev polynomial.

Procedure parameters:

$$\text{void polshtchs }(n,a)$$
n: int; the degree of the polynomial;
a: double *a[0:n]*;
            entry: the coefficients of the power sum;
            exit: the coefficients of the shifted Chebyshev sum.

Procedures used: lintfmpol, polchs.

Remark: See the procedure *polchs*.

```
public static void polshtchs(int n, double a[])
{
  lintfmpol(0.5,0.5,n,a);
  polchs(n,a);
}
```

D. shtchspol

Given the b_k, derives the a_k occurring in the relationship

$$\sum_{k=0}^{n} b_k S_k(x) = \sum_{k=0}^{n} a_k x^k \; ;$$

the $S_k(x)$ being shifted Chebyshev polynomials defined by $S_k(x)=T_k(2x-1)$, and $T_k(x)$ being a Chebyshev polynomial.

Procedure parameters:
$$\text{void shtchspol }(n,a)$$
n: int; the degree of the polynomial;
a: double *a[0:n]*;
            entry: the coefficients of the shifted Chebyshev sum;
            exit: the coefficients of the power sum.

Procedures used: lintfmpol, chspol.

Remark: See the procedure *polchs*.

```
public static void shtchspol(int n, double a[])
{
  chspol(n,a);
  lintfmpol(2.0,-1.0,n,a);
}
```

E. grnnew

Given the coefficients a_i occurring in the polynomial

$$P(x) = \sum_{i=0}^{n} a_i x^i$$

and the tabulation points x_v, computes the divided differences δ_j in the equivalent Newton series representation

$$P(x) = \sum_{j=0}^{n} \left(\prod_{v=0}^{j-1} (x - x_v) \right) \delta_j.$$

Procedure parameters:

$$\text{void grnnew } (n,x,a)$$

n: int; the degree of the polynomial;

x: double $x[0:n-1]$;

     entry: the interpolation points, values of x_v above;

a: double $a[0:n]$;

     entry: the coefficients of the power sum;

     exit: the coefficients of the Newton sum, values of δ_j above.

Remark: See the procedure *polchs*.

```
public static void grnnew(int n, double x[], double a[])
{
   int k,l;

   for (k=n-1; k>=0; k--)
     for (l=n-1; l>=n-1-k; l--) a[l] += a[l+1]*x[n-1-k];
}
```

F. newgrn

Given the coefficients $\delta_j f(x_0)$, together with the values of the arguments x_i from which they are formed, in the truncated Newton interpolation series

$$f_n(x) = \sum_{j=0}^{n} \left(\prod_{v=0}^{j-1} (x - x_v) \right) \delta_j f(x_0)$$

computes the coefficients c_i, $i=0,...,n$, in the equivalent polynomial form

$$f_n(x) = \sum_{i=0}^{n} c_i x^i.$$

Procedure parameters:

$$\text{void newgrn } (n,x,a)$$

n: int; the degree of the polynomial;

x: double $x[0:n-1]$;

     entry: the interpolation points, values of x_i above;

a: double $a[0:n]$;

     entry: the coefficients of the Newton sum, values of $\delta_j f(x_0)$;

     exit: the coefficients of the power sum, values of c_i above.

Procedure used: elmvec.

Remark: See the procedure *polchs*.

```
public static void newgrn(int n, double x[], double a[])
{
  int k;

  for (k=n-1; k>=0; k--)
    Basic.elmvec(k,n-1,1,a,a,-x[k]);
}
```

G. lintfmpol

Given the a_i occurring in the polynomial expression

$$P(x) = \sum_{i=0}^{n} a_i x^i$$

and p, q, derives the b_j occurring in the equivalent expression

$$P(x) = \sum_{j=0}^{n} b_j y^j$$

where $x=py+q$.

Procedure parameters:
$$\text{void lintfmpol } (p,q,n,a)$$

p,q: double;
      entry: defining the linear transformation of the independent variable $x=py+q$; $p=0$ gives the value of the polynomial with argument q;

n: int; the degree of the polynomial;
a: double $a[0:n]$;
      entry: the coefficients of the power sum in x, values of a_i above;
      exit: the coefficients of the power sum in y, values of b_j above.

Procedure used: norderpol.

Remark: See the procedure *polchs*.

```
public static void lintfmpol(double p, double q, int n, double a[])
{
  int k;
  double ppower;

  norderpol(n,n,q,a);
  ppower=p;
  for (k=1; k<=n; k++) {
```

```
    a[k] *= ppower;
    ppower *= p;
  }
}
```

2.7 Operations on orthogonal polynomials

intchs

Given the real coefficients a_j in the expansion

$$f(x) = \sum_{j=0}^{n} a_j T_j(x)$$

where $T_j(x)$ is the Chebyshev polynomial of the first kind of degree j, those in the expansion

$$\int_0^x f(x)dx = \sum_{j=1}^{n+1} b_j T_j(x)$$

are derived.

Procedure parameters:

$$\text{void intchs } (n, a, b)$$

n: int; the degree of the polynomial represented by the Chebyshev series;
a,b: double $a[0:n]$, $b[1:n+1]$;
           entry: the coefficients of the Chebyshev series, values of a_j above;
           exit: the coefficients of the integral Chebyshev series, values of b_j above.

Remark: For a description of the algorithm see [Cle62, FoP68].

```
public static void intchs(int n, double a[], double b[])
{
  int i;
  double h,l,dum;

  if (n == 0) {
    b[1]=a[0];
    return;
  }
  if (n == 1) {
    b[2]=a[1]/4.0;
    b[1]=a[0];
    return;
  }
  h=a[n];
```

```
dum=a[n-1];
b[n+1]=h/((n+1)*2);
b[n]=dum/(n*2);
for (i=n-1; i>=2; i--) {
   l=a[i-1];
   b[i]=(l-h)/(2*i);
   h=dum;
   dum=l;
}
b[1]=a[0]-h/2.0;
}
```

3. Linear Algebra

3.1 Full real general matrices

All procedures in this chapter are class methods of the following *Linear_algebra* class.

```
package numal;

import numal.*;

public class Linear_algebra extends Object {
  // all procedures in this chapter are to be inserted here
}
```

3.1.1 Preparatory procedures

A. dec

Decomposes the $n \times n$ matrix A in the form $LU=PA$, where L is lower triangular, U is unit upper triangular, and P is a permutation matrix.

Procedure parameters:
$$\text{void dec } (a,n,aux,p)$$

a: double $a[1:n,1:n]$;
       entry: the matrix to be decomposed;
       exit: the calculated lower triangular matrix and unit upper triangular matrix with its unit diagonal omitted;
n: int; the order of the matrix;
aux: double $aux[1:3]$;
       entry: *aux[2]:*   a relative tolerance: a reasonable choice for this value is an estimate of the relative precision of the matrix elements; however, it should not be chosen smaller than the machine precision;
       exit: *aux[1]:*   if R is the number of elimination steps performed (see *aux[3]*), then *aux[1]* equals 1 if the determinant of the principal submatrix of order R is positive, else *aux[1]* equals -1;
                  aux[3]:   the number of elimination steps performed; if *aux[3]*<n then the process has been broken off because the selected pivot is too small relative to the maximum of the Euclidean norms of the rows of the given matrix;

p: int *p[1:n]*;

          exit: the pivot indices; row *i* and row *p[i]* are interchanged in the *i*-th iteration.

Procedures used: matmat, mattam, ichrow.

Remark: The decomposition uses only partial pivoting [Dek68, Wi63, Wi65]. Since, in exceptional cases, partial pivoting may yield useless results, even for well-conditioned matrices, the user is advised to use the procedure *gsselm*. However, if the number of variables is small relative to the number of binary digits in the mantissa of the number representation then the procedure *dec* may also be used. Refer to the procedure *gsselm* for more details.

```java
public static void dec(double a[][], int n, double aux[], int p[])
{
  int i,k,k1,pk,d;
  double r,s,eps;
  double v[] = new double[n+1];

  pk=0;
  r = -1.0;
  for (i=1; i<=n; i++) {
    s=Math.sqrt(Basic.mattam(1,n,i,i,a,a));
    if (s > r) r=s;
    v[i]=1.0/s;
  }
  eps=aux[2]*r;
  d=1;
  for (k=1; k<=n; k++) {
    r = -1.0;
    k1=k-1;
    for (i=k; i<=n; i++) {
      a[i][k] -= Basic.matmat(1,k1,i,k,a,a);
      s=Math.abs(a[i][k])*v[i];
      if (s > r) {
        r=s;
        pk=i;
      }
    }
    p[k]=pk;
    v[pk]=v[k];
    s=a[pk][k];
    if (Math.abs(s) < eps) break;
    if (s < 0.0) d = -d;
    if (pk != k) {
      d = -d;
      Basic.ichrow(1,n,k,pk,a);
    }
    for (i=k+1; i<=n; i++)
```

```
        a[k][i]=(a[k][i]-Basic.matmat(1,k1,k,i,a,a))/s;
    }
    aux[1]=d;
    aux[3]=k-1;
}
```

B. gsselm

Decomposes the $n \times n$ matrix A in the form $LU=P_1AP_2$, where L is lower triangular, U is unit upper triangular, and P_1 and P_2 are permutation matrices. This procedure uses partial pivoting, with complete pivoting if the former does not yield stable results.

Procedure parameters:
$$\text{void gsselm } (a,n,aux,ri,ci)$$

a: double $a[1:n,1:n]$;

        entry: the matrix to be decomposed;

        exit: the calculated lower triangular matrix and unit upper triangular matrix with its unit diagonal omitted;

n: int; the order of the matrix;

aux: double $aux[1:7]$;

        entry: $aux[2]$: a relative tolerance: a reasonable choice for this value is an estimate of the relative precision of the matrix elements; however, it should not be chosen smaller than the machine precision;

                  $aux[4]$: a value which is used for controlling pivoting, usually $aux[4]=8$ will give good results;

        exit: $aux[1]$: if R is the number of elimination steps performed (see $aux[3]$), then $aux[1]$ equals 1 if the determinant of the principal submatrix of order R is positive, else $aux[1]$ equals -1;

                  $aux[3]$: the number of elimination steps performed; if $aux[3]<n$ then the process has been broken off because the selected pivot is too small relative to the maximum of the moduli of elements of the given matrix;

                  $aux[5]$: the modulus of an element which is of maximum absolute value for the matrix which had been given in a;

                  $aux[7]$: an upper bound for the growth (the modulus of an element which is of maximum absolute value for the matrices occurring during elimination);

ri: int $ri[1:n]$;

        exit: the pivotal row indices; see ci below;

ci: int $ci[1:n]$;

        exit: the pivotal column indices; in the i-th iteration, row i is interchanged with row $ri[i]$, and column j is interchanged with column $ci[j]$.

Procedures used: rowcst, elmrow, maxelmrow, ichcol, ichrow, absmaxmat.

Remark: The process of Gaussian elimination is performed in at most n steps, where n denotes the order of the matrix. Partial pivoting is used as long as the calculated upper bound for the growth [Busi71] is less than a critical value that equals $aux[4]*n$ times the modulus of an element which is of maximum absolute value for the given matrix. In the partial pivoting strategy, that element is chosen as pivot in the k-th step whose absolute value is maximal for the k-th column of the lower triangular matrix L.   However, if the upper bound for the growth exceeds this critical value in the k-th step then a pivot is selected in the j-th step, j=k,...,n, in such a way that its absolute value is maximal for the remaining submatrix of order n-k+1 (complete pivoting). Since in practice, if we choose $aux[4]$ properly, the upper bound for the growth rarely exceeds this critical value [Busi71, Wi63], we usually take advantage of the greater speed of partial pivoting (order n-k+1 in the k-th step), while in a few doubtful cases numerical difficulties will be recognized, and the process will switch to complete pivoting (order $(n-k+1)^2$ in the k-th step). Using the procedure $gsselm$, the upper bound for the relative error in the solution of a linear system [Wi63, Wi65] will be at most $aux[4]*n$ times the upper bound using Gaussian elimination with complete pivoting only. Usually this will be a crude overestimate. The choice $aux[4] < 1/n$ will result in complete pivoting only, while partial pivoting will be used in every step if we choose $aux[4] > 2^{n-1}/n$. Usually $aux[4]=8$ will give good results [Busi71]. The process will also switch to complete pivoting if the modulus of the pivot obtained with partial pivoting is less than a certain tolerance which equals the given relative tolerance $aux[2]$ times the modulus of an element which is of maximum absolute value for the given matrix. If all elements in the remaining submatrix are smaller in absolute value than this tolerance, then the process is broken off and the previous step number is delivered in $aux[3]$. In contrast with the method used in the procedure dec, no equilibrating is done in this pivoting strategy. The user has to take care for a reasonable scaling of the matrix elements.

```
public static void gsselm(double a[][], int n, double aux[], int ri[], int ci[])
{
  int ii[] = new int[1];
  int jj[] = new int[1];
  int i,j,p,q,r,r1,jpiv,rank,signdet;
  boolean partial;
  double crit,pivot,rgrow,max,aid,max1,eps;

  aux[5]=rgrow=Basic.absmaxmat(1,n,1,n,ii,jj,a);
  i = ii[0];
  j = jj[0];
  crit=n*rgrow*aux[4];
  eps=rgrow*aux[2];
  max=0.0;
  rank=n;
```

```
signdet=1;
partial = rgrow != 0;
for (q=1; q<=n; q++)
  if (q != j) {
    aid=Math.abs(a[i][q]);
    if (aid > max) max=aid;
  }
rgrow += max;
for (r=1; r<=n; r++) {
  r1=r+1;
  if (i != r) {
    signdet = -signdet;
    Basic.ichrow(1,n,r,i,a);
  }
  if (j != r) {
    signdet = -signdet;
    Basic.ichcol(1,n,r,j,a);
  }
  ri[r]=i;
  ci[r]=j;
  pivot=a[r][r];
  if (pivot < 0.0) signdet = -signdet;
  if (partial) {
    max=max1=0.0;
    j=r1;
    Basic.rowcst(r1,n,r,a,1.0/pivot);
    for (p=r1; p<=n; p++) {
      Basic.elmrow(r1,n,p,r,a,a,-a[p][r]);
      aid=Math.abs(a[p][r1]);
      if (max < aid) {
        max=aid;
        i=p;
      }
    }
    for (q=r1+1; q<=n; q++) {
      aid=Math.abs(a[i][q]);
      if (max1 < aid) max1=aid;
    }
    aid=rgrow;
    rgrow += max1;
    if ((rgrow > crit) || (max < eps)) {
      partial=false;
      rgrow=aid;
      max=Basic.absmaxmat(r1,n,r1,n,ii,jj,a);
      i = ii[0];
      j = jj[0];
    }
  } else {
    if (max <= eps) {
      rank=r-1;
```

```
        if (pivot < 0.0) signdet = -signdet;
        break;
      }
      max = -1.0;
      Basic.rowcst(r1,n,r,a,1.0/pivot);
      for (p=r1; p<=n; p++) {
        jpiv=Basic.maxelmrow(r1,n,p,r,a,a,-a[p][r]);
        aid=Math.abs(a[p][jpiv]);
        if (max < aid) {
          max=aid;
          i=p;
          j=jpiv;
        }
      }
      if (rgrow < max) rgrow=max;
    }
  }
  aux[1]=signdet;
  aux[3]=rank;
  aux[7]=rgrow;
}
```

C. onenrminv

Computes $\|a^{-1}\|_1$, a being an $n{\times}n$ matrix assumed to have been triangularly decomposed by the procedure *dec* or *gsselm*. a^{-1} is first computed by forward and back substitution [Dek68, Wi63, Wi65], and $\|a^{-1}\|_1$ is then determined.

Procedure parameters:
$$\text{double onenrminv } (a,n)$$

onenrminv: given the 1-norm of the calculated inverse of the matrix whose triangularly decomposed form is given in array *a*;

a: double *a[1:n, 1:n]*;

    entry: the triangularly decomposed form of a matrix, as delivered by the procedure *dec* or *gsselm*; the elements of *a* remain unaltered on exit of the procedure;

n: int; the order of the matrix.

Procedure used: matvec.

```
public static double onenrminv(double a[][], int n)
{
  int i,j;
  double norm,max,aid;
  double y[] = new double[n+1];

  norm=0.0;
  for (j=1; j<=n; j++) {
```

```
    for (i=1; i<=n; i++)
      y[i]=(i < j) ? 0 : ((i == j) ? 1.0/a[i][i] :
                          -Basic.matvec(j,i-1,i,a,y)/a[i][i]);
    max=0.0;
    for (i=n; i>=1; i--) {
      aid = y[i] -= Basic.matvec(i+1,n,i,a,y);
      max += Math.abs(aid);
    }
    if (norm < max) norm=max;
  }
  return (norm);
}
```

D. erbelm

Computes a rough error bound r for the relative error $\|\Delta x\|_1 / \|x\|$ in the solution of a linear system of n-equations $Ax = b$; it is assumed that A has been triangularly decomposed by the procedure $gsselm$ [Dek68, Wi63, Wi65], and that an upper bound for $\|A^{-1}\|_1$ (possibly determined by the procedure $onenrminv$) is available (r is a function of A alone). With eps the machine precision, n the order of A, $epsa$ an upper bound for the relative error in the elements of A, g the upper bound for the growth of auxiliary numbers produced during Gauss elimination (see the procedure $gsselm$: $aux[7]$), C the calculated inverse of A^{-1}, and
$$q = g * (0.75*n^3 + 4.5*n^2) * eps + epsa$$
$$p = q * \|C\|_1 / (1 - q*\|C\|_1)$$
r is given by $r = p/(1-p)$. If A is very badly conditioned, it may occur that the above formula becomes unusable.

Procedure parameters:

                    void erbelm ($n,aux,nrminv$)

n:	int; the order of the linear system in consideration;	
aux:	double $aux[0:11]$;	

       entry: $aux[0]$:   the machine precision;

                   $aux[5]$:   the modulus of an element which is of maximum absolute value for the matrix of the linear system, this value is delivered by the procedure $gsselm$;

                   $aux[6]$:   an upper bound for the relative error in the elements of the matrix of the linear system;

                   $aux[7]$:   an upper bound for the growth during Gaussian elimination, this value is delivered in $aux[7]$ by the procedure $gsselm$;

       exit: $aux[9]$:   the value of $nrminv$;

                   $aux[11]$:   a rough upper bound for the relative error in the solution of a linear system when Gaussian elimination is used for the calculation of this solution, if no use can be made of the formula for the error bound because of a very bad condition of the matrix then $aux[11] = -1$;

nrminv: double;
     entry: the 1-norm of the inverse of the matrix of the linear system must
             be given in *nrminv*, this value may be obtained by the procedure
             onenrminv.

```
public static void erbelm(int n, double aux[], double nrminv)
{
  double aid,eps;

  eps=aux[0];
  aid=(1.06*eps*(0.75*n+4.5)*(n*n)*aux[7]+aux[5]*aux[6])*nrminv;
  aux[11]=(2.0*aid >= (1.0-eps)) ? -1.0 : aid/(1.0-2.0*aid);
  aux[9]=nrminv;
}
```

E. gsserb

Performs a triangular decomposition of the $n \times n$ matrix A and calculates an upper
bound for the relative error in the solution of linear systems of the form $Ax = b$.

Procedure parameters:
                        void gsserb (*a,n,aux,ri,ci*)
a: double *a[1:n,1:n]*;
           entry: the matrix to be decomposed;
           exit: the calculated lower triangular matrix and unit upper triangular
                   matrix with its unit diagonal omitted;
n: int; the order of the matrix;
aux: double *aux[0:11]*;
           entry: *aux[0]*:   the machine precision;
                   aux[2]:   a relative tolerance;
                   aux[4]:   a value which is used for controlling pivoting;
                   aux[6]:   an upper bound for the relative precision of the matrix
                               elements;
           exit: *aux[1]*:   if R is the number of elimination steps performed, then
                               aux[1] equals 1 if the determinant of the principal
                               submatrix of order R is positive, otherwise *aux[1]* = -1;
                   aux[3]:   the number of elimination steps performed;
                   aux[5]:   the modulus of an element which is of maximum absolute
                               value for the matrix which has been given in array *a*;
                   aux[7]:   an upper bound for the growth;
                   aux[9]:   if *aux[3]=n* then *aux[9]* will be equal to the 1-norm of the
                               inverse matrix, else *aux[9]* will be undefined;
                   aux[11]: if *aux[3]=n* then the value of *aux[11]* will be a rough upper
                               bound for the relative error in the solution of linear
                               systems with a matrix as given in array *a*, else *aux[11]* will
                               be undefined. If no use can be made of the formula for the
                               error bound as given above because of a very bad
                               condition of the matrix then *aux[11]* = -1;

ri: int *ri[1:n]*;
        exit: the pivotal row indices; see *gsselm*;
ci: int *ci[1:n]*;
        exit: the pivotal column indices.

Procedures used: gsselm, onenrminv, erbelm.

```
public static void gsserb(double a[][], int n, double aux[], int ri[], int ci[])
{
  gsselm(a,n,aux,ri,ci);
  if (aux[3] == n) erbelm(n,aux,onenrminv(a,n));
}
```

F. gssnri

Performs a triangular decomposition of the $n \times n$ matrix A and calculates $\|A^{-1}\|_1$.

Procedure parameters:
                    void gssnri *(a,n,aux,ri,ci)*
a: double *a[1:n,1:n]*;
        entry: the matrix to be decomposed;
        exit: the calculated lower triangular matrix and unit upper triangular
                matrix with its unit diagonal omitted;
n: int; the order of the matrix;
aux: double *aux[1:9]*;
        entry: *aux[2]:*   a relative tolerance;
                  aux[4]:   a value used for controlling pivoting;
        exit: *aux[1]:*   if R is the number of elimination steps performed, then
                            aux[1] equals 1 if the determinant of the principal
                            submatrix of order R is positive, otherwise *aux[1]* = -1;
                aux[3]:   the number of elimination steps performed;
                aux[5]:   the modulus of an element which is of maximum absolute
                            value for the matrix which has been given in array *a*;
                aux[7]:   an upper bound for the growth;
                aux[9]:   if *aux[3]=n* then *aux[9]* will be equal to the 1-norm of the
                            inverse matrix, else *aux[9]* will be undefined;
ri: int *ri[1:n]*;
        exit: the pivotal row indices; see *gsselm*;
ci   int *ci[1:n]*;
        exit: the pivotal column indices.

Procedures used: gsselm, onenrminv.

```
public static void gssnri(double a[][], int n, double aux[], int ri[], int ci[])
{
  gsselm(a,n,aux,ri,ci);
  if (aux[3] == n) aux[9]=onenrminv(a,n);
}
```

3.1.2 Calculation of determinant

determ

Calculates the determinant of a triangularly decomposed matrix. The calculation of the determinant is done directly by calculating the product of the diagonal elements of the lower triangular matrix given in array *a*. The user is warned that overflow may occur if the order of the matrix is large.

Procedure parameters:

$$\text{double determ } (a, n, sign)$$

determ: delivers the calculated value of the determinant of the matrix;

a: double $a[1{:}n, 1{:}n]$;

entry: the diagonal elements of the lower triangular matrix, obtained by triangular decomposition of the matrix, has to be given in $a_{i,i}$, $i=1,...,n$;

n: int; the order of the matrix whose determinant has to be calculated;

sign: int;

entry: if the determinant of the matrix is positive then the value of *sign* should be +1, else -1; this value is delivered by the procedure *gsselm* or *dec* in *aux[1]*.

```
public static double determ(double a[][], int n, int sign)
{
    int i;
    double det;

    det=1.0;
    for (i=1; i<=n; i++) det *= a[i][i];
    return (sign*Math.abs(det));
}
```

3.1.3 Solution of linear equations

A. sol

Solves the linear system whose matrix has been triangularly decomposed. *sol* should be called after the procedure *dec* and solves the linear system with a matrix whose triangularly decomposed form as produced by *dec* is given in array *a* and a right-hand side as given in array *b*. *sol* leaves the array *a* and the permutation array *p* unaltered. After one call of *dec*, several calls of *sol* may follow for solving several systems having the same matrix but different right-hand sides.

Procedure parameters:

$$\text{void sol } (a,n,p,b)$$

a: double *a[1:n, 1:n]*;

        entry: the triangularly decomposed form of the matrix of the linear system
                as produced by the procedure *dec*;

n: int; the order of the matrix;

p: int *p[1:n]*;

        exit: the pivotal indices, as produced by *dec*;

b: double *b[1:n]*;

        entry: the right-hand side of the linear system;

        exit: the solution of the linear system.

Procedure used: matvec.

```
public static void sol(double a[][], int n, int p[], double b[])
{
  int k,pk;
  double r;

  for (k=1; k<=n; k++) {
    r=b[k];
    pk=p[k];
    b[k]=(b[pk]-Basic.matvec(1,k-1,k,a,b))/a[k][k];
    if (pk != k) b[pk]=r;
  }
  for (k=n; k>=1; k--) b[k] -= Basic.matvec(k+1,n,k,a,b);
}
```

B. decsol

Solves a well-conditioned linear system of equations $Ax = b$ whose order is small
relative to the number of binary digits in the number representation.

Procedure parameters:

$$\text{void decsol } (a,n,aux,b)$$

a: double *a[1:n, 1:n]*;

        entry: the *n*-th order matrix;

        exit: the calculated lower triangular matrix and unit upper triangular
                matrix with its unit diagonal omitted;

n: int; the order of the matrix;

aux: double *aux[1:3]*;

        entry: *aux[2]*:   a relative tolerance; a reasonable choice for this value is
                            an estimate of the relative precision of the matrix
                            elements; however, it should not be chosen smaller than
                            the machine precision;

        exit: *aux[1]*:   if R is the number of elimination steps performed
                            (see *aux[3]*), then *aux[1]* equals 1 if the determinant of the
                            principal submatrix of order R is positive, else *aux[1]=-1*;

aux[3]:   the number of elimination steps performed, if *aux[3]* < *n* then the process is terminated and no solution will be calculated;

b: double *b[1:n]*;
   entry: the right-hand side of the linear system;
   exit: if *aux[3]* = *n*, then the calculated solution of the linear system is overwritten on *b*, else *b* remains unaltered.

Procedures used: dec, sol.

```
public static void decsol(double a[][], int n, double aux[], double b[])
{
  int p[] = new int[n+1];

  dec(a,n,aux,p);
  if (aux[3] == n) sol(a,n,p,b);
}
```

C. solelm

Solves the *n*×*n* system of equations *Ax* = *b* whose matrix has been triangularly decomposed by the procedure *gsselm* or *gsserb*. *solelm* leaves the matrix *a* and permutation arrays *ri* and *ci* unaltered. After one call of *gsselm* or *gsserb*, several calls of *solelm* may follow for solving several systems having the same matrix but different right-hand sides.

Procedure parameters:
                            void solelm (*a,n,ri,ci,b*)
a: double *a[1:n, 1:n]*;
   entry: the triangularly decomposed form of the matrix of the linear system as produced by *gsselm*;
n: int; the order of the matrix;
ri: int *ri[1:n]*;
   entry: the pivotal row indices, as produced by *gsselm*;
ci: int *ci[1:n]*;
   entry: the pivotal column indices, as produced by *gsselm*;
b: double *b[1:n]*;
   entry: the right-hand side of the linear system;
   exit: the solution of the linear system.

Procedure used: sol.

```
public static void solelm(double a[][], int n, int ri[], int ci[], double b[])
{
  int r,cir;
  double w;
```

```
  sol(a,n,ri,b);
  for (r=n; r>=1; r--) {
    cir=ci[r];
    if (cir != r) {
      w=b[r];
      b[r]=b[cir];
      b[cir]=w;
    }
  }
}
```

D. gsssol

Solves a linear system. *gsssol* first calls *gsselm* to decompose the matrix and then *solelm* to solve the linear system.

Procedure parameters:
$$\text{void gsssol } (a,n,aux,b)$$

a: double $a[1:n,1:n]$;
          entry: the n-th order matrix;
          exit: the calculated lower triangular matrix and unit upper triangular matrix with its unit diagonal omitted;
n: int; the order of the matrix;
aux: double $aux[1:7]$;
          entry: $aux[2]$:   a relative tolerance; a reasonable choice for this value is an estimate of the relative precision of the matrix elements; however, it should not be chosen smaller than the machine precision;
                  $aux[4]$:   a value used for controlling pivoting, see *gsselm*;
          exit: $aux[1]$:   if R is the number of elimination steps performed (see $aux[3]$), then $aux[1]$ equals 1 if the determinant of the principal submatrix of order R is positive, else $aux[1]=-1$;
                  $aux[3]$:   the number of elimination steps performed, if $aux[3] < n$ then the process is terminated and no solution will be calculated;
                  $aux[5]$:   the modulus of an element which is of maximum absolute value for the matrix given in array a;
                  $aux[7]$:   an upper bound for the growth, see *gsselm*;
b: double $b[1:n]$;
          entry: the right-hand side of the linear system;
          exit: if $aux[3] = n$, then the calculated solution of the linear system is overwritten on b, else b remains unaltered.

Procedures used: solelm, gsselm.

```
public static void gsssol(double a[][], int n, double aux[], double b[])
{
```

```
    int ri[] = new int[n+1];
    int ci[] = new int[n+1];

    gsselm(a,n,aux,ri,ci);
    if (aux[3] == n) solelm(a,n,ri,ci,b);
}
```

E. gsssolerb

Solves the $n \times n$ system of equation $Ax = b$, and provides an upper bound for the relative error in x. gsssolerb calls gsserb to perform the triangular decomposition and to calculate an upper bound for the relative error, and then calls solelm to calculate x.

Procedure parameters:
$$\text{void gsssolerb } (a,n,aux,b)$$

a: double $a[1:n, 1:n]$;
      entry: the n-th order matrix;
      exit: the calculated lower triangular matrix and unit upper triangular matrix with its unit diagonal omitted;
n: int;
      entry: the order of the matrix;
aux: double $aux[0:11]$;
      entry: $aux[0]$:   the machine precision;
                  $aux[2]$:   a relative tolerance; a reasonable choice for this value is an estimate of the relative precision of the matrix elements; however, it should not be chosen smaller than the machine precision;
                  $aux[4]$:   a value used for controlling pivoting, see gsselm;
                  $aux[6]$:   an upper bound for the relative precision of the given matrix elements;
      exit: $aux[1]$:   if R is the number of elimination steps performed (see $aux[3]$), then $aux[1]$ equals 1 if the determinant of the principal submatrix of order R is positive, else $aux[1]=-1$;
                  $aux[3]$:   the number of elimination steps performed, if $aux[3] < n$ then the process is terminated and no solution will be calculated;
                  $aux[5]$:   the modulus of an element which is of maximum absolute value for the matrix given in array a;
                  $aux[7]$:   an upper bound for the growth, see gsselm;
                  $aux[9]$:   if $aux[3] = n$ then $aux[9]$ will be equal to the 1-norm of the inverse matrix, else $aux[9]$ will be undefined;
                  $aux[11]$: if $aux[3] = n$ then the value of $aux[11]$ will be a rough upper bound for the relative error in the calculated solution of the given linear system, else $aux[11]$ will be undefined; if no use can be made of the formula for the error bound because of a very bad condition of the matrix, otherwise $aux[11] = -1$.

b: double *b[1:n]*;
              entry: the right-hand side of the linear system;
              exit: if *aux[3]* = *n*, then the calculated solution of the linear system is
                       overwritten on *b*, else *b* remains unaltered.

Procedures used: solelm, gsserb.

```
public static void gsssolerb(double a[] [], int n, double aux[], double b[])
{
  int ri[] = new int[n+1];
  int ci[] = new int[n+1];

  gsserb(a,n,aux,ri,ci);
  if (aux[3] == n) solelm(a,n,ri,ci,b);
}
```

3.1.4 Matrix inversion

A. inv

Calculates the inverse of a matrix that has been triangularly decomposed by *dec*.

Procedure parameters:
$$\text{void inv } (a,n,p)$$

a: double *a[1:n,1:n]*;
              entry: the triangularly decomposed form of the matrix as produced by the
                       procedure *dec*;
              exit: the calculated inverse matrix;
n: int; the order of the matrix;
p: int *p[1:n]*;
              entry: the pivotal indices, as produced by *dec*.

Procedures used: matmat, ichcol, dupcolvec.

```
public static void inv(double a[] [], int n, int p[])
{
  int j,k,kl;
  double r;
  double v[] = new double[n+1];

  for (k=n; k>=1; k--) {
    kl=k+1;
    for (j=n; j>=kl; j--) {
      a[j][kl]=v[j];
      v[j] = -Basic.matmat(kl,n,k,j,a,a);
```

```
    }
    r=a[k][k];
    for (j=n; j>=k1; j--) {
      a[k][j]=v[j];
      v[j] = -Basic.matmat(k1,n,j,k,a,a)/r;
    }
    v[k]=(1.0-Basic.matmat(k1,n,k,k,a,a))/r;
  }
  Basic.dupcolvec(1,n,1,a,v);
  for (k=n-1; k>=1; k--) {
    k1=p[k];
    if (k1 != k) Basic.ichcol(1,n,k,k1,a);
  }
}
```

B. decinv

Obtains the inverse of the $n \times n$ matrix by partial pivoting using successive calls of the procedures *dec* and *inv*.

Procedure parameters:
$$\text{void decinv } (a,n,aux)$$

a: double *a[1:n, 1:n]*;
      entry: the matrix whose inverse has to be calculated;
      exit: if *aux[3]* = *n*, then the calculated inverse matrix;
n: int; the order of the matrix;
aux: double *aux[1:3]*;
      entry: *aux[2]*: a relative tolerance; a reasonable choice for this value is an estimate of the relative precision of the matrix elements; however, it should not be chosen smaller than the machine precision;
      exit: *aux[1]*: if *R* is the number of elimination steps performed (see *aux[3]*), then *aux[1]* equals 1 if the determinant of the principal submatrix of order *R* is positive, else *aux[1]*=-1;
             aux[3]: the number of elimination steps performed, if *aux[3]* < *n* then the process is terminated and no inverse will be calculated.

Procedures used: dec, inv.

```
public static void decinv(double a[][], int n, double aux[])
{
  int p[] = new int[n+1];

  dec(a,n,aux,p);
  if (aux[3] == n) inv(a,n,p);
}
```

C. inv1

Calculates the inverse of a matrix that has been triangularly decomposed by the procedure *gsselm* or *gsserb*. The 1-norm of the inverse matrix might also be calculated.

Procedure parameters:
$$\text{double inv1 } (a,n,ri,ci,withnorm)$$

inv1:   if the value of *withnorm* is true then the value of *inv1* will be equal to the 1-norm of the calculated inverse matrix, else *inv1* = 0;

a: double $a[1:n,1:n]$;
             entry: the triangularly decomposed form of the matrix as produced by *gsselm*;
             exit: the calculated inverse matrix;

n: int; the order of the matrix;

ri: int $ri[1:n]$;
             entry: the pivotal row indices, as produced by *gsselm*;

ci: int $ci[1:n]$;
             entry: the pivotal column indices, as produced by *gsselm*;

withnorm: boolean;
             entry: if the value of *withnorm* is true then the 1-norm of the inverse matrix will be calculated and assigned to *inv1*, else *inv1* = 0.

Procedures used: ichrow, inv.

```
public static double inv1(double a[][], int n, int ri[], int ci[],
                    boolean withnorm)
{
  int l,k,k1;
  double aid,nrminv;

  inv(a,n,ri);
  nrminv=0.0;
  if (withnorm)
    for (l=1; l<=n; l++) nrminv += Math.abs(a[l][n]);
  for (k=n-1; k>=1; k--) {
    if (withnorm) {
      aid=0.0;
      for (l=1; l<=n; l++) aid += Math.abs(a[l][k]);
      if (nrminv < aid) nrminv=aid;
    }
    k1=ci[k];
    if (k1 != k) Basic.ichrow(1,n,k,k1,a);
  }
  return (nrminv);
}
```

D. gssinv

Uses the procedure *gsselm* to perform a triangular decomposition of the matrix and the procedure *inv1* to calculate the inverse matrix.

Procedure parameters:
$$\text{void gssinv } (a,n,aux)$$

a: double *a[1:n, 1:n]*;
  entry: the matrix whose inverse has to be calculated;
  exit: if *aux[3] = n* then the calculated inverse matrix;

n: int;
  entry: the order of the matrix;

aux: double *aux[1:9]*;
  entry: *aux[2]*: a relative tolerance; a reasonable choice for this value is an estimate of the relative precision of the matrix elements; however, it should not be chosen smaller than the machine precision;
  aux[4]: a value used for controlling pivoting, see *gsselm*;
  exit: *aux[1]*: if *R* is the number of elimination steps performed (see *aux[3]*), then *aux[1]* equals 1 if the determinant of the principal submatrix of order *R* is positive, else *aux[1]=-1*;
  aux[3]: the number of elimination steps performed, if *aux[3] < n* then the process is terminated and no solution will be calculated;
  aux[5]: the modulus of an element which is of maximum absolute value for the matrix given in array *a*;
  aux[7]: an upper bound for the growth, see *gsselm*;
  aux[9]: if *aux[3] = n* then *aux[9]* will be equal to the 1-norm of the calculated inverse matrix, else *aux[9]* will be undefined.

Procedures used: inv1, gsselm.

```
public static void gssinv(double a[][], int n, double aux[])
{
   int ri[] = new int[n+1];
   int ci[] = new int[n+1];

   gsselm(a,n,aux,ri,ci);
   if (aux[3] == n) aux[9]=inv1(a,n,ri,ci,true);
}
```

E. gssinverb

Uses the procedure *gsselm* to perform the triangular decomposition of the matrix *a*, the procedure *inv1* to calculate the inverse matrix and its 1-norm and the procedure *erbelm* to calculate an upper bound for the relative error in the calculated inverse.

A Numerical Library in Java for Scientists and Engineers

Procedure parameters:
$$\text{void gssinverb } (a,n,aux)$$

a: double $a[1:n,1:n]$;
       entry: the matrix whose inverse has to be calculated;
       exit: if $aux[3] = n$ then the calculated inverse matrix;

n: int;
       entry: the order of the matrix;

aux: double $aux[0:11]$;
       entry: $aux[0]$:   the machine precision;
               $aux[2]$:   a relative tolerance; a reasonable choice for this value is an estimate of the relative precision of the matrix elements; however, it should not be chosen smaller than the machine precision;
               $aux[4]$:   a value used for controlling pivoting, see *gsselm*;
               $aux[6]$:   an upper bound for the relative precision of the given matrix elements;
       exit: $aux[1]$:   if R is the number of elimination steps performed (see $aux[3]$), then $aux[1]$ equals 1 if the determinant of the principal submatrix of order R is positive, else $aux[1]=-1$;
               $aux[3]$:   the number of elimination steps performed, if $aux[3] < n$ then the process is terminated and no solution will be calculated;
               $aux[5]$:   the modulus of an element which is of maximum absolute value for the matrix given in array a;
               $aux[7]$:   an upper bound for the growth, see *gsselm*;
               $aux[9]$:   if $aux[3] = n$ then $aux[9]$ will be equal to the 1-norm of the inverse matrix, else $aux[9]$ will be undefined;
               $aux[11]$: if $aux[3] = n$ then the value of $aux[11]$ will be a rough upper bound for the relative error in the calculated inverse matrix, else $aux[11]$ will be undefined; if no use can be made of the formula for the error bound because of a very bad condition of the matrix, otherwise $aux[11]=-1$.

Procedures used: inv1, gsselm, erbelm.

```
public static void gssinverb(double a[][], int n, double aux[])
{
  int ri[] = new int[n+1];
  int ci[] = new int[n+1];

  gsselm(a,n,aux,ri,ci);
  if (aux[3] == n) erbelm(n,aux,inv1(a,n,ri,ci,true));
}
```

3.1.5 Iteratively improved solution

A. itisol

Solves a linear system $Ax = b$ whose matrix has been triangularly decomposed by the procedure *gsselm* or *gsserb*. This solution will be refined iteratively until the calculated relative correction to this solution will be less than a prescribed value (see *aux[10]*). Each iteration of the refinement process consists of the following three steps [Bus72a, Dek71, Wi65]:

(a) calculate in double precision, the residual vector r defined by $r = Ax^k - b$, where x^k denotes the solution obtained in the k-th iteration;

(b) calculate the solution c of the linear system $Ac = r$, with the aid of the triangularly decomposed matrix as given in the array *lu*;

(c) calculate the new solution $x^{k+1} = x^k - c$.

Procedure parameters:

$$\text{void itisol } (a,lu,n,aux,ri,ci,b)$$

a: double $a[1:n, 1:n]$;
       entry: the matrix of the linear system;

lu: double $lu[1:n, 1:n]$;
       entry: the triangularly decomposed form of the matrix given in *a*, as delivered by *gsselm*;

n: int; the order of the matrix;

aux: double $aux[10:13]$;
       entry: *aux[10]*: a relative tolerance for the solution vector; if the 1-norm of the vector of corrections to the solution divided by the 1-norm of the calculated solution is smaller than *aux[10]*, then the process will stop; the user should not choose the value of *aux[10]* smaller than the relative precision of the elements of the matrix and the right-hand side of the linear system;

                   aux[12]: the maximum number of iterations allowed for the refinement of the solution; if the number of iterations exceeds the value of *aux[12]* then the process will be broken off; usually *aux[12]* = 5 will give good results;

       exit: *aux[11]*: the 1-norm of the vector of corrections to the solution in the last iteration step divided by the 1-norm of the calculated solution;

                   if *aux[11]* > *aux[10]* then the process has been broken off because the number of iterations exceeded the value given in *aux[12]*;

                  aux[13]: the 1-norm of the residual vector r above;

ri: int $ri[1:n]$;
       entry: the pivotal row indices, as produced by *gsselm*;

ci: int $ci[1:n]$;
       entry: the pivotal column indices, as produced by *gsselm*;

b: double $b[1:n]$;
       entry: the right-hand side of the linear system;
       exit: the calculated solution of the linear system.

Procedures used: solelm, inivec, dupvec.

Remark: If the condition of the matrix is not too bad then the precision of the
calculated solution will be of the order of the precision asked for in
aux[10]. If the condition of the matrix is very bad then this process will
possibly not converge or, in exceptional cases, converge to a useless
result. If the user wants to make certain about the precision of the
calculated solution then the procedure *itisolerb* should be used. *itisol*
leaves *a, lu, ri,* and *ci* unaltered, so after one call of *gsselm* several calls of
itisol may follow to calculate the solution of several linear systems with
the same matrix but different right-hand sides.

```
public static void itisol(double a[][], double lu[][], int n, double aux[],
                          int ri[], int ci[], double b[])
{
  int i,j,iter,maxiter;
  double maxerx,erx,nrmres,nrmsol,r,rr,dtemp;
  double res[] = new double[n+1];
  double sol[] = new double[n+1];

  maxerx=erx=aux[10];
  maxiter=(int) aux[12];
  Basic.inivec(1,n,sol,0.0);
  Basic.dupvec(1,n,0,res,b);
  iter=1;
  do {
    solelm(lu,n,ri,ci,res);
    erx=nrmsol=nrmres=0.0;
    for (i=1; i<=n; i++) {
      r=res[i];
      erx += Math.abs(r);
      rr=sol[i]+r;
      sol[i]=rr;
      nrmsol += Math.abs(rr);
    }
    erx /= nrmsol;
    for (i=1; i<=n; i++) {
      dtemp = -(double)b[i];
      for (j=1; j<=n; j++)
        dtemp += (double)a[i][j]*(double)sol[j];
      r = -dtemp;
      res[i]=r;
      nrmres += Math.abs(r);
    }
    iter++;
  } while ((iter <= maxiter) && (maxerx < erx));
  Basic.dupvec(1,n,0,b,sol);
  aux[11]=erx;
  aux[13]=nrmres;
}
```

B. gssitisol

Uses the procedure *gsselm* to perform a triangular decomposition of the matrix and the procedure *itisol* to calculate an iteratively refined solution of the given linear system.

Procedure parameters:
$$\text{void gssitisol } (a,n,aux,b)$$

a: double $a[1:n, 1:n]$;

    entry: the n-th order matrix;

    exit: the calculated lower triangular matrix and unit upper triangular matrix with its unit diagonal omitted;

n: int;   entry: the order of the matrix;

aux: double $aux[1:13]$;

    entry: $aux[2]$:   a relative tolerance for the process of triangular decomposition; a reasonable choice for this value is an estimate of the relative precision of the matrix elements; however, it should not be chosen smaller than the machine precision;

    $aux[4]$:   a value used for controlling pivoting, see *gsselm*;

    $aux[10]$:   a relative tolerance for the solution vector; if the 1-norm of the vector of corrections to the solution divided by the 1-norm of the calculated solution is smaller than $aux[10]$, then the process will stop; the user should not choose the value of $aux[10]$ smaller than the relative precision of the elements of the matrix and the right-hand side of the linear system;

    $aux[12]$:   the maximum number of iterations allowed for the refinement of the solution; if the number of iterations exceeds the value of $aux[12]$ then the process will be broken off; usually $aux[12] = 5$ will give good results;

    exit: $aux[1]$:   if R is the number of elimination steps performed (see $aux[3]$), then $aux[1]$ equals 1 if the determinant of the principal submatrix of order R is positive, else $aux[1]=-1$;

    $aux[3]$:   the number of elimination steps performed, if $aux[3] < n$ then the process has been broken off and no solution will have been calculated;

    $aux[5]$:   modulus of an element which is of maximum absolute value for array a;

    $aux[7]$:   an upper bound for the growth, see *gsselm*;

    $aux[11]$:   if $aux[3] < n$ then $aux[11]$ will be undefined, else $aux[11]$ will be equal to the 1-norm of the vector of corrections to the solution in the last step divided by the 1-norm of the calculated solution; if $aux[11] > aux[10]$ then the process has been broken off because the number of iterations exceeded the value given in $aux[12]$;

    $aux[13]$:   if $aux[3] = n$ then the value of $aux[13]$ will be equal to the 1-norm of the residual vector (see *itisol*), else $aux[13]$ will be undefined;

b: double $b[1:n]$;
          entry: the right-hand side of the linear system;
          exit: if $aux[3] = n$, then the calculated solution of the linear system is
                   overwritten on b, else b remains unaltered.

Procedures used: dupmat, gsselm, itisol.

```
public static void gssitisol(double a[][], int n, double aux[], double b[])
{
   int ri[] = new int[n+1];
   int ci[] = new int[n+1];
   double aa[][] = new double[n+1][n+1];

   Basic.dupmat(1,n,1,n,aa,a);
   gsselm(a,n,aux,ri,ci);
   if (aux[3] == n) itisol(aa,a,n,aux,ri,ci,b);
}
```

C. itisolerb

Solves a linear system $Ax = b$ whose matrix has been triangularly decomposed by
the procedure *gssnri* (the procedure *gssnri* also delivers the proper values for the
odd elements of the array *aux*). *itisolerb* calculates, with the use of the procedure
itisol, an iteratively improved solution of the linear system. Moreover, with values
of the machine precision ε and of $\|A^{-1}\|_1$ available, and also with upper bounds η_A
for $\|\Delta A\|_1/\|A\|_1$, ΔA being the matrix of errors in A, and η_b for $\|\Delta b\|_1/\|b\|_1$,
Δb being the vector of errors in b available, *itisolerb* calculates a realistic upper
bound for the relative error $\|\Delta x\|_1/\|x\|_1$ in the calculated solution [Bus72a,
Dek71]. The latter bound is given by

$$\frac{\|\Delta x\|_1}{\|x\|_1} = \frac{p}{1-p}$$

where

$$p = \left(\frac{\|Ax - b\|_1}{\|x\|_1} + \frac{\eta_b}{\|x\|_1} + \eta_A \right) \frac{\|C\|_1}{1 - q\|C\|_1}$$

C being the computed inverse of A, and

$$q = \alpha \left(0.75n^3 + 4.5n^2 \right) \varepsilon + \eta_A$$

α being the upper bound for the growth of auxiliary numbers during Gauss
elimination (the value allocated to $aux[7]$ at exit from the procedure *gssnri*).
itisolerb leaves a, lu, ri, and ci unaltered, so after one call of *gssnri* several calls of
itisolerb may follow, to calculate the solution of several linear systems with the
same matrix but different right-hand sides.

Procedure parameters:

void itisolerb (a,lu,n,aux,ri,ci,b)

a: double a[1:n,1:n];
          entry: the matrix of the linear system;
lu: double lu[1:n,1:n];
          entry: the triangularly decomposed form of the matrix given in a, as
                  delivered by gssnri;
n: int;
          entry: the order of the matrix;
aux: double aux[0:13];
          entry: aux[0]:   the machine precision;
                  aux[5]:   the modulus of an element which is of maximum absolute
                            value for the matrix of the linear system, this value is
                            delivered by gssnri in aux[5];
                  aux[6]:   an upper bound for the relative error in the elements of
                            the matrix of the linear system; the value of η_A above;
                  aux[7]:   an upper bound for the growth during Gaussian
                            elimination, this value is delivered by gssnri in aux[7]; the
                            value of α above;
                  aux[8]:   an upper bound for the relative error in the elements of
                            the right-hand side of the linear system; the value of η_b
                            above;
                  aux[9]:   the 1-norm of the inverse matrix, this value is delivered
                            by gssnri in aux[9]; the value of $\|C\|_1$ above;
                  aux[10]: a relative tolerance for the solution vector; if the 1-norm
                            of the vector of corrections to the solution divided by the
                            1-norm of the calculated solution is smaller than aux[10],
                            then the process will stop; the user should not choose the
                            value of aux[10] smaller than the relative precision of the
                            elements of the matrix and the right-hand side of the
                            linear system, given in aux[6] and aux[8];
                  aux[12]: the maximum number of iterations allowed for the
                            refinement of the solution; if the number of iterations
                            exceeds the value of aux[12], then the process will be
                            broken off; usually aux[12] = 5 will give good results;
          exit: aux[11]: a realistic upper bound (the value of p/(1-p) above) for the
                            relative error in the calculated solution; if no use can be
                            made of the error formula then aux[11] = -1;
                  aux[13]: the 1-norm of the residual vector; the value of $\|Ax\text{-}b\|_1$ for
                            which the last iterative scheme operates.
ri: int ri[1:n];
          entry: the pivotal row indices, as produced by gssnri;
ci: int ci[1:n];
          entry: the pivotal column indices, as produced by gssnri;
b: double b[1:n];
          entry: the right-hand side of the linear system;
          exit: the calculated solution of the linear system.

Procedure used: itisol.

```
public static void itisolerb(double a[][], double lu[][], int n, double aux[],
                             int ri[], int ci[], double b[])
{
  int i;
  double nrmsol,nrminv,nrmb,alfa,tola,eps;

  eps=aux[0];
  nrminv=aux[9];
  tola=aux[5]*aux[6];
  nrmb=nrmsol=0.0;
  for (i=1; i<=n; i++) nrmb += Math.abs(b[i]);
  itisol(a,lu,n,aux,ri,ci,b);
  for (i=1; i<=n; i++) nrmsol += Math.abs(b[i]);
  alfa=1.0-(1.06*eps*aux[7]*(0.75*n+4.5)*n*n+tola)*nrminv;
  if (alfa < eps)
    aux[11] = -1.0;
  else {
    alfa=((aux[13]+aux[8]*nrmb)/nrmsol+tola)*nrminv/alfa;
    aux[11]=(1.0-alfa < eps) ? -1.0 : alfa/(1.0-alfa);
  }
}
```

D. gssitisolerb

Uses the procedure *gssnri* to perform a triangular decomposition of the matrix and
the procedure *itisolerb* to calculate an iteratively refined solution of the given linear
system and a realistic upper bound for the relative error in the solution.

Procedure parameters:
$$\text{void gssitisolerb } (a,n,aux,b)$$

a: double *a[1:n, 1:n]*;
          entry: the *n*-th order matrix;
          exit: the calculated lower triangular matrix and unit upper triangular
                  matrix with its unit diagonal omitted;
n: int;
          entry: the order of the matrix;
aux: double *aux[0:13]*;
          entry: *aux[0]*:   the machine precision;
                  aux[2]:   a relative tolerance; a reasonable choice for this value is
                              an estimate of the relative precision of the matrix
                              elements; however, it should not be chosen smaller than
                              the machine precision;
                  aux[4]:   a value used for controlling pivoting, see *gsselm*;
                  aux[6]:   an upper bound for the relative error in the matrix
                              elements of the linear system;
                  aux[8]:   an upper bound for the relative error in the elements of
                              the right-hand side;

$aux[10]$:   a relative tolerance for the solution vector; if the 1-norm of the vector of corrections to the solution divided by the 1-norm of the calculated solution is smaller than $aux[10]$, then the process will stop; the user should not choose the value of $aux[10]$ smaller than the relative precision of the elements of the matrix and the right-hand side of the linear system ($aux[10] \geq aux[2]$);

$aux[12]$:   the maximum number of iterations allowed for the refinement of the solution; if the number of iterations exceeds the value of $aux[12]$ then the process will be broken off; usually $aux[12] = 5$ will give good results;

exit: $aux[1]$:   if R is the number of elimination steps performed (see $aux[3]$), then $aux[1]$ equals 1 if the determinant of the principal submatrix of order R is positive, else $aux[1]=-1$;

$aux[3]$:   the number of elimination steps performed, if $aux[3] < n$ then the process has been broken off and no solution will have been calculated;

$aux[5]$:   the modulus of an element which is of maximum absolute value for the matrix which had been given in array a;

$aux[7]$:   an upper bound for the growth, see $gsselm$;

$aux[9]$:   if $aux[3] = n$ then $aux[9]$ will be equal to the 1-norm of the calculated inverse matrix, else $aux[9]$ will be undefined;

$aux[11]$:   if $aux[3] < n$ then $aux[11]$ will be undefined, else $aux[11]$ will be equal to a realistic upper bound for the relative error in the calculated solution; however, if no use can be made of the error formula (see $itisolerb$) then $aux[11]=-1$;

$aux[13]$:   if $aux[3] = n$ then the value of $aux[13]$ will be equal to the 1-norm of the residual vector (see $itisol$), else $aux[13]$ will be undefined;

b: double $b[1:n]$;

entry: the right-hand side of the linear system;

exit: if $aux[3] = n$, then the calculated solution of the linear system is overwritten on b, else b remains unaltered.

Procedures used: dupmat, gssnri, itisolerb.

```
public static void gssitisolerb(double a[][], int n, double aux[], double b[])
{
    int ri[] = new int[n+1];
    int ci[] = new int[n+1];
    double aa[][] = new double[n+1][n+1];

    Basic.dupmat(1,n,1,n,aa,a);
    gssnri(a,n,aux,ri,ci);
    if (aux[3] == n) itisolerb(aa,a,n,aux,ri,ci,b);
}
```

3.2 Real symmetric positive definite matrices

3.2.1 Preparatory procedures

A. chldec2

Calculates the Cholesky decomposition of a positive definite symmetric matrix whose upper triangle is given in a two-dimensional array. For a given symmetric positive definite matrix A, it computes the upper triangular matrix U for which $U^TU=A$. The columns $u_{(k)}$ $(k=1,...,n)$ of U are determined in succession. The decomposition process is broken off at stage k if

$$a_{k,k} - u_{(k)}^T u_{(k)} < \tau * B, \quad \text{with } B = \max a_{i,i}, \quad 1 \le i \le n,$$

where τ is a tolerance supplied by the user. In this case the matrix, possibly modified by rounding errors, is not positive definite.

Procedure parameters:
$$\text{void chldec2 } (a,n,aux)$$

a: double $a[1:n,1:n]$;
            entry: the upper triangle of the positive definite symmetric matrix must be given in the upper triangular part of a (the elements $a[i,j]$, $i \le j$);
            exit: the Cholesky decomposition of the matrix is delivered in the upper triangle of a;
n: int;
            entry: the order of the matrix;
aux: double $aux[2:3]$;
            entry: $aux[2]$:   a relative tolerance used to control the calculation of the diagonal elements;
            normal exit: $aux[3] = n$;
            abnormal exit: if the decomposition fails because the matrix is not positive definite then $aux[3] = k-1$, where k is the last stage number.

Procedure used: tammat.

Remark: The method used is Cholesky's square root method without pivoting [Dek68, Wi65].

```
public static void chldec2(double a[][], int n, double aux[])
{
  int k,j;
  double r,epsnorm;

  r=0.0;
  for (k=1; k<=n; k++)
    if (a[k][k] > r) r=a[k][k];
```

```
  epsnorm=aux[2]*r;
  for (k=1; k<=n; k++) {
    r=a[k][k]-Basic.tammat(1,k-1,k,k,a,a);
    if (r <= epsnorm) {
      aux[3]=k-1;
      return;
    }
    a[k][k]=r=Math.sqrt(r);
    for (j=k+1; j<=n; j++)
      a[k][j]=(a[k][j]-Basic.tammat(1,k-1,j,k,a,a))/r;
  }
  aux[3]=n;
}
```

B. chldec1

Performs the Cholesky decomposition of a positive definite symmetric matrix whose upper triangle is given columnwise in a one-dimensional array, by Cholesky's square root method without pivoting (see the description of the procedure *chldec2*).

Procedure parameters:
$$\text{void chldec1 } (a,n,aux)$$

a: double $a[1:(n+1)n/2]$;

      entry: the upper triangular part of the positive definite symmetric matrix must be given columnwise in array a (the (i,j)-th element of the matrix should be given in array $a[(j-1)j/2+i]$, $1 \le i \le j \le n$);

      exit: the Cholesky decomposition of the matrix is delivered columnwise in a;

n: int;

      entry: the order of the matrix;

aux: double $aux[2:3]$;

      entry: $aux[2]$:   a relative tolerance used to control the calculation of the diagonal elements;

      normal exit: $aux[3] = n$;

      abnormal exit: if the decomposition fails because the matrix is not positive definite then $aux[3] = k-1$, where k is the last stage number.

Procedure used: vecvec.

```
public static void chldec1(double a[], int n, double aux[])
{
  int j,k,kk,kj,low,up;
  double r,epsnorm;

  r=0.0;
```

```
  kk=0;
  for (k=1; k<=n; k++) {
    kk += k;
    if (a[kk] > r) r=a[kk];
  }
  epsnorm=aux[2]*r;
  kk=0;
  for (k=1; k<=n; k++) {
    kk += k;
    low=kk-k+1;
    up=kk-1;
    r=a[kk]-Basic.vecvec(low,up,0,a,a);
    if (r <= epsnorm) {
      aux[3]=k-1;
      return;
    }
    a[kk]=r=Math.sqrt(r);
    kj=kk+k;
    for (j=k+1; j<=n; j++) {
      a[kj]=(a[kj]-Basic.vecvec(low,up,kj-kk,a,a))/r;
      kj +=j;
    }
  }
  aux[3]=n;
}
```

3.2.2 Calculation of determinant

A. chldeterm2

Calculates the determinant of a symmetric positive definite matrix whose Cholesky matrix is given in the upper triangle of a two-dimensional array. *chldeterm2* should be called after a successful call of *chldec2* or *chldecsol2*. *chldeterm2* should not be called if overflow is to be expected.

Procedure parameters:

$$\text{double chldeterm2 } (a,n)$$

a: double $a[1{:}n, 1{:}n]$;

        entry: the upper triangular part of the Cholesky matrix as produced by *chldec2* or *chldecsol2* must be given in the upper triangle of *a*;

        exit: the contents of *a* are not changed;

n: int;

        entry: the order of the matrix.

```
public static double chldeterm2(double a[][], int n)
{
```

```
    int k;
    double d;

    d=1.0;
    for (k=1; k<=n; k++) d *= a[k][k];
    return (d*d);
}
```

B. chldeterm1

Calculates the determinant of a symmetric positive definite matrix whose Cholesky matrix is given columnwise in a one-dimensional array. *chldeterm1* should be called after a successful call of *chldec1* or *chldecsol1*. *chldeterm1* should not be called if overflow is to be expected.

Procedure parameters:
$$\text{double chldeterm1 } (a,n)$$
a: double *a[1:(n+1)n/2]*;
          entry: the upper triangular part of the Cholesky matrix as produced by
                  chldec1 or *chldecsol1* must be given columnwise in array *a*;
          exit: the contents of *a* are not changed;
n: int;
          entry: the order of the matrix.

```
public static double chldeterm1(double a[], int n)
{
    int k,kk;
    double d;

    d=1.0;
    kk=0;
    for (k=1; k<=n; k++) {
        kk += k;
        d *= a[kk];
    }
    return (d*d);
}
```

3.2.3 Solution of linear equations

A. chlsol2

Calculates the solution of a system of linear equations, provided that the coefficient matrix has been decomposed by a successful call of the procedure *chldec2* or *chldecsol2*. The solution is obtained by carrying out the forward and back substitution with the Cholesky matrix and the right-hand side. The elements of the Cholesky matrix are not changed. Several systems of linear equations with the same coefficient matrix but different right-hand sides can be solved by successive calls of *chlsol2* [Dek68].

Procedure parameters:
$$\text{void chlsol2 } (a,n,b)$$

a: double *a[1:n,1:n]*;
            entry: the upper triangular part of the Cholesky matrix as produced by
                     chldec2 or *chldecsol2* must be given in the upper triangle of *a*;
            exit: the contents of *a* are not changed;
n: int;
            entry: the order of the matrix;
b: double *b[1:n]*;
            entry: the right-hand side of the system of linear equations;
            exit: the solution of the system.

Procedures used: matvec, tamvec.

```
public static void chlsol2(double a[][], int n, double b[])
{
  int i;

  for (i=1; i<=n; i++)
    b[i]=(b[i]-Basic.tamvec(1,i-1,i,a,b))/a[i][i];
  for (i=n; i>=1; i--)
    b[i]=(b[i]-Basic.matvec(i+1,n,i,a,b))/a[i][i];
}
```

B. chlsol1

Calculates the solution of a system of linear equations, provided that the coefficient matrix has been decomposed by a successful call of the procedure *chldec1* or *chldecsol1*. The solution is obtained by carrying out the forward and back substitution with the Cholesky matrix and the right-hand side. The elements of the Cholesky matrix are not changed. Several systems of linear equations with the same coefficient matrix but different right-hand sides can be solved by successive calls of *chlsol1*.

Procedure parameters:
$$\text{void chlsol1 } (a,n,b)$$

a: double *a[1:(n+1)n/2]*;
            entry: the upper triangular part of the Cholesky matrix as produced by
                     chldec1 or *chldecsol1* must be given columnwise in array *a*;
            exit: the contents of *a* are not changed;

n: int;
            entry: the order of the matrix;
b: double *b[1:n]*;
            entry: the right-hand side of the system of linear equations;
            exit: the solution of the system.

Procedures used: vecvec, seqvec.

```
public static void chlsol1(double a[], int n, double b[])
{
  int i,ii;

  ii=0;
  for (i=1; i<=n; i++) {
    ii += i;
    b[i] = (b[i]-Basic.vecvec(1,i-1,ii-i,b,a))/a[ii];
  }
  for (i=n; i>=1; i--) {
    b[i] = (b[i]-Basic.seqvec(i+1,n,ii+i,0,a,b))/a[ii];
    ii -= i;
  }
}
```

C. chldecsol2

Solves a system of linear equations with a symmetric positive definite coefficient
matrix by calling *chldec2* and, if the call is successful, *chlsol2*. The coefficient
matrix must be given in the upper triangle of a two-dimensional array.

Procedure parameters:
$$\text{void chldecsol2 } (a,n,aux,b)$$

a: double *a[1:n, 1:n]*;
            entry: the upper triangle of the symmetric positive definite matrix must be
                    given in the upper triangular part of *a* (the elements $a[i,j]$, $i \le j$);
            exit: the Cholesky decomposition of the matrix is delivered in the upper
                    triangle of *a*;
n: int;
            entry: the order of the matrix;
aux: double *aux[2:3]*;
            entry: *aux[2]*:   a relative tolerance used to control the calculation of the
                                diagonal elements;
            normal exit: $aux[3] = n$;
            abnormal exit: if the decomposition fails because the matrix is not
                                positive definite then $aux[3] = k-1$, where k is the last
                                stage number.
b: double *b[1:n]*;
            entry: the right-hand side of the system of linear equations;
            exit: the solution of the system.

Procedures used: chldec2, chlsol2.

```
public static void chldecsol2(double a[][], int n, double aux[], double b[])
{
   chldec2(a,n,aux);
   if (aux[3] == n) chlsol2(a,n,b);
}
```

D. chldecsol1

Solves a system of linear equations with a symmetric positive definite coefficient matrix by calling *chldec1* and, if the call is successful, *chlsol1*. The upper triangle of the coefficient matrix must be stored columnwise in a one-dimensional array.

Procedure parameters:
$$\text{void chldecsol1 } (a,n,aux,b)$$
a: double $a[1:(n+1)n/2]$;
           entry: the upper triangular part of the symmetric positive definite matrix
                   must be given columnwise in array *a* (the elements (i,j)-th element
                   of the matrix must be given in $a[(j-1)*j/2+i]$, $1 \le i \le j \le n$);
           exit: the Cholesky decomposition of the matrix is delivered columnwise
                   in *a*;
n: int;
           entry: the order of the matrix;
aux: double $aux[2:3]$;
           entry: $aux[2]$:   a relative tolerance used to control the calculation of the
                              diagonal elements;
           normal exit: $aux[3] = n$;
           abnormal exit: if the decomposition fails because the matrix is not
                             positive definite then $aux[3] = k-1$, where k is the last
                             stage number.
b: double $b[1:n]$;
           entry: the right-hand side of the system of linear equations;
           exit: the solution of the system.

Procedures used: chldec1, chlsol1.

```
public static void chldecsol1(double a[], int n, double aux[], double b[])
{
   chldec1(a,n,aux);
   if (aux[3] == n) chlsol1(a,n,b);
}
```

3.2.4 Matrix inversion

A. chlinv2

Calculates the inverse X of a symmetric positive definite matrix A, provided that the matrix has been decomposed ($A = U^TU$, where U is the Cholesky matrix) by a successful call of *chldec2* or *chldecsol2*. The Cholesky matrix must be given in the upper triangle of a two-dimensional array. The inverse X is obtained from the conditions that X be symmetric and UX be a lower triangular matrix whose main diagonal elements are the reciprocals of the diagonal elements of U. The upper triangular elements of X are calculated by back substitution [Dek68].

Procedure parameters:
$$\text{void chlinv2 } (a,n)$$

a: double *a[1:n, 1:n]*;
      entry: the upper triangular part of the Cholesky matrix as produced by *chldec2* or *chldecsol2* must be given in the upper triangle of *a*;
      exit: the upper triangular part of the inverse matrix is delivered in the upper triangle of *a*;
n: int;
      entry: the order of the matrix.

Procedures used: matvec, tamvec, dupvecrow.

```
public static void chlinv2(double a[][], int n)
{
  int i,j,i1;
  double r;
  double u[] = new double[n+1];

  for (i=n; i>=1; i--) {
    r=1.0/a[i][i];
    i1=i+1;
    Basic.dupvecrow(i1,n,i,u,a);
    for (j=n; j>=i1; j--)
      a[i][j] = -(Basic.tamvec(i1,j,j,a,u)+Basic.matvec(j+1,n,j,a,u))*r;
    a[i][i]=(r-Basic.matvec(i1,n,i,a,u))*r;
  }
}
```

B. chlinv1

Calculates the inverse X of a symmetric positive definite matrix A, provided that the matrix has been decomposed ($A = U^TU$, where U is the Cholesky matrix) by a successful call of *chldec1* or *chldecsol1*. The upper triangular part of the Cholesky matrix must be given columnwise in a one-dimensional array. The inverse X is obtained from the conditions that X be symmetric and UX be a lower triangular matrix whose main diagonal elements are the reciprocals of the diagonal elements of U. The upper triangular elements of X are calculated by back substitution.

Procedure parameters:
$$\text{void chlinv1 } (a,n)$$

a: double *a[1:(n+1)n/2]*;

          entry: the upper triangular part of the Cholesky matrix as produced by *chldec1* or *chldecsol1* must be given columnwise in array *a*;

          exit: the upper triangular part of the inverse matrix is delivered columnwise in array *a*;

n: int;

          entry: the order of the matrix.

Procedures used: seqvec, symmatvec.

```
public static void chlinv1(double a[], int n)
{
  int i,ii,i1,j,ij;
  double r;
  double u[] = new double[n+1];

  ii=((n+1)*n)/2;
  for (i=n; i>=1; i--) {
    r=1.0/a[ii];
    i1=i+1;
    ij=ii+i;
    for (j=i1; j<=n; j++) {
      u[j]=a[ij];
      ij += j;
    }
    for (j=n; j>=i1; j--) {
      ij -= j;
      a[ij] = -Basic.symmatvec(i1,n,j,a,u)*r;
    }
    a[ii]=(r-Basic.seqvec(i1,n,ii+i,0,a,u))*r;
    ii -= i;
  }
}
```

C. chldecinv2

Calculates the inverse of a symmetric positive definite coefficient matrix by calling *chldec2* and, if the call is successful, *chlinv2*. The coefficient matrix must be given in the upper triangle of a two-dimensional array.

Procedure parameters:

$$\text{void chldecinv2 } (a,n,aux)$$

a: double *a[1:n, 1:n]*;

      entry: the upper triangle of the symmetric positive definite matrix must be given in the upper triangular part of *a* (the elements *a[i,j]*, *i≤j*);

      exit: the upper triangular part of the inverse matrix is delivered in the upper triangle of *a*;

n: int;

      entry: the order of the matrix;

aux: double *aux[2:3]*;

      entry: *aux[2]*:   a relative tolerance used to control the calculation of the diagonal elements;

      normal exit: *aux[3] = n*;

      abnormal exit: if the decomposition fails because the matrix is not positive definite then *aux[3] = k-1*, where *k* is the last stage number.

Procedures used: chldec2, chlinv2.

```
public static void chldecinv2 (double a[][], int n, double aux[])
{
  chldec2 (a,n,aux);
  if (aux[3] == n) chlinv2 (a,n);
}
```

D. chldecinv1

Calculates the inverse of a symmetric positive definite matrix by calling *chldec1* and, if the call is successful, *chlinv1*. The upper triangle of the coefficient matrix must be stored columnwise in a one-dimensional array.

Procedure parameters:

$$\text{void chldecinv1 } (a,n,aux)$$

a: double *a[1:(n+1)n/2]*;

      entry: the upper triangular part of the symmetric positive definite matrix must be given columnwise in array *a* (the elements (i,j)-th element of the matrix must be given in *a[(j-1)j/2+i]*, *1≤i≤j≤n*);

      exit: the upper triangular part of the inverse matrix is delivered columnwise in *a*;

n: int;

      entry: the order of the matrix;

aux: double *aux[2:3]*;

      entry: *aux[2]*:   a relative tolerance used to control the calculation of the diagonal elements;

      normal exit: *aux[3] = n*;

      abnormal exit: if the decomposition fails because the matrix is not positive definite then *aux[3] = k-1*, where *k* is the last stage number.

Procedures used: chldec1, chlinv1.

```
public static void chldecinv1(double a[], int n, double aux[])
{
  chldec1(a,n,aux);
  if (aux[3] == n) chlinv1(a,n);
}
```

3.3 General real symmetric matrices

3.3.1 Preparatory procedure

decsym2

Calculates the LDL^T decomposition of a symmetric matrix which may be indefinite and/or singular. Before the decomposition is performed, a check is made to see whether the matrix is symmetric. If the matrix is asymmetric then no decomposition is performed.

Given a real symmetric $n \times n$ matrix A, it attempts to obtain the pivot reference integers $p(i)$ associated with a permutation matrix P, a unit lower triangular matrix L and a block diagonal matrix D with 1×1 or 2×2 blocks of elements on its principal diagonal and zeros elsewhere, such that $P^T AP = LDL^T$. The decomposition process used involves the recursive construction of a system of $i \times i$ matrices $A^{(i)}$ $(i=n,...,k)$ with $A^{(n)} = AA^{(i-s)}$, where $s=1$ or $s=2$ is determined from $A^{(i)}$, by use of the decomposition

$$P^{(i)^T} A^{(i)} P^{(i)} = \begin{pmatrix} E & C^T \\ C & Y \end{pmatrix}$$

where $P^{(i)}$ is an $i \times i$ permutation matrix, and E is $s \times s$ nonsingular and of the reduction formula

$$A^{(i-s)} = Y - CE^{-1}C^T.$$

In the following $I_{m,r}$ is the permutation matrix obtained by interchanging rows m and r of the $i \times i$ identity matrix I; $\alpha = (1 + 17^{1/2})/8 \approx 0.6404$ is a fixed parameter; the determinations of $P^{(i)}$ and s at each stage result from computations of a sequence of steps which may terminate at the stages a2, b2, c1, or d1 (thus if termination at stage a2 does not take place, b1 and b2 are embarked upon, etc.).

a1: obtain $\lambda = \max | A^{(i)}_{k,1}|$ $(2 \le k \le i)$ and r for which $\lambda = | A^{(i)}_{r,1}|$;

a2: if $| A^{(i)}_{1,1}| \ge \alpha\lambda$, set $p^{(i)} = I, s=1$;

b1: obtain $\sigma = \max | A^{(i)}_{k,r}|$ $(1 \le k, r \le i; k \ne r)$;

b2: if $| A^{(i)}_{1,1}| \sigma \ge \alpha \lambda^2$, set $p^{(i)} = I, s = 1$;

c1: if $|A_{r,r}^{(i)}| \geq \alpha\sigma$, set $P^{(i)} = I_{1,r}$, s=1;

d1: set $P^{(i)} = I_{2,r}$, s=2.

With

$$\mu^{(i)} = \max|A_{j,k}^{(i)}| \ (1 \leq j,k \leq i),$$

the choice of $\alpha = (1 + 17^{1/2})/8$ above leads to the element growth inequality

$$\max(\mu^{(1)}, \mu^{(2)}) \leq 2.57^{n-1} \mu^{(n)}.$$

When $D_{i,i+1}=0$, $L_{i+1,i}=0$, so that if the block structure of D is known, the elements of D and L may be stored in the closed upper triangular part of the two dimensional array in which the elements of A are stored at entry. At exit from *decsym2*, the successive locations of the one-dimensional integer array p in the parameter list of *decsym2* contain not only the pivot reference integers associated with the $P^{(i)}$ above, but also information concerning the block structure of D: if $p[i]>0$ and $p[i+1]=0$, $D_{i,i+1}\neq0$, and $L_{i+1,i}=0$. Upon successful exit from *decsym2*, the successive locations of the one-dimensional real array *detaux* contains numbers which are of use in computing the determinant of A. If $p[i]$, $p[i+1] > 0$, $detaux[i] = D_{i,i}$; if $p[i] > 0$, $p[i+1] = 0$, $detaux[i] = 1$ and $detaux[i+1]$ is the determinant of

$$\begin{pmatrix} D_{i,i} & D_{i,i+1} \\ D_{i+1,i} & D_{i+1,i+1} \end{pmatrix}.$$

If g is the number of 1×1 blocks in D, so that $h=(n-g)/2$ is the number of 2×2 blocks, and u, v, and w are respectively the numbers of positive, negative, and zero 1×1 blocks, then $u+h$ is the number of positive eigenvalues of A, $v+h$ the number of negative eigenvalues, and w the number of zero eigenvalues (these numbers are delivered in successive locations of *aux* at exit from *decsym2*). The decision as to whether a 1×1 block $D_{i,i}$ is zero is governed by the small real number τ allocated to the real variable *tol* upon call of *decsym2*: if $|D_{i,i}| < \tau$, $D_{i,i}$ is taken to be zero.

Procedure parameters:

void decsym2 (a,n,tol,aux,p,detaux)

a: double a[1:n,1:n];

          entry: the symmetric coefficient matrix;

          exit: the elements of the LDL^T decomposition of a are stored in the upper triangular part of a; D is a block diagonal matrix with blocks of order 1 or 2, for a block of order 2 we always have $D_{i,i+1}\neq0$ and $L_{i+1,i}=0$, so that D and L^T fit in the upper triangular part of a, the strictly lower triangular part of a is left undisturbed;

n: int;

          entry: the order of the matrix;

tol: double;

          entry: a relative tolerance used to control the calculation of the block diagonal elements, the value of τ above;

aux: int aux[2:5];

          exit: aux[2]: if the matrix is symmetric then 1; otherwise 0, and no decomposition of a is performed;

aux[3]:	if the matrix is symmetric then the number of its positive eigenvalues, otherwise 0; if *aux[3]=n* then the matrix is positive definite;
aux[4]:	if the matrix is symmetric then the number of its negative eigenvalues, otherwise 0; if *aux[4]=n* then the matrix is negative definite;
aux[5]:	if the matrix is symmetric then the number of its zero eigenvalues, otherwise 0; if *aux[5]=0* then the matrix is symmetric and nonsingular;
p:	int *p[1:n]*;
exit:	a vector recording (1) the interchanges performed on array *a* during the computation of the decomposition, and (2) the block structure of *D*; if *p[i] > 0* and *p[i+1] = 0* a *2×2* block has been found ($D_{i,i+1} \neq 0$ and $L_{i+1,i}=0$);
detaux:	double *detaux[1:n]*;
exit:	if *p[i]>0* and *p[i+1]>0* then *detaux[i]* equals the exit value of *a[i,i]*; if *p[i]>0* and *p[i+1]=0* then *detaux[i]=1* and *detaux[i+1]* equals the value of the determinant of the corresponding *2×2* diagonal block as determined by *decsym2*.

Procedures used: elmrow, ichrow, ichrowcol.

Remark: The procedure *decsym2* computes the *LDL^T* decomposition of a symmetric matrix according to a method due to Bunch, Kaufman, and Parlett [BunK77, BunKP76]. For the inertia problem it is important that *decsym2* can accept singular matrices. However, in order to find the number of zero eigenvalues of singular matrices, the singular value decomposition might be preferred.

```
public static void decsym2(double a[][], int n, double tol, int aux[], int p[],
                           double detaux[])
{
  int i,j,m,ip1,ip2;
  boolean onebyone,sym;
  double det,s,t,alpha,lambda,sigma,aii,aip1,aip1i,temp;

  aux[3]=aux[4]=0;
  sym=true;
  i=0;
  while (sym && (i < n)) {
    i++;
    j=i;
    while (sym && (j < n)) {
      j++;
      sym = sym && (a[i][j] == a[j][i]);
    }
  }
  if (sym)
    aux[2]=1;
  else {
```

```
    aux[2]=0;
    aux[5]=n;
    return;
  }
  alpha=(1.0+Math.sqrt(17.0))/8.0;
  p[n]=n;
  i=1;
  while (i < n) {
    ip1=i+1;
    ip2=i+2;
    aii=Math.abs(a[i][i]);
    p[i]=i;
    lambda=Math.abs(a[i][ip1]);
    j=ip1;
    for (m=ip2; m<=n; m++)
      if (Math.abs(a[i][m]) > lambda) {
        j=m;
        lambda=Math.abs(a[i][m]);
      }
    t=alpha*lambda;
    onebyone=true;
    if (aii < t) {
      sigma=lambda;
      for (m=ip1; m<=j-1; m++)
        if (Math.abs(a[m][j]) > sigma) sigma=Math.abs(a[m][j]);
      for (m=j+1; m<=n; m++)
        if (Math.abs(a[j][m]) > sigma) sigma=Math.abs(a[j][m]);
      if (sigma*aii < lambda) {
        if (alpha*sigma < Math.abs(a[j][j])) {
          Basic.ichrow(j+1,n,i,j,a);
          Basic.ichrowcol(ip1,j-1,i,j,a);
          t=a[i][i];
          a[i][i]=a[j][j];
          a[j][j]=t;
          p[i]=j;
        } else {
          if (j > ip1) {
            Basic.ichrow(j+1,n,ip1,j,a);
            Basic.ichrowcol(ip2,j-1,ip1,j,a);
            t=a[i][i];
            a[i][i]=a[j][j];
            a[j][j]=t;
            t=a[i][j];
            a[i][j]=a[i][ip1];
            a[i][ip1]=t;
          }
          temp=a[i][ip1];
          det=a[i][i]*a[ip1][ip1]-temp*temp;
          aip1i=a[i][ip1]/det;
          aii=a[i][i]/det;
```

```
            aip1=a[ip1][ip1]/det;
            p[i]=j;
            p[ip1]=0;
            detaux[i]=1.0;
            detaux[ip1]=det;
            for (j=ip2; j<=n; j++) {
              s=aip1i*a[ip1][j]-aip1*a[i][j];
              t=aip1i*a[i][j]-aii*a[ip1][j];
              Basic.elmrow(j,n,j,i,a,a,s);
              Basic.elmrow(j,n,j,ip1,a,a,t);
              a[i][j]=s;
              a[ip1][j]=t;
            }
            aux[3]++;
            aux[4]++;
            i=ip2;
            onebyone=false;
          }
        }
      }
      if (onebyone) {
        if (tol < Math.abs(a[i][i])) {
          aii=a[i][i];
          detaux[i]=a[i][i];
          if (aii > 0.0)
            aux[3]++;
          else
            aux[4]++;
          for (j=ip1; j<=n; j++) {
            s = -a[i][j]/aii;
            Basic.elmrow(j,n,j,i,a,a,s);
            a[i][j]=s;
          }
        }
        i=ip1;
      }
    }
    if (i == n) {
      if (tol < Math.abs(a[n][n])) {
        if (a[n][n] > 0.0)
          aux[3]++;
        else
          aux[4]++;
      }
      detaux[n]=a[n][n];
    }
    aux[5]=n-aux[3]-aux[4];
}
```

3.3.2 Calculation of determinant

determsym2

Calculates the determinant of a symmetric matrix A, det(A). The procedure *decsym2* should be called to perform the LDL^T decomposition of the symmetric matrix. Given the values of the determinants of the blocks of the $n \times n$ block diagonal matrix D (which has 1×1 or 2×2 blocks of elements on its principal diagonal and zeros elsewhere) and the number m of zero eigenvalues of A, *determsym2* evaluates det(A). If $m \neq 0$ then det(A)=0, otherwise det(A) is simply the product of the above determinants of blocks.

Procedure parameters:

$$\text{double determsym2 } (detaux, n, aux)$$

determsym2: delivers the calculated value of the determinant of the matrix;
detaux: double *detaux[1:n]*;
        entry: the array *detaux* as produced by *decsym2*;
n: int; the order of the array *detaux* (= the order of the matrix);
aux: int *aux[2:5]*;
        entry: the array *aux* as produced by *decsym2*.

```
public static double determsym2(double detaux[], int n, int aux[])
{
  int i;
  double det;

  if (aux[5] > 0)
    det=0.0;
  else {
    det=1.0;
    for (i=1; i<=n; i++) det *= detaux[i];
  }
  return (det);
}
```

3.3.3 Solution of linear equations

A. solsym2

Solves a symmetric system of linear equations, assuming that the matrix has been decomposed into LDL^T form by a call of *decsym2*.

Procedure parameters:

void solsym2 (a,n,b,p,detaux)

a: double a[1:n, 1:n];
          entry: the LDL^T decomposition of A as produced by decsym2;

n: int;
          entry: the order of the matrix;

b: double b[1:n];
          entry: the right-hand side of the system of linear equations;
          exit: the solution of the system;

p: int p[1:n];
          entry: a vector recording the interchanges performed on the matrix by decsym2, p also contains information on the block structure of the matrix as decomposed by decsym2;

detaux: double detaux[1:n];
          entry: the array detaux as produced by decsym2.

Procedures used: matvec, elmvecrow.

```
public static void solsym2(double a[][], int n, double b[], int p[],
                           double detaux[])
{
  int i,ii,k,ip1,pi,pii;
  double det,temp,save;

  save=0.0;
  i=1;
  while (i < n) {
    ip1=i+1;
    pi=p[i];
    save=b[pi];
    if (p[ip1] > 0) {
      b[pi]=b[i];
      b[i]=save/a[i][i];
      Basic.elmvecrow(ip1,n,i,b,a,save);
      i=ip1;
    } else {
      temp=b[i];
      b[pi]=b[ip1];
      det=detaux[ip1];
      b[i]=(temp*a[ip1][ip1]-save*a[i][ip1])/det;
      b[ip1]=(save*a[i][i]-temp*a[i][ip1])/det;
      Basic.elmvecrow(i+2,n,i,b,a,temp);
      Basic.elmvecrow(i+2,n,ip1,b,a,save);
      i += 2;
    }
  }
  if (i == n) {
    b[i] /= a[i][i];
    i=n-1;
  } else
    i=n-2;
```

```
   while (i > 0) {
     if (p[i] == 0)
       ii=i-1;
     else
       ii=i;
     for (k=ii; k<=i; k++) {
       save=b[k];
       save += Basic.matvec(i+1,n,k,a,b);
       b[k]=save;
     }
     pii=p[ii];
     b[i]=b[pii];
     b[pii]=save;
     i=ii-1;
   }
 }
}
```

B. decsolsym2

Computes the solution of a symmetric system of linear equations by first calling
decsym2 to compute the *LDL^T* decomposition of the symmetric matrix. If the
matrix is found to be nonsingular, then the procedure *solsym2* is called to
compute the solution vector, otherwise the procedure *solsym2* is not called.

Procedure parameters:
$$\text{void decsolsym2 } (a,n,b,tol,aux)$$

a: double $a[1:n, 1:n]$;
         entry: see *decsym2*;
         exit: see *decsym2*;
n: int;
         entry: the order of the matrix;
b: double $b[1:n]$;
         entry: the right-hand side of the system of linear equations;
         exit: if the matrix *a* is nonsingular, then *b* contains the solution of the
                 system, otherwise *b* is left undisturbed;
tol: double;
         entry: see *decsym2*;
aux: int $aux[2:5]$;
         exit: see *decsym2*.

Procedures used: decsym2, solsym2.

```
public static void decsolsym2(double a[][], int n, double b[], double tol,
                              int aux[])
{
   int p[] = new int[n+1];
   double detaux[] = new double[n+1];
```

```
  decsym2(a,n,tol,aux,p,detaux);
  if (aux[5] == 0) solsym2(a,n,b,p,detaux);
}
```

3.4 Real full rank overdetermined systems

3.4.1 Preparatory procedures

A. lsqortdec

Reduces an $n \times m$ matrix A $(n \geq m)$ to the column permuted form A', where $R=QA'$, R being an $n \times m$ upper triangular matrix, and Q being an $n \times m$ orthogonal matrix having the form

$$\prod_{\upsilon=1}^{p} \left(1 - 2u^{(\upsilon)^T} u^{(\upsilon)} / u^{(\upsilon)} u^{(\upsilon)^T}\right) \quad (p \leq m).$$

Q is the product of at most m Householder matrices which are represented by their generating vectors. A is reduced to R in at most m stages. At the k-th stage the k-th column of the (already modified) matrix is interchanged with the column of maximum Euclidean norm (the pivot column). Then the matrix is multiplied by a Householder matrix such that the subdiagonal elements of the k-th column become zero while the first k-1 columns remain unchanged. The process terminates prematurely if at some stage the Euclidean norm of the pivotal column is less than some tolerance (a given tolerance $aux[2]$ times the maximum of the Euclidean norms of the columns of the given matrix).

Procedure parameters:
$$\text{void lsqortdec } (a,n,m,aux,aid,ci)$$

a: double $a[1:n,1:m]$;
        entry: the coefficient matrix of the linear least square problem;
        exit: in the upper triangle of a (the elements $a[i,j]$ with $i<j$) the superdiagonal elements of the upper triangular matrix (matrix R above) produced by the Householder transformation; in the other part of the columns of a the significant elements of the generating vectors of the Householder matrices used for the Householder triangularization (values of $u^{(\upsilon)}$ above);

n: int;
        entry: the number of rows of the matrix;
m: int;
        entry: the number of columns of the matrix $(n \geq m)$;
aux: double $aux[2:5]$;
        entry: $aux[2]$:   contains a relative tolerance used for calculating the diagonal elements of the upper triangular matrix;

exit: *aux[3]*:   delivers the number of the diagonal elements of the upper triangular matrix which are found not negligible, normal exit *aux[3]* = *m*;

aux[5]:   the maximum of the Euclidean norms of the columns of the given matrix;

aid: double *aid[1:m]*;

normal exit: *aux[3]* = *m*, *aid* contains the diagonal elements of the upper triangular matrix produced by the Householder triangularization;

ci: int *ci[1:m]*;

exit: contains the pivotal indices of the interchanges of the columns of the given matrix.

Procedures used: tammat, elmcol, ichcol.

Remark: The method is Householder triangularization with column interchanges. It is a modification of [Dek68] where a derivation is given by [BusiG65].

```
public static void lsqortdec(double a[][], int n, int m, double aux[], double aid[],
                             int ci[])
{
  int j,k,kpiv;
  double beta,sigma,norm,w,eps,akk,aidk,temp;
  double sum[] = new double[m+1];

  norm=0.0;
  aux[3]=m;
  for (k=1; k<=m; k++) {
    w=sum[k]=Basic.tammat(1,n,k,k,a,a);
    if (w > norm) norm=w;
  }
  w=aux[5]=Math.sqrt(norm);
  eps=aux[2]*w;
  for (k=1; k<=m; k++) {
    sigma=sum[k];
    kpiv=k;
    for (j=k+1; j<=m; j++)
      if (sum[j] > sigma) {
        sigma=sum[j];
        kpiv=j;
      }
    if (kpiv != k) {
      sum[kpiv]=sum[k];
      Basic.ichcol(1,n,k,kpiv,a);
    }
    ci[k]=kpiv;
    akk=a[k][k];
    sigma=Basic.tammat(k,n,k,k,a,a);
    w=Math.sqrt(sigma);
    aidk=aid[k]=((akk < 0.0) ? w : -w);
```

```
    if (w < eps) {
      aux[3]=k-1;
      break;
    }
    beta=1.0/(sigma-akk*aidk);
    a[k][k]=akk-aidk;
    for (j=k+1; j<=m; j++) {
      Basic.elmcol(k,n,j,k,a,a,-beta*Basic.tammat(k,n,k,j,a,a));
      temp=a[k][j];
      sum[j] -= temp*temp;
    }
  }
}
}
```

B. lsqdglinv

Computes the principal diagonal elements of the inverse of A^TA, where A is the coefficient matrix of a linear least squares problem. It is assumed that A has been decomposed after calling *lsqortdec* successfully. These values can be used for the computation of the standard deviations of least squares solutions.

Procedure parameters:
$$\text{void lsqdglinv } (a,m,aux,ci,diag)$$
a,m,aid,ci: see *lsqortdec*; the contents of *a, aid,* and *ci* should be produced by a successful call of *lsqortdec*;
diag: double *diag[1:m]*;
                  exit: the diagonal elements of the inverse of A^TA, where A is the matrix of the linear least squares problem.

Procedures used: vecvec, tamvec.

```
public static void lsqdglinv(double a[][], int m, double aid[], int ci[],
                             double diag[])
{
  int j,k,cik;
  double w;

  for (k=1; k<=m; k++) {
    diag[k]=1.0/aid[k];
    for (j=k+1; j<=m; j++)
      diag[j] = -Basic.tamvec(k,j-1,j,a,diag)/aid[j];
    diag[k]=Basic.vecvec(k,m,0,diag,diag);
  }
  for (k=m; k>=1; k--) {
    cik=ci[k];
    if (cik != k) {
      w=diag[k];
```

```
        diag[k]=diag[cik];
        diag[cik]=w;
      }
    }
  }
}
```

3.4.2 Least squares solution

A. lsqsol

Determines the least squares solution of the overdetermined system $Ax = b$, where A is an $n \times m$ matrix ($n \geq m$). It is assumed that A has been decomposed by calling *lsqortdec* successfully. The least squares solutions of several overdetermined systems with the same coefficient matrix can be solved by successive calls of *lsqsol* with different right-hand sides.

Procedure parameters:
$$\text{void lsqsol } (a,n,m,aid,ci,b)$$
a,n,m,aid,ci: see *lsqortdec*; the contents of the arrays *a, aid*, and *ci* should be produced by a successful call of *lsqortdec*;

b: double $b[1:n]$;

         entry: contains the right-hand side of a linear least squares problem;

         exit: $b[1:m]$ contains the solution of the problem;

                $b[m+1:n]$ contains a vector with Euclidean length equal to the Euclidean length of the residual vector.

Procedures used: matvec, tamvec, elmveccol.

```
public static void lsqsol(double a[][], int n, int m, double aid[], int ci[],
                          double b[])
{
  int k,cik;
  double w;

  for (k=1; k<=m; k++)
    Basic.elmveccol(k,n,k,b,a,Basic.tamvec(k,n,k,a,b)/(aid[k]*a[k][k]));
  for (k=m; k>=1; k--)
    b[k]=(b[k]-Basic.matvec(k+1,m,k,a,b))/aid[k];
  for (k=m; k>=1; k--) {
    cik=ci[k];
    if (cik != k) {
      w=b[k];
      b[k]=b[cik];
      b[cik]=w;
```

```
        }
    }
}
```

B. lsqortdecsol

Computes the least squares solution of an overdetermined system $Ax = b$ (n linear equations in m unknowns), and computes the principal diagonal elements of the inverse of A^TA. The matrix A is first reduced to the column permuted form A', where $R = QA'$ (see *lsqortdec*) by calling *lsqortdec* and, if this call is successful, the least squares solutions are determined by *lsqsol* and the principal diagonal elements are calculated by *lsqdglinv*.

Procedure parameters:

                  void lsqortdecsol (a,n,m,aux,diag,b)
a: double *a[1:n,1:m]*;
           entry: the coefficient matrix of the linear least square problem;
           exit: in the upper triangle of *a* (the elements *a[i,j]* with *i<j*) the
                   superdiagonal elements of the upper triangular matrix (matrix *R*
                   above) produced by the Householder transformation; in the other
                   part of the columns of *a* the significant elements of the generating
                   vectors of the Householder matrices used for the Householder
                   triangularization;
n: int;
           entry: the number of rows of the matrix;
m: int;
           entry: the number of columns of the matrix ($n \geq m$);
aux: double *aux[2:5]*;
           entry: *aux[2]:* contains a relative tolerance used for calculating the
                              diagonal elements of the upper triangular matrix;
           exit: *aux[3]:* delivers the number of the diagonal elements of the upper
                              triangular matrix which are found not negligible, normal
                              exit *aux[3] = m*;
                   aux[5]: the maximum of the Euclidean norms of the columns of
                              the given matrix;
diag: double *diag[1:m]*;
           exit: the diagonal elements of the inverse of A^TA, where A is the matrix
                   of the linear least squares problem;
b: double *b[1:n]*;
           entry: contains the right-hand side of a linear least squares problem;
           exit: *b[1:m]* contains the solution of the problem;
                   b[m+1:n] contains a vector with Euclidean length equal to the
                   Euclidean length of the residual vector.

Procedures used: lsqortdec, lsqdglinv, lsqsol.

```
public static void lsqortdecsol(double a[][], int n, int m, double aux[],
                                double diag[], double b[])
{
  int ci[] = new int[m+1];
  double aid[] = new double[m+1];

  lsqortdec(a,n,m,aux,aid,ci);
  if (aux[3] == m) {
    lsqdglinv(a,m,aid,ci,diag);
    lsqsol(a,n,m,aid,ci,b);
  }
}
```

3.4.3 Inverse matrix of normal equations

lsqinv

Calculates the inverse of the matrix A^TA, where A is the coefficient matrix of a linear least squares problem. *lsqinv* can be used for the calculation of the covariance matrix of a linear least squares problem. Given the $m{\times}m$ upper triangular matrix U and the pivot reference integers p_i, $i=1,...,m$, associated with the permutation matrix P, both occurring in the decomposition $Q^TU=AP$ of the $n{\times}m$ matrix A $(n{\geq}m)$, where Q is $n{\times}n$ orthogonal, *lsqinv* constructs the $m{\times}m$ inverse matrix $X = (A^TA)^{-1}$. *lsqinv* is to be called after a successful call of *lsqortdec* which obtains the decomposition in question. Since $U^TU = P^TA^TAP$, X may be obtained by inverting U^TU in a call of *chlinv2*. Afterwards the covariance matrix is obtained by interchanges of the columns and rows of the inverse matrix.

Procedure parameters:
$$\text{void lsqinv } (a,m,aid,c)$$

a: double $a[1{:}m, 1{:}m]$;

          entry: in the upper triangle of a (the elements $a[i,j]$ with $1{\leq}i{<}j{\leq}m$) the superdiagonal elements should be given of the upper triangular matrix (U above) that is produced by the Householder triangularization in a call of the procedure *lsqortdec* with normal exit $(aux[3]{=}m)$;

          exit: the upper triangle of the symmetric inverse matrix is delivered in the upper triangular elements of the array a ($a[i,j]$, $1{\leq}i{\leq}j{\leq}m$);

m: int;

          entry: the number of columns of the matrix of the linear least squares problem;

aid: double $aid[1{:}m]$;

          entry: contains the diagonal elements of the upper triangular matrix produced by *lsqortdec*;

ci: int $ci[1{:}m]$;

          entry: contains the pivotal indices produced by a call of *lsqortdec*.

Procedures used: chlinv2, ichcol, ichrow, ichrowcol.

```
public static void lsqinv(double a[][], int m, double aid[], int c[])
{
  int i,ci;
  double w;

  for (i=1; i<=m; i++) a[i][i]=aid[i];
  chlinv2(a,m);
  for (i=m; i>=1; i--) {
    ci=c[i];
    if (ci != i) {
      Basic.ichcol(1,i-1,i,ci,a);
      Basic.ichrow(i+1,ci-1,i,ci,a);
      Basic.ichrow(ci+1,m,i,ci,a);
      w=a[i][i];
      a[i][i]=a[ci][ci];
      a[ci][ci]=w;
    }
  }
}
```

3.4.4 Least squares with linear constraints

A. lsqdecomp

Let A denote the given matrix. *lsqdecomp* produces an n-th order orthogonal matrix Q and an $n \times m$ upper triangular matrix U such that $U=QA$ with permuted columns.

The constrained least squares problem considered is that of determining $x \in R^m$ which minimizes $\|r_2\|_E$ where

$$r_2 = b_2 - A_2 x, \quad b_2 \in R^{n_2},$$

and A_2 is an $n_2 \times m$ matrix, subject to the condition that

$$A_1 x = b \quad \text{where} \quad b_1 \in R^{n_1},$$

A_1 being $n_1 \times m$ ($n_1 + n_2 = n$). The required solution satisfies the equation

$$\begin{pmatrix} 0 & 0 & A_1 \\ 0 & I & A_2 \\ A_1^T & A_2^T & 0 \end{pmatrix} \begin{pmatrix} \lambda \\ r_2 \\ x \end{pmatrix} = \begin{pmatrix} b_1 \\ b_2 \\ c \end{pmatrix} \quad \text{where } \lambda \in R^{n_1}, \tag{1}$$

λ is a vector of Lagrange multipliers, and $c \in R^m$ is zero. In the following,

$(A,B)^t$ denotes the compound matrix $\begin{bmatrix} A \\ B \end{bmatrix}$.

Let P be an $m{\times}m$ permutation matrix such that $(A_1,A_2)^t = (E,F)^t$ where $E=(G_{1,1},G_{1,2})$, $F=(G_{2,1},G_{2,2})$, $G_{1,1}$ being an $n_1{\times}n_1$ nonsingular matrix and, for example, $G_{2,2}$ being $(n-n_1){\times}(m-n_1)$. Determine an $n_1{\times}n_1$ orthogonal matrix Q_1 such that $Q_1E=(U_1,R_1)$ where U_1 is an $n_1{\times}n_1$ upper triangular matrix, and R_1 is $n_1{\times}(m-n_1)$. Determine a further $(n-n_1){\times}(n-n_1)$ orthogonal matrix Q_2 such that

$$Q_2\left(G_{2,2} - G_{2,1}U_1^{-1}R_1\right) = \left(U_2,0\right)^t$$

where U_2 is an $(m-n_1){\times}(m-n_1)$ upper triangular matrix, and 0 represents an $(n-m){\times}(m-n_1)$ matrix of zeros. With

$$B = U_1^{-T}G_{2,1} \quad \text{and} \quad R = \left(I,0^T\right)Q_2$$

I representing the $(m-n_1){\times}(m-n_1)$ unit matrix, the two formulae

$$R = \begin{pmatrix} U_1 & R_1 \\ 0 & U_2 \end{pmatrix}, \quad APR^{-1} = \begin{pmatrix} Q_1 & B \\ 0 & R_2 \end{pmatrix}^T$$

are valid. Thus, setting

$$y = (y_1,y_2)^t, \quad d = (d_1,d_2)^t \quad \text{where } y_1,d_1 \in R^{n_1}, \quad y_2,d_2 \in R^{m-n_1}$$

with

$$x = PR^{-1}y, \quad d = P^T R^{-T}c \tag{2}$$

equation (1) may be rewritten as

$$\begin{pmatrix} 0 & 0 & Q_1^T & 0 \\ 0 & I & B^T & R^T \\ Q_1 & B & 0 & 0 \\ 0 & R_2 & 0 & 0 \end{pmatrix} \begin{pmatrix} \lambda \\ r_2 \\ y_1 \\ y_2 \end{pmatrix} = \begin{pmatrix} b_1 \\ b_2 \\ d_1 \\ d_2 \end{pmatrix}.$$

These equations may be solved: successively

$$y_1 = Q_1b_1, \quad y_2 = g_1 - d_1$$

where

$$g = (g_1,g_2)^t = Q_2(b_2 - B^T y_1),$$
$$r_2 = Q_2^T(d_2,g_2)^t, \quad \lambda = Q_1^T(d_1 - Br_2).$$

x may be obtained from y by use of the first of relationships (2).

The orthogonal matrices Q_1 and Q_2 and permutation matrix P are obtained recursively by means of a process which may be explained as follows. Starting with a matrix G and setting $G=A^{(1)}$, $A^{(k+1)}$ is obtained from $A^{(k)}$ by use of the relationship

$$A^{(k+1)} = (I - \beta_k u^{(k)}u^{(k)^T})A^{(k)}$$

where, with $n(k)=n_1$ if $k<n_1$ and $n(k)=n$ if $k>n_1$ and j so chosen that

$$s_j = \sum_{i=k}^{n(k)} \left(A_{i,j}^{(k)}\right)^2, \quad k \le j \le n$$

is a maximum,

$$A_{i,j}^{(k+1)} = 0 \quad \text{for} \quad i = k+1,\ldots,n(k).$$

(The k-th pivot reference integer $p(k)$ associated with P is j.)

$A_{i,j}^{(k+1)}$ is rendered zero as desired by setting

$$\sigma_k = \sqrt{s_j}, \quad \beta_k = \left(\sigma_k\left(\sigma_k + | A_{k,j}^{(k)} |\right)\right)^{-1}$$

$$u_i^{(k)} = 0 \quad \text{for} \quad i < k, \ i > n(k)$$

$$u_k^{(k)} = \text{sign}(A_{k,j}^{(k)})(\sigma_k + | A_{k,j}^{(k)} |)$$

$$u_i^{(k)} = A_{i,j}^{(k)} \quad k < i \le n(k)$$

With $G = A_1$ in the above, Q_2 is the product

$$\left(I - \beta_{n_1} u^{(n_1)} u^{(n_1)^T}\right) \cdots \left(I - \beta_1 u^{(1)} u^{(1)^T}\right) \quad \text{and} \quad A_1^{(n_1)} = (U_1, R_1).$$

Setting

$$A^{(n_1+1)} = \begin{pmatrix} U_1 & R_1 \\ 0 & R' \end{pmatrix}$$

where

$$R' = G_{2,2} - Q_1^T R_1, \quad A^{(m+1)} = (R,0)'.$$

The user must prescribe a small real tolerance *tol*. If in the above $(s_j)^{1/2}$ is less than *tol* multiplied by the maximum Euclidean norm over the columns of $A^{(k)}$, the decomposition process is broken off.

Procedure parameters:

            void lsqdecomp (a,n,m,n1,aux,aid,ci)

a: double *a[1:n,1:m]*;

       entry: the original least squares matrix, where the first *n1* rows should form the constraint matrix (i.e., the first *n1* equations are to be strictly satisfied);

       exit: in the upper triangle of *a* (the elements *a[i,j]* with *i<j*) the superdiagonal part of the upper triangular matrix produced by the Householder transformation; in the other part of the columns of *a* the significant elements of the generating vectors of the Householder matrices used for the Householder triangularization;

n: int;

       entry: the number of rows of the matrix;

m: int;

       entry: the number of columns of the matrix;

n1:   int;

       entry: number of linear constraints, i.e., the first *n1* rows of *a* set up a system of *n1* linear equations that must be strictly satisfied (if there are no constraints, *n1* must be chosen zero);

aux: double *aux[2:7]*;

       entry: *aux[2]*:   contains a relative tolerance used for calculating the diagonal elements of the upper triangular matrix;

       exit: *aux[3]*:   delivers the number of the diagonal elements which are not negligible, normal exit *aux[3] = m*;

> *aux[5]*: the maximum of the Euclidean norms of the columns of the given matrix;

aid: double *aid[1:m]*;
normal exit: *aux[3]* = *m*, *aid* contains the diagonal elements of the upper triangular matrix produced by the Householder transformation;

ci: int *ci[1:m]*;
exit: contains the pivotal indices of the interchanges of the columns of the given matrix.

Procedures used: matmat, tammat, elmcol, ichcol.

Remark: See [BjG67] for the QR decomposition of a least squares matrix.

```
public static void lsqdecomp(double a[][], int n, int m, int n1, double aux[],
                            double aid[], int ci[])
{
  int j,k,kpiv,nr,s;
  boolean fsum;
  double beta,sigma,norm,aidk,akk,w,eps,temp;
  double sum[] = new double[m+1];

  norm=0.0;
  aux[3]=m;
  nr=n1;
  fsum=true;
  for (k=1; k<=m; k++) {
    if (k == n1+1) {
      fsum=true;
      nr=n;
    }
    if (fsum)
      for (j=k; j<=m; j++) {
        w=sum[j]=Basic.tammat(k,nr,j,j,a,a);
        if (w > norm) norm=w;
      }
    fsum=false;
    eps=aux[2]*Math.sqrt(norm);
    sigma=sum[k];
    kpiv=k;
    for (j=k+1; j<=m; j++)
      if (sum[j] > sigma) {
        sigma=sum[j];
        kpiv=j;
      }
    if (kpiv != k) {
      sum[kpiv]=sum[k];
      Basic.ichcol(1,n,k,kpiv,a);
    }
    ci[k]=kpiv;
```

```
akk=a[k][k];
sigma=Basic.tammat(k,nr,k,k,a,a);
w=Math.sqrt(sigma);
aidk=aid[k]=((akk < 0.0) ? w : -w);
if (w < eps) {
  aux[3]=k-1;
  break;
}
beta=1.0/(sigma-akk*aidk);
a[k][k]=akk-aidk;
for (j=k+1; j<=m; j++) {
  Basic.elmcol(k,nr,j,k,a,a,-beta*Basic.tammat(k,nr,k,j,a,a));
  temp=a[k][j];
  sum[j] -= temp*temp;
}
if (k == n1)
  for (j=n1+1; j<=n; j++)
    for (s=1; s<=m; s++) {
      nr = (s > n1) ? n1 : s-1;
      w=a[j][s]-Basic.matmat(1,nr,j,s,a,a);
      a[j][s] = (s > n1) ? w : w/aid[s];
    }
  }
}
```

B. lsqrefsol

Solves a constrained least squares problem consisting of the determination that $x \in R^m$ which minimizes
$$\| r_2 \|_E \quad \text{where} \quad r_2 = b_2 - A_2 x, \quad b_2 \in R^{n_2},$$
and A_2 is an $n_2 \times m$ matrix, subject to the condition that
$$A_1 x = b_1 \quad \text{where} \quad b_1 \in R^{n_1},$$
A_1 being $n_1 \times m$ $(n_1+n_2=n)$. The required solution satisfies the equation $Bz=h$ where

$$B = \begin{pmatrix} 0 & 0 & A_1 \\ 0 & I & A_2 \\ A_1^T & A_2^T & 0 \end{pmatrix}$$

$$z = (\lambda, r_2, x)^T, \quad h = (b_1, b_2, 0)^T, \quad \lambda \in R^{n_1},$$

λ being a vector of Lagrange multipliers. It is assumed that the components of the vectors $u^{(k)}$ associated with the elementary reflectors defining orthogonal matrices Q_1 and Q_2 and those of an upper triangular matrix R, together with pivot reference integers associated with a permutation matrix P, have all been obtained by means of a successful call of *lsqdecomp*.

lsqrefsol first obtains a numerical solution $z^{(1)}$ of the equation $Bz=h$, and then uses an iterative scheme of the form $f^{(s)}=h-Bz^{(s)}$, $B\delta z^{(s)}=f^{(s)}$, $z^{(s+1)}=z^{(s)}+\delta z^{(s)}$, $s=1,2,...$ to obtain a refined solution to this equation, and in so doing a refined estimate of x, $z^{(1)}$ and, at each state, $\delta z^{(s)}$ are derived by the solution process outlined in the documentation to *lsqdecomp*. The above iterative scheme [BjG67] is terminated if either (a) $\| \delta z^{(s)} \|_E \leq \varepsilon \| z^{(s)} \|_E$ where ε is a small real tolerance prescribed by the user or (b) $s=smax$ where the integer *smax* is also prescribed by the user. The least squares solutions of several overdetermined systems with the same constraints and coefficient matrix can be solved by successive calls of *lsqrefsol* with different right-hand sides.

Procedure parameters:

> void lsqrefsol (a,qr,n,m,n1,aux,aid,ci,b,ldx,x,res)

a: double $a[1:n,1:m]$;
> entry: the original least squares matrix, where the first *n1* rows should form the constraint matrix (i.e., the first *n1* equations are to be strictly satisfied);

qr: double $qr[1:n,1:m]$;
> entry: the QR decomposition of the original least squares matrix as delivered by a successful call of *lsqdecomp*;

n: int;
> entry: the number of rows of the matrices *a* and *qr*;

m: int;
> entry: the number of columns of the matrices *a* and *qr*;

n1: int;
> entry: number of linear constraints;

aux: double $aux[2:7]$;
> entry: $aux[2]$: contains a relative tolerance (value of ε above) as a criterion to stop iterative refining, if the Euclidean norm of the correction is smaller than $aux[2]$ times the current approximation of the solution then the iterative refining is stopped;
>
> $aux[6]$: maximum number of iterations allowed (value of *smax* above), usually $aux[6]=5$ will be sufficient;
>
> exit: $aux[7]$: the number of iterations performed (the last value of *s* for which a correction term $\delta z^{(s)}$ is determined in the above);

aid: double $aid[1:m]$;
> entry: the diagonal elements of the upper triangular matrix as delivered by a successful call of *lsqdecomp*;

ci: int $ci[1:m]$;
> entry: the pivotal indices as produced by *lsqdecomp*;

b: double $b[1:n]$;
> entry: the right-hand side of the least squares problem; first *n1* elements form the right-hand sides of the constraints;

ldx: double $ldx[0:0]$;
> exit: the Euclidean norm of the last correction of the solution (the value of $\| \delta x^{(s)} \|_E$ for the last $\delta x^{(s)}$ determined in the above, $x^{(s)}$ being formed from the last *m* components of $z^{(s)}$);

x: double $x[1:m]$;
> exit: the solution vector;

res: double *res[1:n]*;
            exit: the residual vector ($f^{(s)}$ in the above) corresponding to the solution.

Procedures used: vecvec, matvec, tamvec, elmveccol, ichcol.

```
public static void lsqrefsol(double a[][], double qr[][], int n, int m, int n1,
                             double aux[], double aid[], int ci[], double b[],
                             double ldx[], double x[], double res[])
{
  boolean startup;
  int i,j,k,s;
  double c1,nexve,ndx,ndr,d,corrnorm,dtemp;
  double f[] = new double[n+1];
  double g[] = new double[m+1];

  for (j=1; j<=m; j++) {
    s=ci[j];
    if (s != j) Basic.ichcol(1,n,j,s,a);
  }
  for (j=1; j<=m; j++) x[j]=g[j]=0.0;
  for (i=1; i<=n; i++) {
    res[i]=0.0;
    f[i]=b[i];
  }
  k=0;
  do {
    startup = (k <= 1);
    ndx=ndr=0.0;
    if (k != 0) {
      for (i=1; i<=n; i++) res[i] += f[i];
      for (s=1; s<=m; s++) {
        x[s] += g[s];
        dtemp=0.0;
        for (i=1; i<=n; i++)
          dtemp += (double)a[i][s]*(double)res[i];
        d=dtemp;
        g[s]=(-d-Basic.tamvec(1,s-1,s,qr,g))/aid[s];
      }
      for (i=1; i<=n; i++) {
        dtemp = (i > n1) ? res[i] : 0.0;
        for (s=1; s<=m; s++)
          dtemp += (double)a[i][s]*(double)x[s];
        f[i]=(double)b[i]-dtemp;
      }
    }
    nexve=Math.sqrt(Basic.vecvec(1,m,0,x,x)+Basic.vecvec(1,n,0,res,res));
    for (s=1; s<=n1; s++)
      Basic.elmveccol(s,n1,s,f,qr,Basic.tamvec(s,n1,s,qr,f)/(qr[s][s]*aid[s]));
    for (i=n1+1; i<=n; i++)
      f[i] -= Basic.matvec(1,n1,i,qr,f);
```

```
    for (s=n1+1; s<=m; s++)
      Basic.elmveccol(s,n,s,f,qr,Basic.tamvec(s,n,s,qr,f)/(qr[s][s]*aid[s]));
    for (i=1; i<=m; i++) {
      c1=f[i];
      f[i]=g[i];
      g[i] = (i > n1) ? c1-g[i] : c1;
    }
    for (s=m; s>=1; s--) {
      g[s]=(g[s]-Basic.matvec(s+1,m,s,qr,g))/aid[s];
      ndx += g[s]*g[s];
    }
    for (s=m; s>=n1+1; s--)
      Basic.elmveccol(s,n,s,f,qr,Basic.tamvec(s,n,s,qr,f)/(qr[s][s]*aid[s]));
    for (s=1; s<=n1; s++)
      f[s] -= Basic.tamvec(n1+1,n,s,qr,f);
    for (s=n1; s>=1; s--)
      Basic.elmveccol(s,n1,s,f,qr,Basic.tamvec(s,n1,s,qr,f)/(qr[s][s]*aid[s]));
    aux[7]=k;
    for (i=1; i<=n; i++) ndr += f[i]*f[i];
    corrnorm=Math.sqrt(ndx+ndr);
    k++;
  } while (startup || (corrnorm>aux[2]*nexve && k<=aux[6]));
  ldx[0]=Math.sqrt(ndx);
  for (s=m; s>=1; s--) {
    j=ci[s];
    if (j != s) {
      c1=x[j];
      x[j]=x[s];
      x[s]=c1;
      Basic.ichcol(1,n,j,s,a);
    }
  }
}
```

3.5 Other real matrix problems

3.5.1 Solution of overdetermined systems

A. solsvdovr

Solves an overdetermined system of linear equations. *solsvdovr* determines that $x \in R^n$ with minimum $\|x\|_2$ which minimizes $\|Ax-b\|_2$, where A is a real $m \times n$ matrix, $b \in R^m$ ($m \geq n$), the matrices U, Λ, V occurring in the singular value decomposition $A = U \Lambda V^T$ being available, where U is an $m \times n$ column orthogonal matrix ($U^T U = I$), Λ is an $n \times n$ diagonal matrix whose diagonal elements are the singular values of A (λ_i, i=1,...,n), and V is an $n \times n$ orthogonal matrix ($V^TV=VV^T=I$). The analytic solution of the above problem is $x=A^+b$, where A^+ is the pseudo-inverse of A: numerically, $x=V\Lambda^{(-1)}U^Tb$, where $\Lambda^{(-1)}$ is a diagonal matrix whose successive diagonal elements are $(\lambda_i)^{-1}$ if $\lambda_i > \delta$, and 0 otherwise, δ being a small positive real number prescribed by the user. The two stages in the determination of x are the formation of $b'=\Lambda^{(-1)}U^Tb'$ and that of $x=Vb'$.

Procedure parameters:
$$\text{void solsvdovr } (u,val,v,m,n,x,em)$$
u: double $u[1:m,1:n]$;
            entry: the matrix U in the singular values decomposition $U\Lambda V^T$;
val: double $val[1:n]$;
            entry: the singular values (diagonal elements λ_i of Λ);
v: double $v[1:n,1:n]$;
            entry: the matrix V in the singular values decomposition;
m: int;
            entry: the length of the right-hand side vector;
n: int;
            entry: the number of unknowns, n should satisfy $n \leq m$;
x: double $x[1:m]$;
            entry: the right-hand side vector;
            exit: the solution vector;
em: double $em[6:6]$;
            entry: the minimal nonneglectable singular value (value of δ in the above).

Procedures used: matvec, tamvec.

Remark: See [WiR71] for the solution of an overdetermined system of linear equations.

```
public static void solsvdovr(double u[][], double val[], double v[][], int m,
                             int n, double x[], double em[])
{
  int i;
  double min;
  double x1[] = new double[n+1];

  min=em[6];
  for (i=1; i<=n; i++)
    x1[i] = (val[i] <= min) ? 0.0 : Basic.tamvec(1,m,i,u,x)/val[i];
  for (i=1; i<=n; i++) x[i]=Basic.matvec(1,n,i,v,x1);
}
```

B. solovr

Solves an overdetermined system of linear equations. *solovr* determines that $x \in R^n$ with minimum $\|x\|_2$ which minimizes $\|Ax-b\|_2$, where A is a real $m \times n$ matrix, $b \in R^m$ ($m \geq n$). *solovr* first calls *qrisngvaldec* to obtain the matrices U, Λ, V occurring in the singular value decomposition $A = U \Lambda V^T$ and, if all singular values of A can be determined by use of this procedure, calls *solsvdovr* to obtain x.

Procedure parameters:
$$\text{int solovr } (a, m, n, x, em)$$
solovr: given the number of singular values of a which cannot be found by *qrisngvaldec*;

a: double $a[1:m, 1:n]$;
entry: the matrix of the system;

m: int;
entry: the number of rows of a;

n: int;
entry: the number of columns of a, $n \leq m$;

x: double $x[1:m]$;
entry: the right-hand side vector;
exit: the solution vector;

em: double $em[0:7]$;
entry: $em[0]$: the machine precision;
$em[2]$: the relative precision of the singular values;
$em[4]$: the maximal number of iterations to be performed in the singular values decomposition;
$em[6]$: the minimal nonneglected singular value;
exit: $em[1]$: the infinity norm of the matrix;
$em[3]$: the maximal neglected superdiagonal element;
$em[5]$: the number of iterations performed in the singular values decomposition;
$em[7]$: the numerical rank of the matrix, i.e., the number of singular values $\geq em[6]$.

Procedures used: qrisngvaldec, solsvdovr.

```
public static int solovr(double a[][], int m, int n, double x[], double em[])
{
  int i;

  double val[] = new double[n+1];
  double v[][] = new double[n+1][n+1];
  i=qrisngvaldec(a,m,n,val,v,em);
  if (i == 0) solsvdovr(a,val,v,m,n,x,em);
  return i;
}
```

3.5.2 Solution of underdetermined systems

A. solsvdund

Solves an underdetermined system of linear equations [WiR71]. *solsvdund* determines that $x \in R^m$ with minimum $\|x\|_2$ which minimizes $\|A^T x-b\|$, where A is a real $m \times n$ matrix, $b \in R^n$ ($m \geq n$), the matrices U,Λ,V as described in the documentation to *solsvdovr* being available. The analytic solution of the above problem is $x=(A^+)^T b$, where A^+ is the pseudo-inverse of A. The two stages in the determination of x are the formation of $b'=\Lambda^{(-1)}V^T b$ and that of $x=Ub'$.

Procedure parameters:
$$\text{void solsvdund } (u,val,v,m,n,x,em)$$
u: double *u[1:m,1:n]*;
      entry: the matrix U in the singular values decomposition $U\Lambda V^T$;
val: double *val[1:n]*;
      entry: the singular values (diagonal elements of Λ);
v: double *v[1:n,1:n]*;
      entry: the matrix V in the singular values decomposition;
m: int;
      entry: the number of unknowns;
n: int;
      entry: the length of the right-hand side vector, n should satisfy $n \leq m$;
x: double *x[1:m]*;
      entry: the right-hand side vector in *x[1:n]*;
      exit: the solution vector in *x[1:m]*;
em: double *em[6:6]*;
      entry: the minimal nonneglectable singular value.

Procedures used: matvec, tamvec.

```
public static void solsvdund(double u[][], double val[], double v[][], int m,
                      int n, double x[], double em[])
{
  int i;
  double min;
  double x1[] = new double[n+1];

  min=em[6];
  for (i=1; i<=n; i++)
    x1[i] = (val[i] <= min) ? 0.0 : Basic.tamvec(1,n,i,v,x)/val[i];
  for (i=1; i<=m; i++)  x[i] = Basic.matvec(1,n,i,u,x1);
}
```

B. solund

Solves an underdetermined system of linear equations. *solund* determines that $x \in R^m$ with minimum $\|x\|_2$ which minimizes $\|A^T x - b\|_2$, where A is a real $m \times n$ matrix, $b \in R^n$ ($m \geq n$). *solund* first calls *qrisngvaldec* to obtain the matrices U, Λ, V occurring in the singular value decomposition $A = U \Lambda V^T$ and, if all singular values of A may be obtained by use of this procedure, calls *solsvdund* to obtain x.

Procedure parameters:
$$\text{int solund } (a, m, n, x, em)$$
solund: given the number of singular values of a which cannot be found by *qrisngvaldec*;
a: double $a[1:m, 1:n]$;
 entry: the transpose of the matrix;
m: int;
 entry: the number of rows of a;
n: int;
 entry: the number of columns of a, $n \leq m$;
x: double $x[1:m]$;
 entry: the right-hand side vector in $x[1:n]$;
 exit: the solution vector;
em: double $em[0:7]$;
 entry: $em[0]$: the machine precision;
 $em[2]$: the relative precision for the singular values;
 $em[4]$: the maximal number of iterations to be performed in the singular values decomposition;
 $em[6]$: the minimal nonneglected singular value;
 exit: $em[1]$: the infinity norm of the matrix;
 $em[3]$: the maximal neglected superdiagonal element;
 $em[5]$: the number of iterations performed in the singular values decomposition;
 $em[7]$: the numerical rank of the matrix,
 i.e., the number of singular values $\geq em[6]$.

Procedures used: qrisngvaldec, solsvdund.

```
public static int solund(double a[][], int m, int n, double x[], double em[])
{
  int i;
  double val[] = new double[n+1];
  double v[][] = new double[n+1][n+1];

  i=qrisngvaldec(a,m,n,val,v,em);
  if (i == 0) solsvdund(a,val,v,m,n,x,em);
  return i;
}
```

3.5.3 Solution of homogeneous equation

A. homsolsvd

Given the matrices U, Λ, V occurring in a real singular value decomposition $A = U\Lambda V^T$ where A is an $m \times n$ matrix $(m \geq n)$, the diagonal elements λ_i of Λ are reordered in such a way that, for the new order, $\lambda_i \geq \lambda_{i+1}$; the columns of U and V are simultaneously reordered to preserve the correspondence between the λ_i and the columns.

With u_i being the successive columns of U, v_i those of V, and λ_i the diagonal elements of Λ (all in the new ordering)

$$Av_i = \lambda_i u_i \qquad u_i^T A = \lambda_i v_i^T \qquad i = 1, \ldots, n.$$

If, for some $r<n$, $\lambda_i<\delta$ $(i=r+1,\ldots,n)$, δ being a small positive real number, then approximately

$$Av_i = u_i^T A = 0 \qquad i = r+1, \ldots, n$$

for any $x \in R^n$,

$$x = \sum_{i=1}^{n} x_i v_i, \qquad Ax = \sum_{i=1}^{r} x_i \lambda_i u_i,$$

i.e., Ax lies in the column subspace in R^m spanned by u_i $(i=1,\ldots,r)$; any vector x in the complementary column subspace in R^n spanned by $v_{r+1},\ldots,v_n$ satisfies the equation $Ax \approx 0$; for any $x^T \in R^m$, $x^T A$ lies in the column subspace in R^n spanned by v_i $(i=1,\ldots,r)$; any vector x in the complementary row subspace in R^m spanned by $(u_{r+1})^T, \ldots, (u_n)^T$ satisfies the equation $x^T A = 0$.

After a call of *homsolsvd*, the reordered columns of U and V and reordered λ_i are available in the two dimensional arrays u and v, and the one dimensional array *val*, respectively. The latter elements λ_i may thus be inspected, and the vector sets $v_1,\ldots,v_r$, $v_{r+1},\ldots,v_n$, $u_1,\ldots,u_r$, and $u_{r+1},\ldots,u_n$ defining the above four subspaces may be extracted [WiR71].

Procedure parameters:

void homsolsvd (u, val, v, m, n)

u: double $u[1:m, 1:n]$;
           entry: the matrix U in the singular values decomposition $U\Lambda V^T$;
           exit: the components $u[i,j]$ of U occurring in a reordered decomposition $A = U\Lambda V^T$ in which the diagonal elements of Λ are set in nonascending order;
val: double $val[1:n]$;
           entry: the singular values;
           exit: the array will be ordered in such a way that $val[i] < val[j]$ if $j<i$;
v: double $v[1:n, 1:n]$;
           entry: the matrix V in the singular values decomposition;
           exit: the components $v[i,j]$ of V in the reordered decomposition above;
m: int;
           entry: the number of rows of u;
n: int;
           entry: the number of columns of u;

Procedure used: ichcol.

```
public static void homsolsvd(double u[][], double val[], double v[][], int m, int n)
{
   int i,j;
   double x;

   for (i=n; i>=2; i--)
     for (j=i-1; j>=1; j--)
       if (val[i] > val[j]) {
         x=val[i];
         val[i]=val[j];
         val[j]=x;
         Basic.ichcol(1,m,i,j,u);
         Basic.ichcol(1,n,i,j,v);
       }
}
```

B. homsol

Given an $m \times n$ matrix A $(m \geq n)$, capable of real singular value decomposition, *homsol* determines the vectors $u_j \in R^m$ $(j=r,...,n)$ defining the column subspace in which Ax lies for all $x \in R^n$, the vectors $v_j \in R^n$ $(j=r+1,...,n)$ defining the complementary column subspace in which $x \in R^n$ must lie for the condition $Ax=0$ to hold, the vectors v_j $(j=1,...,r)$ defining the column subspace in which x^TA lies for all $x \in R^m$ and the vectors $(u_j)^T$ $(j=r+1,...,n)$ defining the complementary row subspace in which $x^T \in R^m$ must lie for the condition $x^TA=0$ to hold.

   homsol first calls *qrisngvaldec* to obtain the matrices U, Λ, V occurring in the singular value decomposition $A=U\Lambda V^T$ and, if all singular values of A can be determined by use this procedure, calls *homsolsvd* to reorder the columns of U and V and the diagonal elements of Λ. The numerical rank (r in the above) of A is determined by the conditions $\lambda_i > \delta$ $(i=1,...,r)$ $\lambda_i \leq \delta$ $(i=r+1,...,n)$ for the reordered λ_i, where δ is a small positive real number supplied by the user.

Procedure parameters:
                            int homsol (a,m,n,v,em)
homsol: given the number of singular values of a which cannot be found by
           qrisngvaldec;
a: double $a[1:m,1:n]$;
           entry: the matrix of the system;
           exit: the value of the components $(u_j)^{(i)}$ of u_j, $i=1,...,m$; $j=1,...,n$;
m: int;
           entry: the number of rows of a;
n: int;
           entry: the number of columns of a;
v: double $v[1:n,1:n]$;
           exit: the value of the components $(v_j)^{(i)}$ of v_j, $i,j=1,...,n$;

em: double *em[0:7]*;
            entry: *em[0]*:   the machine precision;
                     em[2]:   the relative precision of the singular values;
                     em[4]:   the maximal number of iterations to be performed in the
                                singular values decomposition;
                     em[6]:   the minimal nonneglected singular value;
            exit: *em[1]*:   the infinity norm of the matrix;
                     em[3]:   the maximal neglected superdiagonal element;
                     em[5]:   the number of iterations performed in the singular values
                                decomposition;
                     em[7]:   the numerical rank of the matrix,
                                i.e., the number of singular values $\geq$ *em[6]*.

Procedures used: qrisngvaldec, homsolsvd.

```
public static int homsol(double a[][], int m, int n, double v[][], double em[])
{
  int i;
  double val[] = new double[n+1];

  i=qrisngvaldec(a,m,n,val,v,em);
  if (i == 0) homsolsvd(a,val,v,m,n);
  return i;
}
```

3.5.4 Pseudo-inversion

A. psdinvsvd

Calculates the pseudo-inverse of a matrix [WiR71]. Given the matrices U,Λ,V
occurring in a real singular value decomposition $A=U\Lambda V^T$ of the $m{\times}n$ matrix A
$(m{\geq}n)$, *psdinvsvd* determines the generalized inverse A^+ of A (which satisfies and is
defined by the relationships $AA^+A=A$, $A^+AA^+=A^+$, AA^+ and A^+A symmetric in the real
case). With λ_i (i=1,...,n) being the diagonal elements of Λ, the successive diagonal
elements of the diagonal matrix $\Lambda^{(-1)}$ are determined by taking them to be $(\lambda_i)^{-1}$ if
$\lambda_i{>}\delta$, and 0 otherwise, δ being a small positive real number prescribed by the user.
Thereafter $A^+=V\Lambda^{(-1)}U^T$. The matrix $X=\Lambda^{(-1)}U^T$ is first constructed, and then $A^+=VX$.

Procedure parameters:
                        void psdinvsvd (*u,val,v,m,n,em*)
u: double *u[1:m, 1:n]*;
           entry: the matrix U in the singular values decomposition $U\Lambda V^T$;
           exit: the transpose of the pseudo-inverse;
val: double *val[1:n]*;
           entry: the singular values;

v:	double *v[1:n, 1:n]*;
	entry: the matrix *V* in the singular values decomposition;
m:	int;
	entry: the number of rows of *U*;
n:	int;
	entry: the number of columns of *V*;
em:	double *em[6:6]*;
	entry: the minimal nonneglectable singular value.

Procedure used: matvec.

```
public static void psdinvsvd(double u[][], double val[], double v[][], int m,
                    int n, double em[])
{
  int i,j;
  double min,vali;
  double x[] = new double[n+1];

  min=em[6];
  for (i=1; i<=n; i++)
    if (val[i] > min) {
      vali=1.0/val[i];
      for (j=1; j<=m; j++) u[j][i] *= vali;
    } else
      for (j=1; j<=m; j++) u[j][i]=0.0;
  for (i=1; i<=m; i++) {
    for (j=1; j<=n; j++) x[j]=u[i][j];
    for (j=1; j<=n; j++) u[i][j]=Basic.matvec(1,n,j,v,x);
  }
}
```

B. psdinv

Determines the generalized inverse A^+ of the *m×n* matrix *A* (*m≥n*). The singular value decomposition $A=U\Lambda V^T$ is first carried out by a call of *qrisngvaldec* and, if all singular values have been obtained, *psdinvsvd* is then called to construct A^+.

Procedure parameters:

$$\text{int psdinv } (a,m,n,em)$$

psdinv:	given the number of singular values of *a* which cannot be found by *qrisngvaldec*;
a:	double *a[1:m, 1:n]*;
	entry: the given matrix;
	exit: the transpose of the pseudo-inverse;
m:	int;
	entry: the number of rows of *a*;
n:	int;
	entry: the number of columns of *a*, *n≤m*;

em: double *em[0:7]*;
            entry: *em[0]*:   the machine precision;
                     em[2]:   the relative precision of the singular values;
                     em[4]:   the maximal number of iterations to be performed;
                     em[6]:   the minimal nonneglected singular value;
            exit: *em[1]*:   the infinity norm of the matrix;
                     em[3]:   the maximal neglected superdiagonal element;
                     em[5]:   the number of iterations performed in the singular values
                                decomposition;
                     em[7]:   the numerical rank of the matrix,
                                i.e., the number of singular values $\geq$ *em[6]*.

Procedures used: qrisngvaldec, psdinvsvd.

```
public static int psdinv(double a[][], int m, int n, double em[])
{
  int i;
  double val[] = new double[n+1];
  double v[][] = new double[n+1][n+1];

  i=qrisngvaldec(a,m,n,val,v,em);
  if (i == 0) psdinvsvd(a,val,v,m,n,em);
  return i;
}
```

3.6 Real sparse nonsymmetric band matrices

3.6.1 Preparatory procedure

decbnd

Performs the decomposition of a matrix whose nonzero elements are in band form, and whose band elements are stored rowwise in a one-dimensional array [Dek68].
Given an $n \times n$ matrix A for whose elements $A_{i,j}=0$ when $i>j+lw$ or $j>i+rw$ $(i,j=1,...,n)$, *decbnd* obtains (a) a unit lower triangular band matrix M for whose elements $M_{i,j}=0$ when $i>j+lw$ or $j>i$, and $M_{i,i}=1$; (b) an upper triangular band matrix U with $U_{i,j}=0$ when $j>i+lw+rw$ or $i>j$; and (c) a sequence of pivot reference integers $p(i)$ $(i=1,...,n)$ associated with a permutation matrix P, such that $MU=PA$, by Gauss elimination using partial pivoting.
The method used involves the recursive construction of matrices $A^{(i)}$, with $A^{(1)}=A$. At the i-th stage the elements $(A_{k,j})^{(i)}$ $(j=1,...,i-1; k=j+1,...,n)$ are zero and
$$(A_{k,j})^{(i)}=(U_{k,j})^{(i)} \ (k=1,...,i-1; j=k,k+1,...,n).$$
Then (a) the smallest integer I for which $|(A_{l,i})^{(i)}| \geq |(A_{k,i})^{(i)}|$ for $k \geq i$ is determined; (b) $p(i)$ is set equal to I; (c) rows i and I of $A^{(i)}$ are interchanged (the i-th row of $A^{(i+1)}$ has

now been determined); (d) with $M_{k,i}=(A_{k,i})^{(i)}/(A_{i,i})^{(i+1)}$, row($k$) of $A^{(i)}$ is replaced by row(k) - $M_{k,i}$row(i) to form row(k) of $A^{(i+1)}$ ($k=i+1,...,\min(n,i+lw)$). (The elements $(A_{k,i})^{(i+1)}$, $k>i$, are thus zero.) The process is arrested if $\delta>\delta_i$, where

$$\delta_i=|(A_{i,i})^{(i+1)}|/\|i\text{-th row of }A^{(i)}\|_2,$$

and δ is a small positive real number prescribed by the user.

Procedure parameters:

$$\text{void decbnd }(a,n,lw,rw,aux,m,pi)$$

a: double $a[1:(lw+rw)(n-1)+n]$;

     entry: *a* contains rowwise the band elements of the band matrix in such a way that the (i,j)-th element of the matrix is given in $a[(lw+rw)(i-1)+j]$, $i=1,...,n$ and $j=\max(1,i-lw),...,\min(n,i+rw)$, the values of the remaining elements of *a* are irrelevant;

     exit: the band elements of the Gaussian eliminated matrix, which is an upper triangular band matrix U with $(lw+rw)$ codiagonals, are rowwise delivered in *a* as follows: the (i,j)-th element of U is $a[(lw+rw)(i-j)+j]$, $i=1,...,n$ and $j=i,...,\min(n,i+lw+rw)$;

n: int;

     entry: the order of the band matrix;

lw: int;

     entry: number of left codiagonals of *a*;

rw: int;

     entry: number of right codiagonals of *a*;

aux: double $aux[1:5]$;

     entry: $aux[2]$:   a relative tolerance to control the elimination process (value of δ above);

     exit: $aux[1]$:   if successful, given the sign of the determinant of the matrix (+1 or -1);

         $aux[3]$:   if successful then $aux[3]=n$; otherwise $aux[3]=i-1$, where the reduction process was terminated at stage i;

         $aux[5]$:   if successful, given the minimum absolute value of pivot(i) divided by the Euclidean norm of the i-th row (value of $\min \delta_j$, $1\le j\le n$, in the above);

m: double $m[1:lw(n-2)+1]$;

     exit: the Gaussian multipliers (values of $M_{i,j}$ above) of all eliminations in such a way that the i-th multiplier of the j-th step is $m[lw(j-1)+i-j]$;

p: int $p[1:n]$;

     exit: the pivotal indices.

Procedures used: vecvec, elmvec, ichvec.

```
public static void decbnd(double a[], int n, int lw, int rw, double aux[],
                    double m[], int p[])
{
  int i,j,k,kk,kk1,pk,mk,ik,lw1,f,q,w,w1,w2,nrw,iw,sdet;
  double r,s,eps,min;
  double v[] = new double[n+1];
```

```
f=lw;
w1=lw+rw;
w=w1+1;
w2=w-2;
iw=0;
sdet=1;
nrw=n-rw;
lw1=lw+1;
q=lw-1;
for (i=2; i<=lw; i++) {
  q--;
  iw += w1;
  for (j=iw-q; j<=iw; j++) a[j]=0.0;
}
iw = -w2;
q = -lw;
for (i=1; i<=n; i++) {
  iw += w;
  if (i <= lw1) iw--;
  q += w;
  if (i > nrw) q--;
  v[i]=Math.sqrt(Basic.vecvec(iw,q,0,a,a));
}
eps=aux[2];
min=1.0;
kk = -w1;
mk = -lw;
if (f > nrw) w2 += nrw-f;
for (k=1; k<=n; k++) {
  if (f < n) f++;
  ik = kk += w;
  mk += lw;
  s=Math.abs(a[kk])/v[k];
  pk=k;
  kk1=kk+1;
  for (i=k+1; i<=f; i++) {
    ik += w1;
    m[mk+i-k]=r=a[ik];
    a[ik]=0.0;
    r=Math.abs(r)/v[i];
    if (r > s) {
      s=r;
      pk=i;
    }
  }
  if (s < min) min=s;
  if (s < eps) {
    aux[3]=k-1;
    aux[5]=s;
    aux[1]=sdet;
```

```
        return;
    }
    if (k+w2 >= n) w2--;
    p[k]=pk;
    if (pk != k) {
        v[pk]=v[k];
        pk -= k;
        Basic.ichvec(kk1,kk1+w2,pk*w1,a);
        sdet = -sdet;
        r=m[mk+pk];
        m[mk+pk]=a[kk];
        a[kk]=r;
    } else
        r=a[kk];
    if (r < 0.0) sdet = -sdet;
    iw=kk1;
    lw1=f-k+mk;
    for (i=mk+1; i<=lw1; i++) {
        s = m[i] /= r;
        iw += w1;
        Basic.elmvec(iw,iw+w2,kk1-iw,a,a,-s);
    }
  }
}
aux[3]=n;
aux[5]=min;
aux[1]=sdet;
}
```

3.6.2 Calculation of determinant

determbnd

Calculates the determinant of the Gaussian eliminated upper triangular matrix provided with the correct sign that is delivered by *decbnd* or *decsolbnd*. *determbnd* should not be called when overflow can be expected.

Procedure parameters:
$$\text{double determbnd } (a,n,lw,rw,sgndet)$$
determbnd: delivers the determinant of the band matrix;
a: double $a[1:(lw+rw)*(n-1)+n]$;
                entry: the contents of *a* are produced by *decbnd* or *decsolbnd*;
n: int;
                entry: the order of the band matrix;
lw: int; number of left codiagonals of *a*;
rw: int; number of right codiagonals of *a*;

sgndet: int;
                   entry: the sign of the determinant as delivered in *aux[1]* by *decbnd*,
                           if the elimination was successful.

```
public static double determbnd(double a[], int n, int lw, int rw, int sgndet)
{
  int i,l;
  double p;

  l=1;
  p=1.0;
  lw += rw+1;
  for (i=1; i<=n; i++) {
    p=a[l]*p;
    l += lw;
  }
  return (Math.abs(p)*sgndet);
}
```

3.6.3 Solution of linear equations

A. solbnd

Calculates the solution of a system of linear equations, provided that the matrix
has been decomposed by a successful call of *decbnd*. The solution of the linear
system is obtained by carrying out the elimination, for which the Gaussian
multipliers are saved, on the right-hand side, and by solving the new system with
the upper triangular band matrix, as produced by *decbnd*, by back substitution.
The solutions of several systems with the same coefficient matrix can be obtained
by successive calls of *solbnd*.

Procedure parameters:
                        void solbnd (*a,n,lw,rw,m,p,b*)
a,n,lw,rw,m,p: see *decbnd*;
                  entry: the contents of the arrays *a,m,*and *p* are as produced by
                           decbnd;
b: double *b[1:n]*;
                  entry: the right-hand side of the system of linear equations.

Procedures used: vecvec, elmvec.

```
public static void solbnd(double a[], int n, int lw, int rw, double m[], int p[],
                          double b[])
```

```
{
  int f,i,k,kk,w,w1,w2,shift;
  double s;

  f=lw;
  shift = -lw;
  w1=lw-1;
  for (k=1; k<=n; k++) {
    if (f < n) f++;
    shift += w1;
    i=p[k];
    s=b[i];
    if (i != k) {
      b[i]=b[k];
      b[k]=s;
    }
    Basic.elmvec(k+1,f,shift,b,m,-s);
  }
  w1=lw+rw;
  w=w1+1;
  kk=(n+1)*w-w1;
  w2 = -1;
  shift=n*w1;
  for (k=n; k>=1; k--) {
    kk -= w;
    shift -= w1;
    if (w2 < w1) w2++;
    b[k]=(b[k]-Basic.vecvec(k+1,k+w2,shift,b,a))/a[kk];
  }
}
```

B. decsolbnd

Calculates the solution of a system of linear equations by Gaussian elimination
with partial pivoting if the coefficient matrix is in band form and is stored rowwise
in a one-dimensional array. *decsolbnd* performs Gaussian elimination in the same
way as *decbnd*, meanwhile also carrying out the elimination with the given right-
hand side. The solution of the eliminated system is obtained by back substitution.

Procedure parameters:
$$\text{void decsolbnd } (a,n,lw,rw,aux,b)$$
a,n,lw,rw,aux: see *decbnd*;
b: see *solbnd*.

Procedures used: vecvec, elmvec, ichvec.

```
public static void decsolbnd(double a[], int n, int lw, int rw, double aux[],
                             double b[])
{
  int i,j,k,kk,kk1,pk,ik,lw1,f,q,w,w1,w2,iw,nrw,shift,sdet;
  double r,s,eps,min;
  double m[] = new double[lw+1];
  double v[] = new double[n+1];

  f=lw;
  sdet=1;
  w1=lw+rw;
  w=w1+1;
  w2=w-2;
  iw=0;
  nrw=n-rw;
  lw1=lw+1;
  q=lw-1;
  for (i=2; i<=lw; i++) {
    q--;
    iw += w1;
    for (j=iw-q; j<=iw; j++) a[j]=0.0;
  }
  iw = -w2;
  q = -lw;
  for (i=1; i<=n; i++) {
    iw += w;
    if (i <= lw1) iw--;
    q += w;
    if (i > nrw) q--;
    v[i]=Math.sqrt(Basic.vecvec(iw,q,0,a,a));
  }
  eps=aux[2];
  min=1.0;
  kk = -w1;
  if (f > nrw) w2 += nrw-f;
  for (k=1; k<=n; k++) {
    if (f < n) f++;
    ik = kk += w;
    s=Math.abs(a[kk])/v[k];
    pk=k;
    kk1=kk+1;
    for (i=k+1; i<=f; i++) {
      ik += w1;
      m[i-k]=r=a[ik];
      a[ik]=0.0;
      r=Math.abs(r)/v[i];
      if (r > s) {
        s=r;
        pk=i;
      }
```

```
      }
      if (s < min) min=s;
      if (s < eps) {
         aux[3]=k-1;
         aux[5]=s;
         aux[1]=sdet;
         return;
      }
      if (k+w2 >= n) w2--;
      if (pk != k) {
         v[pk]=v[k];
         pk -= k;
         Basic.ichvec(kk1,kk1+w2,pk*w1,a);
         sdet = -sdet;
         r=b[k];
         b[k]=b[pk+k];
         b[pk+k]=r;
         r=m[pk];
         m[pk]=a[kk];
         a[kk]=r;
      } else
         r=a[kk];
      iw=kk1;
      lw1=f-k;
      if (r < 0.0) sdet = -sdet;
      for (i=1; i<=lw1; i++) {
         s = m[i] /= r;
         iw += w1;
         Basic.elmvec(iw,iw+w2,kk1-iw,a,a,-s);
         b[k+i] -= b[k]*s;
      }
   }
   aux[3]=n;
   aux[5]=min;
   kk=(n+1)*w-w1;
   w2 = -1;
   shift=n*w1;
   for (k=n; k>=1; k--) {
      kk -= w;
      shift -= w1;
      if (w2 < w1) w2++;
      b[k]=(b[k]-Basic.vecvec(k+1,k+w2,shift,b,a))/a[kk];
   }
   aux[1]=sdet;
}
```

3.7 Real sparse nonsymmetric tridiagonal matrices

3.7.1 Preparatory procedures

A. dectri

Given the $n \times n$ tridiagonal matrix T ($T_{i,j}=0$ for $|i-j|>1$) obtains a lower bidiagonal matrix L ($L_{i,j}=0$ for $i>j+1$ and $j>i$) and unit upper bidiagonal matrix U ($U_{i,i}=1$, $U_{i,j}=0$ for $i>j$ and $j>i+1$) such that $T=LU$. The columns of L and rows of U are determined in succession. If at stage k, $|L_{k,k}|<tol^*\|t_k\|_1$, where tol is a tolerance prescribed by the user and t_k is the k-th row of T, the decomposition process is discontinued, and an integer K is given the value $k-1$; otherwise K is given the value n at exit.

Procedure parameters:
$$\text{void dectri } (sub, diag, supre, n, aux)$$
sub: double *sub[1:n-1]*;
  entry: the subdiagonal of the given matrix T, $T_{i+1,i}$ should be given in *sub[i]*, *i=1,...,n-1*;
  exit: the lower bidiagonal matrix, the value $L_{i+1,i}$ will be delivered in *sub[i]*, *i=1,...,aux[3]-1*;
diag: double *diag[1:n]*;
  entry: the diagonal of T, the value of $T_{i,i}$ in *diag[i]*;
  exit: $L_{i,i}$ will be delivered in *diag[i]*, *i=1,...,aux[3]*;
supre: double *supre[1:n-1]*;
  entry: the superdiagonal of T, value of $T_{i,i+1}$ in *supre[i]*, *i=1,...,n-1*;
  exit: $U_{i,i+1}$ will be delivered in *supre[i]*, *i=1,...,aux[3]-1*;
n: int;
  entry: the order of the matrix;
aux: double *aux[2:5]*;
  entry: *aux[2]*: a relative tolerance; a reasonable choice for this value is an estimate of the relative precision of the matrix elements; however, it should not be chosen smaller than the machine precision (the value of *tol* above);
  exit: *aux[3]*: the number of elimination steps performed (value of K above);
     aux[5]: if *aux[3]=n* then *aux[5]* will be equal to the infinity norm of the matrix, else *aux[5]* is set equal to the value of that element which causes the breakdown of the decomposition (value of $T_{K+1,K+1}$).

```
public static void dectri(double sub[], double diag[], double supre[], int n,
                          double aux[])
{
  int i,n1;
  double d,r,s,u,norm,norm1,tol;
```

```
tol=aux[2];
d=diag[1];
r=supre[1];
norm=norm1=Math.abs(d)+Math.abs(r);
if (Math.abs(d) <= norm1*tol) {
   aux[3]=0.0;
   aux[5]=d;
   return;
}
u=supre[1]=r/d;
s=sub[1];
n1=n-1;
for (i=2; i<=n1; i++) {
   d=diag[i];
   r=supre[i];
   norm1=Math.abs(s)+Math.abs(d)+Math.abs(r);
   diag[i] = d -= u*s;
   if (Math.abs(d) <= norm1*tol) {
      aux[3]=i-1;
      aux[5]=d;
      return;
   }
   u=supre[i]=r/d;
   s=sub[i];
   if (norm1 > norm) norm=norm1;
}
d=diag[n];
norm1=Math.abs(d)+Math.abs(s);
diag[n] = d -= u*s;
if (Math.abs(d) <= norm1*tol) {
   aux[3]=n1;
   aux[5]=d;
   return;
}
if (norm1 > norm) norm=norm1;
aux[3]=n;
aux[5]=norm;
}
```

B. dectripiv

Given the $n \times n$ tridiagonal matrix T ($T_{i,j}=0$ for $|i{-}j|>1$) attempts to obtain a lower bidiagonal matrix L ($L_{i,j}=0$ for $i>j+1$) and unit upper band matrix (with at most two codiagonals) U ($U_{i,i}=1$, $U_{i,j}=0$ for $i>j$ and $j>i+2$) such that $LU=PT$ where P is a permutation matrix resulting from either no interchange or interchange of consecutive rows at each stage in the LU decomposition. The columns of $l^{(k)}$ ($k=1,...,n$) of L are determined in succession. If $|(l_i)^{(k)}|<tol^*\|t^{(i)}\|_1$, $(l_i)^{(k)}$ is the i-th

component of $l^{(k)}$ and $t^{(i)}$ is the i-th row of T and *tol* is a tolerance prescribed by the user, then the decomposition process is discontinued, and an integer K is given the value k-1; otherwise K is given the value n at exit.

Procedure parameters:
$$\text{void dectripiv } (sub,diag,supre,n,aid,aux,piv)$$

sub: double *sub[1:n-1]*;
         entry: the subdiagonal of the given matrix T, $T_{i+1,i}$ should be given in *sub[i]*, i=1,...,n-1;
         exit: the lower bidiagonal matrix, the value $L_{i+1,i}$ will be delivered in *sub[i]*, i=1,...,$aux[3]$-1;
diag: double *diag[1:n]*;
         entry: the diagonal of T, value of $T_{i,i}$ in *diag[i]*;
         exit: $L_{i,i}$ will be delivered in *diag[i]*, i=1,...,$aux[3]$;
supre: double *supre[1:n-1]*;
         entry: the superdiagonal of T, value of $T_{i,i+1}$ in *supre[i]*, i=1,...,n-1;
         exit: $U_{i,i+1}$ will be delivered in *supre[i]*, i=1,...,$aux[3]$-1;
n: int;
         entry: the order of the matrix;
aid: double *aid[1:n-2]*;
         exit: the value of $U_{i,i+2}$ will be delivered in *aid[i]*, i=1,...,$aux[3]$-2;
aux: double *aux[2:5]*;
         entry: *aux[2]*:   a relative tolerance; a reasonable choice for this value is an estimate of the relative precision of the matrix elements, however, it should not be chosen smaller than the machine precision (value of *tol* above);
         exit: *aux[3]*:   the number of elimination steps performed (value of K above);
                    aux[5]:   if *aux[3]*=n then *aux[5]* will be equal to the infinity norm of the matrix, else *aux[5]* is set equal to the value of that element which causes the breakdown of the decomposition (value of $T_{K+1,K+1}$);
piv: boolean *piv[1:n-1]*;
         exit: the value of *piv[i]* will be true if the i-th and (i+1)-th row are interchanged, i=1,...,$\min(aux[3],n$-$1)$, else *piv[i]* will be false.

```
public static void dectripiv(double sub[], double diag[], double supre[], int n,
                       double aid[], double aux[], boolean piv[])
{
  int i,i1,n1,n2;
  double d,r,s,u,t,q,v,w,norm,norm1,norm2,tol;

  tol=aux[2];
  d=diag[1];
  r=supre[1];
  norm=norm2=Math.abs(d)+Math.abs(r);
  n2=n-2;
  for (i=1; i<=n2; i++) {
    i1=i+1;
    s=sub[i];
```

```
    t=diag[i1];
    q=supre[i1];
    norm1=norm2;
    norm2=Math.abs(s)+Math.abs(t)+Math.abs(q);
    if (norm2 > norm) norm=norm2;
    if (Math.abs(d)*norm2 < Math.abs(s)*norm1) {
      if (Math.abs(s) <= tol*norm2) {
        aux[3]=i-1;
        aux[5]=s;
        return;
      }
      diag[i]=s;
      u=supre[i]=t/s;
      v=aid[i]=q/s;
      sub[i]=d;
      w = supre[i1] = -v*d;
      d=diag[i1]=r-u*d;
      r=w;
      norm2=norm1;
      piv[i]=true;
    } else {
      if (Math.abs(d) <= tol*norm1) {
        aux[3]=i-1;
        aux[5]=d;
        return;
      }
      u=supre[i]=r/d;
      d=diag[i1]=t-u*s;
      aid[i]=0.0;
      piv[i]=false;
      r=q;
    }
  }
  n1=n-1;
  s=sub[n1];
  t=diag[n];
  norm1=norm2;
  norm2=Math.abs(s)+Math.abs(t);
  if (norm2 > norm) norm=norm2;
  if (Math.abs(d)*norm2 < Math.abs(s)*norm1) {
    if (Math.abs(s) <= tol*norm2) {
      aux[3]=n2;
      aux[5]=s;
      return;
    }
    diag[n1]=s;
    u=supre[n1]=t/s;
    sub[n1]=d;
    d=diag[n]=r-u*d;
    norm2=norm1;
```

```
    piv[n1]=true;
  } else {
    if (Math.abs(d) <= tol*norm1) {
      aux[3]=n2;
      aux[5]=d;
      return;
    }
    u=supre[n1]=r/d;
    d=diag[n]=t-u*s;
    piv[n1]=false;
  }
  if (Math.abs(d) <= tol*norm2) {
    aux[3]=n1;
    aux[5]=d;
    return;
  }
  aux[3]=n;
  aux[5]=norm;
}
```

3.7.2 Solution of linear equations

A. soltri

Given the $n \times n$ lower bidiagonal matrix L ($L_{i,j}=0$ for $i>j+1$ or $j>i$) and unit upper bidiagonal matrix U ($U_{i,i}=1$, $U_{i,j}=0$ for $i>j$ or $j>i+1$) solves the system of equations $Tx-b$ by forward and back substitution, where $T=LU$ (this decomposition results from successful exit from *dectri*). One call of *dectri* followed by several calls of *soltri* may be used to solve several linear systems having the same tridiagonal matrix, but different right-hand sides.

Procedure parameters:

$$\text{void soltri } (sub,diag,supre,n,b)$$

sub: double *sub[1:n-1]*;
           entry: the subdiagonal of the lower bidiagonal matrix, as delivered by
                   dectri;
diag: double *diag[1:n]*;
           entry: the diagonal of the lower bidiagonal matrix, as delivered by *dectri*;
supre: double *supre[1:n-1]*;
           entry: the superdiagonal of the upper bidiagonal matrix, as delivered by
                   dectri;
n: int;
           entry: the order of the matrix;
b: double *b[1:n]*;
           entry: the right-hand side of the linear system;
           exit: the calculated solution of the linear system.

```
public static void soltri(double sub[], double diag[], double supre[], int n,
                          double b[])
{
   int i;
   double r;

   r = b[1] /= diag[1];
   for (i=2; i<=n; i++) r=b[i]=(b[i]-sub[i-1]*r)/diag[i];
   for (i=n-1; i>=1; i--) r = b[i] -= supre[i]*r;
}
```

B. decsoltri

Given the $n \times n$ tridiagonal matrix T, attempts to obtain a lower bidiagonal matrix L and unit upper bidiagonal matrix U such that $T=LU$ and, if successful in so doing, solves the system of equations $Tx=b$. The triangular decomposition of the given matrix is done by calling *dectri* and the forward and back substitution by calling *soltri*.

Procedure parameters:
$$\text{void decsoltri } (sub, diag, supre, n, aux, b)$$

sub: double *sub[1:n-1]*;
        entry: the subdiagonal of the given matrix T, $T_{i+1,i}$ should be given in *sub[i]*, *i=1,...,n-1*;
        exit: the lower bidiagonal matrix, the value $L_{i+1,i}$ will be delivered in *sub[i]*, *i=1,...,aux[3]-1*;

diag: double *diag[1:n]*;
        entry: the diagonal of T, value of $T_{i,i}$ in *diag[i]*;
        exit: $L_{i,i}$ will be delivered in *diag[i]*, *i=1,...,aux[3]*;

supre: double *supre[1:n-1]*;
        entry: the superdiagonal of T, value of $T_{i,i+1}$ in *supre[i]*, *i=1,...,n-1*;
        exit: $U_{i,i+1}$ will be delivered in *supre[i]*, *i=1,...,aux[3]-1*;

n: int;
        entry: the order of the matrix;

aux: double *aux[2:5]*;
        entry: *aux[2]*:   a relative tolerance; a reasonable choice for this value is an estimate of the relative precision of the matrix elements; however, it should not be chosen smaller than the machine precision;
        exit: *aux[3]*:   the number of elimination steps performed;
                aux[5]:   if *aux[3]=n* then *aux[5]* will be equal to the infinity norm of the matrix, else *aux[5]* is set equal to the value of that element which causes the breakdown of the decomposition;

b: double *b[1:n]*;
        entry: the right-hand side of the linear system;
        exit: if *aux[3]=n* then the solution of the linear system is overwritten on *b*, else *b* remains unaltered.

Procedures used: dectri, soltri.

```
public static void decsoltri(double sub[], double diag[], double supre[], int n,
                             double aux[], double b[])
{
   dectri(sub,diag,supre,n,aux);
   if (aux[3] == n) soltri(sub,diag,supre,n,b);
}
```

C. soltripiv

Given the $n{\times}n$ lower bidiagonal matrix L ($L_{i,j}=0$ for $i>j+1$ and $j>i$) and the unit upper band matrix (with at most two codiagonals) U ($U_{i,i}=1$, $U_{i,j}=0$ for $i>j$ and $j>i+2$) such that $LU=PT$ (where P is a permutation matrix resulting from either no interchange or interchange of consecutive rows at each stage in the LU decomposition of T), solves the system of equations $Tx=b$ (the above decomposition is produced at successful exit from *dectripiv*). One call of *dectripiv* followed by several calls of *soltripiv* may be used to solve several linear systems having the same tridiagonal matrix, but different right-hand sides.

Procedure parameters:
$$\text{void soltripiv } (sub,diag,supre,n,aid,piv,b)$$
sub: double *sub[1:n-1]*;
           entry: the subdiagonal of the lower-bidiagonal matrix, as delivered by
                   dectripiv;
diag: double *diag[1:n]*;
           entry: the diagonal of the lower bidiagonal matrix, as delivered by
                   dectripiv;
supre: double *supre[1:n-1]*;
           entry: the first codiagonal of the upper triangular matrix, as delivered by
                   dectripiv;
n: int;
           entry: the order of the matrix;
aid: double *aid[1:n-2]*;
           entry: the second codiagonal of the upper triangular matrix, as delivered
                   by *dectripiv*;
piv: boolean *piv[1:n-1]*;
           entry: the pivot information as delivered by *dectripiv*;
b: double *b[1:n]*;
           entry: the right-hand side of the linear system;
           exit: the calculated solution of the linear system.

```
public static void soltripiv(double sub[], double diag[], double supre[], int n,
                             double aid[], boolean piv[], double b[])
{
   int i,n1;
   double bi,bi1,r,s,t;
```

```
n1=n-1;
for (i=1; i<=n1; i++) {
  if (piv[i]) {
     bi=b[i+1];
     bi1=b[i];
  } else {
     bi=b[i];
     bi1=b[i+1];
  }
  r=b[i]=bi/diag[i];
  b[i+1]=bi1-sub[i]*r;
}
r = b[n] /= diag[n];
t = b[n1] -= supre[n1]*r;
for (i=n-2; i>=1; i--) {
  s=r;
  r=t;
  t = b[i] -= supre[i]*r + ((piv[i]) ? aid[i]*s : 0.0);
}
}
```

D. decsoltripiv

Given the $n \times n$ tridiagonal matrix T ($T_{i,j}=0$ for $|i\text{-}j|>1$) attempts to obtain a lower bidiagonal matrix L ($L_{i,j}=0$ for $i>j+1$ and $j>i$) and unit upper band matrix (with at most two codiagonals) U ($U_{i,i}=1$, $U_{i,j}=0$ for $i>j$ and $j>i+2$) such that $LU=PT$ where P is a permutation matrix resulting from either no interchange or interchange of consecutive rows at each stage in the LU decomposition and, if successful in so doing, solves the system of equations $Tx=b$. One call of *decsoltripiv* is equivalent to calling consecutively *dectripiv* and *soltripiv*. However, *decsoltripiv* does not make use of *dectripiv* and *soltripiv*, to save memory space and time. This is only true in the case that linear systems with different matrices have to be solved.

Procedure parameters:
$$\text{void decsoltripiv } (sub,diag,supre,n,aux,b)$$
sub: double *sub[1:n-1]*;
      entry: the subdiagonal of the given matrix T, $T_{i+1,i}$ should be given in *sub[i]*, $i=1,...,n-1$;
      exit: the elements of *sub* will be destroyed;
diag: double *diag[1:n]*;
      entry: the diagonal of T, value of $T_{i,i}$ in *diag[i]*;
      exit: the elements of *diag* will be destroyed;
supre: double *supre[1:n-1]*;
      entry: the superdiagonal of T, value of $T_{i,i+1}$ in *supre[i]*, $i=1,...,n-1$;
      exit: the elements of *supre* will be destroyed;
n: int;
      entry: the order of the matrix;

aux: double *aux[2:5]*;
         entry: *aux[2]*:   a relative tolerance; a reasonable choice for this value is
                              an estimate of the relative precision of the matrix
                              elements; however, it should not be chosen smaller than
                              the machine precision;
         exit: *aux[3]*:   the number of elimination steps performed;
                  aux[5]:   if *aux[3]=n* then *aux[5]* will be equal to the infinity norm of
                              the matrix, else *aux[5]* is set equal to the value of that
                              element   which   causes   the   breakdown   of   the
                              decomposition;
b: double *b[1:n]*;
         entry: the right-hand side of the linear system;
         exit: if *aux[3]=n* then the solution of the linear system is overwritten on
                  b, else *b* remains unaltered.

```
public static void decsoltripiv(double sub[], double diag[], double supre[], int n,
                               double aux[], double b[])
{
  int i,i1,n1,n2;
  double d,r,s,u,t,q,v,w,norm,norm1,norm2,tol,bi,bi1,bi2;
  boolean piv[] = new boolean[n+1];

  tol=aux[2];
  d=diag[1];
  r=supre[1];
  bi=b[1];
  norm=norm2=Math.abs(d)+Math.abs(r);
  n2=n-2;
  for (i=1; i<=n2; i++) {
    i1=i+1;
    s=sub[i];
    t=diag[i1];
    q=supre[i1];
    bi1=b[i1];
    norm1=norm2;
    norm2=Math.abs(s)+Math.abs(t)+Math.abs(q);
    if (norm2 > norm) norm=norm2;
    if (Math.abs(d)*norm2 < Math.abs(s)*norm1) {
      if (Math.abs(s) <= tol*norm2) {
        aux[3]=i-1;
        aux[5]=s;
        return;
      }
      u=supre[i]=t/s;
      b[i] = bi1 /= s;
      bi -= bi1*d;
      v=sub[i]=q/s;
      w = supre[i1] = -v*d;
      d=diag[i1]=r-u*d;
      r=w;
```

```
      norm2=norm1;
      piv[i]=true;
    } else {
      if (Math.abs(d) <= tol*norm1) {
         aux[3]=i-1;
         aux[5]=d;
         return;
      }
      u=supre[i]=r/d;
      b[i] = bi /= d;
      bi=bi1-bi*s;
      d=diag[i1]=t-u*s;
      piv[i]=false;
      r=q;
   }
}
n1=n-1;
s=sub[n1];
t=diag[n];
norm1=norm2;
bi1=b[n];
norm2=Math.abs(s)+Math.abs(t);
if (norm2 > norm) norm=norm2;
if (Math.abs(d)*norm2 < Math.abs(s)*norm1) {
   if (Math.abs(s) <= tol*norm2) {
      aux[3]=n2;
      aux[5]=s;
      return;
   }
   u=supre[n1]=t/s;
   b[n1] = bi1 /= s;
   bi -= bi1*d;
   d=r-u*d;
   norm2=norm1;
} else {
   if (Math.abs(d) <= tol*norm1) {
      aux[3]=n2;
      aux[5]=d;
      return;
   }
   u=supre[n1]=r/d;
   b[n1] = bi /= d;
   bi=bi1-bi*s;
   d=t-u*s;
}
if (Math.abs(d) <= tol*norm2) {
   aux[3]=n1;
   aux[5]=d;
   return;
}
```

```
aux[3]=n;
aux[5]=norm;
bi1=b[n]=bi/d;
bi = b[n1] -= supre[n1]*bi1;
for (i=n-2; i>=1; i--) {
  bi2=bi1;
  bi1=bi;
  bi = b[i] -= supre[i]*bi1 + ((piv[i]) ? sub[i]*bi2 : 0.0);
}
}
```

3.8 Sparse symmetric positive definite band matrices

3.8.1 Preparatory procedure

chldecbnd

Decomposes a symmetric positive definite band matrix A ($A_{i,j}=0$ for $|i-j|>w$, $i,j=1,...,n$) into the form $A=U^TU$, where U is an upper triangular band matrix ($U_{i,j}=0$ for $j<i$ and $j>i+w$) by use of Cholesky's square root method. For $j=1,...,n$, with $j(w)=\max(i,j-w)$

$$\delta_j = A_{j,j} - \sum_{k=j(w)}^{j-1} U_{k,j}^2$$

$$U_{j,j} = \sqrt{\delta_j}$$

$$U_{i,j} = \left(A_{i,j} - \sum_{k=j(w)}^{i-1} U_{k,j}U_{k,i} \right) \Big/ U_{i,i} \qquad i = j(w),...,j-1$$

If when $j=k$, either $\delta_j<0$ or $\delta_j < \delta \max|a_{i,i}|$ for $1\leq i\leq j$, δ being a small positive real number prescribed by the user, then the above process is terminated, and the value of $k-1$ is allocated to *aux[3]* as a failure indication [Dek68].

Procedure parameters:

$$\text{void chldecbnd } (a,n,w,aux)$$

a: double *a[1:w(n-1)+n]*;

      entry: *a* contains columnwise the upper triangular band elements of the symmetric band matrix (the (i,j)-th element of the matrix is in location *a[(j-1)w+i]*, $j=1,...,n$, $i=\max(1,j-w),...,j$);

      exit: the band elements of the Cholesky matrix, which is an upper triangular band matrix with w superdiagonals, are delivered columnwise in *a*;

n: int;

      entry: the order of the band matrix;

w: int;
        entry: number of superdiagonals of the matrix;
aux: double *aux[2:3]*;
        entry: *aux[2]*: a relative tolerance to control the calculation of the diagonal elements of the Cholesky matrix (value of δ above);
        exit: *aux[3]*: if successful then *aux[3]=n*; otherwise *aux[3]=k-1*, where k is the index of the diagonal element of the Cholesky matrix that cannot be calculated.

Procedure used: vecvec.

```
public static void chldecbnd(double a[], int n, int w, double aux[])
{
  int j,k,jmax,kk,kj,w1,start;
  double r,eps,max;

  max=0.0;
  kk = -w;
  w1=w+1;
  for (j=1; j<=n; j++) {
    kk += w1;
    if (a[kk] > max) max=a[kk];
  }
  jmax=w;
  w1=w+1;
  kk = -w;
  eps=aux[2]*max;
  for (k=1; k<=n; k++) {
    if (k+w > n) jmax--;
    kk += w1;
    start=kk-k+1;
    r=a[kk]-Basic.vecvec(((k <= w1) ? start : kk-w),kk-1,0,a,a);
    if (r <= eps) {
      aux[3]=k-1;
      return;
    }
    a[kk]=r=Math.sqrt(r);
    kj=kk;
    for (j=1; j<=jmax; j++) {
      kj += w;
      a[kj]=(a[kj]-Basic.vecvec(((k+j <= w1) ? start : kk-w+j),kk-1,kj-kk,a,a))/r;
    }
  }
  aux[3]=n;
}
```

3.8.2 Calculation of determinant

chldetermbnd

With the Cholesky decomposition $A=U^TU$ of the symmetric positive definite band matrix A for whose elements $A_{i,j}=0$ for $|i\text{-}j|>w$, $i,j=1,...,n$) available, chldetermbnd computes the determinant Δ of A,

$$\Delta = \left(\prod_{j=1}^{n} U_{j,j} \right)^{2}$$

chldetermbnd should not be called when overflow can be expected.

Procedure parameters:
$$\text{double chldetermbnd } (a,n,w)$$

chldetermbnd: delivers the determinant of symmetric positive definite band matrix whose Cholesky matrix is stored in a (value of Δ above);

a: double $a[1:w(n-1)+n]$;

        entry: the contents of a are as produced by chldecbnd or chldecsolbnd;

n: int;

        entry: the order of the band matrix;

w: int;

        entry: number of superdiagonals of the matrix.

```
public static double chldetermbnd(double a[], int n, int w)
{
  int j,kk,w1;
  double p;

  w1=w+1;
  kk = -w;
  p=1.0;
  for (j=1; j<=n; j++) {
    kk += w1;
    p *= a[kk];
  }
  return (p*p);
}
```

3.8.3 Solution of linear equations

A. chlsolbnd

With the Cholesky decomposition $A=U^TU$ of the symmetric positive definite band matrix A for whose elements $A_{i,j}=0$ for $|i\text{-}j|>w$, $i,j=1,...,n$) available, *chlsolbnd* carries out the solution of the systems of equations $Ax=b$. The relationships

$$y_i = \left(b_j - \sum_{k=j(w)}^{j-1} U_{k,j} y_k \right) \Big/ U_{j,j}$$

$$x_j = \left(y_j - \sum_{k=j+1}^{J(w)} U_{j,k} x_k \right) \Big/ U_{j,j}$$

for $j=1,...,n$, where $j(w)=\max(1,j\text{-}w)$, $J(w)=\min(n,j+w)$, are used (they follow from $U^Ty=b$, $Ux=y$).

Procedure parameters:

$$\text{void chlsolbnd } (a,n,w,b)$$

a: double *a[1:w(n-1)+n]*;
          entry: the contents of *a* are as produced by *chldecbnd*;
n: int;
          entry: the order of the band matrix;
w: int;
          entry: number of superdiagonals of the matrix;
b: double *b[1:n]*;
          entry: the right-hand side of the system of linear equations;
          exit: the solution of the system.

Procedures used: vecvec, scaprd1.

```
public static void chlsolbnd(double a[], int n, int w, double b[])
{
  int k,imax,kk,w1;

  kk = -w;
  w1=w+1;
  for (k=1; k<=n; k++) {
    kk += w1;
    b[k]=(b[k]-Basic.vecvec(((k <= w1) ? 1 : k-w),k-1,kk-k,b,a))/a[kk];
  }
  imax = -1;
  for (k=n; k>=1; k--) {
    if (imax < w) imax++;
    b[k]=(b[k]-Basic.scaprd1(kk+w,w,k+1,1,imax,a,b))/a[kk];
    kk -= w1;
  }
}
```

B. chldecsolbnd

Solves the system of equations $Ax=b$ where A is an $n \times n$ symmetric positive definite band matrix for whose elements $A_{i,j}=0$ for $|i-j|>w$, $i,j=1,...,n$), by use of Cholesky's square root method. The decomposition $A=U^TU$ is first obtained by means of *chldecbnd* and, if this was successful, *chlsolbnd* is then called to obtain x.

Procedure parameters:
$$\text{void chldecsolbnd } (a,n,w,aux,b)$$
a,n,w,aux: see *chldecbnd*;
b: double $b[1:n]$;
                 entry: the right-hand side of the system of linear equations;
                 exit: the solution of the system.

Procedures used: chldecbnd, chlsolbnd.

```
public static void chldecsolbnd(double a[], int n, int w, double aux[], double b[])
{
   chldecbnd(a,n,w,aux);
   if (aux[3] == n) chlsolbnd(a,n,w,b);
}
```

3.9 Symmetric positive definite tridiagonal matrices

3.9.1 Preparatory procedure

decsymtri

Given the $n \times n$ symmetric tridiagonal matrix T ($T_{i,j}=0$ for $|i-j|>1$) attempts to obtain a unit upper diagonal matrix U ($U_{i,i}=1$, $U_{i,j}=0$ for $i>j$ and $j>i+1$) and unit upper bidiagonal matrix U ($U_{i,i}=1$, $U_{i,j}=0$ for $i>j$ and $j>i+1$) and a diagonal matrix D such that $U^TDU=T$. The rows of U and diagonal elements of D are determined in succession. If at stage k, $|T_{k,k}|<tol*\|t^{(k)}\|_1$, where *tol* is a tolerance prescribed by the user and $t^{(k)}$ is the k-th row of T, the decomposition process is discontinued, and an integer K is given the value $k-1$; otherwise K is given the value n at exit.

Procedure parameters:
$$\text{void decsymtri } (diag,co,n,aux)$$
$diag$:   double $diag[1:n]$;
           entry: the diagonal of T, value of $T_{i,i}$ in $diag[i]$, $i=1,...,n$;
           exit: $D_{i,i}$ will be delivered in $diag[i]$, $i=1,...,aux[3]$;
co: double $co[1:n-1]$;
           entry: the codiagonal of T, value of $T_{i,i+1}$ in $co[i]$, $i=1,...,n-1$;
           exit: $U_{i,i+1}$ will be delivered in $co[i]$, $i=1,...,aux[3]-1$;

n: int;
            entry: the order of the matrix;
aux: double *aux[2:5]*;
            entry: *aux[2]:*   a relative tolerance; a reasonable choice for this value is
                                 an estimate of the relative precision of the matrix
                                 elements; however, it should not be chosen smaller than
                                 the machine precision (the value of *tol* above);
            exit: *aux[3]:*   the number of elimination steps performed (value of K
                                 above);
                    aux[5]:   if *aux[3]=n* then *aux[5]* will be equal to the infinity norm of
                                 the matrix, else *aux[5]* is set equal to the value of that
                                 element which causes the breakdown of the
                                 decomposition (value of $T_{K+1,K+1}$).

```
public static void decsymtri(double diag[], double co[], int n, double aux[])
{
  int i,n1;
  double d,r,s,u,tol,norm,normr;

  s=0.0;
  tol=aux[2];
  d=diag[1];
  r=co[1];
  norm=normr=Math.abs(d)+Math.abs(r);
  if (Math.abs(d) <= normr*tol) {
     aux[3]=0.0;
     aux[5]=d;
     return;
  }
  u=co[1]=r/d;
  n1=n-1;
  for (i=2; i<=n1; i++) {
     s=r;
     r=co[i];
     d=diag[i];
     normr=Math.abs(s)+Math.abs(d)+Math.abs(r);
     diag[i] = d -= u*s;
     if (Math.abs(d) <= normr*tol) {
        aux[3]=i-1;
        aux[5]=d;
        return;
     }
     u=co[i]=r/d;
     if (normr > norm) norm=normr;
  }
  d=diag[n];
  normr=Math.abs(d)+Math.abs(r);
  diag[n] = d -= u*s;
  if (Math.abs(d) <= normr*tol) {
     aux[3]=n1;
```

```
      aux[5]=d;
      return;
   }
   if (normr > norm) norm=normr;
   aux[3]=n;
   aux[5]=norm;
}
```

3.9.2 Solution of linear equations

A. solsymtri

Given the $n \times n$ unit upper bidiagonal matrix U ($U_{i,i}=1$, $U_{i,j}=0$ for $i>j$ and $j>i+1$) and diagonal matrix D, solves the system of equations $Tx=b$, where $T=U^T D U$ (this decomposition results from successful exit from *decsymtri*). One call of *decsymtri* followed by several calls of *solsymtri* may be used to solve several linear systems having the same symmetric tridiagonal matrix, but different right-hand sides.

Procedure parameters:
$$\text{void solsymtri } (diag,co,n,b)$$
diag: double *diag[1:n]*;
           entry: the diagonal matrix, as delivered by *decsymtri*;
co: double *co[1:n-1]*;
           entry: the codiagonal of the unit upper bidiagonal matrix, as delivered by
                   decsymtri;
n: int;
           entry: the order of the matrix;
b: double *b[1:n]*;
           entry: the right-hand side of the linear system;
           exit: the calculated solution of the linear system.

```
public static void solsymtri(double diag[], double co[], int n, double b[])
{
   int i;
   double r,s;

   r=b[1];
   b[1]=r/diag[1];
   for (i=2; i<=n; i++) {
      r=b[i]-co[i-1]*r;
      b[i]=r/diag[i];
   }
   s=b[n];
   for (i=n-1; i>=1; i--) s = b[i] -= co[i]*s;
}
```

B. decsolsymtri

Given the $n \times n$ symmetric tridiagonal matrix T ($T_{i,j}=0$ for $|i-j|>1$) attempts to obtain a unit bidiagonal matrix U ($U_{i,i}=1$, $U_{i,j}=0$ for $i>j$ and $j>i+1$) and a diagonal matrix D such that $U^T DU=T$ by calling *decsymtri* and, if successful in so doing, solves the system of equations $Tx=b$ by calling *solsymtri*.

Procedure parameters:
$$\text{void decsolsymtri } (diag,co,n,aux,b)$$

diag: double *diag[1:n]*;
      entry: the diagonal of T, value of $T_{i,i}$ in *diag[i]*, $i=1,...,n$;
      exit: $D_{i,i}$ will be delivered in *diag[i]*, $i=1,...,aux[3]$;

co: double *co[1:n-1]*;
      entry: the codiagonal of T, value of $T_{i,i+1}$ in *co[i]*, $i=1,...,n-1$;
      exit: $U_{i,i+1}$ will be delivered in *co[i]*, $i=1,...,aux[3]-1$;

n: int;
      entry: the order of the matrix;

aux: double *aux[2:5]*;
      entry: *aux[2]*:   a relative tolerance; a reasonable choice for this value is an estimate of the relative precision of the matrix elements; however, it should not be chosen smaller than the machine precision;
      exit: *aux[3]*:   the number of elimination steps performed;
                   aux[5]:   if *aux[3]=n* then *aux[5]* will be equal to the infinity norm of the matrix, else *aux[5]* is set equal to the value of that element which causes the breakdown of the decomposition;

b: double *b[1:n]*;
      entry: the right-hand side of the linear system;
      exit: if *aux[3]=n* then the solution of the linear system is overwritten on b, else b remains unaltered.

Procedures used: decsymtri, solsymtri.

```
public static void decsolsymtri(double diag[], double co[], int n, double aux[],
                              double b[])
{
  decsymtri(diag,co,n,aux);
  if (aux[3] == n) solsymtri(diag,co,n,b);
}
```

3.10 Sparse real matrices — Iterative methods

conjgrad

Solves the system of equations $Ax=b$, where A is an $n{\times}n$ positive definite symmetric matrix, by the method of conjugate gradients [R71]. This method involves the construction of sequences of approximations x_k to the solution, differences $\Delta x_k=x_{k+1}-x_k$, and the associated residual vectors $r^{(k)}=b-Ax^{(k)}$. With x_0 an initial approximation to x, and

$$\rho_k = (r_k,r_k) \qquad \eta_k = (r_k,Ar_k) \qquad (k=0,1,...)$$

the recursion is

$$q_0 = \eta_0/\rho_0, \qquad \Delta x_0 = r_0/q_0$$

and for $k=1,2,...$

$$x_k = x_{k-1} + \Delta x_{k-1}$$
$$e_{k-1} = q_{k-1}\rho_k/\rho_{k-1}$$
$$q_k = (\eta_k/\rho_k) + e_{k-1}$$
$$\Delta x_k = (r_k + e_{k-1}\Delta x_{k-1})/q_k .$$

Setting $a_k=1/q_k$, $b_k=e_{k-1}/q_{k-1}$, $p_k=r_k+e_{k-1}\Delta x_{k-1}$, the above formulae may be written as

$$b_k = \rho_k/\rho_{k-1}$$
$$p_k = r_k + b_kp_{k-1}, \qquad q_k = Ap_k$$
$$a_k = \rho_k/(p_k,q_k)$$
$$x_{k+1} = x_k + a_kp_k, \qquad r_{k+1} = r_k - a_kq_k$$
$$\rho_{k+1} = (r_{k+1},r_{k+1}).$$

The computations are terminated if $\max(\|r_k\|_2, \|\Delta x_{k-1}\|_2){\leq}tol$, where *tol* is a tolerance prescribed by the user.

Procedure parameters:

$$\text{void conjgrad } (method,x,r,l,n,iterate,norm2)$$

method: a class that defines two procedures *matvec* and *goon*, this class must implement the LA_conjgrad_methods interface;

void matvec(double p[ ], double q[ ])

defines the coefficient matrix A of the system as follows: at each call *matvec* delivers in q the matrix-vector product Ap, p, and q are one-dimensional arrays, double $p[l{:}n]$, $q[l{:}n]$;

boolean goon(int iterate[ ], double norm2[ ])

goon indicates the continuation of the process depending on the current values of *iterate* and *norm2*, if *goon* equals false then the iteration process is stopped;

x: double $x[l{:}n]$;

entry: an initial approximation to the solution;

exit: the solution;

r: double $r[l{:}n]$;

entry: the right-hand side of the system;

exit: the residual $b-Ax$;

l,n: int;

entry: l and n are respectively the lower and upper bound of the arrays $x,r,p,$and q;

iterate: int *iterate[0:0]*;
             exit: delivers the number of iteration steps already performed;
norm2:   double *norm2[0:0]*;
             exit: delivers the squared Euclidean norm of the residual.

Procedures used: vecvec, elmvec.

```
public interface LA_conjgrad_methods {

  void matvec(double p[], double q[]);
  boolean goon(int iterate[], double norm2[]);
}



public static void conjgrad(LA_conjgrad_methods method, double x[], double r[],
                       int l, int n, int iterate[], double norm2[])
{
  int i;
  double a,b,prr,rrp;
  double p[] = new double[n+1];
  double ap[] = new double[n+1];

  rrp=0.0;
  prr=1.0;
  iterate[0]=0;
  do {
    if (iterate[0] == 0) {
      method.matvec(x,p);
      for (i=1; i<=n; i++) p[i] = r[i] -= p[i];
      prr=Basic.vecvec(1,n,0,r,r);
    } else {
      b=rrp/prr;
      prr=rrp;
      for (i=1; i<=n; i++) p[i]=r[i]+b*p[i];
    }
    method.matvec(p,ap);
    a=prr/Basic.vecvec(1,n,0,p,ap);
    Basic.elmvec(1,n,0,x,p,a);
    Basic.elmvec(1,n,0,r,ap,-a);
    norm2[0]=rrp=Basic.vecvec(1,n,0,r,r);
    (iterate[0])++;
  } while (method.goon(iterate,norm2));
}
```

3.11 Similarity transformation

3.11.1 Equilibration - real matrices

A. eqilbr

Equilibrates an $n \times n$ real matrix A by determining a real diagonal matrix D whose diagonal elements are integer powers of two, and a permutation matrix P such that the Euclidean norms of the k-th row and the k-th column of $A'=PDAD^{-1}P^{-1}$ are approximately equal $(k=1,...,n)$. With $j(k)=1,2,...,n,1,2,...,$ (i.e., $j(k)=(k-1 \bmod(n))+1$, $k=1,2,...$) a sequence of diagonal matrices D_k, where the $j(k)$-th diagonal element of D_k is ρ_k^{-1} (ρ_k being an integer power of two) and all others one, together with a sequence of matrices $A_k=D_kA_{k-1}D_k^{-1}$, $A_0=A$ are obtained. ρ_k is determined by the condition that the Euclidean norms of the $j(k)$-th column and $j(k)$-th row of A_k have approximately equal values. If all of diagonal elements of either the $j(k)$-th row or the $j(k)$-th column of A_{k-1} is nearly zero, the row and column in question are interchanged with the next pair for which this is not so. The process is terminated if either (a) $1/2 < \rho_k < 2$ for one whole cycle of the $j(k)$ or (b) $k=(n+1)n^2$. D is the product of the D_k.

Procedure parameters:
$$\text{void eqilbr } (a,n,em,d,inter)$$

a: double $a[1:n,1:n]$;
          entry: the matrix to be equilibrated;
          exit: the equilibrated matrix;
n: int;
          entry: the order of the matrix;
em: double $em[0:0]$;
          entry: the machine precision;
d: double $d[1:n]$;
          exit: the main diagonal of the transforming diagonal matrix;
$inter$: int $inter[1:n]$;
          exit: information defining the possible interchanging of some rows and
                   the corresponding columns.

Procedures used: tammat, mattam, ichcol, ichrow.

Remark: The procedure is equilibrated by means of Osborne's diagonal
                       similarity transformation possibly with interchanges [DekHo68,
                       Os60].

```
public static void eqilbr(double a[][], int n, double em[], double d[], int inter[])
{
  int i,im,i1,p,q,j,t,count,exponent,ni;
  double c,r,eps,omega,factor,di;
```

```
factor=1.0/(2.0*Math.log(2.0));
eps=em[0];
omega=1.0/eps;
t=p=1;
q=ni=i=n;
count=((n+1)*n)/2;
for (j=1; j<=n; j++) {
  d[j]=1.0;
  inter[j]=0;
}
i = (i < q) ? i+1 : p;
while (count > 0 && ni > 0) {
  count--;
  im=i-1;
  i1=i+1;
  c=Math.sqrt(Basic.tammat(p,im,i,i,a,a)+Basic.tammat(i1,q,i,i,a,a));
  r=Math.sqrt(Basic.mattam(p,im,i,i,a,a)+Basic.mattam(i1,q,i,i,a,a));
  if (c*omega <= r*eps) {
    inter[t]=i;
    ni=q-p;
    t++;
    if (p != i) {
      Basic.ichcol(1,n,p,i,a);
      Basic.ichrow(1,n,p,i,a);
      di=d[i];
      d[i]=d[p];
      d[p]=di;
    }
    p++;
  } else
    if (r*omega <= c*eps) {
      inter[t] = -i;
      ni=q-p;
      t++;
      if (q != i) {
        Basic.ichcol(1,n,q,i,a);
        Basic.ichrow(1,n,q,i,a);
        di=d[i];
        d[i]=d[q];
        d[q]=di;
      }
      q--;
    } else {
      exponent=(int) (Math.log(r/c)*factor);
      if (Math.abs(exponent) > 1.0) {
        ni=q-p;
        c=Math.pow(2.0,exponent);
        r=1.0/c;
        d[i] *= c;
        for (j=1; j<=im; j++) {
```

```
                a[j][i] *=c;
                a[i][j] *= r;
            }
            for (j=i1; j<=n; j++) {
                a[j][i] *=c;
                a[i][j] *= r;
            }
        } else
            ni--;
    }
    i = (i < q) ? i+1 : p;
  }
}
```

B. baklbr

Given the diagonal elements of the $n \times n$ diagonal matrix D, the pivot reference integers associated with a permutation matrix P, and a set of column vectors $v^{(j)}$ ($j=n1,...,n2$), constructs the sequence of vectors $u^{(j)}=PDv^{(j)}$ ($j=n1,...,n2$). (If $v^{(j)}$ is an eigenvector of $PDAD^{-1}P^{-1}$, $u^{(j)}$ is an eigenvector of A.)

Procedure parameters:
$$\text{void baklbr } (n,n1,n2,d,inter,vec)$$

n: int;
            entry: the length of the vectors to be transformed;
n1,n2:   int;
            entry: the serial numbers of the first and last vector to be transformed;
d: double $d[1:n]$;
            entry: the main diagonal of the transforming diagonal matrix of order n,
                     as produced by *eqilbr*;
inter: int $inter[1:n]$;
            entry: information defining the possible interchanging of some rows and
                     columns, as produced by *eqilbr*;
vec: double $vec[1:n,n1:n2]$;
            entry: the $n2-n1+1$ vectors of length n to be transformed;
            exit: the $n2-n1+1$ vectors of length n resulting from the back
                     transformation.

Procedure used: ichrow.

```
public static void baklbr(int n, int n1, int n2, double d[], int inter[],
                          double vec[][])
{
    int i,j,k,p,q;
    double di;

    p=1;
```

```
q=n;
for (i=1; i<=n; i++) {
  di=d[i];
  if (di != 1)
    for (j=n1; j<=n2; j++) vec[i][j] *= di;
  k=inter[i];
  if (k > 0)
    p++;
  else
    if (k < 0) q--;
}
for (i=p-1+n-q; i>=1; i--) {
  k=inter[i];
  if (k > 0) {
    p--;
    if (k != p) Basic.ichrow(n1,n2,k,p,vec);
  } else {
    q++;
    if (-k != q) Basic.ichrow(n1,n2,-k,q,vec);
  }
}
}
```

3.11.2 Equilibration - complex matrices

A. eqilbrcom

Equilibrates an $n\times n$ complex matrix A by determining a real diagonal matrix D whose diagonal elements are integer powers of two, and a permutation matrix P such that if $C=PDAD^{-1}P^{-1}$, the diagonal elements of $C'C-CC'$ (C' denotes the conjugate transpose of C) are approximately zero. With $j(k)=1,2,...,n,1,2,...,$ (i.e., $j(k)=(k-1 \bmod(n))+1$, $k=1,2,...$) a sequence of diagonal matrices D_k, where the $j(k)$-th diagonal element of D_k is ρ_k^{-1} (ρ_k being an integer power of two) and all others one, together with a sequence of matrices C_k for which $C_k=D_kC_{k-1}D_{k-1}^{-1}$, $C_0=A$ are determined. ρ_k is determined by the condition that the Euclidean norms of the $j(k)$-th column and $j(k)$-th row of C_k have approximately equal values. If all off-diagonal elements of either the $j(k)$-th column or the $j(k)$-th row of C_{k-1} are nearly zero, the row and column in question are interchanged with the next pair for which this is not so. The process is terminated if either (a) $1/2 < \rho_k < 2$ for one whole cycle of the $j(k)$, or (b) $k=kmax+1$, where $kmax$ is an integer prescribed by the user [DekHo68, Os60, PaR69].

Procedure parameters:

$$\text{void eqilbrcom } (a1,a2,n,em,d,inter)$$

a1,a2:   double *a1[1:n,1:n], a2[1:n,1:n]*;

       entry: the real part and imaginary part of the matrix to be equilibrated must be given in the arrays *a1* and *a2*, respectively;

       exit: the real part and the imaginary part of the equilibrated matrix are delivered in the arrays *a1* and *a2*, respectively;

n: int;

       entry: the order of the given matrix;

em: double *em[0:7]*;

       entry: *em[0]*:   the machine precision;

              em[6]:   the maximum allowed number of iterations (value of *kmax* above);

       exit: *em[7]*:   the number of iterations performed;

d: double *d[1:n]*;

       exit: the scaling factors of the diagonal similarity transformation;

inter: int *inter[1:n]*;

       exit: information defining the possible interchanging of some rows and the corresponding columns.

Procedures used: ichcol, ichrow, tammat, mattam.

```
public static void eqilbrcom(double a1[][], double a2[][], int n, double em[],
                             double d[], int inter[])
{
  int i,p,q,j,t,count,exponent,ni,im,i1;
  double c,r,eps,di;

  eps=em[0]*em[0];
  t=p=1;
  q=ni=i=n;
  count=(int) em[6];
  for (j=1; j<=n; j++) {
    d[j]=1.0;
    inter[j]=0;
  }
  i = (i < q) ? i+1 : p;
  while (count > 0 && ni > 0) {
    count--;
    im=i-1;
    i1=i+1;
    c=Basic.tammat(p,im,i,i,a1,a1)+Basic.tammat(i1,q,i,i,a1,a1)+
      Basic.tammat(p,im,i,i,a2,a2)+Basic.tammat(i1,q,i,i,a2,a2);
    r=Basic.mattam(p,im,i,i,a1,a1)+Basic.mattam(i1,q,i,i,a1,a1)+
      Basic.mattam(p,im,i,i,a2,a2)+Basic.mattam(i1,q,i,i,a2,a2);
    if (c/eps <= r) {
      inter[t]=i;
      ni=q-p;
      t++;
      if (p != i) {
```

```
        Basic.ichcol(1,n,p,i,a1);
        Basic.ichrow(1,n,p,i,a1);
        Basic.ichcol(1,n,p,i,a2);
        Basic.ichrow(1,n,p,i,a2);
        di=d[i];
        d[i]=d[p];
        d[p]=di;
      }
      p++;
    } else
      if (r/eps <= c) {
        inter[t] = -i;
        ni=q-p;
        t++;
        if (q != i) {
          Basic.ichcol(1,n,q,i,a1);
          Basic.ichrow(1,n,q,i,a1);
          Basic.ichcol(1,n,q,i,a2);
          Basic.ichrow(1,n,q,i,a2);
          di=d[i];
          d[i]=d[q];
          d[q]=di;
        }
        q--;
      } else {
        exponent=(int) Math.ceil(Math.log(r/c)*0.36067);
        if (Math.abs(exponent) > 1) {
          ni=q-p;
          c=Math.pow(2.0,exponent);
          d[i] *= c;
          for (j=1; j<=im; j++) {
            a1[j][i] *= c;
            a1[i][j] /= c;
            a2[j][i] *= c;
            a2[i][j] /= c;
          }
          for (j=i1; j<=n; j++) {
            a1[j][i] *= c;
            a1[i][j] /= c;
            a2[j][i] *= c;
            a2[i][j] /= c;
          }
        } else
          ni--;
      }
    i = (i < q) ? i+1 : p;
  }
  em[7]=em[6]-count;
}
```

B. baklbrcom

Given a real $n{\times}n$ diagonal matrix D, a permutation matrix P, and a sequence of complex vectors $v^{(k)}$ $(k=n1,...,n2)$, constructs the vectors $u^{(k)}=DPv^{(k)}$ $(k=n1,...,n2)$. (If $v^{(k)}$, $k=n1,...,n2$, are eigenvectors of $P^{-1}D^{-1}ADP$, $u^{(k)}$, $k=n1,...,n2$, are corresponding eigenvectors of A.)

Procedure parameters:
$$\text{void baklbrcom } (n,n1,n2,d,inter,vr,vi)$$

n: int;
    entry: the order of the matrix of which the eigenvectors are calculated;
n1,n2: int;
    entry: the eigenvectors corresponding to the eigenvalues with indices $n1,...,n2$ are to be transformed;
d: double $d[1:n]$;
    entry: the scaling factors of the diagonal similarity transformation as delivered by *eqilbrcom*;
inter: int $inter[1:n]$;
    entry: information defining the interchanging of some rows and columns, as produced by *eqilbrcom*;
vr,vi: double $vr[1:n,n1:n2]$, $vi[1:n,n1:n2]$;
    entry: the back transformation is performed on the eigenvectors with the real parts given in array *vr*, and the imaginary parts given in array *vi*;
    exit: the real parts and imaginary parts of the resulting eigenvectors are delivered in the columns of the arrays *vr* and *vi*, respectively.

Procedure used: baklbr.

```
public static void baklbrcom(int n, int n1, int n2, double d[], int inter[],
                            double vr[][], double vi[][])
{
  baklbr(n,n1,n2,d,inter,vr);
  baklbr(n,n1,n2,d,inter,vi);
}
```

3.11.3 To Hessenberg form - real symmetric

A. tfmsymtri2

Reduces an $n{\times}n$ symmetric matrix A to tridiagonal form T by setting $A^{(1)}=A$ and $A^{(i+1)}=P^{(i)}AP^{(i)}$ $(i=1,...,n-2)$ when $T=A^{(n-1)}$ (setting
$$(v^{(i)})^{\mathrm{T}} = (A_{k,1}^{(i)},...,A_{k,k-2}^{(i)},A_{k,k-1}^{(i)},0,0,...,0),$$

and
$$(u^{(i)})^{\mathrm{T}}=(v^{(i)})^{\mathrm{T}}\pm((v^{(i)})^{\mathrm{T}}v^{(i)})e_k,$$
where $k=n-i+1$ and e_k is the k-th unit vector, and $\pm$ is the sign of $A_{k,k-1}$, $P^{(i)} = I - 2u^{(i)}(u^{(i)})^{\mathrm{T}}/(u^{(i)})^{\mathrm{T}}u^{(i)}$ in the above). The i-th transformation reduces the elements in the $(n-i+1)$-th row and column of $A^{(i)}$ which do not belong to the principal diagonal and the two adjacent diagonals, to zero; if all of these elements are already less in absolute value than $\|A\|_\infty$ times the machine precision, the i-th transformation is skipped [DekHo68, Wi65].

Procedure parameters:
$$\text{void tfmsymtri2 } (a,n,d,b,bb,em)$$
a: double $a[1:n,1:n]$;
     entry: the upper triangle of the symmetric matrix must be given in the upper triangular part of a (the elements $a[i,j]$, $i{\leq}j$);
     exit: the data for Householder's back transformation is delivered in the upper triangular part of a, the elements $a[i,j]$, $i{>}j$ are neither used nor changed;
n: int;
     entry: the order of the given matrix;
d: double $d[1:n]$;
     exit: the main diagonal of the symmetric tridiagonal matrix T, produced by Householder's transformation;
b: double $b[1:n]$;
     exit: the codiagonal elements of T are delivered in $b[1:n-1]$, $b[n]$ is set equal to zero;
bb: double $bb[1:n]$;
     exit: the squares of the codiagonal elements of T are delivered in $bb[1:n-1]$, $bb[n]$ is set equal to zero;
em: double $em[0:1]$;
     entry: $em[0]$: the machine precision;
     exit: $em[1]$: the infinity norm of the original matrix.

Procedures used: tamvec, matmat, tammat, elmveccol, elmcolvec, elmcol.

```
public static void tfmsymtri2(double a[][], int n, double d[], double b[],
                         double bb[], double em[])
{
  int i,j,r,r1;
  double w,x,a1,b0,bb0,machtol,norm;

  norm=0.0;
  for (j=1; j<=n; j++) {
    w=0.0;
    for (i=1; i<=j; i++) w += Math.abs(a[i][j]);
    for (i=j+1; i<=n; i++) w += Math.abs(a[j][i]);
    if (w > norm) norm=w;
  }
  machtol=em[0]*norm;
  em[1]=norm;
  r=n;
```

```
for (r1=n-1; r1>=1; r1--) {
  d[r]=a[r][r];
  x=Basic.tammat(1,r-2,r,r,a,a);
  a1=a[r1][r];
  if (Math.sqrt(x) <= machtol) {
    b0=b[r1]=a1;
    bb[r1]=b0*b0;
    a[r][r]=1.0;
  } else {
    bb0=bb[r1]=a1*a1+x;
    b0 = (a1 > 0.0) ? -Math.sqrt(bb0) : Math.sqrt(bb0);
    a1=a[r1][r]=a1-b0;
    w=a[r][r]=1.0/(a1*b0);
    for (j=1; j<=r1; j++)
      b[j]=(Basic.tammat(1,j,j,r,a,a)+Basic.matmat(j+1,r1,j,r,a,a))*w;
    Basic.elmveccol(1,r1,r,b,a,Basic.tamvec(1,r1,r,a,b)*w*0.5);
    for (j=1; j<=r1; j++) {
      Basic.elmcol(1,j,j,r,a,a,b[j]);
      Basic.elmcolvec(1,j,j,a,b,a[j][r]);
    }
    b[r1]=b0;
  }
  r=r1;
}
d[1]=a[1][1];
a[1][1]=1.0;
b[n]=bb[n]=0.0;
}
```

B. baksymtri2

Performs the back substitutions upon the intermediate numbers produced by *tfmsymtri2*.

Procedure parameters:

$$\text{void baksymtri2 } (a,n,n1,n2,vec)$$

a: double $a[1:n,1:n]$;

          entry: the data for the back transformation, as produced by *tfmsymtri2*,
                  must be given in the upper triangular part of a;

n: int;

          entry: the order of the given matrix;

n1,n2: int;

          entry: the lower and upper bound, respectively, of the column numbers of
                  vec;

vec: double $vec[1:n, n1:n2]$;

          entry: the vectors on which the back transformation has to be performed;

          exit: the transformed vectors.

Procedures used: tammat, elmcol.

```
public static void baksymtri2(double a[][], int n, int n1, int n2, double vec[][])
{
  int j,k;
  double w;

  for (j=2; j<=n; j++) {
    w=a[j][j];
    if (w < 0.0)
      for (k=n1; k<=n2; k++)
        Basic.elmcol(1,j-1,k,j,vec,a,Basic.tammat(1,j-1,j,k,a,vec)*w);
  }
}
```

C. tfmprevec

Computes the matrix P for which $PAP^T=T$ where A is an $n{\times}n$ matrix, and the $n{\times}n$ matrix T is of tridiagonal form, using the intermediate results generated by *tfmsymtri2*.

Procedure parameters:
$$\text{void tfmprevec } (a,n)$$

a: double *a[1:n, 1:n]*;
         entry: the data for the back transformation, as produced by *tfmsymtri2*, must be given in the upper triangular part of *a*;
         exit: the matrix which transforms the original matrix into a similar tridiagonal one;
n: int;
         entry: the order of the given matrix.

Procedures used: tammat, elmcol.

```
public static void tfmprevec(double a[][], int n)
{
  int i,j,j1,k;
  double ab;

  j1=1;
  for (j=2; j<=n; j++) {
    for (i=1; i<=j1-1; i++) a[i][j1]=0.0;
    for (i=j; i<=n; i++) a[i][j1]=0.0;
    a[j1][j1]=1.0;
    ab=a[j][j];
    if (ab < 0)
      for (k=1; k<=j1; k++)
```

```
        Basic.elmcol(1,j1,k,j,a,a,Basic.tammat(1,j1,j,k,a,a)*ab);
      j1=j;
    }
  for (i=n-1; i>=1; i--) a[i][n]=0.0;
  a[n][n]=1.0;
}
```

D. tfmsymtri1

Reduces an $n \times n$ symmetric matrix A to tridiagonal form T as in the implementation of *tfmsymtri2*. It is assumed that the upper triangular elements of A are given in a linear array.

Procedure parameters:
$$\text{void tfmsymtri1 } (a,n,d,b,bb,em)$$

a: double $a[1:(n+1)n/2]$;
          entry: the upper triangle of the given matrix must be given in such a way
                  that the (i,j)-th element of the matrix is $a[(j-1)j/2+i]$, $1 \le i \le j \le n$;
          exit: the data for Householder's back transformation as used by
                  baksymtri1;

n: int;
          entry: the order of the given matrix;
d: double $d[1:n]$;
          exit: the main diagonal of the symmetric tridiagonal matrix T, produced
                  by Householder's transformation;
b: double $b[1:n]$;
          exit: the codiagonal elements of T are delivered in $b[1:n-1]$, $b[n]$ is set
                  equal to zero;
bb: double $bb[1:n]$;
          exit: the squares of the codiagonal elements of T are delivered in
                  $bb[1:n-1]$, $bb[n]$ is set equal to zero;
em: double $em[0:1]$;
          entry: $em[0]$: the machine precision;
          exit: $em[1]$: the infinity norm of the original matrix.

Procedures used: vecvec, seqvec, elmvec.

```
public static void tfmsymtri1(double a[], int n, double d[], double b[],
                              double bb[], double em[])
{
  int i,j,r,r1,p,q,ti,tj;
  double s,w,x,a1,b0,bb0,norm,machtol;

  norm=0.0;
  tj=0;
  for (j=1; j<=n; j++) {
    w=0.0;
```

```
    for (i=1; i<=j; i++) w += Math.abs(a[i+tj]);
    tj += j;
    ti=tj+j;
    for (i=j+1; i<=n; i++) {
      w += Math.abs(a[ti]);
      ti += i;
    }
    if (w > norm) norm=w;
  }
  machtol=em[0]*norm;
  em[1]=norm;
  q=((n+1)*n)/2;
  r=n;
  for (r1=n-1; r1>=1; r1--) {
    p=q-r;
    d[r]=a[q];
    x=Basic.vecvec(p+1,q-2,0,a,a);
    a1=a[q-1];
    if (Math.sqrt(x) <= machtol) {
      b0=b[r1]=a1;
      bb[r1]=b0*b0;
      a[q]=1.0;
    } else {
      bb0=bb[r1]=a1*a1+x;
      b0 = (a1 > 0.0) ? -Math.sqrt(bb0) : Math.sqrt(bb0);
      a1=a[q-1]=a1-b0;
      w=a[q]=1.0/(a1*b0);
      tj=0;
      for (j=1; j<=r1; j++) {
        ti=tj+j;
        s=Basic.vecvec(tj+1,ti,p-tj,a,a);
        tj=ti+j;
        b[j]=(Basic.seqvec(j+1,r1,tj,p,a,a)+s)*w;
        tj=ti;
      }
      Basic.elmvec(1,r1,p,b,a,Basic.vecvec(1,r1,p,b,a)*w*0.5);
      tj=0;
      for (j=1; j<=r1; j++) {
        ti=tj+j;
        Basic.elmvec(tj+1,ti,p-tj,a,a,b[j]);
        Basic.elmvec(tj+1,ti,-tj,a,b,a[j+p]);
        tj=ti;
      }
      b[r1]=b0;
    }
    q=p;
    r=r1;
  }
  d[1]=a[1];
  a[1]=1.0;
```

```
  b[n]=bb[n]=0.0;
}
```

E. baksymtri1

Performs the back substitutions upon the intermediate numbers produced by
tfmsymtri1.

Procedure parameters:
$$\text{void baksymtri1 } (a,n,n1,n2,vec)$$
a: double *a[1:(n+1)n/2]*;
        entry: the data for the back transformation, as produced by *tfmsymtri1*;
n: int;
        entry: the order of the given matrix;
n1,n2: int;
        entry: the lower and upper bound, respectively, of the column numbers of
               vec;
vec: double *vec[1:n, n1:n2]*;
        entry: the vectors on which the back transformation has to be performed;
        exit: the transformed vectors.

Procedures used: vecvec, elmvec.

```
public static void baksymtri1(double a[], int n, int n1, int n2, double vec[][])
{
  int j,j1,k,ti,tj;
  double w;
  double auxvec[] = new double[n+1];

  for (k=n1; k<=n2; k++) {
    for (j=1; j<=n; j++) auxvec[j]=vec[j][k];
    tj=j1=1;
    for (j=2; j<=n; j++) {
      ti=tj+j;
      w=a[ti];
      if (w < 0.0)
        Basic.elmvec(1,j1,tj,auxvec,a,Basic.vecvec(1,j1,tj,auxvec,a)*w);
      j1=j;
      tj=ti;
    }
    for (j=1; j<=n; j++) vec[j][k]=auxvec[j];
  }
}
```

3.11.4 To Hessenberg form - real asymmetric

A. tfmreahes

Given a real $n \times n$ matrix A, obtains a sequence of pivot reference integers $1 \leq p(i) \leq n$ $(i=1,...,n)$ associated with a permutation matrix P, a unit lower triangular matrix L with $L_{i,1}=0$ $(i=2,...,n)$, $|L_{i,j}|<1$ $(j=2,...,n; i=j+1,...,n)$ such that $PLH=APL$, where H is an upper Hessenberg matrix ($H_{i,j}=0$ for $i>j+1$). For further details see [DekHof68, Wi65].

Procedure parameters:
$$\text{void tfmreahes } (a,n,em,index)$$

a: double $a[1:n,1:n]$;
           entry: the matrix to be transformed;
           exit: the upper Hessenberg matrix is delivered in the upper triangle and
                   the first subdiagonal of *a*, the transforming matrix *L*, in the
                   remaining part of *a*, i.e., $a[i,j]=L_{i,j+1}$, for $i=3,...,n$ and $j=1,...,i-2$;

n: int;
           entry: the order of the given matrix;
em: double $em[0:1]$;
           entry: $em[0]$: the machine precision;
           exit: $em[1]$: the infinity norm of the original matrix;
index: int $inter[1:n]$;
           exit: the pivotal indices defining the stabilizing row and column
                   interchanges.

Procedures used: matvec, matmat, ichcol, ichrow.

```
public static void tfmreahes(double a[][], int n, double em[], int index[])
{
  int i,j,j1,k,l;
  double s,t,machtol,macheps,norm;
  double b[] = new double[n];

  macheps=em[0];
  norm=0.0;
  for (i=1; i<=n; i++) {
    s=0.0;
    for (j=1; j<=n; j++) s += Math.abs(a[i][j]);
    if (s > norm) norm=s;
  }
  em[1]=norm;
  machtol=norm*macheps;
  index[1]=0;
  for (j=2; j<=n; j++) {
    j1=j-1;
    l=0;
```

```
    s=machtol;
    for (k=j+1; k<=n; k++) {
      t=Math.abs(a[k][j1]);
      if (t > s) {
        l=k;
        s=t;
      }
    }
    if (l != 0) {
      if (Math.abs(a[j][j1]) < s) {
        Basic.ichrow(1,n,j,l,a);
        Basic.ichcol(1,n,j,l,a);
      } else
        l=j;
      t=a[j][j1];
      for (k=j+1; k<=n; k++) a[k][j1] /=t;
    } else
      for (k=j+1; k<=n; k++) a[k][j1]=0.0;
    for (i=1; i<=n; i++)
      b[i-1] = a[i][j] += ((l == 0) ? 0.0 : Basic.matmat(j+1,n,i,j1,a,a)) -
                         Basic.matvec(1,(j1 < i-2) ? j1 : i-2,i,a,b);
    index[j]=l;
  }
}
```

B. bakreahes1

Given a sequence of pivot reference integers $p(i)$ ($1 \leq p(i) \leq n$, $i=1,...,n$) associated with an $n \times n$ permutation matrix P, a unit lower triangular matrix L with $L_{i,1}=0$ ($i=2,...,n$) and a single column vector v, forms the vector $u=PLv$. (If v is an eigenvector of H, where $PLH=APL$, u is an eigenvector of A.)

Procedure parameters:

                        void bakreahes1 $(a,n,index,v)$

a: double $a[1:n,1:n]$;
            entry: the transforming matrix L, as produced by *tfmreahes* must be given
                    in the part below the first subdiagonal of a, i.e., $a[i,j]=L_{i,j+1}$, for
                    $i=3,...,n$ and $j=1,...,i-2$;
n: int;
            entry: the length of the vector to be transformed;
index: int *index[1:n]*;
            entry: the pivotal indices defining the stabilizing row and column
                    interchanges as produced by *tfmreahes*;
v: double $v[1:n]$;
            entry: the vector to be transformed;
            exit: the transformed vector.

Procedure used: matvec.

```
public static void bakreahes1(double a[][], int n, int index[], double v[])
{
  int i,l;
  double w;
  double x[] = new double[n+1];

  for (i=2; i<=n; i++) x[i-1]=v[i];
  for (i=n; i>=2; i--) {
    v[i] += Basic.matvec(1,i-2,i,a,x);
    l=index[i];
    if (l > i) {
      w=v[i];
      v[i]=v[l];
      v[l]=w;
    }
  }
}
```

C. bakreahes2

Given a sequence of pivot reference integers $p(i)$ $(1 \leq p(i) \leq n,\ i=1,...,n)$ associated with an $n \times n$ permutation matrix P, a unit lower triangular matrix L with $L_{i,1}=0$ $(i=2,...,n)$ and a system of column vectors $b^{(k)}$ $(k=n1,...,n2)$ forms the vectors $v^{(k)}=PLb^{(k)}$ $(k=n1,...,n2)$. (If $b^{(k)}$ is an eigenvector of H, where $PLH=APL$, $v^{(k)}$ is an eigenvector of A.)

Procedure parameters:
$$\text{void bakreahes2 } (a,n,n1,n2,index,vec)$$

a: double $a[1{:}n,1{:}n]$;
            entry: the transforming matrix L, as produced by *tfmreahes* must be given in the part below the first subdiagonal of a, i.e., $a[i,j]=L_{i,j+1}$, for $i=3,...,n$ and $j=1,...,i{-}2$;
n: int;
            entry: the length of the vectors to be transformed;
n1,n2: int;
            the column numbers of the first and last vector to be transformed;
index: int $index[1{:}n]$;
            entry: the pivotal indices defining the stabilizing row and column interchanges as produced by *tfmreahes*;
vec: double $vec[1{:}n,n1{:}n2]$;
            entry: the $n2{-}n1{+}1$ vectors of length n to be transformed;
            exit: the $n2{-}n1{+}1$ vectors of length n resulting from the back transformation.

Procedures used: tamvec, ichrow.

```
public static void bakreahes2(double a[][], int n, int n1, int n2, int index[],
                              double vec[][])
{
  int i,l,k;
  double u[] = new double[n+1];

  for (i=n; i>=2; i--) {
    for (k=i-2; k>=1; k--) u[k+1]=a[i][k];
    for (k=n1; k<=n2; k++)
      vec[i][k] += Basic.tamvec(2,i-1,k,vec,u);
    l=index[i];
    if (l > i) Basic.ichrow(n1,n2,i,l,vec);
  }
}
```

3.11.5 To Hessenberg form - complex Hermitian

A. hshhrmtri

Given an $n{\times}n$ Hermitian matrix A, obtains a complex diagonal matrix C and k ($k{\leq}n{-}2$) vectors $u^{(j)}$ such that with

$$E = \prod_{j=1}^{k}\left(I - 2u^{(j)^{T}}u^{(j)}/(u^{(j)}u^{(j)^{T}})\right)$$

$S=C^{-1}E'AEC$ (E' denotes the conjugate transpose of E) is a real symmetric tridiagonal matrix ($S_{i,j}=0$ if $|i\text{-}j|>1$ and $S_{i,i+1}=S_{i+1,i}$).

Procedure parameters:

void hshhrmtri (a,n,d,b,bb,em,tr,ti)

a: double $a[1:n,1:n]$;
         entry: the real part of the upper triangle of the Hermitian matrix must be
                 given in the upper triangular part of a (the elements $a[i,j]$, $i{\leq}j$); the
                 imaginary part of the strict lower triangle of the Hermitian matrix
                 must be given in the strict lower part of a (the elements $a[i,j]$, $i{>}j$);
         exit: the components of the $u^{(j)}$ and C;
n: int;
         entry: the order of the matrix;
d: double $d[1:n]$;
         exit: the main diagonal of the resulting symmetric tridiagonal matrix
                 ($S_{i,i}$ above);
b: double $b[1:n-1]$;
         exit: the codiagonal elements of the resulting symmetric tridiagonal
                 matrix (the values $S_{i,i+1}$, $i=1,...,n-1$, in the above);

bb: double *bb[1:n-1]*;
            exit: the squares of the moduli of the codiagonal elements of the
                    resulting   symmetric   tridiagonal   matrix   (the   values
                    $(S_{i,i+1})^2$, *i=1,...,n-1*, in the above);
em: double *em[0:1]*;
            entry: *em[0]*: the machine precision;
            exit: *em[1]*: an estimate for a norm of the original matrix;
tr,ti: double *tr[1:n-1]*, *ti[1:n-1]*;
            exit: data for subsequent back transformations.

Procedures used: matvec,   tamvec,   matmat,   tammat,   mattam,   elmveccol,
                    elmcolvec, elmcol, elmrow, elmvecrow, elmrowvec, elmrowcol,
                    elmcolrow, carpol.

Remark: See *hshhrmtrival*.

```
public static void hshhrmtri(double a[][], int n, double d[], double b[],
                        double bb[], double em[], double tr[], double ti[])
{
  int i,j,j1,jm1,r,rm1;
  double nrm,w,tol2,x,ar,ai,h,k,t,q,ajr,arj,bj,bbj;
  double mod[] = new double[1];
  double c[] = new double[1];
  double s[] = new double[1];

  nrm=0.0;
  for (i=1; i<=n; i++) {
    w=Math.abs(a[i][i]);
    for (j=i-1; j>=1; j--)
      w += Math.abs(a[i][j])+Math.abs(a[j][i]);
    for (j=i+1; j<=n; j++)
      w += Math.abs(a[i][j])+Math.abs(a[j][i]);
    if (w > nrm) nrm=w;
  }
  t=em[0]*nrm;
  tol2=t*t;
  em[1]=nrm;
  r=n;
  for (rm1=n-1; rm1>=1; rm1--) {
    x=Basic.tammat(1,r-2,r,r,a,a)+Basic.mattam(1,r-2,r,r,a,a);
    ar=a[rm1][r];
    ai = -a[r][rm1];
    d[r]=a[r][r];
    Basic.carpol(ar,ai,mod,c,s);
    if (x < tol2) {
      a[r][r] = -1.0;
      b[rm1]=mod[0];
      bb[rm1]=mod[0]*mod[0];
    } else {
      h=mod[0]*mod[0]+x;
```

```
      k=Math.sqrt(h);
      t=a[r][r]=h+mod[0]*k;
      if (ar == 0.0 && ai == 0.0)
        a[rm1][r]=k;
      else {
        a[rm1][r]=ar+c[0]*k;
        a[r][rm1] = -ai-s[0]*k;
        s[0] = -s[0];
      }
      c[0] = -c[0];
      j=1;
      jm1=0;
      for (j1=2; j1<=r; j1++) {
        b[j]=(Basic.tammat(1,j,j,r,a,a)+Basic.matmat(j1,rm1,j,r,a,a)+
              Basic.mattam(1,jm1,j,r,a,a)-Basic.matmat(j1,rm1,r,j,a,a))/t;
        bb[j]=(Basic.matmat(1,jm1,j,r,a,a)-Basic.tammat(j1,rm1,j,r,a,a)-
              Basic.matmat(1,j,r,j,a,a)-Basic.mattam(j1,rm1,j,r,a,a))/t;
        jm1=j;
        j=j1;
      }
      q=(Basic.tamvec(1,rm1,r,a,b)-Basic.matvec(1,rm1,r,a,bb))/t/2.0;
      Basic.elmveccol(1,rm1,r,b,a,-q);
      Basic.elmvecrow(1,rm1,r,bb,a,q);
      j=1;
      for (j1=2; j1<=r; j1++) {
        ajr=a[j][r];
        arj=a[r][j];
        bj=b[j];
        bbj=bb[j];
        Basic.elmrowvec(j,rm1,j,a,b,-ajr);
        Basic.elmrowvec(j,rm1,j,a,bb,arj);
        Basic.elmrowcol(j,rm1,j,r,a,a,-bj);
        Basic.elmrow(j,rm1,j,r,a,a,bbj);
        Basic.elmcolvec(j1,rm1,j,a,b,-arj);
        Basic.elmcolvec(j1,rm1,j,a,bb,-ajr);
        Basic.elmcol(j1,rm1,j,r,a,a,bbj);
        Basic.elmcolrow(j1,rm1,j,r,a,a,bj);
        j=j1;
      }
      bb[rm1]=h;
      b[rm1]=k;
    }
    tr[rm1]=c[0];
    ti[rm1]=s[0];
    r=rm1;
  }
  d[1]=a[1][1];
}
```

B. hshhrmtrival

Given an $n \times n$ Hermitian matrix A, obtains m complex vectors $c^{(k)}$ ($m \leq n-2$) such that with $E=E_m E_{m-1}...E_1$, where

$$E_k = I - 2c^{(k)}\overline{c}^{(k)^T} / \overline{c}^{(k)} c^{(k)}, \quad EA\overline{E}^T = S, \quad (\overline{E}^T: \text{conjugate and transpose of } E)$$

where S is a Hermitian tridiagonal matrix and delivers the values $S_{i,i}$ ($i=1,...,n$) and of $|T_{i,i+1}|_2$ ($i=1,...,n-1$).

Procedure parameters:
$$\text{void hshhrmtrival } (a,n,d,bb,em)$$

a: double $a[1:n, 1:n]$;
           entry: the real part of the upper triangle of the Hermitian matrix must be given in the upper triangular part of a (the elements $a[i,j]$, $i \leq j$); the imaginary part of the strict lower triangle of the Hermitian matrix must be given in the strict lower part of a (the elements $a[i,j]$, $i>j$);
           exit: the elements of a are altered;
n: int;
           entry: the order of the given matrix;
d: double $d[1:n]$;
           exit: the main diagonal of the resulting Hermitian tridiagonal matrix ($S_{i,i}$ above);
bb: double $bb[1:n-1]$;
           exit: the squares of the moduli of the codiagonal elements of the resulting Hermitian tridiagonal matrix (the values $(S_{i,i+1})^2$, $i=1,...,n-1$, in the above);
em: double $em[0:1]$;
           entry: $em[0]$: the machine precision;
           exit: $em[1]$: an estimate for a norm of the original matrix.

Procedures used: matvec, tamvec, matmat, tammat, mattam, elmveccol, elmcolvec, elmcol, elmrow, elmvecrow, elmrowvec, elmrowcol, elmcolrow.

Remark: *hshhrmtrival* transforms a Hermitian matrix into a similar Hermitian tridiagonal matrix by means of Householder's transformation. *hshhrmtri* transforms a Hermitian matrix into a similar real tridiagonal matrix by means of Householder's transformation followed by a complex diagonal unitary similarity transformation in order to make the resulting tridiagonal matrix real symmetric. Householder's transformation for complex Hermitian matrices is a unitary similarity transformation, transforming a Hermitian matrix into a similar complex tridiagonal one [Mu66, Wi65].

```
public static void hshhrmtrival(double a[][], int n, double d[], double bb[],
                                double em[])
{
  int i,j,j1,jm1,r,rm1;
  double nrm,w,tol2,x,ar,ai,h,t,q,ajr,arj,dj,bbj,mod2;
```

```
nrm=0.0;
for (i=1; i<=n; i++) {
  w=Math.abs(a[i][i]);
  for (j=i-1; j>=1; j--)
    w += Math.abs(a[i][j])+Math.abs(a[j][i]);
  for (j=i+1; j<=n; j++)
    w += Math.abs(a[i][j])+Math.abs(a[j][i]);
  if (w > nrm) nrm=w;
}
t=em[0]*nrm;
tol2=t*t;
em[1]=nrm;
r=n;
for (rm1=n-1; rm1>=1; rm1--) {
  x=Basic.tammat(1,r-2,r,r,a,a)+Basic.mattam(1,r-2,r,r,a,a);
  ar=a[rm1][r];
  ai = -a[r][rm1];
  d[r]=a[r][r];
  if (x < tol2)
    bb[rm1]=ar*ar+ai*ai;
  else {
    mod2=ar*ar+ai*ai;
    if (mod2 == 0.0) {
      a[rm1][r]=Math.sqrt(x);
      t=x;
    } else {
      x += mod2;
      h=Math.sqrt(mod2*x);
      t=x+h;
      h=1.0+x/h;
      a[r][rm1] = -ai*h;
      a[rm1][r]=ar*h;
    }
    j=1;
    jm1=0;
    for (j1=2; j1<=r; j1++) {
      d[j]=(Basic.tammat(1,j,j,r,a,a)+Basic.matmat(j1,rm1,j,r,a,a)+
           Basic.mattam(1,jm1,j,r,a,a)-Basic.matmat(j1,rm1,r,j,a,a))/t;
      bb[j]=(Basic.matmat(1,jm1,j,r,a,a)-Basic.tammat(j1,rm1,j,r,a,a)-
            Basic.matmat(1,j,r,j,a,a)-Basic.mattam(j1,rm1,j,r,a,a))/t;
      jm1=j;
      j=j1;
    }
    q=(Basic.tamvec(1,rm1,r,a,d)-Basic.matvec(1,rm1,r,a,bb))/t/2.0;
    Basic.elmveccol(1,rm1,r,d,a,-q);
    Basic.elmvecrow(1,rm1,r,bb,a,q);
    j=1;
    for (j1=2; j1<=r; j1++) {
      ajr=a[j][r];
```

```
        arj=a[r][j];
        dj=d[j];
        bbj=bb[j];
        Basic.elmrowvec(j,rm1,j,a,d,-ajr);
        Basic.elmrowvec(j,rm1,j,a,bb,arj);
        Basic.elmrowcol(j,rm1,j,r,a,a,-dj);
        Basic.elmrow(j,rm1,j,r,a,a,bbj);
        Basic.elmcolvec(j1,rm1,j,a,d,-arj);
        Basic.elmcolvec(j1,rm1,j,a,bb,-ajr);
        Basic.elmcol(j1,rm1,j,r,a,a,bbj);
        Basic.elmcolrow(j1,rm1,j,r,a,a,dj);
        j=j1;
      }
      bb[rm1]=x;
    }
    r=rm1;
  }
  d[1]=a[1][1];
}
```

C. bakhrmtri

Given a complex $n \times n$ diagonal matrix C and $m \le n-2$ complex vectors $u^{(k)}$ (as produced, for example, at exit from *hshhrmtri*) and a set of complex vectors $v^{(j)}$ ($j=n1,...,n2$) forms the vectors

$$u^{(j)} = \prod_{k=1}^{m}\left(I - 2u^{(k)^T}u^{(k)} / u^{(k)}u^{(k)^T}\right)Cv^{(j)} \quad (j = n1,...,n2).$$

Procedure parameters:

	void bakhrmtri (*a,n,n1,n2,vecr,veci,tr,ti*)
a:	double *a[1:n,1:n]*;
	entry: the data for the back transformation as produced by *hshhrmtri*;
n:	int;
	entry: the order of the matrix of which the eigenvectors are calculated;
n1,n2:	int;
	entry: the eigenvectors corresponding to the eigenvalues with indices *n1,...,n2* are to be transformed;
vecr,veci:	double *vecr[1:n,n1:n2]*, *veci[1:n,n1:n2]*;
	entry: the back transformation is performed on the real eigenvectors given in the columns of array *vecr*,
	exit: *vecr* and *veci* contain the real part and imaginary part of the transformed eigenvectors;
tr,ti:	double *tr[1:n-1]*, *ti[1:n-1]*;
	entry: data at exit from *hshhrmtri*.

Procedures used: matmat, tammat, elmcol, elmcolrow, commul, comrowcst.

Remark: See hshhrmtrival.

```
public static void bakhrmtri(double a[][], int n, int n1, int n2, double vecr[][],
                              double veci[][], double tr[], double ti[])
{
  int i,j,r,rm1;
  double c,s,t,qr,qi;
  double tmp1[] = new double[1];
  double tmp2[] = new double[1];

  for (i=1; i<=n; i++)
    for (j=n1; j<=n2; j++) veci[i][j]=0.0;
  c=1.0;
  s=0.0;
  for (j=n-1; j>=1; j--) {
    Basic.commul(c,s,tr[j],ti[j],tmp1,tmp2);
    c=tmp1[0];
    s=tmp2[0];
    Basic.comrowcst(n1,n2,j,vecr,veci,c,s);
  }
  rm1=2;
  for (r=3; r<=n; r++) {
    t=a[r][r];
    if (t > 0.0)
      for (j=n1; j<=n2; j++) {
        qr=(Basic.tammat(1,rm1,r,j,a,vecr)-Basic.matmat(1,rm1,r,j,a,veci))/t;
        qi=(Basic.tammat(1,rm1,r,j,a,veci)+Basic.matmat(1,rm1,r,j,a,vecr))/t;
        Basic.elmcol(1,rm1,j,r,vecr,a,-qr);
        Basic.elmcolrow(1,rm1,j,r,vecr,a,-qi);
        Basic.elmcolrow(1,rm1,j,r,veci,a,qr);
        Basic.elmcol(1,rm1,j,r,veci,a,-qi);
      }
    rm1=r;
  }
}
```

3.11.6 To Hessenberg form — Complex non-Hermitian

A. hshcomhes

Given an $n \times n$ complex matrix A, determines m complex vectors $u^{(j)}$ ($m \le n-2$) and an $n \times n$ complex diagonal matrix D such that with $E = E_m E_{m-1} \ldots E_1$, where

$$E_j = I - 2u^{(j)}u^{(j)^T}\Big/u^{(j)^T}\overline{u}^{(j)}$$

$D^{-1}EAED = H$ where H is a complex upper Hessenberg matrix ($H_{i,j} = 0$ for $i > j+1$) with real nonnegative subdiagonal elements $H_{j+1,j}$. The $u^{(j)}$ are determined by imposing the conditions that the first j elements $u^{(j)}$ are zero and that with $A^{(1)}=A$, $A^{(j+1)}A^{(j)}E^{(j)}$ has zeros in positions $i=j+2,...,n$ of the j-th column $(j=1,...,n-2)$. D is determined by imposing the condition that the elements of the subdiagonal of $D^{-1}A^{(n-1)}D$ are absolute values of those of $A^{(n-1)}$. For further details see [Mu66, Wi65].

Procedure parameters:

                         void hshcomhes (ar,ai,n,em,b,tr,ti,del)

ar,ai: double $ar[1:n,1:n]$, $ai[1:n,1:n]$;

        entry: the real part and the imaginary part of the matrix to be transformed must be given in the arrays *ar* and *ai*, respectively;

        exit: the real part and the imaginary part of the upper triangle of the resulting upper Hessenberg matrix are delivered in the corresponding parts of the arrays *ar* and *ai*, respectively; data for the Householder back transformation are delivered in the strict lower triangles of the arrays *ar* and *ai*;

n: int;

        entry: the order of the given matrix;

em: double $em[0:1]$;

        entry: $em[0]$: the machine precision;

                   $em[1]$: an estimate of the norm of the complex matrix (for example, the sum of the infinity norms of the real part and imaginary part of the matrix);

b: double $b[1:n-1]$;

        exit: the real nonnegative subdiagonal of the resulting upper Hessenberg matrix;

tr,ti: double $tr[1:n]$, $ti[1:n]$;

        exit: the real part and the imaginary part of the diagonal elements of a diagonal similarity transformation are delivered in the arrays *tr* and *ti*, respectively;

del: double $del[1:n-2]$;

        exit: information concerning the sequence of Householder matrices.

Procedures used: hshcomcol, matmat, elmrowcol, hshcomprd, carpol, commul, comcolcst, comrowcst.

```
public static void hshcomhes(double ar[][], double ai[][], int n, double em[],
                    double b[], double tr[], double ti[], double del[])
{
  int r,rm1,i,nm1;
  double tol,t,xr,xi;
  double tmp1[] = new double[1];
  double tmp2[] = new double[1];
  double tmp3[] = new double[1];
  double tmp4[] = new double[1];
  boolean temp;

  nm1=n-1;
  t=em[0]*em[1];
  tol=t*t;
```

```
 rm1=1;
 for (r=2; r<=nm1; r++) {
   temp=Basic.hshcomcol(r,n,rm1,ar,ai,tol,tmp1,tmp2,tmp3,tmp4);
   b[rm1]=tmp1[0];
   tr[r]=tmp2[0];
   ti[r]=tmp3[0];
   t=tmp4[0];
   if (temp) {
     for (i=1; i<=n; i++) {
       xr=(Basic.matmat(r,n,i,rm1,ai,ai)-Basic.matmat(r,n,i,rm1,ar,ar))/t;
       xi=(-Basic.matmat(r,n,i,rm1,ar,ai)-Basic.matmat(r,n,i,rm1,ai,ar))/t;
       Basic.elmrowcol(r,n,i,rm1,ar,ar,xr);
       Basic.elmrowcol(r,n,i,rm1,ar,ai,xi);
       Basic.elmrowcol(r,n,i,rm1,ai,ar,xi);
       Basic.elmrowcol(r,n,i,rm1,ai,ai,-xr);
     }
     Basic.hshcomprd(r,n,r,n,rm1,ar,ai,ar,ai,t);
   }
   del[rm1]=t;
   rm1=r;
 }
 if (n > 1) {
   Basic.carpol(ar[n][nm1],ai[n][nm1],tmp1,tmp2,tmp3);
   b[nm1]=tmp1[0];
   tr[n]=tmp2[0];
   ti[n]=tmp3[0];
 }
 rm1=1;
 tr[1]=1.0;
 ti[1]=0.0;
 for (r=2; r<=n; r++) {
   Basic.commul(tr[rm1],ti[rm1],tr[r],ti[r],tmp1,tmp2);
   tr[r]=tmp1[0];
   ti[r]=tmp2[0];
   Basic.comcolcst(1,rm1,r,ar,ai,tr[r],ti[r]);
   Basic.comrowcst(r+1,n,r,ar,ai,tr[r],-ti[r]);
   rm1=r;
 }
}
```

B. bakcomhes

Given m complex vectors $u^{(j)}$ ($m \le n-2$), an $n \times n$ complex diagonal matrix D and a sequence of complex column vectors $v^{(k)}$ ($k=n1,...,n2$) computes the vectors $w^{(k)}=EDv^{(k)}$ ($k=n1,...,n2$) where

$$E = \prod_{j=1}^{m} \left(I - 2u^{(j)}\overline{u}^{(j)^T} \Big/ \overline{u}^{(j)^T} u^{(j)} \right).$$

(If $v^{(k)}$ ($k=n1,...,n2$) are eigenvectors of $H=D^{-1}EAED$, $w^{(k)}$ ($k=n1,...,n2$) are corresponding eigenvectors of A.) For further details see [Mu66, Wi65].

Procedure parameters:

$\qquad$ void bakcomhes ($ar,ai,tr,ti,del,vr,vi,n,n1,n2$)

ar,ai,tr,ti,del: $\quad$ double $ar[1:n,1:n]$, $ai[1:n,1:n]$, $tr[1:n]$, $ti[1:n]$, $del[1:n-2]$;

$\qquad$ entry: the data for the back transformation as produced by *hshcomhes*;

vr,vi: $\quad$ double $vr[1:n,n1:n2]$;

$\qquad$ entry: the back transformation is performed on the eigenvectors with the real parts given in vr and the imaginary parts given in vi;

$\qquad$ exit: the real parts and imaginary parts of the resulting eigenvectors are delivered in the columns of the vr and vi, respectively;

n: $\quad$ int;

$\qquad$ entry: the order of the matrix of which the eigenvectors are calculated;

$n1,n2$: $\quad$ int;

$\qquad$ entry: the eigenvectors corresponding to the eigenvalues with indices $n1,...,n2$ are to be transformed.

Procedures used: $\quad$ comrowcst, hshcomprd.

```
public static void bakcomhes(double ar[][], double ai[][], double tr[], double ti[],
                    double del[], double vr[][], double vi[][], int n,
                    int n1, int n2)
{
  int i,r,rm1;
  double h;

  for (i=2; i<=n; i++)
    Basic.comrowcst(n1,n2,i,vr,vi,tr[i],ti[i]);
  r=n-1;
  for (rm1=n-2; rm1>=1; rm1--) {
    h=del[rm1];
    if (h > 0.0) Basic.hshcomprd(r,n,n1,n2,rm1,vr,vi,ar,ai,h);
    r=rm1;
  }
}
```

3.12 Other transformations

3.12.1 To bidiagonal form - real matrices

A. hshreabid

Reduces an $m \times n$ symmetric matrix A to bidiagonal form B. With $A = A_1$, $u^{(1)}$ is so chosen that all elements but the first in the first column of

$$A_1^{'} = \left(I - 2u^{(1)}u^{(1)^T} / u^{(1)^T} u^{(1)}\right) A_1$$

are zero; $v^{(1)}$ is so chosen that all elements but the first two of the first row in

$$A_1^{''} = A_1^{'} \left(I - 2v^{(1)}v^{(1)^T} / v^{(1)^T} v^{(1)}\right)$$

are zero; the first row and column are stripped from $A_1^{''}$ to produce the $(m-1) \times (n-1)$ matrix A_2, and the process is repeated.

Procedure parameters:
$$\text{void hshreabid } (a,m,n,d,b,em)$$

a: double $a[1:m,1:n]$;
   entry: the given matrix;
   exit: data concerning the premultiplying and postmultiplying matrices;
m: int;
   entry: the number of rows of the given matrix;
n: int;
   entry: the number of columns of the given matrix;
d: double $d[1:n]$;
   exit: the diagonal of the bidiagonal matrix (diagonal of B above);
b: double $b[1:n]$;
   exit: the superdiagonal of the bidiagonal matrix is delivered in $b[1:n-1]$;
em: double $em[0:1]$;
   entry: $em[0]$: the machine precision;
   exit: $em[1]$: the infinity norm of the original matrix.

Procedures used: tammat, mattam, elmcol, elmrow.

Remark: *hshreabid* slightly improves a part of a procedure (svd) of Golub and Reinsch [WiR71] by skipping a transformation if the column or row is already in the desired form, (i.e., if the sum of the squares of the elements that ought to be zero is smaller than a certain constant). In svd the transformation is skipped only if the norm of the full row or column is small enough. As a result, some ill-defined transformations are skipped in *hshreabid*. Moreover, if a transformation is skipped, a zero is not stored in the diagonal or superdiagonal, but the value that would have been found if the column or row were in the desired form already is stored.

```
public static void hshreabid(double a[][], int m, int n, double d[], double b[],
                                double em[])
{
  int i,j,i1;
  double norm,machtol,w,s,f,g,h;

  norm=0.0;
  for (i=1; i<=m; i++) {
    w=0.0;
    for (j=1; j<=n; j++) w += Math.abs(a[i][j]);
    if (w > norm) norm=w;
  }
  machtol=em[0]*norm;
  em[1]=norm;
  for (i=1; i<=n; i++) {
    i1=i+1;
    s=Basic.tammat(i1,m,i,i,a,a);
    if (s < machtol)
      d[i]=a[i][i];
    else {
      f=a[i][i];
      s += f*f;
      d[i] = g = (f < 0.0) ? Math.sqrt(s) : -Math.sqrt(s);
      h=f*g-s;
      a[i][i]=f-g;
      for (j=i1; j<=n; j++)
        Basic.elmcol(i,m,j,i,a,a,Basic.tammat(i,m,i,j,a,a)/h);
    }
    if (i < n) {
      s=Basic.mattam(i1+1,n,i,i,a,a);
      if (s < machtol)
        b[i]=a[i][i1];
      else {
        f=a[i][i1];
        s += f*f;
        b[i] = g = (f < 0.0) ? Math.sqrt(s) : -Math.sqrt(s);
        h=f*g-s;
        a[i][i1]=f-g;
        for (j=i1; j<=m; j++)
          Basic.elmrow(i1,n,j,i,a,a,Basic.mattam(i1,n,i,j,a,a)/h);
      }
    }
  }
}
```

B. psttfmmat

Computes the postmultiplying matrix from the intermediate results generated by *hshreadbid*.

Procedure parameters:

$$\text{void psttfmmat } (a,n,v,b)$$

a: double *a[1:n, 1:n]*;
           entry: the data concerning the postmultiplying matrix, as generated by *hshreabid*;
n: int;
           entry: the number of columns and rows of *a*;
v: double *v[1:n, 1:n]*;
           exit: the postmultiplying matrix;
b: double *b[1:n]*;
           exit: the superdiagonal as generated by *hshreabid*.

Procedures used: matmat, elmcol.

```
public static void psttfmmat(double a[][], int n, double v[][], double b[])
{
  int i,i1,j;
  double h;

  i1=n;
  v[n][n]=1.0;
  for (i=n-1; i>=1; i--) {
    h=b[i]*a[i][i1];
    if (h < 0.0) {
      for (j=i1; j<=n; j++) v[j][i]=a[i][j]/h;
      for (j=i1; j<=n; j++)
        Basic.elmcol(i1,n,j,i,v,v,Basic.matmat(i1,n,i,j,a,v));
    }
    for (j=i1; j<=n; j++) v[i][j]=v[j][i]=0.0;
    v[i][i]=1.0;
    i1=i;
  }
}
```

C. pretfmmat

Computes the premultiplying matrix from the intermediate results generated by *hshreadbid*.

Procedure parameters:

void pretfmmat (a,m,n,d)

a: double a[1:m,1:n];
        entry: the data concerning the premultiplying matrix, as generated by
               hshreabid;
        exit: the premultiplying matrix;
m: int;
        entry: the number of rows of <i>a</i>;
n: int;
        entry: the number of columns of <i>a</i>, <i>n</i> should satisfy $n{\leq}m$;
d: double d[1:n];
        entry: the diagonal as generated by <i>hshreabid</i>.

Procedures used: tammat, elmcol.

```
public static void pretfmmat(double a[][], int m, int n, double d[])
{
  int i,i1,j;
  double g,h;

  for (i=n; i>=1; i--) {
    i1=i+1;
    g=d[i];
    h=g*a[i][i];
    for (j=i1; j<=n; j++) a[i][j]=0.0;
    if (h < 0.0) {
      for (j=i1; j<=n; j++)
        Basic.elmcol(i,m,j,i,a,a,Basic.tammat(i1,m,i,j,a,a)/h);
      for (j=i; j<=m; j++) a[j][i] /= g;
    } else
      for (j=i; j<=m; j++) a[j][i]=0.0;
    a[i][i] += 1.0;
  }
}
```

3.13 The (ordinary) eigenvalue problem

3.13.1 Real symmetric tridiagonal matrices

A. valsymtri

Calculates all or some consecutive eigenvalues in descending order of magnitude, of a symmetric $n{\times}n$ tridiagonal matrix T. With d_i ($i=1,...,n$) the values of the elements of the principal diagonal of T, b_i ($i=1,...,n-1$) the squares of the values of the codiagonal elements of T, δ the product of the machine precision and $\|T\|$, the sequence $f(k,x)$ given by

$$f(1,x)=d_1-x, \; f(k,x)=(d_k-xb_{k-1})/v_k$$

where $v_k=f(k-1,x)$ if $|f(k-1,x)|>\delta$, $-\delta$ if $f(k-1,x)\leq0$, and δ otherwise $(k=2,...,n)$, is determined,

$$f(n,x)=p(n,x)/p(n-1,x)$$

where $p(n,x)$ = determinant of $T-xI$, and the $f(k,x)$ form a Sturm sequence. The zeros of $f(n,x)$ (i.e., the eigenvalues of T) are computed by a sophisticated mixture of linear interpolation and bisection (see *zeroin*). It is assumed that the values of d_i, b_i, and $\|T\|$ above are available; these are generated as output from the tridiagonalization of a symmetric $n\times n$ matrix by both *tfmsymtri2* and *tfmsymtri1*. For further details see [DekHo68, Wi65].

Procedure parameters:

$$\text{void valsymtri } (d,bb,n,n1,n2,val,em)$$

d: double $d[1:n]$;
           entry: the main diagonal of the symmetric tridiagonal matrix (values of d_i above);
bb: double $bb[1:n-1]$;
           entry: the squares of the codiagonal elements of the symmetric tridiagonal matrix (the values of b_i in the above);
n: int;
           entry: the order of the given matrix;
n1,n2:   int;
           entry: the serial number of the first and last eigenvalue to be calculated, respectively;
val: double $val[n1:n2]$;
           exit: the $n2-n1+1$ calculated consecutive eigenvalues in nonincreasing order;
em: double $em[0:3]$;
           entry: $em[0]$: the machine precision;
                   $em[1]$: an upper bound for the moduli of the eigenvalues of the given matrix;
                   $em[2]$: a relative tolerance for the eigenvalues;
           exit: $em[3]$: the total number of iterations used for calculating the eigenvalues.

```
public static void valsymtri(double d[], double bb[], int n, int n1, int n2,
                     double val[], double em[])
{
  boolean extrapolate;
  int k,ext;
  double max,x,y,macheps,norm,re,machtol,lambda, c,fc,b,fb,a,fa,dd,fd,fdb,fda,w,
         mb,tol,m,p,q;
  int count[] = new int[1];
  double lb[] = new double[1];
  double ub[] = new double[1];

  fd=dd=0.0;
  macheps=em[0];
  norm=em[1];
```

```
re=em[2];
machtol=norm*macheps;
max=norm/macheps;
count[0]=0;
ub[0]=1.1*norm;
lb[0] = -ub[0];
lambda=ub[0];
for (k=n1; k<=n2; k++) {
  y=ub[0];
  lb[0] = -1.1*norm;
  x=lb[0];

  /* look for the zero of the polynomial function */

  b=x;
  fb=sturm(d,bb,n,x,k,machtol,max,count,lb,ub);
  a=x=y;
  fa=sturm(d,bb,n,x,k,machtol,max,count,lb,ub);
  c=a;
  fc=fa;
  ext=0;
  extrapolate=true;
  while (extrapolate) {
    if (Math.abs(fc) < Math.abs(fb)) {
      if (c != a) {
        dd=a;
        fd=fa;
      }
      a=b;
      fa=fb;
      b=x=c;
      fb=fc;
      c=a;
      fc=fa;
    }
    tol=Math.abs(x)*re+machtol;
    m=(c+b)*0.5;
    mb=m-b;
    if (Math.abs(mb) > tol) {
      if (ext > 2)
        w=mb;
      else {
        if (mb == 0.0)
          tol=0.0;
        else
          if (mb < 0.0) tol = -tol;
        p=(b-a)*fb;
        if (ext <= 1)
          q=fa-fb;
        else {
```

```
                fdb=(fd-fb)/(dd-b);
                fda=(fd-fa)/(dd-a);
                p *= fda;
                q=fdb*fa-fda*fb;
              }
              if (p < 0.0) {
                p = -p;
                q = -q;
              }
              w=(p<Double.MIN_VALUE || p<=q*tol) ? tol : ((p<mb*q) ? p/q : mb);
            }
            dd=a;
            fd=fa;
            a=b;
            fa=fb;
            x = b += w;
            fb=sturm(d,bb,n,x,k,machtol,max,count,lb,ub);
            if ((fc >= 0.0) ? (fb >= 0.0) : (fb <= 0.0)) {
              c=a;
              fc=fa;
              ext=0;
            } else
              ext = (w == mb) ? 0 : ext+1;
          } else
            break;
        }
        y=c;

        /* end of the zero finding procedure */

        val[k] = lambda = (x > lambda) ? lambda : x;
        if (ub[0] > x)
          ub[0] = (x > y) ? x : y;
      }
      em[3]=count[0];
    }



static private double sturm(double d[], double bb[], int n, double x, int k,
                            double machtol, double max, int count[], double lb[],
                            double ub[])
{
  /* this sturm procedure is used internally by VALSYMTRI */

  int p,i;
  double f;

  (count[0])++;
  p=k;
  f=d[1]-x;
```

```
for (i=2; i<=n; i++) {
  if (f <= 0.0) {
    p++;
    if (p > n) return ((p == n) ? f : (n-p)*max);
  } else
    if (p < i-1) {
      lb[0] = x;
      return ((p == n) ? f : (n-p)*max);
    }
  if (Math.abs(f) < machtol)
    f = (f <= 0.0) ? -machtol : machtol;
  f=d[i]-x-bb[i-1]/f;
}
if (p == n || f <= 0.0)
  if (x < ub[0]) ub[0] = x;
else
  lb[0] = x;
return ((p == n) ? f : (n-p)*max);
}
```

B. vecsymtri

Calculates, by use of inverse iteration, the eigenvectors x_i corresponding to the given consecutive eigenvalues λ_i (in nonincreasing order) ($i=n1,...,n2$) of the $n \times n$ symmetric tridiagonal matrix T. The inverse iteration procedure used involves the determination of the vectors $x_i^{(k)}$ from the relationships $(T-\lambda_i I)x_i^{(k+1)}=Y^{(k)}x_i^{(k)}$ ($k=0,1,...$), where $Y^{(k)}$ is so chosen that $\|x_i^{(k+1)}\|=1$. The values of the machine precision m_0, $\|T\|$ the norm of the given matrix m_1, an orthogonalization parameter m_4, the relative tolerance of the eigenvectors m_6, and the maximum number of iterations allowed for the calculation of each eigenvector m_8 must be provided. If $|\lambda_i - \lambda_j| < m_0 * m_1$ for any j in $1 \leq j \leq i-1$, then λ_i is modified so that $|\lambda_i - \lambda_j| = m_0 * m_1$.

If $|\lambda_i - \lambda_j| < m_4 * m_1$ for any j in the range $1 \leq j \leq i-1$, then a Gram-Schmidt process is carried out at each iteration step, so that the eigenvectors are orthogonal to within working precision. The iteration terminates when either $\|x_n\|_2 < m_1 * m_6$ or $k > m_8$. vecsymtri may be called with $n1=1$ (so that eigenvectors are already known). However, it may occur that from previous use of vecsymtri or some other procedure, the eigenvectors corresponding to the $n1-1$ largest eigenvalues $\lambda_1,...,\lambda_{m+1}$ are already known. In this case the eigenvalues $\lambda_1,...,\lambda_{n1-1}$ must be provided (in array val) as well as the corresponding eigenvector (in array vec). Moreover, if vecsymtri has been called previously, then the number m_5 of eigenvectors involved in the last Gram-Schmidt process used must be provided. For further details see [DekHo68].

Procedure parameters:

void vecsymtri (*d,b,n,n1,n2,val,vec,em*)

d: double *d[1:n]*;
         entry: the main diagonal of the symmetric tridiagonal matrix (values of $T_{i,i}$
                  above);
b: double *b[1:n]*;
         entry: the codiagonal of the symmetric tridiagonal matrix (the values of
                  $T_{i,i+1} = T_{i+1,i}$ in the above) followed by an additional element 0;
n: int;
         entry: the order of the given matrix;
n1,n2: int;
         entry: lower and upper bound of the array *val*;
val: double *val[n1:n2]*;
         entry: a row of nonincreasing eigenvalues (values of $\lambda_1, \lambda_2, ..., \lambda_{n2}$ above) as
                  delivered by *valsymtri*;
vec: double *vec[1:n,n1:n2]*;
         entry: if *n1*>1 then the components of the eigenvectors corresponding to
                  $\lambda_1, ..., \lambda_{n1-1}$ must be given in *vec[1:n,1:n1-1]*;
         exit: the eigenvectors corresponding with the given eigenvalues $\lambda_1, ..., \lambda_{n2}$;
em: double *em[0:9]*;
         entry: *em[0]*:   the machine precision (value of m_0 above);
                  em[1]:   a norm of the given matrix (value of m_1 above);
                  em[4]:   the orthogonalization parameter (value of m_4 above);
                  em[5]:   if *n1*>1 then *em[5]* should be given the value of the number
                            of eigenvectors involved in the last Gram-Schmidt process
                            used in the determination of the eigenvectors corresponding
                            to $\lambda_1, ..., \lambda_{m-1}$ (value of m_5 above);
                  em[6]:   a relative tolerance for the eigenvectors;
                  em[8]:   maximum number of iterations allowed for calculating each
                            eigenvector;
         exit: *em[5]*:   the   number   of   eigenvectors   involved   in   the   last
                            Gram-Schmidt orthogonalization;
                  em[7]:   the maximum Euclidean norm of the residues;
                  em[9]:   the largest number of iterations performed for the
                            calculation of some eigenvector.

Procedures used: vecvec, tamvec, elmveccol.

```
public static void vecsymtri(double d[], double b[], int n, int n1, int n2,
                             double val[], double vec[][], double em[])
{
  boolean iterate;
  int i,j,k,count,maxcount,countlim,orth,ind;
  double bi,bi1,u,w,y,mi1,lambda,oldlambda,ortheps,valspread,spr,res,maxres,norm,
         newnorm,oldnorm,machtol,vectol;
  boolean index[] = new boolean[n+1];
  double m[] = new double[n+1];
  double p[] = new double[n+1];
  double q[] = new double[n+1];
  double r[] = new double[n+1];
  double x[] = new double[n+1];
```

```
oldlambda=res=0.0;
norm=em[1];
machtol=em[0]*norm;
valspread=em[4]*norm;
vectol=em[6]*norm;
countlim=(int) em[8];
ortheps=Math.sqrt(em[0]);
maxcount=ind=0;
maxres=0.0;
if (n1 > 1) {
  orth=(int) em[5];
  oldlambda=val[n1-orth];
  for (k=n1-orth+1; k<=n1-1; k++) {
    lambda=val[k];
    spr=oldlambda-lambda;
    if (spr < machtol) lambda=oldlambda-machtol;
    oldlambda=lambda;
  }
} else
  orth=1;
for (k=n1; k<=n2; k++) {
  lambda=val[k];
  if (k > 1) {
    spr=oldlambda-lambda;
    if (spr < valspread) {
      if (spr < machtol) lambda=oldlambda-machtol;
      orth++;
    } else
      orth=1;
  }
  count=0;
  u=d[1]-lambda;
  bi=w=b[1];
  if (Math.abs(bi) < machtol) bi=machtol;
  for (i=1; i<=n-1; i++) {
    bi1=b[i+1];
    if (Math.abs(bi1) < machtol) bi1=machtol;
    if (Math.abs(bi) >= Math.abs(u)) {
      mi1=m[i+1]=u/bi;
      p[i]=bi;
      y=q[i]=d[i+1]-lambda;
      r[i]=bi1;
      u=w-mi1*y;
      w = -mi1*bi1;
      index[i]=true;
    } else {
      mi1=m[i+1]=bi/u;
      p[i]=u;
      q[i]=w;
```

```
      r[i]=0.0;
      u=d[i+1]-lambda-mi1*w;
      w=bi1;
      index[i]=false;
    }
    x[i]=1.0;
    bi=bi1;
  } /* transform */
  p[n] = (Math.abs(u) < machtol) ? machtol : u;
  q[n]=r[n]=0.0;
  x[n]=1.0;
  iterate=true;
  while (iterate) {
    u=w=0.0;
    for (i=n; i>=1; i--) {
      y=u;
      u=x[i] = (x[i]-q[i]*u-r[i]*w)/p[i];
      w=y;
    } /* next iteration */
    newnorm=Math.sqrt(Basic.vecvec(1,n,0,x,x));
    if (orth > 1) {
      oldnorm=newnorm;
      for (j=k-orth+1; j<=k-1; j++)
        Basic.elmveccol(1,n,j,x,vec,-Basic.tamvec(1,n,j,vec,x));
      newnorm=Math.sqrt(Basic.vecvec(1,n,0,x,x));
      if (newnorm < ortheps*oldnorm) {
        ind++;
        count=1;
        for (i=1; i<=ind-1; i++) x[i]=0.0;
        for (i=ind+1; i<=n; i++) x[i]=0.0;
        x[ind]=1.0;
        if (ind == n) ind=0;
        w=x[1];
        for (i=2; i<=n; i++) {
          if (index[i-1]) {
            u=w;
            w=x[i-1]=x[i];
          } else
            u=x[i];
          w=x[i]=u-m[i]*w;
        }
        continue; /* iterate on */
      } /* new start */
    } /* orthogonalization */
    res=1.0/newnorm;
    if (res > vectol || count == 0) {
      count++;
      if (count <= countlim) {
        for (i=1; i<=n; i++) x[i] *= res;
        w=x[1];
```

```
          for (i=2; i<=n; i++) {
             if (index[i-1]) {
                u=w;
                w=x[i-1]=x[i];
             } else
                u=x[i];
             w=x[i]=u-m[i]*w;
          }
       } else
          break;
    } else
       break;
  }
  for (i=1; i<=n; i++) vec[i][k]=x[i]*res;
  if (count > maxcount) maxcount=count;
  if (res > maxres) maxres=res;
  oldlambda=lambda;
  }
  em[5]=orth;
  em[7]=maxres;
  em[9]=maxcount;
}
```

C. qrivalsymtri

Calculates all eigenvalues of a symmetric tridiagonal matrix T using square-root-free QR iteration. Values of m_0, the machine precision, a relative tolerance m_2, and $\|T\|$ must be provided. The relative error in the computed eigenvalues is less than $m_2*\|T\|$, and the absolute error is less than $m_0*\|T\|$. If some eigenvalues λ_i are very small, the absolute error in their determination may be made smaller by prescribing a value of m_0 which is less than its true value. When this is done, the calculation of b_i/m_0^2 should cause no overflow, and that of $(m_0*\|T\|)^2$ no underflow, where b_i is the i-th element of the codiagonal of T. For further details see [Re71, Wi65].

Procedure parameters:

$$\text{int qrivalsymtri } (d,bb,n,em)$$

qrivalsymtri: given the number of eigenvalues not calculated;
d: double $d[1:n]$;
  entry: the main diagonal of the symmetric tridiagonal matrix;
bb: double $bb[1:n]$;
  entry: the squares of the codiagonal elements of the symmetric tridiagonal matrix followed by an additional element 0;
  exit: the squares of the codiagonal elements of the symmetric tridiagonal matrix resulting from the QR iteration;
n: int;
  entry: the order of the given matrix;

em: double *em[0:5]*;
                      entry: *em[0]*:   the machine precision;
                               em[1]:   a norm of the given matrix;
                               em[2]:   a relative tolerance for the eigenvalues;
                               em[4]:   the maximum allowed number of iterations;
                      exit: *em[3]*:   the maximum absolute value of the codiagonal
                                          elements neglected;
                               em[5]:   the number of iterations performed.

```
public static int qrivalsymtri(double d[], double bb[], int n, double em[])
{
  int i,i1,low,oldlow,n1,count,max;
  double bbtol,bbmax,bbi,bbn1,machtol,dn,delta,f,num,shift,g,h,t,p,r,s,c,oldg;

  t=em[2]*em[1];
  bbtol=t*t;
  machtol=em[0]*em[1];
  max=(int) em[4];
  bbmax=0.0;
  count=0;
  oldlow=n;
  n1=n-1;
  while (n > 0) {
    i=n;
    do {
      low=i;
      i--;
    } while ((i >= 1) ? bb[i] > bbtol : false);
    if (low > 1)
      if (bb[low-1] > bbmax) bbmax=bb[low-1];
    if (low == n)
      n=n1;
    else {
      dn=d[n];
      delta=d[n1]-dn;
      bbn1=bb[n1];
      if (Math.abs(delta) < machtol)
        r=Math.sqrt(bbn1);
      else {
        f=2.0/delta;
        num=bbn1*f;
        r = -num/(Math.sqrt(num*f+1.0)+1.0);
      }
      if (low == n1) {
        d[n]=dn+r;
        d[n1] -= r;
        n -= 2;
      } else {
        count++;
        if (count > max) break;
```

```
              if (low < oldlow) {
                shift=0.0;
                oldlow=low;
              } else
                shift=dn+r;
              h=d[low]-shift;
              if (Math.abs(h) < machtol)
                  h = (h <= 0.0) ? -machtol : machtol;
              g=h;
              t=g*h;
              bbi=bb[low];
              p=t+bbi;
              i1=low;
              for (i=low+1; i<=n; i++) {
                s=bbi/p;
                c=t/p;
                h=d[i]-shift-bbi/h;
                if (Math.abs(h) < machtol)
                    h = (h <= 0.0) ? -machtol : machtol;
                oldg=g;
                g=h*c;
                t=g*h;
                d[i1]=oldg-g+d[i];
                bbi = (i == n) ? 0.0 : bb[i];
                p=t+bbi;
                bb[i1]=s*p;
                i1=i;
              }
              d[n]=g+shift;
            }
          }
        n1=n-1;
      }
    em[3]=Math.sqrt(bbmax);
    em[5]=count;
    return n;
}
```

D. qrisymtri

Calculates the eigenvalues and eigenvectors of a symmetric tridiagonal matrix T of order n simultaneously by QR iteration. The process requires input of an auxiliary matrix A. If T results from tridiagonalization by use of *tfmprevec* then A has precisely the form produced as output by that procedure. If T is simply given as a symmetric tridiagonal matrix then A must be set to be the unit matrix. The maximum allowed number of iterations must be prescribed. If this number is exceeded, and the eigenvalues $\lambda_1,...,\lambda_{n-k}$ and their corresponding eigenvectors have been determined, *qrisymtri* is given the value k, $\lambda_1,...,\lambda_{n-k}$ are to be found in locations $k+1,...,n$ of an output array D, their eigenvectors are to be found in columns $k+1,...,n$ of an output array A, and the numbers found in the remaining positions of these arrays are rough approximations only. For further details see [DekHo68, Wi65].

Procedure parameters:
$$\text{int qrisymtri } (a,n,d,b,bb,em)$$

qrisymtri: given the number (k above) of eigenvalues and eigenvectors not calculated;

a: double $a[1:n,1:n]$;

     entry: some input matrix, possibly the identity matrix;

     exit: the eigenvectors of the original symmetric tridiagonal matrix, premultiplied by the input matrix a;

n: int;

     entry: the order of the given matrix;

d: double $d[1:n]$;

     entry: the main diagonal of the symmetric tridiagonal matrix;

     exit: the eigenvalues of the matrix in some arbitrary order;

b: double $b[1:n]$;

     entry: the codiagonal of the symmetric tridiagonal matrix followed by an additional element 0;

     exit: the codiagonal of the symmetric tridiagonal matrix resulting from the QR iteration, followed by an additional element 0;

bb: double $bb[1:n]$;

     entry: the squared codiagonal elements of the symmetric tridiagonal matrix, followed by an additional element 0;

     exit: the squared codiagonal elements of the symmetric tridiagonal matrix resulting from the QR iteration;

em: double $em[0:5]$;

     entry: $em[0]$: the machine precision;

              $em[1]$: a norm of the given matrix;

              $em[2]$: a relative tolerance for the QR iteration;

              $em[4]$: the maximum allowed number of iterations;

     exit: $em[3]$: the maximum absolute value of the codiagonal elements neglected;

              $em[5]$: the number of iterations performed.

Procedure used: rotcol.

```
public static int qrisymtri(double a[][], int n, double d[], double b[],
                        double bb[], double em[])
{
   int j,j1,k,m,m1,count,max;
   double bbmax,r,s,sin,t,cos,oldcos,g,p,w,tol,tol2,lambda,dk1;

   g=0.0;
   tol=em[2]*em[1];
   tol2=tol*tol;
   count=0;
```

```
bbmax=0.0;
max=(int) em[4];
m=n;
do {
  k=m;
  m1=m-1;
  while (true) {
    k--;
    if (k <= 0) break;
    if (bb[k] < tol2) {
      if (bb[k] > bbmax) bbmax=bb[k];
      break;
    }
  }
  if (k == m1)
    m=m1;
  else {
    t=d[m]-d[m1];
    r=bb[m1];
    if (Math.abs(t) < tol)
      s=Math.sqrt(r);
    else {
      w=2.0/t;
      s=w*r/(Math.sqrt(w*w*r+1.0)+1.0);
    }
    if (k == m-2) {
      d[m] += s;
      d[m1] -= s;
      t = -s/b[m1];
      r=Math.sqrt(t*t+1.0);
      cos=1.0/r;
      sin=t/r;
      Basic.rotcol(1,n,m1,m,a,cos,sin);
      m -= 2;
    } else {
      count++;
      if (count > max) break;
      lambda=d[m]+s;
      if (Math.abs(t) < tol) {
        w=d[m1]-s;
        if (Math.abs(w) < Math.abs(lambda)) lambda=w;
      }
      k++;
      t=d[k]-lambda;
      cos=1.0;
      w=b[k];
      p=Math.sqrt(t*t+w*w);
      j1=k;
      for (j=k+1; j<=m; j++) {
        oldcos=cos;
```

```
          cos=t/p;
          sin=w/p;
          dk1=d[j]-lambda;
          t *= oldcos;
          d[j1]=(t+dk1)*sin*sin+lambda+t;
          t=cos*dk1-sin*w*oldcos;
          w=b[j];
          p=Math.sqrt(t*t+w*w);
          g=b[j1]=sin*p;
          bb[j1]=g*g;
          Basic.rotcol(1,n,j1,j,a,cos,sin);
          j1=j;
        }
        d[m]=cos*t+lambda;
        if (t < 0.0) b[m1] = -g;
      }
    }
  } while (m > 0);
  em[3]=Math.sqrt(bbmax);
  em[5]=count;
  return m;
}
```

3.13.2 Real symmetric full matrices

A. eigvalsym2

Computes the m largest eigenvalues of an $n \times n$ symmetric matrix A whose upper triangular part is stored in a two dimensional array, using linear interpolation on a function derived from a Sturm sequence. A is first reduced to similar tridiagonal form T by a call of *tfmsymtri2*, and *valsymtri* is then used to calculate the eigenvalues.

Procedure parameters:
$$\text{void eigvalsym2 } (a,n,numval,val,em)$$

a: double $a[1:n,1:n]$;
            entry: the upper triangle of the symmetric matrix must be given in the upper triangular part of a (the elements $a[i,j]$, $i \leq j$);
            exit: the data for the Householder's back transformation (which is not used by this procedure) is delivered in the upper triangular part of a;

n: int;
            entry: the order of the given matrix;
numval: int;
            entry: the serial number of the last eigenvalue to be calculated;

val: double *val[1:numval]*;
         exit: the *numval* largest eigenvalues in monotonically nonincreasing order;
em: double *em[0:3]*;
         entry: *em[0]*:   the machine precision;
                  em[2]:   a relative tolerance for the eigenvalues;
         exit: *em[1]*:   the infinity norm of the original matrix;
                  em[3]:   the number of iterations used for calculating the *numval* eigenvalues.

Procedures used: tfmsymtri2, valsymtri.

```
public static void eigvalsym2(double a[][], int n, int numval, double val[],
                              double em[])
{
  double b[]  = new double[n+1];
  double bb[] = new double[n+1];
  double d[]  = new double[n+1];

  tfmsymtri2(a,n,d,b,bb,em);
  valsymtri(d,bb,n,1,numval,val,em);
}
```

B. eigsym2

Computes the *m* largest eigenvalues and corresponding eigenvectors of an *n×n* symmetric matrix *A* whose upper triangular part is stored in a two-dimensional array, by means of inverse iteration. *A* is first reduced to similar tridiagonal form *T* by a call of *tfmsymtri2*, *valsymtri* is used to calculate the eigenvalues, *vecsymtri* is used to calculate the corresponding eigenvectors, and *baksymtri2* is then used to perform the necessary back transformation. *eigsym2* requires, as does *vecsymtri*, the value of an orthogonalization parameter m_4. When the distance between two consecutive eigenvalues is less than m_4, the eigenvectors are orthogonalized by a Gram-Schmidt process. The number m_5 of eigenvectors involved in the last Gram-Schmidt process is given by *eigsym2* as part of its output. If *eigsym2* is called more than once (to compute successive sets of eigenvalues and eigenvectors), the value m_5 arising from the preceding call is required as input for the current call.

Procedure parameters:
                  void eigsym2 (*a,n,numval,val,vec,em*)
a: double *a[1:n, 1:n]*;
         entry: the upper triangle of the symmetric matrix must be given in the upper triangular part of *a* (the elements *a[i,j]*, *i≤j*);
         exit: the data for Householder's back transformation is delivered in the upper triangular part of *a*; the elements *a[i,j]* for *i>j* are neither used nor changed;

n: int;
             entry: the order of the given matrix;
numval: int;
             entry: the serial number of the last eigenvalue to be
                      calculated;
val: double *val[1:numval]*;
             exit: the *numval* largest eigenvalues in monotonically nonincreasing
                      order;
vec: double *vec[1:n, 1:numval]*;
             exit: the *numval* calculated eigenvectors, stored columnwise,
                      corresponding to the calculated eigenvalues;
em: double *em[0:9]*;
             entry: *em[0]*: the machine precision;
                    em[2]: the relative tolerance for the eigenvalues;
                    em[4]: the orthogonalization parameter (value of m_4 above);
                    em[6]: the relative tolerance for the eigenvectors;
                    em[8]: the maximum number of inverse iterations allowed for
                              the calculation of each eigenvector;
             exit: *em[1]*: the infinity norm of the matrix;
                    em[3]: the number of iterations used for calculating the *numval*
                              eigenvalues;
                    em[5]: the number of eigenvectors involved in the last
                              Gram-Schmidt orthogonalization;
                    em[7]: the maximum Euclidean norm of the residues of the
                              calculated eigenvectors;
                    em[9]: the largest number of inverse iterations performed for
                              the calculation of some eigenvector.

Procedures used: tfmsymtri2, valsymtri, vecsymtri, baksymtri2.

```
public static void eigsym2(double a[][], int n, int numval, double val[],
                           double vec[][], double em[])
{
  double b[] = new double[n+1];
  double bb[] = new double[n+1];
  double d[] = new double[n+1];

  tfmsymtri2(a,n,d,b,bb,em);
  valsymtri(d,bb,n,1,numval,val,em);
  vecsymtri(d,b,n,1,numval,val,vec,em);
  baksymtri2(a,n,1,numval,vec);
}
```

C. eigvalsym1

Computes the *m* largest eigenvalues of an *n×n* symmetric matrix *A* whose upper triangular part is stored in a one dimensional array, using linear interpolation on a function derived from a Sturm sequence. *A* is first reduced to similar tridiagonal form *T* by a call of *tfmsymtri1*, and *valsymtri* is then used to calculate the eigenvalues.

Procedure parameters:
$$\text{void eigvalsym1 } (a,n,numval,val,em)$$
a: double *a[1:(n+1)n/2]*;
             entry: the upper triangle of the symmetric matrix must be given in such a way that the (*i,j*)-th element of the matrix is *a[(j-1)j/2+i]*, *1≤i≤j≤n*;
             exit: the data for the Householder's back transformation (which is not used by this procedure) is delivered in *a*;
n: int;
             entry: the order of the given matrix;
numval: int;
             entry: the serial number of the last eigenvalue to be calculated;
val: double *val[1:numval]*;
             exit: the *numval* largest eigenvalues in monotonically nonincreasing order;
em: double *em[0:3]*;
             entry: *em[0]*: the machine precision;
                     em[2]: a relative tolerance for the eigenvalues;
             exit: *em[1]*: the infinity norm of the original matrix;
                     em[3]: the number of iterations used for calculating the *numval* eigenvalues.

Procedures used: tfmsymtri1, valsymtri.

```
public static void eigvalsym1(double a[], int n, int numval, double val[],
                               double em[])
{
  double b[]  = new double[n+1];
  double bb[] = new double[n+1];
  double d[]  = new double[n+1];

  tfmsymtri1(a,n,d,b,bb,em);
  valsymtri(d,bb,n,1,numval,val,em);
}
```

D. eigsym1

Computes the *m* largest eigenvalues and corresponding eigenvectors of an *n×n* symmetric matrix *A* whose upper triangular part is stored in a one-dimensional array, by means of inverse iteration. *A* is first reduced to similar tridiagonal form *T* by a call of *tfmsymtri1*, *valsymtri* is used to calculate the eigenvalues, *vecsymtri* is used to calculate the corresponding eigenvectors, and *baksymtri1* is then used to calculate the necessary back transformation. The significances of an orthogonalization parameter m_4 and a number m_5 of eigenvectors involved in a Gram-Schmidt process are as for *eigsym2*.

Procedure parameters:
$$\text{void eigsym1 } (a,n,numval,val,vec,em)$$

a: double *a[1:(n+1)n/2]*;
    entry: the upper triangle of the symmetric matrix must be given in such a way that the (i,j)-th element of the matrix is $a[(j-1)j/2+i]$, $1 \le i \le j \le n$;
    exit: the data for Householder's back transformation;
n: int;
    entry: the order of the given matrix;
numval: int;
    entry: the serial number of the last eigenvalue to be calculated;
val: double *val[1:numval]*;
    exit: the *numval* largest eigenvalues in monotonically nonincreasing order;
vec: double *vec[1:n,1:numval]*;
    exit: the *numval* calculated eigenvectors, stored columnwise, corresponding to the calculated eigenvalues;
em: double *em[0:9]*;
    entry: *em[0]*: the machine precision;
        em[2]: the relative tolerance for the eigenvalues;
        em[4]: the orthogonalization parameter (value of m_4 above);
        em[6]: the relative tolerance for the eigenvectors;
        em[8]: the maximum number of inverse iterations allowed for the calculation of each eigenvector;
    exit: *em[1]*: the infinity norm of the matrix;
        em[3]: the number of iterations used for calculating the *numval* eigenvalues;
        em[5]: the number of eigenvectors involved in the last Gram-Schmidt orthogonalization;
        em[7]: the maximum Euclidean norm of the residues of the calculated eigenvectors;
        em[9]: the largest number of inverse iterations performed for the calculation of some eigenvector.

Procedures used: tfmsymtri1, valsymtri, vecsymtri, baksymtri1.

```
public static void eigsym1(double a[], int n, int numval, double val[],
                           double vec[][], double em[])
{
  double b[]  = new double[n+1];
  double bb[] = new double[n+1];
  double d[]  = new double[n+1];

  tfmsymtri1(a,n,d,b,bb,em);
  valsymtri(d,bb,n,1,numval,val,em);
  vecsymtri(d,b,n,1,numval,val,vec,em);
  baksymtri1(a,n,1,numval,vec);
```

}

E. qrivalsym2

Computes all eigenvalues of an $n \times n$ symmetric matrix A whose upper triangular part is stored in a two-dimensional array, by means of QR iteration. A is first reduced to similar tridiagonal form by a call of *tfmsymtri2*, and *qrivalsymtri* is then used to calculate the eigenvalues.

Procedure parameters:
$$\text{int qrivalsym2 } (a,n,val,em)$$
qrivalsym2: given the number of eigenvalues not calculated;
a: double *a[1:n,1:n]*;
       entry: the upper triangle of the symmetric matrix must be given in the upper triangular part of *a* (the elements *a[i,j]*, $i \leq j$);
       exit: the data for Householder's back transformation (which is not used by this procedure) is delivered in the upper triangular part of *a*; the elements *a[i,j]* for *i>j* are neither used nor changed;
n: int;
       entry: the order of the given matrix;
val: double *val[1:n]*;
       exit: the eigenvalues of the matrix in some arbitrary order;
em: double *em[0:5]*;
       entry: *em[0]*: the machine precision;
              em[2]: the relative tolerance for the eigenvalues;
              em[4]: the maximum allowed number of iterations;
       exit: *em[1]*: the infinity norm of the matrix;
              em[3]: the maximum absolute value of the codiagonal elements neglected;
              em[5]: the number of iterations performed.

Procedures used: tfmsymtri2, qrivalsymtri.

```
public static int qrivalsym2(double a[][], int n, double val[], double em[])
{
  int i;
  double b[] = new double[n+1];
  double bb[] = new double[n+1];

  tfmsymtri2(a,n,val,b,bb,em);
  i=qrivalsymtri(val,bb,n,em);
  return i;
}
```

F. qrisym

Computes all eigenvalues and corresponding eigenvectors of an $n \times n$ symmetric matrix A whose upper triangular part is stored in a two-dimensional array, by means of QR iteration. A is first reduced to similar tridiagonal form by a call of *tfmsymtri2*, the back transformation on the eigenvectors is prepared by a call of *tfmprevec*, and the eigenvalues and eigenvectors are then calculated by *qrisymtri*.

Procedure parameters:
$$\text{int qrisym } (a,n,val,em)$$

qrisym: given the number of eigenvalues and eigenvectors not calculated;
a: double $a[1:n, 1:n]$;
           entry: the upper triangle of the symmetric matrix must be given in the upper triangular part of a (the elements $a[i,j]$, $i \leq j$);
           exit: the eigenvectors of the symmetric matrix, stored columnwise;
n: int;
           entry: the order of the given matrix;
val: double $val[1:n]$;
           exit: the eigenvalues of the matrix corresponding to the calculated eigenvectors;
em: double $em[0:5]$;
           entry: $em[0]$: the machine precision;
                   $em[2]$: the relative tolerance for the QR iteration;
                   $em[4]$: the maximum allowed number of iterations;
           exit: $em[1]$: the infinity norm of the matrix;
                   $em[3]$: the maximum absolute value of the codiagonal elements neglected;
                   $em[5]$: the number of iterations performed.

Procedures used: tfmsymtri2, tfmprevec, qrisymtri.

```
public static int qrisym(double a[][], int n, double val[], double em[])
{
   int i;
   double b[] = new double[n+1];
   double bb[] = new double[n+1];

   tfmsymtri2(a,n,val,b,bb,em);
   tfmprevec(a,n);
   i=qrisymtri(a,n,val,b,bb,em);
   return i;
}
```

G. qrivalsym1

Computes all eigenvalues of an $n \times n$ symmetric matrix A whose upper triangular part is stored in a one-dimensional array, by means of QR iteration. A is first reduced to similar tridiagonal form by a call of *tfmsymtri1*, and *qrivalsymtri* is then used to calculate the eigenvalues.

Procedure parameters:
$$\text{int qrivalsym1 } (a,n,val,em)$$

qrivalsym1: given the number of eigenvalues not calculated;

a: double $a[1:(n+1)n/2]$;

entry: the upper triangle of the symmetric matrix must be given in such a way that the (i,j)-th element of the matrix is $a[(j-1)j/2+i]$, $1 \leq i \leq j \leq n$;

exit: the data for Householder's back transformation (which is not used by this procedure);

n: int;

entry: the order of the given matrix;

val: double $val[1:n]$;

exit: the eigenvalues of the matrix in some arbitrary order;

em: double $em[0:5]$;

entry: $em[0]$: the machine precision;

$em[2]$: the relative tolerance for the eigenvalues;

$em[4]$: the maximum allowed number of iterations;

exit: $em[1]$: the infinity norm of the matrix;

$em[3]$: the maximum absolute value of the codiagonal elements neglected;

$em[5]$: the number of iterations performed.

Procedures used: tfmsymtri1, qrivalsymtri.

```
public static int qrivalsym1(double a[], int n, double val[], double em[])
{
  int i;
  double b[]  = new double[n+1];
  double bb[] = new double[n+1];

  tfmsymtri1(a,n,val,b,bb,em);
  i=qrivalsymtri(val,bb,n,em);
  return i;
}
```

3.13.3 Symmetric matrices - Auxiliary procedures

A. mergesort

Determines the permutation of p integers in the range $l \leq i \leq u$, which rearranges the real numbers a_i $(i=l,...,u)$ in nonincreasing order of magnitude, so that $|a_{p(i)}| \geq |a_{p(i+1)}|$, $i=l,...,u-1$. For sorting by merging see [AHU74, K73].

Procedure parameters:

$$\text{void mergesort } (a,p,low,up)$$

a: double *a[low:up]*;
          entry: the vector to be stored into nondecreasing order;
          exit: the contents of *a* are left invariant;
p: int *p[low:up]*;
          exit: the permutation of indices corresponding to sorting the elements of
                   vec1 into nondecreasing order;
low: int;
          entry: the lower index of the arrays *a* and *p* (value of *l* above);
up: int;
          entry: the upper index of the arrays *a* and *p* (value of *u* above).

```
public static void mergesort(double a[], int p[], int low, int up)
{
  int i,lo,step,stap,umlp1,umsp1,rest,restv;
  int hp[] = new int[up+1];

  for (i=low; i<=up; i++) p[i]=i;
  restv=0;
  umlp1=up-low+1;
  step=1;
  do {
    stap=2*step;
    umsp1=up-stap+1;
    for (lo=low; lo<=umsp1; lo += stap)
      merge(lo,step,step,p,a,hp);
    rest=up-lo+1;
    if (rest > restv && restv > 0)
      merge(lo,rest-restv,restv,p,a,hp);
    restv=rest;
    step *= 2;
  } while (step < umlp1);
}


static private void merge(int lo, int ls, int rs, int p[], double a[], int hp[])
{
  /* this procedure is used internally by MERGESORT */

  int l,r,i,pl,pr;
  boolean lout,rout;

  l=lo;
  r=lo+ls;
  lout=rout=false;
  i=lo;
  do {
    pl=p[l];
```

```
    pr=p[r];
    if (a[pl] > a[pr]) {
      hp[i]=pr;
      r++;
      rout = (r == lo+ls+rs);
    } else {
      hp[i]=pl;
      l++;
      lout = (l == lo+ls);
    }
    i++;
  } while (!(lout || rout));
  if (rout) {
    for (i=lo+ls-1; i>=l; i--) p[i+rs]=p[i];
    r=l+rs;
  }
  for (i=r-1; i>=lo; i--) p[i]=hp[i];
}
```

B. vecperm

Given a sequence of distinct integers $p(i)$, $i=l,...,u$, in the range $l{\leq}p(i){\leq}u$ and the real numbers a_i ($i=l,...,u$), constructs the sequence of real numbers $a_{p(i)}$, $i=l,...,u$.

Procedure parameters:
$$\text{void vecperm } (perm,low,upp,vector)$$
perm: int *perm[low:upp]*;
        entry: a given permutation (for example, as produced by *mergesort*) of the
                 numbers in the array vector;
        exit: the contents of *a* are left invariant;
low: int;
        entry: the lower index of the arrays *perm* and *vector* (value of *l* above);
upp: int;
        entry: the upper index of the arrays *perm* and *vector* (value of *u* above);
vector: double *vector[low:upp]*;
        entry: the real vector to be permuted;
        exit: the permuted vector elements.

```
public static void vecperm(int perm[], int low, int upp, double vector[])
{
  int t,j,k;
  double a;
  boolean todo[] = new boolean[upp+1];

  for (t=low; t<=upp; t++) todo[t]=true;
  for (t=low; t<=upp; t++)
    if (todo[t]) {
```

```
            k=t;
            a=vector[k];
            j=perm[k];
            while (j != t) {
               vector[k]=vector[j];
               todo[k]=false;
               k=j;
               j=perm[k];
            }
            vector[k]=a;
            todo[k]=false;
         }
   }
```

C. rowperm

Given a sequence of distinct integers $p(j)$, $j=l,...,u$, in the range $l \leq p(j) \leq u$ and the elements $M_{i,j}$, $j=l,...,u$, belonging to the i-th row of a real matrix, constructs the corresponding row elements $M_{i,p(j)}$, $j=l,...,u$.

Procedure parameters:
$$\text{void rowperm } (perm, low, upp, i, mat)$$

perm: int *perm[low:upp]*;

         entry: a given permutation (for example, as produced by *mergesort*) of the numbers in the array vector;

low: int;

         entry: the lower index of the arrays *perm*;

upp: int;

         entry: the upper index of the arrays *perm*;

i: int;

         entry: the row index of the matrix elements;

mat: double *mat[i:i,low:upp]*;

         entry: *mat[i,low:upp]* should contain the elements to be permuted;

         exit: *mat[i,low:upp]* contains the row of permuted elements.

```
public static void rowperm(int perm[], int low, int upp, int i, double mat[][])
{
  int t,j,k;
  double a;
  boolean todo[] = new boolean[upp+1];

  for (t=low; t<=upp; t++) todo[t]=true;
  for (t=low; t<=upp; t++)
    if (todo[t]) {
      k=t;
      a=mat[i][k];
      j=perm[k];
```

```
      while (j != t) {
        mat[i][k]=mat[i][j];
        todo[k]=false;
        k=j;
        j=perm[k];
      }
      mat[i][k]=a;
      todo[k]=false;
    }
}
```

3.13.4 Symmetric matrices - Orthogonalization

orthog

Given a system of linearly independent columns x_j, $j=l,...,u$, of an n-rowed matrix X, determines an orthonormal system of columns

$$x'_j = \sum_{k=l}^{u} c_{j,k} x_k, \qquad j = l,...,u$$

$((x_i',x_j)$ equals 0 when $i{\neq}j$ and equals 1 with $i{=}j)$ of a corresponding matrix X' by the modified Gram-Schmidt orthogonalization [Wi65].

Procedure parameters:
$$\text{void orthog } (n,lc,uc,x)$$

n: int;
           entry: the order of the matrix x;
lc: int;
           entry: the lower column index of the matrix x;
uc: int;
           entry: the upper column index of the matrix x;
x: double $x[1{:}n,lc{:}uc]$;
           entry: the matrix columns to be orthogonalized;
           exit: the orthogonalized matrix columns.

Procedures used: tammat, elmcol.

```
public static void orthog(int n, int lc, int uc, double x[][])
{
  int i,j,k;
  double normx;

  for (j=lc; j<=uc; j++) {
    normx=Math.sqrt(Basic.tammat(1,n,j,j,x,x));
    for (i=1; i<=n; i++) x[i][j] /=normx;
    for (k=j+1; k<=uc; k++)
```

```
        Basic.elmcol(1,n,k,j,x,x,-Basic.tammat(1,n,k,j,x,x));
    }
}
```

3.13.5 Symmetric matrices - Iterative improvement

symeigimp

Given single precision estimates λ_j' of the eigenvalues λ_j and single precision estimates u_j' of the eigenvectors u_j (j=1,...,n) of an $n{\times}n$ real symmetric matrix A, derives (a) ranges of the form $[\lambda_j''-\delta_j, \lambda_n''+\eta_j]$ within which the λ_j lie, λ_j'' being an improved double precision estimate of λ_j, and (b) improved single precision estimates u_j'' of the corresponding eigenvector u_j (j=1,...,n). For further details see [GrK69].

Procedure parameters:

	void symeigimp (*n,a,vec,val,lbound,ubound,aux*)
n:	int;
	entry: the order of the matrix *a*;
a:	double *a[1:n,1:n]*;
	entry: contains a real symmetric matrix whose eigensystem has to be improved;
vec:	double *vec[1:n,1:n]*;
	entry: contains a matrix whose columns are a system of approximate eigenvectors of matrix *a*; initial approximations;
	exit: improved approximations;
val:	double *val[1:n]*;
	entry: initial approximations of the eigenvalues of *a*;
	exit: improved eigenvalues of *a*;
lbound, ubound:	double *lbound[1:n], ubound[1:n]*;
	exit: the lower and upper error bounds respectively for the eigenvalue approximations in *val* such that the *i*-th exact eigenvalue lies between *val[i]-lbound[i]* and *val[i]+ubound[i]* (values of δ_j and η_j, j=1,...,n, respectively, in the above);
aux:	double *aux[0:5]*;
	entry: *aux[0]*: the relative precision of the elements of *a*;
	aux[2]: the relative tolerance for the residual matrix; the iteration ends when the maximum absolute value of the residual elements is smaller than *aux[2]*aux[1]*;
	aux[4]: the maximum number of iterations allowed;
	exit: *aux[1]*: infinity norm of the matrix *a*;
	aux[3]: maximum absolute element of the residual matrix;
	aux[5]: number of iterations.

Procedures used: vecvec, matmat, tammat, mergesort, vecperm, rowperm, orthog, qrisym, infnrmmat.

```
public static void symeigimp(int n, double a[][], double vec[][], double val[],
                             double lbound[], double ubound[], double aux[])
{
  boolean stop;
  int k,i,j,i0,i1,i01,iter,maxitp1,n1,i0m1,i1p1;
  double s,max,tol,mateps,relerra,reltolr,norma,eps2,dl,dr,m1,dtemp;
  int itmp[] = new int[1];
  int perm[] = new int[n+1];
  double em[] = new double[6];
  double rq[] = new double[n+1];
  double eps[] = new double[n+1];
  double z[] = new double[n+1];
  double val3[] = new double[n+1];
  double eta[] = new double[n+1];
  double r[][] = new double[n+1][n+1];
  double p[][] = new double[n+1][n+1];
  double y[][] = new double[n+1][n+1];

  max=0.0;
  norma=Basic.infnrmmat(1,n,1,n,itmp,a);
  i=itmp[0];
  relerra=aux[0];
  reltolr=aux[2];
  maxitp1=(int) (aux[4]+1.0);
  mateps=relerra*norma;
  tol=reltolr*norma;
  for (iter=1; iter<=maxitp1; iter++) {
    if (iter == 1)
      stop=false;
    else
      stop=true;
    max=0.0;
    for (j=1; j<=n; j++)
      for (i=1; i<=n; i++) {
        dtemp = -(double)(vec[i][j])*(double)(val[j]);
        for (k=1; k<=n; k++)
          dtemp += (double)(a[i][k])*(double)(vec[k][j]);
        r[i][j]=dtemp;
        if (Math.abs(r[i][j]) > max) max=Math.abs(r[i][j]);
      }
    if (max > tol) stop=false;
    if ((!stop) && (iter < maxitp1)) {
      for (i=1; i<=n; i++) {
        dtemp=(double)(val[i]);
        for (k=1; k<=n; k++)
          dtemp += (double)(vec[k][i])*(double)(r[k][i]);
```

```
      rq[i]=dtemp;
    }
    for (j=1; j<=n; j++) {
      for (i=1; i<=n; i++)
        eta[i]=r[i][j]-(rq[j]-val[j])*vec[i][j];
      z[j]=Math.sqrt(Basic.vecvec(1,n,0,eta,eta));
    }
    mergesort(rq,perm,1,n);
    vecperm(perm,1,n,rq);
    for (i=1; i<=n; i++) {
      eps[i]=z[perm[i]];
      val3[i]=val[perm[i]];
      rowperm(perm,1,n,i,vec);
      rowperm(perm,1,n,i,r);
    }
    for (i=1; i<=n; i++)
      for (j=i; j<=n; j++)
        p[i][j]=p[j][i]=Basic.tammat(1,n,i,j,vec,r);
  }
  i0=1;
  do {
    j=i1=i0;
    j++;
    while ((j > n) ? false :
                    (rq[j]-rq[j-1] <= Math.sqrt((eps[j]+eps[j-1])*norma))) {
      i1=j;
      j++;
    }
    if (stop || (iter == maxitp1)) {
      i=i0;
      do {
        j=i01=i;
        j++;
        while ((j>i1) ? false : rq[j]-rq[j-1] <= eps[j]+eps[j-1]) {
          i01=j;
          j++;
        }
        if (i == i01) {
          if (i < n) {
            if (i == 1)
              dl=dr=rq[i+1]-rq[i]-eps[i+1];
            else {
              dl=rq[i]-rq[i-1]-eps[i-1];
              dr=rq[i+1]-rq[i]-eps[i+1];
            }
          } else
            dl=dr=rq[i]-rq[i-1]-eps[i-1];
          eps2=eps[i]*eps[i];
          lbound[i]=eps2/dr+mateps;
          ubound[i]=eps2/dl+mateps;
```

```
      } else
        for (k=i; k<=i01; k++)
          lbound[k]=ubound[k]=eps[k]+mateps;
      i01++;
      i=i01;
    } while (i <= i1);  /* bounds */
  } else {
    if (i0 == i1) {
      for (k=1; k<=n; k++)
        if (k == i0)
          y[k][i0]=1.0;
        else
          r[k][i0]=p[k][i0];
      val[i0]=rq[i0];
    } else {
      n1=i1-i0+1;
      em[0]=em[2]=Double.MIN_VALUE;
      em[4]=10*n1;
      double val4[] = new double[n1+1];
      double pp[][] = new double[n1+1][n1+1];
      m1=0.0;
      for (k=i0; k<=i1; k++) m1 += val3[k];
      m1 /= n1;
      for (i=1; i<=n1; i++)
        for (j=1; j<=n1; j++) {
          pp[i][j]=p[i+i0-1][j+i0-1];
          if (i == j) pp[i][j] += val3[j+i0-1]-m1;
        }
      for (i=i0; i<=i1; i++) {
        val3[i]=m1;
        val[i]=rq[i];
      }
      qrisym(pp,n1,val4,em);
      mergesort(val4,perm,1,n1);
      for (i=1; i<=n1; i++)
        for (j=1; j<=n1; j++)
          p[i+i0-1][j+i0-1]=pp[i][perm[j]];
      i0m1=i0-1;
      i1p1=i1+1;
      for (j=i0; j<=i1; j++) {
        for (i=1; i<=i0m1; i++) {
          s=0.0;
          for (k=i0; k<=i1; k++) s += p[i][k]*p[k][j];
          r[i][j]=s;
        }
        for (i=i1p1; i<=n; i++) {
          s=0.0;
          for (k=i0; k<=i1; k++) s += p[i][k]*p[k][j];
          r[i][j]=s;
        }
```

```
              for (i=i0; i<=i1; i++) y[i][j]=p[i][j];
            }
        } /* innerblock */
      } /* not stop */
      i0=i1+1;
    } while (i0 <= n);   /* while i0 loop */
    if ((!stop) && (iter < maxitp1)) {
      for (j=1; j<=n; j++)
        for (i=1; i<=n; i++)
          if (val3[i] != val3[j])
            y[i][j]=r[i][j]/(val3[j]-val3[i]);
      for (i=1; i<=n; i++) {
        for (j=1; j<=n; j++) z[j]=Basic.matmat(1,n,i,j,vec,y);
        for (j=1; j<=n; j++) vec[i][j]=z[j];
      }
      orthog(n,1,n,vec);
    } else {
      aux[5]=iter-1;
      break;
    }
  } /* for iter loop */
  aux[1]=norma;
  aux[3]=max;
}
```

3.13.6 Asymmetric matrices in Hessenberg form

A. reavalqri

Determines the eigenvalues λ_j ($j=1,...,n$) (all assumed to be real) of an $n{\times}n$ real upper Hessenberg matrix A ($A_{i,j}=0$ for $i>j+1$) by single QR iteration. The method used is the single QR iteration of Francis [DekHo68, Fr61, Wi65]. The eigenvalues of a real upper-Hessenberg matrix are calculated provided that the matrix has real eigenvalues only.

Procedure parameters:

$$\text{int reavalqri } (a,n,em,val)$$

reavalqri: given the value 0 provided that the process is completed within *em[4]* iterations; otherwise *reavalqri* is given the value k, of the number of eigenvalues not calculated;

a: double *a[1:n,1:n]*;
            entry: the elements of the real upper Hessenberg matrix must be given in the upper triangle and the first subdiagonal of array *a*;
            exit: the Hessenberg part of array *a* is altered;

n: int;
            entry: the order of the given matrix;

em: double *em[0:5]*;
              entry: *em[0]*: the machine precision;
                      em[1]: a norm of the given matrix;
                      em[2]: the relative tolerance used for the QR iteration
                              (*em[2]>em[0]*); if the absolute value of some subdiagonal
                              element is smaller than *em[1]*em[2]* then this element is
                              neglected and the matrix is partitioned;
                      em[4]: the   maximum   allowed   number   of   iterations
                              (for example, *em[4]=10*n*);
              exit: *em[3]*: the maximum absolute value of the subdiagonal
                              elements neglected;
                      em[5]: the number of QR iterations performed; if the iteration
                              process is not completed within *em[4]* iterations then
                              the value *em[4]+1* is delivered and in this case only the
                              last *n-k* elements of *val* are approximate eigenvalues of
                              the given matrix, where *k* is delivered in *reavalqri*;
val: double *val[1:n]*;
              exit: the eigenvalues of the given matrix are delivered in *val*.

Procedures used: rotcol, rotrow.

```
public static int reavalqri(double a[][], int n, double em[], double val[])
{
  int n1,i,i1,q,max,count;
  double det,w,shift,kappa,nu,mu,r,tol,delta,machtol,s;

  nu=mu=0.0;
  machtol=em[0]*em[1];
  tol=em[1]*em[2];
  max=(int) em[4];
  count=0;
  r=0.0;
  do {
    n1=n-1;
    i=n;
    do{
       q=i;
       i--;
    } while ((i >= 1) ? (Math.abs(a[i+1][i]) > tol) : false);
    if (q > 1)
       if (Math.abs(a[q][q-1]) > r) r=Math.abs(a[q][q-1]);
    if (q == n) {
       val[n]=a[n][n];
       n=n1;
    } else {
       delta=a[n][n]-a[n1][n1];
       det=a[n][n1]*a[n1][n];
       if (Math.abs(delta) < machtol)
         s=Math.sqrt(det);
       else {
```

```
          w=2.0/delta;
          s=w*w*det+1.0;
          s = (s <= 0.0) ? -delta*0.5 : w*det/(Math.sqrt(s)+1.0);
        }
        if (q == n1) {
          val[n]=a[n][n]+s;
          val[n1]=a[n1][n1]-s;
          n -= 2;
        } else {
          count++;
          if (count > max) break;
          shift=a[n][n]+s;
          if (Math.abs(delta) < tol) {
            w=a[n1][n1]-s;
            if (Math.abs(w) < Math.abs(shift)) shift=w;
          }
          a[q][q] -= shift;
          for (i=q; i<=n-1; i++) {
            i1=i+1;
            a[i1][i1] -= shift;
            kappa=Math.sqrt(a[i][i]*a[i][i]+a[i1][i]*a[i1][i]);
            if (i > q) {
              a[i][i-1]=kappa*nu;
              w=kappa*mu;
            } else
              w=kappa;
            mu=a[i][i]/kappa;
            nu=a[i1][i]/kappa;
            a[i][i]=w;
            Basic.rotrow(i1,n,i,i1,a,mu,nu);
            Basic.rotcol(q,i,i,i1,a,mu,nu);
            a[i][i] += shift;
          }
          a[n][n-1]=a[n][n]*nu;
          a[n][n]=a[n][n]*mu+shift;
        }
      }
    }
  } while (n > 0);
  em[3]=r;
  em[5]=count;
  return n;
}
```

B. reaveches

Determines the eigenvector corresponding to a single real eigenvalue λ of an $n \times n$ real Hessenberg matrix A ($A_{i,j}=0$ for $i>j+1$) by inverse iteration

[DekHo68, Vr66, Wi65]. Starting with $x^{(0)}=(1,1,...,1)^T$, vectors $y^{(k)}$ and $x^{(k)}$ are produced by use of the scheme

$$(A-\lambda I)y^{(k)}=x^{(k)}, \quad x^{(k+1)}=y^{(k)}/\|y^{(k)}\|_E \quad (k=0,1,...).$$

The process is terminated if either

(a) $\|(A-\lambda I)x^{(k)}\|_E \le \|A\|*tol$,

where tol is a tolerance prescribed by the user, or

(b) $k=kmax+1$,

where $kmax$ is an integer prescribed by the user.

Procedure parameters:

$$\text{void reaveches } (a,n,lambda,em,v)$$

a: double $a[1:n,1:n]$;

               entry: the elements of the real upper Hessenberg matrix must be given in the upper triangle and the first subdiagonal of array a;

               exit: the Hessenberg part of array a is altered;

n: int;

               entry: the order of the given matrix;

$lambda$:   double;

               the given real eigenvalue of the upper Hessenberg matrix (value of λ above);

em: double $em[0:9]$;

               entry: $em[0]$: the machine precision;

                        $em[1]$: a norm of the given matrix;

                        $em[6]$: the tolerance used for eigenvector (value of tol above, $em[6] > em[0]$); the inverse iteration ends if the Euclidean norm of the residue vector is smaller than $em[1]*em[6]$;

                        $em[8]$: the maximum allowed number of iterations (value of $kmax$ above, for example, $em[8]=5$);

               exit: $em[7]$: the Euclidean norm of the residue vector of the calculated eignenvector;

                        $em[9]$: the number of inverse iterations performed; if $em[7]$ remains larger than $em[1]*em[6]$ during $em[8]$ iterations then the value $em[8]+1$ is delivered;

v: double $v[1:n]$;

               exit: the calculated eigenvector is delivered in v.

Procedures used: vecvec, matvec.

```
public static void reaveches(double a[][], int n, double lambda, double em[],
                        double v[])
{
  int i,i1,j,count,max;
  double m,r,norm,machtol,tol;
  boolean p[] = new boolean[n+1];

  r=0.0;
  norm=em[1];
  machtol=em[0]*norm;
  tol=em[6]*norm;
```

```
  max=(int) em[8];
  a[1][1]  -= lambda;
  for (i=1; i<=n-1; i++) {
    i1=i+1;
    r=a[i][i];
    m=a[i1][i];
    if (Math.abs(m) < machtol) m=machtol;
    p[i] = (Math.abs(m) <= Math.abs(r));
    if (p[i]) {
      a[i1][i] = m /= r;
      for (j=i1; j<=n; j++)
        a[i1][j]=((j > i1) ? a[i1][j] : a[i1][j]-lambda)-m*a[i][j];
    } else {
      a[i][i]=m;
      a[i1][i] = m = r/m;
      for (j=i1; j<=n; j++) {
        r = (j > i1) ? a[i1][j] : a[i1][j]-lambda;
        a[i1][j]=a[i][j]-m*r;
        a[i][j]=r;
      }
    }
  }
  if (Math.abs(a[n][n]) < machtol) a[n][n]=machtol;
  for (j=1; j<=n; j++) v[j]=1.0;
  count=0;
  do {
    count++;
    if (count > max) break;
    for (i=1; i<=n-1; i++) {
      i1=i+1;
      if (p[i])
        v[i1]  -= a[i1][i]*v[i];
      else {
        r=v[i1];
        v[i1]=v[i]-a[i1][i]*r;
        v[i]=r;
      }
    }
    for (i=n; i>=1; i--)
      v[i]=(v[i]-Basic.matvec(i+1,n,i,a,v))/a[i][i];
    r=1.0/Math.sqrt(Basic.vecvec(1,n,0,v,v));
    for (j=1; j<=n; j++) v[j] *= r;
  } while (r > tol);
  em[7]=r;
  em[9]=count;
}
```

C. reaqri

Computes all eigenvalues λ_j (assumed to be real) and corresponding eigenvectors u_j ($j=1,...,n$) of the $n \times n$ real upper Hessenberg matrix A ($A_{i,j}=0$ for $i>j+1$) by single QR iteration. The eigenvectors are calculated by a direct method [see DekHo68], in contrast with *reaveches* which uses inverse iteration. If the Hessenberg matrix is not too ill-conditioned with respect to its eigenvalue problem then this method yields numerically independent eigenvectors and is competitive with inverse iteration as to accuracy and computation time.

Procedure parameters:
$$\text{int reaqri } (a,n,em,val,vec)$$

reaqri: given the value 0 provided that the process is completed within *em[4]* iterations; otherwise *reaqri* is given the value *k*, of the number of eigenvalues and eigenvectors not calculated;

a: double *a[1:n,1:n]*;
  entry: the elements of the real upper Hessenberg matrix must be given in the upper triangle and the first subdiagonal of array *a*;
  exit: the Hessenberg part of array *a* is altered;

n: int;
  entry: the order of the given matrix;

em: double *em[0:5]*;
  entry: *em[0]*: the machine precision;
    em[1]: a norm of the given matrix;
    em[2]: the relative tolerance used for the QR iteration (*em[2]>em[0]*); if the absolute value of some subdiagonal element is smaller than *em[1]*em[2]* then this element is neglected and the matrix is partitioned;
    em[4]: the maximum allowed number of iterations (for example, *em[4]=10*n*);
  exit: *em[3]*: the maximum absolute value of the subdiagonal elements neglected;
    em[5]: the number of QR iterations performed; if the iteration process is not completed within *em[4]* iterations then the value *em[4]+1* is delivered and in this case only the last *n-k* elements of *val* and the last *n-k* columns of *vec* are approximated eigenvalues and eigenvectors of the given matrix, where *k* is delivered in *reaqri*;

val: double *val[1:n]*;
  exit: the eigenvalues of the given matrix are delivered in *val*;

vec: double *vec[1:n,1:n]*;
  exit: the calculated eigenvectors corresponding to the eigenvalues in *val[1:n]* are delivered in the columns of *vec*.

Procedures used: matvec, rotcol, rotrow.

```
public static int reaqri(double a[][], int n, double em[], double val[],
                double vec[][])
```

```
{
  int m1,i,i1,m,j,q,max,count;
  double w,shift,kappa,nu,mu,r,tol,s,machtol,elmax,t,delta,det;
  double tf[] = new double[n+1];

  nu=mu=0.0;
  machtol=em[0]*em[1];
  tol=em[1]*em[2];
  max=(int) em[4];
  count=0;
  elmax=0.0;
  m=n;
  for (i=1; i<=n; i++) {
    vec[i][i]=1.0;
    for (j=i+1; j<=n; j++) vec[i][j]=vec[j][i]=0.0;
  }
  do {
    m1=m-1;
    i=m;
    do {
      q=i;
      i--;
    } while ((i >= 1) ? (Math.abs(a[i+1][i]) > tol) : false);
    if (q > 1)
      if (Math.abs(a[q][q-1]) > elmax) elmax=Math.abs(a[q][q-1]);
    if (q == m) {
      val[m]=a[m][m];
      m=m1;
    } else {
      delta=a[m][m]-a[m1][m1];
      det=a[m][m1]*a[m1][m];
      if (Math.abs(delta) < machtol)
        s=Math.sqrt(det);
      else {
        w=2.0/delta;
        s=w*w*det+1.0;
        s = (s <= 0.0) ? -delta*0.5 : w*det/(Math.sqrt(s)+1.0);
      }
      if (q == m1) {
        val[m] = a[m][m] += s;
        val[q] = a[q][q] -= s;
        t = (Math.abs(s) < machtol) ? (s+delta)/a[m][q] : a[q][m]/s;
        r=Math.sqrt(t*t+1.0);
        nu=1.0/r;
        mu = -t*nu;
        a[q][m] -= a[m][q];
        Basic.rotrow(q+2,n,q,m,a,mu,nu);
        Basic.rotcol(1,q-1,q,m,a,mu,nu);
        Basic.rotcol(1,n,q,m,vec,mu,nu);
        m -= 2;
```

```
      } else {
        count++;
        if (count > max) {
          em[3]=elmax;
          em[5]=count;
          return m;
        }
        shift=a[m][m]+s;
        if (Math.abs(delta) < tol) {
          w=a[m1][m1]-s;
          if (Math.abs(w) < Math.abs(shift)) shift=w;
        }
        a[q][q] -= shift;
        for (i=q; i<=m1; i++) {
          i1=i+1;
          a[i1][i1] -= shift;
          kappa=Math.sqrt(a[i][i]*a[i][i]+a[i1][i]*a[i1][i]);
          if (i > q) {
            a[i][i-1]=kappa*nu;
            w=kappa*mu;
          } else
            w=kappa;
          mu=a[i][i]/kappa;
          nu=a[i1][i]/kappa;
          a[i][i]=w;
          Basic.rotrow(i1,n,i,i1,a,mu,nu);
          Basic.rotcol(1,i,i,i1,a,mu,nu);
          a[i][i] += shift;
          Basic.rotcol(1,n,i,i1,vec,mu,nu);
        }
        a[m][m1]=a[m][m]*nu;
        a[m][m]=a[m][m]*mu+shift;
      }
    }
  } while (m > 0);
  for (j=n; j>=2; j--) {
    tf[j]=1.0;
    t=a[j][j];
    for (i=j-1; i>=1; i--) {
      delta=t-a[i][i];
      tf[i]=Basic.matvec(i+1,j,i,a,tf)/
            ((Math.abs(delta) < machtol) ? machtol : delta);
    }
    for (i=1; i<=n; i++) vec[i][j]=Basic.matvec(1,j,i,vec,tf);
  }
  em[3]=elmax;
  em[5]=count;
  return m;
}
```

D. comvalqri

Determines the real and complex eigenvalues λ_j (j=1,...,n) of an $n \times n$ real upper Hessenberg matrix A ($A_{i,j}=0$ for $i>j+1$) by the double QR iteration of Francis [DekHo68, Fr61, Wi65].

Procedure parameters:
$$\text{int comvalqri } (a,n,em,re,im)$$

comvalqri: given the value 0 provided that the process is completed within *em[4]* iterations; otherwise *comvalqri* is given the value *k*, of the number of eigenvalues not calculated;

a: double *a[1:n,1:n]*;
       entry: the elements of the real upper Hessenberg matrix must be given in the upper triangle and the first subdiagonal of array *a*;
       exit: the Hessenberg part of array *a* is altered;

n: int; entry: the order of the given matrix;

em: double *em[0:5]*;
       entry: *em[0]*: the machine precision;
                 em[1]: a norm of the given matrix;
                 em[2]: the relative tolerance used for the QR iteration (*em[2]>em[0]*); if the absolute value of some subdiagonal element is smaller than *em[1]*em[2]* then this element is neglected and the matrix is partitioned;
                 em[4]: the maximum allowed number of iterations (for example, *em[4]=10*n*);
       exit: *em[3]*: the maximum absolute value of the subdiagonal elements neglected;
                 em[5]: the number of QR iterations performed; if the iteration process is not completed within *em[4]* iterations then the value *em[4]+1* is delivered and in this case only the last *n-k* elements of *re* and *im* are approximate eigenvalues of the given matrix, where *k* is delivered in *comvalqri*;

re, im: double *re[1:n]*, *im[1:n]*;
       exit: the real and imaginary parts of the calculated eigenvalues of the given matrix are delivered in *re, im[1:n]*, the members of each nonreal complex conjugate pair being consecutive.

```
public static int comvalqri(double a[][], int n, double em[], double re[],
                            double im[])
{
  boolean b;
  int i,j,p,q,max,count,n1,p1,p2,imin1,i1,i2,i3;
  double disc,sigma,rho,g1,g2,g3,psi1,psi2,aa,e,k,s,norm,machtol2,tol,w;

  norm=em[1];
  w=em[0]*norm;
  machtol2=w*w;
  tol=em[2]*norm;
```

```
max=(int) em[4];
count=0;
w=0.0;
do {
  i=n;
  do {
    q=i;
    i--;
  } while ((i >= 1) ? (Math.abs(a[i+1][i]) > tol) : false);
  if (q > 1)
    if (Math.abs(a[q][q-1]) > w) w=Math.abs(a[q][q-1]);
  if (q >= n-1) {
    n1=n-1;
    if (q == n) {
      re[n]=a[n][n];
      im[n]=0.0;
      n=n1;
    } else {
      sigma=a[n][n]-a[n1][n1];
      rho = -a[n][n1]*a[n1][n];
      disc=sigma*sigma-4.0*rho;
      if (disc > 0.0) {
        disc=Math.sqrt(disc);
        s = -2.0*rho/(sigma+((sigma >= 0.0) ? disc : -disc));
        re[n]=a[n][n]+s;
        re[n1]=a[n1][n1]-s;
        im[n]=im[n1]=0.0;
      } else {
        re[n]=re[n1]=(a[n1][n1]+a[n][n])/2.0;
        im[n1]=Math.sqrt(-disc)/2.0;
        im[n] = -im[n1];
      }
      n -= 2;
    }
  } else {
    count++;
    if (count > max) break;
    n1=n-1;
    sigma=a[n][n]+a[n1][n1]+Math.sqrt(Math.abs(a[n1][n-2]*a[n][n1])*em[0]);
    rho=a[n][n]*a[n1][n1]-a[n][n1]*a[n1][n];
    i=n-1;
    do {
      p1=i1=i;
      i--;
    } while ((i-1 >= q) ? (Math.abs(a[i][i-1]*a[i1][i]*
        (Math.abs(a[i][i]+a[i1][i1]-sigma)+Math.abs(a[i+2][i1]))) >
        Math.abs(a[i][i]*((a[i][i]-sigma)+a[i][i1]*a[i1][i]+rho))*tol) : false);
    p=p1-1;
    p2=p+2;
    for (i=p; i<=n-1; i++) {
```

```
        imin1=i-1;
        i1=i+1;
        i2=i+2;
        if (i == p) {
          g1=a[p][p]*(a[p][p]-sigma)+a[p][p1]*a[p1][p]+rho;
          g2=a[p1][p]*(a[p][p]+a[p1][p1]-sigma);
          if (p1 <= n1) {
            g3=a[p1][p]*a[p2][p1];
            a[p2][p]=0.0;
          } else
            g3=0.0;
        } else {
          g1=a[i][imin1];
          g2=a[i1][imin1];
          g3 = (i2 <= n) ? a[i2][imin1] : 0.0;
        }
        k = (g1 >= 0.0) ? Math.sqrt(g1*g1+g2*g2+g3*g3) :
                          -Math.sqrt(g1*g1+g2*g2+g3*g3);
        b = (Math.abs(k) > machtol2);
        aa = (b ? g1/k+1.0 : 2.0);
        psi1 = (b ? g2/(g1+k) : 0.0);
        psi2 = (b ? g3/(g1+k) : 0.0);
        if (i != q)
          a[i][imin1] = (i == p) ? -a[i][imin1] : -k;
        for (j=i; j<=n; j++) {
          e=aa*(a[i][j]+psi1*a[i1][j]+((i2 <= n) ? psi2*a[i2][j] : 0.0));
          a[i][j] -= e;
          a[i1][j] -= psi1*e;
          if (i2 <= n) a[i2][j] -= psi2*e;
        }
        for (j=q; j<=((i2 <= n) ? i2 : n); j++) {
          e=aa*(a[j][i]+psi1*a[j][i1]+((i2 <= n) ? psi2*a[j][i2] : 0.0));
          a[j][i] -= e;
          a[j][i1] -= psi1*e;
          if (i2 <= n) a[j][i2] -= psi2*e;
        }
        if (i2 <= n1) {
          i3=i+3;
          e=aa*psi2*a[i3][i2];
          a[i3][i] = -e;
          a[i3][i1] = -psi1*e;
          a[i3][i2] -= psi2*e;
        }
      }
    }
  } while (n > 0);
  em[3]=w;
  em[5]=count;
  return n;
}
```

E. comveches

Determines the eigenvector corresponding to a single complex eigenvalue $\lambda+\mu i$ of an $n \times n$ real upper Hessenberg matrix A ($A_{i,j}=0$ for $i>j+1$) by inverse iteration [DekHo68, Vr66, Wi65]. Starting with $x^{(0)}=(1,1,...,1)^{T}$, vectors $y^{(k)}$ and $x^{(k)}$ are produced by use of the scheme

$$(A-\lambda I)y^{(k)}=x^{(k)}, \quad x^{(k+1)}=y^{(k)}/\|y^{(k)}\|_{E} \quad (k=0,1,...).$$

The process is terminated if either (a) $\|(A-\lambda I)x^{(k)}\|_{E} \leq \|A\|*tol$, where tol is a tolerance prescribed by the user, or (b) $k=kmax+1$, where $kmax$ is an integer prescribed by the user.

Procedure parameters:

> void comveches $(a,n,lambda,mu,em,u,v)$

a: double $a[1:n,1:n]$;

> entry: the elements of the real upper Hessenberg matrix must be given in the upper triangle and the first subdiagonal of array a;
>
> exit: the Hessenberg part of array a is altered;

n: int;

> entry: the order of the given matrix;

lambda, mu: double;

> real and imaginary part of the given eigenvalue (values of λ and μ above);

em: double $em[0:9]$;

> entry: $em[0]$: the machine precision;
>
> $em[1]$: a norm of the given matrix;
>
> $em[6]$: the tolerance used for eigenvector (value of tol above, $em[6]>em[0]$); the inverse iteration ends if the Euclidean norm of the residue vector is smaller than $em[1]*em[6]$;
>
> $em[8]$: the maximum allowed number of iterations (value of $kmax$ above, for example, $em[8]=5$);
>
> exit: $em[7]$: the Euclidean norm of the residue vector of the calculated eigenvector;
>
> $em[9]$: the number of inverse iterations performed; if $em[7]$ remains larger than $em[1]*em[6]$ during $em[8]$ iterations then the value $em[8]+1$ is delivered;

u, v: double $u[1:n]$, $v[1:n]$;

> exit: the real and imaginary parts of the calculated eigenvector are delivered in the arrays u and v.

Procedures used: vecvec, matvec, tamvec.

```
public static void comveches(double a[][], int n, double lambda, double mu,
                        double em[], double u[], double v[])
{
  int i,i1,j,count,max;
  double aa,bb,d,m,r,s,w,x,y,norm,machtol,tol;
  boolean p[] = new boolean[n+1];
```

```
double g[] = new double[n+1];
double f[] = new double[n+1];

w=0.0;
norm=em[1];
machtol=em[0]*norm;
tol=em[6]*norm;
max=(int) em[8];
for (i=2; i<=n; i++) {
  f[i-1]=a[i][i-1];
  a[i][1]=0.0;
}
aa=a[1][1]-lambda;
bb = -mu;
for (i=1; i<=n-1; i++) {
  i1=i+1;
  m=f[i];
  if (Math.abs(m) < machtol) m=machtol;
  a[i][i]=m;
  d=aa*aa+bb*bb;
  p[i] = (Math.abs(m) < Math.sqrt(d));
  if (p[i]) {
    f[i]=r=m*aa/d;
    g[i] = s = -m*bb/d;
    w=a[i1][i];
    x=a[i][i1];
    a[i1][i]=y=x*s+w*r;
    a[i][i1]=x=x*r-w*s;
    aa=a[i1][i1]-lambda-x;
    bb = -(mu+y);
    for (j=i+2; j<=n; j++) {
      w=a[j][i];
      x=a[i][j];
      a[j][i]=y=x*s+w*r;
      a[i][j]=x=x*r-w*s;
      a[j][i1] = -y;
      a[i1][j] -= x;
    }
  } else {
    f[i]=r=aa/m;
    g[i]=s=bb/m;
    w=a[i1][i1]-lambda;
    aa=a[i][i1]-r*w-s*mu;
    a[i][i1]=w;
    bb=a[i1][i]-s*w+r*mu;
    a[i1][i] = -mu;
    for (j=i+2; j<=n; j++) {
      w=a[i1][j];
      a[i1][j]=a[i][j]-r*w;
      a[i][j]=w;
```

```
      a[j][i1]=a[j][i]-s*w;
      a[j][i]=0.0;
    }
  }
}
p[n]=true;
d=aa*aa+bb*bb;
if (d < machtol*machtol) {
  aa=machtol;
  bb=0.0;
  d=machtol*machtol;
}
a[n][n]=d;
f[n]=aa;
g[n] = -bb;
for (i=1; i<=n; i++) {
  u[i]=1.0;
  v[i]=0.0;
}
count=0;
do {
  if (count > max) break;
  for (i=1; i<=n; i++)
    if (p[i]) {
      w=v[i];
      v[i]=g[i]*u[i]+f[i]*w;
      u[i]=f[i]*u[i]-g[i]*w;
      if (i < n) {
        v[i+1] -= v[i];
        u[i+1] -= u[i];
      }
    } else {
      aa=u[i+1];
      bb=v[i+1];
      u[i+1]=u[i]-(f[i]*aa-g[i]*bb);
      u[i]=aa;
      v[i+1]=v[i]-(g[i]*aa+f[i]*bb);
      v[i]=bb;
    }
  for (i=n; i>=1; i--) {
    i1=i+1;
    u[i]=(u[i]-Basic.matvec(i1,n,i,a,u)+(p[i] ? Basic.tamvec(i1,n,i,a,v) :
                a[i1][i]*v[i1]))/a[i][i];
    v[i]=(v[i]-Basic.matvec(i1,n,i,a,v)-(p[i] ? Basic.tamvec(i1,n,i,a,u) :
                a[i1][i]*u[i1]))/a[i][i];
  }
  w=1.0/Math.sqrt(Basic.vecvec(1,n,0,u,u)+Basic.vecvec(1,n,0,v,v));
  for (j=1; j<=n; j++) {
    u[j] *= w;
    v[j] *= w;
```

```
      }
      count++;
   } while (w > tol);
   em[7]=w;
   em[9]=count;
}
```

3.13.7 Real asymmetric full matrices

A. reaeigval

Determines the eigenvalues λ_j $(j=1,...,n)$ (all assumed to be real) of a real $n{\times}n$ matrix A. A is equilibrated to the form $A'=PDAD^{-1}P^{-1}$ (by means of a call of *eqilbr*) and transformed to similar real upper Hessenberg form H (by means of a call of *tfmreahes*). The eigenvalues of H are then computed by QR iteration (by means of a call of *reavalqri*). The procedure *reaeigval* should be used only if all eigenvalues are real. For further details see [DekHo68, Fr61, Wi65].

Procedure parameters:
$$\text{int reaeigval } (a,n,em,val)$$
reaeigval: given the value 0 provided that the process is completed within *em[4]* iterations; otherwise *reaeigval* is given the value k, of the number of eigenvalues not calculated;
a: double *a[1:n,1:n]*;
   entry: the matrix whose eigenvalues are to be calculated;
   exit: the array elements are altered;
n: int;
   entry: the order of the given matrix;
em: double *em[0:5]*;
   entry: *em[0]*: the machine precision;
       em[2]: the relative tolerance used for the QR iteration ($em[2] > em[0]$);
       em[4]: the maximum allowed number of iterations (for example, $em[4]=10n$);
   exit: *em[1]*: the infinity norm of the equilibrated matrix;
       em[3]: the maximum absolute value of the subdiagonal elements neglected;
       em[5]: the number of QR iterations performed; if the iteration process is not completed within *em[4]* iterations then the value *em[4]*+1 is delivered and in this case only the last n-k elements of *val* are approximate eigenvalues of the given matrix, where k is delivered in *reaeigval*;
val: double *val[1:n]*;
   exit: the eigenvalues of the given matrix are delivered in monotonically nonincreasing order.

Procedures used: eqilbr, tfmreahes, reavalqri.

```
public static int reaeigval(double a[][], int n, double em[], double val[])
{
    int i,j,k;
    double r;
    int ind[]  = new int[n+1];
    int ind0[] = new int[n+1];
    double d[] = new double[n+1];

    eqilbr(a,n,em,d,ind0);
    tfmreahes(a,n,em,ind);
    k=reavalqri(a,n,em,val);
    for (i=k+1; i<=n; i++)
      for (j=i+1; j<=n; j++)
        if (val[j] > val[i]) {
          r=val[i];
          val[i]=val[j];
          val[j]=r;
        }
    return k;
}
```

B. reaeig1

Determines the eigenvalues λ_j (all assumed to be real) and corresponding eigenvectors $u^{(j)}$ ($j=1,...,n$) of a real $n \times n$ matrix A. A is equilibrated to the form $A'=PDAD^{-1}P^{-1}$ (by means of a call of *eqilbr*) and transformed to similar real upper Hessenberg form H (by means of a call of *tfmreahes*). The eigenvalues of H are then computed by single QR iteration (by means of a call of *reavalqri*). The eigenvectors of H are determined either by direct use of an iterative scheme of the form

$$\left(H - \lambda_j I\right)y_k^{(j)} = x_k^{(j)}, \quad x_{k+1}^{(j)} = y_k^{(j)} / \left\|y_k^{(j)}\right\|_E \tag{1}$$

or by delayed application of such a scheme. If $\min |\lambda_j-\lambda_i| \leq \varepsilon \|H\|$ ($j \neq i$) for λ_i ranging over the previously determined eigenvalues, λ_j is replaced in (1) by μ_j, where $\min |\mu_j-\lambda_i| = \varepsilon \|H\|$, ε being the value of the machine precision supplied by the user. The inverse iteration scheme is terminated if (a) $\|(H-\lambda_j I)x_k^{(j)}\| \leq \|H\| \tau$, where τ is a relative tolerance prescribed by the user, or (b) $k=kmax+1$, where the integer value of *kmax*, the maximum permitted number of inverse iterations, is also prescribed by the user. The above inverse iteration is performed by *reaveches*. The eigenvectors $v^{(j)}$ of the equilibrated matrix A' are then obtained from those, $x^{(j)}$, of H by back transformation (by means of a call of *bakreahes2*), and the eigenvectors $w^{(j)}$ of the original matrix A are recovered from the $v^{(j)}$ by means of a call of *baklbr*. Finally, the $w^{(j)}$ are scaled to $u^{(j)}$ by imposing the condition that the largest element of $u^{(j)}$ is one (by means of a call of *reascl*). The procedure *reaeig1* should be used only if all eigenvalues are real.

Procedure parameters:
$$\text{int reaeig1 } (a,n,em,val,vec)$$

reaeig1: given the value zero provided that the process is completed within *em[4]* iterations; otherwise *reaeig1* is given the value *k*, of the number of eigenvalues and eigenvectors not calculated;

a: double *a[1:n, 1:n]*;

  entry: the matrix whose eigenvalues and eigenvectors are to be calculated;

  exit: the array elements are altered;

n: int;

  entry: the order of the given matrix;

em: double *em[0:9]*;

  entry: *em[0]*: the machine precision (the value of ε above);

  em[2]: the relative tolerance used for the QR iteration (*em[2] > em[0]*);

  em[4]: the maximum allowed number of iterations (for example, *em[4]=10n*);

  em[6]: the tolerance used for the eigenvectors (value of τ above; *em[6]>em[2]*); for each eigenvector the inverse iteration ends if the Euclidean norm of the residue vector is smaller than *em[1]*em[6]*;

  em[8]: the maximum allowed number of inverse iterations for the calculation of each eigenvector (value of *kmax* above; for example, *em[8]=5*);

  exit: *em[1]*: the infinity norm of the equilibrated matrix;

  em[3]: the maximum absolute value of the subdiagonal elements neglected;

  em[5]: the number of QR iterations performed; if the iteration process is not completed within *em[4]* iterations then the value *em[4]+1* is delivered and in this case only the last *n-k* elements of *val* and the last *n-k* columns of *vec* are approximate eigenvalues and eigenvectors of the given matrix, where *k* is delivered in *reaeig1*;

  em[7]: the maximum Euclidean norm of the residues of the calculated eigenvectors of the transformed matrix;

  em[9]: the largest number of inverse iterations performed for the calculation of some eigenvector; if, for some eigenvector the Euclidean norm of the residue remains larger than *em[1]*em[6]*, then the value *em[8]+1* is delivered; nevertheless, the eigenvectors may then very well be useful, this should be judged from the value delivered in *em[7]* or from some other test;

val: double *val[1:n]*;

  exit: the eigenvalues of the given matrix are delivered in monotonically decreasing order;

vec: double *vec[1:n, 1:n]*;

  exit: the calculated eigenvectors corresponding to the eigenvalues in *val[1:n]* are delivered in the columns of *vec*.

Procedures used: eqilbr, tfmreahes, bakreahes2, baklbr, reavalqri, reaveches, reascl.

```java
public static int reaeig1(double a[][], int n, double em[], double val[],
                         double vec[][])
{
  int i,k,max,j,l;
  double residu,r,machtol;
  int ind[] = new int[n+1];
  int ind0[] = new int[n+1];
  double d[] = new double[n+1];
  double v[] = new double[n+1];
  double b[][] = new double[n+1][n+1];

  residu=0.0;
  max=0;
  eqilbr(a,n,em,d,ind0);
  tfmreahes(a,n,em,ind);
  for (i=1; i<=n; i++)
    for (j=((i == 1) ? 1 : i-1); j<=n; j++) b[i][j]=a[i][j];
  k=reavalqri(b,n,em,val);
  for (i=k+1; i<=n; i++)
    for (j=i+1; j<=n; j++)
      if (val[j] > val[i]) {
        r=val[i];
        val[i]=val[j];
        val[j]=r;
      }
  machtol=em[0]*em[1];
  for (l=k+1; l<=n; l++) {
    if (l > 1)
      if (val[l-1]-val[l] < machtol) val[l]=val[l-1]-machtol;
    for (i=1; i<=n; i++)
      for (j=((i == 1) ? 1 : i-1); j<=n; j++) b[i][j]=a[i][j];
    reaveches(b,n,val[l],em,v);
    if (em[7] > residu) residu=em[7];
    if (em[9] > max) max=(int) em[9];
    for (j=1; j<=n; j++) vec[j][l]=v[j];
  }
  em[7]=residu;
  em[9]=max;
  bakreahes2(a,n,k+1,n,ind,vec);
  baklbr(n,k+1,n,d,ind0,vec);
  Basic.reascl(vec,n,k+1,n);
  return k;
}
```

C. reaeig3

Determines all eigenvalues λ_j (assumed to be real) and corresponding eigenvectors $u^{(j)}$ (j=1,...,n) of a real $n{\times}n$ matrix A by equilibration to the form $A'=PDAD^{-1}P^{-1}$ (calling *eqilbr*), transformation to similar real upper Hessenberg form H (calling *tfmreahes*), computation of the eigenvalues of H by QR iteration and direct determination of the eigenvectors of H (calling *reaqri*); if all eigenvalues of H have been determined, there follow back transformation from the eigenvectors of H to the eigenvectors of A' (calling *bakreahes2*) further back transformation from the eigenvectors of A' to those of A (calling *baklbr*) and finally scaling of the eigenvectors of A in such a way that the element of maximum size in each eigenvector is one (calling *reascl*). The procedure *reaeig3* should be used only if all eigenvalues are real.

Procedure parameters:
$$\text{int reaeig3 } (a,n,em,val,vec)$$

reaeig3: given the value zero provided that the process is completed within *em[4]* iterations; otherwise *reaeig3* is given the value k, of the number of eigenvalues not calculated;

a: double $a[1{:}n,1{:}n]$;
   entry: the matrix whose eigenvalues and eigenvectors are to be calculated;
   exit: the array elements are altered;

n: int;
   entry: the order of the given matrix;

em: double $em[0{:}5]$;
   entry: *em[0]*: the machine precision;
       em[2]: the relative tolerance used for the QR iteration   (*em[2]* > *em[0]*);
       em[4]: maximum   allowed   number   of   QR   iterations (for example, *em[4]*=10n);
   exit: *em[1]*: the infinity norm of the equilibrated matrix;
       em[3]: the maximum absolute value of the subdiagonal elements neglected;
       em[5]: the number of QR iterations performed; if the iteration process is not completed within *em[4]* iterations then the value *em[4]*+1 is delivered and in this case only the last n-k elements of *val* are approximate eigenvalues of the given matrix and no useful eigenvectors are delivered, where k is delivered in *reaeig3*;

val: double $val[1{:}n]$;
   exit: the eigenvalues of the given matrix are delivered;

vec: double $vec[1{:}n,1{:}n]$;
   exit: the calculated eigenvectors corresponding to the eigenvalues in $val[1{:}n]$ are delivered in the columns of *vec*.

Procedures used: eqilbr, tfmreahes, bakreahes2, baklbr, reaqri, reascl.

```
public static int reaeig3(double a[][], int n, double em[], double val[],
                          double vec[][])
{
  int i;
  int ind[] = new int[n+1];
  int ind0[] = new int[n+1];
  double d[] = new double[n+1];

  eqilbr(a,n,em,d,ind0);
  tfmreahes(a,n,em,ind);
  i=reaqri(a,n,em,val,vec);
  if (i == 0) {
    bakreahes2(a,n,1,n,ind,vec);
    baklbr(n,1,n,d,ind0,vec);
    Basic.reascl(vec,n,1,n);
  }
  return i;
}
```

D. comeigval

Determines the real and complex eigenvalues λ_i ($i=1,...,n$) of a real $n \times n$ matrix A by equilibration to the form $A'=PDAD^{-1}P^{-1}$ (calling *eqilbr*) transformation to similar real upper Hessenberg form H (calling *tfmreahes*) and computation of the eigenvalues of H by double QR iteration (calling *comvalqri*).

Procedure parameters:

$$\text{int comeigval } (a,n,em,re,im)$$

comeigval: given the value zero provided that the process is completed within *em[4]* iterations; otherwise, *comeigval* is given the value k, of the number of eigenvalues not calculated;

a: double *a[1:n,1:n]*;

       entry: the matrix whose eigenvalues are to be calculated;

       exit: the array elements are altered;

n: int;

       entry: the order of the given matrix;

em: double *em[0:5]*;

       entry: *em[0]*:   the machine precision;

                  em[2]:   the relative tolerance used for the QR iteration (*em[2] > em[0]*);

                  em[4]:   the maximum allowed number of iterations (for example, *em[4]=10n*);

       exit: *em[1]*:   the infinity norm of the equilibrated matrix;

                  em[3]:   the maximum absolute value of the subdiagonal elements neglected;

                  em[5]:   the number of QR iterations performed; if the iteration process is not completed within *em[4]* iterations then the value *em[4]+1* is delivered and in this case only the last *n-k* elements of *re* and *im* are approximate eigenvalues of the given matrix, where *k* is delivered in *comeigval*;

re, im: double *re[1:n], im[1:n]*;

  exit: the real and imaginary parts of the calculated eigenvalues of the given matrix are delivered in *re, im[1:n]*, the members of each nonreal complex conjugate pair being consecutive.

Procedures used: eqilbr, tfmreahes, comvalqri.

```
public static int comeigval(double a[][], int n, double em[], double re[],
                            double im[])
{
  int i;
  int ind[]  = new int[n+1];
  int ind0[] = new int[n+1];
  double d[] = new double[n+1];

  eqilbr(a,n,em,d,ind0);
  tfmreahes(a,n,em,ind);
  i=comvalqri(a,n,em,re,im);
  return i;
}
```

E. comeig1

Determines the real and complex eigenvalues λ_j and corresponding eigenvectors $u^{(j)}$ ($j=1,...,n$) of a real $n \times n$ matrix A. A is equilibrated to the form $A'=PDAD^{-1}P^{-1}$ (by means of a call of *eqilbr*) and transformed to similar real upper Hessenberg form H (by means of a call of *tfmreahes*). The eigenvalues of H are then computed by double QR iteration (by means of a call of *comvalqri*). The real eigenvectors and complex eigenvectors of H are determined either by direct use of an iterative scheme of the form

$$\left(H - \lambda_j I\right)y_k^{(j)} = x_k^{(j)}, \qquad x_{k+1}^{(j)} = y_k^{(j)} / \left\|y_k^{(j)}\right\|_E \tag{1}$$

or by delayed application of such a scheme. If $\min|\lambda_j-\lambda_i| \leq \varepsilon \|H\|$ ($j \neq i$) for λ_i ranging over the previously determined eigenvalues, λ_j is replaced in (1) by μ_j, where $\min|\mu_j-\lambda_i| = \varepsilon \|H\|$, ε being the value of the machine precision supplied by the user. The inverse iteration scheme is terminated if (a) $\|(H-\lambda_j I)x_k^{(j)}\| \leq \|H\| \tau$, where τ is a relative tolerance prescribed by the user, or (b) $k=kmax+1$, where the integer value of *kmax*, the maximum permitted number of inverse iterations, is also prescribed by the user. The above inverse iteration is performed, when λ_j is real, by *reaveches*, and when λ_j is complex, by *comveches*. The eigenvectors $v^{(j)}$ of the equilibrated matrix A' are then obtained from those, $x^{(j)}$, of H by back transformation (by means of a call of *bakreahes2*) and the eigenvectors $w^{(j)}$ of the original matrix A are recovered from the $v^{(j)}$ by means of a call of *baklbr*. Finally, the $w^{(j)}$ are scaled to $u^{(j)}$ by imposing the condition that the largest element of $u^{(j)}$ is one (by means of a call of *comscl*).

Procedure parameters:
$$\text{int comeig1 } (a,n,em,re,im,vec)$$

comeig1: given the value zero provided that the process is completed within *em[4]* iterations; otherwise *comeig1* is given the value *k*, of the number of eigenvalues and eigenvectors not calculated;

a: double *a[1:n,1:n]*;
  entry: the matrix whose eigenvalues and eigenvectors are to be calculated;
  exit: the array elements are altered;

n: int;
  entry: the order of the given matrix;

em: double *em[0:9]*;
  entry: *em[0]*: the machine precision (the value of ε above);
    em[2]: the relative tolerance used for the QR iteration (*em[2] > em[0]*);
    em[4]: the maximum allowed number of iterations (for example, *em[4]=10n*);
    em[6]: the tolerance used for the eigenvectors (value of τ above; *em[6]>em[2]*); for each eigenvector the inverse iteration ends if the Euclidean norm of the residue vector is smaller than *em[1]*em[6]*;
    em[8]: the maximum allowed number of inverse iterations for the calculation of each eigenvector (value of *kmax* above; for example, *em[8]=5*);
  exit: *em[1]*: the infinity norm of the equilibrated matrix;
    em[3]: the maximum absolute value of the subdiagonal elements neglected;
    em[5]: the number of QR iterations performed; if the iteration process is not completed within *em[4]* iterations then the value *em[4]+1* is delivered and in this case only the last *n-k* elements of *re, im* and columns of *vec* are approximate eigenvalues and eigenvectors of the given matrix, where *k* is delivered in *comeig1*;
    em[7]: the maximum Euclidean norm of the residues of the calculated eigenvectors of the transformed matrix;
    em[9]: the largest number of inverse iterations performed for the calculation of some eigenvector; if the Euclidean norm of the residue for one or more eigenvectors remains larger than *em[1]*em[6]*, then the value *em[8]+1* is delivered; nevertheless, the eigenvectors may then very well be useful, this should be judged from the value delivered in *em[7]* or from some other test;

re, im: double *re[1:n], im[1:n]*;
  exit: the real and imaginary parts of the calculated eigenvalues of the given matrix are delivered in arrays *re[1:n]* and *im[1:n]*, the members of each nonreal complex conjugate pair being consecutive;

vec: double *vec[1:n, 1:n]*;
              exit: the calculated eigenvectors are delivered in the columns of *vec*;
                        an eigenvector corresponding to a real eigenvalue given in array
                        re is delivered in the corresponding column of array *vec*; the
                        real and imaginary part of an eigenvector corresponding to the
                        first member of a nonreal complex conjugate pair of eigenvalues
                        given in the arrays *re, im* are delivered in the two consecutive
                        columns of array *vec* corresponding to this pair (the
                        eigenvectors corresponding to the second members of nonreal
                        complex conjugate pairs are not delivered, since they are
                        simply the complex conjugate of those corresponding to the
                        first member of such pairs).

Procedures used: eqilbr, tfmreahes, bakreahes2, baklbr, reaveches, comvalqri,
                    comveches, comscl.

```
public static int comeig1(double a[][], int n, double em[], double re[],
                          double im[], double vec[][])
{
  boolean again;
  int i,j,k,ii,pj,itt;
  double x,y,max,neps,temp1,temp2;
  int ind[] = new int[n+1];
  int ind0[] = new int[n+1];
  double d[] = new double[n+1];
  double u[] = new double[n+1];
  double v[] = new double[n+1];
  double ab[][] = new double[n+1][n+1];

  eqilbr(a,n,em,d,ind0);
  tfmreahes(a,n,em,ind);
  for (i=1; i<=n; i++)
    for (j=((i == 1) ? 1 : i-1); j<=n; j++) ab[i][j]=a[i][j];
  k=comvalqri(ab,n,em,re,im);
  neps=em[0]*em[1];
  max=0.0;
  itt=0;
  for (i=k+1; i<=n; i++) {
    x=re[i];
    y=im[i];
    pj=0;
    again=true;
    do {
      for (j=k+1; j<=i-1; j++) {
        temp1=x-re[j];
        temp2=y-im[j];
        if (temp1*temp1+temp2*temp2 <= neps*neps) {
          if (pj == j)
            neps=em[2]*em[1];
          else
```

```
            pj=j;
        x += 2.0*neps;
        again = (!again);
        break;
      }
    }
  again = (!again);
  } while (again);
  re[i]=x;
  for (ii=1; ii<=n; ii++)
      for (j=((ii == 1) ? 1 : ii-1); j<=n; j++) ab[ii][j]=a[ii][j];
  if (y != 0.0) {
    comveches(ab,n,re[i],im[i],em,u,v);
    for (j=1; j<=n; j++) vec[j][i]=u[j];
    i++;
    re[i]=x;
  } else
    reaveches(ab,n,x,em,v);
  for (j=1; j<=n; j++) vec[j][i]=v[j];
  if (em[7] > max) max=em[7];
  if (itt < em[9]) itt=(int) em[9];
  }
  em[7]=max;
  em[9]=itt;
  bakreahes2(a,n,k+1,n,ind,vec);
  baklbr(n,k+1,n,d,ind0,vec);
  Basic.comscl(vec,n,k+1,n,im);
  return k;
}
```

3.13.8 Complex Hermitian matrices

A. eigvalhrm

Determines the n' largest eigenvalues λ_j of the $n \times n$ Hermitian matrix A. A is first reduced by a similarity transformation to real tridiagonal form T by calling *hshhrmtrival*. The n' largest eigenvalues λ_j of T are then determined by means of a call of *valsymtri*.

Procedure parameters:
$$\text{void eigvalhrm } (a,n,numval,val,em)$$
a: double $a[1:n,1:n]$;
                entry: the real part of the upper triangle of the Hermitian matrix must
                        be given in the upper triangular part of a (the elements
                        $a[i,j]$, $i \leq j$); the imaginary part of the strict lower triangle of the

Hermitian matrix must be given in the strict lower part of *a*
(the elements *a[i,j]*, *i>j*);

exit:　　the array elements are altered;

n:　　　int;

entry:　the order of the given matrix;

numval:　int;

entry:　*eigvalhrm* calculates the largest *numval* eigenvalues of the
Hermitian matrix (the value of *n'* above);

val:　　double *val[1:numval]*;

exit:　　in array *val* the largest *numval* eigenvalues are delivered in
monotonically nonincreasing order;

em:　　double *em[0:3]*;

entry:　*em[0]*: the machine precision (value of ε above);

em[2]: the relative tolerance used for the eigenvalues (value of
μ above); more precisely, the tolerance for each
eigenvalue λ is $|\lambda|*em[2]+em[1]*em[0]$;

exit:　　*em[1]*: an estimate of a norm of the original matrix;

em[3]: the number of iterations performed.

Procedures used:　　hshhrmtrival, valsymtri.

```
public static void eigvalhrm(double a[][], int n, int numval, double val[],
                    double em[])
{
  double d[] = new double[n+1];
  double bb[] = new double[n];

  hshhrmtrival(a,n,d,bb,em);
  valsymtri(d,bb,n,1,numval,val,em);
}
```

B. eighrm

Determines the *n'* largest eigenvalues λ_j (j=1,...,n) and corresponding eigenvectors
$u^{(j)}$ (j=1,...,n) of the *n×n* Hermitian matrix *A*. *A* is first reduced to similar real
tridiagonal form *T* by a call of *hshhrmtri*. The required eigenvalues of this
tridiagonal matrix are then obtained by a call of *valsymtri*. The corresponding
eigenvectors of *T* are determined by inverse iteration and (possibly) Gram-Schmidt
orthogonalization by a call of *vecsymtri*; the required eigenvectors of *A* are then
recovered from those of *T* by means of a call of *bakhrmtri*.

Procedure parameters:

void eighrm (*a,n,numval,val,vecr,veci,em*)

a:　　　double *a[1:n, 1:n]*;

entry:　the real part of the upper triangle of the Hermitian matrix must
be given in the upper triangular part of *a* (the elements
a[i,j], *i≤j*); the imaginary part of the strict lower triangle of the

Hermitian matrix must be given in the strict lower part of *a* (the elements *a[i,j]*, *i>j*);

    exit: the array elements are altered;

n: int;

    entry: the order of the given matrix;

numval: int;

    entry: *eighrm* calculates the largest *numval* eigenvalues of the Hermitian matrix (the value of *n'* above);

val: double *val[1:numval]*;

    exit: in array *val* the largest *numval* eigenvalues are delivered in monotonically nonincreasing order;

vecr,veci: double *vecr[1:n,1:numval]*, *veci[1:n,1:numval]*;

    exit: the calculated eigenvectors; the complex eigenvector with real part *vecr[1:n,i]* and the imaginary part *veci[1:n,i]* corresponds to the eigenvalue *val[i]*, *i=1,...,numval*;

em: double *em[0:9]*;

    entry: *em[0]*: the machine precision;

              em[2]: the relative tolerance used for the eigenvalues; more precisely, the tolerance for each eigenvalue λ is $|\lambda|*em[2]+em[1]*em[0]$;

              em[4]: the orthogonalization parameter (for example, *em[4]*=0.01);

              em[6]: the tolerance for the eigenvectors;

              em[8]: the maximum number of inverse iterations allowed for the calculation of each eigenvector;

    exit: *em[1]*: an estimate of a norm of the original matrix;

              em[3]: the number of iterations performed;

              em[5]: the number of eigenvectors involved in the last Gram-Schmidt orthogonalization;

              em[7]: the maximum Euclidean norm of the residues of the calculated eigenvectors;

              em[9]: the largest number of inverse iterations performed for the calculation of some eigenvector; if, however, for some calculated eigenvector, the Euclidean norm of the residues remains greater than *em[1]*em[6]*, then *em[9]=em[8]+1*.

Procedures used: hshhrmtri, valsymtri, vecsymtri, bakhrmtri.

```
public static void eighrm(double a[][], int n, int numval, double val[],
                       double vecr[][], double veci[][], double em[])
{
  double bb[] = new double[n];
  double tr[] = new double[n];
  double ti[] = new double[n];
  double d[] = new double[n+1];
  double b[] = new double[n+1];

  hshhrmtri(a,n,d,b,bb,em,tr,ti);
  valsymtri(d,bb,n,1,numval,val,em);
```

```
  b[n]=0.0;
  vecsymtri(d,b,n,1,numval,val,vecr,em);
  bakhrmtri(a,n,1,numval,vecr,veci,tr,ti);
}
```

C. qrivalhrm

Determines all eigenvalues λ_j of the $n \times n$ Hermitian matrix A. A is first reduced by a similarity transformation to real tridiagonal form T by calling *hshhrmtrival*. The eigenvalues λ_j of T are then determined QR iteration using a call of *qrivalsymtri*.

Procedure parameters:
$$\text{int qrivalhrm } (a,n,val,em)$$

qrivalhrm: given the value zero provided the QR iteration is completed within *em[4]* iterations; otherwise, *qrivalhrm* is given the number of eigenvalues, k, not calculated and only the last n-k elements of *val* are approximate eigenvalues of the original Hermitian matrix;

a: double *a[1:n,1:n]*;

      entry: the real part of the upper triangle of the Hermitian matrix must be given in the upper triangular part of *a* (the elements *a[i,j]*, $i \le j$); the imaginary part of the strict lower triangle of the Hermitian matrix must be given in the strict lower part of *a* (the elements *a[i,j]*, $i > j$);

      exit: the array elements are altered;

n: int;

      entry: the order of the given matrix;

val: double *val[1:n]*;

      exit: the calculated eigenvalues;

em: double *em[0:5]*;

      entry: *em[0]*:   the machine precision;

               em[2]:   the relative tolerance used for the QR iteration;

               em[4]:   the maximum allowed number of iterations;

      exit: *em[1]*:   an estimate of a norm of the original matrix;

               em[3]:   the maximum absolute value of the codiagonal elements neglected;

               em[5]:   number of iterations performed; *em[5]=em[4]+1* when *qrivalhrm*$\neq$0.

Procedures used: hshhrmtrival, qrivalsymtri.

```
public static int qrivalhrm(double a[][], int n, double val[], double em[])
{
  int i;
  double bb[] = new double[n+1];

  hshhrmtrival(a,n,val,bb,em);
  bb[n]=0.0;
```

```
i=qrivalsymtri(val,bb,n,em);
return i;
}
```

D. qrihrm

Determines all eigenvalues λ_j and corresponding eigenvectors $u^{(j)}$ of the $n \times n$ Hermitian matrix A. A is first reduced to similar real tridiagonal form T by a call of *hshhrmtri*. The eigenvalues λ_j and corresponding eigenvectors $v^{(j)}$ of T are then determined QR iteration, using a call of *qrisymtri*. The eigenvectors $u^{(j)}$ of A are then obtained from the $v^{(j)}$ by means of a call of *bakhrmtri*.

Procedure parameters:
$$\text{int qrihrm } (a,n,val,vr,vi,em)$$

qrihrm: *qrihrm*=0, provided the process is completed within *em[4]* iterations; otherwise, *qrihrm* is given the number of eigenvalues, k, not calculated and only the last n-k elements of *val* are approximate eigenvalues and the columns of the arrays *vr, vi[1:n,n-k:n]* are approximate eigenvectors of the original Hermitian matrix;

a: double *a[1:n,1:n]*;
            entry: the real part of the upper triangle of the Hermitian matrix must be given in the upper triangular part of *a* (the elements *a[i,j]*, $i \leq j$); the imaginary part of the strict lower triangle of the Hermitian matrix must be given in the strict lower part of *a* (the elements *a[i,j]*, $i > j$);
            exit: the array elements are altered;

n: int;
            entry: the order of the given matrix;

val: double *val[1:n]*;
            exit: the calculated eigenvalues;

vr,vi: double *vr[1:n,1:n], vi[1:n,1:n]*;
            exit: the calculated eigenvectors; the complex eigenvector with real part *vr[1:n,i]* and the imaginary part *vi[1:n,i]* corresponds to the eigenvalue *val[i]*, $i=1,...,n$;

em: double *em[0:5]*;
            entry: *em[0]*: the machine precision;
                    em[2]: the relative tolerance for the QR iteration;
                    em[4]: maximum allowed number of iterations (for example, *em[4]*=10n);
            exit: *em[1]*: an estimate of a norm of the original matrix;
                    em[3]: the maximum absolute value of the codiagonal elements neglected;
                    em[5]: number of iterations performed; *em[5]*=*em[4]*+1 when *qrihrm*≠0.

Procedures used: hshhrmtri, qrisymtri, bakhrmtri.

```
public static int qrihrm(double a[][], int n, double val[], double vr[][],
                        double vi[][], double em[])
{
    int i,j;
    double b[]  = new double[n+1];
    double bb[] = new double[n+1];
    double tr[] = new double[n];
    double ti[] = new double[n];

    hshhrmtri(a,n,val,b,bb,em,tr,ti);
    for (i=1; i<=n; i++) {
        vr[i][i]=1.0;
        for (j=i+1; j<=n; j++) vr[i][j]=vr[j][i]=0.0;
    }
    b[n]=bb[n]=0.0;
    i=qrisymtri(vr,n,val,b,bb,em);
    bakhrmtri(a,n,i+1,n,vr,vi,tr,ti);
    return i;
}
```

3.13.9 Complex upper-Hessenberg matrices

A. valqricom

Determines the eigenvalues λ_j (j=1,...,n) of an $n \times n$ complex upper Hessenberg matrix A ($A_{i,j}=0$ for i>j+1) with real subdiagonal elements $A_{j+1,j}$, by QR iteration. For further details see the documentation of the procedure *qricom*.

Procedure parameters:

$$\text{int valqricom } (a1,a2,b,n,em,val1,val2)$$

valqricom: given the value zero provided the process is computed within *em[4]* iterations; otherwise, given the number, *k*, of eigenvalues is not calculated and only the last *n-k* elements of the arrays *val1* and *val2* are approximate eigenvalues of the upper Hessenberg matrix;

a1,a2: double *a1[1:n,1:n], a2[1:n,1:n]*;

    entry: the real part and the imaginary part of the upper triangle of the upper Hessenberg matrix must be given in the corresponding parts of the arrays *a1* and *a2*;

    exit: the array elements in the upper triangle of *a1* and *a2* are altered;

b: double *b[1:n-1]*;

    entry: the subdiagonal of the upper Hessenberg matrix;

    exit: the elements of *b* are altered;

n: int;

    entry: the order of the given matrix;

em: double *em[0:5]*;
                entry: *em[0]*: the machine precision;
                        em[1]: an estimate of the norm of the upper Hessenberg matrix
                                (e.g., the sum of the infinity norms of the real and
                                imaginary parts of the matrix);
                        em[2]: the relative tolerance for the QR iteration;
                        em[4]: the maximum allowed number of iterations (e.g., 10*n*);
                exit: *em[3]*: the maximum absolute value of the subdiagonal
                                elements neglected;
                        em[5]: the number of iterations performed; *em[5]=em[4]+1* in
                                the case *valqricom≠0*;
val1,val2: double *val1[1:n]*, *val2[1:n]*;
                exit: the real part and the imaginary part of the calculated
                        eigenvalues are delivered in *val1* and *val2*, respectively.

Procedures used: comkwd, rotcomrow, rotcomcol, comcolcst.

```
public static int valqricom(double a1[][], double a2[][], double b[], int n,
                            double em[], double val1[], double val2[])
{
  int nm1,i,i1,q,q1,max,count;
  double r,z1,z2,dd1,dd2,cc,hc,a1nn,a2nn,aij1,aij2,ai1i,kappa,nui,mui1,mui2,muim11,
         muim12,nuim1,tol;
  double g1[] = new double[1];
  double g2[] = new double[1];
  double k1[] = new double[1];
  double k2[] = new double[1];

  hc=0.0;
  tol=em[1]*em[2];
  max=(int) em[4];
  count=0;
  r=0.0;
  if (n > 1) hc=b[n-1];
  do {
    nm1=n-1;
    i=n;
    do {
      q=i;
      i--;
    } while ((i >= 1) ? (Math.abs(b[i]) > tol) : false);
    if (q > 1)
      if (Math.abs(b[q-1]) > r) r=Math.abs(b[q-1]);
    if (q == n) {
      val1[n]=a1[n][n];
      val2[n]=a2[n][n];
      n=nm1;
      if (n > 1) hc=b[n-1];
    } else {
      dd1=a1[n][n];
```

```
dd2=a2[n][n];
cc=b[nm1];
comkwd((a1[nm1][nm1]-dd1)/2.0,(a2[nm1][nm1]-dd2)/2.0,cc*a1[nm1][n],
        cc*a2[nm1][n],g1,g2,k1,k2);
if (q == nm1) {
  val1[nm1]=g1[0]+dd1;
  val2[nm1]=g2[0]+dd2;
  val1[n]=k1[0]+dd1;
  val2[n]=k2[0]+dd2;
  n -= 2;
  if (n > 1) hc=b[n-1];
} else {
  count++;
  if (count > max) break;
  z1=k1[0]+dd1;
  z2=k2[0]+dd2;
  if (Math.abs(cc) > Math.abs(hc)) z1 += Math.abs(cc);
  hc=cc/2.0;
  i=q1=q+1;
  aij1=a1[q][q]-z1;
  aij2=a2[q][q]-z2;
  ai1i=b[q];
  kappa=Math.sqrt(aij1*aij1+aij2*aij2+ai1i*ai1i);
  mui1=aij1/kappa;
  mui2=aij2/kappa;
  nui=ai1i/kappa;
  a1[q][q]=kappa;
  a2[q][q]=0.0;
  a1[q1][q1] -= z1;
  a2[q1][q1] -= z2;
  Basic.rotcomrow(q1,n,q,q1,a1,a2,mui1,mui2,nui);
  Basic.rotcomcol(q,q,q,q1,a1,a2,mui1,-mui2,-nui);
  a1[q][q] += z1;
  a2[q][q] += z2;
  for (i1=q1+1; i1<=n; i1++) {
    aij1=a1[i][i];
    aij2=a2[i][i];
    ai1i=b[i];
    kappa=Math.sqrt(aij1*aij1+aij2*aij2+ai1i*ai1i);
    muim11=mui1;
    muim12=mui2;
    nuim1=nui;
    mui1=aij1/kappa;
    mui2=aij2/kappa;
    nui=ai1i/kappa;
    a1[i1][i1] -= z1;
    a2[i1][i1] -= z2;
    Basic.rotcomrow(i1,n,i,i1,a1,a2,mui1,mui2,nui);
    a1[i][i]=muim11*kappa;
    a2[i][i] = -muim12*kappa;
```

```
            b[i-1]=nuim1*kappa;
            Basic.rotcomcol(q,i,i,i1,a1,a2,mui1,-mui2,-nui);
            a1[i][i] += z1;
            a2[i][i] += z2;
            i=i1;
        }
        aij1=a1[n][n];
        aij2=a2[n][n];
        kappa=Math.sqrt(aij1*aij1+aij2*aij2);
        if ((kappa < tol) ? true : (aij2*aij2 <= em[0]*aij1*aij1)) {
            b[nm1]=nui*aij1;
            a1[n][n]=aij1*mui1+z1;
            a2[n][n] = -aij1*mui2+z2;
        } else {
            b[nm1]=nui*kappa;
            a1nn=mui1*kappa;
            a2nn = -mui2*kappa;
            mui1=aij1/kappa;
            mui2=aij2/kappa;
            Basic.comcolcst(q,nm1,n,a1,a2,mui1,mui2);
            a1[n][n]=mui1*a1nn-mui2*a2nn+z1;
            a2[n][n]=mui1*a2nn+mui2*a1nn+z2;
        }
      }
    }
  }
  } while (n > 0);
  em[3]=r;
  em[5]=count;
  return n;
}
```

B. qricom

Computes the eigenvalues λ_j and corresponding eigenvectors $v^{(j)}$ (j=1,...,n) of an $n \times n$ complex upper Hessenberg matrix A ($A_{i,j}=0$ for $i>j+1$) with real subdiagonal elements $A_{j+1,j}$. A is transformed into a complex upper triangular matrix U by means of the Francis QR iteration [Fr61, Wi65]. The diagonal elements of U are its eigenvalues; the eigenvectors of U are obtained by solving the associated upper triangular system, and the eigenvectors of A are recovered by back transformation.

Procedure parameters:

              int qricom (a1,a2,b,n,em,val1,val2,vec1,vec2)

qricom: given the value zero provided the process is computed within *em[4]* iterations; otherwise given the number, *k*, of eigenvalues not calculated and only the last *n-k* elements of the arrays, *val1* and *val2* are approximate eigenvalues of the upper Hessenberg matrix and no useful eigenvectors are delivered;

a1,a2:   double *a1[1:n,1:n], a2[1:n,1:n]*;
            entry: the real part and the imaginary part of the upper triangle of the
                    upper Hessenberg matrix must be given in the corresponding
                    parts of the arrays *a1* and *a2*;
            exit: the array elements in the upper triangle of *a1* and *a2* are
                    altered;
b: double *b[1:n-1]*;
            entry: the real subdiagonal of the upper Hessenberg matrix;
            exit: the elements of *b* are altered;
n: int;
            entry: the order of the given matrix;
em: double *em[0:5]*;
            entry: *em[0]*: the machine precision;
                    em[1]: an estimate of the norm of the upper Hessenberg matrix
                              (e.g., the sum of the infinity norms of the real and
                              imaginary parts of the matrix);
                    em[2]: the relative tolerance for the QR iteration;
                    em[4]: the maximum allowed number of iterations (e.g., 10n);
            exit: *em[3]*: the maximum absolute value of the subdiagonal
                              elements neglected;
                    em[5]: the number of iterations performed; *em[5]=em[4]+1* in
                              the case *qricom*$\neq$0;
val1,val2: double *val1[1:n], val2[1:n]*;
            exit: the real part and the imaginary part of the calculated
                    eigenvalues are delivered in *val1* and *val2*, respectively.
vec1,vec2: double *vec1[1:n,1:n], vec2[1:n,1:n]*;
            exit: the eigenvectors of the upper Hessenberg matrix; the
                    eigenvector with real part *vec1[1:n,j]* and imaginary part
                    vec2[1:n,j] corresponds to the eigenvalue *val1[j]+val2[j]*i,
                    j=1,...,*n*.

Procedures used: comkwd, rotcomrow, rotcomcol, comcolcst, comrowcst, matvec,
                    commatvec, comdiv.

```
public static int qricom(double a1[][], double a2[][], double b[], int n,
                         double em[], double val1[], double val2[], double vec1[][],
                         double vec2[][])
{
  int m,nm1,i,i1,j,q,q1,max,count;
  double r,z1,z2,dd1,dd2,cc,p1,p2,t1,t2,delta1,delta2,mv1,mv2,h,h1,h2,hc,aij12,
         aij22,a1nn,a2nn,aij1,aij2,aili,kappa,nui,mui1,mui2,muim11,muim12,nuim1,
         tol,machtol;
  double tf1[] = new double[n+1];
  double tf2[] = new double[n+1];
  double g1[] = new double[1];
  double g2[] = new double[1];
  double k1[] = new double[1];
  double k2[] = new double[1];
  double tmp1[] = new double[1];
  double tmp2[] = new double[1];
```

```
hc=0.0;
tol=em[1]*em[2];
machtol=em[0]*em[1];
max=(int) em[4];
count=0;
r=0.0;
m=n;
if (n > 1) hc=b[n-1];
for (i=1; i<=n; i++) {
  vec1[i][i]=1.0;
  vec2[i][i]=0.0;
  for (j=i+1; j<=n; j++)
    vec1[i][j]=vec1[j][i]=vec2[i][j]=vec2[j][i]=0.0;
}
do {
  nm1=n-1;
  i=n;
  do {
    q=i;
    i--;
  } while ((i >= 1) ? (Math.abs(b[i]) > tol) : false);
  if (q > 1)
    if (Math.abs(b[q-1]) > r) r=Math.abs(b[q-1]);
  if (q == n) {
    val1[n]=a1[n][n];
    val2[n]=a2[n][n];
    n=nm1;
    if (n > 1) hc=b[n-1];
  } else {
    dd1=a1[n][n];
    dd2=a2[n][n];
    cc=b[nm1];
    p1=(a1[nm1][nm1]-dd1)*0.5;
    p2=(a2[nm1][nm1]-dd2)*0.5;
    comkwd(p1,p2,cc*a1[nm1][n],cc*a2[nm1][n],g1,g2,k1,k2);
    if (q == nm1) {
      a1[n][n]=val1[n]=g1[0]+dd1;
      a2[n][n]=val2[n]=g2[0]+dd2;
      a1[q][q]=val1[q]=k1[0]+dd1;
      a2[q][q]=val2[q]=k2[0]+dd2;
      kappa=Math.sqrt(k1[0]*k1[0]+k2[0]*k2[0]+cc*cc);
      nui=cc/kappa;
      mui1=k1[0]/kappa;
      mui2=k2[0]/kappa;
      aij1=a1[q][n];
      aij2=a2[q][n];
      h1=mui1*mui1-mui2*mui2;
      h2=2.0*mui1*mui2;
      h = -nui*2.0;
```

```
        a1[q][n]=h*(p1*mui1+p2*mui2)-nui*nui*cc+aij1*h1+aij2*h2;
        a2[q][n]=h*(p2*mui1-p1*mui2)+aij2*h1-aij1*h2;
        Basic.rotcomrow(q+2,m,q,n,a1,a2,mui1,mui2,nui);
        Basic.rotcomcol(1,q-1,q,n,a1,a2,mui1,-mui2,-nui);
        Basic.rotcomcol(1,m,q,n,vec1,vec2,mui1,-mui2,-nui);
        n -= 2;
        if (n > 1) hc=b[n-1];
        b[q]=0.0;
      } else {
        count++;
        if (count > max) {
          em[3]=r;
          em[5]=count;
          return n;
        }
        z1=k1[0]+dd1;
        z2=k2[0]+dd2;
        if (Math.abs(cc) > Math.abs(hc)) z1 += Math.abs(cc);
        hc=cc/2.0;
        q1=q+1;
        aij1=a1[q][q]-z1;
        aij2=a2[q][q]-z2;
        aili=b[q];
        kappa=Math.sqrt(aij1*aij1+aij2*aij2+aili*aili);
        mui1=aij1/kappa;
        mui2=aij2/kappa;
        nui=aili/kappa;
        a1[q][q]=kappa;
        a2[q][q]=0.0;
        a1[q1][q1] -= z1;
        a2[q1][q1] -= z2;
        Basic.rotcomrow(q1,m,q,q1,a1,a2,mui1,mui2,nui);
        Basic.rotcomcol(1,q,q,q1,a1,a2,mui1,-mui2,-nui);
        a1[q][q] += z1;
        a2[q][q] += z2;
        Basic.rotcomcol(1,m,q,q1,vec1,vec2,mui1,-mui2,-nui);
        for (i=q1; i<=nm1; i++) {
          i1=i+1;
          aij1=a1[i][i];
          aij2=a2[i][i];
          aili=b[i];
          kappa=Math.sqrt(aij1*aij1+aij2*aij2+aili*aili);
          muim11=mui1;
          muim12=mui2;
          nuim1=nui;
          mui1=aij1/kappa;
          mui2=aij2/kappa;
          nui=aili/kappa;
          a1[i1][i1] -= z1;
          a2[i1][i1] -= z2;
```

```
        Basic.rotcomrow(i1,m,i,i1,a1,a2,mui1,mui2,nui);
        a1[i][i]=muim11*kappa;
        a2[i][i] = -muim12*kappa;
        b[i-1]=nuim1*kappa;
        Basic.rotcomcol(1,i,i,i1,a1,a2,mui1,-mui2,-nui);
        a1[i][i] += z1;
        a2[i][i] += z2;
        Basic.rotcomcol(1,m,i,i1,vec1,vec2,mui1,-mui2,-nui);
      }
      aij1=a1[n][n];
      aij2=a2[n][n];
      aij12=aij1*aij1;
      aij22=aij2*aij2;
      kappa=Math.sqrt(aij12+aij22);
      if ((kappa < tol) ? true : (aij22 <= em[0]*aij12)) {
        b[nm1]=nui*aij1;
        a1[n][n]=aij1*mui1+z1;
        a2[n][n] = -aij1*mui2+z2;
      } else {
        b[nm1]=nui*kappa;
        a1nn=mui1*kappa;
        a2nn = -mui2*kappa;
        mui1=aij1/kappa;
        mui2=aij2/kappa;
        Basic.comcolcst(1,nm1,n,a1,a2,mui1,mui2);
        Basic.comcolcst(1,nm1,n,vec1,vec2,mui1,mui2);
        Basic.comrowcst(n+1,m,n,a1,a2,mui1,-mui2);
        Basic.comcolcst(n,m,n,vec1,vec2,mui1,mui2);
        a1[n][n]=mui1*a1nn-mui2*a2nn+z1;
        a2[n][n]=mui1*a2nn+mui2*a1nn+z2;
      }
    }
  }
} while (n > 0);
for (j=m; j>=2; j--) {
  tf1[j]=1.0;
  tf2[j]=0.0;
  t1=a1[j][j];
  t2=a2[j][j];
  for (i=j-1; i>=1; i--) {
    delta1=t1-a1[i][i];
    delta2=t2-a2[i][i];
    Basic.commatvec(i+1,j,i,a1,a2,tf1,tf2,tmp1,tmp2);
    mv1=tmp1[0];
    mv2=tmp2[0];
    if (Math.abs(delta1) < machtol && Math.abs(delta2) < machtol) {
      tf1[i]=mv1/machtol;
      tf2[i]=mv2/machtol;
    } else {
      Basic.comdiv(mv1,mv2,delta1,delta2,tmp1,tmp2);
```

```
            tf1[i]=tmp1[0];
            tf2[i]=tmp2[0];
        }
    }
    for (i=1; i<=m; i++) {
        Basic.commatvec(1,j,i,vec1,vec2,tf1,tf2,tmp1,tmp2);
        vec1[i][j]=tmp1[0];
        vec2[i][j]=tmp2[0];
    }
}
em[3]=r;
em[5]=count;
return n;
}
```

3.13.10 Complex full matrices

A. eigvalcom

Computes the eigenvalues λ_j (j=1,...,n) of an $n \times n$ complex matrix A. A is first transformed to equilibrated form A' (by means of a call of *eqilbrcom*); A' is then transformed to complex upper Hessenberg form H with real subdiagonal elements (by means of a call of *hshcomhes*). The eigenvalues λ_j (j=1,...,n) of H are then determined by QR iteration (by means of a call of *valqricom*).

Procedure parameters:
$$\text{int eigvalcom } (ar,ai,n,em,valr,vali)$$
eigvalcom: given the value zero provided the process is computed within *em[4]* iterations; otherwise given the number, k, of eigenvalues not calculated and only the last $n-k$ elements of the arrays, *valr* and *vali* are approximate eigenvalues of the original matrix;

ar,ai: double *ar[1:n,1:n]*, *ai[1:n,1:n]*;
  entry: the real part and the imaginary part of the matrix must be given in the arrays *ar* and *ai*, respectively;
  exit: the array elements of *ar* and *ai* are altered;

n: int;
  entry: the order of the given matrix;

em: double *em[0:7]*;
  entry: *em[0]*: the machine precision;
    em[2]: the relative tolerance for the QR iteration;
    em[4]: the maximum allowed number of QR iterations (e.g., $10n$);
    em[6]: the maximum allowed number of iterations for equilibrating the original matrix (e.g., *em[6]*=$n^*n/2$);
  exit: *em[1]*: the Euclidean norm of the equilibrated matrix;

em[3]:   the maximum absolute value of the subdiagonal elements neglected in the QR iteration;

em[5]:   the number of QR iterations performed; *em[5]=em[4]*+1 in the case *eigvalcom≠0*;

em[7]:   the number of iterations performed for equilibrating the original matrix;

valr,vali: double *valr[1:n]*, *vali[1:n]*;

exit: the real part and the imaginary part of the calculated eigenvalues are delivered in *valr* and *vali*, respectively.

Procedures used: eqilbrcom, comeucnrm, hshcomhes, valqricom.

```
public static int eigvalcom(double ar[][], double ai[][], int n, double em[],
                           double valr[], double vali[])
{
  int i;
  int ind[] = new int[n+1];
  double d[] = new double[n+1];
  double b[] = new double[n+1];
  double del[] = new double[n+1];
  double tr[] = new double[n+1];
  double ti[] = new double[n+1];

  eqilbrcom(ar,ai,n,em,d,ind);
  em[1]=Basic.comeucnrm(ar,ai,n-1,n);
  hshcomhes(ar,ai,n,em,b,tr,ti,del);
  i=valqricom(ar,ai,b,n,em,valr,vali);
  return i;
}
```

B. eigcom

Computes the eigenvalues λ_j and eigenvectors $u^{(j)}$ (j=1,...,n) of an $n \times n$ complex matrix A. A is first transformed to equilibrated form A' (by means of a call of *eqilbrcom*); A' is then transformed to complex upper Hessenberg form H with real subdiagonal elements (by means of a call of *hshcomhes*). The eigenvalues λ_j and eigenvectors $v^{(j)}$ (j=1,...,n) of H are then determined by QR iteration (by means of a call of *qricom*), and the $u^{(j)}$ are then recovered from the $v^{(j)}$ by two successive back transformations (by means of calls of *bakcomhes* and *baklbrcom*).

Procedure parameters:

$$\text{int eigcom } (ar,ai,n,em,valr,vali,vr,vi)$$

eigcom: given the value zero provided the process is computed within *em[4]* iterations; otherwise given the number, k, of eigenvalues not calculated and only the last $n\text{-}k$ elements of the arrays, *valr* and *vali* are approximate eigenvalues of the original matrix and no useful eigenvectors are delivered;

ar,ai: double *ar[1:n, 1:n]*, *ai[1:n, 1:n]*;
        entry: the real part and the imaginary part of the matrix must be given in the arrays *ar* and *ai*, respectively;
        exit: the array elements of *ar* and *ai* are altered;
n: int;
        entry: the order of the given matrix;
em: double *em[0:7]*;
        entry: *em[0]*: the machine precision;
                em[2]: the relative tolerance for the QR iteration;
                em[4]: the maximum allowed number of QR iterations (e.g., $10n$);
                em[6]: the maximum allowed number of iterations for equilibrating the original matrix (e.g., *em[6]*=$n*n/2$);
        exit: *em[1]*: the Euclidean norm of the equilibrated matrix;
                em[3]: the maximum absolute value of the subdiagonal elements neglected in the QR iteration;
                em[5]: the number of QR iterations performed; *em[5]*=*em[4]*+1 in the case *eigcom*≠0;
                em[7]: the number of iterations performed for equilibrating the original matrix;
valr,vali: double *valr[1:n]*, *vali[1:n]*;
        exit: the real part and the imaginary part of the calculated eigenvalues are delivered in *valr* and *vali*, respectively;
vr,vi: double *vr[1:n, 1:n]*, *vi[1:n, 1:n]*;
        exit: the eigenvectors of the matrix; the normalized eigenvector with real part *vr[1:n,j]* and imaginary part *vi[1:n,j]* corresponds to the eigenvalue *valr[j]*+*vali[j]**i, j=1,...,n.

Procedures used: eqilbrcom, comeucnrm, hshcomhes, qricom, bakcomhes, baklbrcom, sclcom.

```
public static int eigcom(double ar[][], double ai[][], int n, double em[],
                         double valr[], double vali[], double vr[][], double vi[][])
{
  int i;
  int ind[]  = new int[n+1];
  double d[]  = new double[n+1];
  double b[]  = new double[n+1];
  double del[] = new double[n+1];
  double tr[]  = new double[n+1];
  double ti[]  = new double[n+1];

  eqilbrcom(ar,ai,n,em,d,ind);
  em[1]=Basic.comeucnrm(ar,ai,n-1,n);
  hshcomhes(ar,ai,n,em,b,tr,ti,del);
  i=qricom(ar,ai,b,n,em,valr,vali,vr,vi);
  if (i == 0) {
    bakcomhes(ar,ai,tr,ti,del,vr,vi,n,1,n);
    baklbrcom(n,1,n,d,ind,vr,vi);
    Basic.sclcom(vr,vi,n,1,n);
```

```
  }
  return i;
}
```

3.14 The generalized eigenvalue problem

3.14.1 Real asymmetric matrices

A. qzival

Solves the generalized matrix eigenvalue problem $Ax=\lambda Bx$ by means of QZ iteration [MS71]. Given two $n \times n$ matrices A and B, *qzival* determines complex $\alpha^{(j)}$ and real $\text{ß}^{(j)}$ such that $\text{ß}^{(j)}A - \alpha^{(j)}B$ is singular $(j=1,...,n)$.

QZ iteration may fail to converge in the determination of a sequence $\alpha^{(i)}$, $\text{ß}^{(i)}$ $(i=1,...,m)$. Such failure is signalled at return from *qzival* by the allocation of the value -1 to the array *iter[i]*, $(i=1,...,m)$ and allocation of a nonnegative value, that of the number of iterations required in each case, to the array *iter[j]* $(j=m+1,...,n)$ to signal success in the determination of the remaining pairs of numbers $\alpha^{(j)}$, $\text{ß}^{(j)}$. In particular, if *iter[1]* contains the value zero at exit, then the computations have been completely successful.

If QZ iteration is completely successful (i.e., $m=0$ above), A and B are both reduced by unitary transformations to quasi-upper triangular form U and upper triangular form V, respectively: U has 1×1 or 2×2 blocks on the principal diagonal, but $U_{i,j}=0$ for $i>j+1$ and $U_{i+1,i}=0$ for those elements not belonging to the blocks; $V_{i,j}=0$ for $i>j$.

The sets $\alpha^{(j)}$, $\text{ß}^{(j)}$ $(j=m+1,...,n)$ may contain complex conjugate pairs in the sense that for some $k \geq m+1$, $\alpha^{(k)}/\text{ß}^{(k)}$ = conjugate of $\{\alpha^{(k+1)}/\text{ß}^{(k+1)}\}$; the components of such pairs are stored consecutively in the output arrays *alfr*, *alfi*, and *beta*.

Procedure parameters:
$$\text{void qzival } (n,a,b,alfr,alfi,beta,iter,em)$$

n: int;
        entry: the number of rows and columns of the matrices *a* and *b*;
a: double *a[1:n,1:n]*;
        entry: the given matrix;
        exit: a quasi upper triangular matrix (value of *U* above);
b: double *b[1:n,1:n]*;
        entry: the given matrix;
        exit: an upper triangular matrix (value of *V* above);
alfr: double *alfr[1:n]*;
        exit: the real part of $\alpha^{(j)}$ given in *alfr[j]*, $j=m+1,...,n$, in the above;
alfi: double *alfi[1:n]*;
        exit: the imaginary part of $\alpha^{(j)}$ given in *alfi[j]*, $j=m+1,...,n$, in the above;

beta: double *beta[1:n]*;
   exit: the value of ß$^{(j)}$ given in *beta[j]*, j=*m*+1,...,*n*, in the above;
iter: int *iter[1:n]*;
   exit: trouble indicator and iteration counter, see above; if *iter[1]*=0 then
      no trouble is signalized;
em: double *em[0:1]*;
   entry: *em[0]*: the smallest positive machine number;
      em[1]: the relative precision of elements of *a* and *b*.

Procedures used: elmcol, hshdecmul, hestgl2, hsh2col, hsh3col, hsh2row2, hsh3row2, chsh2, hshvecmat, hshvectam.

```
public static void qzival(int n, double a[][], double b[][], double alfr[],
                    double alfi[], double beta[], int iter[], double em[])
{
  boolean stationary,goon,out;
  int i,q,m,m1,q1,j,k,k1,k2,k3,km1,l;
  double dwarf,eps,epsa,epsb,anorm,bnorm,ani,bni,constt,a10,a20,a30,b11,b22,b33,
         b44,a11,a12,a21,a22,a33,a34,a43,a44,b12,b34,old1,old2,an,bn,e,c,d,er,ei,
         a11r,a11i,a12r,a12i,a21r,a21i,a22r,a22i,cz,szr,szi,cq,sqr,sqi,ssr,ssi,
         tr,ti,bdr,bdi,r;
  double tmp1[] = new double[1];
  double tmp2[] = new double[1];
  double tmp3[] = new double[1];

  old1=old2=0.0;
  dwarf=em[0];
  eps=em[1];
  hshdecmul(n,a,b,dwarf);
  hestgl2(n,a,b);
  anorm=bnorm=0.0;
  for (i=1; i<=n; i++) {
    bni=0.0;
    iter[i]=0;
    ani = (i > 1) ? Math.abs(a[i][i-1]) : 0.0;
    for (j=i; j<=n; j++) {
      ani += Math.abs(a[i][j]);
      bni += Math.abs(b[i][j]);
    }
    if (ani > anorm) anorm=ani;
    if (bni > bnorm) bnorm=bni;
  }
  if (anorm == 0.0) anorm=eps;
  if (bnorm == 0.0) bnorm=eps;
  epsa=eps*anorm;
  epsb=eps*bnorm;
  m=n;
  out=false;
  do {
    i=q=m;
```

```
while ((i > 1) ? Math.abs(a[i][i-1]) > epsa : false) {
  q=i-1;
  i--;
}
if (q > 1) a[q][q-1]=0.0;
goon=true;
while (goon) {
  if (q >= m-1) {
    m=q-1;
    goon=false;
  } else {
    if (Math.abs(b[q][q]) <= epsb) {
      b[q][q]=0.0;
      q1=q+1;
      hsh2col(q,q,n,q,a[q][q],a[q1][q],a,b);
      a[q1][q]=0.0;
      q=q1;
    } else {
      goon=false;
      m1=m-1;
      q1=q+1;
      constt=0.75;
      (iter[m])++;
      stationary = (iter[m] == 1) ? true :
                                    (Math.abs(a[m][m-1]) >= constt*old1 &&
                                     Math.abs(a[m-1][m-2]) >= constt*old2);
      if (iter[m] > 30 && stationary) {
        for (i=1; i<=m; i++) iter[i] = -1;
        out=true;
        break;
      }
      if (iter[m] == 10 && stationary) {
        a10=0.0;
        a20=1.0;
        a30=1.1605;
      } else {
        b11=b[q][q];
        b22 = (Math.abs(b[q1][q1]) < epsb) ? epsb : b[q1][q1];
        b33 = (Math.abs(b[m1][m1]) < epsb) ? epsb : b[m1][m1];
        b44 = (Math.abs(b[m][m]) < epsb) ? epsb : b[m][m];
        a11=a[q][q]/b11;
        a12=a[q][q1]/b22;
        a21=a[q1][q]/b11;
        a22=a[q1][q1]/b22;
        a33=a[m1][m1]/b33;
        a34=a[m1][m]/b44;
        a43=a[m][m1]/b33;
        a44=a[m][m]/b44;
        b12=b[q][q1]/b22;
        b34=b[m1][m]/b44;
```

```
          a10=((a33-a11)*(a44-a11)-a34*a43+a43*b34*a11)/a21+a12-a11*b12;
          a20=(a22-a11-a21*b12)-(a33-a11)-(a44-a11)+a43*b34;
          a30=a[q+2][q1]/b22;
        }
        old1=Math.abs(a[m][m-1]);
        old2=Math.abs(a[m-1][m-2]);
        for (k=q; k<=m1; k++) {
          k1=k+1;
          k2=k+2;
          k3 = (k+3 > m) ? m : k+3;
          km1 = (k-1 < q) ? q : k-1;
          if (k != m1) {
            if (k == q)
              hsh3col(km1,km1,n,k,a10,a20,a30,a,b);
            else {
              hsh3col(km1,km1,n,k,a[k][km1],a[k1][km1],a[k2][km1],a,b);
              a[k1][km1]=a[k2][km1]=0.0;
            }
            hsh3row2(1,k3,k,b[k2][k2],b[k2][k1],b[k2][k],a,b);
            b[k2][k]=b[k2][k1]=0.0;
          } else {
            hsh2col(km1,km1,n,k,a[k][km1],a[k1][km1],a,b);
            a[k1][km1]=0.0;
          }
          hsh2row2(1,k3,k3,k,b[k1][k1],b[k1][k],a,b);
          b[k1][k]=0.0;
        }
      }
    }
  } /* goon loop */
  if (out) break;
} while (m >= 3);

m=n;
do {
  if ((m > 1) ? (a[m][m-1] == 0) : true) {
    alfr[m]=a[m][m];
    beta[m]=b[m][m];
    alfi[m]=0.0;
    m--;
  } else {
    l=m-1;
    if (Math.abs(b[l][l]) <= epsb) {
      b[l][l]=0.0;
      hsh2col(1,1,n,l,a[l][l],a[m][l],a,b);
      a[m][l]=b[m][l]=0.0;
      alfr[l]=a[l][l];
      alfr[m]=a[m][m];
      beta[l]=b[l][l];
      beta[m]=b[m][m];
```

```
      alfi[m]=alfi[l]=0.0;
  } else
    if (Math.abs(b[m][m]) <= epsb) {
      b[m][m]=0.0;
      hsh2row2(1,m,m,l,a[m][m],a[m][l],a,b);
      a[m][l]=b[m][l]=0.0;
      alfr[l]=a[l][l];
      alfr[m]=a[m][m];
      beta[l]=b[l][l];
      beta[m]=b[m][m];
      alfi[m]=alfi[l]=0.0;
    } else {
      an=Math.abs(a[l][l])+Math.abs(a[l][m])+
         Math.abs(a[m][l])+Math.abs(a[m][m]);
      bn=Math.abs(b[l][l])+Math.abs(b[l][m])+Math.abs(b[m][m]);
      a11=a[l][l]/an;
      a12=a[l][m]/an;
      a21=a[m][l]/an;
      a22=a[m][m]/an;
      b11=b[l][l]/bn;
      b12=b[l][m]/bn;
      b22=b[m][m]/bn;
      e=a11/b11;
      c=((a22-e*b22)/b22-(a21*b12)/(b11*b22))/2.0;
      d=c*c+(a21*(a12-e*b12))/(b11*b22);
      if (d >= 0.0) {
        e += ((c < 0.0) ? c-Math.sqrt(d) : c+Math.sqrt(d));
        a11 -= e*b11;
        a12 -= e*b12;
        a22 -= e*b22;
        if (Math.abs(a11)+Math.abs(a12) >= Math.abs(a21)+Math.abs(a22))
          hsh2row2(1,m,m,l,a12,a11,a,b);
        else
          hsh2row2(1,m,m,l,a22,a21,a,b);
        if (an >= Math.abs(e)*bn)
          hsh2col(l,l,n,l,b[l][l],b[m][l],a,b);
        else
          hsh2col(l,l,n,l,a[l][l],a[m][l],a,b);
        a[m][l]=b[m][l]=0.0;
        alfr[l]=a[l][l];
        alfr[m]=a[m][m];
        beta[l]=b[l][l];
        beta[m]=b[m][m];
        alfi[m]=alfi[l]=0.0;
      } else {
        er=e+c;
        ei=Math.sqrt(-d);
        a11r=a11-er*b11;
        a11i=ei*b11;
        a12r=a12-er*b12;
```

```
    a12i=ei*b12;
    a21r=a21;
    a21i=0.0;
    a22r=a22-er*b22;
    a22i=ei*b22;
    if (Math.abs(a11r)+Math.abs(a11i)+Math.abs(a12r)+
        Math.abs(a12i) >= Math.abs(a21r)+Math.abs(a22r)+Math.abs(a22i)) {
      Basic.chsh2(a12r,a12i,-a11r,-a11i,tmp1,tmp2,tmp3);
      cz=tmp1[0];
      szr=tmp2[0];
      szi=tmp3[0];
    }
    else {
      Basic.chsh2(a22r,a22i,-a21r,-a21i,tmp1,tmp2,tmp3);
      cz=tmp1[0];
      szr=tmp2[0];
      szi=tmp3[0];
    }
    if (an >= (Math.abs(er)+Math.abs(ei))*bn) {
      Basic.chsh2(cz*b11+szr*b12,szi*b12,szr*b22,szi*b22,tmp1,tmp2,tmp3);
      cq=tmp1[0];
      sqr=tmp2[0];
      sqi=tmp3[0];
    }
    else {
      Basic.chsh2(cz*a11+szr*a12,szi*a12,cz*a21+szr*a22,szi*a22,
                  tmp1,tmp2,tmp3);
      cq=tmp1[0];
      sqr=tmp2[0];
      sqi=tmp3[0];
    }
    ssr=sqr*szr+sqi*szi;
    ssi=sqr*szi-sqi*szr;
    tr=cq*cz*a11+cq*szr*a12+sqr*cz*a21+ssr*a22;
    ti=cq*szi*a12-sqi*cz*a21+ssi*a22;
    bdr=cq*cz*b11+cq*szr*b12+ssr*b22;
    bdi=cq*szi*b12+ssi*b22;
    r=Math.sqrt(bdr*bdr+bdi*bdi);
    beta[l]=bn*r;
    alfr[l]=an*(tr*bdr+ti*bdi)/r;
    alfi[l]=an*(tr*bdi-ti*bdr)/r;
    tr=ssr*a11-sqr*cz*a12-cq*szr*a21+cq*cz*a22;
    ti = -ssi*a11-sqi*cz*a12+cq*szi*a21;
    bdr=ssr*b11-sqr*cz*b12+cq*cz*b22;
    bdi = -ssi*b11-sqi*cz*b12;
    r=Math.sqrt(bdr*bdr+bdi*bdi);
    beta[m]=bn*r;
    alfr[m]=an*(tr*bdr+ti*bdi)/r;
    alfi[m]=an*(tr*bdi-ti*bdr)/r;
  }
```

```
      }
    m -= 2;
  }
} while (m > 0);
}
```

B. qzi

Solves the generalized matrix eigenvalue problem $Ax=\lambda Bx$ by means of QZ iteration [MS71]. The procedure *qzi* applies the same method as *qzival*. Given two $n \times n$ matrices A and B, *qzi* determines complex $\alpha^{(i)}$ and real $\beta^{(i)}$ such that $\beta^{(i)}A-\alpha^{(i)}B$ is singular and vectors $x^{(i)}$ such that $\beta^{(i)}Ax^{(i)}=\alpha^{(i)}Bx^{(i)}$ (j=1,...,n), the latter being normalized by the condition that
$$\max(|\,\text{real}(x_i^{(i)})\,|,\ |\,\text{imag}(x_i^{(i)})\,|)=1\ (1 \le i \le n) \text{ for each } x^{(i)} \text{ and}$$
$$\text{either real}(x_i^{(i)})=1 \text{ or imag}(x_i^{(i)}) \text{ for some } x_i^{(i)}.$$
With regard to the determination of the $\alpha^{(i)}$, $\beta^{(i)}$, the remarks made in the documentation to *qzival* apply with equal force here. In particular, (a) QZ iteration may fail to converge in the determination of a sequence $\alpha^{(i)}$, $\beta^{(i)}$ (i=1,...,m); this failure is signalled by the insertion of -1 in the array *iter[i]*, (i=1,...,m) and for those $\alpha^{(i)}$, $\beta^{(i)}$ that are determined, the required number of QZ iterations required in each case is allocated to the array *iter[j]* (j=m+1,...,n), (b) a quasi-upper triangular matrix U and an upper triangular matrix V are produced, and (c) the sets $\alpha^{(i)}$, $\beta^{(i)}$ (j=m+1,...,n) may contain complex conjugate pairs in the sense that for some $k \ge m+1$, $\alpha^{(k)}/\beta^{(k)}$ = conjugate of $\{\alpha^{(k+1)}/\beta^{(k+1)}\}$, and the components of such pairs are stored consecutively in the output arrays *alfr*, *alfi*, and *beta*.

Procedure parameters:
$$\text{void qzi } (n,a,b,x,alfr,alfi,beta,iter,em)$$

n: int;
         entry: the number of rows and columns of the matrices *a*, *b*, and *x*;
a: double *a[1:n,1:n]*;
         entry: the given matrix;
         exit: a quasi upper triangular matrix (value of *U* above);
b: double *b[1:n,1:n]*;
         entry: the given matrix;
         exit: an upper triangular matrix (value of *V* above);
x: double *x[1:n,1:n]*;
         entry: the $n \times n$ unit matrix;
         exit: the matrix of eigenvectors (components of $x^{(i)}$ above); the eigenvectors are stored in *x* as follows: if *alfi[m]=0* then *x[.,m]* is the *m*-th real eigenvector; otherwise, for each pair of consecutive columns *x[.,m]* and *x[.,m+1]* are the real and imaginary parts of the *m*-th complex eigenvector, *x[.,m]* and *-x[.,m+1]* are the real and imaginary parts of the (*m+1*)-st complex eigenvector; the eigenvectors are normalized such that the largest component is one or 1+0*i;

alfr: double *alfr[1:n]*;

        exit: the real part of $\alpha^{(j)}$ given in *alfr[j]*, j=m+1,...,n, in the above;

alfi: double *alfi[1:n]*;

        exit: the imaginary part of $\alpha^{(j)}$ given in *alfi[j]*, j=m+1,...,n, in the above;

beta: double *beta[1:n]*;

        exit: the value of $ß^{(j)}$ given in *beta[j]*, j=m+1,...,n, in the above;

iter: int *iter[1:n]*;

        exit: trouble indicator and iteration counter, see *qzival*; if *iter[1]*=0 then no trouble is signalized;

em: double *em[0:1]*;

        entry: *em[0]*: the smallest positive machine number;

               em[1]: the relative precision of elements of *a* and *b*.

Procedures used: matmat, hshdecmul, hestgl3, hsh2col, hsh2row3, hsh3row3, hsh3col, chsh2, comdiv.

```
public static void qzi(int n, double a[][], double b[][], double x[][],
                       double alfr[], double alfi[], double beta[], int iter[],
                       double em[])
{
  boolean stationary,goon,out;
  int i,q,m,m1,q1,j,k,k1,k2,k3,km1,l,mr,mi,ll;
  double dwarf,eps,epsa,epsb,anorm,bnorm,ani,bni,constt,a10,a20,a30,b11,b22,b33,
         b44,a11,a12,a21,a22,a33,a34,a43,a44,b12,b34,old1,old2,an,bn,e,c,d,er,ei,
         a11r,a11i,a12r,a12i,a21r,a21i,a22r,a22i,cz,szr,szi,cq,sqr,sqi,ssr,ssi,
         tr,ti,bdr,bdi,r,betm,alfm,sl,sk,tkk,tkl,tlk,tll,almi,almr,slr,sli,skr,
         ski,dr,di,tkkr,tkki,tklr,tkli,tlkr,tlki,tllr,tlli,s;
  double tmp1[] = new double[1];
  double tmp2[] = new double[1];
  double tmp3[] = new double[1];

  old1=old2=d=dr=di=0.0;
  dwarf=em[0];
  eps=em[1];
  hshdecmul(n,a,b,dwarf);
  hestgl3(n,a,b,x);
  anorm=bnorm=0.0;
  for (i=1; i<=n; i++) {
    bni=0.0;
    iter[i]=0;
    ani = (i > 1) ? Math.abs(a[i][i-1]) : 0.0;
    for (j=i; j<=n; j++) {
      ani += Math.abs(a[i][j]);
      bni += Math.abs(b[i][j]);
    }
    if (ani > anorm) anorm=ani;
    if (bni > bnorm) bnorm=bni;
  }
  if (anorm == 0.0) anorm=eps;
  if (bnorm == 0.0) bnorm=eps;
```

```
epsa=eps*anorm;
epsb=eps*bnorm;
m=n;
out=false;
do {
  i=q=m;
  while ((i > 1) ? Math.abs(a[i][i-1]) > epsa : false) {
    q=i-1;
    i--;
  }
  if (q > 1) a[q][q-1]=0.0;
  goon=true;
  while (goon) {
    if (q >= m-1) {
      m=q-1;
      goon=false;
    } else {
      if (Math.abs(b[q][q]) <= epsb) {
        b[q][q]=0.0;
        q1=q+1;
        hsh2col(q,q,n,q,a[q][q],a[q1][q],a,b);
        a[q1][q]=0.0;
        q=q1;
      } else {
        goon=false;
        m1=m-1;
        q1=q+1;
        constt=0.75;
        (iter[m])++;
        stationary = (iter[m] == 1) ? true :
                                   (Math.abs(a[m][m-1]) >= constt*old1 &&
                                    Math.abs(a[m-1][m-2]) >= constt*old2);
        if (iter[m] > 30 && stationary) {
          for (i=1; i<=m; i++) iter[i] = -1;
          out=true;
          break;
        }
        if (iter[m] == 10 && stationary) {
          a10=0.0;
          a20=1.0;
          a30=1.1605;
        } else {
          b11=b[q][q];
          b22 = (Math.abs(b[q1][q1]) < epsb) ? epsb : b[q1][q1];
          b33 = (Math.abs(b[m1][m1]) < epsb) ? epsb : b[m1][m1];
          b44 = (Math.abs(b[m][m]) < epsb) ? epsb : b[m][m];
          a11=a[q][q]/b11;
          a12=a[q][q1]/b22;
          a21=a[q1][q]/b11;
          a22=a[q1][q1]/b22;
```

```
            a33=a[m1][m1]/b33;
            a34=a[m1][m]/b44;
            a43=a[m][m1]/b33;
            a44=a[m][m]/b44;
            b12=b[q][q1]/b22;
            b34=b[m1][m]/b44;
            a10=((a33-a11)*(a44-a11)-a34*a43+a43*b34*a11)/a21+a12-a11*b12;
            a20=(a22-a11-a21*b12)-(a33-a11)-(a44-a11)+a43*b34;
            a30=a[q+2][q1]/b22;
          }
        old1=Math.abs(a[m][m-1]);
        old2=Math.abs(a[m-1][m-2]);
        for (k=q; k<=m1; k++) {
          k1=k+1;
          k2=k+2;
          k3 = (k+3 > m) ? m : k+3;
          km1 = (k-1 < q) ? q : k-1;
          if (k != m1) {
            if (k == q)
              hsh3col(km1,km1,n,k,a10,a20,a30,a,b);
            else {
              hsh3col(km1,km1,n,k,a[k][km1],a[k1][km1],a[k2][km1],a,b);
              a[k1][km1]=a[k2][km1]=0.0;
            }
            hsh3row3(1,k3,n,k,b[k2][k2],b[k2][k1],b[k2][k],a,b,x);
            b[k2][k]=b[k2][k1]=0.0;
          } else {
            hsh2col(km1,km1,n,k,a[k][km1],a[k1][km1],a,b);
            a[k1][km1]=0.0;
          }
          hsh2row3(1,k3,k3,n,k,b[k1][k1],b[k1][k],a,b,x);
          b[k1][k]=0.0;
        }
      }
    }
  } /* goon loop */
  if (out) break;
} while (m >= 3);

m=n;
do {
  if ((m > 1) ? (a[m][m-1] == 0) : true) {
    alfr[m]=a[m][m];
    beta[m]=b[m][m];
    alfi[m]=0.0;
    m--;
  } else {
    l=m-1;
    if (Math.abs(b[l][l]) <= epsb) {
      b[l][l]=0.0;
```

```
          hsh2col(1,1,n,1,a[l][l],a[m][l],a,b);
        a[m][l]=b[m][l]=0.0;
        alfr[l]=a[l][l];
        alfr[m]=a[m][m];
        beta[l]=b[l][l];
        beta[m]=b[m][m];
        alfi[m]=alfi[l]=0.0;
      } else
        if (Math.abs(b[m][m]) <= epsb) {
          b[m][m]=0.0;
          hsh2row3(1,m,m,n,1,a[m][m],a[m][l],a,b,x);
          a[m][l]=b[m][l]=0.0;
          alfr[l]=a[l][l];
          alfr[m]=a[m][m];
          beta[l]=b[l][l];
          beta[m]=b[m][m];
          alfi[m]=alfi[l]=0.0;
        } else {
          an=Math.abs(a[l][l])+Math.abs(a[l][m])+
             Math.abs(a[m][l])+Math.abs(a[m][m]);
          bn=Math.abs(b[l][l])+Math.abs(b[l][m])+Math.abs(b[m][m]);
          a11=a[l][l]/an;
          a12=a[l][m]/an;
          a21=a[m][l]/an;
          a22=a[m][m]/an;
          b11=b[l][l]/bn;
          b12=b[l][m]/bn;
          b22=b[m][m]/bn;
          e=a11/b11;
          c=((a22-e*b22)/b22-(a21*b12)/(b11*b22))/2.0;
          d=c*c+(a21*(a12-e*b12))/(b11*b22);
          if (d >= 0.0) {
            e += ((c < 0.0) ? c-Math.sqrt(d) : c+Math.sqrt(d));
            a11 -= e*b11;
            a12 -= e*b12;
            a22 -= e*b22;
            if (Math.abs(a11)+Math.abs(a12) >= Math.abs(a21)+Math.abs(a22))
              hsh2row3(1,m,m,n,1,a12,a11,a,b,x);
            else
              hsh2row3(1,m,m,n,1,a22,a21,a,b,x);
            if (an >= Math.abs(e)*bn)
              hsh2col(1,1,n,1,b[l][l],b[m][l],a,b);
            else
              hsh2col(1,1,n,1,a[l][l],a[m][l],a,b);
            a[m][l]=b[m][l]=0.0;
            alfr[l]=a[l][l];
            alfr[m]=a[m][m];
            beta[l]=b[l][l];
            beta[m]=b[m][m];
            alfi[m]=alfi[l]=0.0;
```

```
} else {
  er=e+c;
  ei=Math.sqrt(-d);
  a11r=a11-er*b11;
  a11i=ei*b11;
  a12r=a12-er*b12;
  a12i=ei*b12;
  a21r=a21;
  a21i=0.0;
  a22r=a22-er*b22;
  a22i=ei*b22;
  if (Math.abs(a11r)+Math.abs(a11i)+Math.abs(a12r)+
      Math.abs(a12i) >= Math.abs(a21r)+Math.abs(a22r)+Math.abs(a22i)) {
    Basic.chsh2(a12r,a12i,-a11r,-a11i,tmp1,tmp2,tmp3);
    cz=tmp1[0];
    szr=tmp2[0];
    szi=tmp3[0];
  }
  else {
    Basic.chsh2(a22r,a22i,-a21r,-a21i,tmp1,tmp2,tmp3);
    cz=tmp1[0];
    szr=tmp2[0];
    szi=tmp3[0];
  }
  if (an >= (Math.abs(er)+Math.abs(ei))*bn) {
    Basic.chsh2(cz*b11+szr*b12,szi*b12,szr*b22,szi*b22,tmp1,tmp2,tmp3);
    cq=tmp1[0];
    sqr=tmp2[0];
    sqi=tmp3[0];
  }
  else {
    Basic.chsh2(cz*a11+szr*a12,szi*a12,cz*a21+szr*a22,szi*a22,
                tmp1,tmp2,tmp3);
    cq=tmp1[0];
    sqr=tmp2[0];
    sqi=tmp3[0];
  }
  ssr=sqr*szr+sqi*szi;
  ssi=sqr*szi-sqi*szr;
  tr=cq*cz*a11+cq*szr*a12+sqr*cz*a21+ssr*a22;
  ti=cq*szi*a12-sqi*cz*a21+ssi*a22;
  bdr=cq*cz*b11+cq*szr*b12+ssr*b22;
  bdi=cq*szi*b12+ssi*b22;
  r=Math.sqrt(bdr*bdr+bdi*bdi);
  beta[l]=bn*r;
  alfr[l]=an*(tr*bdr+ti*bdi)/r;
  alfi[l]=an*(tr*bdi-ti*bdr)/r;
  tr=ssr*a11-sqr*cz*a12-cq*szr*a21+cq*cz*a22;
  ti = -ssi*a11-sqi*cz*a12+cq*szi*a21;
  bdr=ssr*b11-sqr*cz*b12+cq*cz*b22;
```

```
               bdi = -ssi*b11-sqi*cz*b12;
               r=Math.sqrt(bdr*bdr+bdi*bdi);
               beta[m]=bn*r;
               alfr[m]=an*(tr*bdr+ti*bdi)/r;
               alfi[m]=an*(tr*bdi-ti*bdr)/r;
            }
         }
      m -= 2;
   }
} while (m > 0);

for (m=n; m>=1; m--)
   if (alfi[m] == 0.0) {
      alfm=alfr[m];
      betm=beta[m];
      b[m][m]=1.0;
      ll=m;
      for (l=m-1; l>=1; l--) {
         sl=0.0;
         for (j=ll; j<=m; j++)
            sl += (betm*a[l][j]-alfm*b[l][j])*b[j][m];
         if ((l != 1) ? (betm*a[l][l-1] == 0.0) : true) {
            d=betm*a[l][l]-alfm*b[l][l];
            if (d == 0.0) d=(epsa+epsb)/2.0;
            b[l][m] = -sl/d;
         } else {
            k=l-1;
            sk=0.0;
            for (j=ll; j<=m; j++)
               sk += (betm*a[k][j]-alfm*b[k][j])*b[j][m];
            tkk=betm*a[k][k]-alfm*b[k][k];
            tkl=betm*a[k][l]-alfm*b[k][l];
            tlk=betm*a[l][k];
            tll=betm*a[l][l]-alfm*b[l][l];
            d=tkk*tll-tkl*tlk;
            if (d == 0.0) d=(epsa+epsb)/2.0;
            b[l][m]=(tlk*sk-tkk*sl)/d;
            b[k][m] = (Math.abs(tkk) >= Math.abs(tlk)) ?
                        -(sk+tkl*b[l][m])/tkk : -(sl+tll*b[l][m])/tlk;
            l--;
         }
         ll=l;
      }
   } else {
      almr=alfr[m-1];
      almi=alfi[m-1];
      betm=beta[m-1];
      mr=m-1;
      mi=m;
      b[m-1][mr]=almi*b[m][m]/(betm*a[m][m-1]);
```

```
b[m-1][mi]=(betm*a[m][m]-almr*b[m][m])/(betm*a[m][m-1]);
b[m][mr]=0.0;
b[m][mi] = -1.0;
ll=m-1;
for (l=m-2; l>=1; l--) {
  slr=sli=0.0;
  for (j=ll; j<=m; j++) {
    tr=betm*a[l][j]-almr*b[l][j];
    ti = -almi*b[l][j];
    slr += tr*b[j][mr]-ti*b[j][mi];
    sli += tr*b[j][mi]+ti*b[j][mr];
  }
  if ((l != 1) ? (betm*a[l][l-1] == 0.0) : true) {
    dr=betm*a[l][l]-almr*b[l][l];
    di = -almi*b[l][l];
    Basic.comdiv(-slr,-sli,dr,di,tmp1,tmp2);
    b[l][mr]=tmp1[0];
    b[l][mi]=tmp2[0];
  } else {
    k=l-1;
    skr=ski=0.0;
    for (j=ll; j<=m; j++) {
      tr=betm*a[k][j]-almr*b[k][j];
      ti = -almi*b[k][j];
      skr += tr*b[j][mr]-ti*b[j][mi];
      ski += tr*b[j][mi]+ti*b[j][mr];
    }
    tkkr=betm*a[k][k]-almr*b[k][k];
    tkki = -almi*b[k][k];
    tklr=betm*a[k][l]-almr*b[k][l];
    tkli = -almi*b[k][l];
    tlkr=betm*a[l][k];
    tlki=0.0;
    tllr=betm*a[l][l]-almr*b[l][l];
    tlli = -almi*b[l][l];
    dr=tkkr*tllr-tkki*tlli-tklr*tlkr;
    di=tkkr*tlli+tkki*tllr-tkli*tlkr;
    if (dr == 0.0 && di == 0.0) dr=(epsa+epsb)/2.0;
    Basic.comdiv(tlkr*skr-tkkr*slr+tkki*sli,
                 tlkr*ski-tkkr*sli-tkki*slr,dr,di,tmp1,tmp2);
    b[l][mr]=tmp1[0];
    b[l][mi]=tmp2[0];
    if (Math.abs(tkkr)+Math.abs(tkki) >= Math.abs(tlkr)) {
      Basic.comdiv(-skr-tklr*b[l][mr]+tkli*b[l][mi],
                   -ski-tklr*b[l][mi]-tkli*b[l][mr],tkkr,tkki,tmp1,tmp2);
      b[k][mr]=tmp1[0];
      b[k][mi]=tmp2[0];
    }
    else {
      Basic.comdiv(-slr-tllr*b[l][mr]+tlli*b[l][mi],
```

```
                                    -sli-tllr*b[l][mi]-tlli*b[l][mr],tlkr,tlki,tmp1,tmp2);
                  b[k][mr]=tmp1[0];
                  b[k][mi]=tmp2[0];
              }
            l--;
          }
          ll=l;
        }
      m--;
    }
  for (m=n; m>=1; m--)
    for (k=1; k<=n; k++) x[k][m]=Basic.matmat(1,m,k,m,x,b);
  for (m=n; m>=1; m--) {
    s=0.0;
    if (alfi[m] == 0.0) {
      for (k=1; k<=n; k++) {
        r=Math.abs(x[k][m]);
        if (r >= s) {
          s=r;
          d=x[k][m];
        }
      }
      for (k=1; k<=n; k++) x[k][m] /= d;
    } else {
      for (k=1; k<=n; k++) {
        r=Math.abs(x[k][m-1])+Math.abs(x[k][m]);
        an=x[k][m-1]/r;
        bn=x[k][m]/r;
        r *= Math.sqrt(an*an+bn*bn);
        if (r >= s) {
          s=r;
          dr=x[k][m-1];
          di=x[k][m];
        }
      }
      for (k=1; k<=n; k++) {
        Basic.comdiv(x[k][m-1],x[k][m],dr,di,tmp1,tmp2);
        x[k][m-1]=tmp1[0];
        x[k][m]=tmp2[0];
      }
      m--;
    }
  }
}
```

C. hshdecmul

Given an $n \times n$ matrix A, determines vectors $u^{(i)}$ ($i=1,...,n'-1$; $n' \leq n$) such that with

$$E = \prod_{i=1}^{n'-1} \left(I - 2u^{(i)^T} u^{(i)} / u^{(i)} u^{(i)^T} \right),$$

$A'=EA$ is of upper triangular form and, also given the $n \times n$ matrix B, forms $B'=EB$. *hshdecmul* is used in *qzi* and *qzival*.

Procedure parameters:
$$\text{void hshdecmul } (n,a,b,dwarf)$$

n: int;
          entry: the order of the given matrices;
a: double *a[1:n, 1:n]*;
          entry: the given matrix;
          exit: the transformed matrix (value of A' above);
b: double *b[1:n, 1:n]*;
          entry: the given matrix;
          exit: the upper triangular matrix (value of B' above);
dwarf: double;
          entry: the smallest positive machine number.

Procedures used: tammat, hshvecmat.

```
public static void hshdecmul(int n, double a[][], double b[][], double dwarf)
{
  int j,k,k1,n1;
  double r,t,c;
  double v[] = new double[n+1];

  k=1;
  n1=n+1;
  for (k1=2; k1<=n1; k1++) {
    r=Basic.tammat(k1,n,k,k,b,b);
    if (r > dwarf) {
      r = (b[k][k] < 0.0) ? -Math.sqrt(r+b[k][k]*b[k][k]) :
                            Math.sqrt(r+b[k][k]*b[k][k]);
      t=b[k][k]+r;
      c = -t/r;
      b[k][k] = -r;
      v[k]=1.0;
      for (j=k1; j<=n; j++) v[j]=b[j][k]/t;
      Basic.hshvecmat(k,n,k1,n,c,v,b);
      Basic.hshvecmat(k,n,1,n,c,v,a);
    }
    k=k1;
  }
}
```

D. hestgl3

Given an $n{\times}n$ matrix A and an $n{\times}n$ upper triangular matrix U, obtains vectors $u_1^{(i)}$, $u_2^{(i)}$ such that with

$$Q_1 = \prod_{i=1}^{n'-1} \left(I - 2u_1^{(i)}u_1^{(i)^T} / u_1^{(i)^T} u_1^{(i)}\right), \quad (n' \le n)$$

and Q_2 similarly defined, $Q_1AQ_2{=}H$ is an upper Hessenberg matrix and $Q_1UQ_2{=}U'$ is an upper triangular matrix and, also given an $n{\times}n$ matrix X, forms $X'{=}Q_1XQ_2$. *hestgl3* is used in *qzi*.

Procedure parameters:
$$\text{void hestgl3 } (n,a,b,x)$$

n: int;
        entry: the order of the given matrices;
a: double $a[1{:}n, 1{:}n]$;
        entry: the given matrix;
        exit: the upper Hessenberg matrix (value of H above);
b: double $b[1{:}n, 1{:}n]$;
        entry: the given upper triangular matrix (value of U above);
        exit: the upper triangular matrix (value of U' above);
x: double $x[1{:}n, 1{:}n]$;
        entry: the given matrix (value of X above);
        exit: the transformed matrix (value of X' above).

Procedures used: hsh2col, hsh2row3.

```
public static void hestgl3(int n, double a[][], double b[][], double x[][])
{
  int nm1,k,l,k1,l1;

  if (n > 2) {
    for (k=2; k<=n; k++)
      for (l=1; l<=k-1; l++) b[k][l]=0.0;
    nm1=n-1;
    k=1;
    for (k1=2; k1<=nm1; k1++) {
      l1=n;
      for (l=n-1; l>=k1; l--) {
        hsh2col(k,l,n,l,a[l][k],a[l1][k],a,b);
        a[l1][k]=0.0;
        hsh2row3(1,n,l1,n,l,b[l1][l1],b[l1][l],a,b,x);
        b[l1][l]=0.0;
        l1=l;
      }
      k=k1;
    }
  }
}
```

E. hestgl2

Given an $n{\times}n$ matrix A and an $n{\times}n$ upper triangular matrix U, obtains vectors $u_1^{(i)}$, $u_2^{(i)}$ such that with

$$Q_1 = \prod_{i=1}^{n'-1}\left(I - 2u_1^{(i)}u_1^{(i)^T} / u_1^{(i)^T} u_1^{(i)}\right), \quad (n' \le n)$$

and Q_2 similarly defined, $Q_1AQ_2=H$ is an upper Hessenberg matrix, and $Q_1UQ_2=U'$ is an upper triangular matrix. *hestgl2* is used in *qzival*.

Procedure parameters:
$$\text{void hestgl2 } (n,a,b)$$

n: int;
       entry: the order of the given matrices;
a: double a[1:n,1:n];
       entry: the given matrix;
       exit: the upper Hessenberg matrix (value of H above);
b: double b[1:n,1:n];
       entry: the given upper triangular matrix (value of U above);
       exit: the upper triangular matrix (value of U' above).

Procedures used: hsh2col, hsh2row2.

```
public static void hestgl2(int n, double a[][], double b[][])
{
  int nm1,k,l,k1,l1;

  if (n > 2) {
    for (k=2; k<=n; k++)
      for (l=1; l<=k-1; l++) b[k][l]=0.0;
    nm1=n-1;
    k=1;
    for (k1=2; k1<=nm1; k1++) {
      l1=n;
      for (l=n-1; l>=k1; l--) {
        hsh2col(k,l,n,l,a[l][k],a[l1][k],a,b);
        a[l1][k]=0.0;
        hsh2row2(1,n,l1,l,b[l1][l1],b[l1][l],a,b);
        b[l1][l]=0.0;
        l1=l;
      }
      k=k1;
    }
  }
}
```

F. hsh2col

(a) Given the values of two elements $M_{k,j}$ $(k=i,i+1)$ belonging to a certain column of a rectangular matrix M, determines a vector v such that all rows except the i-th and $(i+1)$-th of M and $M'=EM$ agree, where $E=I-2vv^T/v^Tv$ and $M_{i,j}=0$, and (b) given the elements $A_{k,j}$ $(k=i,i+1;\ j=la,...,u)$ of the rectangular matrix A, determines the corresponding elements of $A'=EA$, and (c) given the elements $B_{k,j}$ $(k=i,i+1;\ j=lb,...,u)$ of the rectangular matrix B, determines the corresponding elements of $B'=EB$. hsh2col is used in *qzival* and *qzi*.

Procedure parameters:
$$\text{void hsh2col } (la,lb,u,i,a1,a2,a,b)$$

la: int;
      entry: the lower bound of the running column subscript of *a* (value of *la* above);

lb: int;
      entry: the lower bound of the running column subscript of *b* (value of *lb* above);

u: int;
      entry: the upper bound of the running column subscript of *a* and *b* (value of *u* above);

i: int;
      entry: the lower bound of the running row subscript of *a* and *b* (value of *i* above);
                $i+1$ is the upper bound;

a1,a2: double;
      entry: *a1* and *a2* are the *i*-th and (*i*+1)-th component of the vector to be transformed, respectively (values of $M_{k,j}$ $(k=i,i+1)$ above);

a: double a[i:i+1,la:u];
      entry: the given matrix (value of *A* above);
      exit: the transformed matrix (value of *A'* above);

b: double b[i:i+1,lb:u];
      entry: the given matrix (value of *B* above);
      exit: the transformed matrix (value of *B'* above).

Procedure used: hshvecmat.

```
public static void hsh2col(int la, int lb, int u, int i, double a1,double a2,
                           double a[][], double b[][])
{
  double d1,d2,s1,s2,r,d,c;

  if (a2 != 0.0) {
    double v[] = new double[i+2];
    d1=Math.abs(a1);
    d2=Math.abs(a2);
    s1 = (a1 >= 0.0) ? 1.0 : -1.0;
    s2 = (a2 >= 0.0) ? 1.0 : -1.0;
    if (d2 <= d1) {
```

```
     r=d2/d1;
     d=Math.sqrt(1.0+r*r);
     c = -1.0-1.0/d;
     v[i+1]=s1*s2*r/(1.0+d);
   } else {
     r=d1/d2;
     d=Math.sqrt(1.0+r*r);
     c = -1.0-r/d;
     v[i+1]=s1*s2/(r+d);
   }
   v[i]=1.0;
   Basic.hshvecmat(i,i+1,la,u,c,v,a);
   Basic.hshvecmat(i,i+1,lb,u,c,v,b);
  }
}
```

G. hsh3col

(a) Given the values of three elements $M_{k,j}$ ($k=i,i+1,i+2$) belonging to a certain column of a rectangular matrix M, determines a vector v such that all rows except the i-th, ($i+1$)-th, and   ($i+2$)-th of M and $M'=EM$ agree, where
$$E = I - 2vv^T/v^Tv \text{ and } M_{i,j} = M_{i,j+1} = 0,$$
and (b) given the elements $A_{k,j}$ ($k=i,i+1,i+2; j=la,...,u$) of the rectangular matrix A, determines the corresponding elements of $A'=EA$, and (c) given the elements $B_{k,j}$ ($k=i,i+1,i+2; j=lb,...,u$) of the rectangular matrix B, determines the corresponding elements of $B'=EB$. hsh3col is used in *qzival* and *qzi*.

Procedure parameters:
$$\text{void hsh3col } (la,lb,u,i,a1,a2,a3,a,b)$$

la: int;
                 entry: the lower bound of the running column subscript of a
                         (value of *la* above);
lb: int;
                 entry: the lower bound of the running column subscript of b
                         (value of *lb* above);
u: int;
                 entry: the upper bound of the running column subscript of a and b
                         (value of *u* above);
i: int;
                 entry: the lower bound of the running row subscript of a and b
                         (value of *i* above);
                         $i+2$ is the upper bound;
a1,a2,a3: double;
                 entry: $a1$, $a2$, and $a3$ are the i-th, ($i+1$)-th, and ($i+2$)-th component of
                         the vector to be transformed, respectively (values of $M_{k,j}$
                         ($k=i,i+1,i+2$) above);
a: double $a[i:i+2,la:u]$;
                 entry: the given matrix (value of A above);

	exit:	the transformed matrix (value of A' above);
b:		double $b[i:i+2,lb:u]$;
	entry:	the given matrix (value of B above);
	exit:	the transformed matrix (value of B' above).

Procedure used: hshvecmat.

```
public static void hsh3col(int la, int lb, int u, int i, double a1, double a2,
                           double a3, double a[][], double b[][])
{
  double c,d1,d2,d3,s1,s2,s3,r1,r2,r3,d;

  if (a2 != 0.0 || a3 != 0.0) {
    double v[] = new double[i+3];
    d1=Math.abs(a1);
    d2=Math.abs(a2);
    d3=Math.abs(a3);
    s1 = (a1 >= 0.0) ? 1.0 : -1.0;
    s2 = (a2 >= 0.0) ? 1.0 : -1.0;
    s3 = (a3 >= 0.0) ? 1.0 : -1.0;
    if (d1 >= d2 && d1 >= d3) {
      r2=d2/d1;
      r3=d3/d1;
      d=Math.sqrt(1.0+r2*r2+r3*r3);
      c = -1.0-(1.0/d);
      d=1.0/(1.0+d);
      v[i+1]=s1*s2*r2*d;
      v[i+2]=s1*s3*r3*d;
    } else if (d2 >= d1 && d2 >= d3) {
      r1=d1/d2;
      r3=d3/d2;
      d=Math.sqrt(1.0+r1*r1+r3*r3);
      c = -1.0-(s1*r1/d);
      d=1.0/(r1+d);
      v[i+1]=s1*s2*d;
      v[i+2]=s1*s3*r3*d;
    } else {
      r1=d1/d3;
      r2=d2/d3;
      d=Math.sqrt(1.0+r1*r1+r2*r2);
      c = -1.0-(s1*r1/d);
      d=1.0/(r1+d);
      v[i+1]=s1*s2*r2*d;
      v[i+2]=s1*s3*d;
    }
    v[i]=1.0;
    Basic.hshvecmat(i,i+2,la,u,c,v,a);
    Basic.hshvecmat(i,i+2,lb,u,c,v,b);
  }
}
```

H. hsh2row3

(a) Given the values of two elements $M_{i,k}$ ($k=j,j+1$) belonging to a certain row of a rectangular matrix M, determines a vector v such that all columns except the j-th, ($j+1$)-th of M and $M'=ME$ agree, where $E=I-2vv^T/v^Tv$ and $M_{i,j}=0$, and (b) given the elements $A_{i,k}$ ($i=l,...,ua$; $k=j,j+1$) of the rectangular matrix A, determines the corresponding elements of $A'=AE$, and (c) given the elements $B_{i,k}$ ($i=l,...,ub$; $k=j,j+1$) of the rectangular matrix B, determines the corresponding elements of $B'=BE$, and (d) given the elements $X_{i,k}$ ($i=l,...,ux$; $k=j,j+1$) of the rectangular matrix X, determines the corresponding elements of $X'=XE$. hsh2row3 is used in qzi.

Procedure parameters:
$$\text{void hsh2row3 } (l,ua,ub,ux,j,a1,a2,a,b,x)$$

l: int;
    entry: the lower bound of the running row subscript of a, b, and x (value of l above);
ua: int;
    entry: the upper bound of the running row subscript of a (value of ua above);
ub: int;
    entry: the upper bound of the running row subscript of b (value of ub above);
ux: int;
    entry: the upper bound of the running row subscript of x (value of ux above);
j: int;
    entry: the lower bound of the running column subscript of a, b, and x (value of j above);
        $j+1$ is the upper bound;
$a1,a2$: double;
    entry: $a1$ and $a2$ are the j-th and ($j+1$)-th component of the vector to be transformed, respectively (values of $M_{i,k}$ ($k=j,j+1$) above);
a: double $a[l:ua,j:j+1]$;
    entry: the given matrix (value of A above);
    exit: the transformed matrix (value of A' above);
b: double $b[l:ub,j:j+1]$;
    entry: the given matrix (value of B above);
    exit: the transformed matrix (value of B' above);
x: double $x[l:ux,j:j+1]$;
    entry: the given matrix (value of X above);
    exit: the transformed matrix (value of X' above).

Procedure used: hshvectam.

```
public static void hsh2row3(int l, int ua, int ub, int ux, int j, double a1,
                    double a2, double a[][], double b[][], double x[][])
{
    double d1,d2,s1,s2,r,d,c;
```

```
if (a2 != 0.0) {
  double v[] = new double[j+2];
  d1=Math.abs(a1);
  d2=Math.abs(a2);
  s1 = (a1 >= 0.0) ? 1.0 : -1.0;
  s2 = (a2 >= 0.0) ? 1.0 : -1.0;
  if (d2 <= d1) {
    r=d2/d1;
    d=Math.sqrt(1.0+r*r);
    c = -1.0-1.0/d;
    v[j]=s1*s2*r/(1.0+d);
  } else {
    r=d1/d2;
    d=Math.sqrt(1.0+r*r);
    c = -1.0-r/d;
    v[j]=s1*s2/(r+d);
  }
  v[j+1]=1.0;
  Basic.hshvectam(l,ua,j,j+1,c,v,a);
  Basic.hshvectam(l,ub,j,j+1,c,v,b);
  Basic.hshvectam(l,ux,j,j+1,c,v,x);
  }
}
```

I. hsh2row2

(a) Given the values of two elements $M_{i,k}$ ($k=j,j+1$) belonging to a certain row of a rectangular matrix M, determines a vector v such that all columns except the j-th, ($j+1$)-th of M, and $M'=ME$ agree, where $E=I-2vv^T/v^Tv$ and $M_{i,j}=0$, and (b) given the elements $A_{i,k}$ ($i=l,...,ua$; $k=j,j+1$) of the rectangular matrix A, determines the corresponding elements of $A'=AE$, and (c) given the elements $B_{i,k}$ ($i=l,...,ub$; $k=j,j+1$) of the rectangular matrix B, determines the corresponding elements of $B'=BE$. *hsh2row2* is used in *qzival*.

Procedure parameters:

$$\text{void hsh2row2 } (l,ua,ub,j,a1,a2,a,b)$$

l: int;
       entry: the lower bound of the running row subscript of *a* and *b* (value of *l* above);

ua: int;
       entry: the upper bound of the running row subscript of *a* (value of *ua* above);

ub: int;
       entry: the upper bound of the running row subscript of *b* (value of *ub* above);

j: int;
            entry: the lower bound of the running column subscript of *a* and *b*
                    (value of *j* above);
                    j+1 is the upper bound;
a1,a2: double;
            entry: *a1* and *a2* are the *j*-th and (*j*+1)-th component of the vector to be
                    transformed, respectively (values of $M_{i,k}$ (k=*j,j*+1) above);
a: double *a[la:ua,j:j+1]*;
            entry: the given matrix (value of *A* above);
            exit: the transformed matrix (value of *A'* above);
b: double *b[lb:ub,j:j+1]*;
            entry: the given matrix (value of *B* above);
            exit: the transformed matrix (value of *B'* above).

Procedure used: hshvectam.

```
public static void hsh2row2(int l, int ua, int ub, int j, double a1, double a2,
                            double a[][], double b[][])
{
  double d1,d2,s1,s2,r,d,c;

  if (a2 != 0.0) {
    double v[] = new double[j+2];
    d1=Math.abs(a1);
    d2=Math.abs(a2);
    s1 = (a1 >= 0.0) ? 1.0 : -1.0;
    s2 = (a2 >= 0.0) ? 1.0 : -1.0;
    if (d2 <= d1) {
      r=d2/d1;
      d=Math.sqrt(1.0+r*r);
      c = -1.0-1.0/d;
      v[j]=s1*s2*r/(1.0+d);
    } else {
      r=d1/d2;
      d=Math.sqrt(1.0+r*r);
      c = -1.0-r/d;
      v[j]=s1*s2/(r+d);
    }
    v[j+1]=1.0;
    Basic.hshvectam(l,ua,j,j+1,c,v,a);
    Basic.hshvectam(l,ub,j,j+1,c,v,b);
  }
}
```

J. hsh3row3

(a) Given the values of three elements $M_{i,k}$ ($k=j,j+1,j+2$) belonging to a certain row of a rectangular matrix M, determines a vector v such that all columns except the j-th, ($j+1$)-th, and ($j+2$)-th of M and $M'=ME$ agree, where $E=I-2vv^T/v^Tv$ and $M_{i,j}=M_{i+1,j}=0$, and (b) given the elements $A_{i,k}$ ($i=l,...,u$; $k=j,j+1,j+2$) of the rectangular matrix A, determines the corresponding elements of $A'=AE$, and (c) given the elements $B_{i,k}$ ($i=l,...,u$; $k=j,j+1,j+2$) of the rectangular matrix B, determines the corresponding elements of $B'=BE$, and (d) given the elements $X_{i,k}$ ($i=l,...,ux$; $k=j,j+1,j+2$) of the rectangular matrix X, determines the corresponding elements of $X'=XE$. hsh3row3 is used in qzi.

Procedure parameters:
$$\text{void hsh3row3 } (l,u,ux,j,a1,a2,a3,a,b,x)$$

l:	int;
	entry: the lower bound of the running row subscript of a, b, and x (value of l above);
u:	int;
	entry: the upper bound of the running row subscript of a and b (value of u above);
ux:	int;
	entry: the upper bound of the running row subscript of x (value of ux above);
j:	int;
	entry: the lower bound of the running column subscript of a, b, and x (value of j above); $j+2$ is the upper bound;
$a1,a2,a3$:	double;
	entry: $a1$, $a2$, and $a3$ are the j-th, ($j+1$)-th and ($j+2$)-th component of the vector to be transformed, respectively (values of $M_{i,k}$ ($k=j,j+1,j+2$) above);
a:	double $a[l:u,j:j+2]$;
	entry: the given matrix (value of A above);
	exit: the transformed matrix (value of A' above);
b:	double $b[l:u,j:j+2]$;
	entry: the given matrix (value of B above);
	exit: the transformed matrix (value of B' above);
x:	double $x[l:ux,j:j+2]$;
	entry: the given matrix (value of X above);
	exit: the transformed matrix (value of X' above).

Procedure used: hshvectam.

```
public static void hsh3row3(int l, int u, int ux, int j, double a1, double a2,
                          double a3, double a[][], double b[][], double x[][])
{
  double c,d1,d2,d3,s1,s2,s3,r1,r2,r3,d;

  if (a2 != 0.0 || a3 != 0.0) {
    double v[] = new double[j+3];
```

```
d1=Math.abs(a1);
d2=Math.abs(a2);
d3=Math.abs(a3);
s1 = (a1 >= 0.0) ? 1.0 : -1.0;
s2 = (a2 >= 0.0) ? 1.0 : -1.0;
s3 = (a3 >= 0.0) ? 1.0 : -1.0;
if (d1 >= d2 && d1 >= d3) {
  r2=d2/d1;
  r3=d3/d1;
  d=Math.sqrt(1.0+r2*r2+r3*r3);
  c = -1.0-(1.0/d);
  d=1.0/(1.0+d);
  v[j+1]=s1*s2*r2*d;
  v[j]=s1*s3*r3*d;
} else if (d2 >= d1 && d2 >= d3) {
  r1=d1/d2;
  r3=d3/d2;
  d=Math.sqrt(1.0+r1*r1+r3*r3);
  c = -1.0-(s1*r1/d);
  d=1.0/(r1+d);
  v[j+1]=s1*s2*d;
  v[j]=s1*s3*r3*d;
} else {
  r1=d1/d3;
  r2=d2/d3;
  d=Math.sqrt(1.0+r1*r1+r2*r2);
  c = -1.0-(s1*r1/d);
  d=1.0/(r1+d);
  v[j+1]=s1*s2*r2*d;
  v[j]=s1*s3*d;
}
v[j+2]=1.0;
Basic.hshvectam(l,u,j,j+2,c,v,a);
Basic.hshvectam(l,u,j,j+2,c,v,b);
Basic.hshvectam(l,ux,j,j+2,c,v,x);
  }
}
```

K. hsh3row2

(a) Given the values of three elements $M_{i,k}$ ($k=j,j+1,j+2$) belonging to a certain row of a rectangular matrix M, determines a vector v such that all columns except the j-th, ($j+1$)-th, and ($j+2$)-th of M and $M'=ME$ agree, where $E=I-2vv^T/v^Tv$ and $M_{i,j}=M_{i+1,j}=0$, and (b) given the elements $A_{i,k}$ ($i=l,...,u$; $k=j,j+1,j+2$) of the rectangular matrix A, determines the corresponding elements of $A'=AE$, and (c) given the elements $B_{i,k}$ ($i=l,...,u$; $k=j,j+1,j+2$) of the rectangular matrix B, determines the corresponding elements of $B'=BE$. hsh3row2 is used in *qzival*.

Procedure parameters:

$$\text{void hsh3row2 } (l,u,j,a1,a2,a3,a,b)$$

l: int;

      entry: the lower bound of the running row subscript of *a* and *b* (value of *l* above);

u: int;

      entry: the upper bound of the running row subscript of *a* and *b* (value of *u* above);

j: int;

      entry: the lower bound of the running column subscript of *a* and *b* (value of *j* above); *j+2* is the upper bound;

a1,a2,a3: double;

      entry: *a1, a2,* and *a3* are the *j*-th, *(j+1)*-th and *(j+2)*-th component of the vector to be transformed, respectively (values of $M_{i,k}$ $(k=j,j+1,j+2)$ above);

a: double *a[l:u,j:j+2]*;

      entry: the given matrix (value of *A* above);

      exit: the transformed matrix (value of *A'* above);

b: double *b[l:u,j:j+2]*;

      entry: the given matrix (value of *B* above);

      exit: the transformed matrix (value of *B'* above).

Procedure used: hshvectam.

```
public static void hsh3row2(int l, int u, int j, double a1, double a2, double a3,
                            double a[][], double b[][])
{
  double c,d1,d2,d3,s1,s2,s3,r1,r2,r3,d;

  if (a2 != 0.0 || a3 != 0.0) {
    double v[] = new double[j+3];
    d1=Math.abs(a1);
    d2=Math.abs(a2);
    d3=Math.abs(a3);
    s1 = (a1 >= 0.0) ? 1.0 : -1.0;
    s2 = (a2 >= 0.0) ? 1.0 : -1.0;
    s3 = (a3 >= 0.0) ? 1.0 : -1.0;
    if (d1 >= d2 && d1 >= d3) {
      r2=d2/d1;
      r3=d3/d1;
      d=Math.sqrt(1.0+r2*r2+r3*r3);
      c = -1.0-(1.0/d);
      d=1.0/(1.0+d);
      v[j+1]=s1*s2*r2*d;
      v[j]=s1*s3*r3*d;
    } else if (d2 >= d1 && d2 >= d3) {
      r1=d1/d2;
      r3=d3/d2;
      d=Math.sqrt(1.0+r1*r1+r3*r3);
```

```
    c = -1.0-(s1*r1/d);
    d=1.0/(r1+d);
    v[j+1]=s1*s2*d;
    v[j]=s1*s3*r3*d;
  } else {
    r1=d1/d3;
    r2=d2/d3;
    d=Math.sqrt(1.0+r1*r1+r2*r2);
    c = -1.0-(s1*r1/d);
    d=1.0/(r1+d);
    v[j+1]=s1*s2*r2*d;
    v[j]=s1*s3*d;
  }
  v[j+2]=1.0;
  Basic.hshvectam(1,u,j,j+2,c,v,a);
  Basic.hshvectam(1,u,j,j+2,c,v,b);
 }
}
```

3.15 Singular values

3.15.1 Real bidiagonal matrices

A. qrisngvalbid

Computes, by use of a variant of the QR algorithm the singular values of a
bidiagonal $n{\times}n$ matrix A, i.e., the elements $d_1,...,d_n$ of the diagonal matrix D for
which $A=UDV^T$, where $U^TU=V^TV=I$ ($n{\times}n$ unit matrix).

Procedure parameters:

$$\text{int qrisngvalbid } (d,b,n,em)$$

qrisngvalbid: given the number of singular values not found, i.e., a number not
                  equal to zero if the number of iterations exceeds *em[4]*;
d: double *d[1:n]*;
                  entry: the diagonal of the bidiagonal matrix;
                  exit: the singular values;
b: double *b[1:n]*;
                  entry: the super diagonal of the bidiagonal matrix in *b[1:n-1]*;
n: int;
                  entry: the length of *b* and *d*;
em: double *em[1:7]*;
                  entry: *em[1]*: the infinity norm of the matrix;
                          em[2]: the relative precision in the singular values;
                          em[4]: the maximal number of iterations to be performed;
                          em[6]: the minimal nonneglectable singular value;

exit: *em[3]*:   the maximal neglected superdiagonal element;
           em[5]:   the number of iterations performed;
           em[7]:   the numerical rank of the matrix; i.e., the number of
                      singular values greater than or equal to *em[6]*.

Remark: The method is described in detail in [WiR71]. *qrisngvalbid* is a rewriting
                of part of the procedure SVD published there.

```
public static int qrisngvalbid(double d[], double b[], int n, double em[])
{
  int n1,k,k1,i,i1,count,max,rnk;
  double tol,bmax,z,x,y,g,h,f,c,s,min;

  tol=em[2]*em[1];
  count=0;
  bmax=0.0;
  max=(int) em[4];
  min=em[6];
  rnk=n;
  do {
    k=n;
    n1=n-1;
    while (true) {
      k--;
      if (k <= 0) break;
      if (Math.abs(b[k]) >= tol) {
        if (Math.abs(d[k]) < tol) {
          c=0.0;
          s=1.0;
          for (i=k; i<=n1; i++) {
            f=s*b[i];
            b[i] *= c;
            i1=i+1;
            if (Math.abs(f) < tol) break;
            g=d[i1];
            d[i1]=h=Math.sqrt(f*f+g*g);
            c=g/h;
            s = -f/h;
          }
          break;
        }
      } else {
        if (Math.abs(b[k]) > bmax) bmax=Math.abs(b[k]);
        break;
      }
    }
    if (k == n1) {
      if (d[n] < 0.0) d[n] = -d[n];
      if (d[n] <= min) rnk--;
      n=n1;
```

```
    } else {
      count++;
      if (count > max) break;
      k1=k+1;
      z=d[n];
      x=d[k1];
      y=d[n1];
      g = (n1 == 1) ? 0.0 : b[n1-1];
      h=b[n1];
      f=((y-z)*(y+z)+(g-h)*(g+h))/(2.0*h*y);
      g=Math.sqrt(f*f+1.0);
      f=((x-z)*(x+z)+h*(y/((f < 0.0) ? f-g : f+g)-h))/x;
      c=s=1.0;
      for (i=k1+1; i<=n; i++) {
        i1=i-1;
        g=b[i1];
        y=d[i];
        h=s*g;
        g *= c;
        z=Math.sqrt(f*f+h*h);
        c=f/z;
        s=h/z;
        if (i1 != k1) b[i1-1]=z;
        f=x*c+g*s;
        g=g*c-x*s;
        h=y*s;
        y *= c;
        d[i1]=z=Math.sqrt(f*f+h*h);
        c=f/z;
        s=h/z;
        f=c*g+s*y;
        x=c*y-s*g;
      }
      b[n1]=f;
      d[n]=x;
    }
  } while (n > 0);
  em[3]=bmax;
  em[5]=count;
  em[7]=rnk;
  return n;
}
```

B. qrisngvaldecbid

Computes by use of a variant of the QR algorithm the singular value decomposition of an $m \times n$ matrix A $(m \geq n)$, i.e., the $m \times n$ matrix U, the $n \times n$ diagonal

matrix D, and the $n{\times}n$ matrix V for which $A{=}UDV^T$, where $U^TU{=}V^TV{=}I$ ($n{\times}n$ unit matrix). It is assumed that A has been reduced to bidiagonal form by preliminary pre- and post-multiplication by matrices of the form
$$I - 2uu^T/u^Tu,$$
by use of *hshreabid*.

Procedure parameters:

int qrisngvaldecbid (d,b,m,n,u,v,em)

qrisngvaldecbid:	given the number of singular values not found, i.e., a number not equal to zero if the number of iterations exceeds *em[4]*;
d:	double $d[1{:}n]$;
	entry: the diagonal of the bidiagonal matrix;
	exit: the singular values;
b:	double $b[1{:}n]$;
	entry: the super diagonal of the bidiagonal matrix in $b[1{:}n{-}1]$;
m:	int;
	entry: the number of rows of the matrix u;
n:	int;
	entry: the length of b and d, the number of columns of u and the number of columns and rows of v;
u:	double $u[1{:}m,1{:}n]$;
	entry: the premultiplying matrix as produced by *pretfmmat*;
	exit: the premultiplying matrix U of the singular value decomposition UDV^T;
v:	double $v[1{:}n,1{:}n]$;
	entry: the transpose of the postmultiplying matrix as produced by *psttfmmat*;
	exit: the transpose of the postmultiplying matrix V of the singular value decomposition;
em:	double $em[1{:}7]$;
	entry: *em[1]*: the infinity norm of the matrix;
	em[2]: the relative precision in the singular values;
	em[4]: the maximal number of iterations to be performed;
	em[6]: the minimal nonneglectable singular value;
	exit: *em[3]*: the maximal neglected superdiagonal element;
	em[5]: the number of iterations performed;
	em[7]: the numerical rank of the matrix; i.e., the number of singular values greater than or equal to *em[6]*.

Procedure used: rotcol.

Remark: The method is described in detail in [WiR71]. *qrisngvaldecbid* is a rewriting of part of the procedure SVD published there.

```
public static int qrisngvaldecbid(double d[], double b[], int m, int n,
                       double u[][], double v[][], double em[])
{
  int n0,n1,k,k1,i,i1,count,max,rnk;
```

```
double tol,bmax,z,x,y,g,h,f,c,s,min;

tol=em[2]*em[1];
count=0;
bmax=0.0;
max=(int) em[4];
min=em[6];
rnk=n0=n;
do {
  k=n;
  n1=n-1;
  while (true) {
    k--;
    if (k <= 0) break;
    if (Math.abs(b[k]) >= tol) {
      if (Math.abs(d[k]) < tol) {
        c=0.0;
        s=1.0;
        for (i=k; i<=n1; i++) {
          f=s*b[i];
          b[i] *= c;
          i1=i+1;
          if (Math.abs(f) < tol) break;
          g=d[i1];
          d[i1]=h=Math.sqrt(f*f+g*g);
          c=g/h;
          s = -f/h;
          Basic.rotcol(1,m,k,i1,u,c,s);
        }
        break;
      }
    } else {
      if (Math.abs(b[k]) > bmax) bmax=Math.abs(b[k]);
      break;
    }
  }
  if (k == n1) {
    if (d[n] < 0.0) {
      d[n] = -d[n];
      for (i=1; i<=n0; i++) v[i][n] = -v[i][n];
    }
    if (d[n] <= min) rnk--;
    n=n1;
  } else {
    count++;
    if (count > max) break;
    k1=k+1;
    z=d[n];
    x=d[k1];
    y=d[n1];
```

```
    g = (n1 == 1) ? 0.0 : b[n1-1];
    h=b[n1];
    f=((y-z)*(y+z)+(g-h)*(g+h))/(2.0*h*y);
    g=Math.sqrt(f*f+1.0);
    f=((x-z)*(x+z)+h*(y/((f < 0.0) ? f-g : f+g)-h))/x;
    c=s=1.0;
    for (i=k1+1; i<=n; i++) {
       i1=i-1;
       g=b[i1];
       y=d[i];
       h=s*g;
       g *= c;
       z=Math.sqrt(f*f+h*h);
       c=f/z;
       s=h/z;
       if (i1 != k1) b[i1-1]=z;
       f=x*c+g*s;
       g=g*c-x*s;
       h=y*s;
       y *= c;
       Basic.rotcol(1,n0,i1,i,v,c,s);
       d[i1]=z=Math.sqrt(f*f+h*h);
       c=f/z;
       s=h/z;
       f=c*g+s*y;
       x=c*y-s*g;
       Basic.rotcol(1,m,i1,i,u,c,s);
    }
    b[n1]=f;
    d[n]=x;
  }
} while (n > 0);
em[3]=bmax;
em[5]=count;
em[7]=rnk;
return n;
}
```

3.15.2 Real full matrices

A. qrisngval

Computes the singular values of a given matrix. The matrix is first transformed to
bidiagonal form by calling *hshreabid*, and then the singular values are calculated
by *qrisngvalbid*.

Procedure parameters:

$$\text{int qrisngval } (a,m,n,val,em)$$

qrisngval: given the number of singular values not found, i.e., a number not equal to zero if the number of iterations exceeds *em[4]*;

a: double *a[1:m,1:n]*;
entry: the input matrix;
exit: data concerning the transformation to bidiagonal form;

m: int;
entry: the number of rows of the matrix *a*;

n: int;
entry: the number of columns of *a*, *n* should satisfy $n \le m$;

val: double *val[1:n]*;
exit: the singular values;

em: double *em[0:7]*;
entry: *em[0]*: the machine precision;
em[2]: the relative precision in the singular values;
em[4]: the maximal number of iterations to be performed;
em[6]: the minimal nonneglectable singular value;
exit: *em[1]*: the infinity norm of the matrix;
em[3]: the maximal neglected superdiagonal element;
em[5]: the number of iterations performed;
em[7]: the numerical rank of the matrix; i.e., the number of singular values greater than or equal to *em[6]*.

Procedures used: hshreabid, qrisngvalbid.

```
public static int qrisngval(double a[][], int m, int n, double val[], double em[])
{
  int i;
  double b[] = new double[n+1];

  hshreabid(a,m,n,val,b,em);
  i=qrisngvalbid(val,b,n,em);
  return i;
}
```

B. qrisngvaldec

Calculates the singular value decomposition $A=UDV^T$ of a given matrix *A*. The matrix is first transformed to bidiagonal form by calling *hshreabid*, the two transforming matrices are calculated by calling *psttfmmat* and *pretfmmat*, and finally the singular value decomposition is calculated by *qrisngvaldecbid*.

Procedure parameters:

$$\text{int qrisngvaldec } (a,m,n,val,v,em)$$

qrisngvaldec: given the number of singular values not found, i.e., a number not equal to zero if the number of iterations exceeds *em[4]*;

a:	double *a[1:m, 1:n]*;
	entry: the given matrix;
	exit: the matrix *U* in the singular value decomposition *UDV*ᵀ;
m:	int;
	entry: the number of rows of the matrix *a*;
n:	int;
	entry: the number of columns of *a, n* should satisfy *n≤m*;
val:	double *val[1:n]*;
	exit: the singular values;
v:	double *v[1:n, 1:n]*;
	exit: the transpose of matrix *V* in the singular value decomposition;
em:	double *em[0:7]*;
	entry: *em[0]*: the machine precision;
	em[2]: the relative precision in the singular values;
	em[4]: the maximal number of iterations to be performed;
	em[6]: the minimal nonneglectable singular value;
	exit: *em[1]*: the infinity norm of the matrix;
	em[3]: the maximal neglected superdiagonal element;
	em[5]: the number of iterations performed;
	em[7]: the numerical rank of the matrix; i.e., the number of singular values greater than or equal to *em[6]*.

Procedures used: hshreabid, psttfmmat, pretfmmat, qrisngvaldecbid.

```
public static int qrisngvaldec(double a[][], int m, int n, double val[],
                               double v[][], double em[])
{
  int i;
  double b[] = new double[n+1];

  hshreabid(a,m,n,val,b,em);
  psttfmmat(a,n,v,b);
  pretfmmat(a,m,n,val);
  i=qrisngvaldecbid(val,b,m,n,a,v,em);
  return i;
}
```

3.16 Zeros of polynomials

3.16.1 Zeros of general real polynomials

A. zerpol

Attempts to determine, by use of Laguerre's method [Dek66] and composite deflation [Ad67, PeW71, Rey77], the zeros of the polynomial

$$P(z) = \sum_{i=0}^{n} a_i z^i$$

where the a_i are real and $a_n \neq 0$.

For real zeros, Laguerre's method makes use of the recursion

$$z_{k+1} = z_k - \frac{nf_k}{f_k' \pm \left((n-1)^2 (f_k')^2 - n(n-1) f_k f_k''\right)}.$$

Again for real zeros, the deflation used may be described as follows: with α a derived approximation to a zero of $P(z)$, and r determined from the condition that $|a_i \alpha^i|$ takes its maximum value for $0 \leq i \leq n$ when $i=r$, p_i and q_i are obtained by use of the recursions

$$p_{n-1} = a_n, \quad p_{i-1} = a_i + \alpha p_i \qquad (i=n-1,...,r+1)$$
$$q_0 = -a_0/\alpha, \quad q_i = (a_i - q_{i-1})/\alpha \qquad (i=1,...,r-1)$$

(with obvious modifications if $r=0$ or $r=n$), the polynomial

$$\sum_{i=0}^{r-1} q_i z^i + \sum_{i=r}^{n-1} p_i z^i$$

is formed; its roots are those of $P(z)$ with α omitted. If all zeros of $P(z)$ are real, the above deflation may be continued.

Procedure parameters:
$$\text{int zerpol } (n,a,em,re,im,d)$$
zerpol: given the number, k, of zeros not found;
n: int;
          entry: the degree of the polynomial;
a: double $a[0:n]$;
          entry: the coefficients of the polynomial (values of a_i above);
em: double $em[0:4]$;
          entry: $em[0]$: the machine precision;
                  $em[1]$: the maximal number of iterations allowed for each zero (e.g., 40);
          exit: $em[2]$: fail indication;
                  0   successful call;
                  1   upon entry degree $n \leq 0$;
                  2   upon entry leading coefficient $a[n]=0$;
                  3   number of iterations exceeded $em[1]$;

> *em[3]*: number of new starts in the last iteration; if upon exit, *em[2]*=3 and *em[3]*<5 then it may be useful to start again with a higher value of *em[1]*;
>
> *em[4]*: total number of iterations performed;

re,im: double *re[1:n]*, *im[1:n]*;

> exit: the real and imaginary parts of the zeros of the polynomial; the members of each nonreal complex conjugate pair are consecutive;

d: double *d[0:n]*;

> exit: if the call is unsuccessful and only *n-k* zeros have been found, then *d[0:k]* contains the coefficients of the (deflated) polynomial; furthermore the zeros found are delivered in *re[k+1:n]*, *im[k+1:n]*, whereas the remaining parts of *re* and *im* contain no information.

Procedures used: comabs, comsqrt.

```
public static int zerpol(int n, double a[], double em[], double re[], double im[],
                    double d[])
{
  int i,totit,it,fail,start,up,max,giex,itmax,ih,m,split;
  double x,y,newf,oldf,maxrad,ae,tol,h1,h2,ln2,h,side,s1re,s1im,s2re,s2im,dx,dy,
        h3,h4,h5,h6;
  boolean btmp,control;
  int itmp[] = new int[1];
  double f[] = new double[6];
  double tries[] = new double[11];
  double tmp1[] = new double[1];
  double tmp2[] = new double[1];

  oldf=maxrad=0.0;
  totit=it=fail=up=start=max=0;
  ln2=Math.log(2.0);
  newf=Double.MAX_VALUE;
  ae=Double.MIN_VALUE;
  giex=(int) (Math.log(newf)/ln2-40.0);
  tol=em[0];
  itmax=(int) em[1];
  for (i=0; i<=n; i++) d[i]=a[n-i];
  if (n <= 0)
    fail=1;
  else
    if (d[0] == 0.0) fail=2;
  if (fail > 0) {
    em[2]=fail;
    em[3]=start;
    em[4]=totit;
    for (i=(n-1)/2; i>=0; i--) {
      tol=d[i];
      d[i]=d[n-i];
      d[n-i]=tol;
```

```
    }
    return n;
}
while (d[n] == 0.0 && n > 0) {
    re[n]=im[n]=0.0;
    n--;
}
x=y=0.0;
while (n > 2) {
    /* control */
    if (it > itmax) {
        totit += it;
        fail=3;
        em[2]=fail;
        em[3]=start;
        em[4]=totit;
        for (i=(n-1)/2; i>=0; i--) {
            tol=d[i];
            d[i]=d[n-i];
            d[n-i]=tol;
        }
        return n;
    } else
        if (it == 0) {
            maxrad=0.0;
            max=(int) ((giex-Math.log(Math.abs(d[0]))/ln2)/n);
            for (i=1; i<=n; i++) {
                h1 = (d[i] == 0.0) ? 0.0 : Math.exp(Math.log(Math.abs(d[i]/d[0]))/i);
                if (h1 > maxrad) maxrad=h1;
            }
            for (i=1; i<=n-1; i++)
                if (d[i] != 0.0) {
                    ih=(int) ((giex-Math.log(Math.abs(d[i]))/ln2)/(n-i));
                    if (ih < max) max=ih;
                }
            max=max*(int) (ln2/Math.log(n));
            side = -d[1]/d[0];
            side = (Math.abs(side) < tol) ? 0.0 : ((side > 0.0) ? 1.0 : -1.0);
            if (side == 0.0) {
                tries[7]=tries[2]=maxrad;
                tries[9] = -maxrad;
                tries[6]=tries[4]=tries[3]=maxrad/Math.sqrt(2.0);
                tries[5] = -tries[3];
                tries[10]=tries[8]=tries[1]=0.0;
            } else {
                tries[8]=tries[4]=maxrad/Math.sqrt(2.0);
                tries[1]=side*maxrad;
                tries[3]=tries[4]*side;
                tries[6]=maxrad;
                tries[7] = -tries[3];
```

```
        tries[9] = -tries[1];
        tries[2]=tries[5]=tries[10]=0.0;
      }
      if (Basic.comabs(x,y) > 2.0*maxrad) x=y=0.0;
      control=false;
    } else {
      if (it > 1 && newf >= oldf) {
        up++;
        if (up == 5 && start < 5) {
          start++;
          up=0;
          x=tries[2*start-1];
          y=tries[2*start];
          control=false;
        } else
          control=true;
      } else
        control=true;
    } /* end of control */
  if (control) {
    /* laguerre */
    if (Math.abs(f[0]) > Math.abs(f[1])) {
      h1=f[0];
      h6=f[1]/h1;
      h2=f[2]+h6*f[3];
      h3=f[3]-h6*f[2];
      h4=f[4]+h6*f[5];
      h5=f[5]-h6*f[4];
      h6=h6*f[1]+h1;
    } else {
      h1=f[1];
      h6=f[0]/h1;
      h2=h6*f[2]+f[3];
      h3=h6*f[3]-f[2];
      h4=h6*f[4]+f[5];
      h5=h6*f[5]-f[4];
      h6=h6*f[0]+f[1];
    }
    s1re=h2/h6;
    s1im=h3/h6;
    h2=s1re*s1re-s1im*s1im;
    h3=2.0*s1re*s1im;
    s2re=h2-h4/h6;
    s2im=h3-h5/h6;
    h1=s2re*s2re+s2im*s2im;
    h1 = (h1 != 0.0) ? (s2re*h2+s2im*h3)/h1 : 1.0;
    m = (h1 > n-1) ? ((n > 1) ? n-1 : 1) : ((h1 > 1.0) ? (int)h1 : 1);
    h1=(double)(n-m)/(double) m;
    Basic.comsqrt(h1*(n*s2re-h2),h1*(n*s2im-h3),tmp1,tmp2);
    h2=tmp1[0];
```

```
        h3=tmp2[0];
        if (s1re*h2+s1im*h3 < 0.0) {
          h2 = -h2;
          h3 = -h3;
        }
        h2 += s1re;
        h3 += s1im;
        h1=h2*h2+h3*h3;
        if (h1 == 0.0) {
          dx = -n;
          dy=n;
        } else {
          dx = -n*h2/h1;
          dy=n*h3/h1;
        }
        h1=Math.abs(x)*tol+ae;
        h2=Math.abs(y)*tol+ae;
        if (Math.abs(dx) < h1 && Math.abs(dy) < h2) {
          dx = (dx == 0.0) ? h1 : ((dx > 0.0) ? h1 : -h1);
          dy = (dy == 0.0) ? h2 : ((dy > 0.0) ? h2 : -h2);
        }
        x += dx;
        y += dy;
        if (Basic.comabs(x,y) > 2.0*maxrad) {
          h1 = (Math.abs(x) > Math.abs(y)) ? Math.abs(x) : Math.abs(y);
          h2=Math.log(h1)/ln2+1.0-max;
          if (h2 > 0.0) {
            h2=Math.pow(2.0,h2);
            x /= h2;
            y /= h2;
          }
        } /* end of laguerre */
      }
      oldf=newf;
      itmp[0]=it;
      tmp1[0]=newf;
      btmp=zerpolfunction(n,d,f,x,y,tol,itmp,tmp1);
      it=itmp[0];
      newf=tmp1[0];
      if (btmp) {
        if (y != 0.0 && Math.abs(y) < 0.1) {
          h=y;
          y=0.0;
          itmp[0]=it;
          tmp1[0]=newf;
          btmp=zerpolfunction(n,d,f,x,y,tol,itmp,tmp1);
          it=itmp[0];
          newf=tmp1[0];
          if (!btmp) y=h;
        }
```

```
       re[n]=x;
       im[n]=y;
       if (y != 0.0) {
         re[n-1]=x;
         im[n-1] = -y;
       }
       /* deflation */
       if (x == 0.0 && y == 0.0)
         n--;
       else {
         double b[] = new double[n];
         if (y == 0.0) {
           n--;
           b[n] = -d[n+1]/x;
           for (i=1; i<=n; i++) b[n-i]=(b[n-i+1]-d[n-i+1])/x;
           for (i=1; i<=n; i++) d[i] += d[i-1]*x;
         } else {
           h1 = -2.0*x;
           h2=x*x+y*y;
           n -= 2;
           b[n]=d[n+2]/h2;
           b[n-1]=(d[n+1]-h1*b[n])/h2;
           for (i=2; i<=n; i++)
             b[n-i]=(d[n-i+2]-h1*b[n-i+1]-b[n-i+2])/h2;
           d[1] -= h1*d[0];
           for (i=2; i<=n; i++) d[i] -= h1*d[i-1]+h2*d[i-2];
         }
         split=n;
         h2=Math.abs(d[n]-b[n])/(Math.abs(d[n])+Math.abs(b[n]));
         for (i=n-1; i>=0; i--) {
           h1=Math.abs(d[i])+Math.abs(b[i]);
           if (h1 > tol) {
             h1=Math.abs(d[i]-b[i])/h1;
             if (h1 < h2) {
               h2=h1;
               split=i;
             }
           }
         }
         for (i=split+1; i<=n; i++) d[i]=b[i];
         d[split]=(d[split]+b[split])/2.0;
       } /* end of deflation */
       totit += it;
       up=start=it=0;
     }
   }
   if (n == 1) {
     re[1] = -d[1]/d[0];
     im[1]=0.0;
   } else {
```

```
    h1 = -0.5*d[1]/d[0];
    h2=h1*h1-d[2]/d[0];
    if (h2 >= 0.0) {
      re[2] = (h1 < 0.0) ? h1-Math.sqrt(h2) : h1+Math.sqrt(h2);
      re[1]=d[2]/(d[0]*re[2]);
      im[2]=im[1]=0.0;
    } else {
      re[2]=re[1]=h1;
      im[2]=Math.sqrt(-h2);
      im[1] = -im[2];
    }
  }
  em[2]=fail;
  em[3]=start;
  em[4]=totit;
  return 0;
}


static private boolean zerpolfunction(int n, double d[], double f[], double x,
                                double y, double tol, int it[], double newf[])
{
  /* this procedure is used internally by ZERPOL */

  int k,m1,m2;
  double p,q,qsqrt,f01,f02,f03,f11,f12,f13,f21,f22,f23,stop;

  (it[0])++;
  p=2.0*x;
  q = -(x*x+y*y);
  qsqrt=Math.sqrt(-q);
  f01=f11=f21=d[0];
  f02=f12=f22=0.0;
  m1=n-4;
  m2=n-2;
  stop=Math.abs(f01)*0.8;
  for (k=1; k<=m1; k++) {
    f03=f02;
    f02=f01;
    f01=d[k]+p*f02+q*f03;
    f13=f12;
    f12=f11;
    f11=f01+p*f12+q*f13;
    f23=f22;
    f22=f21;
    f21=f11+p*f22+q*f23;
    stop=qsqrt*stop+Math.abs(f01);
  }
  if (m1 < 0) m1=0;
  for (k=m1+1; k<=m2; k++) {
```

```
      f03=f02;
      f02=f01;
      f01=d[k]+p*f02+q*f03;
      f13=f12;
      f12=f11;
      f11=f01+p*f12+q*f13;
      stop=qsqrt*stop+Math.abs(f01);
    }
    if (n == 3) f21=0.0;
    f03=f02;
    f02=f01;
    f01=d[n-1]+p*f02+q*f03;
    f[0]=d[n]+x*f01+q*f02;
    f[1]=y*f01;
    f[2]=f01-2.0*f12*y*y;
    f[3]=2.0*y*(-x*f12+f11);
    f[4]=2.0*(-x*f12+f11)-8.0*y*y*(-x*f22+f21);
    f[5]=y*(6.0*f12-8.0*y*y*f22);
    stop=qsqrt*(qsqrt*stop+Math.abs(f01))+Math.abs(f[0]);
    newf[0]=f02=Basic.comabs(f[0],f[1]);
    return (f02 < (2.0*Math.abs(x*f01)-8.0*(Math.abs(f[0])+Math.abs(f01)*qsqrt)+
                   10.0*stop)*tol*Math.pow(1.0+tol,4*n+3.0)));
  }
```

B. bounds

Calculates upper bounds for the absolute error in given approximated zeros of a polynomial with real coefficients.

Given approximations α_i (i=1,...,n) to the zeros of a polynomial

$$P(z) = \sum_{i=0}^{n} a_i z^i$$

with real coefficients, determines centers γ_l and radii r_l of disjoint discs D_l together with positive integers m_l such that m_l zeros of $P(z)$ are contained in D_l (l=1,...,n'; $m_1+...+m_{n'}=n$).

The results upon which the method used is based may be described as follows. Let

$$R(z) = a_n \prod_{i=1}^{n}(z - \alpha_i) \quad \text{and} \quad Q(z) = R(z) - P(z) = \sum_{k=1}^{n-1} \varepsilon_k z^k .$$

Let m of the α_i, namely $\alpha_{k(i)}$ (i=1,...,m) lie near to each other; let

$$\prod_{i=1}^{n} {}^{(k)} \quad \text{denote a product} \quad \prod_{i=1}^{n}$$

from which the terms with suffix $k(i)$ (i=1,...,m) have been removed; set

$$\gamma = \left(\sum_{i=1}^{m} \alpha_{k(i)} \right) / m$$

and $ß_i = \alpha_i - \gamma$ $(i=1,...,n)$. For z on the circle C with center γ and suitable radius r,

$$|R(z)| \geq |a_n| \prod_{i=1}^{m}\left(r - |\beta_{k(i)}|\right) \prod_{i=1}^{n}{}^{(k)}\left(|\beta_i| - r\right)$$

$$|Q(z)| \leq \sum_{k=0}^{n-1}|\varepsilon_k|\left(|\gamma| + r\right)^k .$$

If $|Q(z)| < |R(z)|$ for all z on C, then $R(z)$ and $R(z)-Q(z)=P(z)$ have the same number of zeros in C, i.e., if

$$\frac{\sum_{k=0}^{m-1}|\varepsilon_k|\left(|\gamma| + r\right)^k}{|a_n| \prod_{i=1}^{n}{}^{(k)}\left(|\beta_i| - r\right)} < \prod_{i=1}^{m}\left(r - |\beta_{k(i)}|\right)$$

C contains m zeros of $P(z)$. A suitable value of r may be estimated by iteration: r_1 is determined from the condition

$$\left(r_1 - d\right)^m = \frac{\sum_{k=0}^{n-1}\varepsilon_k|\gamma|^k}{|a_n| \prod_{i=1}^{n}{}^{(k)}|\beta_i|}$$

where $d = \max|ß_{k(i)}|$ $(1 \leq i \leq m)$, and r from the condition that $(r-d)^m = 1.1(r_1-d)^m$. For a more detailed description see [PeW71, Rey77].

Procedure parameters:

  void bounds $(n,a,re,im,rele,abse,recentre,imcentre,bound)$

n: int;
             entry: the degree of the polynomial;
a: double $a[0:n]$;
             entry: the coefficients of the polynomial of which $re[j]+im[j]*i$ are the approximated zeros (values of a_i above);
re,im: double $re[1:n]$, $im[1:n]$;
             entry: real and imaginary parts of approximated zeros of a polynomial such that the members of each nonreal complex conjugate pair are consecutive (values of α_i above);
             exit: a permutation of the input data;
rele: double;
             entry: relative error in the nonvanishing coefficients $a[j]$ of the given polynomial;
abse: double;
             entry: absolute error in the vanishing coefficients $a[j]$ of the given polynomial; if there are no vanishing coefficients, abse should be zero;
recentre,imcentre: double $recentre[1:n]$, $imcentre[1:n]$;
             exit: real and imaginary parts of the centers of disks in which some number of zeros of the polynomial given by a are situated; the number of identical centers denotes

the number of zeros in that disk;

bound: double *bound[1:n]*;

exit: radius of the disks whose centers are given correspondingly in *recentre* and *imcentre*.

```
public static void bounds(int n, double a[], double re[], double im[], double rele,
                    double abse, double recentre[], double imcentre[],
                    double bound[])
{
  boolean goon;
  int i,j,k,index1,index2,place,clustin;
  double h,min,recent,imcent,xk,yk,zk,corr,boundin,temp1,temp2;
  double rc[] = new double[n+1];
  double c[] = new double[n+1];
  double rce[] = new double[n+1];
  double clust[] = new double[n+1];

  rc[0]=c[0]=a[n];
  rce[0]=Math.abs(c[0]);
  k=0;
  for (i=1; i<=n; i++) {
    rc[i]=rce[i]=0.0;
    c[i]=a[n-i];
  }
  while (k < n) {
    k++;
    xk=re[k];
    yk=im[k];
    zk=xk*xk+yk*yk;
    for (j=k; j>=1; j--) rce[j] += rce[j-1]*Math.sqrt(zk);
    if (yk == 0.0)
      for (j=k; j>=1; j--) rc[j] -= xk*rc[j-1];
    else {
      k++;
      if (k <= n && xk == re[k] && yk == -im[k]) {
        xk = -2.0*xk;
        for (j=k; j>=1; j--) rce[j] += rce[j-1]*Math.sqrt(zk);
        for (j=k; j>=2; j--) rc[j] += xk*rc[j-1]+zk*rc[j-2];
        rc[1] += xk*rc[0];
      }
    }
  }
  rc[0]=rce[0];
  corr=1.06*Double.MIN_VALUE;
  for (i=1; i<=n-1; i++)
    rc[i]=Math.abs(rc[i]-c[i])+rce[i]*corr*(n+i-2)+rele*Math.abs(c[i])+abse;
  rc[n]=Math.abs(rc[n]-c[n])+rce[n]*corr*(n-1)+rele*Math.abs(c[n])+abse;
  for (i=1; i<=n; i++)
    kcluster(1,i,n,rc,re,im,recentre,imcentre,bound,clust);
  goon=true;
```

```
while (goon) {
  index1=index2=0;
  min=Double.MAX_VALUE;
  i=n-(int)(clust[n])+1;
  while (i >= 2) {
    j=i;
    recent=recentre[i];
    imcent=imcentre[i];
    while (j >= 2) {
      j -= clust[j-1];
      temp1=recent-recentre[j];
      temp2=imcent-imcentre[j];
      h=Math.sqrt(temp1*temp1+temp2*temp2);
      if (h < bound[i]+bound[j] && h <= min) {
        index1=j;
        index2=i;
        min=h;
      }
    }
    i -= clust[i-1];
  }
  if (index1 == 0)
    goon=false;
  else {
    if (imcentre[index1] == 0.0) {
      if (imcentre[index2] != 0.0) clust[index2] *= 2.0;
    }
    else
      if (imcentre[index2] == 0.0) clust[index1] *= 2.0;
    k=index1+(int)(clust[index1]);
    if (k != index2) {
      /*  shift */
      double wa1[] = new double[(int)(clust[index2])+1];
      double wa2[] = new double[(int)(clust[index2])+1];
      clustin=(int) clust[index2];
      boundin=bound[index2];
      imcent=imcentre[index2];
      recent=recentre[index2];
      for (j=1; j<=clustin; j++) {
        place=index2+j-1;
        wa1[j]=re[place];
        wa2[j]=im[place];
      }
      for (j=index2-1; j>=k; j--) {
        place=j+clustin;
        re[place]=re[j];
        im[place]=im[j];
        clust[place]=clust[j];
        bound[place]=bound[j];
        recentre[place]=recentre[j];
```

```
               imcentre[place]=imcentre[j];
            }
          for (j=k+clustin-1; j>=k; j--) {
            place=j+1-k;
            re[j]=wa1[place];
            im[j]=wa2[place];
            bound[j]=boundin;
            clust[j]=clustin;
            recentre[j]=recent;
            imcentre[j]=imcent;
          }
       } /* end of shift */
       k=(int) (clust[index1]+clust[k]);
       kcluster(k,index1,n,rc,re,im,recentre,imcentre,bound,clust);
    }
  }
}


static private void kcluster(int k, int m, int n, double rc[], double re[],
                             double im[], double recentre[], double imcentre[],
                             double bound[], double clust[])
{
  /* this procedure is used internally by BOUNDS */

  boolean nonzero;
  int i,stop,l;
  double recent,imcent,d,prod,rad,gr,r,s,h1,h2,temp1,temp2;
  double dist[] = new double[m+k];

  recent=re[m];
  imcent=im[m];
  stop=m+k-1;
  l = (imcent == 0.0) ? 0 : ((imcent > 0.0) ? 1 : -1);
  nonzero = (l != 0);
  for (i=m+1; i<=stop; i++) {
    recent += re[i];
    if (nonzero) {
      nonzero=(l == ((im[i] == 0.0) ? 0 : ((im[i]>0.0) ? 1 : -1)));
      imcent += im[i];
    }
  }
  recent /= k;
  imcent = (nonzero ? imcent/k : 0.0);
  d=0.0;
  rad=0.0;
  for (i=m; i<=stop; i++) {
    recentre[i]=recent;
    imcentre[i]=imcent;
    temp1=re[i]-recent;
```

```
    temp2=im[i]-imcent;
    dist[i]=Math.sqrt(temp1*temp1+temp2*temp2);
    if (d < dist[i]) d=dist[i];
  }
  s=Math.sqrt(recent*recent+imcent*imcent);
  h1=rc[1];
  h2=rc[0];
  for (i=2; i<=n; i++) h1=h1*s+rc[i];
  for (i=1; i<=m-1; i++) {
    temp1=re[i]-recent;
    temp2=im[i]-imcent;
    h2 *= Math.abs(Math.sqrt(temp1*temp1+temp2*temp2));
  }
  for (i=m+k; i<=n; i++) {
    temp1=re[i]-recent;
    temp2=im[i]-imcent;
    h2 *= Math.abs(Math.sqrt(temp1*temp1+temp2*temp2));
  }
  gr=Math.abs((h1 == 0.0) ? 0.0 : ((h2 == 0.0) ? 10.0 : h1/h2));
  if (gr > 0.0)
    do {
      r=rad;
      rad=d+Math.exp(Math.log(1.1*gr)/k);
      if (rad == r) rad *= Math.exp(Math.log(1.1)/k);
      s=Math.sqrt(recent*recent+imcent*imcent)+rad;
      h1=rc[1];
      h2=rc[0];
      for (i=2; i<=n; i++) h1=h1*s+rc[i];
      for (i=1; i<=m-1; i++) {
        temp1=re[i]-recent;
        temp2=im[i]-imcent;
        h2 *= Math.abs(Math.sqrt(temp1*temp1+temp2*temp2)-rad);
      }
      for (i=m+k; i<=n; i++) {
        temp1=re[i]-recent;
        temp2=im[i]-imcent;
        h2 *= Math.abs(Math.sqrt(temp1*temp1+temp2*temp2)-rad);
      }
      gr=(h1 == 0.0) ? 0.0 : ((h2 == 0.0) ? -10.0 : h1/h2);
      prod=1.0;
      for (i=m; i<=stop; i++) prod *= (rad-dist[i]);
    } while (prod <= gr);
  for (i=m; i<=stop; i++) {
    bound[i]=rad;
    clust[i]=k;
  }
}
```

3.16.2 Zeros of orthogonal polynomials

A. allzerortpol

Calculates all zeros of an orthogonal polynomial given by the coefficients of their recurrence relation. *allzerortpol* determines the roots of the polynomial $p_n(x)$, where
$$p_0(x) = 1, \quad p_1(x) = x - b_0$$
$$p_{k+1}(x) = (x-b_k)p_k(x) - c_k p_{k-1}(x) \qquad (k=1,...,n-1)$$
the b_k and c_k are being given ($c_k>0$, $k=1,...,n-1$).

The roots of $p_n(x)$ are the eigenvalues of the $n \times n$ tridiagonal matrix T for which
$$T_{i,i}=b_{i-1} \ (i=1,...,n), \quad T_{i,i+1}=T_{i+1,i}=c_i^{1/2} \ (i=1,...,n-1).$$
These eigenvalues are obtained by QR iteration, using a call of *qrivalsymtri*.

Procedure parameters:
$$\text{void allzerortpol } (n,b,c,zer,em)$$

n: int;
  entry: the degree of the orthogonal polynomial of which the zeros are to be calculated;

b,c: double $b[0:n-1]$, $c[0:n-1]$;
  entry: the elements $b[i]$ and $c[i]$, $i=0,1,...,n-1$, contain the coefficients of the recurrence relation $p_{i+1}(x)=(x-b[i])*p_i(x)-c[i]*p_{i-1}(x)$, $i=0,1,...,n-1$, assuming $c[0]=0$;
  exit: the contents of the arrays b and c are not altered;

zer: double $zer[1:n]$;
  exit: the zeros of the n-th degree orthogonal polynomial;

em: double $em[0:5]$;
  entry: $em[0]$: the machine precision;
    $em[2]$: the relative tolerance of the zeros;
    $em[4]$: the maximal allowed number of iterations (e.g., $5*n$);
  exit: $em[1]$: the value of
    $$\max(|b[0]|+1, c[i]+|b[i]|+1 \ (i=1,...,n-2), c[n-1]+|b[n-1]|);$$
    $em[3]$: the maximum absolute value of the codiagonal elements neglected;
    $em[5]$: the number of iterations performed.

Procedures used: qrivalsymtri, dupvec.

Remark: See the procedure *selzerortpol*.

```
public static void allzerortpol(int n, double b[], double c[], double zer[],
                                double em[])
{
  int i;
  double nrm;
  double bb[] = new double[n+1];
```

```
nrm=Math.abs(b[0]);
for (i=1; i<=n-2; i++)
  if (c[i]+Math.abs(b[i]) > nrm) nrm=c[i]+Math.abs(b[i]);
if (n > 1)
  nrm = (nrm+1 >= c[n-1]+Math.abs(b[n-1])) ? nrm+1.0 : (c[n-1]+Math.abs(b[n-1]));
em[1]=nrm;
for (i=n; i>=1; i--) zer[i]=b[i-1];
Basic.dupvec(1,n-1,0,bb,c);
qrivalsymtri(zer,bb,n,em);
}
```

B. lupzerortpol

Calculates a number of adjacent upper or lower zeros of an orthogonal polynomial given by the coefficients of their recurrence relation. *lupzerortpol* determines either the m smaller roots or the m larger roots of the polynomial $p_n(x)$, where

$$p_0(x) = 1, \quad p_1(x) = x - b_0$$
$$p_{k+1}(x) = (x-b_k)p_k(x) - c_k p_{k-1}(x) \qquad (k=1,...,n-1)$$

the b_k and c_k are being given ($c_k>0$, $k=1,...,n-1$).

The roots of $p_n(x)$ are the eigenvalues of the $n \times n$ tridiagonal matrix T for which

$$T_{i,i}=b_{i-1} \ (i=1,...,n), \quad T_{i,i+1}=T_{i+1,i}=c_i^{1/2} \ (i=1,...,n-1).$$

The m smaller eigenvalues of this matrix are determined by use of a rational variant of the QR algorithm. The m larger roots of $p_n(x)$ may be obtained by supplying $-b_k$ in place of b_k in the above ($k=0,...,n-1$) so that the roots of $p_n(-x)$ are found, and reversing the signs attached to the values of the roots determined in this way.

Procedure parameters:

$$\text{void lupzerortpol } (n,m,b,c,zer,em)$$

n: int;
      entry: the degree of the orthogonal polynomial of which the zeros are to be calculated;

m: int;
      entry: the number of zeros to be calculated;

b,c: double $b[0:n-1]$, $c[0:n-1]$;
      entry: the elements $b[i]$ and $c[i]$, $i=0,1,...,n-1$, contain the coefficients of the recurrence relation $p_{i+1}(x)=(x-b[i])^*p_i(x)-c[i]^*p_{i-1}(x)$, $i=0,1,...,n-1$, assuming $c[0]=0$;
      exit: the contents of the arrays b and c are altered;

zer: double $zer[1:m]$;
      exit: the m lowest zeros are delivered; if the array $b[0:n-1]$ contained the opposite values of the corresponding recurrence coefficients then the opposite values of the m upper zeros are delivered; in either case $zer[i] < zer[i+1]$, $i=1,...,m-1$;

em: double $em[0:6]$;
      entry: $em[0]$: the machine precision;

em[2]: the relative tolerance of the zeros;

em[4]: the maximal allowed number of iterations (e.g., $15*m$);

em[6]: if all zeros are known to be positive then 1 else 0;

exit: em[1]: the value of
$$\max(|b[0]|+1, c[i]+|b[i]|+1 \ (i=1,...,n-2), c[n-1]+|b[n-1]|);$$

em[3]: the maximum absolute value of the theoretical errors of the zeros;

em[5]: the number of iterations performed.

Procedures used: dupvec, infnrmvec.

Remark: See the procedure *selzerortpol*.

```
public static void lupzerortpol(int n, int m, double b[], double c[], double zer[],
                                double em[])
{
  boolean posdef,converge;
  int i,j,k,t;
  double nrm,dlam,eps,delta,e,ep,err,p,q,qp,r,s,tot;
  int itmp[] = new int[1];

  qp=0.0;
  nrm=Math.abs(b[0]);
  for (i=1; i<=n-2; i++)
    if (c[i]+Math.abs(b[i]) > nrm) nrm=c[i]+Math.abs(b[i]);
  if (n > 1)
    nrm = (nrm+1 >= c[n-1]+Math.abs(b[n-1])) ? nrm+1.0 : (c[n-1]+Math.abs(b[n-1]));
  em[1]=nrm;
  for (i=n; i>=1; i--) b[i]=b[i-1];
  for (i=n; i>=2; i--) c[i]=c[i-1];
  posdef = (em[6] == 1.0);
  dlam=em[2];
  eps=em[0];
  c[1]=err=q=s=0.0;
  tot=b[1];
  for (i=n; i>=1; i--) {
    p=q;
    q=Math.sqrt(c[i]);
    e=b[i]-p-q;
    if (e < tot) tot=e;
  }
  if (posdef && (tot < 0.0))
    tot=0.0;
  else
    for(i=1; i<=n; i++) b[i] -= tot;
  t=0;
  for (k=1; k<=m; k++) {
    converge=false;
    /* next qr transformation */
    do {
```

```
      t++;
      tot += s;
      delta=b[n]-s;
      i=n;
      e=Math.abs(eps*tot);
      if (dlam < e) dlam=e;
      if (delta <= dlam) {
        converge=true;
        break;
      }
      e=c[n]/delta;
      qp=delta+e;
      p=1.0;
      for (i=n-1; i>=k; i--) {
        q=b[i]-s-e;
        r=q/qp;
        p=p*r+1.0;
        ep=e*r;
        b[i+1]=qp+ep;
        delta=q-ep;
        if (delta <= dlam) {
          converge=true;
          break;
        }
        e=c[i]/q;
        qp=delta+e;
        c[i+1]=qp*ep;
      }
      if (converge) break;
      b[k]=qp;
      s=qp/p;
    } while (tot+s > tot);  /* end of qr transformation */
    if (!converge) {
      /* irregular end of iteration, deflate minimum diagonal element */
      s=0.0;
      i=k;
      delta=qp;
      for (j=k+1; j<=n; j++)
        if (b[j] < delta) {
          i=j;
          delta=b[j];
        }
    }
    /* convergence */
    if (i < n) c[i+1]=c[i]*e/qp;
    for (j=i-1; j>=k; j--) {
      b[j+1]=b[j]-s;
      c[j+1]=c[j];
    }
    b[k]=tot;
```

```
    c[k] = err += Math.abs(delta);
  }
  em[5]=t;
  em[3]=Basic.infnrmvec(1,m,itmp,c);
  Basic.dupvec(1,m,0,zer,b);
}
```

C. selzerortpol

Calculates a number of adjacent zeros of an orthogonal polynomial given by the coefficients of their recurrence relation [GolW69, L57, St72]. *selzerortpol* determines the subset λ_k ($k=n1,n1+1,...,n2$) of the total set of roots λ_k ($k=1,...,n$) (arranged in descending order) of the polynomial $p_n(x)$, where

$$p_0(x) = 1, \quad p_1(x) = x - b_0$$
$$p_{k+1}(x) = (x-b_k)p_k(x) - c_k p_{k-1}(x) \qquad (k=1,...,n-1)$$

the b_k and c_k are being given ($c_k>0$, $k=1,...,n-1$).

The roots of $p_n(x)$ are the eigenvalues of the $n \times n$ tridiagonal matrix T for which

$$T_{i,i}=b_{i-1} \ (i=1,...,n), \quad T_{i,i+1}=T_{i+1,i}=c_i^{1/2} \ (i=1,...,n-1).$$

These eigenvalues are obtained by means of a call of *valsymtri*.

It is efficient to use *allzerortpol* if more than 50 percent of extreme zeros or more than 25 percent of selected zeros are wanted.

Procedure parameters:

void selzerortpol ($n,n1,n2,b,c,zer,em$)

$n,n1,n2$:   int;

entry: the degree of the orthogonal polynomial of which the $n1$-th up to and including $n2$-th zeros are to be calculated ($zer[n1] \geq zer[n2]$);

b,c: double $b[0:n-1]$, $c[0:n-1]$;

entry: the elements $b[i]$ and $c[i]$, $i=0,1,...,n-1$, contain the coefficients of the recurrence relation $p_{i+1}(x)=(x-b[i])*p_i(x)-c[i]*p_{i-1}(x)$, $i=0,1,...,n-1$, assuming $c[0]=0$;

exit: the contents of the arrays b and c are not altered;

zer: double $zer[n1:n2]$;

exit: the $n2-n1+1$ calculated zeros in decreasing order;

em: double $em[0:5]$;

entry: $em[0]$: the machine precision;

         $em[2]$: the relative tolerance of the zeros;

exit: $em[1]$: the value of

$$\max(|b[0]|+1, \ c[i]+|b[i]|+1 \ (i=1,...,n-2), \ c[n-1]+|b[n-1]|);$$

         $em[5]$: the number of iterations performed.

Procedure used: valsymtri.

```
public static void selzerortpol(int n, int n1, int n2, double b[], double c[],
                        double zer[], double em[])
```

```
{
  int i;
  double nrm;
  double d[] =new double[n+1];

  nrm=Math.abs(b[0]);
  for (i=n-2; i>=1; i--)
    if (c[i]+Math.abs(b[i]) > nrm) nrm=c[i]+Math.abs(b[i]);
  if (n > 1)
    nrm = (nrm+1 >= c[n-1]+Math.abs(b[n-1])) ? nrm+1.0 : (c[n-1]+Math.abs(b[n-1]));
  em[1]=nrm;
  for (i=n; i>=1; i--) d[i]=b[i-1];
  valsymtri(d,c,n,n1,n2,zer,em);
  em[5]=em[3];
}
```

D. alljaczer

Calculates all zeros of the n-th Jacobi polynomial $P_n^{(\alpha,\beta)}(x)$, see [AbS65]. The Jacobi polynomials satisfy the recursion

$$P_0^{(\alpha,\beta)}(x) = 1, \qquad P_1^{(\alpha,\beta)}(x) = \tfrac{1}{2}(\alpha+\beta+2)x+\tfrac{1}{2}(\alpha-\beta),$$
$$2(k+1)(k+\alpha+\beta+1)(2k+\alpha+\beta)\, P_{k+1}^{(\alpha,\beta)}(x) =$$
$$(2k+\alpha+\beta+1)((2k+\alpha+\beta)(2k+\alpha+\beta+2)x+\alpha^2-\beta^2)\, P_k^{(\alpha,\beta)}(x) -$$
$$2k(k+\alpha)(k+\beta)(2k+\alpha+\beta+2)\, P_{k-1}^{(\alpha,\beta)}(x) \qquad\qquad (k=2,3,...)$$

and the coefficient of x^k in $P_k^{(\alpha,\beta)}(x)$ is

$$c_k = 2^{-k}\binom{2k+\alpha+\beta}{k}.$$

The polynomials $p_k(x) = P_k^{(\alpha,\beta)}(x)/c_k$ satisfy the recursion

$$p_0(x)=1, \qquad p_1(x) = x - \frac{\beta-\alpha}{\alpha+\beta+2},$$

$$p_{k+1}(x) = \left(x - \frac{\beta^2-\alpha^2}{(2k+\alpha+\beta)(2k+\alpha+\beta+2)}\right)p_k(x) -$$

$$\frac{4k(k+\alpha)(k+\beta)(k+\alpha+\beta)}{(2k+\alpha+\beta)^2(2k+\alpha+\beta-1)(2k+\alpha+\beta+1)}p_{k-1}(x)$$

The roots of $p_n(x)$ (i.e., those of $P_n^{(\alpha,\beta)}(x)$) may be obtained by a call of *allzerortpol*. However, for the special case in which $\alpha=\beta$,

$$P_{2m}^{(\alpha,\alpha)}(x) = c_m P_m^{(\alpha,-1/2)}(2x^2-1)$$
$$P_{2m-1}^{(\alpha,\alpha)}(x) = d_m x P_m^{(\alpha,1/2)}(2x^2-1)$$

where c_m and d_m are independent of x. Thus, in this special case, the determination of the roots of $P_n^{(\alpha,\beta)}(x)$ may slightly be simplified.

Procedure parameters:

    void alljaczer (n,alfa,beta,zer)

n:	int;
	entry: the upper bound of the array *zer*, $n \geq 1$;
alfa,beta:	double;
	entry: the parameters of the Jacobi polynomial (values of α and ß above);
	alfa, beta > -1;
zer:	double *zer[1:n]*;
	exit: the zeros of the *n*-th Jacobi polynomial with parameters *alfa* and *beta*.

Procedure used: allzerortpol.

```
public static void alljaczer(int n, double alfa, double beta, double zer[])
{
  int i,m;
  double sum,min,gamma,zeri;
  double em[] = new double[6];

  if (alfa == beta) {
    double a[] = new double[n/2+1];
    double b[] = new double[n/2+1];
    m=n/2;
    if (n != 2*m) {
      gamma=0.5;
      zer[m+1]=0.0;
    } else
      gamma = -0.5;
    min=0.25-alfa*alfa;
    sum=alfa+gamma+2.0;
    a[0] = (gamma-alfa)/sum;
    a[1] =min/sum/(sum+2.0);
    b[1] =4.0*(1.0+alfa)*(1.0+gamma)/sum/sum/(sum+1.0);
    for (i=2; i<=m-1; i++) {
      sum=i+i+alfa+gamma;
      a[i] =min/sum/(sum+2.0);
      sum *= sum;
      b[i] =4.0*i*(i+alfa+gamma)*(i+alfa)*(i+gamma)/sum/(sum-1.0);
    }
    em[0] =Double.MIN_VALUE;
    em[2] =Double.MIN_VALUE;
    em[4] =6*m;
    allzerortpol(m,a,b,zer,em);
    for (i=1; i<=m; i++) {
      zer[i] = zeri = -Math.sqrt((1.0+zer[i])/2.0);
      zer[n+1-i] = -zeri;
    }
  } else {
    double a[] = new double[n+1];
    double b[] = new double[n+1];
    min=(beta-alfa)*(beta+alfa);
```

```
      sum=alfa+beta+2.0;
      b[0]=0.0;
      a[0]=(beta-alfa)/sum;
      a[1]=min/sum/(sum+2.0);
      b[1]=4.0*(1.0+alfa)*(1.0+beta)/sum/sum/(sum+1.0);
      for (i=2; i<=n-1; i++) {
        sum=i+i+alfa+beta;
        a[i]=min/sum/(sum+2.0);
        sum *= sum;
        b[i]=4.0*i*(i+alfa+beta)*(i+alfa)*(i+beta)/(sum-1.0)/sum;
      }
      em[0]=Double.MIN_VALUE;
      em[2]=Double.MIN_VALUE;
      em[4]=6*n;
      allzerortpol(n,a,b,zer,em);
    }
}
```

E. alllagzer

Calculates all zeros of the n-th Laguerre polynomial $L_n^{(\alpha)}(x)$, see [AbS65]. The Laguerre polynomials satisfy the recursion

$$L_0^{(\alpha)}(x) = 1, \qquad L_1^{(\alpha)}(x) = \alpha+1-x,$$
$$(k+1)\, L_{k+1}^{(\alpha)}(x) = (2k+\alpha+1-x)\, L_k^{(\alpha)}(x) - (k+\alpha)\, L_{k-1}^{(\alpha)}(x)$$

and the coefficient of x^k in $L_k^{(\alpha)}(x)$ is $c_k=(-1)^k/k!$. The polynomials $p_k(x) = L_k^{(\alpha)}(x)/c_k$ satisfy the recursion

$$p_0(x) = 1, \qquad p_1(x) = x-\alpha-1,$$
$$p_{k+1}(x) = (x-2k-\alpha-1)\, p_k(x) - k(k+\alpha)\, p_{k-1}(x) \qquad (k=2,3,...).$$

The roots of $p_n(x)$ (i.e., those of $L_n^{(\alpha)}(x)$) are obtained by means of a call of *allzerortpol*.

Procedure parameters:

$$\text{void alllagzer } (n, alfa, zer)$$

n: int;
         entry: the upper bound of the array *zer*, $n \geq 1$;
alfa: double;
         entry: the parameter of the Laguerre polynomial (value of α above);
              alfa > -1;
zer: double *zer[1:n]*;
         exit: the zeros of the n-th Laguerre polynomial with parameters *alfa*.

Procedure used: allzerortpol.

```
public static void alllagzer(int n, double alfa, double zer[])
{
  int i;
  double em[] = new double[6];
```

```
double a[] = new double[n+1];
double b[] = new double[n+1];

b[0]=0.0;
a[n-1]=n+n+alfa-1.0;
for (i=1; i<=n-1; i++) {
  a[i-1]=i+i+alfa-1.0;
  b[i]=i*(i+alfa);
}
em[0]=Double.MIN_VALUE;
em[2]=Double.MIN_VALUE;
em[4]=6*n;
allzerortpol(n,a,b,zer,em);
}
```

3.16.3 Zeros of complex polynomials

comkwd

Determines the roots g and k of the quadratic equation $z^2-2pz-q=0$ with complex p and q.

Procedure parameters:

$$\text{void comkwd } (pr,pi,qr,qi,gr,gi,kr,ki)$$

pr,pi,qr,qi: double;

         entry: *pr, qr* are the real parts and *pi, qi* are the imaginary parts of the coefficients of the quadratic equation:

            $z^2 - 2(pr+pi*i)z - (qr+qi*i) = 0$;

gr,gi,kr,ki: double *gr[0:0], gi[0:0], kr[0:0], ki[0:0]*;

         exit: the real parts and the imaginary parts of the dinomial are delivered in *gr, kr* and *gi, ki*, respectively; moreover, the modulus of *gr+gi*i* is greater than or equal to the modulus of *kr+ki*i*.

Procedures used: commul, comdiv, comsqrt.

```
public static void comkwd(double pr, double pi, double qr, double qi, double gr[],
                          double gi[], double kr[], double ki[])
{
  double tmp1,tmp2;
  double hr[] = new double[1];
  double hi[] = new double[1];

  if (qr == 0.0 && qi == 0.0) {
    kr[0] = ki[0] = 0.0;
```

```
      gr[0] = pr*2.0;
      gi[0] = pi*2.0;
      return;
    }
    if (pr == 0.0 && pi == 0.0) {
      Basic.comsqrt(qr,qi,gr,gi);
      kr[0] = -gr[0];
      ki[0] = -gi[0];
      return;
    }
    if (Math.abs(pr) > 1.0 || Math.abs(pi) > 1.0) {
      Basic.comdiv(qr,qi,pr,pi,hr,hi);
      Basic.comdiv(hr[0],hi[0],pr,pi,hr,hi);
      Basic.comsqrt(1.0+hr[0],hi[0],hr,hi);
      Basic.commul(pr,pi,hr[0]+1.0,hi[0],gr,gi);
    } else {
      Basic.comsqrt(qr+(pr+pi)*(pr-pi),qi+pr*pi*2.0,hr,hi);
      if (pr*hr[0]+pi*hi[0] > 0.0) {
        gr[0] = pr+hr[0];
        gi[0] = pi+hi[0];
      } else {
        gr[0] = pr-hr[0];
        gi[0] = pi-hi[0];
      }
    }
    tmp1=gr[0];
    tmp2=gi[0];
    Basic.comdiv(-qr,-qi,tmp1,tmp2,kr,ki);
}
```

4. Analytic Evaluations

All procedures in this chapter are class methods of the following *Analytic_eval* class.

```
package numal;

import numal.*;

public class Analytic_eval extends Object {
   // all procedures in this chapter are to be inserted here
}
```

4.1 Evaluation of an infinite series

A. euler

Applies the Euler transformation to the series

$$\sum_{i=1}^{\infty} a_i .$$

The course of the computations is determined by two parameters, both prescribed by the user: *eps*, a real tolerance; *tim*, a positive integer specifying the number of consecutive transformed sums whose agreement to within the tolerance *eps* is taken to imply that the last of them is an acceptable approximation to the Euler sum of the original series.

   A set of numbers $M_{i,j}$, a sequence of integers $J(i)$, and sequences of partial sums S_i and terms t_i for which $S_{i+1}=S_i+t_{i+1}$ are computed as follows. Initially $M_{1,1}=a_1$, $J(1)=0$, and $S_1=\frac{1}{2}a_1$. For $i\geq1$, and with $M_{i+1,1}=a_{i+1}$, the numbers $M_{i+1,j+1} = \frac{1}{2}(M_{i,j}+M_{i+1,j})$ $(j=1,...,J(i))$ are computed. If $|M_{i+1,J(i)+1}|<|M_{i,J(i)}|$ then $J(i+1)=J(i)+1$ and $t_{i+1}=\frac{1}{2}M_{i+1,J(i)+1}$; otherwise $J(i+1)=J(i)$ and $t_{i+1}=M_{i+1,J(i)+1}$. If $|t_{i+1}|<eps$ for $i=I,...,I+tim-1$ for some $I>0$, the process terminated.

Procedure parameters:
$$\text{double euler } (method,eps,tim)$$
euler: delivers the computed sum of the infinite series;
method: a class that defines a procedure *ai*, this class must implement the AE_euler_method interface;
            double ai(int i)
                this procedure is the *i*-th term of the series, $i \geq 0$;
eps: double;
            entry: summation criterion, see *tim* below;
tim: int;
            entry: the summation is continued until *tim* successive terms of the transformed series are in absolute value less than *eps*.

```
package numal;

public interface AE_euler_method {

  double ai(int i);
}


public static double euler(AE_euler_method method, double eps, int tim)
{
    int i,k,n,t;
    double mn,mp,ds,sum;
    double m[] = new double[16];

    n=t=i=0;
    m[0]=method.ai(i);
    sum=m[0]/2.0;
    do {
        i++;
        mn=method.ai(i);
        for (k=0; k<=n; k++) {
            mp=(mn+m[k])/2.0;
            m[k]=mn;
            mn=mp;
        }
        if (Math.abs(mn) < Math.abs(m[n]) && n < 15) {
            ds=mn/2.0;
            n++;
            m[n]=mn;
        } else
            ds=mn;
        sum += ds;
        t = (Math.abs(ds) < eps) ? t+1 : 0;
    } while (t < tim);
    return sum;
}
```

B. sumposseries

Performs the summation of a convergent series with positive monotonically decreasing terms using the Van Wijngaarden transformation [Dan69, Vnw65] of the series to an alternating series.

$sumposseries$ estimates the sum of a series of real terms

$$\sum_{i=1}^{\infty} v(i) \tag{1}$$

by use of the transformation

$$\sum_{i=1}^{\infty} v(i) = \sum_{j=1}^{\infty} (-1)^{j-1} \sum_{k=1}^{\infty} 2^{k-1} v(2^{k-1} j).$$ (2)

The above transformation may be derived from the identity

$$(1-x)^{-1} = \prod_{k=1}^{\infty} \left(1 + x^{2^{k-1}}\right) \qquad (|x| < 1)$$

Differentiating throughout, and dividing the resulting relationship by the original version

$$x(1-x)^{-1} = \sum_{k=1}^{\infty} \frac{2^{k-1} x^{2^{k-1}}}{1 + x^{2^{k-1}}}$$

$$= \sum_{j=1}^{\infty} (-1)^{j-1} \sum_{k=1}^{\infty} 2^{k-1} x^{2^{k-1} j}$$

Replacing x by the displacement operator D $(Dv(i) = v(i+1))$ and applying both sides of the derived relationship to $v(0)$, the transformation (2) is obtained.

Euler's transformation may be applied to the series on the right-hand side of relationship (2): denoting the Euler sum of a series by $E\{...\}$, relationship (2) yields

$$\sum_{i=1}^{\infty} v(i) = E\left\{ \sum_{j=1}^{\infty} (-1)^{j-1} v'(j) \right\}$$

where

$$v'(j) = \sum_{k=1}^{\infty} 2^{k-1} v(2^{k-1} j).$$ (3)

Each series (3) involves a subsequent of the terms of the series (1), and may converge far more rapidly than the latter. For example, when $v(i) = i^{\alpha}$ $(\alpha > 1)$,

$$\sum_{i=1}^{\infty} v(i) = \sum_{i=1}^{n} v(i) = O\left(n^{1-\alpha}\right)$$

but

$$v'(j) = \sum_{k=1}^{n} 2^{k-1} v(2^{k-1} j) + O\left(2^{(1-\alpha)n}\right)$$

(e.g., when $\alpha = 2$, 2^{-n} decreases to zero more rapidly than n^{-1}.) Again when $v(i) = x^i$ $(|x| < 1)$ the two remainder terms are $O(x^n)$ and $O(x^l)$ (where $l = 2^n$) respectively. The auxiliary sums (3) with even values of j may be obtained from values of sums previously determined:

$$v'(2j) = \tfrac{1}{2} \left(v'(j) - v(j) \right)$$ (4)

Thus the sum of the component series (3) may often be evaluated directly or indirectly by direct summation; the sum of the alternating series of sums on the right-hand side of relationship (2) may then be obtained by use of Euler's transformation. In this way the numerical sum of a large number of terms of the series (1) is effectively approximated by a weighted sum which involves relatively few of them.

It may occur that the auxiliary series (3) do not converge with sufficient rapidity. In such a case, application of the transformation (2) may be repeated:

$$v'(j) = E\left\{\sum_{l=1}^{\infty}(-1)^{l-1}v''(j,l)\right\} \tag{5}$$

where

$$v''(j,l) = \sum_{m=1}^{\infty}2^{m-1}2^{2^{m-1}l-1}v(2^{2^{m-1}l-1}j).$$

Again, for fixed j,

$$v''(j,2l) = \tfrac{1}{2}\left(v''(j,l) - 2^{l-1}v(2^{l-1}j)\right)$$

and relationship (3) still holds. For odd j, the terms $v''(j,l)$ are obtained, directly or indirectly, by summation; they are inserted into an Euler process to yield terms $v'(j)$ given by formula (5); the $v'(j)$ with even j are derived by use of relationship (4); the $v'(j)$ are inserted into a further Euler process to yield the sum (1).

Clearly the above process of recursion may be extended: letting i be an integer vector with components $i(1)$, $i(2)$,..., and $v^{(j)}(i)$ be a function of the first j components of i,

$$\sum_{i=1}^{\infty}v(i) = E\left\{\sum_{i(1)=1}^{\infty}(-1)^{i(1)-1}v^{(1)}(i)\right\}$$

where

$$v^{(j)}(i) = E\left\{\sum_{i(j+1)=1}^{\infty}(-1)^{i(j+1)-1}v^{j+1}(i)\right\} \qquad (j = 1,\ldots,k-1),$$

and $v^{(k)}(i)$ is obtained by direct summation:

$$v^{(k)}(i) = \sum_{i(k+1)=1}^{\infty}f^{(k+1)}(i)v(s^{(k+1)}(i)) \tag{6}$$

where $f^{(k+1)}(i)$, $s^{(k+1)}(i)$ are integer functions of the first $k+1$ components if i, and may be derived by use of the following recursion:

$f(1) = 1, \qquad s(1) = 2^{i(k+1)-1}$
$f(j+1) = f(j)s(j)$
$s(j+1) = 2^{s(j)i(k-j+1)-1} \qquad\qquad (j=1,\ldots,k-1)$
$f^{(k+1)}(i) = f(k)s(k), \qquad s^{(k+1)}(i) = i(1)s(k).$

For fixed $i(1)$, $i(2)$,..., $i(j-1)$, $v^{(j)}(i)$ with even values of $i(j)$ may be determined without summation or transformation: writing $v^{(j)}(i)$ as $v^{(j)}(i(1),i(2),\ldots,i(j))$, etc.

$$v^{(j)}(i(1),i(2),\ldots,i(j-1),2i(j)) = \tfrac{1}{2}(v^{(j)}(i(1),i(2),\ldots,i(j-1),i(j)) -$$
$$f^{(j+1)}(i(1),i(2),\ldots,i(j),1)v(s^{(j+1)}(i(1),i(2),\ldots,i(j),1))) \tag{7}$$

for $j=1,\ldots,k$.

For many series

$$\sum_{i=1}^{\infty}u(i)$$

of real terms with consistent sign

$$\left|\sum_{i=1}^{\infty}u(i)\right| \le R(n) = \left|\frac{2u(n)}{1-u(n+1)/u(n)}\right|$$

If $R(n)\le\varepsilon$ for $n=m,m+1,\ldots,m+tim-1$, ε being a real tolerance, m and tim being positive integers, it may reasonably be inferred that the partial sum

$$\sum_{i=1}^{n-1} u(i)$$

represents the sum of the series in question to the stated tolerance. This test is used to establish the level k at which the recursion functions during the computation. With integer values of m and tim supplied by the user, the above test is first applied to the series (1) itself, and if the test is passed, no transformation is carried out. If the test fails, then with $i(1)=1$, the test is applied to the series (6) with $k=1$; if this test fails, it is applied to the series (6) with $k=2$ and $i(1)=i(2)=1$, and so on; acceptance of the test determines the value of k for which $v^{(k)}(i)$ is to be determined, directly or indirectly, by summation.

With $i(1)=i(2)=...=i(k)=1$, $v^{(k)}(i)$ is determined by summation, and inserted into the Euler process to determine $v^{(k-1)}(i)$. With $i(1)=i(2)=...=i(k-1)=1$, $i(k)=2$, $v^{(k)}(i)$ is determined by use of relationship (7) with $j=k$, and $-v^{(k)}(i)$ is inserted into the above Euler process. With $i(1)=i(2)=...=i(k-1)=1$, $i(k)=3$, $v^{(k)}(i)$ is determined by summation, and so on. When the Euler process produces a transformed sum (namely $v^{(k-1)}(1,1,...,1)$) it is inserted into the penultimate Euler process. $v^{(k-1)}(i)$ with $i(1)=i(2)=...=i(k-1)$, $i(k-1)=2$ is determined by use of relationship (7) with $j=k-1$. The rate of convergence of the series (6) with k replaced by $k-1$ is now tested as described above; if the test is passed, the $v^{(k-1)}(i)$ are determined by direct summation; otherwise k is increased to a level at which convergence is sufficiently rapid, and the process described above is continued; testing rate of convergence takes place after the termination of each Euler process. Finally the Euler process involving the terms $v^{(1)}(i)$ terminates, and the transformed sum of the series (1) is obtained.

Procedure parameters:
        double sumposseries (*method,maxaddup,maxzero,maxrecurs,machexp,tim*)
sumposseries: delivers the computed sum of the infinite series;
method: a class that defines a procedure *ai*, this class must implement the
                AE_sumposseries_method interface;
                double ai(double x)
                    this procedure is the x-th term of the series, $x \geq 1$;
maxaddup: int;
                entry: upper limit for the number of straightforward additions
                       (value of m above);
maxzero: double;
                entry: tolerance in the Euler summation, see *tim* below; *maxzero* is
                       also used as a tolerance for *maxaddup* straightforward
                       additions;
maxrecurs: int;
                entry: upper limit for the recursion depth of the Van Wijngaarden
                       transformation;
machexp: int;
                entry: in order to avoid overflow and evaluation of those terms
                       which can be neglected, *machexp* has to be the largest
                       admissible value for which terms with index $k*(2^{machexp})$ can
                       be computed (k is small); otherwise, overflow might occur in
                       computing a value for the parameter I (in procedure *ai*),
                       which can be an unusually high power of 2;

tim: int;
                    entry: tolerance in the Euler summation; the summation is
                    continued until *tim* successive terms of the transformed
                    series are in absolute value less than *maxzero*.

```
package numal;

public interface AE_sumposseries_method {

  double ai(double x);
}


public static double sumposseries(AE_sumposseries_method method, int maxaddup,
                      double maxzero, int maxrecurs, int machexp,
                      int tim)
{
    int recurs,vl,vl2,vl4;

    recurs=0;
    vl=1000;
    vl2=2*vl;
    vl4=2*vl2;
    return sumposseriessumup(false,method,maxaddup,maxzero,maxrecurs,machexp,
                      tim,recurs,vl,vl2,vl4,0);
}


static private double sumposseriessumup(boolean bjk, AE_sumposseries_method method,
                                int maxaddup, double maxzero, int maxrecurs,
                                int machexp, int tim, int recurs, int vl,
                                int vl2, int vl4, int jj)
{
    /* this procedure is internally used by SUMPOSSERIES */

    boolean transform,jodd;
    int j,j2,k,n,t;
    double i,sum,nextterm,mn,mp,ds,esum,temp,vj;
    double m[] = new double[16];

    i=maxaddup+1;
    j=1;
    transform=false;
    while (true) {
        temp = (bjk) ? sumposseriesbjk(jj,i,machexp,method) : method.ai(i);
        if (temp <= maxzero) {
            if (j >= tim) break;
            j++;
            i++;
        } else {
```

```
            if (recurs != maxrecurs) transform=true;
            break;
        }
    }
    if (!transform) {
        sum=i=0.0;
        j=0;
        do {
            i++;
            nextterm = (bjk) ? sumposseriesbjk(jj,i,machexp,method) : method.ai(i);
            j = (nextterm <= maxzero) ? j+1 : 0;
            sum += nextterm;
        } while (j < tim);
        return sum;
    }
    /* transform series */
    double v[] = new double[vl+1];
    j2=0;
    jodd=true;
    /* euler */
    n=t=j=0;
    jj=j+1;
    if (jodd) {
        jodd=false;
        recurs++;
        temp=vj=sumposseriessumup(true,method,maxaddup,maxzero,maxrecurs,machexp,
                                  tim,recurs,vl,vl2,vl4,jj);
        recurs--;
        if (jj <= vl)
            v[jj]=temp;
        else
            if (jj <= vl2) v[jj-vl]=temp;
    } else {
        jodd=true;
        if (jj > vl4) {
            recurs++;
            vj = -sumposseriessumup(true,method,maxaddup,maxzero,maxrecurs,
                                    machexp,tim,recurs,vl,vl2,vl4,jj);
            recurs--;
        } else {
            j2++;
            i=j2;
            if (jj > vl2) {
                temp = (bjk) ? sumposseriesbjk(jj,i,machexp,method) : method.ai(i);
                vj = -(v[j2-vl]-temp)/2.0;
            }
            else {
                temp = (bjk) ? sumposseriesbjk(jj,i,machexp,method) : method.ai(i);
                temp=v[(jj <= vl) ? jj : jj-vl]=(v[j2]-temp)/2.0;
                vj = -temp;
```

```
            }
        }
    }
    m[0]=vj;
    esum=m[0]/2.0;
    do {
        j++;
        jj=j+1;
        if (jodd) {
            jodd=false;
            recurs++;
            temp=vj=sumposseriessumup(true,method,maxaddup,maxzero,maxrecurs,
                            machexp,tim,recurs,vl,vl2,vl4,jj);
            recurs--;
            if (jj <= vl)
                v[jj]=temp;
            else
                if (jj <= vl2) v[jj-vl]=temp;
        } else {
            jodd=true;
            if (jj > vl4) {
                recurs++;
                vj = -sumposseriessumup(true,method,maxaddup,maxzero,maxrecurs,
                                machexp,tim,recurs,vl,vl2,vl4,jj);
                recurs--;
            } else {
                j2++;
                i=j2;
                if (jj > vl2) {
                    temp = (bjk) ?
                            sumposseriesbjk(jj,i,machexp,method) : method.ai(i);
                    vj = -(v[j2-vl]-temp)/2.0;
                } else {
                    temp = (bjk) ?
                            sumposseriesbjk(jj,i,machexp,method) : method.ai(i);
                    temp=v[(jj <= vl) ? jj : jj-vl]=(v[j2]-temp)/2.0;
                    vj = -temp;
                }
            }
        }
        mn=vj;
        for (k=0; k<=n; k++) {
            mp=(mn+m[k])/2.0;
            m[k]=mn;
            mn=mp;
        }
        if (Math.abs(mn) < Math.abs(m[n]) && n < 15) {
            ds=mn/2.0;
            n++;
            m[n]=mn;
```

```
    } else
        ds=mn;
      esum += ds;
      t = (Math.abs(ds) < maxzero) ? t+1 : 0;
    } while (t < tim);
    return esum;
  }


static private double sumposseriesbjk(int j, double i, double machexp,
                                      AE_sumposseries_method method)
{
    /* this procedure is internally used by SUMPOSSERIES */

    double coeff;

    if (i > machexp) return 0.0;
    coeff=Math.pow(2.0,i-1.0);
    return coeff*method.ai(j*coeff);
}
```

4.2 Quadrature

4.2.1 One-dimensional quadrature

A. qadrat

Evaluates the integral

$$I(f;a,b) = \int_a^b f(x)dx$$

where f is real valued, a and b are finite ($b<a$ is permitted).

      qadrat functions recursively [RoF72]. With $f_i=f(a+ih)$, where $h=(b-a)/32$, two sums of the form

$$U = u_1(f_2 + f_{30}) + \sum_{i=2}^{5} u_i(f_{4i-8} + f_{40-4i}) + u_6 f_{16}$$

$$V = v_0(f_1 + f_{31}) + v_1(f_2 + f_{30}) + \sum_{i=2}^{5} v_i(f_{4i-8} + f_{40-4i}) + v_6 f_{16}$$

are computed (U and V are such that Uh and Vh are equal to $I(f:a,b)$ if f is a polynomial of degree ≤ 16). If either $|h|<hmin=|b-a|\eta_a$, where η_a is an absolute tolerance prescribed by the user, or $|U-V|<|V|\eta_r+\eta_a$, where η_r is a relative tolerance also prescribed by the user, then Vh is accepted as an approximation to $I(f;a,b)$. If neither of these conditions is satisfied, a further sum of the form

$$W = w_0(f_1 + f_{31}) + w_1(f_2 + f_{30}) + \sum_{i=2}^{5} w_i(f_{4i-8} + f_{40-4i}) + w_6 f_{16} - w_7(f_6 + f_{26})$$

is computed (again Wh is equal to $I(f;a,b)$ if f is a polynomial of degree ≤ 16). If now $|W-V| < |W| \eta_r + \eta_a$, then Wh is accepted as an approximation to $I(f;a,b)$. If this further condition is not satisfied, $I(f;a,b)$ is then expressed as

$$\int_a^{(a+b)/2} f(x)dx - \int_b^{(a+b)/2} f(x)dx ,$$

and the two integrals are each treated in the above manner (with $hmin$ left unchanged). The first is approximated by linear sums involving the function values

$$f_i' = f(a+ih') \text{ where } h' = (b-a)/64.$$

Since $f_{2i}' = f_i$ $(i=0,1,2,4,6,8,12,16)$, this set of values may immediately be used in an attempt to evaluate the first of the integrals (1). The second integral is approximated by linear sums involving the function values $f_i' = f(b+ih')$, where $h' = (a-b)/64$, since $f_{2i}' = f_{32-i}$ $(i=0,1,2,4,6,8,12,16)$.

Procedure parameters:

$$\text{double qadrat } (a,b,method,e)$$

qadrat:	delivers the computed value of the definite integral from a to b of the function $f(x)$;
a,b:	double;
	entry: (a,b) denotes the interval of integration; $b<a$ is allowed;
method:	a class that defines a procedure fx, this class must implement the AE_qadrat_method interface;
	double fx(double x)
	fx denotes the integrand $f(x)$;
e:	double $e[1:3]$;

        entry: $e[1]$:   the relative accuracy required (value of η_r above);

                  $e[2]$:   the absolute accuracy required (value of η_a above);

        exit: $e[3]$:   the number of elementary integrations with $h < |b-a|*e[1]$.

```
package numal;

public interface AE_qadrat_method {

  double fx(double x);
}


public static double qadrat(double a, double b, AE_qadrat_method method, double e[])
{
    double x,f0,f2,f3,f5,f6,f7,f9,f14,hmin,hmax,re,ae,result;

    hmax=(b-a)/16.0;
    if (hmax == 0.0) return 0.0;
    re=e[1];
    ae=2.0*e[2]/Math.abs(b-a);
```

```
    e[3]=0.0;
    hmin=Math.abs(b-a)*re;
    x=a;
    f0=method.fx(x);
    x=a+hmax;
    f2=method.fx(x);
    x=a+2.0*hmax;
    f3=method.fx(x);
    x=a+4.0*hmax;
    f5=method.fx(x);
    x=a+6.0*hmax;
    f6=method.fx(x);
    x=a+8.0*hmax;
    f7=method.fx(x);
    x=b-4.0*hmax;
    f9=method.fx(x);
    x=b;
    f14=method.fx(x);
    result = lint(method,e,a,b,f0,f2,f3,f5,f6,f7,f9,f14,hmin,hmax,re,ae)*16.0;
    return result;
}


static private double lint(AE_qadrat_method method, double e[], double x0,
                           double xn, double f0, double f2, double f3, double f5,
                           double f6, double f7, double f9, double f14,
                           double hmin, double hmax, double re, double ae)
{
    /* this procedure is internally used by QADRAT */

    double x,v,w,h,xm,f1,f4,f8,f10,f11,f12,f13;

    xm=(x0+xn)/2.0;
    h=(xn-x0)/32.0;
    x=xm+4.0*h;
    f8=method.fx(x);
    x=xn-4.0*h;
    f11=method.fx(x);
    x=xn-2.0*h;
    f12=method.fx(x);
    v=0.330580178199226*f7+0.173485115707338*(f6+f8)+0.321105426559972*(f5+f9)+
        0.135007708341042*(f3+f11)+0.165714514228223*(f2+f12)+
        0.393971460638127e-1*(f0+f14);
    x=x0+h;
    f1=method.fx(x);
    x=xn-h;
    f13=method.fx(x);
    w=0.260652434656970*f7+0.239063286684765*(f6+f8)+0.263062635477467*(f5+f9)+
        0.218681931383057*(f3+f11)+0.275789764664284e-1*(f2+f12)+
        0.105575010053846*(f1+f13)+0.157119426059518e-1*(f0+f14);
```

```
if (Math.abs(h) < hmin) e[3] += 1.0;
if (Math.abs(v-w) < Math.abs(w)*re+ae || Math.abs(h) < hmin)
    return h*w;
else {
    x=x0+6.0*h;
    f4=method.fx(x);
    x=xn-6.0*h;
    f10=method.fx(x);
    v=0.245673430093324*f7+0.255786258286921*(f6+f8)+
        0.228526063690406*(f5+f9)+0.500557131525460e-1*(f4+f10)+
        0.177946487736780*(f3+f11)+0.584014599347449e-1*(f2+f12)+
        0.874830942871331e-1*(f1+f13)+0.189642078648079e-1*(f0+f14);
    return ((Math.abs(v-w) < Math.abs(v)*re+ae) ? h*v :
        (lint(method,e,x0,xm,f0,f1,f2,f3,f4,f5,f6,f7,hmin,hmax,re,ae)-
        lint(method,e,xn,xm,f14,f13,f12,f11,f10,f9,f8,f7,hmin,hmax,re,ae)));
}
}
```

B. integral

Evaluates the integral

$$\int_\alpha^\beta f(x)dx \tag{1}$$

where f is real valued, $-\infty < \alpha < \infty$, and $-\infty \leq \beta \leq \infty$ (when $\beta < \infty$, $\beta < \alpha$ is permitted). This is done by means of Simpson's rule with Richardson correction [DekR71, RoF72]. If the fourth derivative is too large (and thus the correction term), the total interval is split into two equal parts and the integration process is invoked recursively. This is done in such a way that the total amount of Richardson corrections is slightly smaller than or equal to
$$e[1] * |\text{integral of the function from } a \text{ to } b| + e[2].$$
For the integration of a definite integral over a finite interval, the use of *qadrat* is recommended, especially when high accuracy is required.

    integral may be called a number of times to evaluate the integral of f over a sequence of intervals of the form $[\alpha_0, \alpha_1]$, $[\alpha_1, \alpha_2]$,..., $[\alpha_{n-1}, \alpha_n]$ and, when used in this way, may be directed to accumulate the total integral over the range $[\alpha_0, \alpha_n]$.

    The values of α and β in expression (1) are determined partly by the values of a and b occurring in the parameter list of *integral*. If *ua* is true then $\alpha=a$; otherwise $\alpha=a'$. If *ub* is true then $\beta=b$; if *ub* is false, then $\beta=\infty$ if $b>a$, and $\beta=-\infty$ if $b<a$. Immediately before return from a successful call, *integral* sets $a'=b$ if *ub* is true, and in this way the integrals over the above sequence of intervals may be evaluated without adjustment of a.

    If *ub* is false (so that $\beta=\infty$ or $\beta=-\infty$ above) the integral (1) is expressed as

$$\int_\alpha^b f(x)dx - \int_1^0 f(b-1+z^{-1})z^{-2}dz \tag{2}$$

if $\beta=\infty$, and as

$$\int_{\alpha}^{b} f(x)dx - \int_{1}^{0} f(b+1+z^{-1})z^{-2}dz \tag{3}$$

if ß=-∞. *integral* evaluates the two relevant integrals over finite ranges, and forms their difference.

For integration over a finite subrange, use is made of the formula

$$\int_{\gamma}^{\delta} f(x)dx = I_{1} - \frac{I_{2} - I_{1}}{15} + O(h^{7}) \tag{4}$$

where $h = (\delta-\gamma)/4$, and with $f_i = f(\gamma+ih)$

$I_1 = (h/3)(f_0 + 4f_1 + 2f_2 + 4f_3 + f_4)$
$I_2 = (h/3)(2f_0 + 8f_2 + 2f_4)$.

Acceptance of the value of the expression $I_1 - (I_2-I_1)/15$ as an approximation to the value of the intermediate integral (2) is decided by the values of two tolerances η_a and η_r provided by the user, from which a further tolerance τ_a is derived. If in expression (1) ß is finite (i.e., *ub* is true) then $\tau_a = 180/(ß-\alpha)$. If $|ß|$ is infinite, so that the pair (2) or (3) of integrals is to be evaluated, $\tau_a = 90\eta_a/|b-\alpha|$ for the first integral of the pair, and $\tau_a = 90\eta_a$ for the second. A minimum stepsize *hmin* is also determined. If in expression (1) ß is finite, then $hmin=|ß-\alpha|\eta_r$. If $|ß|$ is infinite so that the pair (2) or (3) of integrals is to be evaluated, $hmin = |b-\alpha|\eta_r$ for the first integral, and $hmin = \eta_r$ for the second.

integral functions recursively. Formula (4) is first examined with γ and δ set equal to the endpoints of the relevant integral over a finite range being evaluated. The sums

$v = 15(f_0 + f_4 + 2f_2 + 4(f_1 + f_3))$
$t = f_0 + 6f_2 + f_2 - 4(f_1 + f_3)$

(so that $v = 180I_1/h$, $t = 180(I_2-I_1)/h$) are formed. If $|t| < |v|\eta_r + \tau_a$, or $|\delta-\gamma| < hmin$, where η_r, τ_a and *hmin* are as defined above, then $I_1 - (I_2-I_1)/15$ is accepted as an approximation to the value of the integral (4); otherwise, the formula

$$\int_{\gamma}^{\delta} f(x)dx = \int_{\gamma}^{(\gamma+\delta)/2} f(x)dx + \int_{(\gamma+\delta)/2}^{\delta} f(x)dx$$

is used, each constituent of the right-hand side being treated as above. The computed approximation to the first constituent integral involves function values $f_i' = f(\gamma+ih')$ where $h' = (\delta-\gamma)/8$, so that $f_{2i}' = f_i$ $(i=0,1,2)$; those even order f_i' have already been evaluated, and may be used directly. The computed approximation to the second constituent integral involves function values $f_i' = f(\frac{1}{2}(\gamma+\delta)+ih')$ so that $f_{2i}' = f_{i+2}$ $(i=0,1,2)$.

Procedure parameters:

        double integral (a,b,method,e,ua,ub)

integral: delivers the computed value of the definite integral of the function from a to b; after successive calls of the procedure, the integral over the total interval is delivered, i.e., the value of a in the last call with *ua* equals true is the starting point of the integral;

a,b: double;
        entry: (a,b) denotes the interval of integration; b<a is allowed;

method: a class that defines a procedure *fx*, this class must implement the `AE_integral_method` interface;

            double fx(double x)

                fx denotes the integrand *f(x)*;

e: double *e[1:6]*;

         entry: *e[1]:*   the relative accuracy required (value of η_r above);

               e[2]:   the absolute accuracy required (value of η_a above);

         exit: *e[3]:*   the number of skipped integration steps;

               e[4]:   the value of the integral from *a* to *b*;

               e[5]:   if *ub* is false then *e[5]=0* else *e[5]=b*;

               e[6]:   if *ub* is false then *e[5]=0* else *e[5]=f(b)*;

ua: boolean;

         entry: determines the starting point of the integration; if *ua* is false then the starting point is *e[5]*; otherwise, the starting point is *a*;

ub: boolean;

         entry: determines the final point of the integration;

               if *ub* is true then the final point is *b*; otherwise, if *b > a* then the final point is ∞ else it is $-\infty$.

```
package numal;

public interface AE_integral_method {

    double fx(double x);
}


public static double integral(double a, double b, AE_integral_method method,
                       double e[], boolean ua, boolean ub)
{
    double re,ae,b1,x;
    double x0[] = new double[1];
    double x1[] = new double[1];
    double x2[] = new double[1];
    double f0[] = new double[1];
    double f1[] = new double[1];
    double f2[] = new double[1];

    b1=0.0;
    re=e[1];
    if (ub)
        ae=e[2]*180.0/Math.abs(b-a);
    else
        ae=e[2]*90.0/Math.abs(b-a);
    if (ua) {
        e[3]=e[4]=0.0;
        x=x0[0]=a;
        f0[0]=method.fx(x);
    } else {
        x=x0[0]=a=e[5];
```

```
        f0[0]=e[6];
    }
    e[5]=x=x2[0]=b;
    e[6]=f2[0]=method.fx(x);
    e[4] += integralqad(false,method,e,x0,x1,x2,f0,f1,f2,re,ae,b1);
    if (!ub) {
        if (a < b) {
            b1=b-1.0;
            x0[0]=1.0;
        } else {
            b1=b+1.0;
            x0[0] = -1.0;
        }
        f0[0]=e[6];
        e[5]=x2[0]=0.0;
        e[6]=f2[0]=0.0;
        ae=e[2]*90.0;
        e[4] -= integralqad(true,method,e,x0,x1,x2,f0,f1,f2,re,ae,b1);
    }
    return e[4];
}



static private double integralqad(boolean transf, AE_integral_method method,
                                  double e[], double x0[], double x1[], double x2[],
                                  double f0[], double f1[], double f2[], double re,
                                  double ae, double b1)
{
    /* this procedure is internally used by INTEGRAL */

    double hmin,x,z;
    double sum[] = new double[1];

    hmin=Math.abs(x0[0]-x2[0])*re;
    x=x1[0]=(x0[0]+x2[0])*0.5;
    if (transf) {
        z=1.0/x;
        x=z+b1;
        f1[0]=method.fx(x)*z*z;
    } else
        f1[0]=method.fx(x);
    sum[0]=0.0;
    integralint(transf,method,e,x0,x1,x2,f0,f1,f2,sum,re,ae,b1,hmin);
    return sum[0]/180.0;
}



static private void integralint(boolean transf, AE_integral_method method,
                                double e[], double x0[], double x1[], double x2[],
                                double f0[], double f1[], double f2[], double sum[],
```

```
                                double re, double ae, double b1, double hmin)
{
    /* this procedure is internally used by INTEGRALQAD of INTEGRAL */

    boolean anew;
    double x3,x4,f3,f4,h,x,z,v,t;

    x4=x2[0];
    x2[0]=x1[0];
    f4=f2[0];
    f2[0]=f1[0];
    anew=true;
    while (anew) {
        anew=false;
        x=x1[0]=(x0[0]+x2[0])*0.5;
        if (transf) {
            z=1.0/x;
            x=z+b1;
            f1[0]=method.fx(x)*z*z;
        } else
            f1[0]=method.fx(x);
        x=x3=(x2[0]+x4)*0.5;
        if (transf) {
            z=1.0/x;
            x=z+b1;
            f3=method.fx(x)*z*z;
        } else
            f3=method.fx(x);
        h=x4-x0[0];
        v=(4.0*(f1[0]+f3)+2.0*f2[0]+f0[0]+f4)*15.0;
        t=6.0*f2[0]-4.0*(f1[0]+f3)+f0[0]+f4;
        if (Math.abs(t) < Math.abs(v)*re+ae)
            sum[0] += (v-t)*h;
        else if (Math.abs(h) < hmin)
            e[3] += 1.0;
        else {
            integralint(transf,method,e,x0,x1,x2,f0,f1,f2,sum,re,ae,b1,hmin);
            x2[0]=x3;
            f2[0]=f3;
            anew=true;
        }
        if (!anew) {
            x0[0]=x4;
            f0[0]=f4;
        }
    }
}
```

4.2.2 Multidimensional quadrature

tricub

Evaluates the double integral

$$I(g;\Omega) = \iint_\Omega g(x,y)dxdy$$

where Ω is a triangle with vertices (x_i,y_i), (x_j,y_j), and (x_k,y_k), and g is real valued. A nested sequence of cubature rules of order 2, 3, 4, and 5 is applied [He73]. If the difference between the result with the 4-th degree rule and the result with the 5-th degree rule is too large, then the triangle is divided into four congruent triangles. This process is applied recursively in order to obtain an adaptive cubature algorithm. The theory of the method adopted may be presented as follows. The area of Ω is

$$\Delta = \tfrac{1}{2}\left| x_iy_j - x_jy_i + x_ky_i - x_iy_k + x_jy_k - x_ky_j \right|.$$

Denote the argument pair (x_i,y_i) by z_i, and the other pairs similarly. Set

$$z_{j,k} = \tfrac{1}{2}(z_j + z_k),$$

and define $z_{k,i}$, $z_{i,j}$ similarly ($z_{j,k}$ is the midpoint of the line joining z_j and z_k). Set

$$z_c = (z_i + z_j + z_k)/3$$

(z_c is the barycenter is Ω). Set $z_i' = \tfrac{1}{2}(z_i + z_c)$ and define z_j', z_k' similarly. Set

$$z_{i;j,k} = \tfrac{1}{2}(z_i + z_{j,k}),$$

and define $z_{k;i,j}$, $z_{j;k,i}$ similarly. Denote values of g at the above points by corresponding suffices, so that $g_i = g(z_i) = g(x_i,y_i)$, $g_{j,k} = g(z_{j,k}) = g(\tfrac{1}{2}(x_j+x_k), \tfrac{1}{2}(y_j+y_k))$ and so on. Set

$$A = g_i + g_j + g_k \qquad\qquad B = g_{j,k} + g_{k,i} + g_{i,j}$$
$$C = g_i' + g_j' + g_k' \qquad\qquad D = g_{i;j,k} + g_{k;i,j} + g_{j;k,i}$$

and

$$S_2 = 5A + 45g_c \qquad\qquad S_3 = 3A + 27g_c + 8B$$
$$S_4 = A + 9g_c + 4B + 12C$$
$$S_5 = (51A + 2187g_c + 276B + 972C - 768D)/63$$

and

$$I_j = S_j\Delta/60 \ .$$

If g is a polynomial in x and y of degree $\leq j$, then $I_j = I(g;\Omega)$ exactly.

The user is asked to provide values of tolerances η_r, η_a. *tricub* determines from these further tolerances: $\tau_r = 30\eta_r$, $\tau_r' = \Delta\eta_r$, and $\tau_a = 30\eta_a/\Delta$.

tricub functions recursively. S_2 and S_3 are evaluated, and if either $\Delta<\tau_r'$ or $|S_3-S_2|<\tau_r$ then I_3 is accepted as an approximation to $I(g;\Omega)$. If neither of these conditions is satisfied, S_4 is evaluated and if $|S_4-S_3|<\tau_r$, I_4 is accepted as an approximation to $I(g;\Omega)$. If this condition is not satisfied, S_5 is evaluated and if $|S_5-S_4|<\tau_r$, I_5 is accepted as an approximation to $I(g;\Omega)$. If this condition is not satisfied Ω is subdivided into four congruent triangles whose vertices are $z_{i,j}, z_{j,k}, z_{k,i;}$ $z_i, z_{i,j}, z_{i,k};$ $z_j, z_{j,k}, z_{j,i;}$ $z_k, z_{k,j}, z_{k,i}$ and the integrals over these four triangles are individually treated as above. The barycenter of the first triangle is z_c (i.e., that of Ω). The barycenters of the remaining three triangles are z_i', z_j', and z_k', respectively. $z_{i;j,k}$, $z_{j;k,i}$, and $z_{k;i,j}$ are the midpoints of the sides of the first triangle. The values of g at the points just mentioned (and at z_i, z_j, z_k) are required in the treatment of the integrals over the four triangles. Further function values at the midpoints of the lines joining z_i and $z_{i,j}$, z_i and $z_{i,k}$, z_j and $z_{j,k}$, z_j and $z_{j,i}$, z_k, and $z_{k,j}$, and z_k and $z_{k,i}$ are required.

Procedure parameters:

$$\text{double tricub }(xi,yi,xj,yj,xk,yk,method,re,ae)$$

tricub: delivers the computed value of the definite integral of the function $g(x,y)$ over the triangular domain with vertices (xi,yi), (xj,yj) and (xk,yk);

xi,yi: double;
              entry: the coordinates of the first vertex of the triangular domain of integration;

xj,yj: double;
              entry: the coordinates of the second vertex of the triangular domain of integration;

xk,yk: double;
              entry: the coordinates of the third vertex of the triangular domain of integration;
              the algorithm is symmetric in the vertices; this implies that the result of the procedure (on all counts) is invariant for any permutation of the vertices;

method: a class that defines a procedure g, this class must implement the AE_tricub_method interface;
              double g(double x, double y)
                  g denotes the integrand $g(x,y)$;

re: double;
              entry: the required relative error (value of η_r above);

ae: double;
              entry: the required absolute error (value of η_a above);
              one should take for *ae* and *re* values which are greater than the absolute and relative error in the computation of the integrand g.

```
package numal;

public interface AE_tricub_method {

  double g(double x, double y);
}


public static double tricub(double xi, double yi, double xj, double yj, double xk,
                            double yk, AE_tricub_method method, double re, double ae)
{
    double surfmin,xz,yz,gi,gj,gk;
    double surf[] = new double[1];

    surf[0]=0.5*Math.abs(xj*yk-xk*yj+xi*yj-xj*yi+xk*yi-xi*yk);
    surfmin=surf[0]*re;
    re *= 30.0;
    ae=30.0*ae/surf[0];
    xz=(xi+xj+xk)/3.0;
```

```
    yz=(yi+yj+yk)/3.0;
    gi=method.g(xi,yi);
    gj=method.g(xj,yj);
    gk=method.g(xk,yk);
    xi *= 0.5;
    yi *= 0.5;
    xj *= 0.5;
    yj *= 0.5;
    xk *= 0.5;
    yk *= 0.5;
    return tricubint(xi,yi,gi,xj,yj,gj,xk,yk,gk,xj+xk,yj+yk,method.g(xj+xk,yj+yk),
                     xk+xi,yk+yi,method.g(xk+xi,yk+yi),xi+xj,yi+yj,
                     method.g(xi+xj,yi+yj),0.5*xz,0.5*yz,method.g(xz,yz),
                     method,re,ae,surf,surfmin)/60.0;
}


static private double tricubint(double ax1, double ay1, double af1, double ax2,
                                double ay2, double af2, double ax3, double ay3,
                                double af3, double bx1, double by1, double bf1,
                                double bx2, double by2, double bf2, double bx3,
                                double by3, double bf3, double px, double py,
                                double pf, AE_tricub_method method, double re,
                                double ae, double surf[], double surfmin)
{
    /* this procedure is internally used by TRICUB */

    double e,i3,i4,i5,a,b,c,sx1,sy1,sx2,sy2,sx3,sy3,cx1,cy1,cf1,cx2,cy2,cf2,cx3,
           cy3,cf3,dx1,dy1,df1,dx2,dy2,df2,dx3,dy3,df3,result;

    a=af1+af2+af3;
    b=bf1+bf2+bf3;
    i3=3.0*a+27.0*pf+8.0*b;
    e=Math.abs(i3)*re+ae;
    if (surf[0] < surfmin || Math.abs(5.0*a+45.0*pf-i3) < e)
        return i3*surf[0];
    else {
        cx1=ax1+px;
        cy1=ay1+py;
        cf1=method.g(cx1,cy1);
        cx2=ax2+px;
        cy2=ay2+py;
        cf2=method.g(cx2,cy2);
        cx3=ax3+px;
        cy3=ay3+py;
        cf3=method.g(cx3,cy3);
        c=cf1+cf2+cf3;
        i4=a+9.0*pf+4.0*b+12.0*c;
        if (Math.abs(i3-i4) < e)
            return i4*surf[0];
```

```
        else {
            sx1=0.5*bx1;
            sy1=0.5*by1;
            dx1=ax1+sx1;
            dy1=ay1+sy1;
            df1=method.g(dx1,dy1);
            sx2=0.5*bx2;
            sy2=0.5*by2;
            dx2=ax2+sx2;
            dy2=ay2+sy2;
            df2=method.g(dx2,dy2);
            sx3=0.5*bx3;
            sy3=0.5*by3;
            dx3=ax3+sx3;
            dy3=ay3+sy3;
            df3=method.g(dx3,dy3);
            i5=(51.0*a+2187.0*pf+276.0*b+972.0*c-
                        768.0*(df1+df2+df3))/63.0;
            if (Math.abs(i4-i5) < e)
                return i5*surf[0];
            else {
                surf[0] *= 0.25;
                result=tricubint(sx1,sy1,bf1,sx2,sy2,bf2,sx3,sy3,bf3,dx1,dy1,df1,
                            dx2,dy2,df2,dx3,dy3,df3,px,py,pf,method,re,ae,
                            surf,surfmin)+
                    tricubint(ax1,ay1,af1,sx3,sy3,bf3,sx2,sy2,bf2,dx1,dy1,df1,
                            ax1+sx2,ay1+sy2,method.g(ax1+sx2,ay1+sy2),
                            ax1+sx3,ay1+sy3,method.g(ax1+sx3,ay1+sy3),
                            0.5*cx1,0.5*cy1,cf1,method,re,ae,surf,surfmin)+
                    tricubint(ax2,ay2,af2,sx3,sy3,bf3,sx1,sy1,bf1,dx2,dy2,df2,
                            ax2+sx1,ay2+sy1,method.g(ax2+sx1,ay2+sy1),
                            ax2+sx3,ay2+sy3,method.g(ax2+sx3,ay2+sy3),
                            0.5*cx2,0.5*cy2,cf2,method,re,ae,surf,surfmin)+
                    tricubint(ax3,ay3,af3,sx1,sy1,bf1,sx2,sy2,bf2,dx3,dy3,df3,
                            ax3+sx2,ay3+sy2,method.g(ax3+sx2,ay3+sy2),
                            ax3+sx1,ay3+sy1,method.g(ax3+sx1,ay3+sy1),
                            0.5*cx3,0.5*cy3,cf3,method,re,ae,surf,surfmin);
                surf[0] *= 4.0;
                return result;
            }
        }
    }
}
```

4.2.3 Gaussian quadrature - General weights

A. reccof

Calculates from a given weight function on [-1,1] the recurrence coefficients of the corresponding orthogonal polynomials [G68a, G68b].

Given $w(x)$, defined for $-1 \leq x \leq 1$, and n, derives the coefficients b_k and c_k occurring in the recursion

$$p_0(x) = 1, \quad p_1(x) = x - b_0$$
$$p_{k+1}(x) = (x-b_k)p_k(x) - c_k p_{k-1}(x) \quad (k=1,...,n) \tag{1}$$

satisfied by the polynomials

$$p_x(x) = x^k + \sum_{l=0}^{k-1} p_{k,l} x^l$$

defined by the conditions

$$\int_{-1}^{1} p_i(x) p_j(x) w(x) dx = \begin{cases} 1 & (i = j) \\ \\ 0 & (i \neq j) \end{cases}$$

The formulae

$$L_k = \int_{-1}^{1} p_k(x)^2 w(x) dx \tag{2}$$

$$b_k = \left(\int_{-1}^{1} x\, p_k(x)^2 w(x) dx \right) \Big/ L_k \tag{3}$$

are used with $k=0$ to determine b_0 (and hence $p_1(x)$) and then with $c_k = L_k/L_{k-1}$ and (1) with $k=1,...,n-1$ to determine the b_k and c_k. The integrals (2) and (3) are evaluated by use of the formula

$$\int_{-1}^{1} f(x) dx = \frac{\pi}{m} \sum_{j=1}^{m} \sin(\theta_j) f(\cos(\theta_j)).$$

The user must supply the value of m.

When w is symmetric $(w(-x)=w(x))$, the above formulae may be simplified (for example, by b_{2k} are zero). For this reason the user is asked to allocate a value to the variable sym upon call of $reccof$: true if w is symmetric, false otherwise.

Procedure parameters:

                 void reccof (n,m,method,b,c,l,sym)

n: int;
            entry: upper bound for the indices of the arrays b, c, l $(n \geq 0)$;

m: int;
            entry: the number of points used in the Gauss-Chebyshev quadrature rule for calculating the approximation of the integral representations of $b[k]$, $c[k]$;

method: a class that defines a procedure wx, this class must implement the AE_reccof_method interface;
            double wx(double x)
               the weight function ($w(x)$ above);

b,c,l: double *b[0:n], c[0:n], l[0:n];*
     exit: the approximate recurrence coefficients for
       $p_{k+1}(x)=(x-b[k])^*p_k(x)-c[k]^*p_{k-1}(x)$, $k=0,1,2,...,n$,
     and the approximate square lengths
       $l[k]$ = integral from -1 to 1 $(w(x)^*p_k(x)^*p_k(x))$ *dx*;

sym: boolean;
     entry: if *sym* is true then the weight function on [-1,1] is even, otherwise the weight function on [-1,1] is not even.

Procedure used: ortpol.

```
package numal;

public interface AE_reccof_method {

    double wx(double x);
}


public static void reccof(int n, int m, AE_reccof_method method, double b[],
                    double c[], double l[], boolean sym)
{
    int i,j,up;
    double x,r,s,pim,h,hh,arg,sa,temp;

    pim=4.0*Math.atan(1.0)/m;
    if (sym) {
        for (j=0; j<=n; j++) {
            r=b[j]=0.0;
            up=m/2;
            for (i=1; i<=up; i++) {
                arg=(i-0.5)*pim;
                x=Math.cos(arg);
                temp=Algebraic_eval.ortpol(j,x,b,c);
                r += Math.sin(arg)*method.wx(x)*temp*temp;
            }
            if (up*2 == m)
                l[j]=2.0*r*pim;
            else {
                x=0.0;
                temp=Algebraic_eval.ortpol(j,0.0,b,c);
                l[j]=(2.0*r+method.wx(x)*temp*temp)*pim;
            }
            c[j] = (j == 0) ? 0.0 : l[j]/l[j-1];
        }
    } else
        for (j=0; j<=n; j++) {
            r=s=0.0;
            up=m/2;
            for (i=1; i<=up; i++) {
```

```
                    arg=(i-0.5)*pim;
                    sa=Math.sin(arg);
                    x=Math.cos(arg);
                    temp=Algebraic_eval.ortpol(j,x,b,c);
                    h=method.wx(x)*temp*temp;
                    x = -x;
                    temp=Algebraic_eval.ortpol(j,x,b,c);
                    hh=method.wx(x)*temp*temp;
                    r += (h+hh)*sa;
                    s += (hh-h)*x*sa;
                }
                b[j]=s*pim;
                if (up*2 == m)
                    l[j]=r*pim;
                else {
                    x=0.0;
                    temp=Algebraic_eval.ortpol(j,0.0,b,c);
                    l[j]=(r+method.wx(x)*temp*temp)*pim;
                }
                c[j] = (j == 0) ? 0.0 : l[j]/l[j-1];
            }
        }
    }
```

B. gsswts

Calculates from the recurrence coefficients of orthogonal polynomials the Gaussian weights of the corresponding weight function [G70].
Given the coefficients b_k and c_k occurring in the recursion

$$p_0(x) = 1, \quad p_1(x) = x - b_0$$
$$p_{k+1}(x) = (x-b_k)p_k(x) - c_k p_{k-1}(x) \qquad (k=1,...,n-1)$$

satisfied by the polynomials

$$p_x(x) = x^k + \sum_{l=0}^{k-1} p_{k,l} x^l$$

defined by the conditions

$$\int_\alpha^\beta p_i(x)p_j(x)w(x)dx = \begin{cases} 1 & (i = j) \\ \\ 0 & (i \neq j) \end{cases}$$

and the n real roots $z_1,...,z_n$ of p_n, derives the weights $w_1,...,w_n$ which render the quadrature formula

$$\sum_{i=1}^n w_i f(z_i) = \int_\alpha^\beta w(x)f(x)dx \Big/ \int_\alpha^\beta w(x)dx$$

exact for all polynomials of degree $< 2n$. The w_i are given by

$$w_i = \left(1 + \sum_{j=1}^{n-1} p_j(z_i)^2 / c_j\right)^{-1}.$$

Procedure parameters:

$$\text{void gsswts } (n, zer, b, c, w)$$

n: int;

          entry: the number of weights to be computed; upper bound for arrays z
          and w ($n \geq 1$);

zer: double $zer[1:n]$;

          entry: the zeros of the n-th degree orthogonal polynomial;

b,c: double $b[0:n-1]$, $c[1:n-1]$;

          entry: the recurrence coefficients;

w: double $w[1:n]$;

          exit: the Gaussian weights divided by the integral over the weight
          function.

Procedure used: allortpol.

```
public static void gsswts(int n, double zer[], double b[], double c[], double w[])
{
    int j,k;
    double s;
    double p[] = new double[n];

    for (j=1; j<=n; j++) {
        Algebraic_eval.allortpol(n-1,zer[j],b,c,p);
        s=0.0;
        for (k=n-1; k>=1; k--) s=(s+p[k]*p[k])/c[k];
        w[j]=1.0/(1.0+s);
    }
}
```

C. gsswtssym

Calculates from the recurrence coefficients of orthogonal polynomials the
Gaussian weights of the corresponding weight function.
Given the coefficients c_k occurring in the recursion

$$p_0(x) = 1, \quad p_1(x) = x$$
$$p_{k+1}(x) = xp_k(x) - c_k p_{k-1}(x) \quad (k=1,\ldots,n-1)$$

satisfied by the polynomials

$$p_x(x) = x^k + \sum_{l=0}^{k-1} p_{k,l} x^l$$

defined by the conditions

$$\int_{-\infty}^{\alpha} p_i(x)p_j(x)w(x)dx = \begin{cases} 1 & (i = j) \\ \\ 0 & (i \neq j) \end{cases}$$

where w is symmetric ($w(-x)=w(x)$), and $n'=n/2$ nonpositive roots $z_1,...,z_{n'}$ ($z_i<z_{i+1}$) of p_n, derives the weights $w_1,...,w_{n''}$ ($n''=(n+1)/2$) which with $w_{n''+1},...,w_n$ further defined by $w_j=w_{n+1-j}$ ($j=1,...,n''-1$) render the quadrature formula

$$\sum_{i=1}^{n} w_i f(z_i) = \int_{\alpha}^{\beta} w(x)f(x)dx \Big/ \int_{\alpha}^{\beta} w(x)dx$$

exact for all polynomials of degree $< 2n$. (For the roots of p_n, $z_j=-z_{n+1-j}$ ($j=1,...,n'-1$) also; in providing n' negative roots, one provides them all). The w_i above are given by

$$w_i = \left(1 + \sum_{j=1}^{n-1} p_j(z_i)^2 / c_j \right)^{-1}.$$

Procedure parameters:
$$\text{void gsswtssym } (n,zer,c,w)$$

n: int;
         entry: the weights of an n-points Gauss rule are to be computed (because of symmetry only $(n+1)/2$ of the values are delivered);
zer: double $zer[1:n/2]$;
         entry: the negative zeros of the n-th degree orthogonal polynomial ($zer[i]<zer[i+1]$, $i=1,2,...,n/2-1$; if n is odd then 0 is also a zero);
c: double $c[1:n-1]$;
         entry: the recurrence coefficients;
w: double $w[1:(n+1)/2]$;
         exit: part of the Gaussian weights divided by the integral of the weight function (note that $w[n+1-k]=w[k]$, $k=1,2,...,(n+1)/2$).

Procedure used: allortpolsym.

Remark: See the procedure *gsswts*, this procedure is supplied for storage economical reasons.

```
public static void gsswtssym(int n, double zer[], double c[], double w[])
{
    int i,twoi,low,up;
    double s;
    double p[] = new double[n];

    low=1;
    up=n;
    while (low < up) {
        Algebraic_eval.allortpolsym(n-1,zer[low],c,p);
        s=p[n-1]*p[n-1];
        for (i=n-1; i>=1; i--) s=s/c[i]+p[i-1]*p[i-1];
        w[low]=1.0/s;
```

```
        low++;
        up--;
    }
    if (low == up) {
        s=1.0;
        for (twoi=n-1; twoi>=2; twoi -= 2)
        s=s*c[twoi-1]/c[twoi]+1.0;
        w[low]=1.0/s;
    }
}
```

4.2.4 Gaussian quadrature - Special weights

A. gssjacwghts

Computes the n roots, x_i, of the n-th Jacobi polynomial $P_n^{(\alpha,\beta)}$ and corresponding Gauss-Christoffel numbers w_i needed for the n-point Gauss-Jacobi quadrature over [-1,1] with weight function $(1-x)^\alpha(1+x)^\beta$. The quadrature formula

$$\int_{-1}^{1}(1-x)^\alpha(1+x)^\beta f(x)dx = \sum_{i=1}^{n} w_i f(x_i)$$

is exact if f is a polynomial of degree $\leq 2n-1$.

The x_i are first determined by means of a call of *alljaczer*, and the w_i then derived by use of the formula

$$w_i = \left(\sum_{j=0}^{n-1} p_j^{(\alpha,\beta)}(x_i)^2\right)^{-1}.$$

Procedure parameters:

$$\text{void gssjacwghts } (n,alfa,beta,x,w)$$

n:　　　　　int;
　　　　　　entry: the upper bound of the arrays x and w $(n\geq1)$;
alfa,beta:　double;
　　　　　　entry: the parameters of the weight function for the Jacobi polynomials,
　　　　　　　　　alfa, beta > -1;
x:　　　　　double $x[1:n]$;
　　　　　　exit: $x[i]$ is the i-th zero of the n-th Jacobi polynomial;
w:　　　　　double $w[1:n]$;
　　　　　　exit: $w[i]$ is the Gauss-Christoffel number associated with the i-th zero of the n-th Jacobi polynomial.

Procedures used: gamma, alljaczer.

Remark: See [AbS65, St72].

```
public static void gssjacwghts(int n, double alfa, double beta, double x[],
                               double w[])
{
    int i,j,m;
    double r0,r1,r2,s,h0,alfa2,xi,min,sum,alfabeta,temp;

    if (alfa == beta) {
        double b[] = new double[n];
        Linear_algebra.alljaczer(n,alfa,alfa,x);
        alfa2=2.0*alfa;
        temp=Special_functions.gamma(1.0+alfa);
        h0=Math.pow(2.0,alfa2+1.0)*temp*temp/Special_functions.gamma(alfa2+2.0);
        b[1]=1.0/Math.sqrt(3.0+alfa2);
        m=n-n/2;
        for (i=2; i<=n-1; i++)
            b[i]=Math.sqrt(i*(i+alfa2)/(4.0*(i+alfa)*(i+alfa)-1.0));
        for (i=1; i<=m; i++) {
            xi=Math.abs(x[i]);
            r0=1.0;
            r1=xi/b[1];
            s=1.0+r1*r1;
            for (j=2; j<=n-1; j++) {
                r2=(xi*r1-b[j-1]*r0)/b[j];
                r0=r1;
                r1=r2;
                s += r2*r2;
            }
            w[i]=w[n+1-i]=h0/s;
        }
    } else {
        double a[] = new double[n+1];
        double b[] = new double[n+1];
        alfabeta=alfa+beta;
        min=(beta-alfa)*alfabeta;
        b[0]=0.0;
        sum=alfabeta+2.0;
        a[0]=(beta-alfa)/sum;
        a[1]=min/sum/(sum+2.0);
        b[1]=2.0*Math.sqrt((1.0+alfa)*(1.0+beta)/(sum+1.0))/sum;
        for (i=2; i<=n-1; i++) {
            sum=i+i+alfabeta;
            a[i]=min/sum/(sum+2.0);
            b[i]=(2.0/sum)*Math.sqrt(i*(sum-i)*(i+alfa)*(i+beta)/(sum*sum-1.0));
        }
        h0=Math.pow(2.0,alfabeta+1.0)*Special_functions.gamma(1.0+alfa)*
            Special_functions.gamma(1.0+beta)/Special_functions.gamma(2.0+alfabeta);
        Linear_algebra.alljaczer(n,alfa,beta,x);
        for (i=1; i<=n; i++) {
            xi=x[i];
```

```
            r0=1.0;
            r1=(xi-a[0])/b[1];
            sum=1.0+r1*r1;
            for (j=2; j<=n-1; j++) {
                r2=((xi-a[j-1])*r1-b[j-1]*r0)/b[j];
                sum += r2*r2;
                r0=r1;
                r1=r2;
            }
            w[i]=h0/sum;
        }
    }
}
```

B. gsslagwghts

Computes the n roots, x_i, of the n-th Laguerre polynomial $L_n^{(\alpha)}$ and corresponding Gauss-Christoffel numbers w_i needed for the n-point Gauss-Laguerre quadrature of $f(x)$ over $(0,\infty]$ with respect to the weight function $x^\alpha e^x$. The quadrature formula

$$\int_0^\infty x^\alpha e^{-x} f(x)dx = \sum_{i=1}^n w_i f(x_i)$$

is exact if f is a polynomial of degree $\leq 2n$-1.

The x_i are first determined by means of a call of *alllagzer*, and the w_i then derived by use of the formula

$$w_i = \left(\sum_{j=0}^{n-1} L_j^{(\alpha)}(x_i)^2 \right)^{-1}.$$

Procedure parameters:

$$\text{void gsslagwghts } (n, alfa, x, w)$$

n:　　　int;

　　　　entry:　the upper bound of the arrays x and w ($n \geq 1$);

alfa:　　double;

　　　　entry:　the parameter of the weight function for the Laguerre polynomials, *alfa* > -1;

x:　　　double $x[1:n]$;

　　　　exit:　$x[i]$ is the i-th zero of the n-th Laguerre polynomial;

w:　　　double $w[1:n]$;

　　　　exit:　$w[i]$ is the Gaussian weight associated with the i-th zero of the n-th Laguerre polynomial.

Procedures used:　　gamma, alllagzer.

Remark:　　　The zeros and weights are computed in the same way as in the procedure *gssjacwghts*.

```
public static void gsslagwghts(int n, double alfa, double x[], double w[])
{
    int i,j;
    double h0,s,r0,r1,r2,xi;
    double a[] = new double[n+1];
    double b[] = new double[n+1];

    a[0]=1.0+alfa;
    a[1]=3.0+alfa;
    b[1]=Math.sqrt(a[0]);
    for (i=2; i<=n-1; i++) {
        a[i]=i+i+alfa+1.0;
        b[i]=Math.sqrt(i*(i+alfa));
    }
    Linear_algebra.alllagzer(n,alfa,x);
    h0=Special_functions.gamma(1.0+alfa);
    for (i=1; i<=n; i++) {
        xi=x[i];
        r0=1.0;
        r1=(xi-a[0])/b[1];
        s=1.0+r1*r1;
        for (j=2; j<=n-1; j++) {
            r2=((xi-a[j-1])*r1-b[j-1]*r0)/b[j];
            r0=r1;
            r1=r2;
            s += r2*r2;
        }
        w[i]=h0/s;
    }
}
```

4.3 Numerical differentiation

4.3.1 Calculation with difference formulas

A. jacobnnf

Calculates the Jacobian matrix of an n-dimensional function of n variables using forward differences. *jacobnnf* computes first order difference quotient approximations

$$J_{i,j} = (f_i(x_1,...,x_{j-1},x_j+\delta_i,x_{j+1},...,x_n)-f_i(x_1,...,x_{j-1},x_j,x_{j+1},...,x_n))/\delta_i \quad (i,j=1,...,n)$$

to the partial derivatives $J_{i,j} = \partial f_i(x)/\partial x_j$ of the components of the function $f(x)$ $(f,x \in R^n)$.

Procedure parameters:

$$\text{void jacobnnf } (n,x,f,jac,method)$$

n: int;
         entry: the number of independent variables and the dimension of the
                 function;
x: double $x[1:n]$;
         entry: the point at which the Jacobian has to be calculated;
f: double $f[1:n]$;
         entry: the values of the function components at the point given in array
                 x;
jac: double $jac[1:n,1:n]$;
         exit: the Jacobian matrix in such a way that the partial derivative of
                 $f[i]$ with respect to $x[j]$ is given in $jac[i,j]$, $i,j=1,...,n$;
method: a class that defines two procedures di and funct, this class must
         implement the AE_jacobnnf_methods interface;
         double di(int i, int n)
                 the partial derivatives to $x[i]$ are approximated by forward
                 differences, using an increment in the i-th variable equal to the
                 value of di, $i=1,...,n$;
         void funct(int n, double x[ ], double f[ ])
                 the meaning of the parameters of the procedure funct is as follows:
                 n: the number of independent variables of the function f;
                 x: the independent variables are given in $x[1:n]$;
                 f: after a call of funct the function components should be given in
                     $f[1:n]$.

```
public static void jacobnnf(int n, double x[], double f[], double jac[][],
                    AE_jacobnnf_methods method)
{
    int i,j;
    double step,aid;
    double f1[] = new double[n+1];

    for (i=1; i<=n; i++) {
        step=method.di(i,n);
        aid=x[i];
        x[i]=aid+step;
        step=1.0/step;
        method.funct(n,x,f1);
        for (j=1; j<=n; j++) jac[j][i]=(f1[j]-f[j])*step;
        x[i]=aid;
    }
}
```

B. jacobnmf

Calculates the Jacobian matrix of an n-dimensional function of m variables using forward differences. *jacobnmf* computes first order difference quotient approximations

$$J_{i,j} = (f_i(x_1,\ldots,x_{j-1},x_j+\delta_i,x_{j+1},\ldots,x_m)-f_i(x_1,\ldots,x_{j-1},x_j,x_{j+1},\ldots,x_m))/\delta_i \qquad (i,j=1,\ldots,n;j=1,\ldots m)$$

to the partial derivatives $J_{i,j} = \partial f_i(x)/\partial x_j$ of the components of the function $f(x)$ $(f \in R^n, x \in R^m)$.

Procedure parameters:
$$\text{void jacobnmf }(n,m,x,f,jac,method)$$

n: int;
entry: the number of function components;
m: int;
entry: the number of independent variables;
x: double $x[1:m]$;
entry: the point at which the Jacobian has to be calculated;
f: double $f[1:n]$;
entry: the values of the function components at the point given in array x;
jac: double $jac[1:n,1:m]$;
exit: the Jacobian matrix in such a way that the partial derivative of $f[i]$ with respect to $x[j]$ is given in $jac[i,j]$, $i=1,\ldots,n$, $j=1,\ldots,m$;
method: a class that defines two procedures *di* and *funct*, this class must implement the AE_jacobnmf_methods interface;
double di(int i, int m)
the partial derivatives to $x[i]$ are approximated by forward differences, using an increment in the i-th variable equal to the value of *di*, $i=1,\ldots,m$;
void funct(int n, int m, double x[ ], double f[ ])
the meaning of the parameters of the procedure *funct* is as follows:
n: the number of function components;
m: the number of independent variables of the function f;
x: the independent variables are given in $x[1:m]$;
f: after a call of *funct* the function components should be given in $f[1:n]$.

```
package numal;

public interface AE_jacobnmf_methods {

  double di(int i, int m);
  void funct(int n, int m, double x[], double f[]);
}
```

```
public static void jacobnmf(int n, int m, double x[], double f[], double jac[][],
                            AE_jacobnmf_methods method)
{
    int i,j;
    double step,aid;
    double f1[] = new double[n+1];

    for (i=1; i<=m; i++) {
        step=method.di(i,m);
        aid=x[i];
        x[i]=aid+step;
        step=1.0/step;
        method.funct(n,m,x,f1);
        for (j=1; j<=n; j++) jac[j][i]=(f1[j]-f[j])*step;
        x[i]=aid;
    }
}
```

C. jacobnbndf

Calculates the Jacobian matrix of an n-dimensional function of n variables, if this Jacobian is known to be a band matrix and have to be stored rowwise in a one-dimensional array. *jacobnbndf* computes first order difference quotient approximations

$$J_{i,j} = (f_i(x_1,...,x_{j-1},x_j+\delta_i,x_{j+1},...,x_n)-f_i(x_1,...,x_{j-1},x_j,x_{j+1},...,x_n))/\delta i$$

for $i=1,...n$; $max(1,i-lw)\leq j\leq min(n,i+rw)$ to the partial derivatives $J_{i,j} = \partial f_i(x)/\partial x_j$ of the components of the function $f(x)$ $(f,x \in R^n)$.

Procedure parameters:
                    void jacobnbndf ($n,lw,rw,x,f,jac,method$)

n: int;
            entry: the number of independent variables and the dimension of the function;
lw: int;
            entry: the number of codiagonals to the left of the main diagonal of the Jacobian matrix, which is known to be a band matrix;
rw: int;
            entry: the number of codiagonals to the right of the main diagonal of the Jacobian matrix;
x: double $x[1:n]$;
            entry: the point at which the Jacobian has to be calculated;
f: double $f[1:n]$;
            entry: the values of the function components at the point given in array x;
jac: double $jac[1:(lw+rw)*(n-1)+n]$;
            exit: the Jacobian matrix in such a way that the (i,j)-th element of the Jacobian, i.e., the partial derivative of $f[i]$ to $x[j]$ is given in $jac[(lw+rw)*(i-1)+j]$, $i=1,...,n$, $j=max(1,i-lw),...,min(n,i+rw)$;

method: a class that defines two procedures *di* and *funct*, this class must implement the AE_jacobnbndf_methods interface;

    double di(int i, int n)

        the partial derivatives to *x[i]* are approximated with forward differences, using an increment to the *i*-th variable that equals the value of *di*, *i=1,...,n*;

    void funct(int n, int l, int u, double x[ ], double f[ ])

        the meaning of the parameters of the procedure *funct* is as follows:

        n: the number of function components;

        l,u: the lower and upper bound of the function component subscript;

        x: the independent variables are given in *x[1:n]*;

        f: after a call of *funct* the function components *f[i]*, *i=l,...,u*, should be given in *f[l:u]*.

```
package numal;

public interface AE_jacobnbndf_methods {

  double di(int i, int n);
  void funct(int n, int l, int u, double x[], double f[]);
}


public static void jacobnbndf(int n, int lw, int rw, double x[], double f[],
                        double jac[], AE_jacobnbndf_methods method)
{
    int i,j,k,l,u,t,b,ll;
    double aid,stepi;

    l=1;
    u=lw+1;
    t=rw+1;
    b=lw+rw;
    for (i=1; i<=n; i++) {
        ll=l;
        double f1[] = new double[u+1];
        stepi=method.di(i,n);
        aid=x[i];
        x[i]=aid+stepi;
        method.funct(n,l,u,x,f1);
        x[i]=aid;
        k = i+((i <= t) ? 0 : i-t)*b;
        for (j=l; j<=u; j++) {
            jac[k]=(f1[j]-f[j])/stepi;
            k += b;
        }
        if (i >= t) l++;
        if (u < n) u++;

    }
}
```

5. Analytic Problems

5.1 Nonlinear equations

5.1.1 Single equation - No derivative available

A. zeroin

Given the values, x_0 and y_0 say, of the end points of an interval assumed to contain a zero of the function $f(x)$, *zeroin* attempts to find, by use of a combination of linear interpolation, extrapolation, and bisection, values, x_n and y_n say, of the end points of a smaller interval containing the zero. The method used is described in detail in [BusD75]. At successful exit $f(x_n)f(y_n){\leq}0$, $|f(x_n)|{\leq}|f(y_n)|$, and $|x_n{-}y_n|{\leq}2{*}tol(x_n)$ where *tol* is a tolerance function prescribed by the user. The value of $tol(x)$ may be changed during the course of the computation or left constant, as the user requires. For example $tol(x)$ may be the function $|x|{*}re{+}ae$ (where *re*, the permissible relative error, should be greater than the machine precision, and *ae*, the permissible absolute error, should not be set equal to zero if the prescribed interval contains the origin).

The algorithm used results in the construction of five sequences of real numbers x_i, a_i, b_i, c_i, and d_i, and two integer sequences $j(i)$, $k(i)$ $(i=1,...,n)$ as follows. Initially, with $x_1=y_0$, if $|f(x_1)|{\leq}|f(x_0)|$ then $b_1=x_1$, $a_1=c_1=x_0$, otherwise $b_1=x_0$, $a_1=c_1=x_1$. Then, for $i=2,...,n$:

(a) $j(i)$ is taken to be the largest integer in the range $1{<}j(i){<}i$ for which $|b_{j(i)}{-}c_{j(i)}|{\leq}{\frac{1}{2}}|b_{j(i)-1}{-}c_{j(i)-1}|$ with $j(i)=1$ if this condition is never satisfied;

(b) with

$$h(b,c) = b + sign(c-b)tol(b)$$
$$m(b,c) = (b+c)/2$$

$$l(b,a) = b - f(b)(b-a)/(f(b)-f(a)) \qquad (f(a){\neq}f(b))$$
$$l(b,a) = \infty \qquad\qquad\qquad\qquad\quad (f(a)=f(b){\neq}0)$$
$$l(b,a) = b \qquad\qquad\qquad\qquad\quad (f(a)=f(b)=0)$$

and

$$w(l,b,c) = l \qquad\qquad \text{if } l \text{ is between } h(b,c) \text{ and } m(b,c);$$
$$w(l,b,c) = h(b,c) \qquad \text{if } |l{-}b|{\leq}tol(b) \text{ and } l \text{ does not lie outside the interval}$$
$$\text{bounded by } b \text{ and } m(b,c);$$
$$w(l,b,c) = m(b,c) \qquad \text{otherwise}$$

and $r(b,a,d)$ determined by the condition that

$$(x-r(b,a,d))/(px+q) = f(x)$$

when $x=a,b,d$, x_i is determined by use of the formula

$$x_i = w(l(b_{i-1},a_{i-1}), b_{i-1},c_{i-1}) \qquad \text{if } j(i) \geq i{-}2$$
$$x_i = w(r(b_{i-1},a_{i-1},d_{i-1}), b_{i-1},c_{i-1}) \qquad \text{if } j(i) = i{-}3$$
$$x_i = m(b_{i-1},c_{i-1}) \qquad\qquad\qquad\qquad \text{otherwise}$$

(c) $k(i)$ is taken to be the largest integer in the range $0 \le k(i) < i$ for which $f(x_{k(i)})f(x_i) \le 0$, and if $|f(x_i)| \le |f(x_{k(i)})|$, then $b_i = x_i$, $c_i = x_{k(i)}$, $a_i = b_{i-1}$, otherwise $b_i = x_{k(i)}$, $a_i = c_i = x_i$; also, if $b_i = x_i$ or $b_i = b_{i-1}$ then $d_i = a_{i-1}$, otherwise $d_i = b_{i-1}$;

(d) n is the smallest integer for which $|b_n - c_n| \le 2*tol(b_n)$, and then $x_n = b_n$, $y_n = c_n$.

The number of evaluations of *fx* and *tolx* is at most $4*\log_2(|x-y|)/tau$, where x and y are the argument values given upon entry, *tau* is the minimum of the tolerance function *tolx* on the initial interval. If upon entry x and y satisfy $f(x)*f(y) \le 0$, then convergence is guaranteed, and the asymptotic order of convergence is 1.618 for simple zeros.

Procedure parameters:

boolean zeroin (x,y,method)

zeroin: on exit *zeroin* is given the value true when a sufficiently small subinterval of J containing a zero of the function $f(x)$ has been found; otherwise *zeroin* is given the value false;

x: double *x[0:0]*;

entry: one endpoint of interval J in which a zero is searched for;

exit: a value approximating the zero within the tolerance $2*tol(x)$ when *zeroin* has the value true, and a presumably worthless argument value otherwise;

y: double *y[0:0]*;

entry: the other endpoint of interval J in which a zero is searched for; upon entry $x < y$ as well as $y < x$ is allowed;

exit: the other straddling approximation of the zero, i.e., upon exit the values of y and x satisfy

(a) $f(x)*f(y) \le 0$,

(b) $|x-y| \le 2*tol(x)$, and

(c) $|f(x)| \le |f(y)|$ when *zeroin* has a value true;

method: a class that defines two procedures *fx*, and *tolx*, this class must implement the `AP_zeroin_methods` interface;

double fx(double x[ ])

defines function *f* as a function depending on the actual parameter for *x*;

double tolx(double x[ ])

defines the tolerance function *tol* which may depend on the actual parameter for *x*; one should choose *tolx* positive and never smaller than the precision of the machine's arithmetic at *x*, i.e., in this arithmetic *x+tolx* and *x-tolx* should always yield values distinct from *x*; otherwise the procedure may get into a loop.

```
package numal;


public interface AP_zeroin_methods {

    double fx(double x[]);
    double tolx(double x[]);
}
```

```java
public static boolean zeroin(double x[], double y[], AP_zeroin_methods method)
{
    boolean extrapolate;
    int ext;
    double c,fc,b,fb,a,fa,d,fd,fdb,fda,w,mb,tol,m,p,q;

    fd=d=0.0;
    b = x[0];
    fb=method.fx(x);
    a = x[0] = y[0];
    fa=method.fx(x);
    c=a;
    fc=fa;
    ext=0;
    extrapolate=true;
    while (extrapolate) {
        if (Math.abs(fc) < Math.abs(fb)) {
            if (c != a) {
                d=a;
                fd=fa;
            }
            a=b;
            fa=fb;
            b = x[0] =c;
            fb=fc;
            c=a;
            fc=fa;
        }
        tol=method.tolx(x);
        m=(c+b)*0.5;
        mb=m-b;
        if (Math.abs(mb) > tol) {
            if (ext > 2)
                w=mb;
            else {
                if (mb == 0.0)
                    tol=0.0;
                else
                    if (mb < 0.0) tol = -tol;
                p=(b-a)*fb;
                if (ext <= 1)
                    q=fa-fb;
                else {
                    fdb=(fd-fb)/(d-b);
                    fda=(fd-fa)/(d-a);
                    p *= fda;
                    q=fdb*fa-fda*fb;
                }
                if (p < 0.0) {
                    p = -p;
```

```
              q = -q;
          }
          w= (p<Double.MIN_VALUE || p<=q*tol) ? tol : ((p<mb*q) ? p/q : mb);
      }
      d=a;
      fd=fa;
      a=b;
      fa=fb;
      x[0] = b += w;
      fb=method.fx(x);
      if ((fc >= 0.0) ? (fb >= 0.0) : (fb <= 0.0)) {
          c=a;
          fc=fa;
          ext=0;
      } else
          ext = (w == mb) ? 0 : ext+1;
    } else
        break;
  }
  y[0] = c;
  return ((fc >= 0.0) ? (fb <= 0.0) : (fb >= 0.0));
}
```

B. zeroinrat

Given the values, x_0 and y_0 say, of the end points of an interval assumed to contain a zero of the function $f(x)$, *zeroinrat* attempts to find, by use of a combination of linear interpolation, extrapolation, and bisection, values, x_n and y_n say, of the end points of a smaller interval containing the zero. The method used is described in detail in [BusD75]. At successful exit $f(x_n)f(y_n) \leq 0$, $|f(x_n)| \leq |f(y_n)|$, and $|x_n-y_n| \leq 2*tol(x_n)$ where *tol* is a tolerance function prescribed by the user   (see the documentation of *zeroin*).

The algorithm used by *zeroinrat* closely resembles that used by *zeroin*, except that x_i is now determined by use of the formula

$$x_i = w(l(b_{i-1},a_{i-1}),\ b_{i-1},c_{i-1}) \qquad \text{if } j(i) \geq i-2$$
$$x_i = w(r(b_{i-1},a_{i-1},d_{i-1}),\ b_{i-1},c_{i-1}) \qquad \text{if } j(i) = i-3$$
$$x_i = w(2r(b_{i-1},a_{i-1},d_{i-1})-b_{i-1},b_{i-1},c_{i-1}) \qquad \text{if } \geq 3 \text{ and } j(i)=i-4$$
$$x_i = m(b_{i-1},c_{i-1}) \qquad \text{otherwise.}$$

The number of evaluations of *fx* and *tolx* is at most $5*\log_2(|x-y|)/tau$, where x and y are the argument values given upon entry, *tau* is the minimum of the tolerance function *tolx* on the initial interval. If upon entry x and y satisfy $f(x)*f(y) \leq 0$, then convergence is guaranteed and the asymptotic order of convergence is 1.839 for simple zeros.

zeroin is preferable for simple (i.e., cheaply to calculate) functions and/or when no high precision is required. *zeroinrat* is preferable for complicated (i.e., expensive) functions when a zero is required in rather high precision and also for functions having a pole near the zero. When the analytic derivative of the function is easily obtained, then *zeroinder* should be taken into consideration.

Procedure parameters:

$$\text{boolean zeroinrat } (x,y,method)$$

zeroinrat: on exit *zeroinrat* is given a value true when a sufficiently small subinterval of *J* containing a zero of the function *f(x)* has been found; otherwise, *zeroinrat* is given the value false;

x: double *x[0:0]*;

  entry: one endpoint of interval *J* in which a zero is searched for;

  exit: a value approximating the zero within the tolerance 2**tol(x)* when *zeroinrat* has a value true;

y: double *y[0:0]*;

  entry: the other endpoint of interval *J* in which a zero is searched for; upon entry *x < y* as well as *y < x* is allowed;

  exit: the other straddling approximation of the zero, i.e., upon exit the values of *y* and *x* satisfy

  (a) $f(x)^*f(y) \leq 0$,

  (b) $|x-y| \leq 2^*tol(x)$, and

  (c) $|f(x)| \leq |f(y)|$ when *zeroinrat* has a value true;

method: a class that defines two procedures *fx*, and *tolx*, this class must implement the `AP_zeroinrat_methods` interface;

  double fx(double x[ ])

    defines function *f* as a function depending on the actual parameter for *x*;

  double tolx(double x[ ])

    defines the tolerance function *tol* which may depend on the actual parameter for *x*; one should choose *tolx* positive and never smaller than the precision of the machine's arithmetic at *x*, i.e., in this arithmetic *x+tolx* and *x-tolx* should always yield values distinct from *x*; otherwise, the procedure may get into a loop.

```
package numal;

public interface AP_zeroinrat_methods {

    double fx(double x[]);
    double tolx(double x[]);
}



public static boolean zeroinrat(double x[], double y[], AP_zeroinrat_methods method)
{
    boolean extrapolate, first;
    int ext;
    double b,fb,a,fa,d,fd,c,fc,fdb,fda,w,mb,tol,m,p,q;

    fd=d=0.0;
     b = x[0];
    fb=method.fx(x);
     a = x[0] = y[0];
    fa=method.fx(x);
    first=true;
```

```
c=a;
fc=fa;
ext=0;
extrapolate=true;
while (extrapolate) {
    if (Math.abs(fc) < Math.abs(fb)) {
        if (c != a) {
            d=a;
            fd=fa;
        }
        a=b;
        fa=fb;
        b = x[0] =c;
        fb=fc;
        c=a;
        fc=fa;
    }
    tol=method.tolx(x);
    m=(c+b)*0.5;
    mb=m-b;
    if (Math.abs(mb) > tol) {
        if (ext > 3)
            w=mb;
        else {
            if (mb == 0.0)
                tol=0.0;
            else
                if (mb < 0.0) tol = -tol;
            p=(b-a)*fb;
            if (first) {
                q=fa-fb;
                first=false;
            } else {
                fdb=(fd-fb)/(d-b);
                fda=(fd-fa)/(d-a);
                p *= fda;
                q=fdb*fa-fda*fb;
            }
            if (p < 0.0) {
                p = -p;
                q = -q;
            }
            if (ext == 3) p *= 2.0;
            w=(p<Double.MIN_VALUE || p<=q*tol) ? tol : ((p<mb*q) ? p/q : mb);
        }
        d=a;
        fd=fa;
        a=b;
        fa=fb;
        x[0] = b += w;
```

```
            fb=method.fx(x);
            if ((fc >= 0.0) ? (fb >= 0.0) : (fb <= 0.0)) {
                c=a;
                fc=fa;
                ext=0;
            } else
                ext = (w == mb) ? 0 : ext+1;
        } else
            break;
    }
    y[0] = c;
    return ((fc >= 0.0) ? (fb <= 0.0) : (fb >= 0.0));
}
```

5.1.2 Single equation - Derivative available

zeroinder

Given the values, x_0 and y_0 say, of the end points of an interval assumed to contain a zero of the function $f(x)$, zeroinder attempts to find, by use of a combination of interpolation and extrapolation based upon the use of a fractional linear function, and of bisection, values, x_n and y_n say, of the end points of a smaller interval containing the zero. At successful exit $f(x_n)f(y_n) \le 0$, $|f(x_n)| \le |f(y_n)|$, and $|x_n-y_n| \le 2*tol(x_n)$ where tol is a tolerance function prescribed by the user (see the documentation of zeroin).

The algorithm used by zeroinder is similar to that used by zeroin, with the difference that the estimate $l(b,a)$ of the zero of $f(x)$ obtained by linear interpolation between the two function values at $x=a$ and $x=b$ is replaced by an estimate obtained by interpolation based upon use of a fractional linear function $(x-u)/(vx+w)$ whose value agrees with that of $f(x)$ at $x=a$ and $x=b$, and whose derivative is also equal in value to that of $f(x)$ at one of these points.

zeroinder is to prefer to zeroin or zeroinrat if the derivative is (much) cheaper to evaluate than the function.

The number of evaluations of fx, dfx, and tolx is at most $4*\log_2(|x-y|)/tau$, where x and y are the argument values given upon entry, tau is the minimum of the tolerance function tolx on the initial interval (i.e., zeroinder requires at most four times the number of steps required for bisection). If upon entry x and y satisfy $f(x)*f(y) \le 0$, then convergence is guaranteed, and the asymptotic order of convergence is 2.414 for a simple zero of f.

Procedure parameters:
                        boolean zeroinder $(x,y,method)$
zeroinder: on exit zeroinder is given a value true when a sufficiently small subinterval of J containing a zero of the function $f(x)$ has been found; otherwise, zeroin is given the value false;

x: double *x[0:0]*;
              entry: one endpoint of interval *J* in which a zero is searched for;
              exit: a value approximating the zero within the tolerance 2**tol(x)*
                      when *zeroinder* has a value true;
y: double *y[0:0]*;
              entry: the other endpoint of interval *J* in which a zero is searched for;
                      upon entry *x* < *y* as well as *y* < *x* is allowed;
              exit: the other straddling approximation of the zero, i.e., upon exit
                      the values of *y* and *x* satisfy
                      (a) $f(x)^*f(y) \leq 0$,
                      (b) $|x-y| \leq 2^*tol(x)$, and
                      (c) $|f(x)| \leq |f(y)|$ when *zeroinder* has a value true;
method: a class that defines three procedures *fx*, *dfx*, and *tolx*, this class must
              implement the `AP_zeroinder_methods` interface;
              double fx(double x[ ])
                  defines function *f* as a function depending on the actual parameter
                  for *x*;
              double dfx(double x[ ])
                  defines derivative *df* of *f* as a function depending on the actual
                  parameter for *x*;
              double tolx(double x[ ])
                  defines the tolerance function *tol* which may depend on the actual
                  parameter for *x*; one should choose *tolx* positive and never smaller
                  than the precision of the machine's arithmetic at *x*, i.e., in this
                  arithmetic *x+tolx* and *x-tolx* should always yield values distinct from
                  x; otherwise, the procedure may get into a loop.

```
package numal;

public interface AP_zeroinder_methods {

    double fx(double x[]);
    double tolx(double x[]);
    double dfx(double x[]);
}



public static boolean zeroinder(double x[], double y[], AP_zeroinder_methods method)
{
    boolean extrapolate;
    int ext;
    double b,fb,dfb,a,fa,dfa,c,fc,dfc,d,w,mb,tol,m,p,q;

    b = x[0];
    fb=method.fx(x);
    dfb=method.dfx(x);
    a = x[0] = y[0];
    fa=method.fx(x);
    dfa=method.dfx(x);
    c=a;
```

```
fc=fa;
dfc=dfa;
ext=0;
extrapolate=true;
while (extrapolate) {
    if (Math.abs(fc) < Math.abs(fb)) {
        a=b;
        fa=fb;
        dfa=dfb;
        b = x[0] =c;
        fb=fc;
        dfb=dfc;
        c=a;
        fc=fa;
        dfc=dfa;
    }
    tol=method.tolx(x);
    m=(c+b)*0.5;
    mb=m-b;
    if (Math.abs(mb) > tol) {
        if (ext > 2)
            w=mb;
        else {
            if (mb == 0.0)
                tol=0.0;
            else
                if (mb < 0.0) tol = -tol;
            d = (ext == 2) ? dfa : (fb-fa)/(b-a);
            p=fb*d*(b-a);
            q=fa*dfb-fb*d;
            if (p < 0.0) {
                p = -p;
                q = -q;
            }
            w=(p<Double.MIN_VALUE || p<=q*tol) ? tol : ((p<mb*q) ? p/q : mb);
        }
        a=b;
        fa=fb;
        dfa=dfb;
        x[0] = b += w;
        fb=method.fx(x);
        dfb=method.dfx(x);
        if ((fc >= 0.0) ? (fb >= 0.0) : (fb <= 0.0)) {
            c=a;
            fc=fa;
            dfc=dfa;
            ext=0;
        } else
            ext = (w == mb) ? 0 : ext+1;
    } else
```

```
        break;
    }
    y[0] = c;
    return ((fc >= 0.0) ? (fb <= 0.0) : (fb >= 0.0));
}
```

5.1.3 System of equations - No Jacobian matrix

A. quanewbnd

Solves systems of nonlinear equations of which the Jacobian is known to be a band matrix and an approximation of the Jacobian is assumed to be available at the initial guess. The method used is the same as given in [Bro71].

quanewbnd computes an approximation to the solution x of the system of equations $f(x)=0$ $(f,x \in R^n)$ for which the components $J^{(k,j)}=\partial f^{(k)}(x)/\partial x^{(j)}$ of the associated Jacobian matrix are zero when $j>i+rw$ or $i>j+lw$, an approximation x_0 to x and an approximation J_0 to $\partial f(x_0)/\partial x_0$ being assumed to be available.

At the i-th stage of the iterative method used, δ_i is obtained by solving (*decsolbnd* is called for this purpose) the band matrix system of equations

$$J_i \delta_i = -f(x_i) \qquad (1)$$

with $x_{i+1} = x_i + \delta_i$ and with $p_i^{(j)} = (0,...,0,\delta_i^{k(j)},...,\delta_i^{h(j)},0,...,0)$ where $h(j)=\max(1,j-lw)$, $k(j)=\min(n,j+rw)$, and with I_j being the matrix whose (j,j)-th element is unity, all other elements being zero, and (the bars denoting Euclidean norms) with

$\Delta_i^{(j)} = I_j f(x_{i+1})(p_i^{(j)})^T/(p_i^{(j)},p_i^{(j)})$ if $\|p_i^{(j)}\| \geq \varepsilon^2 \|\delta_i\| \|p_i\|$, and 0 otherwise

(ε being the machine precision) J_{i+1} is obtained from the relationship

$$J_{i+1} = J_i + \sum_{j=1}^{n} \Delta_i^{(j)}$$

formula (1) is reapplied and the process continued.

Procedure parameters:

        void quanewbnd (*n,lw,rw,x,f,jac,method,in,out*)

n: int;
      entry: the number of independent variables; the number of equations should also be equal to n;
lw: int;
      entry: the number of codiagonals to the left of the main diagonal of the Jacobian;
rw: int;
      entry: the number of codiagonals to the right of the main diagonal of the Jacobian;
x: double $x[1:n]$;
      entry: an initial estimate of the solution of the system that has to be solved;
      exit: the calculated solution of the system;

f: double *f[1:n]*;

          entry: the values of the function components at the initial guess;

          exit: the values of the function components at the calculated solution;

jac: double *jac[1:(lw+rw)*(n-1)+n]*;

          entry: an approximation of the Jacobian at the initial estimate of the solution;

                  an approximation of the (i,j)-th element of the Jacobian is given in *jac[(lw+rw)*(i-1)+j]*, for $i=1,...,n$ and $j=\max(1,i-lw),...,\min(n,i+rw)$;

          exit: an approximation to the Jacobian at the calculated solution;

method: a class that defines a procedure *funct*, this class must implement the AP_quanewbnd_method interface;

          boolean funct(int n, int l, int u, double x[ ], double f[ ])

          entry: the meaning of the parameters of the procedure *funct* is as follows:

               n: the number of independent variables of the function *f*;

               l,u: int; the lower and upper bound of the function component subscript;

               x: the independent variables are given in *x[1:n]*;

               f: after a call of *funct* the function components *f[i]*, *i=l,...,u*, should be given in *f[l:u]*;

          exit: if the value of *funct* is false then the execution of *quanewbnd* will be terminated, while the value of *out[5]* is set equal to 2;

in: double *in[0:4]*;

          entry: *in[0]*: the machine precision;

                in[1]: the relative precision asked for;

                in[2]: the absolute precision asked for; if the value, delivered in *out[5]* equals zero then the last correction vector *d*, say, which is a measure for the error in the solution, satisfies the inequality $\|d\| \le \|x\| * in[1]+in[2]$, whereby *x* denotes the calculated solution, given in array *x* and $\|.\|$ denotes the Euclidean norm; however, we cannot guarantee that the true error in the solution satisfies this inequality, especially if the Jacobian is (nearly) singular at the solution;

                in[3]: the maximum value of the norm of the residual vector allowed; if *out[5]*=0 then this residual vector *r*, say, satisfies: $\|r\| \le in[3]$;

                in[4]: the maximum number of function component evaluations allowed; *l-u+1* function component evaluations are counted for each call of *funct(n,l,u,x,f)*; if *out[5]*=1 then the process is terminated because the number of evaluations exceeded the value given in *in[4]*;

out: double *out[1:5]*;

          exit: *out[1]*: the Euclidean norm of the last step accepted;

                out[2]: the Euclidean norm of the residual vector at the calculated solution;

                out[3]: the number of function component evaluations performed;

                out[4]: the number of iterations carried out;

out[5]: the integer value delivered in *out[5]* gives some information about the termination of the process;

    out[5]=0: the process is terminated in a normal way; the last step and the norm of the residual vector satisfy the conditions (see *in[2]*, *in[3]*); if *out[5]*≠0 then the process is terminated prematurely;

    out[5]=1: the number of function component evaluations exceeds the value given in *in[4]*;

    out[5]=2: a call of *funct* delivered the value false;

    out[5]=3: the approximation to the Jacobian matrix turns out to be singular.

Procedures used: mulvec, dupvec, vecvec, elmvec, decsolbnd.

```
package numal;


public interface AP_quanewbnd_method {

    boolean funct(int n, int l, int u, double x[], double f[]);
}



public static void quanewbnd(int n, int lw, int rw, double x[], double f[],
                            double jac[], AP_quanewbnd_method method, double in[],
                            double out[])
{
    int l,it,fcnt,fmax,err,b,i,j,k,r,m;
    double macheps,reltol,abstol,tolres,nd,mz,res,mul,crit;
    double aux[] = new double[6];
    double delta[] = new double[n+1];

    nd=0.0;
    macheps=in[0];
    reltol=in[1];
    abstol=in[2];
    tolres=in[3];
    fmax=(int)in[4];
    mz=macheps*macheps;
    it=fcnt=0;
    b=lw+rw;
    l=(n-1)*b+n;
    b++;
    res=Math.sqrt(Basic.vecvec(1,n,0,f,f));
    err=0;
    while (true) {
        if (err != 0 || (res < tolres && Math.sqrt(nd) <
            Math.sqrt(Basic.vecvec(1,n,0,x,x))*reltol+abstol)) break;
        it++;
        if (it != 1) {
```

```java
            /* update jac */
            double pp[] = new double[n+1];
            double s[] = new double[n+1];
            crit=nd*mz;
            for (i=1; i<=n; i++) pp[i]=delta[i]*delta[i];
            r=k=1;
            m=rw+1;
            for (i=1; i<=n; i++) {
                mul=0.0;
                for (j=r; j<=m; j++) mul += pp[j];
                j=r-k;
                if (Math.abs(mul) > crit)
                    Basic.elmvec(k,m-j,j,jac,delta,f[i]/mul);
                k += b;
                if (i > lw)
                    r++;
                else
                    k--;
                if (m < n) m++;
            }
        }
        /* direction */
        double lu[] = new double[l+1];
        aux[2]=macheps;
        Basic.mulvec(1,n,0,delta,f,-1.0);
        Basic.dupvec(1,l,0,lu,jac);
        Linear_algebra.decsolbnd(lu,n,lw,rw,aux,delta);
        if (aux[3] != n) {
            err=3;
            break;
        } else {
            Basic.elmvec(1,n,0,x,delta,1.0);
            nd=Basic.vecvec(1,n,0,delta,delta);
            /* evaluate */
            fcnt += n;
            if (!(method.funct(n,1,n,x,f))) {
                err=2;
                break;
            }
            if (fcnt > fmax) err=1;
            res=Math.sqrt(Basic.vecvec(1,n,0,f,f));
        }
    }
    out[1]=Math.sqrt(nd);
    out[2]=res;
    out[3]=fcnt;
    out[4]=it;
    out[5]=err;
}
```

B. quanewbnd1

Solves systems of nonlinear equations of which the Jacobian is known to be a band matrix and an approximation to the Jacobian at the initial guess is calculated using forward differences.

Computes an approximation to the solution x of the system of equations $f(x)=0$ $(f,x \in R^n)$ for which the components $J^{(k,j)}=\partial f^{(k)}(x)/\partial x^{(j)}$ to the associated Jacobian matrix are zero when $j>i+rw$ or $i>j+lw$, an approximation x_0 to x being assumed to be available.

An approximation J_0 to $\partial f(x_0)/\partial x_0$, based upon the use of first order finite difference approximations using equal increments d for all components, is first obtained by means of a call of *jacobnbndf*, and *quanewbnd* is then called to obtain the desired approximation.

Procedure parameters:
$$\text{void quanewbnd1 } (n,lw,rw,x,f,funct,in,out)$$

n: int;

        entry: the number of independent variables; the number of equations
                should also be equal to *n*;

lw: int;

        entry: the number of codiagonals to the left of the main diagonal of the
                Jacobian;

rw: int;

        entry: the number of codiagonals to the right of the main diagonal of the
                Jacobian;

x: double $x[1:n]$;

        entry: an initial estimate of the solution of the system that has to be
                solved;

        exit: the calculated solution of the system;

f: double $f[1:n]$;

        exit: the values of the function components at the calculated solution;

method: a class that defines a procedure *funct*, this class must implement the
        `AP_quanewbnd_method` interface;

        boolean *funct(int n, int l, int u, double x[ ], double f[ ])*

        entry: the meaning of the parameters of the procedure *funct* is as
                follows:

                n: the number of independent variables of the function *f*;
                l,u: int; the lower and upper bound of the function component
                        subscript;
                x: the independent variables are given in $x[1:n]$;
                f: after a call of *funct* the function components $f[i]$, $i=l,...,u$,
                        should be given in $f[l:u]$;

        exit: if the value of *funct* is false then the execution of *quanewbnd1*
                will be terminated, while the value of *out[5]* is set equal to 2;

in: double $in[0:5]$;

        entry: *in[0]*:   the machine precision;
                 in[1]:   the relative precision asked for;
                 in[2]:   the absolute precision asked for; if the value delivered in
                        out[5] equals zero then the last correction vector *d*, say,

which is a measure for the error in the solution, satisfies the inequality $\|d\| \leq \|x\| * in[1] + in[2]$, whereby x denotes the calculated solution, given in array x and $\|.\|$ denotes the Euclidean norm; however, we cannot guarantee that the true error in the solution satisfies this inequality, especially if the Jacobian is (nearly) singular at the solution;

in[3]:   the maximum value of the norm of the residual vector allowed; if out[5]=0 then this residual vector r, say, satisfies: $\|r\| \leq in[3]$;

in[4]:   the maximum number of function component evaluations allowed; $l-u+1$ function component evaluations are counted each call of funct(n,l,u,x,f); if out[5]=1 then the process is terminated because the number of evaluations exceeded the value given in in[4];

in[5]:   the Jacobian matrix at the initial guess is approximated using forward differences, with a fixed increment to each variable that equals the value given in in[5];

out: double out[1:5];

exit: out[1]: the Euclidean norm of the last step accepted;

out[2]: the Euclidean norm of the residual vector at the calculated solution;

out[3]: the number of function component evaluations performed;

out[4]: the number of iterations carried out;

out[5]: the integer value delivered in out[5] gives some information about the termination of the process;

out[5]=0:   the process is terminated in a normal way; the last step and the norm of the residual vector satisfy the conditions (see in[2], in[3]); if out[5]≠0 then the process is terminated prematurely;

out[5]=1:   the number of function component evaluations exceeds the value given in in[4];

out[5]=2:   a call of funct delivered the value false;

out[5]=3:   the approximation of the Jacobian matrix turns out to be singular.

Procedures used: jacobnbndf, quanewbnd.

```
package numal;


public interface AP_quanewbnd_method {

  boolean funct(int n, int l, int u, double x[], double f[]);

}
```

```
public static void quanewbnd1(int n, int lw, int rw, double x[], double f[],
                      AP_quanewbnd_method method, double in[], double out[])
{
  int kk,i,j,k,l,u,t,b;
  double aid,stepi,s;
  double jac[] = new double[(lw+rw)*(n-1)+n+1];

  method.funct(n,1,n,x,f);
  s=in[5];
  kk=(lw+rw)*(n-1)+n*2-((lw-1)*lw+(rw-1)*rw)/2;
  in[4] -= kk;
  l=1;
  u=lw+1;
  t=rw+1;
  b=lw+rw;
  for (i=1; i<=n; i++) {
    double f1[] = new double[u+1];
    stepi=s;
    aid=x[i];
    x[i]=aid+stepi;
    method.funct(n,l,u,x,f1);
    x[i]=aid;
    k = i+((i <= t) ? 0 : i-t)*b;
    for (j=l; j<=u; j++) {
      jac[k]=(f1[j]-f[j])/stepi;
      k += b;
    }
    if (i >= t) l++;
    if (u < n) u++;
  }
  quanewbnd(n,lw,rw,x,f,jac,method,in,out);
  in[4]  += kk;
  out[3] += kk;
}
```

5.2 Unconstrained optimization

5.2.1 One variable - No derivative

minin

Determines a point $x \in [a,b]$ at which the real valued function $f(x)$ assumes a minimum value. It is assumed that, for some point $\mu \in (a,b)$ either (a) f is strictly monotonically decreasing on $[a,\mu)$ and strictly monotonically increasing on $[\mu,b]$, or

(b) these two intervals may be replaced by $[a,\mu]$ and $(\mu,b]$ respectively. Use is made of function values alone.

The method used involves the determination of five sequences of points a_n, b_n, v_n, w_n, and x_n, and further auxiliary numbers, u_n, p_n, and q_n $(n=0,1,...)$. Initially

$$a_0 = a, \quad b_0 = b, \quad v_0 = w_0 = x_0 = a_0 + \tfrac{1}{2}(3-\sqrt{5})(b_0-a_0).$$

At the n-th stage of the computations a local minimum is known to lie in $[a_n,b_n]$ and $x_n \in [a_n,b_n]$. If $\max(x_n-a_n, b_n-x_n) \le 2*tol(x_n)$, where $tol(x)$ is a tolerance function prescribed by the user, the computations terminate, x_n is such that $f(x_n) < f(x_r)$ either (a) for $r=0,1,...,n-1$, or (b) for $r=0,1,...,n-2$ with $f(x_n) = f(x_{n-1})$; $f(w_n) < f(x_r)$ for $r=0,1,...,n-2$ in case (a) and for $r=0,1,...,n-3$ in case (b). If the computations are not terminated, p_n, q_n are determined such that $x_n + p_n/q_n$ is a turning point of the parabola through $(v_n,f(v_n))$, $(w_n,f(w_n))$ and $(x_n,f(x_n))$:

$$p_n = \pm ((x_n-v_n)^2(f(x_n)-f(w_n))-(x_n-w_n)^2(f(x_n)-f(v_n)))$$
$$q_n = \pm 2((x_n-v_n)^2(f(x_n)-f(w_n))-(x_n-w_n)(f(w_n)-f(v_n))).$$

If either $|p_n/q_n| \le tol(x_n)$ or $q_n=0$ or $x_n+p_n/q_n \notin (a_n,b_n)$ then

$$u_{n+1} = \tfrac{1}{2}(\sqrt{5}-1)x_n+\tfrac{1}{2}(3-\sqrt{5})a_n \quad \text{if } x_n \ge \tfrac{1}{2}(a_n+b_n)$$
$$u_{n+1} = \tfrac{1}{2}(\sqrt{5}-1)x_n+\tfrac{1}{2}(3-\sqrt{5})b_n \quad \text{if } x_n < \tfrac{1}{2}(a_n+b_n);$$

otherwise, $u_{n+1} = x_n + p_n/q_n$. The points $a_n,...,w_n$ are updated as follows: if $f(u_{n+1}) \le f(x_n)$ then (a) $a_{n+1}=a_n$ and $b_{n+1}=x_n$ if $u_{n+1} < x_n$, while if $u_{n+1} \ge x_n$, $a_{n+1}=x_n$, and $b_{n+1}=b_n$, and also (b) $v_{n+1}=w_n$, $w_{n+1}=x_n$, and $x_{n+1}=u_{n+1}$; if however, $f(u_{n+1}) > f(x_n)$ then (a) $a_{n+1}=u_{n+1}$ and $b_{n+1}=b_n$ if $u_{n+1}<x_n$, while if $u_{n+1} \ge x_n$, $a_{n+1}=a_n$, and $b_{n+1}=u_{n+1}$ and also (b) $x_{n+1}=x_n$, $v_{n+1}=x_n$, and $w_{n+1}=u_{n+1}$ if $f(u_{n+1}) \le f(w_n)$ or $w_n=x_n$ while, if neither of these conditions is satisfied and either $f(u_{n+1}) \le f(v_n)$ or $v_n=x_n$ or $v_n=w_n$, then $x_{n+1}=x_n$, $v_{n+1}=u_{n+1}$, and $w_{n+1}=w_n$.

The user should be aware of the fact that the choice of $tolx$ may highly affect the behavior of the algorithm, although convergence to a point for which the given function is minimal on the interval is assured. The asymptotic behavior will usually be fine as long as the numerical function is strictly δ-unimodal on the given interval (see [Bre73]), and the tolerance function satisfies $tol(x) \ge \delta$, for all x in the given interval.

Procedure parameters:

         double minin $(x,a,b,method)$

minin: delivers the calculated minimum value of the function, defined by fx, on the interval with endpoints a and b;

x: double $x[0:0]$;

        entry: the actual parameter for fx and $tolx$ depends on x;

        exit: the calculated approximation of the position of the minimum;

a,b: double $a[0:0]$, $b[0:0]$;

        entry: the endpoints of the interval on which a minimum is searched for;

        exit: the endpoints of the interval with length less than $4*tol(x)$ such that $a<x<b$;

method: a class that defines two procedures, fx and $tolx$; this class must implement the `AP_minin_methods` interface;

        double $fx(double\ x[\ ])$

            the function is given by the actual parameter fx which depends on x;

> *double tolx(double x[ ])*
>> defines the tolerance function which may depend on *x*; a suitable
>> tolerance function is: $|x|*re+ae$, where *re* is the relative precision
>> desired, and *ae* is an absolute precision which should not be
>> chosen equal to zero.

```
package numal;

public interface AP_minin_methods {

    double fx(double x[]);
    double tolx(double x[]);
}


public static double minin(double x[], double a[], double b[],
                            AP_minin_methods method)
{
    double z,c,d,e,m,p,q,r,tol,t,u,v,w,fu,fv,fw,fz;

    d=0.0;
    c=(3.0-Math.sqrt(5.0))/2.0;
    if (a[0] > b[0]) {
        z = a[0];
        a[0] = b[0];
        b[0]=z;
    }
    w = x[0] = a[0];
    fw=method.fx(x);
    z = x[0] = b[0];
    fz=method.fx(x);
    if (fz > fw) {
        z=w;
        w = x[0];
        v=fz;
        fz=fw;
        fw=v;
    }
    v=w;
    fv=fw;
    e=0.0;
    while (true) {
        m=(a[0] + b[0])*0.5;
        tol=method.tolx(x);
        t=tol*2.0;
        if (Math.abs(z-m) <= t-(b[0] - a[0])*0.5) break;
        p=q=r=0.0;
        if (Math.abs(e) > tol) {
            r=(z-w)*(fz-fv);
            q=(z-v)*(fz-fw);
```

```java
                p=(z-v)*q-(z-w)*r;
                q=(q-r)*2.0;
                if (q > 0.0)
                    p = -p;
                else
                    q = -q;
                r=e;
                e=d;
            }
            if (Math.abs(p) < Math.abs(q*r*0.5) && p > (a[0]-z)*q && p < (b[0]-z)*q) {
                d=p/q;
                u=z+d;
                if (u-a[0] < t || b[0]-u < t) d = ((z < m) ? tol : -tol);
            } else {
                e = ((z < m) ? b[0] : a[0]) - z;
                d=c*e;
            }
            u = x[0] = z + ((Math.abs(d) >= tol) ? d : ((d>0.0) ? tol : -tol));
            fu=method.fx(x);
            if (fu <= fz) {
                if (u < z)
                    b[0]=z;
                else
                    a[0]=z;
                v=w;
                fv=fw;
                w=z;
                fw=fz;
                z=u;
                fz=fu;
            } else {
                if (u < z)
                    a[0]=u;
                else
                    b[0]=u;
                if (fu <= fw) {
                    v=w;
                    fv=fw;
                    w=u;
                    fw=fu;
                } else
                    if (fu <= fv || v == w) {
                        v=u;
                        fv=fu;
                    }
            }
        }
        x[0]=z;
        return fz;
    }
```

5.2.2 One variable - Derivative available

mininder

Determines a point $x \in [a,b]$ at which the real valued function $f(x)$ of one variable assumes a minimum value, when the analytical derivative $Df(x)$ of the function is available.

    mininder has almost the same structure as the procedure given in [Bre73]. However, cubic interpolation (see [Dav59]) is used instead of quadratic interpolation to approximate the minimum. The method used involves the determination of a sequence of intervals $[a_n, b_n]$ ($n=0,1,...$) known to contain the required minimum point, and a further sequence of points u_n ($n=1,2,...$). Initially $a_0=a$, $b_0=b$. u_{n+1} is the point at which g has a minimum value, where g is the osculating cubic for which $g(x)=f(x)$ and $Dg(x)=Df(x)$ when $x=a_n$ and $x=b_n$. With $Df(a_n) \leq 0$, $Df(b_n)>0$, $a_{n+1}=u_{n+1}$, and $b_{n+1}=b_n$ if $Df(u_n) \leq 0$, while $a_{n+1}=a_n$, $b_{n+1}=u_{n+1}$ if $Df(u_n) > 0$. Termination occurs when $b_{n+1}-a_{n+1} \leq 3*tol(u_{n+1})$, where $tol(x)$ is a tolerance function prescribed by the user.

    The user should be aware of the fact that the choice of *tolx* may highly affect the behavior of the algorithm, although convergence to a point for which the given function is minimal on the interval is assured. The asymptotic behavior will usually be fine as long as the numerical function is strictly δ-unimodal on the given interval (see [Bre73]), and the tolerance function satisfies $tol(x) \geq \delta$, for all x in the given interval. Let the value of *dfx* at the begin and end point of the initial interval be denoted by *dfa* and *dfb*, respectively, then, finding a global minimum is only guaranteed if the function is convex and $dfa \leq 0$ and $dfb \geq 0$. If these conditions are not satisfied then a local minimum or a minimum at one of the end points might be found.

Procedure parameters:

                        double mininder $(x,y,method)$

mininder: delivers the calculated minimum value of the function, defined by *fx*, on the interval with endpoints a and b;

x: double $x[0:0]$;

          entry: one of the end points of the interval on which the function has to be minimized;

          exit: the calculated approximation of the position of the minimum;

y: double $y[0:0]$;

          entry: the other end point of the interval on which the function has to be minimized;

          exit: a value such that $|x-y| \leq 3*tol(x)$;

method: a class that defines three procedures *fx*, *dfx*, and *tolx*; this class must implement the AP_mininder_method interface;

          double fx(double x[ ])

              the function is given by the actual parameter *fx* which depends on x;

> *double dfx(double x[ ])*
>> the derivative of the function is given by the actual parameter *dfx* which depends on *x*; *fx* and *dfx* are evaluated successively for a certain value of *x*;
>
> *double tolx(double x[ ])*
>> defines the tolerance function which may depend on *x*; a suitable tolerance function is: $|x|*re+ae$, where *re* is the relative precision desired, and *ae* is an absolute precision which should not be chosen equal to zero.

```
package numal;

public interface AP_mininder_methods {

    double fx(double x[]);
    double tolx(double x[]);
    double dfx(double x[]);
}



public static double mininder(double x[], double y[], AP_mininder_methods method)
{
    int sgn;
    double a,b,c,fa,fb,fu,dfa,dfb,dfu,e,d,tol,ba,z,p,q,s;

    if (x[0] <= y[0]) {
        a = x[0];
        fa=method.fx(x);
        dfa=method.dfx(x);
        b = x[0] = y[0];
        fb=method.fx(x);
        dfb=method.dfx(x);
    } else {
        b = x[0];
        fb=method.fx(x);
        dfb=method.dfx(x);
        a = x[0] = y[0];
        fa=method.fx(x);
        dfa=method.dfx(x);
    }
    c=(3.0-Math.sqrt(5.0))/2.0;
    d=b-a;
    e=d*2.0;
    z=e*2.0;
    while (true) {
        ba=b-a;
        tol=method.tolx(x);
        if (ba < tol*3.0) break;
        if (Math.abs(dfa) <= Math.abs(dfb)) {
            x[0]=a;
```

```
            sgn=1;
        } else {
            x[0]=b;
            sgn = -1;
        }
        if (dfa <= 0.0 && dfb >= 0.0) {
            z=(fa-fb)*3.0/ba+dfa+dfb;
            s=Math.sqrt(z*z-dfa*dfb);
            p = (sgn == 1) ? dfa-s-z : dfb+s-z;
            p *= ba;
            q=dfb-dfa+s*2.0;
            z=e;
            e=d;
            d = (Math.abs(p) <= Math.abs(q)*tol) ? tol*sgn : -p/q;
        } else
            d=ba;
        if (Math.abs(d) >= Math.abs(z*0.5) || Math.abs(d) > ba*0.5) {
            e=ba;
            d=c*ba*sgn;
        }
        x[0] += d;
        fu=method.fx(x);
        dfu=method.dfx(x);
        if (dfu >= 0.0 || (fu >= fa && dfa <= 0.0)) {
            b = x[0];
            fb=fu;
            dfb=dfu;
        } else {
            a = x[0];
            fa=fu;
            dfa=dfu;
        }
    }
    if (fa <= fb) {
        x[0]=a;
        y[0]=b;
        return fa;
    } else {
        x[0]=b;
        y[0]=a;
        return fb;
    }
}
```

5.2.3 More variables - Auxiliary procedures

A. linemin

Determines the real value α_{min} (>0) of α for which $f(\alpha) = F(x_0+\alpha d)$ attains a minimum, where $F(x)$ is a real valued function of $x \in R^n$; $x_0 \in R^n$ is fixed, and $d \in R^n$ is a fixed direction. It is assumed that the partial derivatives $g_i(x) = \partial F(x)/\partial x_i$ $(i=1,...,n)$ may easily be computed; we then have $f'(\alpha) = \partial f(\alpha)/\partial \alpha = d^T g(x_0+\alpha d)$.

The method employed utilises an iterative process, based upon cubic interpolation (see [Dav59]), for locating a minimum of the real-valued function f. The process involves an initial approximation α_0, three real sequences u_k, v_k, and y_k, a fixed real parameter $\mu \in (0,\frac{1}{2})$, and relative and absolute tolerances, ε_r and ε_a respectively, used to terminate the process.

Initially $y_0=u_0=0$, $v_0=\alpha_0$. At step k:

(a) if $f'(\alpha_k) \geq 0$, compute $y = v_k - (v_k-u_k)(f'(v_k)+w-z)/(f'(v_k)-f'(u_k)+2w)$
   where $z = 3(f(u_k)-f(v_k))/(v_k-u_k) + f'(u_k) + f'(v_k)$ and $w = (z^2 - f'(u_k)f'(v_k))^{1/2}$;
   set $\varepsilon_k = \|x_0 + y_k d\| \varepsilon_v + \varepsilon_a$; y_{k+1} is determined as follows:
   (i) if $(y-u_k) < \varepsilon_k$ then $y_{k+1} = u_k+\varepsilon_k$, else (ii) if $(v_k-y)<\varepsilon_k$ then $y_{k+1}=v_k-\varepsilon_k$, else (iii) $y_{k+1}=y$; u_{k+1} and v_{k+1} are determined as follows: (iv) if $f'(y_{k+1}) \geq 0$ then $v_{k+1}=y_{k+1}$ and $u_{k+1}=u_k$; else (v) $v_{k+1}=v_k$ and $u_{k+1}=y_{k+1}$.

(b) if $f'(v_k) < 0$ then (i) if $(f(v_k)-f(0))/v_k f'(0) > \mu$ (now $0 < v_k < \varepsilon_{min}$) set $u_{k+1}=v_k$ and $y_{k+1}=v_{k+1}=2v_k$; else (ii) (now $u_k < \varepsilon_{min} < v_k$) set $y_{k+1} = \frac{1}{2}(u_k+v_k)$ and if $f'(v_{k+1}) \geq 0$ or $(f(y_{k+1})-f(0))/y_{k+1}f'(0) > \mu$ then $v_{k+1}=y_{k+1}$ and $u_{k+1}=u_k$, else $v_{k+1}=v_k$ and $u_{k+1}=y_{k+1}$.

The above process is terminated if:

I. $\frac{1}{2}|v_k - u_k| < \|x_0 + y_{k+1}d\| \varepsilon_r + \varepsilon_a$,
II. $\mu \leq (f(y_k)-f(0))/y_k f'(0) \leq 1 - \mu$, or
III. the maximal permitted number of simultaneous evaluations of $f(\varepsilon)$ and $f'(\varepsilon)$ is exceeded.

In cases I and II, y_{k+1} is an acceptable approximation to ε_{min}.

The direction vector d and the initial values, x_0, $f(0)=F(x_0)$, and $f'(0)=d^T g(0)$ must be supplied by the user. The user must also prescribe the value of the variable *strongsearch*: if this value is true then the stopping criteria I and III alone are used; if the value is false then all criteria are used. The stopping criterion used when the value of *strongsearch* is false is described in [Fle63] and [GolP67]. A detailed description of this procedure is given in [Bus72b].

Procedure parameters:
   void linemin $(n,x,d,nd,alfa,g,method,f0,f1,df0,df1,evlmax,strongsearch,in)$
n: int;
             entry: the number of variables of the given function f;
x: double $x[1:n]$;
             entry: a vector x_0, such that f is decreasing in x_0, in the direction given by d;
             exit: the calculated approximation of the vector for which f is minimal on the line defined by: $x_0 + alfa*d$, $(alfa>0)$;
d: double $d[1:n]$;
             entry: the direction of the line on which f has to be minimized;

nd: double;
                 entry: the Euclidean norm of the vector given in *d[1:n]*;
alfa: double *alfa[0:0]*;
                 the independent variable that defines the position on the line on
                 which *f* has to be minimized; this line is defined by x_0 + *alfa*d*,
                 (*alfa*>0);
                 entry: an estimate *alfa0* of the value for which *h(alfa)=F(x_0+alfa*d)*
                          is minimal;
                 exit: the calculated approximation *alfam* of the value for which
                          h(alfa) is minimal;
g: double *g[1:n]*;
                 exit: the gradient of *f* at the calculated approximation of the
                          minimum;
method: a class that defines a procedure *funct*, this class must implement
                 the AP_linemin_method interface;
                 double funct(int n, double x[ ], double g[ ])
                     a call of *funct* should effectuate:
                     1.   *funct=f(x)*;
                     2.   the value of *g[i]*, (*i=1,...,n*), becomes the value of the *i*-th
                          component of the gradient of *f* at *x*;
f0:   double;
                 entry: the value of *h(0)*, (see *alfa*);
f1:   double *f1[0:0]*;
                 entry: the value of *h(alfa0)*, (see *alfa*);
                 exit: the value of *h(alfam)*, (see *alfa*);
df0:   double;
                 entry: the value of the derivative of *h* at *alfa*=0;
df1:   double *df1[0:0]*;
                 entry: the value of the derivative of *h* at *alfa=alfa0*;
                 exit: the value of the derivative of *h* at *alfa=alfam*;
evlmax: int *evlmax[0:0]*;
                 entry: the maximum allowed number of calls of *funct*;
                 exit: the number of times *funct* has been called;
strongsearch: boolean:
                 if the value of *strongsearch* is true then the process makes use of
                 two stopping criteria:
                     A: the number of times *funct* has been called exceeds the given
                          value of *evlmax*;
                     B: an interval is found with length less than two times the
                          prescribed precision, on which a minimum is expected;
                 if the value of *strongsearch* is false then the process makes also use
                 of a third stopping criterion:
                     C: $\mu \leq (h(alfak)-h(alfa0))/(alfak*df0) \leq 1 - \mu$, whereby *alfak* is
                          the current iterate and μ a prescribed constant;
in: double *in[1:3]*;
                 entry: *in[1]*:   relative precision, ε_r, necessary for the stopping
                                    criterion B (see *strongsearch*);
                          in[2]:   absolute precision, ε_a, necessary for the stopping
                                    criterion B (see *strongsearch*);

the prescribed precision, ε, at *alfa=alfak* is given by: $\varepsilon = \| x_0 + alfa*d \| \varepsilon_r + \varepsilon_a$, where $\|.\|$ denotes the Euclidean norm;

in[3]:   the parameter μ necessary for stopping criterion C; this parameter must satisfy: $0 < \mu < 0.5$; in practice, a choice of $\mu = 0.0001$ is advised.

Procedures used: vecvec, elmvec, dupvec.

```
package numal;

public interface AP_linemin_method {

    double funct(int n, double x[], double g[]);
}


public static void linemin(int n, double x[], double d[], double nd, double alfa[],
                    double g[], AP_linemin_method method, double f0,
                    double f1[], double df0, double df1[], int evlmax[],
                    boolean strongsearch, double in[])
{
    boolean notinint;
    int evl;
    double f,oldf,df,olddf,mu,alfa0,q,w,y,z,reltol,abstol,eps,aid;
    double x0[] = new double[n+1];

    reltol=in[1];
    abstol=in[2];
    mu=in[3];
    evl=0;
    alfa0=0.0;
    oldf=f0;
    olddf=df0;
    y = alfa[0];
    notinint=true;
    Basic.dupvec(1,n,0,x0,x);
    eps=(Math.sqrt(Basic.vecvec(1,n,0,x,x))*reltol+abstol)/nd;
    q=(f1[0]-f0)/(alfa[0]*df0);
    while (true) {
        if (notinint) notinint = (df1[0] < 0.0 && q > mu);
        aid = alfa[0];
        if (df1[0] >= 0.0) {
            /* cubic interpolation */
            z=3.0*(oldf-f1[0])/alfa[0]+olddf+df1[0];
            w=Math.sqrt(z*z-olddf*df1[0]);
            alfa[0] = alfa[0]*(1.0-(df1[0]+w-z)/(df1[0]-olddf+w*2.0));
            if (alfa[0] < eps)
                alfa[0]=eps;
```

```
        else
            if (aid-alfa[0] < eps) alfa[0]=aid-eps;
    } else
        if (notinint) {
            alfa0 = alfa[0] = y;
            olddf = df1[0];
            oldf = f1[0];
        } else
            alfa[0] *= 0.5;
    y = alfa[0]+alfa0;
    Basic.dupvec(1,n,0,x,x0);
    Basic.elmvec(1,n,0,x,d,y);
    eps=(Math.sqrt(Basic.vecvec(1,n,0,x,x))*reltol+abstol)/nd;
    f=method.funct(n,x,g);
    evl++;
    df=Basic.vecvec(1,n,0,d,g);
    q=(f-f0)/(y*df0);
    if (!(((notinint || strongsearch) ? true :
            (q < mu || q > 1.0-mu)) && (evl < evlmax[0])))) break;
    if (notinint || df > 0.0 || q < mu) {
        df1[0]=df;
        f1[0]=f;
    } else {
        alfa0=y;
        alfa[0]=aid-alfa[0];
        olddf=df;
        oldf=f;
    }
    if (alfa[0] <= eps*2.0) break;
}
alfa[0]=y;
evlmax[0]=evl;
df1[0]=df;
f1[0]=f;
}
```

B. rnk1upd

Determines the upper triangular part of the symmetric $n \times n$ matrix g by use of the formula $g = h + cvv^T$. h being a given symmetric matrix whose upper triangular part is available, c being a real constant, and $v \in R^n$ a vector.

Procedure parameters:
$$\text{void rnk1upd } (h,n,v,c)$$

h: double $h[1:n*(n+1)/2]$;
            entry: the upper triangle (stored columnwise, i.e., $a_{ij}=h[(j-1)*j/2+i]$, $1 \leq i \leq j \leq n$)
                    of the symmetric matrix that has to be updated;
            exit: the upper triangle (stored columnwise) of the updated matrix;

n: int;
           entry: the order of the symmetric matrix whose upper triangle is stored
                   columnwise in the one-dimensional array *h*;
v: double *v[1:n]*;
           entry: the given matrix is updated (another matrix is added to it) with a
                   symmetric matrix *u*, of rank one, defined by: $u_{i,j}=c*v[i]*v[j]$;
c: double; see *v*.

Procedure used: elmvec.

```
public static void rnk1upd(double h[], int n, double v[], double c)
{
    int j,k;

    k=0;
    j=1;
    do {
        k++;
        Basic.elmvec(j,j+k-1,1-j,h,v,v[k]*c);
        j += k;
    } while (k < n);
}
```

C. davupd

Determines the upper triangular part of the symmetric *n×n* matrix *g* by use of the
formula
$$g = h + c_1 v v^T - c_2 w w^T.$$
h being a given symmetric matrix whose upper triangular part is available,
c_1 and c_2 being real constants, and $v, w \in R^n$ vectors.

Procedure parameters:
                      void davupd (*h,n,v,w,c1,c2*)
h: double *h[1:n*(n+1)/2]*;
           entry: the upper triangle (stored columnwise, i.e., $a_{ij}=h[(j-1)*j/2+i]$, $1 \le i \le j \le n$)
                   of the symmetric matrix that has to be updated;
           exit: the upper triangle (stored columnwise) of the updated matrix;
n: int;
           entry: the order of the symmetric matrix whose upper triangle is stored
                   columnwise in the one-dimensional array *h*;
v,w: double *v[1:n]*, *w[1:n]*;
           entry: the given matrix is updated with a symmetric matrix *u*, of rank two,
                   defined by: $u_{i,j}=c1*v[i]*v[j]-c2*w[i]*w[j]$;
c1,c2: double; see *v* and *w* above.

```
public static void davupd(double h[], int n, double v[], double w[], double c1,
                          double c2)
```

```
{
    int i,j,k;
    double vk,wk;

    k=0;
    j=1;
    do {
        k++;
        vk=v[k]*c1;
        wk=w[k]*c2;
        for (i=0; i<=k-1; i++) h[i+j] += v[i+1]*vk-w[i+1]*wk;
        j += k;
    } while (k < n);
}
```

D. fleupd

Determines the upper triangular part of the symmetric $n \times n$ matrix g by use of the formula

$$g = h - c_1 v w^T - c_2 w v^T + (1 + (c_1/c_2)c_1 v v^T).$$

h being a given symmetric matrix whose upper triangular part is available, c_1 and $c_2 (\neq 0)$ being real constants, and $v, w \in R^n$ vectors.

Procedure parameters:

$$\text{void fleupd } (h, n, v, w, c1, c2)$$

h: double $h[1:n(n+1)/2]$;

        entry: the upper triangle (stored columnwise, i.e., $a_{ij}=h[(j-1)j/2+i]$, $1 \leq i \leq j \leq n$) of the symmetric matrix that has to be updated;

        exit: the upper triangle (stored columnwise) of the updated matrix;

n: int;

        entry: the order of the symmetric matrix whose upper triangle is stored columnwise in the one-dimensional array h;

v, w: double $v[1:n]$, $w[1:n]$;

        entry: the given matrix is updated with a symmetric matrix u, of rank two, defined by: $u_{i,j}=c2*v[i]*v[j]-c1*(v[i]*w[j]+w[i]*v[j])$;

$c1, c2$: double; see v and w above.

```
public static void fleupd(double h[], int n, double v[], double w[], double c1,
                          double c2)
{
    int i,j,k;
    double vk,wk;

    k=0;
    j=1;
    do {
        k++;
```

```
        vk = -w[k]*c1+v[k]*c2;
        wk=v[k]*c1;
        for (i=0; i<=k-1; i++) h[i+j] += v[i+1]*vk-w[i+1]*wk;
        j += k;
    } while (k < n);
}
```

5.2.4 More variables - No derivatives

praxis

Determines a point x at which the function $f(x)$ is a minimum, where $x=(x_1,...,x_n)$, by use of an iterative algorithm due to Brent [Bre73]. An initial approximation to x is to be provided.

Procedure parameters:

$$\text{void praxis } (n,x,method,in,out)$$

n: int;
              entry: the number of variables of the function to be minimized;
x: double $x[1:n]$;
              entry: an approximation of the position of the minimum;
              exit: the calculated position of the minimum;
method: a class that defines a procedure *funct*, this class must implement the
              AP_praxis_method interface;
              double funct(int n, double x[ ])
                   funct should deliver the value of the function to be minimized, at
                   the point given by $x[1:n]$; the meaning of the parameters of the
                   procedure *funct* is as follows:
                   n: the number of variables;
                   x: the values of the variables for which the function has to be
                          evaluated;
in: double $in[0:9]$;
              entry: *in[0]*:   the machine precision;
                       in[1]:   the relative tolerance for the stepvector (relative to the
                                    current estimates of the variables), see *in[2]*;
                       in[2]:   the absolute tolerance for the stepvector (relative to
                                    the current estimates of the variables); the process is
                                    terminated when in *in[8]*+1 successive iteration steps
                                    the Euclidean norm of the step vector is less than
                                    $(in[1]*\|x\|+in[2])*0.5$; *in[1]* should be chosen in
                                    agreement with the precision in which the function is
                                    calculated; usually *in[1]* should be chosen such that
                                    $in[1] \geq \sqrt{(in[0])}$; *in[0]* should be chosen different from
                                    zero;
                       in[3],in[4]: *in[3]* and *in[4]* are neither used nor changed;
```
```

in[5]:   the maximum number of function evaluations allowed (i.e., calls of *funct*);

in[6]:   the maximum step size; *in[6]* should be equal to the maximum expected distance between the guess and the minimum; if *in[6]* is too small or too large then the initial rate of convergence will be slow;

in[7]:   the maximum scaling factor; the value of *in[7]* may be used to obtain automatic scaling of the variables; however, this scaling is worthwhile but may be unreliable; therefore, the user should try to scale his problem as well as possible and set *in[7]*=1; in either case, *in[7]* should not be chosen greater than 10;

in[8]:   the process terminates if no substantial improvement of the values of the variables is obtained in *in[8]+1* successive iteration steps (see *in[1]*, *in[2]*); *in[8]*=4 is very cautious; usually, *in[8]*=1 is satisfactory;

in[9]:   if the problem is known to be ill-conditioned (see [Bre73]) then the value of *in[9]* should be negative, otherwise *in[9]*≥0;

out: double *out[1:6]*;

    exit: *out[1]*: this value gives information about the termination of the process;

                  out[1]=0: normal termination;

                  out[1]=1: the process is broken off because at the end of an iteration step the number of calls of *funct* exceeded the value given in *in[5]*;

                  out[1]=2: the process is broken off because the condition of the problem is too bad;

    out[2]: the calculated minimum of the function;

    out[3]: the value of the function at the initial guess;

    out[4]: the number of function evaluations needed to obtain this result;

    out[5]: the number of line searches (see [Bre73]);

    out[6]: the step size in the last iteration step.

Procedures used: inivec, inimat, dupvec, dupmat, dupcolvec, mulrow, mulcol, vecvec, tammat, mattam, ichrowcol, elmveccol, qrisngvaldec.

```
package numal;

public interface AP_praxis_method {

  double funct(int n, double x[]);
}


public static void praxis(int n, double x[], AP_praxis_method method, double in[],
                  double out[])
{
    boolean illc,emergency;
```

```
int i,j,k,k2,maxf,kl,kt,ktm;
double s,sl,dn,dmin,f1,lds,ldt,sf,df,qf1,qd0,qd1,m2,m4,small,vsmall,large,
       vlarge,scbd,ldfac,t2,macheps,reltol,abstol,h,l;
int nl[] = new int[1];
int nf[] = new int[1];
double em[] = new double[8];
double d[] = new double[n+1];
double y[] = new double[n+1];
double z[] = new double[n+1];
double q0[] = new double[n+1];
double q1[] = new double[n+1];
double v[][] = new double[n+1][n+1];
double a[][] = new double[n+1][n+1];
double qa[] = new double[1];
double qb[] = new double[1];
double qc[] = new double[1];
double fx[] = new double[1];
double tmp1[] = new double[1];
double tmp2[] = new double[1];
double tmp3[] = new double[1];

Random ran = new Random(1);
macheps=in[0];
reltol=in[1];
abstol=in[2];
maxf=(int)in[5];
h=in[6];
scbd=in[7];
ktm=(int)in[8];
illc = in[9] < 0.0;
small=macheps*macheps;
vsmall=small*small;
large=1.0/small;
vlarge=1.0/vsmall;
m2=reltol;
m4=Math.sqrt(m2);
ldfac = (illc ? 0.1 : 0.01);
kt=nl[0]=0;
nf[0]=1;
out[3]=qf1=fx[0]=method.funct(n,x);
abstol=t2=small+Math.abs(abstol);
dmin=small;
if (h < abstol*100.0) h=abstol*100;
ldt=h;
Basic.inimat(1,n,1,n,v,0.0);
for (i=1; i<=n; i++) v[i][i]=1.0;
d[1]=qd0=qd1=0.0;
Basic.dupvec(1,n,0,q1,x);
Basic.inivec(1,n,q0,0.0);
emergency=false;
```

```
while (true) {
    sf=d[1];
    d[1]=s=0.0;
    tmp1[0]=d[1];
    tmp2[0]=s;
    praxismin(1,2,tmp1,tmp2,fx,false,n,x,v,qa,qb,qc,qd0,qd1,q0,q1,nf,nl,fx,m2,
              m4,dmin,ldt,reltol,abstol,small,h,method);
    d[1]=tmp1[0];
    s=tmp2[0];
    if (s <= 0.0) Basic.mulcol(1,n,1,1,v,v,-1.0);
    if (sf <= 0.9*d[1] || 0.9*sf >= d[1])
        Basic.inivec(2,n,d,0.0);
    for (k=2; k<=n; k++) {
        Basic.dupvec(1,n,0,y,x);
        sf=fx[0];
        illc = (illc || kt > 0);
        while (true) {
            kl=k;
            df=0.0;
            if (illc) {
                /* random stop to get off resulting valley */
                for (i=1; i<=n; i++) {
                    s=z[i]=(0.1*ldt+t2*Math.pow(10.0,kt))*(ran.nextDouble()-0.5);
                    Basic.elmveccol(1,n,i,x,v,s);
                }
                fx[0]=method.funct(n,x);
                nf[0]++;
            }
            for (k2=k; k2<=n; k2++) {
                sl=fx[0];
                s=0.0;
                tmp1[0]=d[k2];
                tmp2[0]=s;
                praxismin(k2,2,tmp1,tmp2,fx,false,n,x,v,qa,qb,qc,qd0,qd1,q0,q1,
                          nf,nl,fx,m2,m4,dmin,ldt,reltol,abstol,small,h,method);
                d[k2]=tmp1[0];
                s=tmp2[0];
                s = illc ? d[k2]*(s+z[k2])*(s+z[k2]) : sl-fx[0];
                if (df < s) {
                    df=s;
                    kl=k2;
                }
            }
            if (!illc && df < Math.abs(100.0*macheps*fx[0]))
                illc=true;
            else
                break;
        }
        for (k2=1; k2<=k-1; k2++) {
```

```
            s=0.0;
            tmp1[0]=d[k2];
            tmp2[0]=s;
            praxismin(k2,2,tmp1,tmp2,fx,false,n,x,v,qa,qb,qc,qd0,qd1,q0,q1,nf,
                        nl,fx,m2,m4,dmin,ldt,reltol,abstol,small,h,method);
            d[k2]=tmp1[0];
            s=tmp2[0];
        }
        f1=fx[0];
        fx[0]=sf;
        lds=0.0;
        for (i=1; i<=n; i++) {
            sl=x[i];
            x[i]=y[i];
            y[i] = sl -= y[i];
            lds += sl*sl;
        }
        lds=Math.sqrt(lds);
        if (lds > small) {
            for (i=kl-1; i>=k; i--) {
                for (j=1; j<=n; j++) v[j][i+1]=v[j][i];
                d[i+1]=d[i];
            }
            d[k]=0.0;
            Basic.dupcolvec(1,n,k,v,y);
            Basic.mulcol(1,n,k,k,v,v,1.0/lds);
            tmp1[0]=d[k];
            tmp2[0]=lds;
            tmp3[0]=f1;
            praxismin(k,4,tmp1,tmp2,tmp3,true,n,x,v,qa,qb,qc,qd0,qd1,q0,q1,nf,
                        nl,fx,m2,m4,dmin,ldt,reltol,abstol,small,h,method);
            d[k]=tmp1[0];
            lds=tmp2[0];
            f1=tmp3[0];
            if (lds <= 0.0) {
                lds = -lds;
                Basic.mulcol(1,n,k,k,v,v,-1.0);
            }
        }
        ldt *= ldfac;
        if (ldt < lds) ldt=lds;
        t2=m2*Math.sqrt(Basic.vecvec(1,n,0,x,x))+abstol;
        kt = (ldt > 0.5*t2) ? 0 : kt+1;
        if (kt > ktm) {
            out[1]=0.0;
            emergency=true;
        }
    }
    if (emergency) break;
/* quad */
```

```
s=fx[0];
fx[0]=qf1;
qf1=s;
qd1=0.0;
for (i=1; i<=n; i++) {
    s=x[i];
    x[i]=l=q1[i];
    q1[i]=s;
    qd1 += (s-l)*(s-l);
}
l=qd1=Math.sqrt(qd1);
s=0.0;
if ((qd0*qd1 > Double.MIN_VALUE) && (nl[0] >=3*n*n)) {
    tmp1[0]=s;
    tmp2[0]=l;
    tmp3[0]=qf1;
    praxismin(0,2,tmp1,tmp2,tmp3,true,n,x,v,qa,qb,qc,qd0,qd1,q0,q1,nf,nl,
              fx,m2,m4,dmin,ldt,reltol,abstol,small,h,method);
    s=tmp1[0];
    l=tmp2[0];
    qf1=tmp3[0];
    qa[0]=l*(l-qd1)/(qd0*(qd0+qd1));
    qb[0]=(l+qd0)*(qd1-l)/(qd0*qd1);
    qc[0]=l*(l+qd0)/(qd1*(qd0+qd1));
} else {
    fx[0]=qf1;
    qa[0]=qb[0]=0.0;
    qc[0]=1.0;
}
qd0=qd1;
for (i=1; i<=n; i++) {
    s=q0[i];
    q0[i]=x[i];
    x[i]=qa[0]*s+qb[0]*x[i]+qc[0]*q1[i];
}
/* end of quad */
dn=0.0;
for (i=1; i<=n; i++) {
    d[i]=1.0/Math.sqrt(d[i]);
    if (dn < d[i]) dn=d[i];
}
for (j=1; j<=n; j++) {
    s=d[j]/dn;
    Basic.mulcol(1,n,j,j,v,v,s);
}
if (scbd > 1.0) {
    s=vlarge;
    for (i=1; i<=n; i++) {
        sl=z[i]=Math.sqrt(Basic.mattam(1,n,i,i,v,v));
        if (sl < m4) z[i]=m4;
```

```
                if (s > sl) s=sl;
        }
        for (i=1; i<=n; i++) {
            sl=s/z[i];
            z[i]=1.0/sl;
            if (z[i] > scbd) {
                sl=1.0/scbd;
                z[i]=scbd;
            }
            Basic.mulrow(1,n,i,i,v,v,sl);
        }
    }
}
for (i=1; i<=n; i++) Basic.ichrowcol(i+1,n,i,i,v);
em[0]=em[2]=macheps;
em[4]=10*n;
em[6]=vsmall;
Basic.dupmat(1,n,1,n,a,v);
if (Linear_algebra.qrisngvaldec(a,n,n,d,v,em) != 0) {
    out[1]=2.0;
    emergency=true;
}
if (emergency) break;
if (scbd > 1.0) {
    for (i=1; i<=n; i++) Basic.mulrow(1,n,i,i,v,v,z[i]);
    for (i=1; i<=n; i++) {
        s=Math.sqrt(Basic.tammat(1,n,i,i,v,v));
        d[i] *= s;
        s=1.0/s;
        Basic.mulcol(1,n,i,i,v,v,s);
    }
}
for (i=1; i<=n; i++) {
    s=dn*d[i];
    d[i] = (s > large) ? vsmall : ((s < small) ? vlarge : 1.0/(s*s));
}
/* sort */
for (i=1; i<=n-1; i++) {
    k=i;
    s=d[i];
    for (j=i+1; j<=n; j++)
        if (d[j] > s) {
            k=j;
            s=d[j];
        }
    if (k > i) {
        d[k]=d[i];
        d[i]=s;
        for (j=1; j<=n; j++) {
            s=v[j][i];
            v[j][i]=v[j][k];
```

```
                    v[j][k]=s;
                }
            }
        }
        /* end of sort */
        dmin=d[n];
        if (dmin < small) dmin=small;
        illc = (m2*d[1]) > dmin;
        if (nf[0] >= maxf) {
            out[1]=1.0;
            break;
        }
    }
    out[2]=fx[0];
    out[4]=nf[0];
    out[5]=nl[0];
    out[6]=ldt;
}


static private void praxismin(int j, int nits, double d2[], double x1[],
                         double f1[], boolean fk, int n, double x[],
                         double v[][], double qa[], double qb[], double qc[],
                         double qd0, double qd1, double q0[], double q1[],
                         int nf[], int nl[], double fx[], double m2,
                         double m4, double dmin, double ldt, double reltol,
                         double abstol, double small, double h,
                         AP_praxis_method method)
{
    /* this procedure is internally used by PRAXIS */

    boolean loop,dz;
    int k;
    double x2,xm,f0,f2,fm,d1,t2,s,sf1,sx1;

    f2=x2=0.0;
    sf1 = f1[0];
    sx1 = x1[0];
    k=0;
    xm=0.0;
    f0 = fm = fx[0];
    dz = d2[0] < reltol;
    s=Math.sqrt(Basic.vecvec(1,n,0,x,x));
    t2=m4*Math.sqrt(Math.abs(fx[0])/(dz ? dmin : d2[0])+s*ldt)+m2*ldt;
    s=s*m4+abstol;
    if (dz && (t2 > s)) t2=s;
    if (t2 < small) t2=small;
    if (t2 > 0.01*h) t2=0.01*h;
    if (fk && (f1[0] <= fm)) {
        xm = x1[0];
```

```
        fm = f1[0];
    }
    if (!fk || (Math.abs(x1[0]) < t2)) {
        x1[0] = (x1[0] > 0.0) ? t2 : -t2;
        f1[0]=praxisflin(x1[0],j,n,x,v,qa,qb,qc,qd0,qd1,q0,q1,nf,method);
    }
    if (f1[0] <= fm) {
        xm = x1[0];
        fm = f1[0];
    }
    loop=true;
    while (loop) {
        if (dz) {
            /* evaluate praxisflin at another point and
                estimate the second derivative */
            x2 = (f0 < f1[0]) ? -x1[0] : x1[0]*2.0;
            f2=praxisflin(x2,j,n,x,v,qa,qb,qc,qd0,qd1,q0,q1,nf,method);
            if (f2 <= fm) {
                xm=x2;
                fm=f2;
            }
            d2[0]=(x2*(f1[0]-f0)-x1[0]*(f2-f0))/(x1[0]*x2*(x1[0]-x2));
        }
        /* estimate first derivative at 0 */
        d1=(f1[0]-f0)/x1[0]-x1[0]*d2[0];
        dz=true;
        x2 = (d2[0] <= small) ? ((d1 < 0.0) ? h : -h) : -0.5*d1/d2[0];
        if (Math.abs(x2) > h)   x2 = (x2 > 0.0) ? h : -h;
        while (true) {
            f2=praxisflin(x2,j,n,x,v,qa,qb,qc,qd0,qd1,q0,q1,nf,method);
            if (k < nits && f2 > f0) {
                k++;
                if (f0 < f1[0] && x1[0]*x2 > 0.0) break;
                x2=0.5*x2;
            } else {
                loop=false;
                break;
            }
        }
    }
    nl[0]++;
    if (f2 > fm)
        x2=xm;
    else
        fm=f2;
    d2[0] = (Math.abs(x2*(x2-x1[0])) > small) ?
            ((x2*(f1[0]-f0)-x1[0]*(fm-f0))/(x1[0]*x2*(x1[0]-x2))) :
            ((k > 0) ? 0.0 : d2[0]);
    if (d2[0] <= small) d2[0]=small;
    x1[0]=x2;
```

```
    fx[0]=fm;
    if (sf1 < fx[0]) {
        fx[0]=sf1;
        x1[0]=sx1;
    }
    if (j > 0) Basic.elmveccol(1,n,j,x,v,x1[0]);
}


static private double praxisflin(double l, int j, int n, double x[], double v[][],
                           double qa[], double qb[], double []qc, double qd0,
                           double qd1, double q0[], double q1[], int nf[],
                           AP_praxis_method method)
{
    /* this procedure is internally used by PRAXISMIN */

    int i;
    double result;
    double t[] = new double[n+1];

    if (j > 0)
        for (i=1; i<=n; i++) t[i]=x[i]+l*v[i][j];
    else {
        /* search along parabolic space curve */
        qa[0]=l*(l-qd1)/(qd0*(qd0+qd1));
        qb[0]=(l+qd0)*(qd1-l)/(qd0*qd1);
        qc[0]=l*(l+qd0)/(qd1*(qd0+qd1));
        for (i=1; i<=n; i++)
            t[i]=qa[0]*q0[i]+qb[0]*x[i]+qc[0]*q1[i];
    }
    nf[0]++;
    result=method.funct(n,t);
    return result;
}
```

5.2.5 More variables - Gradient available

A. rnk1min

Determines a vector $x \in R^n$ for which the real valued function $F(x)$ attains a minimum. It is assumed that the partial derivatives $g_i(x) = \partial F(x)/\partial x_i$ $(i=1,...,n)$ may easily be computed. *rnk1min* is suitable, in particular, for use in connection with problems for which the $n \times n$ Hessian matrix $G(x)$, whose components are $G_{i,j}(x) = \partial^2 F(x)/\partial x_i \partial x_j$ $(i,j=1,...,n)$, is almost singular at the minimum.

With $H(x) = G(x)^{-1}$, and the initial vector $x^{(0)}$ prescribed, the sequence of vectors $x^{(k)}$ produced by use of the Newton scheme $x^{(k+1)} = x^{(k)} - H(x^{(k)})g(x^{(k)})^T$ (k=0,1,...) under certain conditions converges quadratically. Use of this scheme requires the evaluation and inversion of a Hessian matrix at each stage, and in order to avoid this *rnk1min* determines a sequence of vectors $x^{(k)}$ by use of a scheme of the form $x^{(k+1)} = x^{(k)} - \alpha^{(k)}H^{(k)}g(x^{(k)})$ in which the metric $H^{(k)}$ is an approximation to $H(x^{(k)})$, and is corrected by use of a simple updating formula of the form $H^{(k+1)}=H^{(k)}+C^{(k)}$ where $C^{(k)}$ is of rank one (or possibly rank two), and the $\alpha^{(k)}$ are suitably determined real numbers.

The user is asked to provide at outset the initial approximation vector $x^{(0)}$ and values of the machine precision ε, the required relative and absolute precisions, ε_r and ε_a respectively, in the determination of the minimum, a descent parameter μ in the range $0<\mu<1/2$ (often μ=0.0001 is suitable), an upper bound ε_g for $\|g(x)\|$ (here and in the sequel all norms are Euclidean) and a lower bound F_{min} for $F(x)$, both holding for all x in a domain containing the minimum and all $x^{(k)}$ produced, either a rough estimate $c>0$ of $\|G(x^{(0)})\|$ (1.0 is often suitable) or approximate values of the elements of the upper triangular part of the inverse Hessian matrix $H(x^{(0)})$, the maximum permitted number of simultaneous evaluations of $F(x)$ and $g(x)$ for the required values of x, and an orthogonality parameter ß for which $(\varepsilon/\varepsilon_r)^{1/2}/n \le ß < 1.0$ (often 0.01 is suitable).

The algorithm used consists first of an initialization, and then in determinations of sequences of directions $d^{(k)}$, virtual steplengths $\alpha^{(k)}$, steps $\delta^{(k)}$, approximations $x^{(k)}$ to the minimum, increments in derivatives $\xi^{(k)}$, and metrics $H^{(k)}$. Initially, either $H^{(0)}=cI$ (I being the $n \times n$ unit matrix) or $H^{(0)}$ is the approximation to the inverse Hessian matrix, whose elements have been supplied by the user. Then, for k=0,1,... steps I to VI as follows are carried out.

I.   If

$$g(x^{(k)})^T H^{(k)} g(x^{(k)}) > 0 \qquad (1)$$

then set $d^{(k)} = -H^{(k)}g(x^{(k)})$. If condition (1) is violated, and the new direction $d^{(k)}$ were to be taken as just given then $(d^{(k)})^T g(x^{(k)}) \ge 0$, i.e., the derivative of $F(x)$ in the direction $d^{(k)}$ at the point $x^{(k)}$ would be decreasing. To avoid this, $H^{(k)}$ is decomposed in the form $H^{(k)}=U\Delta U^T$, where U is a matrix of eigenvectors, and Δ an $n \times n$ diagonal matrix of eigenvalues λ_i (i=1,...,n); defining $|\Delta|$ to be the diagonal matrix with diagonal elements $|\lambda_i|$ (i=1,...,n), and $K^{(k)}$ by $K^{(k)} = U|\Delta|U^T$ (so that $K^{(k)}$ is positive definite) we take $d^{(k)} = -K^{(k)}g(x^{(k)})$ when condition (1) is violated.

II.   If k=0, set $\theta^{(k)} = \min(1, 2(F_{min}-F(x^{(0)}))/(-d^{(0)})^T g(x^{(0)}))$, and determine the smallest integer $r \ge 0$ for which either

$$(d^{(0)})^T g(x^{(0)}+2^r\theta^{(0)}d^{(0)}) \ge 0$$

or

$$(F(x^{(0)}-2^r\theta^{(0)}d^{(0)})-F(x^{(0)}))/2^r\theta^{(0)}(d^{(0)})^T g(x^{(0)}) < \mu$$

and set $\theta^{(0)} = 2^r\theta^{(0)}$ for this r. $F(x)$ now decreases and then increases upon the line $x=x^{(0)}+\alpha d^{(0)}$ ($\alpha>0$). An approximation to the value of α specifying the point at which F is minimum upon this line is determined by cubic interpolation (see the theory of *linemin* and [Dav59]) or if $(d^{(0)})^T g(x^{(0)}+\theta^{(0)}d^{(0)}) < 0$ by bisection. For the value of α so determined

$$\mu \le (F(x^{(k)}+\alpha d^{(k)}) - F(x^{(k)}))/\alpha(d^{(k)})^T g(x^{(k)}) \le 1-\mu. \qquad (2)$$

If $k \ne 0$, set $\theta^{(k)} = \|\delta^{(k-1)}\|/\|d^{(k)}\|$ if $k<n$, and $\theta^{(k)}=1$ otherwise. If

$$(F(x^{(k)}+\theta^{(k)}d^{(k)}) - F(x^{(k)}))/\theta^{(k)}(d^{(k)})^T g(x^{(k)}) \ge \mu \qquad (3)$$

set $\alpha^{(k)} = \theta^{(k)}$. If condition (3) is violated, the distance from $x=x^{(k)}$ to the minimum of $F(x)$ upon the line $x^{(k)}+\alpha d^{(k)}$ ($\alpha > 0$) is overestimated by setting $\alpha = \theta^{(k)}$, and a value of α in the range $0 < \alpha < \theta^{(k)}$ is determined by cubic interpolation or bisection as above, such that condition (2) is satisfied for this α, for which we set $\alpha^{(k)} = \alpha$.

III.   Set $\delta^{(k)} = \alpha^{(k)}d^{(k)}$ and $x^{(k+1)} = x^{(k)} + \delta^{(k)}$.

IV.   If condition (1) is satisfied, set $\xi^{(k)} = -\alpha^{(k)}g(x^{(k)})$. Otherwise set $\xi^{(k)} = -\alpha^{(k)}U\Delta U^T d^{(k)}$ where, with U and Δ being determined as in stage I, Δ is the $n \times n$ diagonal matrix with diagonal elements sign (λ_i) $(i=1,...,n)$.

V.   If, with $\gamma^{(k)} = g(x^{(k+1)}) - g(x^{(k)})$,

$$\left| (\gamma^{(k)}-\xi^{(k)})^T\delta^{(k)} \right| > \text{ß} \left\| \gamma^{(k)}-\xi^{(k)} \right\| \left\| \delta^{(k)} \right\| \tag{4}$$

then $H^{(k+1)}$ is determined by use of the rank-one updating formula

$$H^{(k+1)} = H^{(k)} + c^{(k)}u^{(k)}(u^{(k)})^T \tag{5}$$

where $u^{(k)} = \delta^{(k)} - H^{(k)}\gamma^{(k)}$, $c^{(k)} = 1/((\gamma^{(k)})^Tu^{(k)})$.

If condition (4) is violated, but

$$c^{(k)}(\gamma^{(k)})^T\delta^{(k)} > 0 \tag{6}$$

then the rank-two updating formula

$$H^{(k+1)} = H^{(k)} - c_1^{(k)}v^{(k)}(w^{(k)})^T - c_2^{(k)}w^{(k)}(v^{(k)})^T + (1 + (c_1^{(k)}/c_2^{(k)})c_1^{(k)}v^{(k)}(v^{(k)})^T \tag{7}$$

with

$$c_1^{(k)} = 1/(\delta^{(k)})^T\gamma^{(k)}, \quad c_2^{(k)} = -1/(\gamma^{(k)})^TH^{(k)}\gamma^{(k)}, \quad v_{(k)}=\delta^{(k)}, \quad w^{(k)}=H^{(k)}\gamma^{(k)} \tag{8}$$

is used. If both conditions (4) and (6) are violated, the rank-two updating formula

$$H^{(k+1)} = H^{(k)} + c_1^{(k)}v^{(k)}(v^{(k)})^T + c_2^{(k)}w^{(k)}w^{(k)} \tag{9}$$

with $c_1^{(k)},...,w^{(k)}$ as defined by formulae (8) is used.

VI.   If $\left\| H^{(k+1)}g(x^{(k+1)}) \right\| \le \varepsilon_r \left\| x^{(k+1)} \right\| + \varepsilon_a$ and $\left\| g(x^{(k+1)}) \right\| \le \varepsilon_g$ and $g(x^{(k+1)})^TH^{(k+1)}g(x^{(k+1)}) \ge 0$ and $k \ge n$ then $x^{(k+1)}$ is accepted as an approximation to the minimum. If k is equal to the number specified by the maximum number of function evaluations then the process is terminated (and $x^{(k+1)}$ is then not an acceptable approximation).

The eigen-value/vector determination possibly required at stage I (in most practical cases this requirement is of infrequent occurrence) is carried out by a call of *eigsym1*. The line minimization required at stage II is carried out by *linemin*. The updating formulae (5,7,9) possibly required at stage V are implemented by *rnk1upd*, *fleupd*, and *davupd*, respectively. A detailed description of the algorithm and some results about its convergence is given in [Bus72b].

Procedure parameters:

           double rnk1min $(n,x,g,h,method,in,out)$

rnk1min: delivers the calculated least value of the given function;

n: int;

          entry: the number of variables of the function to be minimized;

x: double $x[1:n]$;

          entry: an approximation of a minimum of the function;

          exit: the calculated minimum of the function;

g: double $g[1:n]$;

          exit: the gradient of the function at the calculated minimum;

h: double $h[1:n(n+1)/2]$;

          the upper triangle of an approximation of the inverse Hessian is stored columnwise in h (i.e., the (i,j)-th element $= h[(j-1)j/2+i]$, $1 \le i \le j \le n$); if $in[6] > 0$ initializing of h will be done automatically and the initial approximation of the inverse Hessian will equal the unit matrix multiplied with the value of *in[6]*; if *in[6]*<0 then no initializing of *h* will be done, and the user should give in *h* an approximation of the inverse Hessian, at the starting point; the upper triangle of an approximation of the inverse Hessian at the calculated minimum is delivered in *h*;

method: a class that defines a procedure *funct*, this class must implement the AP_linemin_method interface;

double funct(int n, double x[ ], double g[ ])

a call of *funct* must have the following results:

1. *funct* becomes the value of the function to be minimized at the point *x*;

2. the value of *g[i]*, (*i*=1,...,*n*), becomes the value of the *i*-th component of the gradient of the function at *x*;

in: double *in[0:9]*;

entry: *in[0]*: the machine precision;

in[1]: the relative tolerance for the solution; this tolerance should not be chosen smaller than *in[0]*;

in[2]: the absolute tolerance for the solution;

in[3]: a parameter used for controlling line minimization (see [Fle63, GolP67]); usually a suitable value is 0.0001;

in[4]: the absolute tolerance for the Euclidean norm of the gradient at the solution;

in[5]: a lower bound for the function value;

in[6]: this parameter controls the initialization of the approximation of the inverse Hessian (metric), see *h*; usually the choice *in[6]*=1 will give good results;

in[7]: the maximum allowed number of calls of *funct*;

in[8]: a parameter used for controlling the updating of the metric; it is used to avoid unboundedness of the metric (see [Po70]); the value of *in[8]* should satisfy: $\sqrt{(in[0]/in[1])}/n < in[8] < 1$; usually a suitable value will be 0.01;

out: double *out[0:4]*;

exit: *out[0]*: the Euclidean norm of the product of the metric and the gradient at the calculated minimum;

out[1]: the Euclidean norm of the gradient at the calculated minimum;

out[2]: the number of calls of *funct* necessary to attain this result;

out[3]: the number of iterations in which a line search was necessary;

out[4]: the number of iterations in which a direction had to be calculated with the method given in [Gr67]; in such an iteration a calculation of the eigenvalues and eigenvectors of the metric is necessary.

Procedures used: vecvec, matvec, tamvec, elmvec, symmatvec, inivec, inisymd, mulvec, dupvec, eigsym1, linemin, rnk1upd, davupd, fleupd.

```
package numal;

public interface AP_linemin_method {

    double funct(int n, double x[], double g[]);
}


public static double rnk1min(int n, double x[], double g[], double h[],
                            AP_linemin_method method, double in[], double out[])
{
    boolean ok;
    int i,it,n2,cntl,cnte,evl,evlmax;
    double f,f0,fmin,mu,dg,dg0,ghg,gs,nrmdelta,alfa,macheps,reltol,abstol,eps,
            tolg,orth,aid,temp1,temp2;
    double em[] = new double[10];
    double xtmp1[] = new double[1];
    double xtmp2[] = new double[1];
    double xtmp3[] = new double[1];
    int itmp[] = new int[1];
    double v[] = new double[n+1];
    double delta[] = new double[n+1];
    double gamma[] = new double[n+1];
    double s[] = new double[n+1];
    double p[] = new double[n+1];
    double vec[][] = new double[n+1][n+1];

    macheps=in[0];
    reltol=in[1];
    abstol=in[2];
    mu=in[3];
    tolg=in[4];
    fmin=in[5];
    it=0;
    alfa=in[6];
    evlmax=(int)in[7];
    orth=in[8];
    n2=(n*(n+1))/2;
    cntl=cnte=0;
    if (alfa > 0.0) {
        Basic.inivec(1,n2,h,0.0);
        Basic.inisymd(1,n,0,h,alfa);
    }
    f=method.funct(n,x,g);
    evl=1;
    dg=Math.sqrt(Basic.vecvec(1,n,0,g,g));
    for (i=1; i<=n; i++) delta[i] = -Basic.symmatvec(1,n,i,h,g);
    nrmdelta=Math.sqrt(Basic.vecvec(1,n,0,delta,delta));
    dg0=Basic.vecvec(1,n,0,delta,g);
    ok = dg0 < 0.0;
```

```
eps=Math.sqrt(Basic.vecvec(1,n,0,x,x))*reltol+abstol;
it++;
while ((nrmdelta > eps || dg > tolg || !ok) && (evl < evlmax)) {
    if (!ok) {
        /* calculating greenstadts direction */
        double th[] = new double[n2+1];
        em[0]=macheps;
        em[2]=aid=Math.sqrt(macheps*reltol);
        em[4]=orth;
        em[6]=aid*n;
        em[8]=5.0;
        cnte++;
        Basic.dupvec(1,n2,0,th,h);
        Linear_algebra.eigsym1(th,n,n,v,vec,em);
        for (i=1; i<=n; i++) {
            aid = -Basic.tamvec(1,n,i,vec,g);
            s[i]=aid*Math.abs(v[i]);
            v[i]=((v[i] == 0.0) ? 0.0 : ((v[i] > 0.0) ? aid : -aid));
        }
        for (i=1; i<=n; i++) {
            delta[i]=Basic.matvec(1,n,i,vec,s);
            p[i]=Basic.matvec(1,n,i,vec,v);
        }
        dg0=Basic.vecvec(1,n,0,delta,g);
        nrmdelta=Math.sqrt(Basic.vecvec(1,n,0,delta,delta));
    }
    Basic.dupvec(1,n,0,s,x);
    Basic.dupvec(1,n,0,v,g);
    if (it > n)
        alfa=1.0;
    else {
        if (it != 1)
            alfa /= nrmdelta;
        else {
            alfa=2.0*(fmin-f)/dg0;
            if (alfa > 1.0) alfa=1.0;
        }
    }
    Basic.elmvec(1,n,0,x,delta,alfa);
    f0=f;
    f=method.funct(n,x,g);
    evl++;
    dg=Basic.vecvec(1,n,0,delta,g);
    if (it == 1 || f0-f < -mu*dg0*alfa) {
        /* line minimization */
        i=evlmax-evl;
        cntl++;
        xtmp1[0]=alfa;
        xtmp2[0]=f;
        xtmp3[0]=dg;
```

```
            itmp[0]=i;
            linemin(n,s,delta,nrmdelta,xtmp1,g,method,f0,xtmp2,dg0,xtmp3,itmp,
                    false,in);
            alfa=xtmp1[0];
            f=xtmp2[0];
            dg=xtmp3[0];
            i=itmp[0];
            evl += i;
            Basic.dupvec(1,n,0,x,s);
        }
        Basic.dupvec(1,n,0,gamma,g);
        Basic.elmvec(1,n,0,gamma,v,-1.0);
        if (!ok) Basic.mulvec(1,n,0,v,p,-1.0);
        dg -= dg0;
        if (alfa != 1.0) {
            Basic.mulvec(1,n,0,delta,delta,alfa);
            Basic.mulvec(1,n,0,v,v,alfa);
            nrmdelta *= alfa;
            dg *= alfa;
        }
        Basic.dupvec(1,n,0,p,gamma);
        Basic.elmvec(1,n,0,p,v,1.0);
        for (i=1; i<=n; i++) v[i]=Basic.symmatvec(1,n,i,h,gamma);
        Basic.dupvec(1,n,0,s,delta);
        Basic.elmvec(1,n,0,s,v,-1.0);
        gs=Basic.vecvec(1,n,0,gamma,s);
        ghg=Basic.vecvec(1,n,0,v,gamma);
        aid=dg/gs;
        temp1=Basic.vecvec(1,n,0,delta,p);
        temp2=orth*nrmdelta;
        if (temp1*temp1 > Basic.vecvec(1,n,0,p,p)*temp2*temp2)
            rnk1upd(h,n,s,1.0/gs);
        else
            if (aid >= 0.0)
                fleupd(h,n,delta,v,1.0/dg,(1.0+ghg/dg)/dg);
            else
                davupd(h,n,delta,v,1.0/dg,1.0/ghg);
        for (i=1; i<=n; i++) delta[i] = -Basic.symmatvec(1,n,i,h,g);
        alfa=nrmdelta;
        nrmdelta=Math.sqrt(Basic.vecvec(1,n,0,delta,delta));
        eps=Math.sqrt(Basic.vecvec(1,n,0,x,x))*reltol+abstol;
        dg=Math.sqrt(Basic.vecvec(1,n,0,g,g));
        dg0=Basic.vecvec(1,n,0,delta,g);
        ok = dg0 <= 0.0;
        it++;
    }
    out[0]=nrmdelta;
    out[1]=dg;
    out[2]=evl;
    out[3]=cnt1;
```

```
    out[4]=cnte;
    return f;
}
```

B. flemin

Determines a vector $x \in \mathbb{R}^n$ for which the real valued function $F(x)$ attains a minimum by means of the variable metric algorithm given in [Fle63], except for some details (see [Bus72b]). It is assumed that the partial derivatives $g_i(x) = \partial F(x) / \partial x_i$ ($i=1,...,n$) may easily be computed. *flemin* is suitable, in particular, for use in connection with problems for which n is relatively large and the computation of $F(x)$ and $g(x)$ is relatively cheap.

The theory underlying the operation of *flemin* is similar to that underlying the operation of *rnk1min*, and the algorithms employed are similar in many respects.

The user is asked to provide at outset the initial approximation vector $x^{(0)}$ and the same parameters ε_r, ε_a, μ, ε_g, F_{min}, (possibly) c, and a number specifying the maximum permitted number of simultaneous evaluations of $F(x)$ and $g(x)$, together with (possibly) approximations to the elements in the upper triangular part of $H(x^{(0)})$, as for *rnk1min*.

The initialization for the algorithm employed is the same as that for *rnk1min*. The stages are as follows.

I. Simply set $d^{(k)} = -H^{(k)}g(x^{(k)})$.

II, III. As for *rnk1min*.

IV. If

$$(\delta^{(k)})^T \gamma^{(k)} < 0 \tag{1}$$

then set $H^{(k+1)} = H^{(k)}$. If condition (1) is violated and

$$(\delta^{(k)})^T \gamma^{(k)} \geq (\gamma^{(k)})^T H^{(k)} \gamma^{(k)} \tag{2}$$

then the rank-two updating formulae (7,8) of *rnk1min* are used. If both conditions (1,2) are violated, the rank-two updating formulae (8,9) of *rnk1min* are used.

V. If $\|H^{(k+1)}g(x^{(k+1)})\| \leq \varepsilon_r \|x^{(k+1)}\| + \varepsilon_a$ and $\|g(x^{(k+1)})\| \leq \varepsilon_g$ and $k \geq n$ then $x^{(k+1)}$ is accepted as an approximation to the minimum. If k is equal to the number specified by the maximum number of function evaluations then the process is terminated (and $x^{(k+1)}$ is then not an acceptable approximation).

The above algorithm differs from one given by Fletcher (see [Fle63]), (which may fail when $F(x)$ is not approximately a quadratic function in the components of x) only in the determination of $\theta^{(k)}$ at stage II (the method adopted by *flemin* to a large extent overcomes the difficulty mentioned).

Procedure parameters:
$$\text{double flemin } (n,x,g,h,funct,in,out)$$

flemin: delivers the calculated least value of the given function;
n: int;
  entry: the number of variables of the function to be minimized;

x: double *x[1:n]*;
           entry: an approximation of a minimum of the function;
           exit: the calculated minimum of the function;
g: double *g[1:n]*;
           exit: the gradient of the function at the calculated minimum;
h: double *h[1:n(n+1)/2]*;
           the upper triangle of an approximation of the inverse Hessian is stored
           columnwise in *h* (i.e., the (i,j)-th element = $h[(j-1)j/2+i]$, $1 \le i \le j \le n$); if *in[6]*>0
           initializing of *h* will be done automatically and the initial approximation
           of the inverse Hessian will equal the unit matrix multiplied with the
           value of *in[6]*; if *in[6]*<0 then no initializing of *h* will be done and the user
           should give in *h* an approximation of the inverse Hessian, at the starting
           point; the upper triangle of an approximation of the inverse Hessian at
           the calculated minimum is delivered in *h*;
method: a class that defines a procedure *funct*, this class must implement the
           AP_linemin_method interface;
           double funct(int n, double x[ ], double g[ ])
               a call of *funct* must have the following results:
               1. *funct* becomes the value of the function to be minimized at the
                   point *x*;
               2. the value of *g[i]*, ($i=1,...,n$), becomes the value of the *i*-th
                   component of the gradient of the function at *x*;
in: double *in[1:7]*;
           entry: *in[1]*:   the relative tolerance for the solution;
                    in[2]:   the absolute tolerance for the solution;
                    in[3]:   a parameter used for controlling line minimization
                              (see *rnk1min*); usually a suitable value is 0.0001;
                    in[4]:   the absolute tolerance for the Euclidean norm of the
                              gradient at the solution;
                    in[5]:   a lower bound for the function value;
                    in[6]:   this parameter controls the initialization of the
                              approximation of the inverse Hessian (metric), see *h*;
                              usually the choice *in[6]*=1 will give good results;
                    in[7]:   the maximum allowed number of calls of *funct*;
out: double *out[0:4]*;
           exit: *out[0]*: the Euclidean norm of the product of the metric and the
                              gradient at the calculated minimum;
                    out[1]: the Euclidean norm of the gradient at the calculated
                              minimum;
                    out[2]: the number of calls of *funct*, necessary to attain this
                              result;
                    out[3]: the number of iterations in which a line search was
                              necessary;
                    out[4]: if *out[4]*=-1 then the process is broken off because no
                              downhill direction could be calculated; the precision
                              asked for may not be attained and is possibly chosen too
                              high; normally *out[4]*=0.

Procedures used: vecvec, elmvec, symmatvec, inivec, inisymd, mulvec, dupvec,
                    linemin, davupd, fleupd.

```
package numal;

public interface AP_linemin_method {

  double funct(int n, double x[], double g[]);
}


public static double flemin(int n, double x[], double g[], double h[],
                      AP_linemin_method method, double in[], double out[])
{
    int i,it,cntl,evl,evlmax;
    double f,f0,fmin,mu,dg,dg0,nrmdelta,alfa,reltol,abstol,eps,tolg,aid;
    double xtmp1[] = new double[1];
    double xtmp2[] = new double[1];
    double xtmp3[] = new double[1];
    int itmp[] = new int[1];
    double v[] = new double[n+1];
    double delta[] = new double[n+1];
    double s[] = new double[n+1];

    reltol=in[1];
    abstol=in[2];
    mu=in[3];
    tolg=in[4];
    fmin=in[5];
    alfa=in[6];
    evlmax=(int)in[7];
    out[4]=0.0;
    it=0;
    f=method.funct(n,x,g);
    evl=1;
    cntl=0;
    if (alfa > 0.0) {
        Basic.inivec(1,(n*(n+1))/2,h,0.0);
        Basic.inisymd(1,n,0,h,alfa);
    }
    for (i=1; i<=n; i++) delta[i] = -Basic.symmatvec(1,n,i,h,g);
    dg=Math.sqrt(Basic.vecvec(1,n,0,g,g));
    nrmdelta=Math.sqrt(Basic.vecvec(1,n,0,delta,delta));
    eps=Math.sqrt(Basic.vecvec(1,n,0,x,x))*reltol+abstol;
    dg0=Basic.vecvec(1,n,0,delta,g);
    it++;
    while ((nrmdelta > eps || dg > tolg) && (evl < evlmax)) {
        Basic.dupvec(1,n,0,s,x);
        Basic.dupvec(1,n,0,v,g);
        if (it >= n)
            alfa=1.0;
        else {
```

```
        if (it != 1)
            alfa /= nrmdelta;
        else {
            alfa=2.0*(fmin-f)/dg0;
            if (alfa > 1.0) alfa=1.0;
        }
    }
    Basic.elmvec(1,n,0,x,delta,alfa);
    f0=f;
    f=method.funct(n,x,g);
    evl++;
    dg=Basic.vecvec(1,n,0,delta,g);
    if (it == 1 || f0-f < -mu*dg0*alfa) {
        /* line minimization */
        i=evlmax-evl;
        cntl++;
        xtmp1[0]=alfa;
        xtmp2[0]=f;
        xtmp3[0]=dg;
        itmp[0]=i;
        linemin(n,s,delta,nrmdelta,xtmp1,g,method,f0,xtmp2,dg0,xtmp3,itmp,
                false,in);
        alfa=xtmp1[0];
        f=xtmp2[0];
        dg=xtmp3[0];
        i=itmp[0];
        evl += i;
        Basic.dupvec(1,n,0,x,s);
    }
    if (alfa != 1.0) Basic.mulvec(1,n,0,delta,delta,alfa);
    Basic.mulvec(1,n,0,v,v,-1.0);
    Basic.elmvec(1,n,0,v,g,1.0);
    for (i=1; i<=n; i++) s[i]=Basic.symmatvec(1,n,i,h,v);
    aid=Basic.vecvec(1,n,0,v,s);
    dg=(dg-dg0)*alfa;
    if (dg > 0.0)
        if (dg >= aid)
            fleupd(h,n,delta,s,1.0/dg,(1.0+aid/dg)/dg);
        else
            davupd(h,n,delta,s,1.0/dg,1.0/aid);
    for (i=1; i<=n; i++) delta[i] = -Basic.symmatvec(1,n,i,h,g);
    alfa *= nrmdelta;
    nrmdelta=Math.sqrt(Basic.vecvec(1,n,0,delta,delta));
    eps=Math.sqrt(Basic.vecvec(1,n,0,x,x))*reltol+abstol;
    dg=Math.sqrt(Basic.vecvec(1,n,0,g,g));
    dg0=Basic.vecvec(1,n,0,delta,g);
    if (dg0 > 0.0) {
        out[4] = -1.0;
        break;
    }
```

```
        it++;
    }
    out[0]=nrmdelta;
    out[1]=dg;
    out[2]=evl;
    out[3]=cntl;
    return f;
}
```

5.3 Overdetermined nonlinear systems

5.3.1 Least squares - With Jacobian matrix

A. marquardt

Calculates the least squares solution of an overdetermined system of nonlinear equations with Marquardt's method [BusDK75, M63].

marquardt computes the values of the components $p_1,...,p_n$ of $p \in R^n$ such that

$$\Phi(p) = \sum_{i=1}^{m} \left(f(t_i, p) - d_i \right)^2$$

is a minimum at p, where $f(t,p)$ is a real-valued function, and d_i $(i=1,...,m)$ are given data readings $(m \geq n)$ by use of an improved version of the Levenberg-Marquardt algorithm (the p_j offers a parameter set of best fit of the function $f(t,p)$ to the data at the points t_i).

An initial approximation $p^{(0)}$ to p, two tolerances ε_{re} and ε_a, and a real number ξ expressing the ratio between the gradient and Gauss-Newton directions must be specified. ξ may be taken to be 0.01 for well-conditioned problem; in any case ξ must satisfy the inequality:

machine precision $< \xi \leq 1/$machine precision.

At the k-th stage of the algorithm, the $m \times n$ Jacobian matrix $J(p^{(k)})$, whose components are given by $J_{i,j}(p) = \partial f(t_i,p)/\partial p_j$ $(i=1,...,m; j=1,...,n)$ with $p=p^{(k)}$, is decomposed in the form $J(p^{(k)})=U^{(k)}D^{(k)}V^{(k)}$ where $U^{(k)}$ is an orthogonal $m \times m$ matrix, $D^{(k)}$ is an $m \times n$ matrix whose principal $n \times n$ submatrix is the diagonal matrix $D^{(k)}$ of singular values of $J(p^{(k)})$ ($D^{(k)}$ having zeros elsewhere), and $V^{(k)}$ is an $n \times n$ orthogonal matrix.

The scalar $\rho^{(p)} = \xi \| D^{(k)} \|^2$ is determined. The further number $\lambda^{(k)}$ is determined by setting $\lambda^{(0)} = \lambda^{(-1)} = \rho^{(0)}$ if $k=0$ and, if $k \geq 1$, $\lambda^{(k)} = w\lambda^{(k-1)}$ if $\lambda^{(k-1)} \leq \lambda^{(k-2)}$ and $\lambda^{(k)} \leq \lambda^{(k-1)}$ if $\lambda^{(k-1)} > \lambda^{(k-2)}$ (w is a real number in the range $0<w<1$; *marquardt* uses the value 0.5). If $h(\lambda^{(k)}) \geq \mu$ (μ being a real number in the range $0<\mu\leq0.5$; *marquardt* uses the value 0.5) where, with $s^{(k)}(\lambda)$ defined by

$$(J(p^{(k)})^T J(p^{(k)}) + \lambda I)s^{(k)}(\lambda) = -J(p^{(k)})^T f(p_{(k)}),$$

$$h^{(k)}(\lambda) = (\Phi(p^{(k)}) - \Phi(p^{(k)}+s^{(k)}(\lambda))) / -s^{(k)}(\lambda)^T J(p^{(k)})^T f(p^{(k)}),$$

the components of $f(p^{(k)})$ being $f(t_i,p^{(k)})-d_i x$ ($i=1...,m$) then $\lambda^{(k)}$ is set equal to $\lambda'^{(k)}$. If, however, $h(\lambda'^{(k)})<\mu$, $\lambda^{(k)}$ is taken to be $v_r\max(\lambda'^{(k)},\rho^{(k)})$ (v being a real number in the range $1<v<\infty$; *marquardt* uses the value 10) where r is the smallest nonnegative integer for which the relationship $h(v^r\max(\lambda'^{(k)},\rho^{(k)})) \geq \mu$, is satisfied, $p^{(k+1)}$ is then determined by $p^{(k+1)}=p^{(k)}+s^{(k)}(\lambda^{(k)})$.

The algorithm is terminated (with $p^{(k)}$ accepted as an approximation to p) when

$$\Phi(p^{(k-1)}) - \Phi(p^{(k)}) \leq \varepsilon_r\Phi(p^{(k)}) + \varepsilon_a, \text{ or } \Phi(p^{(k)}) \leq \varepsilon_a.$$

Procedure parameters:

void marquardt $(m,n,par,g,v,method,in,out)$

m: int;

         entry: the number of equations;

n: int;

         entry: the number of unknown variables; n should satisfy $n \leq m$;

par: double *par[1:n]*;

         entry: an approximation to a least squares solution of the system;

         exit: the calculated least squares solution;

g: double *g[1:m]*;

         exit: the residual vector at the calculated solution;

v: double *v[1:n,1:n]*;

         exit: the inverse of the matrix $J^T J$ where J denotes the transpose of the matrix of partial derivatives $\partial g[i]/\partial par[j]$ ($i=1,...,m$; $j=1,...,n$); i.e., $v[i,j]$ contains the (i,j)-th element of $(J(p)^T J(p))^{-1}$ ($i,j=1,...,n$) where p is the value of $p^{(k)}$ in the above at termination;

               note that the standard deviation of σ_j associated with the computed estimate of p_j ($j=1,...,n$) and the correlation $\rho_{i,j}$ between the estimates of p_i and p_j ($i=1,...,n$; $j=i+1,...,n$) may easily be extracted from the contents of the array v by use of the formula $\sigma = \sqrt{(v[j,j]/m)}$ ($j=1,...,n$), $\rho_{i,j} = v[i,j]/\sqrt{(v[i,i]*v[j,j])}$ ($i=1,...,n$; $j=i+1,...,n$);

method: a class that defines two procedures *funct* and *jacobian*, this class must implement the `AP_marquardt_methods` interface;

         boolean *funct(int m, int n, double par[ ], double g[ ])*

            entry: m, n have the same meaning as in the procedure *marquardt*; array *par[1:n]* contains the current values of the unknowns and should not be altered;

                exit: upon completion of a call of *funct*, the array *g[1:m]* should contain the residual vector obtained with the current values of the unknowns; e.g., in curve fitting problems,

                     g[i] = theoretical value *f(x[i], par)* - observed value *y[i]*;

                after a successful call of *funct*, the function should deliver the value true; however, if *funct* delivers the value false then it is assumed that the current estimates of the unknowns lie outside a feasible region and the process is terminated (see *out[1]*); hence, proper programming of *funct* makes it possible to avoid calculation of a residual vector with values of the unknown variables which make no sense or which even may cause overflow in the computation;

void jacobian(int m, int n, double par[ ], double g[ ], double jac[ ][ ])

entry: *m, n, par.* see *funct*;

  g contains the residual vector obtained with the current values of the unknowns and should not be altered;

exit: double *jac[1:m,1:n]*;

  upon completion of a call of *jacobian*, the array *jac* should contain the partial derivatives $\partial g[i]/\partial par[j]$, obtained with the current values of the unknown variables given in *par[1:n]*;

it is a prerequisite for the proper operation of the procedure *marquardt* that the precision of the elements of the matrix *jac* is at least the precision defined by *in[3]* and *in[4]*;

in: double *in[0:6]*;

entry: *in[0]*: the machine precision;

  in[1], in[2]: these are not used by *marquardt*;

  in[3]: the relative tolerance for the difference between the Euclidean norm of the ultimate and penultimate residual vector (value of ε_{re} above); see *in[4]*;

  in[4]: the absolute tolerance for the difference between the Euclidean norm of the ultimate and penultimate residual vector (value of ε_a above);

  the process is terminated if the improvement of the sum of squares is less than

   in[3]*(sum of squares)+*in[4]***in[4]*;

  these tolerances should be chosen greater than the corresponding errors of the calculated residual vector;

  in[5]: the maximum number of calls of *funct* allowed;

  in[6]: a starting value used for the relation between the gradient and the Gauss-Newton direction (value of ξ above); if the problem is well conditioned then a suitable value for *in[6]* will be 0.01; if the problem is ill conditioned then *in[6]* should be greater, but the value of *in[6]* should satisfy: *in[0]* < *in[6]* ≤ 1/*in[0]*;

out: double *out[1:7]*;

exit: *out[1]*: this value gives information about the termination of the process;

  out[1]=0: normal termination;

  out[1]=1: the process has been broken off because the number of calls of *funct* exceeded the number given in *in[5]*;

  out[1]=2: the process has been broken off because a call of *funct* delivered the value false;

  out[1]=3: *funct* became false when called with the initial estimates of *par[1:n]*; the iteration process was not started and so *v[1:n,1:n]* cannot be used;

  out[1]=4: the process has been broken off because the precision asked for cannot be attained; this precision is possibly chosen too high, relative to the precision in which the residual vector is calculated (see *in[3]*);

out[2]: the Euclidean norm of the residual vector calculated with values of the unknowns delivered;

out[3]: the Euclidean norm of the residual vector calculated with the initial values of the unknown variables;

out[4]: the number of calls of *funct* necessary to attain the calculated result;

out[5]: the total number of iterations performed; note that in each iteration one evaluation of the Jacobian matrix had to be made;

out[6]: the improvement vector in the last iteration step;

out[7]: the condition number of $J^T J$, i.e., the ratio of its largest to smallest eigenvalues.

Procedures used: mulcol, dupvec, vecvec, matvec, tamvec, mattam, qrisngvaldec.

```
package numal;

public interface AP_marquardt_methods {

    boolean funct(int m, int n, double par[], double rv[]);
    void jacobian(int m, int n, double par[], double rv[], double jac[][]);
}


public static void marquardt(int m, int n, double par[], double g[], double v[][],
                      AP_marquardt_methods method, double in[], double out[])
{
    boolean emergency;
    int maxfe,fe,it,i,j,err;
    double vv,ww,w,mu,res,fpar,fparpres,lambda,lambdamin,p,pw,reltolres,abstolres,
            temp;
    double em[] = new double[8];
    double val[] = new double[n+1];
    double b[] = new double[n+1];
    double bb[] = new double[n+1];
    double parpres[] = new double[n+1];
    double jac[][] = new double[m+1][n+1];

    lambda=0.0;
    vv=10.0;
    w=0.5;
    mu=0.01;
    ww = (in[6] < 1.0e-7) ? 1.0e-8 : 1.0e-1*in[6];
    em[0]=em[2]=em[6]=in[0];
    em[4]=10*n;
    reltolres=in[3];
    abstolres=in[4]*in[4];
    maxfe=(int)in[5];
    err=0;
    fe=it=1;
```

```
p=fpar=res=0.0;
pw = -Math.log(ww*in[0])/2.30;
if (!method.funct(m,n,par,g)) {
    err=3;
    out[4]=fe;
    out[5]=it-1;
    out[1]=err;
    return;
}
fpar=Basic.vecvec(1,m,0,g,g);
out[3]=Math.sqrt(fpar);
emergency=false;
it=1;
do {
    method.jacobian(m,n,par,g,jac);
    i=Linear_algebra.qrisngvaldec(jac,m,n,val,v,em);
    if (it == 1)
        lambda=in[6]*Basic.vecvec(1,n,0,val,val);
    else
        if (p == 0.0) lambda *= w;
    for (i=1; i<=n; i++) b[i]=val[i]*Basic.tamvec(1,m,i,jac,g);
    while (true) {
        for (i=1; i<=n; i++) bb[i]=b[i]/(val[i]*val[i]+lambda);
        for (i=1; i<=n; i++)
            parpres[i]=par[i]-Basic.matvec(1,n,i,v,bb);
        fe++;
        if (fe >= maxfe)
            err=1;
        else
            if (!method.funct(m,n,parpres,g)) err=2;
        if (err != 0) {
            emergency=true;
            break;
        }
        fparpres=Basic.vecvec(1,m,0,g,g);
        res=fpar-fparpres;
        if (res < mu*Basic.vecvec(1,n,0,b,bb)) {
            p += 1.0;
            lambda *= vv;
            if (p == 1.0) {
                lambdamin=ww*Basic.vecvec(1,n,0,val,val);
                if (lambda < lambdamin) lambda=lambdamin;
            }
            if (p >= pw) {
                err=4;
                emergency=true;
                break;
            }
        } else {
            Basic.dupvec(1,n,0,par,parpres);
```

```
                fpar=fparpres;
                break;
            }
        }
        if (emergency) break;
        it++;
    } while (fpar > abstolres && res > reltolres*fpar+abstolres);
    for (i=1; i<=n; i++)
        Basic.mulcol(1,n,i,i,jac,v,1.0/(val[i]+in[0]));
    for (i=1; i<=n; i++)
        for (j=1; j<=i; j++)
            v[i][j]=v[j][i]=Basic.mattam(1,n,i,j,jac,jac);
    lambda=lambdamin=val[1];
    for (i=2; i<=n; i++)
        if (val[i] > lambda)
            lambda=val[i];
        else
            if (val[i] < lambdamin) lambdamin=val[i];
    temp=lambda/(lambdamin+in[0]);
    out[7]=temp*temp;
    out[2]=Math.sqrt(fpar);
    out[6]=Math.sqrt(res+fpar)-out[2];
    out[4]=fe;
    out[5]=it-1;
    out[1]=err;
}
```

B. gssnewton

Calculates the least squares solution of an overdetermined system of nonlinear equations with the Gauss-Newton method [BusDK75, Har61, Sp67].

gssnewton computes the values of the components $p_1,...,p_n$ of $p \in R^n$ such that

$$\Phi(p) = \sum_{i=1}^{m}\left(f(t_i,p) - d_i\right)^2$$

is a minimum at p, where $f(t,p)$ is a real-valued function, and d_i $(i=1,...,m)$ are given data readings $(m \geq n)$ by use of an improved version of the Gauss-Newton algorithm (the p_j offer a parameter set of best fit of the function $f(t,p)$ to the data at the points t_i).

An initial approximation $p^{(0)}$ to p, three tolerances ε_{re}, ε_a, and ε_0 must be given.

At the k-th stage of the algorithm, the $m \times n$ Jacobian matrix $J(p^{(k)})$, whose components are given by $J_{i,j}(p) = \partial f(t_i,p)/\partial p_j$ $(i=1,...,m; j=1,...,n)$ with $p=p^{(k)}$, is decomposed in the form $J(p^{(k)})=Q^{(k)}R^{(k)}$ where $Q^{(k)}$ is an $m \times n$ orthogonal matrix, and $R^{(k)}$ is an $m \times n$ upper triangular matrix. The direction vector $d^{(k)} \in R^n$ is determined by solving the $n \times n$ upper triangular system derived from the first n equations of the system $R^{(k)}d^{(k)} = -(Q^{(k)})^T f(p^{(k)})$ where the components of $f(p^{(k)}) \in R^m$ are $f(t_i, p^{(k)}) - d_i$ $(i=1,...,m)$.

The scalar $\alpha^{(p)}$ is determined (a) by setting $\alpha^{(k)}=1$ if $\Phi(p^{(k)}+d^{(k)}) \leq \Phi(p^{(k)})$ or, if this condition is violated, (b) by setting $\alpha^{(k)}=2^{-r}$, where r is the smallest nonnegative integer for which

$$\Phi(p^{(k)} + 2^{-r}d^{(k)}) < \Phi(p^{(k)}),$$
$$\Phi(p^{(k)} + 2^{-r}d^{(k)}) < \Phi(p^{(k)} + 2^{-(r+1)}d^{(k)})$$

(these inequalities are inspected for r replaced by r' $(r'=0,1,...,r)$) $p^{(k+1)}$ is then $p^{(k)} + \alpha^{(k)}d^{(k)}$.

The algorithm is terminated (with $p^{(k)}$ accepted as an approximation to p) when (c) $\Phi(p^{(k)}) \leq \varepsilon_0^2$ or (d) $\|p^{(k+1)} - p^{(k)}\| \leq \|p^{(k+1)}\| \varepsilon_{re} + \varepsilon_a$, and the standard deviations of the elements of p are then calculated.

Procedure parameters:

void gssnewton ($m,n,par,rv,jjinv,method,in,out$)

m: int;
               entry: the number of equations;

n: int;
               entry: the number of unknowns in the m equations; ($n \leq m$);

par: double $par[1:n]$;
               entry: an approximation to a least squares solution of the system;
               exit: the calculated least squares solution;

rv: double $rv[1:m]$;
               exit: the residual vector of the system at the calculated solution;

$jjinv$:   double $jjinv[1:n,1:n]$;
               exit: the inverse of the matrix $J^T J$ where J denotes the transpose of the Jacobian matrix at the solution;

$method$: a class that defines two procedures *funct* and *jacobian*, this class must implement the AP_marquardt_methods interface;

         boolean funct(int m, int n, double par[ ], double rv[ ])

                 entry: m, n have the same meaning as in the procedure *gssnewton*; array $par[1:n]$ contains the current values of the unknowns and should not be altered;

                 exit: upon completion of a call of *funct*, the array $rv[1:m]$ should contain the residual vector obtained with the current values of the unknowns;

              the programmer of *funct* may decide that some current estimates of the unknowns lie outside a feasible region; in this case *funct* should deliver the value false and the process is terminated (see $out[1]$), otherwise *funct* should deliver the value true;

        void jacobian(int m, int n, double par[ ], double rv[ ], double jac[ ][ ])

              the parameters of *jacobian* are:

              m, n: see *gssnewton*;

              par: entry: current estimate of the unknowns, these values should not be changed;

                rv: entry: the residual vector of the system of equations corresponding to the vector of unknowns as given in *par*; on exit the values are not changed;

jac:	exit:	double *jac[1:m, 1:n]*;
		a call of the procedure *jacobian* should deliver the Jacobian matrix evaluated at the current estimates of the unknown variables given in *par* in such a way, that the partial derivative $\partial rv[i]/\partial par[j]$ is delivered in *jac[i,j]*, $i=1,...,m$, $j=1,...,n$;

in: double *in[0:7]*;

	entry: *in[0]*:	the machine precision;
	in[1]:	the relative tolerance for the step vector (relative to the vector of current estimates in *par*) (value of ε_{re} above); see *in[2]*; *in[1]* should not be chosen smaller than *in[0]*;
	in[2]:	the absolute tolerance for the step vector (relative to the vector of current estimates in *par*) (value of ε_a above); the process is terminated if in some iteration (but not the first) the Euclidean norm of the calculated Newton step is less than *in[1]*‖par‖+in[2]*;
	in[3]:	not used by *gssnewton*;
	in[4]:	the absolute tolerance for the Euclidean norm of the residual vector (value of ε_0 above); the process is terminated when this norm is less than *in[4]*;
	in[5]:	the maximum allowed number of function evaluations (i.e., calls of *funct*);
	in[6]:	the maximum allowed number of halvings of a calculated Newton step vector, a suitable value is 15;
	in[7]:	the maximum allowed number of successive *in[6]* times halved step vector, suitable values are 1 and 2;

out: double *out[1:9]*;

	exit: *out[1]*:	the process was terminated because *out[1]* has the value:
		1: the norm of the residual vector is small with respect to *in[4]*;
		2: the calculated Newton step is sufficiently small (see *in[1]*, *in[2]*);
		3: the calculated step was completely damped (halved) in *in[7]* successive iterations;
		4: *out[4]* exceeds *in[5]*, the maximum allowed number of calls of *funct*;
		5: the Jacobian was not full-rank (see *out[8]*);
		6: *funct* delivered false at a new vector of estimates of the unknowns;
		7: *funct* delivered false in a call from *jacobian*;
	out[2]:	the Euclidean norm of the last residual vector;
	out[3]:	the Euclidean norm of the initial residual vector;
	out[4]:	the total number of calls of *funct*; *out[4]* will be less than *in[5]+in[6]*;
	out[5]:	the total number of iterations;
	out[6]:	the Euclidean norm of the last step vector;
	out[7]:	iteration number of the last iteration in which the Newton step was halved;
	out[8],*out[9]*:	rank and maximum column norm of the Jacobian

matrix in the last iteration, as delivered by *lsqortdec*
in *aux[3]* and *aux[5]*.

Procedures used: dupvec, vecvec, elmvec, lsqortdec, lsqsol, lsqinv.

```
package numal;


public interface AP_gssnewton_methods {

    boolean funct(int m, int n, double par[], double rv[]);
    void jacobian(int m, int n, double par[], double rv[], double jac[][]);
}



public static void gssnewton(int m, int n, double par[], double rv[],
                        double jjinv[][], AP_gssnewton_methods method,
                        double in[], double out[])
{
    boolean dampingon,fail,testthf,conv;
    int i,j,inr,mit,text,it,itmax,inrmax,tim,feval,fevalmax;
    double rho,res1,res2,rn,reltolpar,abstolpar,abstolres,stap,normx;
    double aux[] = new double[6];
    int ci[] = new int[n+1];
    double pr[] = new double[n+1];
    double aid[] = new double[n+1];
    double sol[] = new double[n+1];
    double fu2[] = new double[m+1];
    double jac[][] = new double[m+2][n+1];

    res1=rho=0.0;
    conv=true;
    itmax=fevalmax=(int)in[5];
    aux[2]=n*in[0];
    tim=(int)in[7];
    reltolpar=in[1]*in[1];
    abstolpar=in[2]*in[2];
    abstolres=in[4]*in[4];
    inrmax=(int)in[6];
    Basic.dupvec(1,n,0,pr,par);
    if (m < n)
        for (i=1; i<=n; i++) jac[m+1][i]=0.0;
    text=4;
    mit=0;
    testthf=true;
    res2=stap=out[5]=out[6]=out[7]=0.0;
    method.funct(m,n,par,fu2);
    rn=Basic.vecvec(1,m,0,fu2,fu2);
    out[3]=Math.sqrt(rn);
    feval=1;
    dampingon=false;
```

```
fail=false;
it=1;
do {
    out[5]=it;
    method.jacobian(m,n,par,fu2,jac);
    if (!testthf) {
        text=7;
        fail=true;
        break;
    }
    Linear_algebra.lsqortdec(jac,m,n,aux,aid,ci);
    if (aux[3] != n) {
        text=5;
        fail=true;
        break;
    }
    Linear_algebra.lsqsol(jac,m,n,aid,ci,fu2);
    Basic.dupvec(1,n,0,sol,fu2);
    stap=Basic.vecvec(1,n,0,sol,sol);
    rho=2.0;
    normx=Basic.vecvec(1,n,0,par,par);
    if (stap > reltolpar*normx+abstolpar || it == 1 && stap > 0.0) {
        inr=0;
        do {
            rho /= 2.0;
            if (inr > 0) {
                res1=res2;
                Basic.dupvec(1,m,0,rv,fu2);
                dampingon = inr > 1;
            }
            for (i=1; i<=n; i++) pr[i]=par[i]-sol[i]*rho;
            feval++;
            if (!method.funct(m,n,pr,fu2)) {
                text=6;
                fail=true;
                break;
            }
            res2=Basic.vecvec(1,m,0,fu2,fu2);
            conv = inr >= inrmax;
            inr++;
        } while ((inr == 1) ? (dampingon || res2 >= rn) :
                              (!conv && (rn <= res1 || res2 < res1)));
        if (fail) break;
        if (conv) {
            mit++;
            if (mit < tim) conv=false;
        } else
            mit=0;
        if (inr > 1) {
            rho *= 2.0;
```

```
                    Basic.elmvec(1,n,0,par,sol,-rho);
                    rn=res1;
                    if (inr > 2) out[7]=it;
                } else {
                    Basic.dupvec(1,n,0,par,pr);
                    rn=res2;
                    Basic.dupvec(1,m,0,rv,fu2);
                }
                if (rn <= abstolres) {
                    text=1;
                    itmax=it;
                } else
                    if (conv && inrmax > 0) {
                        text=3;
                        itmax=it;
                    } else
                        Basic.dupvec(1,m,0,fu2,rv);
            } else {
                text=2;
                rho=1.0;
                itmax=it;
            }
            it++;
        } while (it <= itmax && feval < fevalmax);
        if (!fail) {
            Linear_algebra.lsqinv(jac,n,aid,ci);
            for (i=1; i<=n; i++) {
                jjinv[i][i]=jac[i][i];
                for (j=i+1; j<=n; j++) jjinv[i][j]=jjinv[j][i]=jac[i][j];
            }
        }
        out[6]=Math.sqrt(stap)*rho;
        out[2]=Math.sqrt(rn);
        out[4]=feval;
        out[1]=text;
        out[8]=aux[3];
        out[9]=aux[5];
    }
```

5.4 Differential equations — Initial value problems

5.4.1 First order - No derivatives right hand side

A. rk1

Solves an initial value problem for a single first order ordinary differential equation $dy/dx = f(x,y)$ by means of a 5-th order Runge-Kutta method [see Z64]. The equation is assumed to be nonstiff.

      rk1 is based on an explicit 5-th order Runge-Kutta method and is provided with step length and error control. The error control is based on the last term of the Taylor series which is taken into account. A step is rejected if the absolute value of this last term is greater than $(|fxy|*e[1]+e[2])*|h|/k$, where $k = |b-($if fi is true then a else $d[3])|$ denotes the length of the integration interval, otherwise, a step is accepted. *rk1* uses as its minimal absolute step length $hmin = e[1]*k+e[2]$. If a step of length $|h| = hmin$ is rejected then the step is skipped.

Procedure parameters:
                      void rk1 $(x,a,b,y,ya,method,e,d,fi)$

x:	double $x[0:0]$;
	entry: the independent variable;
	exit: upon completion of a call, x is equal to b;
a:	double;
	entry: the initial value of x;
b:	double;
	entry: a value parameter, giving the end value of x;
y:	double $y[0:0]$;
	entry: the dependent variable;
ya:	double;
	entry: the value of y at $x=a$;
method:	a class that defines a procedure *fxy*, this class must implement the AP_rk1_method interface;
	double fxy(double x[ ], double y[ ])
	the function giving the value of dy/dx;
e:	double $e[1:2]$;
	entry: $e[1]$: a relative tolerance;
	$e[2]$: an absolute tolerance;
d:	double $d[1:4]$;
	exit: $d[1]$: number of steps skipped;
	$d[2]$: equals the step length;
	$d[3]$: equals b;
	$d[4]$: equals $y(b)$;
fi:	boolean;
	entry: if *fi* is true then *rk1* integrates from $x=a$ to $x=b$ with initial value $y(a)=ya$ and trial step $b-a$; if *fi* is false then *rk1* integrates from $x=d[3]$ to $x=b$ with initial value $y(d[3])=d[4]$ and step length $h=d[2]*sign(b-d[3])$, while a and ya are ignored.

```
package numal;

public interface AP_rk1_method {

    double fxy(double x[], double y[]);
}


public static void rk1(double x[], double a, double b, double y[], double ya,
                       AP_rk1_method method, double e[], double d[], boolean fi)
{
    boolean last,first,reject,test,ta,tb;
    double e1,e2,xl,yl,h,ind,hmin,absh,k0,k1,k2,k3,k4,k5,discr,tol,mu,mu1,fh,hl;

    last=true;
    hl=mu1=0.0;
    if (fi) {
        d[3]=a;
        d[4]=ya;
    }
    d[1]=0.0;
    xl=d[3];
    yl=d[4];
    if (fi) d[2]=b-d[3];
    absh=h=Math.abs(d[2]);
    if (b-xl < 0.0) h = -h;
    ind=Math.abs(b-xl);
    hmin=ind*e[1]+e[2];
    e1=e[1]/ind;
    e2=e[2]/ind;
    first=true;
    test=true;
    if (fi) {
        last=true;
        test=false;
    }
    while (true) {
        if (test) {
            absh=Math.abs(h);
            if (absh < hmin) {
                h = (h > 0.0) ? hmin : -hmin;
                absh=hmin;
            }
            ta=(h >= b-xl);
            tb=(h >= 0.0);
            if ((ta && tb) || (!(ta || tb))) {
                d[2]=h;
                last=true;
                h=b-xl;
                absh=Math.abs(h);
```

```
        } else
            last=false;
}
test=true;
x[0]=xl;
y[0]=yl;
k0=method.fxy(x,y)*h;
x[0]=xl+h/4.5;
y[0]=yl+k0/4.5;
k1=method.fxy(x,y)*h;
x[0]=xl+h/3.0;
y[0]=yl+(k0+k1*3.0)/12.0;
k2=method.fxy(x,y)*h;
x[0]=xl+h*0.5;
y[0]=yl+(k0+k2*3.0)/8.0;
k3=method.fxy(x,y)*h;
x[0]=xl+h*0.8;
y[0]=yl+(k0*53.0-k1*135.0+k2*126.0+k3*56.0)/125.0;
k4=method.fxy(x,y)*h;
x[0] = (last ? b : xl+h);
y[0]=yl+(k0*133.0-k1*378.0+k2*276.0+k3*112.0+k4*25.0)/168.0;
k5=method.fxy(x,y)*h;
discr=Math.abs(k0*21.0-k2*162.0+k3*224.0-k4*125.0+k5*42.0)/14.0;
tol=Math.abs(k0)*e1+absh*e2;
reject = discr > tol;
mu=tol/(tol+discr)+0.45;
if (reject) {
    if (absh <= hmin) {
        d[1] += 1.0;
        y[0]=yl;
        first=true;
        if (b == x[0]) break;
        xl = x[0];
        yl = y[0];
    } else
        h *= mu;
} else {
    if (first) {
        first=false;
        hl=h;
        h *= mu;
    } else {
        fh=mu*h/hl+mu-mu1;
        hl=h;
        h *= fh;
    }
    mu1=mu;
    y[0]=yl+(-k0*63.0+k1*189.0-k2*36.0-k3*112.0+k4*50.0)/28.0;
    k5=method.fxy(x,y)*hl;
    y[0]=yl+(k0*35.0+k2*162.0+k4*125.0+k5*14.0)/336.0;
```

```
            if (b == x[0]) break;
            x1 = x[0];
            y1 = y[0];
        }
    }
    if (!last) d[2]=h;
    d[3] = x[0];
    d[4] = y[0];
}
```

B. rke

Solves an initial value problem for a system of first order ordinary differential equations $dy/dx = f(x,y)$, from $x = x0$ to $x = xe$ where $y(x0) = y0$, by means of a 5-th order Runge-Kutta method. The system is assumed to be nonstiff.

    The method upon which *rke* is based, is a member of a class of 5-th order Runge-Kutta formulas presented in [Eng69]. Automatic stepsize control is implemented in a way proposed in [Z64]. For further information see [Be74].

Procedure parameters:
$$\text{void rke } (x,xe,n,y,method,data,fi)$$

x: double $x[0:0]$;
           entry: the independent variable; the initial value $x0$;
xe: double $xe[0:0]$;
           entry: the final value of x;
n: int;
           entry: the number of equations of the system;
y: double $y[1:n]$;
           entry: the dependent variable; the initial values at $x = x0$;
           exit: the values of the solution at $x = xe$;
method: a class that defines two procedures *der* and *out*, this class must implement the `AP_rke_methods` interface;
           void der(int n, double t, double v[ ])
                this procedure performs an evaluation of the right-hand side of the system with dependent variables $v[1:n]$ and independent variable t; upon completion of *der* the right-hand side should be overwritten on $v[1:n]$;
           void out(int n, double x[ ], double xe[ ], double y[ ], double data[])
                after each integration step performed, *out* can be used to obtain information from the solution process, e.g., the values of x, $y[1:n]$, and $data[3:6]$; *out* can also be used to update *data*, but x and xe remain unchanged.
data: double $data[1:6]$;
           in array *data* one should give:
              data[1]:   the relative tolerance;
              data[2]:   the absolute tolerance;
           after each step *data[3:6]* contains:

         data[3]:   the steplength used for the last step;
         data[4]:   the number of integration steps performed;
         data[5]:   the number of integration steps rejected;
         data[6]:   the number of integration steps skipped;
    if upon completion of *rke data[6]* > 0, then results should be considered most critically;

fi: boolean;
    entry: if *fi* is true then the integration starts at *x0* with a trial step *xe - x0*; if *fi* is false then the integration is continued with a step length *data[3] * sign(xe - x0)*;

```
package numal;

public interface AP_rke_methods {

    void der(int n, double t, double v[]);
    void out(int n, double x[], double xe[], double y[], double data[]);
}


public static void rke(double x[], double xe[], int n, double y[],
                       AP_rke_methods method, double data[], boolean fi)
{
    boolean last,first,reject,test,ta,tb;
    int j;
    double xt,h,hmin,ind,hl,ht,absh,fhm,discr,tol,mu,mu1,fh,e1,e2;
    double k0[] = new double[n+1];
    double k1[] = new double[n+1];
    double k2[] = new double[n+1];
    double k3[] = new double[n+1];
    double k4[] = new double[n+1];

    last=true;
    hl=mu1=0.0;
    if (fi) {
        data[3]=xe[0]-x[0];
        data[4]=data[5]=data[6]=0.0;
    }
    absh=h=Math.abs(data[3]);
    if (xe[0] < x[0]) h = -h;
    ind=Math.abs(xe[0]-x[0]);
    hmin=ind*data[1]+data[2];
    e1=12.0*data[1]/ind;
    e2=12.0*data[2]/ind;
    first=true;
    reject=false;
    test=true;
    if (fi) {
        last=true;
        test=false;
```

```
        }
        while (true) {
            if (test) {
                absh=Math.abs(h);
                if (absh < hmin) {
                    h = (xe[0] == x[0]) ? 0.0 : ((xe[0] > x[0]) ? hmin : -hmin);
                    absh=hmin;
                }
                ta=(h >= xe[0]-x[0]);
                tb=(h >= 0.0);
                if ((ta && tb) || (!(ta || tb))) {
                    last=true;
                    h=xe[0]-x[0];
                    absh=Math.abs(h);
                } else
                    last=false;
            }
            test=true;
            if (!reject) {
                for (j=1; j<=n; j++) k0[j]=y[j];
                method.der(n,x[0],k0);
            }
            ht=0.184262134833347*h;
            xt = x[0]+ht;
            for (j=1; j<=n; j++) k1[j]=k0[j]*ht+y[j];
            method.der(n,xt,k1);
            ht=0.690983005625053e-1*h;
            xt=4.0*ht+x[0];
            for (j=1; j<=n; j++) k2[j]=(3.0*k1[j]+k0[j])*ht+y[j];
            method.der(n,xt,k2);
            xt=0.5*h+x[0];
            ht=0.1875*h;
            for (j=1; j<=n; j++)
                k3[j]=((1.74535599249993*k2[j]-k1[j])*2.23606797749979+k0[j])*ht+y[j];
            method.der(n,xt,k3);
            xt=0.723606797749979*h+x[0];
            ht=0.4*h;
            for (j=1; j<=n; j++)
                k4[j]=(((0.517595468166681*k0[j]-k1[j])*0.927050983124840+
                        k2[j])*1.46352549156242+k3[j])*ht+y[j];
            method.der(n,xt,k4);
            xt = (last ? xe[0] : x[0]+h);
            ht=2.0*h;
            for (j=1; j<=n; j++)
                k1[j]=((((2.0*k4[j]+k2[j])*0.412022659166595+k1[j])*
                        2.23606797749979-k0[j])*0.375-k3[j])*ht+y[j];
            method.der(n,xt,k1);
            reject=false;
            fhm=0.0;
            for (j=1; j<=n; j++) {
```

```
            discr=Math.abs((1.6*k3[j]-k2[j]-k4[j])*5.0+k0[j]+k1[j]);
            tol=Math.abs(k0[j])*e1+e2;
            reject = (discr > tol || reject);
            fh=discr/tol;
            if (fh > fhm) fhm=fh;
        }
        mu=1.0/(1.0+fhm)+0.45;
        if (reject) {
            data[5] += 1.0;
            if (absh <= hmin) {
                data[6] += 1.0;
                hl=h;
                reject=false;
                first=true;
                data[3]=hl;
                data[4] += 1.0;
                x[0]=xt;
                method.out(n,x,xe,y,data);
                if (x[0] == xe[0]) break;
            } else
                h *= mu;
        } else {
            if (first) {
                first=false;
                hl=h;
                h *= mu;
            } else {
                fh=mu*h/hl+mu-mu1;
                hl=h;
                h *= fh;
            }
            mu1=mu;
            ht=hl/12.0;
            for (j=1; j<=n; j++)
                y[j]=((k2[j]+k4[j])*5.0+k0[j]+k1[j])*ht+y[j];
            data[3]=hl;
            data[4] += 1.0;
            x[0]=xt;
            method.out(n,x,xe,y,data);
            if (x[0] == xe[0]) break;
        }
    }
}
}
```

C. rk4a

Solves an initial value problem for a single first order ordinary differential equation $dy/dx = f(x,y)$, where $f(x,y)$ may become large, e.g., in the neighborhood of a singularity, by means of a 5-th order Runge-Kutta method [see Z64]. The equation is assumed to be nonstiff.

$\quad$ *rk4a* integrates the given differential equation as it stands if $|f(x,y)| \leq 1$. If, however, $|f(x,y)| > 1$ then y is selected as the variable of integration, and the equation actually solved is $dx/dy = 1/f(x,y)$. The procedure is provided with step size and error control. *rk4a* integrates the differential equation from $x = a$ ($y(a)$ being given), and the direction of integration (i.e., x increasing or decreasing, or y increasing or decreasing) specified at outset, until to within stated tolerances a zero of the function $b(x,y)$ is encountered.

$\quad$ To explain the use of the function $b(x,y)$, we remark that if it is desired to continue integration from $x = a$ to $x = c$, then $b(x,y)$ is simply taken to be x-c; if it is desired to continue integration until $y(x)$ and a given function $g(x)$ have equal values (so that the graphs of y and g intersect), then $b(x,y)$ is taken to be $y(x) - g(x)$; if it is believed but not known that the graphs of $y(x)$ and $g(x)$ intersect in the range $x = a$ to $x = c$, $b(x,y)$ can be taken to be $(x-c)(y(x)-g(x))$ and by subsequent inspection of x, y, and g upon exit from *rk4a*, it can be decided whether the calculations were terminated because a point of intersection had been found or because the end of the range had been reached. The value of $b(x,y)$ for the current values of x and y upon call of *rk4a* is not inspected: it is thus possible of progress systematically from one zero of b to the next.

Procedure parameters:
$$\text{void rk4a } (x,xa,method,y,ya,e,d,fi,xdir,pos)$$

x:	double *x[0:0]*;
	entry:$\quad$the independent variable;
	exit:$\quad$upon completion of a call, x is equal to the most recent value of the independent variable;
xa:	double;
	entry:$\quad$the initial value of x;
method:	a class that defines two procedures b and *fxy*, this class must implement the AP_rk4a_methods interface;

$\qquad$ *double b(double x[ ], double y[ ])*
$\qquad\qquad$ the equation $b=0$ fulfilled within the tolerances *e[4]* and *e[5]* specifies the end of the integration interval, at the end of each integration step b is evaluated and is tested for change of sign$\quad$(it delivers the value of $b(x,y)$ described above);
$\qquad$ *double fxy(double x[ ], double y[ ])*
$\qquad\qquad$ *fxy* gives the value of dy/dx;

y:	double *y[0:0]*;
	entry:$\quad$the dependent variable;
ya:	double;
	entry:$\quad$the value of y at $x = xa$;
e:	double *e[0:5]*;
	entry:$\quad$*e[0]*, *e[2]*:$\quad$relative tolerances for x and y respectively;
	$\qquad\qquad$*e[1]*, *e[3]*:$\quad$absolute tolerances for x and y respectively;

$e[4]$, $e[5]$:   tolerances used in the determination of the zero of b;

d: double $d[0:4]$;

After completion of each step we have:

if $d[0]>0$ then x is the integration variable;

if $d[0]<0$ then y is the integration variable;

$d[1]$ is the number of steps skipped;

$d[2]$ is the step size;

$d[3]$ is equal to the last value of x;

$d[4]$ is equal to the last value of y;

fi: boolean;

entry: if fi is true then the integration is started with initial values
$x = xa$, $y = ya$;

if fi is false then the integration is started with $x = d[3]$, $y = d[4]$;

xdir,pos: boolean;

entry: if fi is true then the integration starts in such a way that

if pos is true and $xdir$ is true then x increases,

if pos is true and $xdir$ is false then y increases,

if pos is false and $xdir$ is true then x decreases,

if pos is false and $xdir$ is false then y decreases.

```
package numal;

public interface AP_rk4a_methods {

    double b(double x[], double y[]);
    double fxy(double x[], double y[]);
}


public static void rk4a(double x[], double xa, AP_rk4a_methods method, double y[],
                        double ya, double e[], double d[], boolean fi, boolean xdir,
                        boolean pos)
{
    boolean extrapolate,iv,first,next,fir,rej,t;
    int i,ext;
    double fhm,absh,s,xl,cond0,s1,cond1,yl,hmin,h,zl,tol,hl,mu,mu1,fzero,c,fc,bb,
           fb,a,fa,dd,fd,fdb,fda,w,mb,m,p,q;
    double e1[] = new double[3];
    double k0[] = new double[1];
    double k1[] = new double[1];
    double k2[] = new double[1];
    double k3[] = new double[1];
    double k4[] = new double[1];
    double k5[] = new double[1];
    double discr[] = new double[1];

    mu=hl=mu1=cond0=fd=dd=0.0;
    if (fi) {
        d[3]=xa;
        d[4]=ya;
```

```
        d[0]=1.0;
    }
    d[1]=0.0;
    x[0]=xl=d[3];
    y[0]=yl=d[4];
    iv = d[0] > 0.0;
    first=fir=true;
    hmin=e[0]+e[1];
    h=e[2]+e[3];
    if (h < hmin) hmin=h;
    while (true) {
        zl=method.fxy(x,y);
        if (Math.abs(zl) <= 1.0) {
            if (!iv) {
                d[2] = h /= zl;
                d[0]=1.0;
                iv=first=true;
            }
            if (fir) {
                t=(((iv && xdir) || (!(iv || xdir))) ? h : h*zl) < 0.0;
                if (fi ? ((t && pos) || (!(t || pos))) : (h*d[2] < 0))
                    h = -h;
            }
            i=1;
        } else {
            if (iv) {
                if (!fir) d[2] = h *= zl;
                d[0] = -1.0;
                iv=false;
                first=true;
            }
            if (fir) {
                h=e[0]+e[1];
                t=(((iv && xdir) || (!(iv || xdir))) ? h : h*zl) < 0.0;
                if (fi ? ((t && pos) || (!(t || pos))) : (h*d[2] < 0))
                    h = -h;
            }
            i=1;
        }
        next=false;
        while (true) {
            absh=Math.abs(h);
            if (absh < hmin) {
                h = (h == 0.0) ? 0.0 : ((h > 0.0) ? hmin : -hmin);
                absh=hmin;
            }
            if (iv) {
                rk4arkstep(x,xl,h,y,yl,zl,method,i,false,k0,k1,k2,k3,k4,k5,discr,mu);
                tol=e[2]*Math.abs(k0[0])+e[3]*absh;
            } else {
```

```
            rk4arkstep(y,yl,h,x,xl,1.0/zl,method,i,true,k0,k1,k2,k3,k4,k5,
                        discr,mu);
            tol=e[0]*Math.abs(k0[0])+e[1]*absh;
        }
        rej = discr[0] > tol;
        mu=tol/(tol+discr[0])+0.45;
        if (!rej) break;
        if (absh <= hmin) {
            if (iv) {
                x[0]=xl+h;
                y[0]=yl+k0[0];
            } else {
                x[0]=xl+k0[0];
                y[0]=yl+h;
            }
            d[1] += 1.0;
            first=true;
            next=true;
            break;
        }
        h *= mu;
        i=0;
    }
    if (!next) {
        if (first) {
            first=fir;
            hl=h;
            h *= mu;
        } else {
            fhm=mu*h/hl+mu-mu1;
            hl=h;
            h *= fhm;
        }
        if (iv)
            rk4arkstep(x,xl,hl,y,yl,zl,method,2,false,k0,k1,k2,k3,k4,k5,
                        discr,mu);
        else
            rk4arkstep(y,yl,hl,x,xl,zl,method,2,true,k0,k1,k2,k3,k4,k5,discr,mu);
        mu1=mu;
    }
    if (fir) {
        fir=false;
        cond0=method.b(x,y);
        if (!(fi || rej)) h=d[2];
    } else {
        d[2]=h;
        cond1=method.b(x,y);
        if (cond0*cond1 <= 0.0) break;
        cond0=cond1;
    }
```

```
        d[3]=xl=x[0];
        d[4]=yl=y[0];
    }
    e1[1]=e[4];
    e1[2]=e[5];
    s1 = iv ? x[0] : y[0];
    s = iv ? xl : yl;
    /* find zero */
    bb=s;
    if (iv) {
        if (s == xl)
            fzero=cond0;
        else
            if (s == s1)
                fzero=cond1;
            else {
                rk4arkstep(x,xl,s-xl,y,yl,zl,method,3,false,k0,k1,k2,k3,k4,k5,
                        discr,mu);
                fzero=method.b(x,y);
            }
    } else {
        if (s == yl)
            fzero=cond0;
        else
            if (s == s1)
                fzero=cond1;
            else {
                rk4arkstep(y,yl,s-yl,x,xl,zl,method,3,true,k0,k1,k2,k3,k4,k5,
                        discr,mu);
                fzero=method.b(x,y);
            }
    }
    fb=fzero;
    a=s=s1;
    if (iv) {
        if (s == xl)
            fzero=cond0;
        else
            if (s == s1)
                fzero=cond1;
            else {
                rk4arkstep(x,xl,s-xl,y,yl,zl,method,3,false,k0,k1,k2,k3,k4,k5,
                        discr,mu);
                fzero=method.b(x,y);
            }
    } else {
        if (s == yl)
            fzero=cond0;
        else
            if (s == s1)
```

```
                        fzero=cond1;
                else {
                    rk4arkstep(y,yl,s-yl,x,xl,zl,method,3,true,k0,k1,k2,k3,k4,k5,
                                discr,mu);
                    fzero=method.b(x,y);
                }
        }
        fa=fzero;
        c=a;
        fc=fa;
        ext=0;
        extrapolate=true;
        while (extrapolate) {
            if (Math.abs(fc) < Math.abs(fb)) {
                if (c != a) {
                    dd=a;
                    fd=fa;
                }
                a=bb;
                fa=fb;
                bb=s=c;
                fb=fc;
                c=a;
                fc=fa;
            }
            tol=Math.abs(e1[1]*s)+Math.abs(e1[2]);
            m=(c+bb)*0.5;
            mb=m-bb;
            if (Math.abs(mb) > tol) {
                if (ext > 2)
                    w=mb;
                else {
                    if (mb == 0.0)
                        tol=0.0;
                    else
                        if (mb < 0.0) tol = -tol;
                    p=(bb-a)*fb;
                    if (ext <= 1)
                        q=fa-fb;
                    else {
                        fdb=(fd-fb)/(dd-bb);
                        fda=(fd-fa)/(dd-a);
                        p *= fda;
                        q=fdb*fa-fda*fb;
                    }
                    if (p < 0.0) {
                        p = -p;
                        q = -q;
                    }
                    w=(p<Double.MIN_VALUE || p<=q*tol) ? tol : ((p<mb*q) ? p/q : mb);
```

```
                }
                dd=a;
                fd=fa;
                a=bb;
                fa=fb;
                s = bb += w;
                if (iv) {
                    if (s == xl)
                        fzero=cond0;
                    else
                        if (s == sl)
                            fzero=cond1;
                        else {
                            rk4arkstep(x,xl,s-xl,y,yl,zl,method,3,false,k0,k1,k2,k3,
                                        k4,k5,discr,mu);
                            fzero=method.b(x,y);
                        }
                } else {
                    if (s == yl)
                        fzero=cond0;
                    else
                        if (s == sl)
                            fzero=cond1;
                        else {
                            rk4arkstep(y,yl,s-yl,x,xl,zl,method,3,true,k0,k1,k2,k3,
                                        k4,k5,discr,mu);
                            fzero=method.b(x,y);
                        }
                }
                fb=fzero;
                if ((fc >= 0.0) ? (fb >= 0.0) : (fb <= 0.0)) {
                    c=a;
                    fc=fa;
                    ext=0;
                } else
                    ext = (w == mb) ? 0 : ext+1;
            } else
                break;
        }
        /* end of finding zero */
        sl = iv ? x[0] : y[0];
        if (iv)
            rk4arkstep(x,xl,s-xl,y,yl,zl,method,3,false,k0,k1,k2,k3,k4,k5,discr,mu);
        else
            rk4arkstep(y,yl,s-yl,x,xl,zl,method,3,true,k0,k1,k2,k3,k4,k5,discr,mu);
        d[3]=x[0];
        d[4]=y[0];
    }
}
```

```
static private void rk4arkstep(double x[], double xl, double h, double y[],
                               double yl, double zl, AP_rk4a_methods method, int d,
                               boolean invf, double k0[], double k1[], double k2[],
                               double k3[], double k4[], double k5[],
                               double discr[], double mu)
{
    /* this procedure is internally used by RK4A */

    if (d != 2) {
        if (d == 3) {
            x[0]=xl;
            y[0]=yl;
            k0[0]=(invf ? (1.0/method.fxy(x,y)) : method.fxy(x,y))*h;
        } else
            if (d == 1)
                k0[0]=zl*h;
            else
                k0[0] *= mu;
        x[0]=xl+h/4.5;
        y[0]=yl+k0[0]/4.5;
        k1[0]=(invf ? (1.0/method.fxy(x,y)) : method.fxy(x,y))*h;
        x[0]=xl+h/3.0;
        y[0]=yl+(k0[0]+k1[0]*3.0)/12.0;
        k2[0]=(invf ? (1.0/method.fxy(x,y)) : method.fxy(x,y))*h;
        x[0]=xl+h*0.5;
        y[0]=yl+(k0[0]+k2[0]*3.0)/8.0;
        k3[0]=(invf ? (1.0/method.fxy(x,y)) : method.fxy(x,y))*h;
        x[0]=xl+h*0.8;
        y[0]=yl+(k0[0]*53.0-k1[0]*135.0+k2[0]*126.0+k3[0]*56.0)/125.0;
        k4[0]=(invf ? (1.0/method.fxy(x,y)) : method.fxy(x,y))*h;
        if (d <= 1) {
            x[0]=xl+h;
            y[0]=yl+(k0[0]*133.0-k1[0]*378.0+k2[0]*276.0+
                    k3[0]*112.0+k4[0]*25.0)/168.0;
            k5[0]=(invf ? (1.0/method.fxy(x,y)) : method.fxy(x,y))*h;
            discr[0]=Math.abs(k0[0]*21.0-k2[0]*162.0+k3[0]*224.0-
                            k4[0]*125.0+k5[0]*42.0)/14.0;
            return;
        }
    }
    x[0]=xl+h;
    y[0]=yl+(-k0[0]*63.0+k1[0]*189.0-k2[0]*36.0-k3[0]*112.0+k4[0]*50.0)/28.0;
    k5[0] = (invf ? (1.0/method.fxy(x,y)) : method.fxy(x,y))*h;
    y[0]=yl+(k0[0]*35.0+k2[0]*162.0+k4[0]*125.0+k5[0]*14.0)/336.0;
}
```

D. rk4na

Solves an initial value problem for a system of first order ordinary differential equations $dx_j(x_0)/dx_0 = f_j(x)$, $(j=1,...,n)$ of which the derivative components are supposed to become large, e.g., in the neighborhood of a singularities, by means of a 5-th order Runge-Kutta method [Z64]. The system is assumed to be nonstiff.

    rk4na integrates the given system of differential equations from $x_0 = a$ $(x_j(a)$, $j=1,...,n$ being given) and the direction of integration (i.e., with l specified, $0 \le l \le n$, x_l increasing or decreasing) prescribed at outset until, to within stated tolerances, a zero of the function $b(x)$ is encountered. ($b(x)$ is, of course, effectively a function of the single variable x_0).

    The role of the function b is similar to that played in the implementation of *rk4a*; now, for example, we may take $b(x)$ to be $x_1 - x_2$, to determine the point at which $x_1(x_0)$ and $x_2(x_0)$ become equal.

    At each integration step, the quantities $|dx_j(x_0)/dx_0|$ $(j=0,1,...,n)$ are inspected, and the variable of integration is taken to be $x_{j'}$, where j' is that value of j corresponding to the maximum of these quantities. The system of equations actually solved has the form

$$dx_j/dx_{j'} = f_j(x)/f_{j'}(x) \quad (j=0,1,...,n;\ j \ne j').$$

Procedure parameters:

                  void rk4na $(x,xa,method,e,d,fi,n,l,pos)$

x: double $x[0:n]$;
          entry: $x[0]$ is the independent variable, $x[1],...,x[n]$ are the dependent variables;
          exit: the solution at $b=0$;
xa: double $xa[0:n]$;
          entry: the initial values of $x[0],...,x[n]$;
method: a class that defines two procedures b and fxj, this class must implement the AP_rk4na_methods interface;
          double b(int n, double x[ ])
              b depends on $x[0],...,x[n]$; if the equation $b=0$ is satisfied within a certain tolerance (see parameter e), the integration is terminated; b is evaluated and tested for change of sign at the end of each step;
          double fxj(int n, int j, double x[ ])
              fxj depends on $x[0],..,x[n]$ and j, defining the right-hand side of the differential equation; at each call it delivers: $dx[j]/dx[0]$;
e: double $e[0:2n+3]$;
          entry: $e[2j]$ and $e[2j+1]$, $0 \le j \le n$, are the relative and the absolute tolerance, respectively, associated with $x[j]$; $e[2n+2]$ and $e[2n+3]$ are the relative and absolute tolerance used in the determination of the zero of b;
d: double $d[0:n+3]$;
          After completion of each step we have:
          $d[0]$ is the number of steps skipped;
          $d[2]$ is the step length;
          $d[j+3]$ is the last value of $x[j]$, $j=0,...,n$;

fi: boolean;
            entry: if *fi* is true then the integration is started with initial condition
                        $x[j] = xa[j]$;
                        if *fi* is false then the integration is continued with $x[j] = d[j+3]$;

n: int;
            entry: the number of equations;

l: int;
            entry: an integer to be supplied by the user, $0 \le l \le n$ (see *pos*);

pos: boolean;
            entry: if *fi* is true then the integration starts in such a way that $x[l]$
                        increases if *pos* is true and $x[l]$ decreases if *pos* is false;
                        if *fi* is false then *pos* is of no significance.

```
package numal;

public interface AP_rk4na_methods {

    double b(int n, double x[]);
    double fxj(int n, int j, double x[]);
}



public static void rk4na(double x[], double xa[], AP_rk4na_methods method,
                         double e[], double d[], boolean fi, int n, int l,
                         boolean pos)
{
    boolean extrapolate,rej,change,next,first,fir,t;
    int j,i,iv,iv0,ext;
    double h,cond0,cond1,fhm,absh,tol,fh,max,x0,x1,s,hmin,hl,mu,mu1,p,fzero,c,fc,bb,
           fb,a,fa,dd,fd,fdb,fda,w,mb,m,q;
    double e1[] = new double[3];
    double xl[] = new double[n+1];
    double discr[] = new double[n+1];
    double y[] = new double[n+1];
    double k[][] = new double[6][n+1];

    hmin=mu=hl=mu1=cond0=x0=fd=dd=0.0;
    if (fi) {
        for (i=0; i<=n; i++) d[i+3]=xa[i];
        d[0]=d[2]=0.0;
    }
    d[1]=0.0;
    for (i=0; i<=n; i++) x[i]=xl[i]=d[i+3];
    iv=(int)d[0];
    h=d[2];
    first=fir=true;
    y[0]=1.0;
    next=false;
    change=true;
    while (true) {
```

```
if (!change) {
    while (true) {
        absh=Math.abs(h);
        if (absh < hmin) {
            h = (h > 0.0) ? hmin : -hmin;
            absh=Math.abs(h);
        }
        rk4narkstep(h,i,n,iv,mu,method,x,xl,y,discr,k);
        rej=false;
        fhm=0.0;
        for (i=0; i<=n; i++)
            if (i != iv) {
                tol=e[2*i]*Math.abs(k[0][i])+e[2*i+1]*absh;
                rej=(tol < discr[i] || rej);
                fh=discr[i]/tol;
                if (fh > fhm) fhm=fh;
            }
        mu=1.0/(1.0+fhm)+0.45;
        if (!rej) break;
        if (absh <= hmin) {
            for (i=0; i<=n; i++)
                if (i != iv)
                    x[i]=xl[i]+k[0][i];
                else
                    x[i]=xl[i]+h;
            d[1] += 1.0;
            first=true;
            next=true;
            break;
        }
        h *= mu;
        i=0;
    }
    if (!next) {
        if (first) {
            first=fir;
            hl=h;
            h *= mu;
        } else {
            fh=mu*h/hl+mu-mu1;
            hl=h;
            h *= fh;
        }
        rk4narkstep(hl,2,n,iv,mu,method,x,xl,y,discr,k);
        mu1=mu;
    }
    next=false;
    if (fir) {
        fir=false;
        cond0=method.b(n,x);
```

```
                    if (!(fi || rej)) h=d[2];
            } else {
                d[2]=h;
                cond1=method.b(n,x);
                if (cond0*cond1 <= 0.0) break;
                cond0=cond1;
            }
            for (i=0; i<=n; i++) d[i+3]=xl[i]=x[i];
        }
        change=false;
        iv0=iv;
        for (j=1; j<=n; j++) y[j]=method.fxj(n,j,x);
        max=Math.abs(y[iv]);
        for (i=0; i<=n; i++)
            if (Math.abs(y[i]) > max) {
                max=Math.abs(y[i]);
                iv=i;
            }
        if (iv0 != iv) {
            first=true;
            d[0]=iv;
            d[2]=h=y[iv]/y[iv0]*h;
        }
        x0=xl[iv];
        if (fir) {
            hmin=e[0]+e[1];
            for (i=1; i<=n; i++) {
                h=e[2*i]+e[2*i+1];
                if (h < hmin) hmin=h;
            }
            h=e[2*iv]+e[2*iv+1];
            t=y[1]/y[iv]*h < 0.0;
            if ((fi && ((t && pos) || !(t || pos))) || (!fi && d[2]*h < 0.0)) h = -h;
        }
        i=1;
    }
}
el[1]=e[2*n+2];
el[2]=e[2*n+3];
x1=x[iv];
s=x0;
/* find zero */
bb=s;
if (s == x0)
    fzero=cond0;
else
    if (s == x1)
        fzero=cond1;
    else {
        rk4narkstep(s-xl[iv],3,n,iv,mu,method,x,xl,y,discr,k);
        fzero=method.b(n,x);
```

```
        }
    fb=fzero;
    a=s=x1;
    if (s == x0)
        fzero=cond0;
    else
        if (s == x1)
            fzero=cond1;
        else {
            rk4narkstep(s-x1[iv],3,n,iv,mu,method,x,x1,y,discr,k);
            fzero=method.b(n,x);
        }
    fa=fzero;
    c=a;
    fc=fa;
    ext=0;
    extrapolate=true;
    while (extrapolate) {
        if (Math.abs(fc) < Math.abs(fb)) {
            if (c != a) {
                dd=a;
                fd=fa;
            }
            a=bb;
            fa=fb;
            bb=s=c;
            fb=fc;
            c=a;
            fc=fa;
        }
        tol=Math.abs(e1[1]*s)+Math.abs(e1[2]);
        m=(c+bb)*0.5;
        mb=m-bb;
        if (Math.abs(mb) > tol) {
            if (ext > 2)
                w=mb;
            else {
                if (mb == 0.0)
                    tol=0.0;
                else
                    if (mb < 0.0) tol = -tol;
                p=(bb-a)*fb;
                if (ext <= 1)
                    q=fa-fb;
                else {
                    fdb=(fd-fb)/(dd-bb);
                    fda=(fd-fa)/(dd-a);
                    p *= fda;
                    q=fdb*fa-fda*fb;
                }
```

```
            if (p < 0.0) {
                p = -p;
                q = -q;
            }
            w=(p<Double.MIN_VALUE || p<=q*tol) ? tol : ((p<mb*q) ? p/q : mb);
        }
        dd=a;
        fd=fa;
        a=bb;
        fa=fb;
        s = bb += w;
        if (s == x0)
            fzero=cond0;
        else
            if (s == x1)
                fzero=cond1;
            else {
                rk4narkstep(s-xl[iv],3,n,iv,mu,method,x,xl,y,discr,k);
                fzero=method.b(n,x);
            }
        fb=fzero;
        if ((fc >= 0.0) ? (fb >= 0.0) : (fb <= 0.0)) {
            c=a;
            fc=fa;
            ext=0;
        } else
            ext = (w == mb) ? 0 : ext+1;
        } else
            break;
    }
    /* end of finding zero */
    x0=s;
    x1=x[iv];
    rk4narkstep(x0-xl[iv],3,n,iv,mu,method,x,xl,y,discr,k);
    for (i=0; i<=n; i++) d[i+3]=x[i];
}


static private void rk4narkstep(double h, int d, int n, int iv, double mu,
                                AP_rk4na_methods method, double x[], double xl[],
                                double y[], double discr[], double k[][])
{
    /* this procedure is internally used by RK4NA */

    int i,j;
    double p;

    if (d != 2) {
        if (d == 3) {
            for (i=0; i<=n; i++) x[i]=xl[i];
```

```
        for (j=1; j<=n; j++) y[j]=method.fxj(n,j,x);
        p=h/y[iv];
        for (i=0; i<=n; i++)
            if (i != iv) k[0][i]=y[i]*p;
    } else
        if (d == 1) {
            p=h/y[iv];
            for (i=0; i<=n; i++)
                if (i != iv) k[0][i]=p*y[i];
        } else
            for (i=0; i<=n; i++)
                if (i != iv) k[0][i] *= mu;
for (i=0; i<=n; i++)
    x[i]=xl[i]+((i == iv) ? h : k[0][i])/4.5;
for (j=1; j<=n; j++) y[j]=method.fxj(n,j,x);
p=h/y[iv];
for (i=0; i<=n; i++)
    if (i != iv) k[1][i]=y[i]*p;
for (i=0; i<=n; i++)
    x[i]=xl[i]+((i == iv) ? h*4.0 : (k[0][i]+k[1][i]*3.0))/12.0;
for (j=1; j<=n; j++) y[j]=method.fxj(n,j,x);
p=h/y[iv];
for (i=0; i<=n; i++)
    if (i != iv) k[2][i]=y[i]*p;
for (i=0; i<=n; i++)
    x[i]=xl[i]+((i == iv) ? h*0.5 : (k[0][i]+k[2][i]*3.0)/8.0);
for (j=1; j<=n; j++) y[j]=method.fxj(n,j,x);
p=h/y[iv];
for (i=0; i<=n; i++)
    if (i != iv) k[3][i]=y[i]*p;
for (i=0; i<=n; i++)
    x[i]=xl[i]+((i == iv) ? h*0.8 : (k[0][i]*53.0-
        k[1][i]*135.0+k[2][i]*126.0+k[3][i]*56.0)/125.0);
for (j=1; j<=n; j++) y[j]=method.fxj(n,j,x);
p=h/y[iv];
for (i=0; i<=n; i++)
    if (i != iv) k[4][i]=y[i]*p;
if (d <= 1) {
    for (i=0; i<=n; i++)
        x[i]=xl[i]+((i == iv) ? h : (k[0][i]*133.0-k[1][i]*378.0+
            k[2][i]*276.0+k[3][i]*112.0+k[4][i]*25.0)/168.0);
    for (j=1; j<=n; j++) y[j]=method.fxj(n,j,x);
    p=h/y[iv];
    for (i=0; i<=n; i++)
        if (i != iv) k[5][i]=y[i]*p;
    for (i=0; i<=n; i++)
        if (i != iv)
            discr[i]=Math.abs(k[0][i]*21.0-k[2][i]*162.0+k[3][i]*224.0-
                            k[4][i]*125.0+k[5][i]*42.0)/14.0;
    return;
```

```
        }
    }
    for (i=0; i<=n; i++)
        x[i]=x1[i]+((i == iv) ? h : (-k[0][i]*63.0+k[1][i]*189.0-k[2][i]*36.0-
                                    k[3][i]*112.0+k[4][i]*50.0)/28.0);
    for (j=1; j<=n; j++) y[j]=method.fxj(n,j,x);
    p=h/y[iv];
    for (i=0; i<=n; i++)
        if (i != iv) k[5][i]=y[i]*p;
    for (i=0; i<=n; i++)
        if (i != iv)
            x[i]=x1[i]+(k[0][i]*35.0+k[2][i]*162.0+k[4][i]*125.0+k[5][i]*14.0)/336.0;
}
```

E. rk5na

Solves an initial value problem for a system of first order ordinary differential equations
$$dx_j/dx_0 = f_j(x)/f_0(x), \quad (j=1,2,\ldots,n)$$
of which the derivative components are supposed to become large, e.g., in the neighborhood of a singularities, by means of a 5-th order Runge-Kutta method [Z64]. The system is assumed to be nonstiff.

      rk5na integrates the given system of differential equations from $x_0 = a$ $(x_j(a)$, $j=1,\ldots,n$, being given) and the direction of integration (i.e., with l specified, $0 \leq l \leq n$, x_l increasing or decreasing) prescribed at outset until, to within stated tolerances, a zero of the function $b(x)$ is encountered.

      The arc length s is used as the variable of integration. Since $ds^2 = dx_0^2 + \ldots + dx_n^2$, it follows that $ds = (dx_0/f_0(x))a$, where

$$a = \sqrt{\sum_{j=0}^{n} f_j(x)^2}$$

The system of equations actually solved is
$$dx_j/dx_0 = f_j(x)/a \quad (j=0,1,\ldots,n).$$
Thus, points at which $f_0(x) = 0$ and dx_j/dx_0 may become infinite cause no special trouble.

      The role of the function b is similar to that played by the same function in the implementation of *rk4na*.

Procedure parameters:
$$\text{void rk5na } (x,xa,method,e,d,fi,n,l,pos)$$
x: double $x[0:n]$;
            entry: the dependent variables;
xa: double $xa[0:n]$;
            entry: the initial values of $x[0],\ldots,x[n]$;
method: a class that defines two procedures b and *fxj*, this class must implement
            the AP_rk5na_methods interface;

      double b(int n, double x[ ])

          b depends on *x[0],...,x[n]*; if, within some tolerance, *b=0* then the integration is terminated; see parameter *e*;

      double fxj(int n, int j, double x[ ])

          fxj depends on *x[0],..,x[n]* and *j*, giving the value of *dx[j]/dx[0]*;

e: double *e[0:2n+3]*;

      entry: *e[2j]* and *e[2j+1]*, $0 \le j \le n$, are the relative and the absolute tolerance, respectively, associated with *x[j]*; while *e[2n+2]* and *e[2n+3]* are the relative and absolute tolerance used in the determination of the zero of *b*;

d: double *d[1:n+3]*;

      After completion of each step we have:

      $|d[1]|$ is the arc length;

      d[2] is the step length;

      d[j+3] is the latest value of *x[j]*, *j=0,...,n*;

fi: boolean;

      entry: if *fi* is true then the integration is started with initial conditions *x[j]=xa[j]*, *j=0,...,n*; if *fi* is false then the integration is continued with *x[j]=d[j+3]*;

n: int;

      entry: the number of equations;

l: int;

      entry: an integer to be supplied by the user, $1 \le l \le n$ (see *pos*);

pos: boolean;

      entry: if *fi* is true then the integration starts in such a way that *x[l]* increases if *pos* is true and *x[l]* decreases if *pos* is false; if *fi* is false then *pos* is ignored.

```
package numal;

public interface AP_rk5na_methods {

    double b(int n, double x[]);
    double fxj(int n, int j, double x[]);
}



public static void rk5na(double x[], double xa[], AP_rk5na_methods method,
                    double e[], double d[], boolean fi, int n, int l,
                    boolean pos)
{
    boolean extrapolate,first,fir,rej,t;
    int j,i,ext;
    double fhm,s,s0,cond0,s1,cond1,h,absh,tol,fh,hl,mu,mu1,fzero,c,fc,bb,fb,a,
            fa,dd,fd,fdb,fda,w,mb,m,p,q;
    double e1[] = new double[3];
    double y[] = new double[n+1];
    double xl[] = new double[n+1];
    double discr[] = new double[n+1];
    double k[][] = new double[6][n+1];
```

```
mu=hl=mu1=cond0=s0=fd=dd=0.0;
if (fi) {
    for (i=0; i<=n; i++) d[i+3]=xa[i];
    d[1]=d[2]=0.0;
}
for (i=0; i<=n; i++) x[i]=xl[i]=d[i+3];
s=d[1];
first=fir=true;
h=e[0]+e[1];
for (i=1; i<=n; i++) {
    absh=e[2*i]+e[2*i+1];
    if (h > absh) h=absh;
}
if (fi) {
    j=1;
    t=method.fxj(n,j,x)*h < 0.0;
    if ((t && pos) || !(t || pos)) h = -h;
} else
    if (d[2]*h < 0.0) h = -h;
i=0;
while (true) {
    rk5narkstep(h,i,n,mu,method,x,xl,y,discr,k);
    rej=false;
    fhm=0.0;
    absh=Math.abs(h);
    for (i=0; i<=n; i++) {
        tol=e[2*i]*Math.abs(k[0][i])+e[2*i+1]*absh;
        rej=(tol < discr[i] || rej);
        fh=discr[i]/tol;
        if (fh > fhm) fhm=fh;
    }
    mu=1.0/(1.0+fhm)+0.45;
    if (rej) {
        h *= mu;
        i=1;
    } else {
        if (first) {
            first=fir;
            hl=h;
            h *= mu;
        } else {
            fh=mu*h/hl+mu-mu1;
            hl=h;
            h *= fh;
        }
        rk5narkstep(hl,2,n,mu,method,x,xl,y,discr,k);
        mu1=mu;
        s += hl;
        if (fir) {
```

```
            cond0=method.b(n,x);
            fir=false;
            if (!fi) h=d[2];
        } else {
            d[2]=h;
            cond1=method.b(n,x);
            if (cond0*cond1 <= 0.0) break;
            cond0=cond1;
        }
        for (i=0; i<=n; i++) d[i+3]=xl[i]=x[i];
        d[1]=s0=s;
        i=0;
    }
}
e1[1]=e[2*n+2];
e1[2]=e[2*n+3];
s1=s;
s=s0;
/* find zero */
bb=s;
if (s == s0)
    fzero=cond0;
else
    if (s == s1)
        fzero=cond1;
    else {
        rk5narkstep(s-s0,3,n,mu,method,x,xl,y,discr,k);
        fzero=method.b(n,x);
    }
fb=fzero;
a=s=s1;
if (s == s0)
    fzero=cond0;
else
    if (s == s1)
        fzero=cond1;
    else {
        rk5narkstep(s-s0,3,n,mu,method,x,xl,y,discr,k);
        fzero=method.b(n,x);
    }
fa=fzero;
c=a;
fc=fa;
ext=0;
extrapolate=true;
while (extrapolate) {
    if (Math.abs(fc) < Math.abs(fb)) {
        if (c != a) {
            dd=a;
            fd=fa;
```

```
        }
        a=bb;
        fa=fb;
        bb=s=c;
        fb=fc;
        c=a;
        fc=fa;
    }
    tol=Math.abs(e1[1]*s)+Math.abs(e1[2]);
    m=(c+bb)*0.5;
    mb=m-bb;
    if (Math.abs(mb) > tol) {
        if (ext > 2)
            w=mb;
        else {
            if (mb == 0.0)
                tol=0.0;
            else
                if (mb < 0.0) tol = -tol;
            p=(bb-a)*fb;
            if (ext <= 1)
                q=fa-fb;
            else {
                fdb=(fd-fb)/(dd-bb);
                fda=(fd-fa)/(dd-a);
                p *= fda;
                q=fdb*fa-fda*fb;
            }
            if (p < 0.0) {
                p = -p;
                q = -q;
            }
            w=(p<Double.MIN_VALUE || p<=q*tol) ? tol : ((p<mb*q) ? p/q : mb);
        }
        dd=a;
        fd=fa;
        a=bb;
        fa=fb;
        s = bb += w;
        if (s == s0)
            fzero=cond0;
        else
            if (s == s1)
                fzero=cond1;
            else {
                rk5narkstep(s-s0,3,n,mu,method,x,xl,y,discr,k);
                fzero=method.b(n,x);
            }
        fb=fzero;
        if ((fc >= 0.0) ? (fb >= 0.0) : (fb <= 0.0)) {
```

```
                    c=a;
                    fc=fa;
                    ext=0;
                } else
                    ext = (w == mb) ? 0 : ext+1;
            } else
                break;
        }
        /* end of finding zero */
        rk5narkstep(s-s0,3,n,mu,method,x,xl,y,discr,k);
        for (i=0; i<=n; i++) d[i+3]=x[i];
        d[1]=s;
    }


    static private void rk5narkstep(double h, int d, int n, double mu,
                            AP_rk5na_methods method, double x[], double xl[],
                            double y[], double discr[], double k[][])
    {
        /* this procedure is internally used by RK5NA */

        int i,j;
        double p,s;

        if (d != 2) {
            if (d == 1)
                for (i=0; i<=n; i++) k[0][i] *= mu;
            else {
                for (i=0; i<=n; i++) x[i]=xl[i];
                for (j=0; j<=n; j++) y[j]=method.fxj(n,j,x);
                s=0.0;
                for (j=0; j<=n; j++) s += y[j]*y[j];
                p=h/Math.sqrt(s);
                for (i=0; i<=n; i++) k[0][i]=y[i]*p;
            }
            for (i=0; i<=n; i++)    x[i]=xl[i]+k[0][i]/4.5;
            for (j=0; j<=n; j++) y[j]=method.fxj(n,j,x);
            s=0.0;
            for (j=0; j<=n; j++) s += y[j]*y[j];
            p=h/Math.sqrt(s);
            for (i=0; i<=n; i++) k[1][i]=y[i]*p;
            for (i=0; i<=n; i++)    x[i]=xl[i]+(k[0][i]+k[1][i]*3.0)/12.0;
            for (j=0; j<=n; j++) y[j]=method.fxj(n,j,x);
            s=0.0;
            for (j=0; j<=n; j++) s += y[j]*y[j];
            p=h/Math.sqrt(s);
            for (i=0; i<=n; i++) k[2][i]=y[i]*p;
            for (i=0; i<=n; i++)    x[i]=xl[i]+(k[0][i]+k[2][i]*3.0)/8.0;
            for (j=0; j<=n; j++) y[j]=method.fxj(n,j,x);
            s=0.0;
```

```
    for (j=0; j<=n; j++) s += y[j]*y[j];
    p=h/Math.sqrt(s);
    for (i=0; i<=n; i++) k[3][i]=y[i]*p;
    for (i=0; i<=n; i++)
        x[i]=xl[i]+(k[0][i]*53.0-k[1][i]*135.0+k[2][i]*126.0+k[3][i]*56.0)/125.0;
    for (j=0; j<=n; j++) y[j]=method.fxj(n,j,x);
    s=0.0;
    for (j=0; j<=n; j++) s += y[j]*y[j];
    p=h/Math.sqrt(s);
    for (i=0; i<=n; i++) k[4][i]=y[i]*p;
    if (d <= 1) {
        for (i=0; i<=n; i++)
            x[i]=xl[i]+(k[0][i]*133.0-k[1][i]*378.0+k[2][i]*276.0+k[3][i]*112.0+
                k[4][i]*25.0)/168.0;
        for (j=0; j<=n; j++) y[j]=method.fxj(n,j,x);
        s=0.0;
        for (j=0; j<=n; j++) s += y[j]*y[j];
        p=h/Math.sqrt(s);
        for (i=0; i<=n; i++) k[5][i]=y[i]*p;
        for (i=0; i<=n; i++)
            discr[i]=Math.abs(k[0][i]*21.0-k[2][i]*162.0+k[3][i]*224.0-
                k[4][i]*125.0+k[5][i]*42.0)/14.0;
        return;
    }
}
for (i=0; i<=n; i++)
    x[i]=xl[i]+(-k[0][i]*63.0+k[1][i]*189.0-k[2][i]*36.0-k[3][i]*112.0+
        k[4][i]*50.0)/28.0;
for (j=0; j<=n; j++) y[j]=method.fxj(n,j,x);
s=0.0;
for (j=0; j<=n; j++) s += y[j]*y[j];
p=h/Math.sqrt(s);
for (i=0; i<=n; i++) k[5][i]=y[i]*p;
for (i=0; i<=n; i++)
    x[i]=xl[i]+(k[0][i]*35.0+k[2][i]*162.0+k[4][i]*125.0+k[5][i]*14.0)/336.0;
}
```

F. multistep

Solves an initial value problem for a system of first order ordinary differential equations

$$dy_i(x)/dx = f_i(x,y(x)) \quad (i=1,2,\dots,n)$$

from $x = x_0$ to $x = x_{end}$ ($x_0 < x_{end}$), $y(x_0)$ being given, based on two linear multistep methods. For stiff problems it uses the backward differentiation method, and for nonstiff problems the Adams-Bashforth-Moulton method [He71].

During the implementation of a typical stage of the variant of the Adams-Moulton method employed, (a) approximations $f_i^{(p+v)}$ to $f_i(x,y(x))$ at $x = x_{p+v}$

($v=0,...,k$; $i=1,...,n$) are available, together with (b) approximations $y_i^{(p+k,v)}$ to $d^v y_i(x)/dx^v$ at $x = x_{p+k}$ ($v=0,...,k$; $i=1,...,n$), and (c) approximations to the derivatives $d^v y_i(x)/dx^v$ ($v=0,...,k$; $i=1,...,n$) at $x=x_{p+k+1}$ are determined by prediction and iterated correction. If the order k is preserved then the $f_i^{(p)}$ are discarded, and the process is repeated.

At a typical stage in the implementation of the variant of the Curtiss-Hirschfelder method employed, approximations $y_i^{(p+v)}$ ($v=0,...,k$; $i=1,...,n$) to $y_i(x)$ at $x = x_{p+v}$ ($v=0,...,k$; $i=1,...,n$) are available, together with the derivative approximations (b) above, and the derivative approximations (c) above are also determined by prediction and iterated correction.

The above process of iterated correction may, if necessary, be accelerated by use of the Jacobian matrix whose elements are given by $\partial f_i(x,y(x))/\partial y_j(x)$ ($i,j=1,...,n$).

The Adam-Moulton method is a relatively rapid process for high-accuracy computation of the solution of systems of differential equations with slowly varying dependent variables. The Curtiss-Hirschfelder method is suitable for the solution of stiff systems, whose solution vector contains one or more rapidly varying components. *multistep* uses a steplength and order control mechanism due to Gear (the steplength is $x_{p+k+1} - x_{p+k}$ above; k is the order). If so requested, *multistep* begins by attempting an unaccelerated Adams-Moulton method; if this is too time-consuming, acceleration using a Jacobian matrix is embarked upon; if this also fails, *multistep* switches to Gear's implementation of the Curtiss-Hirschfelder method, (after the determination of a Jacobian matrix at the commencement of one step, the same matrix is used to accelerate correction over as many subsequent steps as possible). However, it is possible to request *multistep* to use Gear's method at outset. Tactical considerations concerning steplength and order are reversible; strategic decisions concerning the method are not; once *multistep* has embarked upon Gear's method, it continues with it for the duration of the call.

Procedure parameters:

        boolean multistep (x,xend,y,hmin,hmax,yamx,eps,first,save,
                method,jacobian,stiff,n,btmp,itmp,xtmp)

multistep: if difficulties are encountered during the integration (i.e., *save[37]* ≠ 0 or *save[36]* ≠ 0) then *multistep* is set to false, otherwise *multistep* is set to true;

x: double *x[0:0]*;
        entry: the independent variable; the initial value *x0*;
        exit: the final value *xend*;

xend: double;
        entry: the final value of x (*xend* ≥ *x*);

y: double *y[1:6n]*;
        entry: the dependent variable; *y[1:n]* are the initial values of the solution of the system of differential equations at $x = x0$;
        exit: *y[1:n]* are the final values of the solution at $x = xend$;

hmin,hmax: double;
        entry: the minimum and maximum steplength allowed, respectively;

ymax: double *ymax[1:n]*;
        entry: the absolute local error bound divided by *eps*;

exit: *ymax[i]* gives the maximal value of the entry value of *ymax[i]* and the values of $|y[i]|$ during integration;

eps: double;
entry: the relative local error bound;

first: boolean *first[0:0]*;
if *first* is true then the procedure starts the integration with a first order Adams method and a steplength equal to *hmin*, upon completion of a call, *first* is set to false;
if *first* is false then the procedure continues integration;

save: double *save[0:6n+38]*;
in this array the procedure stores information which can be used in a continuing call with *first* = false; also the following messages are delivered:

save[38]=0: an Adams method has been used;
save[38]=1: the procedure switched to Gear's method;
save[37]=0: no error message;
save37]=1: with the *hmin* specified the procedure cannot handle the nonlinearity (decrease *hmin*!);
save[36]:   number of times that the requested local error bound was exceeded;
save[35]:   if *save[36]* is nonzero then *save[35]* gives an estimate of the maximal local error bound, otherwise *save[35]*=0;

method: a class that defines two procedures *deriv* and *fxj*, this class must implement the AP_multistep_methods interface;
void deriv(double df[ ], int n, double x[ ], double y[ ])
    this procedure should deliver *dy[i]/dx* in *df[i]*;
boolean available(int n, double x[ ], double y[ ], double jacobian[ ][ ])
    if an analytic expression of the Jacobian matrix is not available then *available* is set to false, otherwise *available* is set to true and the evaluation of this procedure must have the following side-effect: the entries of the Jacobian matrix *d(dy[i]/dx)/dy[j]* are delivered in the array elements *jacobian[i,j]*;
void out(double h, int k, int n, double x[ ], double y[ ])
    at the end of each accepted step of integration process this procedure is called, the last steplength used, *h*, and the order of the method, *k*, are delivered;
    at each call of the procedure *out*, the current values of the independent variable *x* and of the solution *y[i](x)* are available for use; moreover, in the neighborhood of the current value of *x*, any value of *y[i](x')* can be computed by means of the following interpolation formula

$$y[i](x') = \sum_{j=0}^{k} y[i + j * n]\left(\frac{x' - x}{h}\right)^{j}$$

jacobian: double *jacobian[1:n, 1:n]*;
at each evaluation of the user supplied procedure *available* with the result *available* equals true, the Jacobian matrix has to be assigned to this array;

stiff: boolean;
       entry: if *stiff* equals true then the procedure skips an attempt to solve
           the problem with Adams-Bashforth or Adams-Moulton
           methods, directly using Gear's method;
n: int;
       entry: the number of equations;
btmp: boolean *btmp[0:2]*;
       working storage for internal static data;
itmp: int *itmp[0:3]*;
       working storage for internal static data;
xtmp: double *xtmp[0:7]*;
       working storage for internal static data.

Procedures used: matvec, dec, sol.

```
package numal;

public interface AP_multistep_methods {

    void deriv(double df[], int n, double x[], double y[]);
    boolean available(int n, double x[], double y[], double jacobian[][]);
    void out(double h, int k, int n, double x[], double y[]);
}


public static boolean multistep(double x[], double xend, double y[], double hmin,
                                double hmax, double ymax[], double eps,
                                boolean first[], double save[],
                                AP_multistep_methods method, double jacobian[][],
                                boolean stiff, int n, boolean btmp[], int itmp[],
                                double xtmp[])
{
    double adams1[] = {1.0, 1.0, 144.0, 4.0, 0.0, 0.5, 1.0, 0.5, 576.0, 144.0, 1.0,
                       5.0/12.0, 1.0, 0.75, 1.0/6.0, 1436.0, 576.0, 4.0, 0.375, 1.0,
                       11.0/12.0, 1.0/3.0, 1.0/24.0, 2844.0, 1436.0, 1.0,
                       251.0/720.0, 1.0, 25.0/24.0, 35.0/72.0, 5.0/48.0, 1.0/120.0,
                       0.0, 2844.0, 0.1};
    double adams2[] = {1.0, 1.0, 9.0, 4.0, 0.0, 2.0/3.0, 1.0, 1.0/3.0, 36.0, 20.25,
                       1.0, 6.0/11.0, 1.0, 6.0/11.0, 1.0/11.0, 84.028, 53.778, 0.25,
                       0.48, 1.0, 0.7, 0.2, 0.02, 156.25, 108.51, 0.027778,
                       120.0/274.0, 1.0, 225.0/274.0, 85.0/274.0, 15.0/274.0,
                       1.0/274.0, 0.0, 187.69, 0.0047361};
    boolean adams,withjacobian,conv;
    int m,same,kold;
    double xold,hold;
    boolean newstart,firsttime,trycurtiss,errortestok;
    int i,j,l,k,fails;
    double error,dfi,ss,aa;
    double a[] = new double[6];
    double aux[] = new double[4];
```

```
double a0[] = new double[1];
double tolup[] = new double[1];
double tol[] = new double[1];
double toldwn[] = new double[1];
double tolconv[] = new double[1];
int knew[] = new int[1];
double c[] = new double[1];
double h[] = new double[1];
double ch[] = new double[1];
double chnew[] = new double[1];

int p[] = new int[n+1];
double delta[] = new double[n+1];
double lastdelta[] = new double[n+1];
double df[] = new double[n+1];
double fixy[] = new double[n+1];
double fixdy[] = new double[n+1];
double dy[] = new double[n+1];
double jac[][] = new double[n+1][n+1];
boolean evaluate[] = new boolean[1];
boolean evaluated[] = new boolean[1];
boolean decompose[] = new boolean[1];
boolean decomposed[] = new boolean[1];

adams=btmp[0];
withjacobian=btmp[1];
m=itmp[0];
same=itmp[1];
kold=itmp[2];
xold=xtmp[0];
hold=xtmp[1];
a0[0]=xtmp[2];
tolup[0]=xtmp[3];
tol[0]=xtmp[4];
toldwn[0]=xtmp[5];
tolconv[0]=xtmp[6];
k=0;
error=0.0;
conv=true;
newstart=true;
trycurtiss=false;
errortestok=true;
if (first[0]) {
    firsttime=true;
    first[0] = false;
    m=n;
    save[37]=save[36]=save[35]=0.0;
    method.out(0.0,0,n,x,y);
    adams=(!stiff);
    withjacobian=(!adams);
```

```
        if (withjacobian)
            multistepjacobian(n,x,y,eps,fixy,fixdy,dy,jacobian,method,evaluate,
                             decompose,evaluated);
        if (adams)
            for (j=0; j<=34; j++) save[j]=adams1[j];
        else
            for (j=0; j<=34; j++) save[j]=adams2[j];
    } else {
        firsttime=false;
        withjacobian=(!adams);
        ch[0]=1.0;
        k=kold;
        multistepreset(y,save,x,ch,c,h,decomposed,hmin,hmax,hold,xold,m,k,n);
        multisteporder(a,save,tolup,tol,toldwn,tolconv,a0,decompose,eps,k,n);
        decompose[0]=withjacobian;
    }

    while (newstart) {
        newstart=false;
        if (firsttime) {
            firsttime=false;
            k=1;
            same=2;
            multisteporder(a,save,tolup,tol,toldwn,tolconv,a0,decompose,eps,k,n);
            method.deriv(df,n,x,y);
            if (!withjacobian)
                h[0]=hmin;
            else {
                ss=Double.MIN_VALUE;
                for (i=1; i<=n; i++) {
                    aa=Basic.matvec(1,n,i,jacobian,df)/ymax[i];
                    ss += aa*aa;
                }
                h[0]=Math.sqrt(2.0*eps/Math.sqrt(ss));
            }
            if (h[0] > hmax)
                h[0]=hmax;
            else
                if (h[0] < hmin) h[0]=hmin;
            xold=x[0];
            hold=h[0];
            kold=k;
            ch[0]=1.0;
            for (i=1; i<=n; i++) {
                save[i+38]=y[i];
                save[m+i+38]=y[m+i]=df[i]*h[0];
            }
            method.out(0.0,0,n,x,y);
        }
        fails=0;
```

```
while (x[0] < xend) {
    if (x[0]+h[0] <= xend)
        x[0] += h[0];
    else {
        h[0]=xend-x[0];
        x[0] = xend;
        ch[0]=h[0]/hold;
        c[0]=1.0;
        for (j=m; j<=k*m; j+=m) {
            c[0] *= ch[0];
            for (i=j+1; i<=j+n; i++) y[i] *= c[0];
        }
        same = ((same < 3) ? 3 : same+1);
    }
    /* prediction */
    for (l=1; l<=n; l++) {
        for (i=1; i<=(k-1)*m+1; i+=m)
            for (j=(k-1)*m+1; j>=i; j-=m) y[j] += y[j+m];
        delta[l]=0.0;
    }
    evaluated[0]=false;
    /* correction and estimation local error */
    for (l=1; l<=3; l++) {
        method.deriv(df,n,x,y);
        for (i=1; i<=n; i++) df[i]=df[i]*h[0]-y[m+i];
        if (withjacobian) {
            if (evaluate[0])
                multistepjacobian(n,x,y,eps,fixy,fixdy,dy,jacobian,method,
                                  evaluate,decompose,evaluated);
            if (decompose[0]) {
                /* decompose jacobian */
                decompose[0]=false;
                decomposed[0]=true;
                c[0] = -a0[0]*h[0];
                for (j=1; j<=n; j++) {
                    for (i=1; i<=n; i++)
                        jac[i][j]=jacobian[i][j]*c[0];
                    jac[j][j] += 1.0;
                }
                aux[2]=Double.MIN_VALUE;
                Linear_algebra.dec(jac,n,aux,p);
            }
            Linear_algebra.sol(jac,n,p,df);
        }
        conv=true;
        for (i=1; i<=n; i++) {
            dfi=df[i];
            y[i] += a0[0]*dfi;
            y[m+i] += dfi;
            delta[i] += dfi;
```

```
              conv=(conv && (Math.abs(dfi) < tolconv[0]*ymax[i]));
        }
        if (conv) {
            ss=Double.MIN_VALUE;
            for (i=1; i<=n; i++) {
                aa=delta[i]/ymax[i];
                ss += aa*aa;
            }
            error=ss;
            break;
        }
    }
    /* acceptance or rejection */
    if (!conv) {
        if (!withjacobian) {
            evaluate[0]=withjacobian=((same >= k) || (h[0] < 1.1*hmin));
            if (!withjacobian) ch[0] /= 4.0;
        }
        else if (!decomposed[0]) decompose[0]=true;
        else if (!evaluated[0]) evaluate[0]=true;
        else if (h[0] > 1.1*hmin) ch[0] /= 4.0;
        else if (adams) {
            trycurtiss=true;
            adams=false;
            for (j=0; j<=34; j++) save[j]=adams2[j];
            k=kold=1;
            multistepreset(y,save,x,ch,c,h,decomposed,hmin,hmax,hold,xold,m,
                        k,n);
            multisteporder(a,save,tolup,tol,toldwn,tolconv,a0,decompose,eps,
                        k,n);
            same=2;
        }
        else {
            save[37]=1.0;
            break;
        }
        if (!trycurtiss)
            multistepreset(y,save,x,ch,c,h,decomposed,hmin,hmax,hold,xold,m,
                        k,n);
        trycurtiss=false;
    } else {
        if (error > tol[0]) {
            errortestok=false;
            fails++;
            if (h[0] > 1.1*hmin) {
                if (fails > 2) {
                    if (adams) {
                        adams=false;
                        for (j=0; j<=34; j++) save[j]=adams2[j];
                    }
```

```
                    kold=0;
                    multistepreset(y,save,x,ch,c,h,decomposed,hmin,hmax,
                                   hold,xold,m,k,n);
                    newstart=true;
                } else {
                    multistepstep(knew,chnew,tolup,tol,toldwn,delta,error,
                                   lastdelta,y,ymax,fails,m,k,n);
                    if (knew[0] != k) {
                        k=knew[0];
                        multisteporder(a,save,tolup,tol,toldwn,tolconv,a0,
                                       decompose,eps,k,n);
                    }
                    ch[0] *= chnew[0];
                    multistepreset(y,save,x,ch,c,h,decomposed,hmin,hmax,
                                   hold,xold,m,k,n);
                }
            } else {
                if (adams) {
                    adams=false;
                    for (j=0; j<=34; j++) save[j]=adams2[j];
                } else {
                    if (k == 1) {
                        /* violate eps criterion */
                        c[0]=eps*Math.sqrt(error/tol[0]);
                        if (c[0] > save[35]) save[35]=c[0];
                        save[36] += 1.0;
                        same=4;
                        errortestok=true;
                    }
                }
                if(!errortestok) {
                    k=kold=1;
                    multistepreset(y,save,x,ch,c,h,decomposed,hmin,hmax,
                                   hold,xold,m,k,n);
                    multisteporder(a,save,tolup,tol,toldwn,tolconv,a0,
                                   decompose,eps,k,n);
                    same=2;
                }
            }
        }
        if (errortestok) {
            fails=0;
            for (i=1; i<=n; i++) {
                c[0]=delta[i];
                for (l=2; l<=k; l++) y[l*m+i] += a[l]*c[0];
                if (Math.abs(y[i]) > ymax[i])
              ymax[i]=Math.abs(y[i]);
            }
            same--;
            if (same == 1)
```

```
                        for (i=1; i<=n; i++) lastdelta[i]=delta[i];
                  else
                     if (same == 0) {
                        multistepstep(knew,chnew,tolup,tol,toldwn,delta,error,
                                      lastdelta,y,ymax,fails,m,k,n);
                        if (chnew[0] > 1.1) {
                           decomposed[0]=false;
                           if (k != knew[0]) {
                              if (knew[0] > k)
                                 for (i=1; i<=n; i++)
                                    y[knew[0]*m+i]=delta[i]*a[k]/knew[0];
                              k=knew[0];
                              multisteporder(a,save,tolup,tol,toldwn,tolconv,a0,
                                             decompose,eps,k,n);
                           }
                           same=k+1;
                           if (chnew[0]*h[0] > hmax)
                              chnew[0]=hmax/h[0];
                           h[0] *= chnew[0];
                           c[0]=1.0;
                           for (j=m; j<=k*m; j+=m) {
                              c[0] *= chnew[0];
                              for (i=j+1; i<=j+n; i++) y[i] *= c[0];
                           }
                        } else
                           same=10;
                     }
                  if (x[0] != xend) {
                     xold=x[0];
                     hold=h[0];
                     kold=k;
                     ch[0]=1.0;
                     for (i=k*m+n; i>=1; i--) save[i+38]=y[i];
                     method.out(h[0],k,n,x,y);
                  }
               }
               errortestok=true;
            }
      } /* while (x[0] < xend) loop */
   } /* while (newstart) loop */
   btmp[0]=adams;
   btmp[1]=withjacobian;
   itmp[0]=m;
   itmp[1]=same;
   itmp[2]=kold;
   xtmp[0]=xold;
   xtmp[1]=hold;
   xtmp[2]=a0[0];
   xtmp[3]=tolup[0];
   xtmp[4]=tol[0];
```

```
    xtmp[5]=toldwn[0];
    xtmp[6]=tolconv[0];
    save[38]=(adams ? 0.0 : 1.0);
    return ((save[37] == 0.0) && (save[36] == 0.0));
}


static private void multistepreset(double y[], double save[], double x[],
                                   double ch[], double c[], double h[],
                                   boolean decomposed[], double hmin, double hmax,
                                   double hold, double xold, int m, int k, int n)
{
    /* this procedure is internally used by MULTISTEP */

    int i,j;

    if (ch[0] < hmin/hold)
        ch[0] = hmin/hold;
    else
        if (ch[0] > hmax/hold) ch[0] = hmax/hold;
    x[0] = xold;
    h[0] = hold*(ch[0]);
    c[0] = 1.0;
    for (j=0; j<=k*m; j+=m) {
        for (i=1; i<=n; i++) y[j+i]=save[j+i+38]*c[0];
        c[0] *= ch[0];
    }
    decomposed[0] = false;
}


static private void multisteporder(double a[], double save[], double tolup[],
                                   double tol[], double toldwn[], double tolconv[],
                                   double a0[], boolean decompose[], double eps,
                                   int k, int n)
{
    /* this procedure is internally used by MULTISTEP */

    int i,j;
    double c;

    c=eps*eps;
    j=(k-1)*(k+8)/2-38;
    for (i=0; i<=k; i++) a[i]=save[i+j+38];
    tolup[0] = c*save[j+k+39];
    tol[0] = c*save[j+k+40];
    toldwn[0] = c*save[j+k+41];
    tolconv[0] = eps/(2*n*(k+2));
    a0[0] = a[0];
    decompose[0] = true;
```

```
}


static private void multistepstep(int knew[], double chnew[], double tolup[],
                                  double tol[], double toldwn[], double delta[],
                                  double error, double lastdelta[], double y[],
                                  double ymax[], int fails, int m, int k, int n)
{
    /* this procedure is internally used by MULTISTEP */

    int i;
    double a1,a2,a3,aa,ss;

    if (k <= 1)
        a1=0.0;
    else {
        ss=Double.MIN_VALUE;
        for (i=1; i<=n; i++) {
            aa=y[k*m+i]/ymax[i];
            ss += aa*aa;
        }
        a1=0.75*Math.pow(toldwn[0]/ss,0.5/k);
    }
    a2=0.80*Math.pow(tol[0]/error,0.5/(k+1));
    if (k >= 5 || fails != 0)
        a3=0.0;
    else {
        ss=Double.MIN_VALUE;
        for (i=1; i<=n; i++) {
            aa=(delta[i]-lastdelta[i])/ymax[i];
            ss += aa*aa;
        }
        a3=0.70*Math.pow(tolup[0]/ss,0.5/(k+2));
    }
    if (a1 > a2 && a1 > a3) {
        knew[0] = k-1;
        chnew[0] = a1;
    } else
        if (a2 > a3) {
            knew[0] = k;
            chnew[0] = a2;
        } else {
            knew[0] = k+1;
            chnew[0] = a3;
        }
}


static private void multistepjacobian(int n, double x[], double y[], double eps,
                                      double fixy[], double fixdy[], double dy[],
```

```
                              double jacobian[][],
                              AP_multistep_methods method,
                              boolean evaluate[], boolean decompose[],
                              boolean evaluated[])
{
    /* this procedure is internally used by MULTISTEP */

    int i,j;
    double d;

    evaluate[0] = false;
    decompose[0] = evaluated[0] = true;
    if (!method.available(n,x,y,jacobian)) {
        for (i=1; i<=n; i++) fixy[i]=y[i];
        method.deriv(fixdy,n,x,y);
        for (j=1; j<=n; j++) {
            d=((eps > Math.abs(fixy[j])) ? eps*eps : eps*Math.abs(fixy[j]));
            y[j] += d;
            method.deriv(dy,n,x,y);
            for (i=1; i<=n; i++) jacobian[i][j]=(dy[i]-fixdy[i])/d;
            y[j]=fixy[j];
        }
    }
}
```

G. diffsys

Integrates the system of first order ordinary differential equations
$$dy_i(x)/dx = f_i(x,y)$$
from $x = x0$ to $x = xe$ by means of a high order extrapolation method based on the modified midpoint rule. The method is suitable for high accuracy problems and is not suited for stiff equations.

 diffsys is a slight modification of the algorithm in [BulS65]. By this modification integration from $x0$ until xe can be performed by one call of diffsys. A number of integration steps are taken, starting with the initial step $h0$, in each integration step a number of solutions are computed by means of the modified midpoint rule. Extrapolation is used to improve these solutions, until the required accuracy is met. An integration step is rejected if the accuracy requirements are not fulfilled after nine extrapolation steps. In these cases the integration step is rejected, and integration is tried again with the integration step halved.

 The algorithm is for each step a variable order method (the highest order is 14), and uses a variable number of function evaluations, depending on the order (minimum is 3, maximum is 217). The algorithm is less sensitive to too small values of the initial stepsize than the original algorithm (see [Fox72, HEFS72]). However, bad guesses require still some more computations.

Procedure parameters:

                          void diffsys (x,xe,n,y,method,aeta,reta,s,h0)

x: double x[0:0];
              entry: the independent variable; the initial value x0;
              exit: the final value xe;
xe: double;
              entry: the final value of x (xe ≥ x);
n: int;
              entry: the number of equations;
y: double y[1:n];
              entry: the dependent variable; y[1:n] are the initial values of the solution
                       of the system of differential equations at x = x0;
              exit: y[1:n] are the final values of the solution at x = xe;
method: a class that defines two procedures *derivative* and *output*, this class must
              implement the AP_diffsys_methods interface;
              void derivative(int n, double x, double y[ ], double dy[ ])
                       this procedure should deliver the right-hand side of the i-th
                       differential equation at the point (x,y) as dy[i], i=1,...,n;
              void output(int n, double x[ ], double xe, double y[ ], double s[ ])
                       this procedure is called at the end of each integration step,
                       the user can ask for output of parameters x, xe, y, and s;
aeta: double;
              entry: required absolute precision in the integration process;
reta: double;
              entry: required relative precision in the integration process;
s: double s[1:n];
              the array s is used to control the accuracy of the computed values of y;
              entry: it is advisable to set s[i]=0, i=1,...,n;
              exit: the maximum value of |y[i]|, encountered during integration,
                       if this value exceeds the value of s[i] on entry;
h0:   double;
              entry: the initial step to be taken.

```
package numal;

public interface AP_diffsys_methods {

    void derivative(int n, double x, double y[], double dy[]);
    void output(int n, double x[], double xe, double y[], double s[]);
}



public static void diffsys(double x[], double xe, int n, double y[],
                           AP_diffsys_methods method, double aeta, double reta,
                           double s[], double h0)
{
    boolean bh,last,next,b0,konv;
    int i,j,k,kk,jj,l,m,r,sr;
    double a,b,b1,c,g,h,u,v,ta,fc;
    double d[] = new double[7];
    double ya[] = new double[n+1];
```

```
double yl[] = new double[n+1];
double ym[] = new double[n+1];
double dy[] = new double[n+1];
double dz[] = new double[n+1];
double dt[][] = new double[n+1][7];
double yg[][] = new double[8][n+1];
double yh[][] = new double[8][n+1];

last=false;
h=h0;
do {
    next=false;
    if (h*1.1 >= xe-x[0]) {
        last=true;
        h0=h;
        h=xe-x[0]+Double.MIN_VALUE;
    }
    method.derivative(n,x[0],y,dz);
    bh=false;
    for (i=1; i<=n; i++) ya[i]=y[i];
    while (true) {
        a=h+x[0];
        fc=1.5;
        b0=false;
        m=1;
        r=2;
        sr=3;
        jj = -1;
        for (j=0; j<=9; j++) {
            if (b0) {
                d[1]=16.0/9.0;
                d[3]=64.0/9.0;
                d[5]=256.0/9.0;
            } else {
                d[1]=9.0/4.0;
                d[3]=9.0;
                d[5]=36.0;
            }
            konv=true;
            if (j > 6) {
                l=6;
                d[6]=64.0;
                fc *= 0.6;
            } else {
                l=j;
                d[l]=m*m;
            }
            m *= 2;
            g=h/m;
            b=g*2.0;
```

```
if (bh && j < 8)
    for (i=1; i<=n; i++) {
        ym[i]=yh[j][i];
        yl[i]=yg[j][i];
    }
else {
    kk=(m-2)/2;
    m--;
    for (i=1; i<=n; i++) {
        yl[i]=ya[i];
        ym[i]=ya[i]+g*dz[i];
    }
    for (k=1; k<=m; k++) {
        method.derivative(n,x[0]+k*g,ym,dy);
        for (i=1; i<=n; i++) {
            u=yl[i]+b*dy[i];
            yl[i]=ym[i];
            ym[i]=u;
            u=Math.abs(u);
            if (u > s[i]) s[i]=u;
        }
        if (k == kk && k != 2) {
            jj++;
            for (i=1; i<=n; i++) {
                yh[jj][i]=ym[i];
                yg[jj][i]=yl[i];
            }
        }
    }
}
method.derivative(n,a,ym,dy);
for (i=1; i<=n; i++) {
    v=dt[i][0];
    ta=c=dt[i][0]=(ym[i]+yl[i]+g*dy[i])/2.0;
    for (k=1; k<=l; k++) {
        b1=d[k]*v;
        b=b1-c;
        u=v;
        if (b != 0.0) {
            b=(c-v)/b;
            u=c*b;
            c=b1*b;
        }
        v=dt[i][k];
        dt[i][k]=u;
        ta += u;
    }
    if (Math.abs(y[i]-ta) > reta*s[i]+aeta) konv=false;
    y[i]=ta;
}
```

```
            if (konv) {
                next=true;
                break;
            }
            d[2]=4.0;
            d[4]=16.0;
            b0 = !b0;
            m=r;
            r=sr;
            sr=m*2;
        }
        if (next) break;
        bh = !bh;
        last=false;
        h /= 2.0;
    }
    h *= fc;
    x[0]=a;
    method.output(n,x,xe,y,s);
  } while (!last);
}
```

H. ark

Solves an initial value problem for a system of first order ordinary differential equations which is obtained from semi-discretization of an initial boundary value problem for a parabolic or hyperbolic equation, based on stabilized, explicit Runge-Kutta methods of low order [Be72, Vh71, VhBDS71, VhK71]. Automatic stepsize control is provided but step rejection has been excluded in order to save storage. Because of its limited storage requirements and adaptive stability facilities the method is well suited for the solution of initial boundary value problems for partial differential equations.

$\quad$ *ark* solves the system of differential equations

$$Du_i(t) = f_i(t,u(t)) \quad (i=m0,m0+1,...,m)$$

where $D \equiv d/dt$, with $u(t_0)$ given, from $t = t0$ to $t = te$. The theory upon which the operation of *ark* is based is, in its early stages, the same as that supporting *efrk*. However, *ark* operates with a fixed stability polynomial

$$p_n(z) = 1 + \beta_1 z + ... + \beta_n z^n$$

all of whose coefficients the user supplies at call of *ark*; furthermore, it is assumed that those eigenvalues of $J(t)$ (the Jacobian matrix whose elements are given by

$$J_{i,j}(t) = \partial f_{i+m0-1}(t,u(t))/\partial u_{j+m-1}(t) \quad (i,j=1,...,m-m0+1)$$

which lie in the closed left half-plane $\text{Real}(\delta) \leq 0$ are either all pure real or pure imaginary for $t0 \leq t \leq te$.

$\quad$ With $\sigma(J)$ the spectral radius of $J(t)$ with respect to its eigenvalues in the closed left half-plane, stability polynomials $p_n(z)$ have been constructed such that the associated Runge-Kutta scheme (2) of *efrk* is stable if the stepsize (in t) h satisfies the inequality $h \leq \beta(n)/\sigma(J)$; $\beta(n)$ is a constant associated with the polynomial $p_n(z)$ in question, and in each case $p_n(z)$ has been so determined that

$ß(n)$ assumes its maximum permissible value. Examples of such polynomials $p_n(z)$ and associated constants $ß(n)$, together with the types of equations (i.e., those for which the eigenvalues in the closed left half-plane Real($δ$) ≤ 0 of the associated Jacobian matrix $J(t)$ are either all pure real or all pure imaginary) to which they refer, and the order r of exactness (see the documentation to *efrk*) of the polynomial in question are given in the following table.

n	$ß_1$	$ß_2$	$ß_3$	$ß_4$	$ß_5$	$ß(n)$	type	r
3	1	1/2	1/4			2	imaginary	2
4	1	1/2	1/6	1/24		$2\sqrt{2}$	imaginary	3
4	1	5/32	1/128	1/8192		32	real	1
4	1	1/2	0.078	0.0036		12	real	2
4	1	1/2	1/6	0.0185		6	real	3
5	1	1/2	3/16	1/32	1.128	4	imaginary	2

Procedure parameters:

$$\text{void ark } (t,te,m0,m,u,method,data)$$

t: double *t[0:0]*;
the independent variable t of the systems of ordinary differential equations

$$du/dt = f(t,u), \ u=u0 \text{ at } t=t0;$$

entry: the initial value $t0$;
exit: the final value te;

te: double *te[0:0]*;
entry: the final value of t ($te \geq t$);

m0,m: int *m0[0:0]*, *m[0:0]*;
indices of the first and last equation of the system;

u: double *u[m0:m]*;
entry: the initial values of the solution of the system of differential equations at $t=t0$;
exit: the values of the solution at $t = te$;

method: a class that defines two procedures *derivative* and *out*, this class must implement the AP_ark_methods interface;
void derivative(int m0[ ], int m[ ], double t[ ], double v[ ])
this procedure performs an evaluation of the right-hand side of the system with dependent variables $v[m0:m]$ and independent variable t, upon completion of *derivative*, the right-hand side should be overwritten on $v[m0:m]$;
void out(int m0[ ], int m[ ], double t[ ], double te[ ], double u[ ], double data[ ])
after each integration step performed information can be obtained or updated by this procedure;

data: double *data[1:10+data[1]]*;
in array *data* one should give:
data[1]: the number of evaluations of $f(t,u)$ per integration step ($data[1] \geq data[2]$);
data[2]: the order of accuracy of the method ($data[2] \leq 3$) (value of r above);
data[3]: stability bound (value of $ß(n)$ above);
data[4]: the spectral radius of the Jacobian matrix with respect to those eigenvalues which are located in the nonpositive half plane (value of $σ(J(x))$ above);

> > *data[5]*: the minimal stepsize;
> > *data[6]*: the absolute tolerance;
> > *data[7]*: the relative tolerance;
> > > if both *data[6]* and *dat[7]* are negative then the integration is performed with a constant step *data[5]*;
> > *data[8]*: *data[8]* should be zero if *ark* is called for a first time, for continued integration *data[8]* should not be changed;
> > *data[11],...,data[10+data[1]]*:
> > > polynomial coefficients (the values of ß$_i$ above in *data[i+10]*, *i*=1,...,*data[1]*);
>
> after each step the following byproducts are delivered:
> > *data[8]*: the number of integration steps performed;
> > *data[9]*: an estimate of the local error last made;
> > *data[10]*: information messages:
> > > *data[10]*=0: no difficulty;
> > > *data[10]*=1: minimal steplength exceeds the steplength prescribed by stability theory, i.e., *data[5]*>*data[3]*/*data[4]*; termination of *ark*; decrease minimal steplength;
>
> if necessary, *data[i]*, *i*=4,...,7, can be updated (after each step by means of procedure *out*).

Procedures used: inivec, mulvec, dupvec, vecvec, elmvec, decsol.

```
package numal;

public interface AP_ark_methods {

    void derivative(int m0[], int m[], double t[], double v[]);
    void out(int m0[], int m[], double t[], double te[], double u[], double data[]);
}


public static void ark(double t[], double te[], int m0[], int m[], double u[],
                    AP_ark_methods method, double data[])
{
    double th1[] = {1.0, 0.5, 1.0/6.0, 1.0/3.0, 1.0/24.0, 1.0/12.0, 0.125, 0.25};
    double ec0,ec1,ec2,tau0,tau1,tau2,taus,t2;
    boolean start,step1,last;
    int p,n,q,i,j,k,l,n1,m00;
    double thetanm1,tau,betan,qinv,eta,ss,theta0,tauacc,taustab,aa,bb,cc,ec,mt,lt;
    double th[] = new double[9];
    double aux[] = new double[4];
    double s[] = new double[1];

    n=(int)data[1];
    m00=m0[0];
    double mu[] = new double[n+1];
    double lambda[] = new double[n+1];
    double thetha[] = new double[n+1];
```

```
double ro[] = new double[m[0]+1];
double r[] = new double[m[0]+1];
double alfa[][] = new double[9][n+2];

step1=true;
tau2=tau1=tau0=taus=t2=ec0=0.0;
p=(int)data[2];
ec1=ec2=0.0;
betan=data[3];
thetanm1 = (p == 3) ? 0.75 : 1.0;
theta0=1.0-thetanm1;
s[0]=1.0;
for (j=n-1; j>=1; j--) {
    s[0] = -s[0]*theta0+data[n+10-j];
    mu[j]=data[n+11-j]/s[0];
    lambda[j]=mu[j]-theta0;
}
for (i=1; i<=8; i++)
    for (j=0; j<=n; j++)
        if (i == 1) alfa[i][j+1]=1.0;
        else if (j == 0) alfa[i][j+1]=0.0;
        else if (i == 2 || i == 4 || i == 8)
            alfa[i][j+1]=Math.pow(arkmui(j,n,p,lambda),(i+2)/3);
        else if ((i == 3 || i == 6) && j > 1) {
            s[0]=0.0;
            for (l=1; l<=j-1; l++)
                s[0] += arklabda(j,l,n,p,lambda)*
                                Math.pow(arkmui(l,n,p,lambda),i/3);
            alfa[i][j+1]=s[0];
        }
        else if (i == 5 && j > 2) {
            s[0]=0.0;
            for (l=2; l<=j-1; l++) {
                ss=0.0;
                for (k=1; k<=l-1; k++)
                    ss += arklabda(l,k,n,p,lambda)*arkmui(k,n,p,lambda);
                s[0] += arklabda(j,l,n,p,lambda)*ss;
            }
            alfa[i][j+1]=s[0];
        }
        else if (i == 7 && j > 1) {
            s[0]=0.0;
            for (l=1; l<=j-1; l++)
                s[0] += arklabda(j,l,n,p,lambda)*arkmui(l,n,p,lambda);
            alfa[i][j+1]=s[0]*arkmui(j,n,p,lambda);
        }
        else alfa[i][j+1]=0.0;
n1 = ((n < 4) ? n+1 : ((n < 7) ? 4 : 8));
for (i=1; i<=8; i++) th[i]=th1[i-1];
if (p == 3 && n < 7) th[1]=th[2]=0.0;
```

```
aux[2]=Double.MIN_VALUE;
Linear_algebra.decsol(alfa,n1,aux,th);
Basic.inivec(0,n,thetha,0.0);
Basic.dupvec(0,n1-1,1,thetha,th);
if (!(p == 3 && n < 7)) {
    thetha[0]  -= theta0;
    thetha[n-1] -= thetanm1;
    q=p+1;
} else
    q=3;
qinv=1.0/q;
start=(data[8] == 0.0);
data[10]=0.0;
last=false;
Basic.dupvec(m0[0],m[0],0,r,u);
method.derivative(m0,m,t,r);
do {
    /* stepsize */
    eta=Math.sqrt(Basic.vecvec(m0[0],m[0],0,u,u))*data[7]+data[6];
    if (eta > 0.0) {
        if (start) {
            if (data[8] == 0) {
                tauacc=data[5];
                step1=true;
            } else
                if (step1) {
                    tauacc=Math.pow(eta/ec2,qinv);
                    if (tauacc > 10.0*tau2)
                        tauacc=10.0*tau2;
                    else
                        step1=false;
                } else {
                    bb=(ec2-ec1)/tau1;
                    cc = -bb*t2+ec2;
                    ec=bb*t[0]+cc;
                    tauacc = (ec < 0.0) ? tau2 : Math.pow(eta/ec,qinv);
                    start=false;
                }
        } else {
            aa=((ec0-ec1)/tau0+(ec2-ec1)/tau1)/(tau1+tau0);
            bb=(ec2-ec1)/tau1-(2.0*t2-tau1)*aa;
            cc = -(aa*t2+bb)*t2+ec2;
            ec=(aa*t[0]+bb)*t[0]+cc;
            tauacc = ((ec < 0.0) ? taus : Math.pow(eta/ec,qinv));
            if (tauacc > 2.0*taus) tauacc=2.0*taus;
            if (tauacc < taus/2.0) tauacc=taus/2.0;
        }
    } else
        tauacc=data[5];
    if (tauacc < data[5]) tauacc=data[5];
```

```
taustab=betan/data[4];
if (taustab < data[5]) {
    data[10]=1.0;
    break;
}
tau = ((tauacc > taustab) ? taustab : tauacc);
taus=tau;
if (tau >= te[0]-t[0]) {
    tau=te[0]-t[0];
    last=true;
}
tau0=tau1;
tau1=tau2;
tau2=tau;
/* difference scheme */
Basic.mulvec(m0[0],m[0],0,ro,r,thetha[0]);
if (p == 3) Basic.elmvec(m0[0],m[0],0,u,r,0.25*tau);
for (i=1; i<=n-1; i++) {
    mt=mu[i]*tau;
    lt=lambda[i]*tau;
    for (j=m0[0]; j<=m[0]; j++) r[j]=lt*r[j]+u[j];
    s[0]=t[0]+mt;
    method.derivative(m0,m,s,r);
    if (thetha[i] != 0.0)
        Basic.elmvec(m0[0],m[0],0,ro,r,thetha[i]);
    if (i == n) {
        data[9]=Math.sqrt(Basic.vecvec(m0[0],m[0],0,ro,ro))*tau;
        ec0=ec1;
        ec1=ec2;
        ec2=data[9]/Math.pow(tau,q);
    }
}
Basic.elmvec(m0[0],m[0],0,u,r,thetanm1*tau);
Basic.dupvec(m0[0],m[0],0,r,u);
s[0]=t[0]+tau;
method.derivative(m0,m,s,r);
if (thetha[n] != 0.0)
    Basic.elmvec(m0[0],m[0],0,ro,r,thetha[n]);
data[9]=Math.sqrt(Basic.vecvec(m0[0],m[0],0,ro,ro))*tau;
ec0=ec1;
ec1=ec2;
ec2=data[9]/Math.pow(tau,q);
t2=t[0];
if (last) {
    last=false;
    t[0]=te[0];
} else
    t[0] += tau;
data[8] += 1.0;
method.out(m0,m,t,te,u,data);
```

```
    } while (t[0] != te[0]);
}


static private double arkmui(int i, int n, int p, double lambda[])
{
    /* this procedure is internally used by ARK */

    return ((i==n) ? 1.0 : ((i<1 || i>n) ? 0.0 :
            ((p<3) ? lambda[i] : ((p==3) ? lambda[i]+0.25 : 0.0)))) ;
}


static private double arklabda(int i, int j, int n, int p, double lambda[])
{
    /* this procedure is internally used by ARK */

    return ((p<3) ? ((j==i-1) ? arkmui(i,n,p,lambda) : 0.0) :
            ((p==3) ? ((i==n) ? ((j==0) ? 0.25 :
            ((j==n-1) ? 0.75 : 0.0)) :
            ((j==0) ? ((i==1) ? arkmui(1,n,p,lambda) : 0.25) :
            ((j==i-1) ? lambda[i] : 0.0))) : 0.0)) ;
}
```

I. efrk

Solves an initial value problem for a system of first order ordinary differential
equations by means of an exponentially fitted explicit Runge-Kutta method of first,
second or third order. *efrk* is a special purpose procedure for stiff equations with a
known, clustered eigenvalue spectrum; automatic error control is not provided. A
detailed description of the method is given in [De72], see also [BeDHKW73,
BeDHV74, DekHH72].

    efrk solves the system of differential equations

$$Du_i(t) = f_i(t,u(t)) \qquad i=m0,m0+1,...,m \tag{1}$$

where $D \equiv d/dt$, with $u(t_0)$ given, from $t = t0$ to $t = te$. Consider the N-th order Runge-
Kutta scheme

$$u^{(n+1)} = u^{(n)} + \sum_{j=0}^{N-1} \theta_j k^{(j)} \tag{2}$$

$$k^{(j)} = h_n u(t_n + \mu_j h_n, u^{(n)} + \sum_{j=0}^{j-1} \lambda_{j,l} k^{(l)}) \qquad j = 0,1,...,N-1$$

where $h_n = t_{n+1}-t_n$. For the purposes of theoretical investigation, we work in a space
whose coordinates are numbered $m0,m0+1,...,m$, and it is assumed that, *f* being a
suitable differentiable function, $u^{(n+1)}$ given by formula (2) may be expressed as the
sum of a power series in the variable h_n, with coefficients involving the derivatives
of $u^{(n)}$:

$$u^{(n+1)} = u^{(n)} + \beta_1(Du^{(n)})h_n + \beta_2(D^2u^{(n)})h_n^2 + \beta_3 J^{(n)}D^2u^{(n)}h_n^3 + \tfrac{1}{2}\beta_{3,1}(D^3u^{(n)}-J^{(n)}D^2u^{(n)})h_n^3 +... \quad (3)$$

where $J^{(n)}$ is the $(m-m0+1)\times(m-m0+1)$ Jacobian matrix whose elements are $\partial f_i(t_n,u^{(n)}(t))/\partial u_j^{(n)}(t)$ $(i,j=m0,m0+1,...,m)$. Expanding each of the $k^{(j)}$ in (2) in ascending powers of h_n, it is found that the θ_j, μ_j and $\lambda_{i,j}$ of (2) and the β_j and $\beta_{i,j}$ of (3) are connected by the system of relationships

$$\beta_1 = \sum_{j=0}^{N-1} \theta_j , \qquad \beta_2 = \sum_{j=1}^{N-1} \theta_j \mu_j$$

$$\beta_3 = \sum_{j=2}^{N-1} \theta_j \sum_{l=1}^{j-1} \lambda_{j,l}\mu_l , \qquad \beta_{3,1} = \sum_{j=1}^{N-1} \theta_j \mu_j^2$$

$$\beta_4 = \sum_{j=3}^{N-1} \theta_j \sum_{l=2}^{j-1} \lambda_{j,l} \sum_{i=1}^{l-1} \lambda_{l,i}\mu_i , \qquad \beta_{4,1} = \sum_{j=2}^{N-1} \theta_j \sum_{l=1}^{j-1} \lambda_{j,l}\mu_l^2$$

$$\beta_{4,2} = \sum_{j=2}^{N-1} \theta_j \mu_j \sum_{l=1}^{j-1} \lambda_{j,l}\mu_l , \qquad \beta_{4,3} = \sum_{j=1}^{N-1} \theta_j \mu_j^2 , \qquad \cdots$$

By imposing appropriate conditions upon the β_i and $\beta_{i,j}$, the θ_j, μ_j, and $\lambda_{i,j}$ are determined, and the Runge-Kutta scheme (2) is established.

One type of condition concerns the extent of agreement between the series (3) and the Taylor series expansion of $u(t_n+h_n)$ in ascending powers of h_n, given that u satisfies equation (1). If such agreement is to persist up to and including the terms in h_n^r (the scheme (2) is then said to be r-th order exact) the β_i and $\beta_{i,j}$ must have the following values

r	β_1	β_2	β_3	$\beta_{3,1}$	β_4	$\beta_{4,1}$	$\beta_{4,2}$	$\beta_{4,3}$
1	1							
2	1	1/2						
3	1	1/2	1/6	1/3				
4	1	1/2	1/6	1/3	1/24	1/12	1/8	1/4

A further type of condition concerns stability. Let $u'^{(n+1)}$ be the vector produced by means of the scheme (2) from the vector $u'^{(n)}$, and $u''^{(n+1)}$ similarly from $u''^{(n)}$, where $e^{(n)} = y'^{(n)} - y''^{(n)}$ is small. Then $\quad y'^{(n+1)} - y''^{(n+1)} = p_N(h_n J^{(n)})e^{(n)} + O(e^{(n)})^2$ as $e^{(n)} \to 0$, where

$$p_N(z) = 1 + \beta_1 z + \beta_2 z^2 + ... + \beta_N z^N .$$

The scheme (2) is stable if $\left| p_N(h_n \delta) \right| \le 1$ for all eigenvalues δ of $J^{(n)}$ in the closed left half-plane Real$(\delta) \le 0$.

If all of the above δ are very small, a relatively large stepsize h_n may be taken without violating the above stability condition. For the treatment of such differential equations, it is appropriate to choose the β_i and $\beta_{i,j}$, and by implication the θ_j, μ_j, and $\lambda_{i,j}$ in (2) so as to ensure r-th order exactness for the highest possible value of r. If the above δ are not small (this phenomenon accompanies stiff equations and many systems arising from the numerical solution of partial differential equations) it is wiser to sacrifice a certain degree of exactness in order to obtain stability. One works with a stability polynomial whose first few coefficients β_i are fixed (thus ensuring a certain degree of exactness) and determines the remaining coefficients β_i in such a way that the stability condition $\left| p_N(h_n \delta) \right| \le 1$ is satisfied for the largest possible stepsize h_n.

The problem of determining the coefficients β_i which satisfy the above criteria has been successfully resolved, by a process of exponential fitting, in two cases. The two-cluster case in which, during an integration step, the above δ are to be found in two circles, one with center at the origin, and the other with center z_1 upon the negative real axis, and the three-cluster case in which three circles at the origin and z_1 and the conjugate of z_1 in the left-half plane, are concerned.

For certain systems of differential equations of the form (1), the Jacobian matrix $J^{(n)}$ (and hence its distribution of eigenvalues) does not vary greatly over the range of integration; it is then possible to prescribe at outset a center z_1 and the diameter of a circle with center at z_1 (or two centers z_1 and conjugate(z_1), and equal diameters of two circles) for which, over the entire range of integration, the above δ remain within the circle in question. For other examples, new estimates of z_1 and the diameter must be determined at the commencement of each integration step, and a new Runge-Kutta scheme must be constructed (*efrk* carries this construction out). *efrk* caters for both cases.

Equation (1) may refer to a system in which the initial and final indices (*mo* and *m*) are fixed over the range of integration. It is, however, possible that these indices should vary during the course of the computation. (An example in which this occurs (it concerns the numerical solution of a partial differential equation) is given to illustrate the use of the procedure *ark*.) In such a case, the procedure which evaluates the derivatives (1) must also modify *m0* and *m* in a suitable manner.

When supplied with appropriate data:

(i)   the values of the fixed first r coefficients $ß_i$, $i=1,...,r$, of the stability polynomial $p_N(z)$ (these may be extracted from the above table);

(ii)   the degree of exponential fitting: if z_1 is negative real, l, the number supplied, is the degree of exponential fitting; if imag(z_1)≠0, l is twice this degree (and must therefore be even; in both cases $r+l$ is the total number of evaluations of the set of f_i in (1));

(iii)   the value of z_1 in the form $z_1 = \sigma e^{i\theta}$, where z_1 is the center of a circular region containing a cluster of eigenvalues δ of the Jacobian matrix $J^{(m)}$ as described above during the first integration step (this is step below);

(iv)   the diameter of the above circular region;

(v)   the value of a boolean variable *thirdorder*: the value of *thirdorder* is true if the user requires a third order exact polynomial $p_N(z)$ to be used (in this case, the supplied value of r must be ≥ 3 and the appropriate coefficients $ß_i$, $i=0,1,2,3$, must be supplied), and *thirdorder* is false otherwise;

(vi)   an upper bound *tol* for the rounding errors in the computations of one Runge-Kutta step;

(vii)   an upper bound step for the stepsize (the stepsize actually used during the computations may be reduced by stability constraints);

(viii) $t0$ and te, the initial and final values of t;

(ix) $m0$ and m, the indices of the first and last equations in the set (2) holding when $t=t0$;

(x)   the initial values $u_i(t0)$ $(i=m0,m0+1,...,m)$;

efrk constructs an appropriate Runge-Kutta scheme of the form (3) at the commencement of the first integration step and integrates the system (1), either retaining the constructed scheme throughout, or constructing a new scheme if the estimates of z_1 and the diameter supplied during the course of the computation change sufficiently.

The user is also asked to provide a procedure *output* which is called by *efrk* at the conclusion of each integration step. *output* is intended partly as a monitoring device, and may also be used to change the values of σ and φ in the representation $z_1 = \sigma e^{i\varphi}$ and of the diameter of the circle with center z_1 enclosing a cluster of eigenvalues during the computation.

Procedure parameters:

   void efrk (t,te,m0,m,u,sigma,phi,diameter,method,k,step,r,l,beta,thirdorder,tol)

t: double t[0:0];
              the independent variable t of the systems of ordinary differential
              equations
$$du/dt = f(t,u);$$
              entry: the initial value $t0$;
              exit: the final value te;
te: double;
              entry: the final value of t ($te \geq t$);
m0,m: int;
              indices of the first and last equation of the system;
u: double u[m0:m];
              the dependent variable;
              entry: the initial values of the solution of the system of differential
                       equations at $t=t0$;
              exit: the values of the solution at $t = te$;
sigma: double sigma[0:0];
              entry: the modulus of the point at which exponential fitting is desired
                       (the value of σ above); for example an approximation of the
                       center of the left hand cluster;
              exit: value of σ for the last step;
phi: double phi[0:0];
              entry: the argument of the center of the left hand cluster (the value of
                       φ above); in the case of two complex conjugated clusters, the
                       argument of the center in the second quadrant should be
                       taken;
              exit: value of φ for the last step;
diameter: double diameter[0:0];
              entry: the diameter of the left hand cluster of eigenvalues of the
                       Jacobian matrix of the system of differential equations; in case
                       of nonlinear equations diameter should have such a value that
                       the variation of the eigenvalues in this cluster in the period
                       ($t, t+step$) is less than half the diameter;
              exit: value of the diameter for the last step;
method: a class that defines two procedures derivative and out, this class must
              implement the AP_efrk_methods interface;
              void derivative(int m0, int m, double t, double u[ ])
                  this procedure should deliver the value of $f(t,u)$ in the point (t,u) in
                  the array u;
              void output(int m0, int m, double t[ ], double te, double u[ ],
                          double sigma[ ], double phi[ ], double diameter[ ], int k[ ],
                          double step, int r, int l)
                  this procedure is called at the end of each integration step; the
                  user can ask for output of some parameters, for example t, k, u,
                  r,and l, and compute new values for sigma, phi, diameter, and step;

k: int *k[0:0]*;

                  counts the number of integration steps taken;

                  entry: an (arbitrarily) chosen value of *k0*, e.g., *k0* = 0;

                  exit: *k0* + the number of integration steps performed;

step: double *step[0:0]*;

                  the stepsize chosen will be at most equal to *step*; this stepsize may be reduced by stability constraints, imposed by a positive diameter, or by considerations of internal stability;

r: double;

                  entry: the degree of exactness of the Runge-Kutta scheme to be employed (value of *r* above); *r+l* is the number of evaluations of *f(t,u)* on which the Runge-Kutta scheme is based; for *r*=1,2,≥3, first, second, and third order accuracy may be obtained by an appropriate choice of the array *beta*;

l: double;

                  entry: if *phi*=4*tan^{-1}(1) then *l* is the order of the exponential fitting, otherwise *l* is twice the order of the exponential fitting; note that *l* should be even in the latter case;

beta: double *beta[0:r+l]*;

                  entry: the elements *beta[i]*, *i*=0,...,*r*, should have the value of the *r*+1 first coefficients of the stability polynomial;

thirdorder: boolean;

                  entry: if third order accuracy is desired then *thirdorder* should have the value true, in combination with appropriate choices of *r* (*r* ≥ 3) and the array *beta* (*beta[i]*=1/*i!*, *i*=0,1,2,3); in all other cases *thirdorder* must have the value false;

tol: double;

                  entry: an upper bound for the rounding errors in the computations in one Runge-Kutta step; in some cases (e.g., large values of *sigma* and *r*) *tol* will cause a decrease of the stepsize.

Procedures used: elmvec, dec, sol.

```
package numal;

public interface AP_efrk_methods {

    void derivative(int m0, int m, double t, double u[]);
    void output(int m0, int m, double t[], double te, double u[], double sigma[],
                double phi[], double diameter[], int k[], double step, int r, int l);
}


public static void efrk(double t[], double te, int m0, int m, double u[],
                        double sigma[], double phi[], double diameter[],
                        AP_efrk_methods method, int k[], double step, int r, int l,
                        double beta[], boolean thirdorder, double tol)
{
    boolean first,last,change,complex;
    int n,i,j,c1,c3;
```

```
double theta0,thetanm1,h,b,b0,phi0,phil,cosphi,sinphi,eps,betar,dd,hstab,
       hstabint,c,temp,c2,e,b1,zi,cosiphi,siniphi,cosphil,bb,mt,lt,tht;
double aux[] = new double[4];
int p[] = new int[l+1];
double mu[] = new double[r+1];
double labda[] = new double[r+1];
double pt[] = new double[r+1];
double fac[] = new double[l];
double betac[] = new double[l];
double rl[] = new double[m+1];
double d[] = new double[l+1];
double a[][] = new double[l+1][l+1];

cosphi=sinphi=theta0=thetanm1=0.0;
n=r+1;
first=true;
b0 = -1.0;
betar=Math.pow(beta[r],1.0/r);
last=false;
eps=Double.MIN_VALUE;
phi0=phil=Math.PI;
do {
    /* stepsize */
    h=step;
    dd=Math.abs(sigma[0]*Math.sin(phi[0]));
    complex=(((l/2)*2 == l) && 2.0*dd > diameter[0]);
    if (diameter[0] > 0.0) {
        temp=sigma[0]*sigma[0]/(diameter[0]*(diameter[0]*0.25+dd));
        hstab=Math.pow(temp,l*0.5/r)/betar/sigma[0];
    } else
        hstab=h;
    dd = (thirdorder ? Math.pow(2.0*tol/eps/beta[r],1.0/(n-1))*
        Math.pow(4.0,(l-1.0)/(n-1.0)) : Math.pow(tol/eps,(1.0/r)/betar));
    hstabint=Math.abs(dd/sigma[0]);
    if (h > hstab) h=hstab;
    if (h > hstabint) h=hstabint;
    if (t[0]+h > te*(1.0-k[0]*eps)) {
        last=true;
        h=te-t[0];
    }
    b=h*sigma[0];
    dd=diameter[0]*0.1*h;
    dd *= dd;
    if (h < t[0]*eps) break;
    change=((b0 == -1.0) ||
        ((b-b0)*(b-b0)+b*b0*(phi[0]-phi0)*(phi[0]-phi0) > dd));
    if (change) {
        /* coefficient */
        b0=b;
        phi0=phi[0];
```

```
    if (b >= 1.0) {
        if (complex) {
            /* solution of complex equations */
            if (phi0 != phil) {
                /* elements of matrix */
                phil=phi0;
                cosphi=Math.cos(phil);
                sinphi=Math.sin(phil);
                cosiphi=1.0;
                siniphi=0.0;
                for (i=0; i<=l-1; i++) {
                    c1=r+1+i;
                    c2=1.0;
                    for (j=l-1; j>=1; j-=2) {
                        a[j][l-i]=c2*cosiphi;
                        a[j+1][l-i]=c2*siniphi;
                        c2 *= c1;
                        c1--;
                    }
                    cosphil=cosiphi*cosphi-siniphi*sinphi;
                    siniphi=cosiphi*sinphi+siniphi*cosphi;
                    cosiphi=cosphil;
                }
                aux[2]=0.0;
                Linear_algebra.dec(a,l,aux,p);
            }
            /* right hand side */
            e=Math.exp(b*cosphi);
            b1=b*sinphi-(r+1)*phil;
            cosiphi=e*Math.cos(b1);
            siniphi=e*Math.sin(b1);
            b1=1.0/b;
            zi=Math.pow(b1,r);
            for (j=l; j>=2; j-=2) {
                d[j]=zi*siniphi;
                d[j-1]=zi*cosiphi;
                cosphil=cosiphi*cosphi-siniphi*sinphi;
                siniphi=cosiphi*sinphi+siniphi*cosphi;
                cosiphi=cosphil;
                zi *= b;
            }
            cosiphi=zi=1.0;
            siniphi=0.0;
            for (i=r; i>=0; i--) {
                c1=i;
                c2=beta[i];
                c3=((2*i > l-2) ? 2 : l-2*i);
                cosphil=cosiphi*cosphi-siniphi*sinphi;
                siniphi=cosiphi*sinphi+siniphi*cosphi;
                cosiphi=cosphil;
```

```
            for (j=1; j>=c3; j-=2) {
                d[j] += zi*c2*siniphi;
                d[j-1] -= zi*c2*cosiphi;
                c2 *=c1;
                c1--;
            }
            zi *= b1;
        }
        Linear_algebra.sol(a,l,p,d);
        for (i=1; i<=l; i++) beta[r+i]=d[l+1-i]*b1;
    } else {
        /* form beta */
        if (first) {
            /* form constants */
            first=false;
            fac[0]=1.0;
            for (i=1; i<=l-1; i++) fac[i]=i*fac[i-1];
            pt[r]=l*fac[l-1];
            for (i=1; i<=r; i++) pt[r-i]=pt[r-i+1]*(l+i)/i;
        }
        if (l == 1) {
            c=1.0-Math.exp(-b);
            for (j=1; j<=r; j++) c=beta[j]-c/b;
            beta[r+1]=c/b;
        } else
            if (b > 40.0)
                for (i=r+1; i<=r+l; i++) {
                    c=0.0;
                    for (j=0; j<=r; j++)
                        c=beta[j]*pt[j]/(i-j)-c/b;
                    beta[i]=c/b/fac[l+r-i]/fac[i-r-1];
                }
            else {
                dd=c=Math.exp(-b);
                betac[l-1]=dd/fac[l-1];
                for (i=1; i<=l-1; i++) {
                    c=b*c/i;
                    dd += c;
                    betac[l-1-i]=dd/fac[l-1-i];
                }
                bb=1.0;
                for (i=r+1; i<=r+l; i++) {
                    c=0.0;
                    for (j=0; j<=r; j++)
                        c=(beta[j]-((j < l) ? betac[j] : 0.0))*
                            pt[j]/(i-j)-c/b;
                    beta[i]=c/b/fac[l+r-i]/fac[i-r-1]+
                            ((i < l) ? bb*betac[i] : 0.0);
                    bb *= b;
                }
```

```
                    }
                }
            }
        labda[0]=mu[0]=0.0;
        if (thirdorder) {
            theta0=0.25;
            thetanm1=0.75;
            if (b < 1.0) {
                c=mu[n-1]=2.0/3.0;
                labda[n-1]=5.0/12.0;
                for (j=n-2; j>=1; j--) {
                    c=mu[j]=c/(c-0.25)/(n-j+1);
                    labda[j]=c-0.25;
                }
            } else {
                c=mu[n-1]=beta[2]*4.0/3.0;
                labda[n-1]=c-0.25;
                for (j=n-2; j>=1; j--) {
                    c=mu[j]=c/(c-0.25)*beta[n-j+1]/beta[n-j]/((j < 1) ? b : 1.0);
                    labda[j]=c-0.25;
                }
            }
        } else {
            theta0=0.0;
            thetanm1=1.0;
            if (b < 1.0)
                for (j=n-1; j>=1; j--) mu[j]=labda[j]=1.0/(n-j+1);
            else {
                labda[n-1]=mu[n-1]=beta[2];
                for (j=n-2; j>=1; j--)
                    mu[j]=labda[j]=beta[n-j+1]/beta[n-j]/((j < 1) ? b : 1.0);
            }
        }
    }
    k[0]++;
    /* difference scheme */
    i = -1;
    do {
        i++;
        mt=mu[i]*h;
        lt=labda[i]*h;
        for (j=m0; j<=m; j++) rl[j]=u[j]+lt*rl[j];
        method.derivative(m0,m,t[0]+mt,rl);
        if (i == 0 || i == n-1) {
            tht=((i == 0) ? theta0*h : thetanm1*h);
            Basic.elmvec(m0,m,0,u,rl,tht);
        }
    } while (i < n-1);
    t[0] += h;
    method.output(m0,m,t,te,u,sigma,phi,diameter,k,step,r,l);
```

```
    } while (!last);
}
```

5.4.2 First order - Jacobian matrix available

A. efsirk

Solves an initial value problem, given as an autonomous system of first order differential equations

$$Dy(x) = f(y) \tag{1}$$

where $D = d/dx$ and $f, y, \in R^m$, with $y(x_0)$ prescribed, over the interval $x_0 \leq x \leq x_e$ by means of an exponentially fitted, A-stable, semi-implicit Runge-Kutta method of third order [BeDHKW73, BeDHV74, Vh72]. This procedure is suitable for the integration of stiff equations. The algorithm uses for each step two function evaluations and if the input parameter *linear* is false one evaluation of the Jacobian matrix. The stepsize is not determined by the accuracy of the numerical solution, but by the amount by which the given differential equation differs from a linear equation. The procedure does not reject integration steps.

The method used has the form

$$y_{n+1} = y_n + (1/4)h_n f_n(y_n) + (3/4)h_n f(y_n + \Lambda_{1,0}{}^{(n)}(h_n J_n)h_n f(y_n)) \tag{2}$$

where $x^{n+1} = x_n + h_n$, and with $y_n \approx y(x_n)$, $J_n = \partial f(y_n)/\partial y_n$,

$$\Lambda_{1,0}^{(n)}(z) = \frac{2}{3}\left(\frac{1 - \frac{1}{6}\left(1 + 3\alpha_1^{(n)}\right)z}{1 - \frac{1}{2}\left(1 + \alpha_1^{(n)}\right)z + \frac{1}{12}\left(1 + 3\alpha_1^{(n)}\right)z^2} \right)$$

The stability function for the above method is

$$R(z) = 1 + z + \frac{3}{4}z_2\Lambda_{1,0}^{(n)}(z) = \frac{1 + \frac{1}{2}\left(1 - \alpha_1^{(n)}\right)z + \frac{1}{12}\left(1 - 3\alpha_1^{(n)}\right)z^2}{1 - \frac{1}{2}\left(1 + \alpha_1^{(n)}\right)z + \frac{1}{12}\left(1 + 3\alpha_1^{(n)}\right)z^2}$$

and the parameter $\alpha_1^{(n)}$ in $\Lambda_{1,0}{}^{(n)}(z)$ is determined by use of the relationship

$$R_n(h_n\delta_n) = e^{h_n\delta_n}$$

where δ_n is the eigenvalue of J_n in the left half plane with largest absolute value, so that

$$\alpha_1^{(n)} = \frac{1}{3z_n}\left(\frac{z_n^2 + 6z_n + 12 - e^{z_n}\left(z_n^2 - 6z_n + 12\right)}{e^{z_n}\left(z_n - 2\right) + z_n + 2} \right)$$

where $z_n = h_n\delta_n$.

The right-hand side vector in equation (1) is computed by means of a user supplied procedure *derivative*. The Jacobian matrix J_n is computed by another user supplied procedure *jacobian*. If the variable *linear* in the parameter list is given the value true upon call then *jacobian* is called upon when $n=0$ in formula

(2) and not otherwise; if *linear* has the value false then *jacobian* is called at the commencement of each step of the integration method. When *linear* has the value false upon call, the stepsize h_n is determined as follows by use of a second method alternative to (2) for the numerical solution of equations (1). This method produces approximations y'_n to $y(x_n)$ and has the form

$$y'_{n+1} = y_n + v_0^{(n)}h_nf(y_n) + v_1^{(n)}h_nf(y_n + \Lambda_{1,0}(h_nJ_n)h_nf(y_n)) +$$
$$v_2^{(n)}\Lambda_{1,0}(h_nJ_n)h_nf(y_n) + v_3^{(n)}h_nf(y_{n+1}).$$

(3)

When

$$v_3^{(n)} = \frac{1}{2}\left(\frac{1+3\alpha_1^{(n)}}{5-3\alpha_1^{(n)}}\right), \qquad v_2^{(n)} = \frac{3}{4}\left(1+2v_3^{(n)}\right)$$

$$v_1^{(n)} = \frac{1}{2} - 2v_3^{(n)}, \qquad v_0^{(n)} = 0$$

the stability function of the method (3), namely

$$R'(z) = 1 + v_0^{(n)}z + v_1^{(n)}z(1 + \Lambda_{1,0}(z)) + v_2^{(n)}\Lambda_{1,0}(z)z + v_3^{(n)}zR(z)$$

is $R(z)$ above. If y_{n+1} *and* y'_{n+1} were to be determined by use of the schemes (2) and (3) respectively with step length h'_n in both cases, then, since the method (3) is second order exact, $\|y'_{n+1} - y_{n+1}\| \approx c_nh'_n{}^3$, where c_n is independent of the variable h'_n. If $c_n \approx c_{n-1}$, an optimal steplength h'_n may be determined by use of the relationships

$$h'_n = (tol/c_n)^{1/3} \approx (tol/c_{n-1})^{1/3} \approx h_{n-1}(tol/\|y'_n-y_n\|)^{1/3}$$

or, approximately,

$$h'_n = ((4/3)tol / (tol + \|y'_n-y_n\|) + 1/3)h_{n-1}$$

(4)

where *tol* is an acceptable local tolerance (taken by *efsirk* to be $tol = \eta_a + \eta_r\|y_n\|$, where η_a and η_r are absolute and relative tolerances prescribed by the user). Thus y'_n is determined by use of formula (4). With two limits *hmin* and *hmax* prescribed by the user, h_n is taken to be h'_n if $hmin \le h'_n \le hmax$, to be *hmin* if $h'_n < hmin$, and to be *hmax* if $h'_n > hmax$.

When *linear* has the value true upon call, a constant value $h_n = hmax$ is used in formula (2), and the reference values y'_n are not determined.

Procedure parameters:

void efsirk $(x,xe,m,y,delta,method,j,n,aeta,reta,hmin,hmax,linear)$

x: double $x[0:0]$;
                entry: the independent variable; the initial value $x0$;
                exit: the final value xe;
xe: double;
                entry: the final value of x;
m: int;
                entry: the number of differential equations;
y: double $y[1:m]$;
                the dependent variable; during the integration process the computed solution at the point x is associated with the array y;
                entry: the initial values of the solution of the system;
                exit: the final values of the solution;
delta: double $delta[0:0]$;
                delta denotes the real part of the point at which exponential fitting is desired; alternatives:

delta=(an estimate of) the real part of the, in absolute value, largest eigenvalue of the Jacobian matrix of the system;

delta<-10^{14}, in order to obtain asymptotic stability;

delta=0, in order to obtain a higher order of accuracy in case of linear or almost linear equations;

method: a class that defines three procedures *derivative*, *jacobian*, and *output*, this class must implement the AP_efsirk_methods interface;

void derivative(int m, double a[ ], double delta[ ])

in *efsirk* when *derivative* is called *a[i]* contains the values of *y[i]*; upon completion of a call of *derivative*, the array *a* should contain the values of *f(y)*;

note that the variable *x* should not be used in *derivative* because the differential equation is supposed to be autonomous;

void jacobian(int m, double j[ ][ ], double y[ ], double delta[ ])

in *efsirk* when *jacobian* is called the array *y* contains the values of the dependent variable;

upon completion of a call of *jacobian*, the array *j* should contain the values of the Jacobian matrix of *f(y)*;

void output(double x[ ], double xe, int m, double y[ ], double delta[ ], double j[ ][ ], int n[ ])

the user can ask for output of parameters;

j: double *j[1:m,1:m]*;

j is an auxiliary array which is used in the procedure *jacobian*;

n: int *n[0:0]*;

counts the integration steps;

aeta,reta: double;

entry: required absolute and relative local accuracy;

hmin,hmax: double;

entry: minimal and maximal stepsize by which the integration is performed;

linear: boolean;

entry: if *linear* is true then the procedure *jacobian* will only be called if $n = 1$; the integration will then be performed with a stepsize *hmax*; the corresponding reduction of computing time can be exploited in case of linear or almost linear equations.

Procedures used: vecvec, matvec, matmat, gsselm, solelm.

```
package numal;

public interface AP_efsirk_methods {

    void derivative(int m, double a[], double delta[]);
    void jacobian(int m, double j[][], double y[], double delta[]);
    void output(double x[], double xe, int m, double y[], double delta[],
            double j[][], int n[]);

}
```

```
public static void efsirk(double x[], double xe, int m, double y[], double delta[],
                          AP_efsirk_methods method, double j[][], int n[],
                          double aeta, double reta, double hmin, double hmax,
                          boolean linear)
{
    boolean lin;
    int k,l;
    double step,h,mu0,mu1,mu2,theta0,theta1,nu1,nu2,nu3,yk,fk,c1,c2,d,discr,eta,
           s,z1,z2,e,alpha1,a,b;
    double aux[] = new double[8];
    int ri[] = new int[m+1];
    int ci[] = new int[m+1];
    double f[] = new double[m+1];
    double k0[] = new double[m+1];
    double labda[] = new double[m+1];
    double j1[][] = new double[m+1][m+1];

    nu3=h=z2=mu1=mu2=mu0=theta0=theta1=nu1=nu2=0.0;
    aux[2]=Double.MIN_VALUE;
    aux[4]=8.0;
    for (k=1; k<=m; k++) f[k]=y[k];
    n[0] = 0;
    method.output(x,xe,m,y,delta,j,n);
    step=0.0;
    do {
        n[0]++;
        /* difference scheme */
        method.derivative(m,f,delta);
        /* step size */
        if (linear)
            s=h=hmax;
        else
            if (n[0] == 1 || hmin == hmax)
                s=h=hmin;
            else {
                eta=aeta+reta*Math.sqrt(Basic.vecvec(1,m,0,y,y));
                c1=nu3*step;
                for (k=1; k<=m; k++) labda[k] += c1*f[k]-y[k];
                discr=Math.sqrt(Basic.vecvec(1,m,0,labda,labda));
                s=h=(eta/(0.75*(eta+discr))+0.33)*h;
                if (h < hmin)
                    s=h=hmin;
                else
                    if (h > hmax) s=h=hmax;
            }
        if (x[0]+s > xe) s=xe-x[0];
        lin=((step == s) && linear);
        step=s;
        if (!linear || n[0] == 1) method.jacobian(m,j,y,delta);
        if (!lin) {
```

```
    /* coefficient */
    z1=step*delta[0];
    if (n[0] == 1) z2=z1+z1;
    if (Math.abs(z2-z1) > 1.0e-6*Math.abs(z1) || z2 > -1.0) {
        a=z1*z1+12.0;
        b=6.0*z1;
        if (Math.abs(z1) < 0.1)
            alpha1=(z1*z1/140.0-1.0)*z1/30.0;
        else if (z1 < 1.0e-14)
            alpha1=1.0/3.0;
        else if (z1 < -33.0)
            alpha1=(a+b)/(3.0*z1*(2.0+z1));
        else {
            e=((z1 < 230.0) ? Math.exp(z1) : Double.MAX_VALUE);
            alpha1=((a-b)*e-a-b)/(((2.0-z1)*e-2.0-z1)*3.0*z1);
        }
        mu2=(1.0/3.0+alpha1)*0.25;
        mu1 = -(1.0+alpha1)*0.5;
        mu0=(6.0*mu1+2.0)/9.0;
        theta0=0.25;
        theta1=0.75;
        a=3.0*alpha1;
        nu3=(1.0+a)/(5.0-a)*0.5;
        a=nu3+nu3;
        nu1=0.5-a;
        nu2=(1.0+a)*0.75;
        z2=z1;
    }
    c1=step*mu1;
    d=step*step*mu2;
    for (k=1; k<=m; k++) {
        for (l=1; l<=m; l++)
            j1[k][l]=d*Basic.matmat(1,m,k,l,j,j)+c1*j[k][l];
        j1[k][k] += 1.0;
    }
    Linear_algebra.gsselm(j1,m,aux,ri,ci);
}
c1=step*step*mu0;
d=step*2.0/3.0;
for (k=1; k<=m; k++) {
    k0[k]=fk=f[k];
    labda[k]=d*fk+c1*Basic.matvec(1,m,k,j,f);
}
Linear_algebra.solelm(j1,m,ri,ci,labda);
for (k=1; k<=m; k++) f[k]=y[k]+labda[k];
method.derivative(m,f,delta);
c1=theta0*step;
c2=theta1*step;
d=nu1*step;
for (k=1; k<=m; k++) {
```

```
        yk=y[k];
        fk=f[k];
        labda[k]=yk+d*fk+nu2*labda[k];
        y[k]=f[k]=yk+c1*k0[k]+c2*fk;
    }
    x[0] += step;
    method.output(x,xe,m,y,delta,j,n);
} while (x[0] < xe);
}
```

B. eferk

Solves an initial value problem, given as an autonomous system of first order differential equations

$$Dy(x) = f(y) \tag{1}$$

where $D \equiv d/dx$ and $f, y, \in R^m$, with $y(x_0)$ prescribed, over the interval $x_0 \le x \le xe$ by means of an exponentially fitted, semi-implicit Runge-Kutta method of third order [BeDHKW73, BeDHV74, DekHH72, Vh72]. This procedure is suitable for the integration of stiff differential equations. The algorithm uses for each step two function evaluations and if the input parameter *linear* is false one evaluation of the Jacobian matrix. The stepsize is determined by an estimation of the local truncation error based on the residual function. Integration steps are not rejected.

The method used is one of a class having the form

$$y_{n+1} = y_n + \theta_0(z)h_nf_n(y_n) + \theta_1(z)h_nf(y_n + \Lambda_{1,0}(z)h_nf(y_n))$$

where, with $x^{n+1} = x_n + h_n$, $y_n \approx y(x_n)$ and $z = h_nJ_n$ where $J_n = \partial f(y_n)/\partial y_n$, θ_0, θ_1 and $\Lambda_{1,0}$ being polynomials. Such a method is first order exact if $\theta_0(0) + \theta_1(0) = 1$, second order exact if also $\theta_0'(0) + \theta'_1(0) + \theta_1(0)\Lambda_{1,0}(0) = 1/2$, and third order consistent if also

$$\theta_0''(0) + \theta_1''(0) + 2\theta_1(0)\Lambda_{1,0}'(0) + 2\theta_1'(0)\Lambda_{1,0}(0) = 1/3$$

and

$$\theta_1(0)\Lambda_{1,0}^2(0) = 1/3.$$

The choice $\theta_0(z) = 1/4$, $\theta_1(z) = 3/4$ is made, and the method used by *eferk* has the form

$$y_{n+1} = y_n + \frac{1}{4}h_nf(y_n) + \frac{3}{4}h_nf(y_n + \frac{4}{3}(\frac{1}{2} + \frac{1}{6}h_nJ_n + \sum_{j=4}^{m}\beta_j(h_nJ_n)^{j-2})h_nf(y_n)) \tag{2}$$

The stability function for this method is

$$R_m(z) = \sum_{j=0}^{p}\frac{z^j}{j!}\sum_{j=p+1}^{m}\beta_jz^j$$

with $p = 3$. The β_j may be determined by exponential fitting of order r_j at the s points z_j ($j=1,...,s$), so that

$$\frac{d^k}{dz_j^k}R_m(z_j) = e^{z_j} \qquad (k = 0,...,r_j; \; j = 1,...,s)$$

where

$$\sum_{j=1}^{s}\left(r_j+1\right)=m-p$$

Approximately

$$R_m(z)\approx\left(\sum_{j=0}^{p}\frac{z^j}{j!}\right)\prod_{j=1}^{s}\left(\frac{z-z_j}{z_j}\right)^{r_j}$$

That part of the stability region (i.e., the domain over which $|R_m(z)| \le 1$) in the neighborhood of z_j is a small circle with z_j as center and radius ρ_j where

$$\left(\rho_j\right)^{r_j}=p!z_j^{(r_j-p)}\prod_{\substack{i=1\\(i\ne j)}}^{s}\left|\frac{z_i}{z_i-z_j}\right|^{r_i}$$

If exponential fitting takes place at one point z_0 alone
$$\rho_0=|z_0|\,|p!z_0^p|^{1/(m-p)}.$$
If fitting takes place at two complex conjugate points z_0 and conjugate(z_0), with $\arg(\Lambda_0)=\varphi$
$$\rho_0=|z_0|\,|(p!z_0^p)/2\sin\varphi|^{2/(m-p)}.$$
Since these stability regions are small, it is necessary to determine z_0 precisely. Furthermore the regions prescribe a maximal step length h_{stab} for stability: if $d\delta_n$ is the distance between the smallest eigenvalue of J_n and the point of fitting δ_n then, when $p=3$,

$$h_{stab}=\frac{1}{\delta_n}6^{1/3}\left|\frac{\delta_n}{d\delta_n}\right|^{\frac{m}{3}-1}.$$

The step-choice strategy is based upon the following considerations. If $y_{n+1}(h)$ is the value of the solution to equation (1) at $x=x_n+h$ with initial values $y_{n+1}(0)=y_n$ produced by use of the scheme (2), and $\xi_n(h)=dy_{n+1}(h)/dh-f(y_{n+1}(h))$ then $h_n\|\xi_n(h)\|$ is a good estimate of the local error $\|y(x_n+h)-y_{n+1}(h)\|$, and $\rho_n'(h)=h|R(hJ_n)f(y_n)-f(y_{n+1})|$ is a good estimate of $h_n\|\xi_n(h)\|$. With h_n, $tol=\eta_a+\eta_r\|y_n\|$, (where η_a and η_r are absolute and relative tolerances prescribed by the user) and $\rho_n'(h_n)$ available, the quantity

$$h_{n+1}'=\left(\frac{4\,tol}{(3+j)tol+\rho_n'(h_n)}+\frac{1}{3-j}\right)h_n$$

with $j=0$ for nonlinear equations, and $j=1$ for approximately linear equations, is determined. With two limits $hmin$ and $hmax$ prescribed by the user, the steplength h_{n+1} is taken to be h_{n+1}' if $hmin \le h_{n+1}' \le hmax$; if h_{n+1}' lies outside this range, h_{n+1} is $hmax$ if $h_{n+1}'>hmax$ and $hmin$ if $h_{n+1}'<hmin$.

It can occur that the system of equations (1) is derived from the system $Dy(x)=g(x,y)$ with $y(x_0)=y_0$, where $y, g \in R^{m-1}$, simply by taking the first $m-1$ components of y in (1) to be those of y, and the m-th component of y to be x. In this case, the m-th component of f in (1) is always 1, and certain simplifications in the formulae can be exploited. If it is known that equations (1) have been derived in the above manner then the variable aut in the parameter list of $eferk$ should be given the value false upon call, otherwise true.

Procedure parameters:

void eferk (*x,xe,m,y,sigma,phi,method,j,k,l,aut,aeta,reta,hmin,hmax,linear*)

x: double *x[0:0]*;

entry: the independent variable; the initial value *x0*;

exit: the final value *xe*;

xe: double;

entry: the final value of x ($xe \geq x$);

m: int;

entry: the number of equations;

y: double *y[1:m]*;

the dependent variable;

entry: the initial values of the system of differential equations: *y[i]* at
$x=x0$;

exit: the final values of the solution: *y[i]* at $x=xe$;

sigma: double *sigma[0:0]*;

the modulus of the point at which exponential fitting is desired, for
example the largest negative eigenvalue of the Jacobian matrix of the
system of differential equations;

phi: double;

the argument of the complex point at which exponential fitting is desired;

method: a class that defines three procedures *derivative, jacobian,* and *output,*
this class must implement the AP_eferk_methods interface;

void derivative(int m, double y[ ])

this procedure should deliver the right-hand side of the *i*-th
differential equation at the point (*y*) as *y[i]*;

void jacobian(int m, double j[ ][ ], double y[ ], double sigma[ ])

in this procedure the Jacobian at the point (*y*) has to be assigned to
the array *j*;

void output(double x[ ], double xe, int m, double y[ ], double j[ ][ ], int k[ ])

this procedure is called at the end of each integration step; the
user can ask for output of some parameters, for example, *x, y, j, k*;

j: double *j[1:m, 1:m]*;

the Jacobian matrix of the system; array *j* should be updated in the
procedure *jacobian*;

k: int *k[0:0]*;

counts the integration steps taken;

l: int;

entry: if *phi*=4*tan⁻¹(1) then *l* is the order of the exponential fitting, else *l*
is twice the order of the exponential fitting;

aut: boolean;

entry: if the system has been written in autonomous form by adding the
equation $dy[m]/dx = 1$ to the system then *aut* may have the value
false, else *aut* should have the value true;

aeta: double;

entry: required absolute precision in the integration process, *aeta* has to
be positive;

reta: double;

entry: required relative precision in the integration process, *reta* has to
be positive;

hmin: double;

entry: the steplength chosen will be at least equal to *hmin*;

hmax: double;
           entry: the steplength chosen will be at most equal to *hmax*;
linear: boolean;
           entry: the procedure *jacobian* is called only if *linear* is false or $k = 0$;
                   so if the system is linear then *linear* may have the value true.

Procedures used: vecvec, matvec, dec, sol.

```
package numal;

public interface AP_eferk_methods {

    void derivative(int m, double y[]);
    void jacobian(int m, double j[][], double y[], double sigma[]);
    void output(double x[], double xe, int m, double y[], double j[][], int k[]);
}



public static void eferk(double x[], double xe, int m, double y[], double sigma[],
                         double phi, AP_eferk_methods method, double j[][], int k[],
                         int l, boolean aut, double aeta, double reta, double hmin,
                         double hmax, boolean linear)
{
    boolean change,last;
    int m1,i,c1,q;
    double h,b,b0,phi0,cosphi,sinphi,eta,discr,fac,s,cos2phi,sina,e,zi,c2,cosiphi,
           siniphi,cosphil,emin1,b1,b2,a0,a1,a2,a3,c,ddd,betai,bethai;
    double aux[] = new double[4];
    int p[] = new int[l+1];
    double beta[] = new double[l+1];
    double betha[] = new double[l+1];
    double betac[] = new double[l+4];
    double k0[] = new double[m+1];
    double d[] = new double[m+1];
    double d1[] = new double[m+1];
    double d2[] = new double[m+1];
    double dd[] = new double[l+1];
    double a[][] = new double[l+1][l+1];

    h=cosphi=sinphi=0.0;
    b0 = phi0 = -1.0;
    betac[l]=betac[l+1]=betac[l+2]=betac[l+3]=0.0;
    beta[0]=1.0/6.0;
    betha[0]=0.5;
    fac=1.0;
    for (i=2; i<=l-1; i++) fac *= i;
    m1=(aut ? m : m-1);
    k[0] = 0;
    last=false;
    do {
```

```
for (i=1; i<=m; i++) d[i]=y[i];
method.derivative(m,d);
if (!linear || k[0] == 0) method.jacobian(m,j,y,sigma);
/* step size */
eta=aeta+reta*Math.sqrt(Basic.vecvec(1,m1,0,y,y));
if (k[0] == 0) {
    discr=Math.sqrt(Basic.vecvec(1,m1,0,d,d));
    h=eta/discr;
} else {
    s=0.0;
    for (i=1; i<=m1; i++) {
        ddd=d[i]-d2[i];
        s += ddd*ddd;
    }
    discr=h*Math.sqrt(s)/eta;
    h *= (linear ? 4.0/(4.0+discr)+0.5 : 4.0/(3.0+discr)+1.0/3.0);
}
if (h < hmin) h=hmin;
if (h > hmax) h=hmax;
b=Math.abs(h*sigma[0]);
change=(Math.abs(1.0-b/b0) > 0.05 || phi != phi0);
if (1.1*h >= xe-x[0]) {
    change=last=true;
    h=xe-x[0];
}
if (!change) h=h*b0/b;
if (change) {
    /* coefficient */
    b0=b=Math.abs(h*sigma[0]);
    if (b >= 0.1) {
        if (phi != Math.PI && l == 2 ||
            Math.abs(phi-Math.PI) > 0.01) {
            /* solution of complex equations */
            if (l == 2) {
                phi0=phi;
                cosphi=Math.cos(phi0);
                sinphi=Math.sin(phi0);
                e=Math.exp(b*cosphi);
                zi=b*sinphi-3.0*phi0;
                sina=((Math.abs(sinphi) < 1.0e-6) ? -e*(b+3.0) :
                                              e*Math.sin(zi)/sinphi);
                cos2phi=2.0*cosphi*cosphi-1.0;
                betha[2]=(0.5+(2.0*cosphi+(1.0+2.0*cos2phi+sina)/b)/b)/b/b;
                sina=((Math.abs(sinphi) < 1.0e-6) ? e*(b+4.0) :
                                              sina*cosphi-e*Math.cos(zi));
                betha[1] = -(cosphi+(1.0+2.0*cos2phi+
                            (4.0*cosphi*cos2phi+sina)/b)/b)/b;
                beta[1]=betha[2]+2.0*cosphi*(betha[1]-1.0/6.0)/b;
                beta[2]=(1.0/6.0-betha[1])/b/b;
            } else {
```

```
    if (phi0 != phi) {
        /* elements of matrix */
        phi0=phi;
        cosphi=Math.cos(phi0);
        sinphi=Math.sin(phi0);
        cosiphi=1.0;
        siniphi=0.0;
        for (i=0; i<=l-1; i++) {
            c1=4+i;
            c2=1.0;
            for (q=l-1; q>=1; q-=2) {
                a[q][l-i]=c2*cosiphi;
                a[q+1][l-i]=c2*siniphi;
                c2 *= c1;
                c1--;
            }
            cosphil=cosiphi*cosphi-siniphi*sinphi;
            siniphi=cosiphi*sinphi+siniphi*cosphi;
            cosiphi=cosphil;
        }
        aux[2]=0.0;
        Linear_algebra.dec(a,l,aux,p);
    }
    /* right hand side */
    e=Math.exp(b*cosphi);
    zi=b*sinphi-4.0*phi0;
    cosiphi=e*Math.cos(zi);
    siniphi=e*Math.sin(zi);
    zi=1.0/b/b/b;
    for (q=l; q>=2; q-=2) {
        dd[q]=zi*siniphi;
        dd[q-1]=zi*cosiphi;
        cosphil=cosiphi*cosphi-siniphi*sinphi;
        siniphi=cosiphi*sinphi+siniphi*cosphi;
        cosiphi=cosphil;
        zi *= b;
    }
    siniphi=2.0*sinphi*cosphi;
    cosiphi=2.0*cosphi*cosphi-1.0;
    cosphil=cosphi*(2.0*cosiphi-1.0);
    dd[l] += sinphi*(1.0/6.0+(cosphi+(1.0+2.0*cosiphi*
                (1.0+2.0*cosphi/b))/b)/b);
    dd[l-1] -= cosphi/6.0+(0.5*cosiphi+(cosphil+
                (2.0*cosiphi*cosiphi-1.0)/b)/b)/b;
    dd[l-2] += sinphi*(0.5+(2.0*cosphi+(2.0*cosiphi+1.0)/b)/b);
    dd[l-3] -= 0.5*cosphi-(cosiphi+cosphil/b)/b;
    if (l >= 5) {
        dd[l-4] += sinphi+siniphi/b;
        dd[l-5] -= cosphi+cosiphi/b;
        if (l >= 7) {
```

```
                            dd[l-6] += sinphi;
                            dd[l-7] -= cosphi;
                        }
                }
                Linear_algebra.sol(a,l,p,dd);
                zi=1.0/b;
                for (i=1; i<=l; i++) {
                    beta[i]=dd[l+1-i]*zi;
                    betha[i]=(i+3)*beta[i];
                    zi /= b;
                }
            }
        } else {
            /* form beta */
            if (l == 1) {
                betha[1]=(0.5-(1.0-(1.0-Math.exp(-b))/b)/b)/b;
                beta[1]=(1.0/6.0-betha[1])/b;
            } else if (l == 2) {
                e=Math.exp(-b);
                emin1=e-1.0;
                betha[1]=(1.0-(3.0+e+4.0*emin1/b)/b)/b;
                betha[2]=(0.5-(2.0+e+3.0*emin1/b)/b)/b/b;
                beta[2]=(1.0/6.0-betha[1])/b/b;
                beta[1]=(1.0/3.0-(1.5-(4.0+e+5.0*emin1/b)/b)/b)/b;
            } else {
                betac[l-1]=c=ddd=Math.exp(-b)/fac;
                for (i=l-1; i>=1; i--) {
                    c=i*b*c/(l-i);
                    betac[i-1]=ddd=ddd*i+c;
                }
                b2=0.5-betac[2];
                b1=(1.0-betac[1])*(l+1)/b;
                b0=(1.0-betac[0])*(l+2)*(l+1)*0.5/b/b;
                a3=1.0/6.0-betac[3];
                a2=b2*(l+1)/b;
                a1=b1*(l+2)*0.5/b;
                a0=b0*(l+3)/3.0/b;
                ddd=1/b;
                for (i=1; i<=l; i++) {
                    beta[i]=(a3/i-a2/(i+1)+a1/(i+2)-a0/(i+3))*ddd+betac[i+3];
                    betha[i]=(b2/i-b1/(i+1)+b0/(i+2))*ddd+betac[i+2];
                    ddd=ddd*(l-i)/i/b;
                }
            }
        }
    } else
        for (i=1; i<=l; i++) {
            betha[i]=beta[i-1];
            beta[i]=beta[i-1]/(i+3);
        }
```

```
        }
        method.output(x,xe,m,y,j,k);
        /* difference scheme */
        if (m1 < m) {
            d2[m]=1.0;
            k0[m]=y[m]+2.0*h/3.0;
            y[m]  += 0.25*h;
        }
        for (q=1; q<=m1; q++) {
            k0[q]=y[q]+2.0*h/3.0*d[q];
            y[q]  += 0.25*h*d[q];
            d1[q]=h*Basic.matvec(1,m,q,j,d);
            d2[q]=d1[q]+d[q];
        }
        for (i=0; i<=1; i++) {
            betai=4.0*beta[i]/3.0;
            bethai=betha[i];
            for (q=1; q<=m1; q++) d[q]=h*d1[q];
            for (q=1; q<=m1; q++) {
                k0[q]  += betai*d[q];
                d1[q]=Basic.matvec(1,m1,q,j,d);
                d2[q]  += bethai*d1[q];
            }
        }
        method.derivative(m,k0);
        for (q=1; q<=m; q++) y[q]  += 0.75*h*k0[q];
        k[0]++;
        x[0] += h;
    } while (!last);
    method.output(x,xe,m,y,j,k);
}
```

C. liniger1vs

Solves an initial value problem, given as an autonomous system of first order differential equations

$$Dy(x) = f(y) \tag{1}$$

where $D \equiv d/dx$ and $f,y, \in R^m$, with $y(x0)$ prescribed, over the interval $x0 \le x \le xe$ by means of an implicit, first order accurate, exponentially fitted one-step method. Automatic stepsize control is provided. This procedure is suitable for the integration of stiff differential equations. The algorithm is based on [LiW70]. The stepsize strategy requires many extra array operations. The user may avoid this extra work by giving the parameters *aeta* and *reta* a negative value and prescribing a stepsize *hmax*.

The method used has the form

$$y_{n+1} = y_n + h \left(\mu_n f_n + (1 - \mu_n) f_{n+1}\right) \tag{2}$$

where, with $x^n = x_0 + nh$ $(n=0,1,...)$, $y_n \approx y(x_n)$ and $f_n = f(y_n)$.

Equation (1) may be written as
$$Dy = f_n + J_n (y - y_n) + \ldots$$
J_n being the Jacobian matrix $\partial f(y_n)/\partial y_n$. The linearized version of (1) is thus $Dy' = f(y')$ where $f(y')=f_n-J_ny_n+J_ny'$. The exact solution of the linearized equation with $y'(x_n)=y_n$ is

$$y'(x) = y_n - (x - x_n) \, e_2 \, (J_n(x - x_n))f_n$$

where $e_2(z) = (e^z-1)/z$ so that $y_{n+1}' = y_n - he_2(hJ_n)f_n$. With y_n'' a computed value of y_n (i.e., of y_n) the scheme (2) yields

$$(I - (1-\mu_n)hJ_n)y_{n+1}'' = (I + \mu_nhJ_n)y_n'' + h(f_n-J_ny_n).$$

Thus

$$y_{n+1}'' - y_{n+1}' = R_n(hJ_n)(y_n'' - y_n') + h(R_n^{(2)}(hJ_n) - e_2(hJ_n))f_n \qquad (3)$$

where R_n is the stability function

$$R_n(z) = (1 + \mu_nz) / (1 - (1-\mu_n)z)$$

and $R_n^{(2)}(z) = (R_n(z) - 1)/z$. If all eigenvalues of J_n not in the neighborhood of the origin are clustered about the single point δ_n, and μ_n is determined by the condition

$$R_n(h\delta_n) = e^{h\delta_n} \, , \qquad i.e., \qquad \mu_n = \left(1 - e^{-h\delta_n}\right)^{-1} - \left(h\delta_n\right)^{-1}$$

the second term on the right-hand side of equation (3) is negligible, and
$$\left\| y_{n+1}'' - y_{n+1}' \right\| \le \left\| R_n(hJ_n) \right\| \, \left\| y_n'' - y_n' \right\|.$$
When $\delta_n \in (-\infty,0)$, $\mu_n \in [0,\frac{1}{2}]$ and $\left| R_n(z) \right| < 1$ for $z \in (-\infty,0)$. Thus, with such δ_n and μ_n determined by exponential fitting in this way, the scheme (2) is stable.

y_{n+1} in formula (2) is determined by setting $y_{n+1}^{(0)} = y_n$ and using the Newton scheme

$$[I - (1-\mu_n)hJ_{n+1}^{(l)}]\Delta_{n+1}^{(l)} = y_n - y_{n+1}^{(l)} + \mu_nhf_n + (1-\mu_n)hf_{n+1}^{(l)} \quad (l=0,1,\ldots) \qquad (4)$$

where $\Delta_{n+1}^{(l)} = y_{n+1}^{(l+1)} - y_{n+1}^{(l)}$, and $J_{n+1}^{(l)}$ is an approximation to the Jacobian matrix $Df(y_{n+1}^{(l)})/Dy_{n+1}^{(l)}$. This scheme is continued until either $\left\| \Delta_{n+1}^{(l)} \right\| \le \eta_a + \left\| y_{n+1}^{(l)} \right\|\eta_r$, where η_a and η_r are absolute and relative tolerances prescribed by the user, or $l = itmax$, where $itmax$ is a similarly prescribed integer.

The Jacobian matrix $J_{n+1}^{(l)}$ occurring in formula (4) is updated at every iteration and every integration step.

Procedure parameters:
      void liniger1vs (x,xe,m,y,sigma,method,j,itmax,hmin,hmax,aeta,reta,info)

x: double x[0:0];
      entry: the independent variable; the initial value $x0$;
      exit: the final value xe;

xe: double;
      entry: the final value of x ($xe \ge x$);

m: int;
      entry: the number of equations;

y: double y[1:m];
      the dependent variable;
      entry: the initial values of the system of differential equations: $y[i]$ at $x=x0$;
      exit: the final values of the solution: $y[i]$ at $x=xe$;

sigma: double sigma[0:0];
      the modulus of the point at which exponential fitting is desired, for

example the largest negative eigenvalue of the Jacobian matrix of the
system of differential equations;

method: a class that defines three procedures *derivative*, *jacobian*, and *output*,
this class must implement the AP_liniger1vs_methods interface;

void derivative(int m, double y[ ], double sigma[ ])

this procedure should deliver the right-hand side of the *i*-th
differential equation at the point (*y*) as *y[i]*;

void jacobian(int m, double j[ ][ ], double y[ ], double sigma[ ])

in this procedure (an approximation to) the Jacobian has to be
assigned to the array *j*;

void output(double x[ ], double xe, int m, double y[ ],
double sigma[ ], double j[ ][ ], double info[ ])

this procedure is called at the end of each integration step; the
user can ask for output of some parameters, for example, *x*, *y*, *j*,
and *info*;

j: double *j[1:m, 1:m]*;
the Jacobian matrix of the system; array *j* should be updated in the
procedure *jacobian*;

itmax: int;
entry: an upper bound for the number of iterations in Newton's process,
used to solve the implicit equations;

hmin: double;
entry: minimal stepsize by which the integration is performed;

hmax: double;
entry: maximal stepsize by which the integration is performed;

aeta: double;
entry: required absolute precision in the integration process;

reta: double;
entry: required relative precision in the integration process;
if both *aeta* and *reta* have negative values then integration will be
performed with a stepsize equal to *hmax*, which may be varied by
the user; in this case the absolute values of *aeta* and *reta* will
control the Newton iteration;

info: double *info[1:9]*;
during integration and upon exit this array contains the following
information:

info[1]:   number of integration steps taken;
info[2]:   number of derivative evaluations used;
info[3]:   number of Jacobian evaluations used;
info[4]:   number of integration steps equal to *hmin* taken;
info[5]:   number of integration steps equal to *hmax* taken;
info[6]:   maximal number of integrations taken in the Newton process;
info[7]:   local error tolerance;
info[8]:   estimated local error;
info[9]:   maximum value of the estimated local error.

Procedures used: inivec, mulvec, mulrow, dupvec, matvec, elmvec, vecvec, dec,
sol.

```
package numal;

public interface AP_liniger1vs_methods {

    void derivative(int m, double y[], double sigma[]);
    void jacobian(int m, double j[][], double y[], double sigma[]);
    void output(double x[], double xe, int m, double y[], double sigma[],
                double j[][], double info[]);
}


public static void liniger1vs(double x[], double xe, int m, double y[],
                              double sigma[], AP_liniger1vs_methods method,
                              double j[][], int itmax, double hmin, double hmax,
                              double aeta, double reta, double info[])
{
    boolean last,first,evaljac,evalcoef;
    int i,st,lastjac,k,q;
    double h,hnew,e,e1,eta,eta1,discr,hl,b;
    double aux[] = new double[4];
    double mu[] = new double[1];
    double mu1[] = new double[1];
    double beta[] = new double[1];
    double p[] = new double[1];
    int pi[] = new int[m+1];
    double dy[] = new double[m+1];
    double yl[] = new double[m+1];
    double yr[] = new double[m+1];
    double f[] = new double[m+1];
    double a[][] = new double[m+1][m+1];

    lastjac=0;
    h=discr=e=e1=0.0;
    first=evaljac=true;
    last=evalcoef=false;
    Basic.inivec(1,9,info,0.0);
    eta=reta*Math.sqrt(Basic.vecvec(1,m,0,y,y))+aeta;
    eta1=eta/Math.sqrt(Math.abs(reta));
    Basic.dupvec(1,m,0,f,y);
    method.derivative(m,f,sigma);
    info[2]=1.0;
    st=1;
    do {
        /* step size */
        if (eta < 0.0) {
            hl=h;
            h=hnew=hmax;
            info[5] += 1.0;
            if (1.1*hnew > xe-x[0]) {
                last=true;
```

```
            h=hnew=xe-x[0];
        }
        evalcoef=(h != hl);
    } else if (first) {
        h=hnew=hmin;
        first=false;
        info[4] += 1.0;
    } else {
        b=discr/eta;
        hl=h;
        if (b < 0.01) b=0.01;
        hnew = (b > 0.0) ? h*Math.pow(b,-1.0/p[0]) : hmax;
        if (hnew < hmin) {
            hnew=hmin;
            info[4] += 1.0;
        } else
            if (hnew > hmax)  {
                hnew=hmax;
                info[5] += 1.0;
            }
        if (1.1*hnew >= xe-x[0]) {
            last=true;
            h=hnew=xe-x[0];
        } else
            if (Math.abs(h/hnew-1.0) > 0.1) h=hnew;
        evalcoef=(h !=hl);
    }
    info[1] += 1.0;
    if (evaljac) {
        method.jacobian(m,j,y,sigma);
        info[3] += 1.0;
        h=hnew;
        liniger1vscoef(m,a,j,aux,pi,h,sigma,mu,mu1,beta,p);
        evaljac=false;
        lastjac=st;
    } else
        if (evalcoef)
            liniger1vscoef(m,a,j,aux,pi,h,sigma,mu,mu1,beta,p);
    i=1;
    do {
        /* iteration */
        if (reta < 0.0) {
            if (i == 1) {
                Basic.mulvec(1,m,0,dy,f,h);
                for (k=1; k<=m; k++) yl[k]=y[k]+mu[0]*dy[k];
                Linear_algebra.sol(a,m,pi,dy);
                e=1.0;
            } else {
                for (k=1; k<=m; k++) dy[k]=yl[k]-y[k]+mu1[0]*f[k];
                if (e*Math.sqrt(Basic.vecvec(1,m,0,y,y)) > e1*e1) {
```

```
                evaljac=(i >= 3);
                if (i > 3) {
                    info[3] += 1.0;
                    method.jacobian(m,j,y,sigma);
                    for (q=1; q<=m; q++) {
                        Basic.mulrow(1,m,q,q,a,j,-mu1[0]);
                        a[q][q] += 1.0;
                    }
                    aux[2]=0.0;
                    Linear_algebra.dec(a,m,aux,pi);
                }
            }
            Linear_algebra.sol(a,m,pi,dy);
        }
        e1=e;
        e=Math.sqrt(Basic.vecvec(1,m,0,dy,dy));
        Basic.elmvec(1,m,0,y,dy,1.0);
        eta=Math.sqrt(Basic.vecvec(1,m,0,y,y))*reta+aeta;
        discr=0.0;
        Basic.dupvec(1,m,0,f,y);
        method.derivative(m,f,sigma);
        info[2] += 1.0;
    } else {
        if (i == 1) {
            /* linearity */
            for (k=1; k<=m; k++) dy[k]=y[k]-mu1[0]*f[k];
            Linear_algebra.sol(a,m,pi,dy);
            Basic.elmvec(1,m,0,dy,y,-1.0);
            e=Math.sqrt(Basic.vecvec(1,m,0,dy,dy));
            if (e*(st-lastjac) > eta) {
                method.jacobian(m,j,y,sigma);
                lastjac=st;
                info[3] += 1.0;
                h=hnew;
                liniger1vscoef(m,a,j,aux,pi,h,sigma,mu,mu1,beta,p);
                /* linearity */
                for (k=1; k<=m; k++) dy[k]=y[k]-mu1[0]*f[k];
                Linear_algebra.sol(a,m,pi,dy);
                Basic.elmvec(1,m,0,dy,y,-1.0);
                e=Math.sqrt(Basic.vecvec(1,m,0,dy,dy));
            }
            evaljac=(e*(st+1-lastjac) > eta);
            Basic.mulvec(1,m,0,dy,f,h);
            for (k=1; k<=m; k++) yl[k]=y[k]+mu[0]*dy[k];
            Linear_algebra.sol(a,m,pi,dy);
            for (k=1; k<=m; k++)
                yr[k]=h*beta[0]*Basic.matvec(1,m,k,j,dy);
            Linear_algebra.sol(a,m,pi,yr);
            Basic.elmvec(1,m,0,yr,dy,1.0);
        } else {
```

```
                for (k=1; k<=m; k++) dy[k]=yl[k]-y[k]+mu1[0]*f[k];
                if (e > eta1 && discr > eta1) {
                    info[3] += 1.0;
                    method.jacobian(m,j,y,sigma);
                    for (q=1; q<=m; q++) {
                        Basic.mulrow(1,m,q,q,a,j,-mu1[0]);
                        a[q][q] += 1.0;
                    }
                    aux[2]=0.0;
                    Linear_algebra.dec(a,m,aux,pi);
                }
                Linear_algebra.sol(a,m,pi,dy);
                e=Math.sqrt(Basic.vecvec(1,m,0,dy,dy));
            }
            Basic.elmvec(1,m,0,y,dy,1.0);
            eta=Math.sqrt(Basic.vecvec(1,m,0,y,y))*reta+aeta;
            eta1=eta/Math.sqrt(reta);
            Basic.dupvec(1,m,0,f,y);
            method.derivative(m,f,sigma);
            info[2] += 1.0;
            for (k=1; k<=m; k++) dy[k]=yr[k]-h*f[k];
            discr=Math.sqrt(Basic.vecvec(1,m,0,dy,dy))/2.0;
        }
        if (i > info[6]) info[6]=i;
        i++;
    } while (e > Math.abs(eta) && discr > 1.3*eta && i <= itmax);
    info[7]=eta;
    info[8]=discr;
    x[0] += h;
    if (discr > info[9]) info[9]=discr;
    method.output(x,xe,m,y,sigma,j,info);
    st++;
    } while (!last);
}


static private void liniger1vscoef(int m, double a[][], double j[][], double aux[],
                                   int pi[], double h, double sigma[], double mu[],
                                   double mu1[], double beta[], double p[])
{
    /* this procedure is internally used by LINIGER1VS */

    int q;
    double b,e;

    b=Math.abs(h*sigma[0]);
    if (b > 40.0) {
        mu[0] = 1.0/b;
        beta[0] = 1.0;
        p[0] = 2.0+2.0/(b-2.0);
```

```
   } else if (b < 0.04) {
       e=b*b/30.0;
       p[0]  = 3.0-e;
       mu[0]  = 0.5-b/12.0*(1.0-e/2.0);
       beta[0]  = 0.5+b/6.0*(1.0-e);
   } else {
       e=Math.exp(b)-1.0;
       mu[0]  = 1.0/b-1.0/e;
       beta[0]  = (1.0-b/e)*(1.0+1.0/e);
       p[0]  = (beta[0]-mu[0])/(0.5-mu[0]);
   }
   mul[0]  = h*(1.0-mu[0]);
   for (q=1; q<=m; q++) {
       Basic.mulrow(1,m,q,q,a,j,-mul[0]);
       a[q][q]  += 1.0;
   }
   aux[2]=0.0;
   Linear_algebra.dec(a,m,aux,pi);
}
```

D. liniger2

Solves an initial value problem, given as an autonomous system of first order differential equations
$$Dy(x) = f(y) \qquad (1)$$
where $D \equiv d/dx$ and $f, y, \in R^m$, with $y(x_0)$ prescribed, over the interval $x0 \leq x \leq xe$ by means of an implicit, second (or possibly third) order accurate, exponentially fitted one-step method [see BeDHKW73, DekHH72, LiW70]. No automatic stepsize control is provided. This procedure is suitable for the integration of stiff differential equations.

The method used has the form
$$y_{n+1} = y_n + (1/2)h[(1-a_n)f_n + (1+a_n)f_{n+1}] + (1/4)h^2[(b_n-a_n)J_nf_n - (b_n+a_n)J_{n+1}f_{n+1}] \qquad (2)$$
where, with $x^n = x_0 + nh$ $(n=0,1,...)$, $y_n \approx y(x_n)$, $f_n = f(y_n)$ and $J_n = \partial f(y_n)/\partial y_n$.
An analysis similar to that given in the documentation of *liniger1vs* reveals that the stability function for the above method has the form
$$R(z) = [1 + (1/2)(1-a_n)z + (1/4)(b_n-a_n)z^2] / [1 - (1/2)(1+a_n)z + (1/4)(b_n+a_n)z^2].$$
Exponential fitting is now possible at two points: with distinct nonzero z_0 and z_1,
$$R(z_0) = e^{z_0} \quad \text{and} \quad R(z_1) = e^{z_1}$$
if $\quad a_n = 2[g(z_1)-g(z_0)]/[z_1g(z_0)-z_0g(z_1)], \quad b_n = 2(z_1-z_0)/[z_1g(z_0)-z_0g(z_1)]$
where $g(z) = z^2(1-e^z)/[e^z(2-z)-(2+z)]$.

It may be shown that the method (2) is stable in three cases. In the first
(a) the eigenvalues of J_n not in the neighborhood of the origin are clustered about two points $\delta_0^{(n)}$ and $\delta_1^{(n)}$ on the negative real axis; a_n and b_n are then obtained as above by setting $z_0=h\delta_0^{(n)}$, $z_1=h\delta_1^{(n)}$;
(b) the above eigenvalues are clustered about two complex conjugate points $\delta_0^{(n)}=\sigma e^{i\varphi}$, $\delta_1^{(n)}=\sigma e^{-i\varphi}$, in the left half plane $\pi/2 < \varphi < 3\pi/2$; a_n and b_n are then determined as in (a); and

(c)   the above eigenvalues are clustered about a single point $\delta_0^{(n)}$ on the negative
real axis; a_n and b_n are then obtained by limiting relationships of the above
form with $z_0 = h\delta_0^{(n)}$ and $z_1=0$.

In cases (a) and (b) above, the method (2) is second order accurate; in case (c) it is
third order accurate.

y_{n+1} in formula (2) is determined by setting $y_{n+1}^{(n)} = y_n$ and using the
modified Newton scheme

$$[4I - 2h(1+a_n)J_{n+1}^{(l)} + h^2(b_n+a_n)(J_{n+1}^{(l)})^2]\Delta y_{n+1}^{(l)} =$$
$$h[2(1+a_n)I - h(b_n+a_n)J_{n+1}^{(l)}]f_{n+1}^{(l)} + 4y_n + 2h(1-a_n)f_n + h^2(b_n-a_n)J_nf_n - 4y_{n+1}^{(l)}$$

where $\Delta_{n+1}^{(l)} = y_{n+1}^{(l+1)} - y_{n+1}^{(l)}$ and $J_{n+1}^{(l)}$ is an approximation to the Jacobian matrix
$\partial f(y_{n+1}^{(l)}/\partial y_{n+1}^{(l)}$. This scheme is continued until conditions analogous to those given
in the documentation to *liniger1vs* are satisfied.

Procedure parameters:
          void liniger2 (*x,xe,m,y,sigma1,sigma2,method,j,k,itmax,step,aeta,reta*)

x: double *x[0:0]*;
          entry: the independent variable; the initial value *x0*;
          exit: the final value *xe*;
xe: double;
          entry: the final value of *x* ($xe \geq x$);
m: int;
          entry: the number of equations;
y: double *y[1:m]*;
          the dependent variable;
          entry: the initial values of the system of differential equations:
                 y[i] at *x=x0*;
          exit: the final values of the solution: *y[i]* at *x=xe*;
sigma1: double *sigma1[0:0]*;
          the modulus of the point at which exponential fitting is desired, this
          point may be complex or real and negative;
sigma2: double *sigma2[0:0]*;
          sigma2 may define three different types of exponential fitting; fitting in
          two complex conjugated points, fitting in two real negative points, or
          fitting in one point combined with third order accuracy;
          if third order is desired then *sigma2* should have the value 0;
          if fitting in a second negative point is desired then *sigma2* should have
          the value of the modulus of this point;
          if fitting in two complex conjugated points is desired then *sigma2* should
          be minus the value of the argument of the point in the second quadrant
          (thus a value between $-\pi$ and $-\pi/2$);
method: a class that defines four procedures *f, evaluate, jacobian*, and *output*, this
          class must implement the AP_liniger2_methods interface;
          double f(int m, double y[ ], int i, double sigma1[ ], double sigma2[ ])
                 this procedure should deliver the right-hand side of the *i*-th
                 differential equation as *f*;
          boolean evaluate (int itnum)
                 evaluate should be assigned the value true if it is desired that the
                 Jacobian of the system is updated in the *itnum*-th iteration step of
                 the Newton process;

void jacobian(int m, double j[ ][ ], double y[ ], double sigma1[ ],
*   double sigma2[ ])*
in this procedure the Jacobian has to be assigned to the array *j*; or an approximation of the Jacobian, if only second order accuracy is required;

void output(double x[ ], double xe, int m, double y[ ], double sigma1[ ],
*   double sigma2[ ], double j[ ][ ], int k[ ])*
this procedure is called at the end of each integration step; the user can ask for output of some parameters, for example, *x*, *y*, *j*, and *k*;

j: double *j[1:m, 1:m]*;
           the Jacobian matrix of the system; array *j* should be updated in the procedure *jacobian*;

k: int *k[0:0]*;
           counts the number of integration steps taken;

itmax: int;
           entry: an upper bound for the number of iterations in Newton's process, used to solve the implicit equations;

step: double;
           entry: the length of the integration step, to be prescribed by the user;

aeta: double;
           entry: required absolute precision in the Newton process, used to solve the implicit equations;

reta: double;
           entry: required relative precision in the Newton process, used to solve the implicit equations.

Procedures used: vecvec, matvec, matmat, dec, sol.

```
package numal;

public interface AP_liniger2_methods {

  double f(int m, double y[], int i, double sigma1[], double sigma2[]);
  boolean evaluate(int itnum);
  void jacobian(int m, double j[][], double y[], double sigma1[], double sigma2[]);
  void output(double x[], double xe, int m, double y[], double sigma1[],
              double sigma2[], double j[][], int k[]);
}


public static void liniger2(double x[], double xe, int m, double y[],
                            double sigma1[], double sigma2[],
                            AP_liniger2_methods method, double j[][], int k[],
                            int itmax, double step, double aeta, double reta)
{
    boolean last;
    int i,itnum;
    double h,hl,jfl,eta,discr;
    double aux[] = new double[4];
```

```
double c0[] = new double[1];
double c1[] = new double[1];
double c2[] = new double[1];
double c3[] = new double[1];
double c4[] = new double[1];
int pi[] = new int[m+1];
double dy[] = new double[m+1];
double yl[] = new double[m+1];
double fl[] = new double[m+1];
double a[][] = new double[m+1][m+1];

last=false;
k[0] = 0;
hl=0.0;
do {
    k[0]++;
    /* step size */
    h=step;
    if (1.1*h >= xe-x[0]) {
        last=true;
        h=xe-x[0];
        x[0]=xe;
    } else
        x[0] += h;
    /* newton iteration */
    itnum=0;
    while (true) {
        itnum++;
        if (method.evaluate(itnum)) {
            method.jacobian(m,j,y,sigma1,sigma2);
            liniger2coef(m,j,a,aux,pi,h,sigma1,sigma2,c0,c1,c2,c3,c4);
        } else
            if (itnum == 1 && h != hl)
                liniger2coef(m,j,a,aux,pi,h,sigma1,sigma2,c0,c1,c2,c3,c4);
        for (i=1; i<=m; i++) fl[i]=method.f(m,y,i,sigma1,sigma2);
        if (itnum == 1)
            for (i=1; i<=m; i++) {
                jfl=Basic.matvec(1,m,i,j,fl);
                dy[i]=h*(fl[i]-c4[0]*jfl);
                yl[i]=y[i]+c2[0]*fl[i]+c3[0]*jfl;
            }
        else
            for (i=1; i<=m; i++)
                dy[i]=yl[i]-y[i]+c1[0]*fl[i]-c0[0]*Basic.matvec(1,m,i,j,fl);
        Linear_algebra.sol(a,m,pi,dy);
        for (i=1; i<=m; i++) y[i] += dy[i];
        if (itnum >= itmax) break;
        eta=Math.sqrt(Basic.vecvec(1,m,0,y,y))*reta+aeta;
        discr=Math.sqrt(Basic.vecvec(1,m,0,dy,dy));
        if (eta >= discr) break;
```

```
        }
        hl=h;
        method.output(x,xe,m,y,sigma1,sigma2,j,k);
    } while (!last);
}



static private void liniger2coef(int m, double j[][], double a[][], double aux[],
                                 int pi[], double h, double sigma1[],
                                 double sigma2[], double c0[], double c1[],
                                 double c2[], double c3[], double c4[])
{
    /* this procedure is internally used by LINIGER2 */

    boolean out,doublefit;
    int i,k;
    double b1,b2,r,r1,r2,ex,zeta,eta,sinl,cosl,sinh,cosh,d,p,q;

    ex=p=q=0.0;
    out=false;
    doublefit=false;
    b1=h*sigma1[0];
    b2=h*sigma2[0];
    if (b1 < 0.1) {
        p=0.0;
        q=1.0/3.0;
        out=true;
    }
    if (!out) {
        if (b2 < 0.0) {
            /* complex */
            eta=Math.abs(b1*Math.sin(sigma2[0]));
            zeta=Math.abs(b1*Math.cos(sigma2[0]));
            if (eta < b1*b1*1.0e-6) {
                b1=b2=zeta;
                doublefit=true;
            }
            if (!doublefit)
                if (zeta > 40.0) {
                    p=1.0-4.0*zeta/b1/b1;
                    q=4.0*(1.0-zeta)/b1/b1+1.0;
                } else {
                    ex=Math.exp(zeta);
                    sinl=Math.sin(eta);
                    cosl=Math.cos(eta);
                    sinh=0.5*(ex-1.0/ex);
                    cosh=0.5*(ex+1.0/ex);
                    d=eta*(cosh-cosl)-0.5*b1*b1*sinl;
                    p=(zeta*sinl+eta*sinh-4.0*zeta*zeta/b1/b1*(cosh-cosl))/d;
                    q=eta*((cosh-cosl-zeta*sinh-eta*sinl)*4.0/b1/b1+cosh+cosl)/d;
```

```
            }
    } else if (b1 < 1.0 || b2 < 0.1) {
        /* third order */
        q=1.0/3.0;
        if (b1 > 40.0)
            r=b1/(b1-2.0);
        else {
            ex=Math.exp(-b1);
            r=b1*(1.0-ex)/(b1-2.0+(b1+2.0)*ex);
        }
        p=r/3.0-2.0/b1;
    } else if (Math.abs(b1-b2) < b1*b1*1.0e-6)
        doublefit=true;
    else {
        if (b1 > 40.0)
            r=b1/(b1-2.0);
        else {
            ex=Math.exp(-b1);
            r=b1*(1.0-ex)/(b1-2.0+(b1+2.0)*ex);
        }
        r1=r*b1;
        if (b2 > 40.0)
            r=b2/(b2-2.0);
        else {
            ex=Math.exp(-b2);
            r=b2*(1.0-ex)/(b2-2.0+(b2+2.0)*ex);
        }
        r2=r*b2;
        d=b2*r1-b1*r2;
        p=2.0*(r2-r1)/d;
        q=2.0*(b2-b1)/d;
    }
    if (doublefit) {
        b1=0.5*(b1+b2);
        if (b1 > 40.0)
            r=b1/(b1-2.0);
        else {
            ex=Math.exp(-b1);
            r=b1*(1.0-ex)/(b1-2.0+(b1+2.0)*ex);
        }
        r1=r;
        if (b1 > 40.0) ex=0.0;
        r2=b1/(1.0-ex);
        r2=1.0-ex*r2*r2;
        q=1.0/(r1*r1*r2);
        p=r1*q-2.0/b1;
    }
}
c0[0] = 0.25*h*h*(p+q);
c1[0] = 0.5*h*(1.0+p);
```

```
c2[0] = h-c1[0];
c3[0] = 0.25*h*h*(q-p);
c4[0] = 0.5*h*p;
for (i=1; i<=m; i++) {
    for (k=1; k<=m; k++)
        a[i][k]=c0[0]*Basic.matmat(1,m,i,k,j,j)-c1[0]*j[i][k];
    a[i][i] += 1.0;
}
aux[2]=0.0;
Linear_algebra.dec(a,m,aux,pi);
}
```

E. gms

Solves an initial value problem, given as an autonomous system of first order differential equations

$$Dy(x) = f(y) \qquad (1)$$

where D denotes differentiation with respect to the indicated argument and y, $f \in R^r$, with $y(x_0)$ prescribed, over the interval $x_0 \le x \le x_e$ by means of a third order, exponentially fitted, generalized multistep method with automatic stepsize control [VhV74, Vw74]. This procedure is suitable for the integration of stiff differential equations.

The procedure *gms* describes an implementation of a third order three-step generalized linear multistep method with quasi-zero parasitic roots and quasi-adaptive stability function. The procedure supplies the additional starting values and performs a stepsize control which is based on the nonlinearity of the differential equation. By this control the Jacobian matrix is incidentally evaluated. It is possible to eliminate the stepsize control. Then, one has to give the number of integration steps per Jacobian evaluation. For linear equations the stepsize control is automatically eliminated, while the procedure performs one evaluation of the Jacobian. However, in this case the three-step scheme is reduced to a one-step scheme. The procedure uses one function evaluation per integration step, and it does not reject integration steps. Each change in the steplength or each reevaluation of the Jacobian costs one LU-decomposition. It is possible to fit exponentially. This fitting is equivalent to fitting in the sense of Liniger and Willoughby, only when the Jacobian matrix is evaluated at each integration step. When the system to be integrated is nonlinear and the Jacobian matrix is not evaluated at each integration step, it is recommended to fit at infinity (input parameter *delta* $\le -10^{15}$).

The method used is one of a class having the form

$$y_{n+1} = R(h_n J_n^*) + h_n \sum_{l=1}^{k} B_l(h_n J_n^*)\left(f(y_{n+1-l}) - J_n^* y_{n+1-l}\right) \qquad (2)$$

where, with $x_{n+1} = x_n + h_n$, $y_n \approx y(x_n)$, J_n^* is an approximation to the Jacobian matrix $J_n = \partial f(y_n)/\partial y_n$, R and the B_l being rational functions.

Considerations of stability demand that $\|B_l(h_n J_n^*)\|$ ($l=1,...,k$) be as small as possible: the stabilizing condition $B_l(z) \to 0$ as $\text{Real}(z) \to -\infty$ ($l=1,...,k$) is imposed.

The method (2) is k-th order accurate if, with
$$q_l = (x_{n-l} - x_n)/h_n \quad (l=0,1,\ldots,k-1)$$
and

$$C_0(z) = R(z) - \sum_{l=1}^{k} z B_l(z)$$

$$C_j(z) = \frac{1}{j!} \sum_{l=1}^{k} (j - z q_{l-1}) q_{l-1}^{j-1} B_l(z)$$

$$C_j^{(i)} = \lim D^i C_j(z) \qquad (z \to 0)$$

the conditions
$$C_j^{(0)} = (1/j!) \ (j=1,\ldots,k) \qquad C_i^{(j-1)} = 0 \ (j=1,\ldots,k; \ i=0,\ldots,j-1)$$
$$\lim D^i R(z) = 1 \quad (z \to 0) \qquad (i=0,\ldots,k) \tag{3}$$
are satisfied or, alternatively, with

$$\sum_{l=1}^{k} q_{l-1}^{j-1} B_l(z) = G_j(z) \qquad (j=1,\ldots,k)$$

$$G_1(z) = [R(z)-1+O(z^{k+1})]/z, \quad G_{j+1}(z) = [jG_j(z)-1-O(z^{k+1-j})]/z \qquad (j=1,\ldots,k-1) \tag{4}$$

as $z \to 0$. The one-step scheme
$$y_{n+1} = y_n + (J^*)^{-1}[R(h_n J_n^*) - I]f(y_n)$$
is second order accurate if $J_n^* = J_n$ exactly, and (3) is satisfied for some $k > 1$.

The local truncation error of the scheme (2) at $x = x_n$ is

$$\sum_{j=1}^{k} \left[\left(\frac{1}{(k+j)!} - c_{k+j}^{(0)} \right) D^{k+j} y(x_n) - \sum_{i=0}^{k+j-1} \frac{c_i^{(k-i+j)}}{(k-i+j)!} (J_n^*)^{k-i+j} D^i y(x_n) \right] h_n^{k+j}$$

The choice of $B_l(z)$ resulting from taking the 0 terms in the expressions (4) for the G_j to be zero minimizes the local truncation error, and in this case
$$c_j(z) = 1/j! \ (j=1,\ldots,k-1), \qquad c_n(z) = 1/k! + O(z)$$
and $c_j^{(i)} = 0 \ (j=0,1,\ldots,k-1; \ i=1,2,\ldots), \ c_k^{(i)} = 0 \ (i=2,3,\ldots)$. The local truncation error then becomes

$$\sum_{j=1}^{k} \left[\left(\frac{1}{(k+j)!} - c_{k+j}^{(0)} \right) D^{k+j} y(x_n) - \sum_{i=1}^{j} c_{k+j-i}^{(i)} (J_n^*)^i D^{k+j-i} y(x_n) \right] h_n^{k+j}$$

The stability function R used by gms is

$$R(z) = \frac{1 + \frac{1}{2}\left(1 - \alpha^{(n)}\right)z + \frac{1}{12}\left(1 - 3\alpha^{(n)}\right)z^2}{1 - \frac{1}{2}\left(1 + \alpha^{(n)}\right)z + \frac{1}{12}\left(1 + 3\alpha^{(n)}\right)z^2}$$

The parameter $\alpha^{(n)}$ is determined by exponential fitting at the point $z = h_n \delta_n$ where δ_n is the eigenvalue of greatest absolute value on the negative real axis of J_n^*:

$$R(h_n \delta_n) = e^{h_n \delta_n} \quad \text{if} \quad \alpha^{(n)} = \frac{1}{3z_n}\left(\frac{e^{z_n}\left(z_n^2 - 6z_n + 12\right) - \left(z_n^2 + 6z_n + 12\right)}{e^{z_n}\left(2 - z_n\right) - \left(2 + z_n\right)} \right)$$

where $z_n = h_n \delta_n$. If $\alpha^{(n)} = 0$, R is fourth order consistent; otherwise third.

Taking the denominator of $R(z)$ to be $Q(z)$, the G_j have, when $k=3$, the form $G_j(z)=(g_{1,j}+g_{2,j}z)/Q(z)$ where $g_{1,j} = 1/j$, $g_{2,1} = -\alpha^{(n)}/2$, $g_{2,2} = g_{2,3} = -(1 + 3\alpha^{(n)})/12$, and then

$$B_l(z) = \sum_{j=1}^{3} b_{l,j} G_j(z) \qquad (l = 1,2,3)$$

where $b_{1,1} = 1$, $\qquad$ $b_{2,1} = b_{3,1} = 0$, $\qquad$ $b_{1,2} = -(q_1+q_2)/q_1q_2$, $\qquad$ $b_{2,2} = -q_2/(q_1^2-q_1q_2)$,
$b_{3,2} = -q_1/(q_2^2-q_1q_2)$, $\qquad$ $b_{1,3} = 1/q_1q_2$, $\qquad$ $b_{2,3} = 1/(q_1^2-q_1q_2)$, $\qquad$ $b_{3,3} = 1/(q_2^2-q_1q_2)$.
The three step, third order scheme of the class (2) with minimized local truncation error then becomes

$$Q\left(h_n J_n^*\right)y_{n+1} = y_n + \sum_{l=1}^{k}\sum_{j=1}^{k} h_n b_{l,j} d_{1,j} f(y_{n+1-l}) +$$

$$h_n^2 (J_n^*)^2 \left[\sum_{l=1}^{k}\sum_{j=1}^{k} \left(h_n b_{l,j} d_{2,j} f(y_{n+1-l}) - b_{l,j} d_{1,j} y_{n+1-l} \right) + \frac{1-\alpha^{(n)}}{2} y_n \right] + \qquad (5)$$

$$h_n^2 (J_n^*)^2 \left[\sum_{l=1}^{k}\sum_{j=1}^{k} \left(-b_{l,j} d_{2,j} y_{n+1-l} \right) + \frac{1-3\alpha^{(n)}}{12} y_n \right]$$

with $k=3$. When $k=2$, the $b_{l,j}$ are given by $b_{1,1} = 1$, $b_{2,1} = 0$, $b_{1,2} = 1 - 1/q_1$, $b_{2,2} = 1/q_1$.

The user is asked to provide an initial stepsize h_0. $J_0^* = J_0$ is then evaluated exactly, and the one-step scheme is used to determine y_1. The two-step scheme (resulting from the substitution $k=2$ above) is used to determine y_2, and thereafter the three-step scheme (5) is used.

The user is also asked to provide two limits $hmin$ and $hmax$ for the minimal and maximal stepsizes, respectively. If these are equal (their particular values are then ignored) the computations are continued using the scheme (5) with constant stepsize h_0. In this case the user must also provide an integer value $nsjev$; the approximation J_n^* is then updated every $nsjev$ steps.

The user is also asked to allocate a value to the parameter $linear$ upon call. If this value is true then a constant stepsize h and a constant Jacobian $J_n^* = J_0$ are used throughout the computation.

If the value allocated to $linear$ is false then the following method of estimating a stepsize is used. If y'_{n+1} were to be determined by use of a two step scheme (with $k=2$ above) and y_{n+1} were to be computed by use of the three-step scheme (5) with stepsize h_n' in both cases, then $\| y_{n+1} - y_{n+1}' \| \approx c_n h_n'^3$ where c_n is independent of h_n'. Thus with a permitted local tolerance tol, and assuming $c_{n-1} = c_n$, the maximal value of h_n' is given by

$h_n' = (tol/c_n)^{1/3}$
$\qquad \approx (tol/c_{n-1})^{1/3}$
$\qquad = h_{n-1}(tol/\| y_n-y_n' \|)^{1/3}$
$\qquad \approx h_{n-1}[((4/3)tol)/(tol + \| y_n-y_n' \|) + 1/3]$

where y_n' has been computed as above (from y_{n-1}) by means of a two-step method with stepsize h_{n-1}. The bound $tol = \eta_a + \eta_r \| y_n \|$ is used, where η_a and η_r are absolute and relative tolerances prescribed by the user. A new stepsize is chosen (i.e., h_n is taken to be h_n' rather than h_{n-1}) if $\left| (h_{n-1}-h_n')/h_{n-1} \right| > 0.1$. A new J_n^* is determined if (i) h_n is to be changed and a new J_n^* was not determined at the preceding step, and (ii) $10\left| h_{n-1}-h_n' \right| < 0.1\left| h_{n-1} \right|$. (Naturally h_n is taken to be $hmin$ or $hmax$ as is appropriate, if h_n' lies outside the range $hmin \le h_n' \le hmax$).

Formula (5) represents a system of linear equations. The matrix $Q(h_n J_n^*)$ is decomposed in LU form, and y_{n+1} is obtained from this decomposition. The parameters jev and lu record the current number of evaluations of J_n^* and of LU decompositions (since h_n may be changed while J_n^* remains unchanged, these two numbers may differ).

Procedure parameters:

> void gms (*x,xe,r,y,h,hmin,hmax,delta, aeta,reta,n,jev,lu,nsjev,linear*)

x: double *x[0:0]*;

>   entry: the independent variable; the initial value *x0*;
>   exit: the final value *xe*;

xe: double;

>   entry: the final value of x $(xe \geq x)$;

r: int;

>   entry: the number of differential equations;

y: double *y[1:r]*;

>   the dependent variable;
>   entry: the initial value of y;
>   exit: the solution y at the point x after each integration step;

h: double;

>   entry: the steplength when the integration has to be performed without the stepsize mechanism; otherwise, the initial steplength (see *hmin* and *hmax* below);

hmin,hmax: double;

>   entry: minimal and maximal steplength by which the integration is allowed to be performed; by putting *hmin=hmax* the stepsize mechanism is eliminated; in this case the given values for *hmin* and *hmax* are irrelevant, while the integration is performed with the steplength given by *h*;

delta: double *delta[0:0]*;

>   entry: the real part of the point at which exponential fitting is desired; alternatives:
>   *delta*=(an estimate of) the real part of the largest eigenvalue in modulus of the Jacobian matrix of the system;
> $delta \leq -10^{15}$, in order to obtain asymptotic stability;
>   *delta*=0, in order to obtain a higher order of accuracy in case of linear equations;

method: a class that defines three procedures *derivative*, *jacobian*, and *out*, this class must implement the `AP_gms_methods` interface;

>   *void derivative(int r, double y[ ], double delta[ ])*
>   this procedure should replace the *i*-th component of the solution y by the *i*-th component of the derivative $f(y)$, $i=1,...,r$;
>   *void jacobian(int r, double j[ ][ ], double y[ ], double delta[ ])*
>   in *gms* when this procedure is called, the array y contains the values of the dependent variable; upon completion of a call of *jacobian* the array j should contain the values of the Jacobian matrix of $f(y)$;
>   *void out(double x[ ], double xe, int r, double y[ ],double delta[ ],*
>   *int n[ ],int jev[ ], int lu[ ])*
>   this procedure is called after each integration step; the user can ask for output of some parameters, for example, x, y, *delta*, n, *jev*, and *lu*;

aeta,reta: double;
            entry: measure of the absolute and relative local accuracy required;
                     these values are irrelevant when the integration is performed
                     without the stepsize mechanism;
n: int *n[0:0]*;
            exit: the number of integration steps;
jev: int *jev[0:0]*;
            exit: the number of Jacobian evaluations;
lu: int *lu[0:0]*;
            exit: the number of LU-decompositions;
nsjev: int;
            entry: number of integration steps per Jacobian evaluation;
                     the value of *nsjev* is relevant only when the integration is
                     performed without the stepsize mechanism and the system to
                     be solved is nonlinear;
linear: boolean;
            entry: *linear* equals true when the system to be integrated is linear,
                     otherwise false;
                     if *linear* is true the stepsize mechanism is automatically
                     eliminated.

Procedures used: vecvec, matvec, matmat, elmrow, elmvec, dupvec, gsselm,
                        solelm, colcst, mulvec.

```
package numal;

public interface AP_gms_methods {

    void derivative(int r, double y[], double delta[]);
    void jacobian(int r, double j[][], double y[], double delta[]);
    void out(double x[], double xe, int r, double y[], double delta[], int n[],
            int jev[], int lu[]);
}


public static void gms(double x[], double xe, int r, double y[], double h,
                        double hmin, double hmax, double delta[],
                        AP_gms_methods method, double aeta, double reta, int n[],
                        int jev[], int lu[], int nsjev, boolean linear)
{
    boolean strategy,change;
    int k,l,count,count1,kchange;
    double a,x0,eta,h0,h1,discr;
    double aux[] = new double[10];
    boolean reeval[] = new boolean[1];
    boolean update[] = new boolean[1];
    int nsjev1[] = new int[1];
    double alfa[] = new double[1];
    double s1[] = new double[1];
    double s2[] = new double[1];
```

```
double xl0[] = new double[1];
double xl1[] = new double[1];
double q1[] = new double[1];
double q2[] = new double[1];
int ri[] = new int[r+1];
int ci[] = new int[r+1];
double y1[] = new double[r+1];
double y0[] = new double[r+1];
double yl[] = new double[3*r+1];
double fl[] = new double[3*r+1];
double bd1[][] = new double[4][4];
double bd2[][] = new double[4][4];
double hjac[][] = new double[r+1][r+1];
double h2jac2[][] = new double[r+1][r+1];
double rqz[][] = new double[r+1][r+1];

/* initialization */
eta=0.0;
lu[0]=jev[0]=n[0]=nsjev1[0]=kchange=0;
x0=x[0];
discr=0.0;
k=1;
h1=h0=h;
count = -2;
aux[2]=Double.MIN_VALUE;
aux[4]=8.0;
Basic.dupvec(1,r,0,yl,y);
reeval[0]=change=true;
strategy=((hmin != hmax) && !linear);
q1[0] = -1.0;
q2[0] = -2.0;
count1=0;
xl0[0]=xl1[0]=0.0;
method.out(x,xe,r,y,delta,n,jev,lu);
x[0] += h1;
/* operator construction */
gmsopconstruct(reeval,update,r,hjac,h2jac2,rqz,y,aux,ri,ci,lu,jev,nsjev1,delta,
               alfa,h1,h0,s1,s2,method);
bd1[1][1]=1.0;
bd2[1][1] = -alfa[0]*0.5;
if (!linear)
    gmscoefficient(xl1,xl0,x0,change,n,q1,q2,h1,alfa,bd1,bd2,strategy);
while (true) {
    gmsdiffscheme(k,count,r,fl,yl,n,nsjev1,y0,alfa,bd1,bd2,h1,y,hjac,h2jac2,
                  rqz,ri,ci,delta,method);
    if (strategy) count++;
    if (count == 1) {
        /* test accuracy */
        k=2;
        Basic.dupvec(1,r,0,yl,y);
```

```
        gmsdiffscheme(k,count,r,fl,yl,n,nsjevl,y0,alfa,bd1,bd2,h1,y,hjac,h2jac2,
                    rqz,ri,ci,delta,method);
    k=3;
    eta=aeta+reta*Math.sqrt(Basic.vecvec(1,r,0,yl,yl));
    Basic.elmvec(1,r,0,y,yl,-1.0);
    discr=Math.sqrt(Basic.vecvec(1,r,0,y,y));
    Basic.dupvec(1,r,0,y,yl);
}
method.out(x,xe,r,y,delta,n,jev,lu);
if (x[0] >= xe) break;
/* step size */
x0=x[0];
h0=h1;
if ((n[0] <= 2) && !linear) (k)++;
if (count == 1) {
    a=eta/(0.75*(eta+discr))+0.33;
    h1 = (a <= 0.9 || a >= 1.1) ? a*h0 : h0;
    count=0;
    reeval[0]=(a <= 0.9 && nsjevl[0] != 1);
    count1 = (a >= 1.0 || reeval[0]) ? 0 : count1+1;
    if (count1 == 10) {
        count1=0;
        reeval[0]=true;
        h1=a*h0;
    }
} else {
    h1=h;
    reeval[0]=((nsjev == nsjevl[0]) && !strategy && !linear);
}
if (strategy)
    h1 = (h1 > hmax) ? hmax : ((h1 < hmin) ? hmin : h1);
x[0] += h1;
if (x[0] >= xe) {
    h1=xe-x0;
    x[0]=xe;
}
if ((n[0] <= 2) && !linear) reeval[0]=true;
if (h1 != h0) {
    update[0]=true;
    kchange=3;
}
if (reeval[0]) update[0]=true;
change=((kchange > 0) && !linear);
kchange--;
if (update[0])
    /* operator construction */
    gmsopconstruct(reeval,update,r,hjac,h2jac2,rqz,y,aux,ri,ci,lu,jev,
                    nsjevl,delta,alfa,h1,h0,s1,s2,method);
if (!linear)
    gmscoefficient(xl1,xl0,x0,change,n,q1,q2,h1,alfa,bd1,bd2,strategy);
```

```
        /* next integration step */
        for (l=2; l>=1; l--) {
            Basic.dupvec(l*r+1,(l+1)*r,-r,yl,yl);
            Basic.dupvec(l*r+1,(l+1)*r,-r,fl,fl);
        }
        Basic.dupvec(1,r,0,yl,y);
    }
}


static private void gmsopconstruct(boolean reeval[], boolean update[], int r,
                                   double hjac[][], double h2jac2[][],
                                   double rqz[][], double y[], double aux[],
                                   int ri[], int ci[], int lu[], int jev[],
                                   int nsjev1[], double delta[], double alfa[],
                                   double h1, double h0, double s1[], double s2[],
                                   AP_gms_methods method)
{
    /* this procedure is internally used by GMS */

    int i,j;
    double a,a1,z1,e,q;

    if (reeval[0]) {
        method.jacobian(r,hjac,y,delta);
        jev[0]++;
        nsjev1[0] = 0;
        if (delta[0] <= 1.0e-15)
            alfa[0]=1.0/3.0;
        else {
            z1=h1*delta[0];
            a=z1*z1+12.0;
            a1=6.0*z1;
            if (Math.abs(z1) < 0.1)
                alfa[0]=(z1*z1/140.0-1.0)*z1/30.0;
            else if (z1 < -33.0)
                alfa[0]=(a+a1)/(3.0*z1*(2.0+z1));
            else {
                e=Math.exp(z1);
                alfa[0]=((a-a1)*e-a-a1)/(((2.0-z1)*e-2.0-z1)*z1*3.0);
            }
        }
        s1[0] = -(1.0+alfa[0])*0.5;
        s2[0]=(alfa[0]*3.0+1.0)/12.0;
    }
    a=h1/h0;
    a1=a*a;
    if (reeval[0]) a=h1;
    if (a != 1.0)
        for (j=1; j<=r; j++) Basic.colcst(1,r,j,hjac,a);
```

```
    for (i=1; i<=r; i++) {
        for (j=1; j<=r; j++) {
            q=h2jac2[i][j]=(reeval[0] ? Basic.matmat(1,r,i,j,hjac,hjac) :
                                        h2jac2[i][j]*a1);
            rqz[i][j]=s2[0]*q;
        }
        rqz[i][i] += 1.0;
        Basic.elmrow(1,r,i,i,rqz,hjac,s1[0]);
    }
    Linear_algebra.gsselm(rqz,r,aux,ri,ci);
    lu[0]++;
    reeval[0]=update[0]=false;
}


static private void gmscoefficient(double xl1[], double xl0[], double x0,
                                   boolean change, int n[], double q1[],
                                   double q2[], double h1, double alfa[],
                                   double bd1[][], double bd2[][], boolean strategy)
{
    /* this procedure is internally used by GMS */

    double a,q12,q22,q1q2,xl2;

    xl2=xl1[0];
    xl1[0]=xl0[0];
    xl0[0]=x0;
    if (change) {
        if (n[0] > 2) {
            q1[0]=(xl1[0]-xl0[0])/h1;
            q2[0]=(xl2-xl0[0])/h1;
        }
        q12=q1[0]*q1[0];
        q22=q2[0]*q2[0];
        q1q2=q1[0]*q2[0];
        a = -(3.0*alfa[0]+1.0)/12.0;
        bd1[1][3]=1.0+(1.0/3.0-(q1[0]+q2[0])*0.5)/q1q2;
        bd1[2][3]=(1.0/3.0-q2[0]*0.5)/(q12-q1q2);
        bd1[3][3]=(1.0/3.0-q1[0]*0.5)/(q22-q1q2);
        bd2[1][3] = -alfa[0]*0.5+a*(1.0-q1[0]-q2[0])/q1q2;
        bd2[2][3]=a*(1.0-q2[0])/(q12-q1q2);
        bd2[3][3]=a*(1.0-q1[0])/(q22-q1q2);
        if (strategy || n[0] <= 2) {
            bd1[2][2]=1.0/(2.0*q1[0]);
            bd1[1][2]=1.0-bd1[2][2];
            bd2[2][2] = -(3.0*alfa[0]+1.0)/(12.0*q1[0]);
            bd2[1][2] = -bd2[2][2]-alfa[0]*0.5;
        }
    }
}
```

```
static private void gmsdiffscheme(int k, int count, int r, double fl[], double yl[],
                        int n[], int nsjev1[], double y0[], double alfa[],
                        double bd1[][], double bd2[][], double h1,
                        double y[], double hjac[][], double h2jac2[][],
                        double rqz[][], int ri[], int ci[],
                        double delta[], AP_gms_methods method)
{
    /* this procedure is internally used by GMS */

    int i,l;

    if (count != 1) {
      Basic.dupvec(1,r,0,fl,yl);
      method.derivative(r,fl,delta);
      n[0]++;
      nsjev1[0]++;
    }
    Basic.mulvec(1,r,0,y0,yl,(1.0-alfa[0])/2.0-bd1[1][k]);
    for (l=2; l<=k; l++)
      Basic.elmvec(1,r,r*(l-1),y0,yl,-bd1[l][k]);
    for (l=1; l<=k; l++)
      Basic.elmvec(1,r,r*(l-1),y0,fl,h1*bd2[l][k]);
    for (i=1; i<=r; i++) y[i]=Basic.matvec(1,r,i,hjac,y0);
    Basic.mulvec(1,r,0,y0,yl,(1.0-3.0*alfa[0])/12.0-bd2[1][k]);
    for (l=2; l<=k; l++)
      Basic.elmvec(1,r,r*(l-1),y0,yl,-bd2[l][k]);
    for (i=1; i<=r; i++) y[i] += Basic.matvec(1,r,i,h2jac2,y0);
    Basic.dupvec(1,r,0,y0,yl);
    for (l=1; l<=k; l++)
      Basic.elmvec(1,r,r*(l-1),y0,fl,h1*bd1[l][k]);
    Basic.elmvec(1,r,0,y,y0,1.0);
    Linear_algebra.solelm(rqz,r,ri,ci,y);
}
```

F. impex

Solves an initial value problem, given as an autonomous system of first order differential equations

$$dy_i(t)/dt = f_i(t,y(t)) \qquad (i=1,...,n)$$

with $y(t_0)$ prescribed, over the interval $t_0 \le t \le tend$ by means of an implicit midpoint rule with smoothing and extrapolation. Automatic stepsize control is provided. This procedure is suitable for the integration of stiff differential equations.

The integration method [see BeDHV74, Li73] is based on the computation of two independent solutions with stepsizes h and $h/2$ by the implicit midpoint rule. Passive smoothing and passive extrapolation are performed to obtain stability and high accuracy. The algorithm uses for each step at least three function evaluations, and on change of stepsize or at slow convergence in the iteration process an approximation of the Jacobian matrix (computed by divided differences or explicitly specified by the user). If the computed local error exceeds tolerance, the last step is rejected. Moreover, two global errors are computed.

Procedure parameters:

    void impex (*n,t0,tend,y0,method, h0,hmax,presch,eps,weights,fail*)

n: int;
    entry: the number of equations;
t0:   double;
    entry: the initial value of the independent variable;
tend: double;
    entry: the final value of the independent variable;
y0:   double *y0[1:n]*;
    entry: the initial values of the system of differential equations:
      y0[i] at *t=t0*;
method: a class that defines four procedures *deriv, available, update*, and *control*,
    this class must implement the AP_impex_methods interface;
    void deriv(double t, double y[ ], double f[ ], int n)
      this procedure should deliver the values of *f(t,y)* in the array *f[1:n]*;
    boolean available(double t, double y[ ], double a[ ][ ], int n)
      if an analytic expression of the Jacobian matrix at the point *(t,y)* is not available then this procedure should deliver the value false; otherwise, this procedure should deliver the value true, and the Jacobian matrix should be assigned to the array *a[1:n,1:n]*;
    void update(double weights[ ], double y2[ ], int n)
      this procedure may change the arrays *weights*, according to the value of an approximation to *y(t)*, given in the array *y2[1:n]*;
    void control(double tp[ ], double t, double h, double hnew,
      double y[ ][ ], double err[ ], int n, double tend)
      control is called on entry to *impex* and further as soon as the inequality *tp ≤ t* holds; during a call of *control*, printing of results and change of stepsize (if *presch* has the value true) is then possible; the meaning of the parameters of *control* is:
    tp: double *tp[0:0]*;
      entry: the value of the independent variable at which a call of *control* is desired;
      exit: a new value *(tp > t)* at which a call of *control* is desired;
    t: double;
      the actual value of the independent variable, up to which integration has been performed;
    h: double;
      halve the actual stepsize;
    hnew: double;
      the new stepsize; if *presch* is true then the user may prescribe a new stepsize by changing *hnew*;
    y: double *y[1:5,1:n]*;
      the value of the dependent variable and its first four divided differences at the point *t* are given in this array;

error: double *error[1:3]*;
the elements of this array contain the following errors:
error[1]:   the local error;
error[2]:   the global error of second order in *h*;
error[3]:   the global error of fourth order in *h*;

n: int;
the number of equations;

tend: double;
the final value of the independent variable;

h0:   double;
entry: the initial stepsize;

hmax: double;
entry: maximal stepsize by which the integration is performed;

presch: boolean;
entry: indicator for choice of stepsize;
the stepsize is automatically controlled if *presch* equals false; otherwise the stepsize has to be prescribed, either only initially or also by the procedure *control*;

eps: double;
entry: bound for the estimate of the local error;

weights: double *weights[1:n]*;
entry: the initial weights for the computation of the weighted Euclidean norm of the errors; note that the choice *weights[i]*=1 implies an estimation of the absolute error, whereas *weights[i]*=*y[i]* defines a relative error;

fail: boolean *fail[0:0]*;
exit: if the procedure fails to solve the system of equations, *fail* will have the value true upon exit; this may occur upon divergence of the iteration process used in the midpoint rule, while integration is performed with a user defined prescribed stepsize.

Procedures used: inivec, inimat, mulvec, mulrow, dupvec, duprowvec, dupmat, vecvec, matvec, matmat, elmvec, elmrow, dec, sol.

```
package numal;

public interface AP_impex_methods {

    void deriv(double t, double y[], double f[], int n);
    boolean available(double t, double y[], double a[][], int n);
    void update(double weights[], double y2[], int n);
    void control(double tp[], double t, double h, double hnew, double y[][],
                double err[], int n, double tend);
}


public static void impex(int n, double t0, double tend, double y0[],
                AP_impex_methods method, double h0, double hmax,
                boolean presch, double eps, double weights[],
                boolean fail[])
```

```
{
    boolean start,two,halv,gotoMstp;
    int i,k,eci;
    double t,h,h2,hnew,alf,lq,alf1,c0,c1,c2,c3,b0,b1,b2,b3,w,sl1,sn,lr;
    double err[] = new double[4];
    double e[] = new double[5];
    double d[] = new double[5];
    double t1[] = new double[1];
    double t2[] = new double[1];
    double t3[] = new double[1];
    double tp[] = new double[1];
    int ps1[] = new int[n+1];
    int ps2[] = new int[n+1];
    double y[] = new double[n+1];
    double z[] = new double[n+1];
    double s1[] = new double[n+1];
    double s2[] = new double[n+1];
    double s3[] = new double[n+1];
    double u1[] = new double[n+1];
    double u3[] = new double[n+1];
    double w1[] = new double[n+1];
    double w2[] = new double[n+1];
    double w3[] = new double[n+1];
    double ehr[] = new double[n+1];
    double r[][] = new double[6][n+1];
    double rf[][] = new double[6][n+1];
    double a1[][] = new double[n+1][n+1];
    double a2[][] = new double[n+1][n+1];
    double kof[][] = new double[5][5];

    eci=0;
    start=halv=two=true;
    if (presch)
        h=h0;
    else {
        if (h0 > hmax)
            h=hmax;
        else
            h=h0;
        if (h > (tend-t0)/4.0) h=(tend-t0)/4.0;
    }
    hnew=h;
    alf=0.0;
    t=tp[0]=t0;
    Basic.inivec(1,3,err,0.0);
    Basic.inivec(1,n,ehr,0.0);
    Basic.duprowvec(1,n,1,r,y0);
    method.control(tp,t,h,hnew,r,err,n,tend);

    Init:  for (;;) {
```

```
/* initialization */
h2=hnew;
h=h2/2.0;
Basic.dupvec(1,n,0,s1,y0);
Basic.dupvec(1,n,0,s2,y0);
Basic.dupvec(1,n,0,s3,y0);
Basic.dupvec(1,n,0,w1,y0);
Basic.duprowvec(1,n,1,r,y0);
Basic.inivec(1,n,u1,0.0);
Basic.inivec(1,n,w2,0.0);
Basic.inimat(2,5,1,n,r,0.0);
Basic.inimat(1,5,1,n,rf,0.0);
t=t1[0]=t0;
t2[0]=t0-2.0*h-1.0e-6;
t3[0]=2.0*t2[0]+1.0;
gotoMstp=false;
if (impexrecomp(a1,h,t,s1,ps1,n,method)) {
    fail[0] = presch;
    if (fail[0]) return;
    if (eci > 1) t -= h2;
    halv=two=false;
    hnew=h2/2.0;
    if (start)
        continue Init;
    else {
        if (tp[0] <= t)
        method.control(tp,t,h,hnew,r,err,n,tend);
        if (start) start=false;
        if (hnew == h2) t += h2;
        eci++;
        if (t >= tend+h2) return;
        gotoMstp=true;
    }
}
if (!gotoMstp) {
    if (impexrecomp(a2,h2,t,w1,ps2,n,method)) {
        fail[0] = presch;
        if (fail[0]) return;
        if (eci > 1) t -= h2;
        halv=two=false;
        hnew=h2/2.0;
        if (start)
            continue Init;
        else {
            if (tp[0] <= t)
            method.control(tp,t,h,hnew,r,err,n,tend);
            if (start) start=false;
            if (hnew == h2) t += h2;
            eci++;
```

```
                         if (t >= tend+h2) return;
                 }
         } else {
             start=true;
             for (eci=0; eci<=3; eci++) {
                 /* one large step */
                 if (impexlargestep(n,y,t,t1,t2,t3,s1,s2,s3,h,h2,z,u1,u3,w1,w2,
                                     w3,ps1,ps2,weights, a1,a2,eps,method)) {
                     fail[0] = presch;
                     if (fail[0]) return;
                     if (eci > 1) t -= h2;
                     halv=two=false;
                     hnew=h2/2.0;
                     if (start)
                         continue Init;
                     else {
                         if (tp[0] <= t)
                             method.control(tp,t,h,hnew,r,err,n,tend);
                         if (start) start=false;
                         if (hnew == h2) t += h2;
                         eci++;
                         if (t >= tend+h2) return;
                         gotoMstp=true;
                         break;
                     }
                 } else {
                     t += h2;
                     if (eci > 0) {
                         /* backward differences */
                         impexbackdiff(n,u1,u3,w1,w2,w3,s1,s2,s3,r,rf);
                         method.update(weights,s2,n);
                     }
                 }
             }
             if (!gotoMstp)
                 eci=4;
         }
     }
Mstp:  for (;;) {
         if (hnew != h2) {
             eci=1;
             /* change of information */
             c1=hnew/h2;
             c2=c1*c1;
             c3=c2*c1;
             kof[2][2]=c1;
             kof[2][3]=(c1-c2)/2.0;
             kof[2][4]=c3/6.0-c2/2.0+c1/3.0;
             kof[3][3]=c2;
             kof[3][4]=c2-c3;
```

```
kof[4][4]=c3;
for (i=1; i<=n; i++)
    u1[i]=r[2][i]+r[3][i]/2.0+r[4][i]/3.0;
alf1=Basic.matvec(1,n,1,rf,u1)/Basic.vecvec(1,n,0,u1,u1);
alf=(alf+alf1)*c1;
for (i=1; i<=n; i++) {
    e[1]=rf[1][i]-alf1*u1[i];
    e[2]=rf[2][i]-alf1*2.0*r[3][i];
    e[3]=rf[3][i]-alf1*4.0*r[4][i];
    e[4]=rf[4][i];
    d[1]=r[1][i];
    rf[1][i] = e[1] *= c2;
    for (k=2; k<=4; k++) {
        r[k][i]=d[k]=Basic.matmat(k,4,k,i,kof,r);
        rf[k][i]=e[k]=c2*Basic.matvec(k,4,k,kof,e);
    }
    s1[i]=d[1]+e[1];
    w1[i]=d[1]+4.0*e[1];
    s2[i]=s1[i]-(d[2]+e[2]/2.0);
    s3[i]=s2[i]-(d[2]+e[2])+(d[3]+e[3]/2.0);
}
t3[0]=t-hnew;
t2[0]=t-hnew/2.0;
t1[0]=t;
h2=hnew;
h=h2/2.0;
err[1]=0.0;
if (halv) {
    for (i=1; i<=n; i++) ps2[i]=ps1[i];
    Basic.dupmat(1,n,1,n,a2,a1);
}
if (two) {
    for (i=1; i<=n; i++) ps1[i]=ps2[i];
    Basic.dupmat(1,n,1,n,a1,a2);
} else
    if (impexrecomp(a1,hnew/2.0,t,s1,ps1,n,method)) {
        fail[0] = presch;
        if (fail[0]) return;
        if (eci > 1) t -= h2;
        halv=two=false;
        hnew=h2/2.0;
        if (start)
            continue Init;
        else {
            if (tp[0] <= t)
                method.control(tp,t,h,hnew,r,err,n,tend);
            if (start) start=false;
            if (hnew == h2) t += h2;
            eci++;
            if (t >= tend+h2) return;
```

```
                          continue Mstp;
                  }
              }
          if (!halv)
              if (impexrecomp(a2,hnew,t,w1,ps2,n,method)) {
                  fail[0] = presch;
                  if (fail[0]) return;
                  if (eci > 1) t -= h2;
                  halv=two=false;
                  hnew=h2/2.0;
                  if (start)
                      continue Init;
                  else {
                      if (tp[0] <= t)
                          method.control(tp,t,h,hnew,r,err,n,tend);
                      if (start) start=false;
                      if (hnew == h2) t += h2;
                      eci++;
                      if (t >= tend+h2) return;
                      continue Mstp;
                  }
              }
          /* one large step */
          if (impexlargestep(n,y,t,t1,t2,t3,s1,s2,s3,h,h2,z,u1,u3,w1,w2,w3,
                            ps1,ps2,weights,a1,a2,eps,method)) {
              fail[0] = presch;
              if (fail[0]) return;
              if (eci > 1) t -= h2;
              halv=two=false;
              hnew=h2/2.0;
              if (start)
                  continue Init;
              else {
                  if (tp[0] <= t)
                      method.control(tp,t,h,hnew,r,err,n,tend);
                  if (start) start=false;
                  if (hnew == h2) t += h2;
                  eci++;
                  if (t >= tend+h2) return;
                  continue Mstp;
              }
          }
          t += h2;
          eci=2;
      }
      /* one large step */
      if (impexlargestep(n,y,t,t1,t2,t3,s1,s2,s3,h,h2,z,u1,u3,w1,w2,w3,
                        ps1,ps2,weights,a1,a2,eps,method)) {
          fail[0] = presch;
          if (fail[0]) return;
```

```
            if (eci > 1) t -= h2;
            halv=two=false;
            hnew=h2/2.0;
            if (start)
              · continue Init;
            else {
                if (tp[0] <= t)
                    method.control(tp,t,h,hnew,r,err,n,tend);
                if (start) start=false;
                if (hnew == h2) t += h2;
                eci++;
                if (t >= tend+h2) return;
                continue Mstp;
            }
        }
        /* backward differences */
        impexbackdiff(n,u1,u3,w1,w2,w3,s1,s2,s3,r,rf);
        method.update(weights,s2,n);
        /* error estimates */
        c0=c1=c2=c3=0.0;
        for (i=1; i<=n; i++) {
            w=weights[i]*weights[i];
            b0=rf[4][i]/36.0;
            c0 += b0*b0*w;
            lr=Math.abs(b0);
            b1=rf[1][i]+alf*r[2][i];
            c1 += b1*b1*w;
            b2=rf[3][i];
            c2 += b2*b2*w;
            sl1=Math.abs(rf[1][i]-rf[2][i]);
            sn = (sl1 < 1.0e-10) ? 1.0 : Math.abs(rf[1][i]-r[4][i]/6.0)/sl1;
            if (sn > 1.0) sn=1.0;
            if (start) {
                sn *= sn*sn*sn;
                lr *= 4.0;
            }
            ehr[i]=b3=sn*ehr[i]+lr;
            c3 += b3*b3*w;
        }
        b0=err[1];
        err[1]=b1=Math.sqrt(c0);
        err[2]=Math.sqrt(c1);
        err[3]=Math.sqrt(c3)+Math.sqrt(c2)/2.0;
        lq=eps/((b0 < b1) ? b1 : b0);
        if (b0 < b1 && lq >= 80.0) lq=10.0;
        if (eci < 4 && lq > 80.0) lq=20.0;
        halv=two=false;
        if (!presch) {
            if (lq < 1.0) {
                /* reject */
```

```
                    if (start) {
                        hnew=Math.pow(lq,1.0/5.0)*h/2.0;
                        continue Init;
                    } else {
                        for (k=1; k<=4; k++) Basic.elmrow(1,n,k,k+1,r,r,-1.0);
                        for (k=1; k<=3; k++) Basic.elmrow(1,n,k,k+1,r,r,-1.0);
                        for (k=1; k<=4; k++) Basic.elmrow(1,n,k,k+1,rf,rf,-1.0);
                        t -= h2;
                        halv=true;
                        hnew=h;
                        continue Mstp;
                    }
                } else {
                    /* step size */
                    if (lq < 2.0) {
                        halv=true;
                        hnew=h;
                    } else {
                        if (lq > 80.0)
                            hnew=((lq > 5120.0) ? Math.pow(lq/5.0,1.0/5.0) : 2.0)*h2;
                        if (hnew > hmax) hnew=hmax;
                        if (tend > t && tend-t < hnew) hnew=tend-t;
                        two=(hnew == 2.0*h2);
                    }
                }
            }
            if (tp[0] <= t) method.control(tp,t,h,hnew,r,err,n,tend);
            if (start) start=false;
            if (hnew == h2) t += h2;
            eci++;
            if (t >= tend+h2) return;
            continue Mstp;
        } /*  Mstp loop */
    } /*  Init loop */
}


static private boolean impexrecomp(double a[][], double h, double t, double y[],
                                   int ps[], int n, AP_impex_methods method)
{
    /* this procedure is internally used by IMPEX */

    int i,j;
    double sl,ss;
    double aux[] = new double[4];

    sl=h/2.0;
    if (!method.available(t,y,a,n)) {
        double f1[] = new double[n+1];
        double f2[] = new double[n+1];
```

```
            method.deriv(t,y,f1,n);
            for (i=1; i<=n; i++) {
                ss=1.0e-6*y[i];
                if (Math.abs(ss) < 1.0e-6) ss=1.0e-6;
                y[i] += ss;
                method.deriv(t,y,f2,n);
                for (j=1; j<=n; j++) a[j][i]=(f2[j]-f1[j])/ss;
                y[i] -= ss;
            }
        }
        for (i=1; i<=n; i++) {
            Basic.mulrow(1,n,i,i,a,a,-sl);
            a[i][i] += 1.0;
        }
        aux[2]=1.0e-14;
        Linear_algebra.dec(a,n,aux,ps);
        if (aux[3] < n)
            return true;
        else
            return false;
    }



    static private boolean impexlargestep(int n, double y[], double t, double t1[],
                                  double t2[], double t3[], double s1[],
                                  double s2[], double s3[], double h, double h2,
                                  double z[], double u1[], double u3[],
                                  double w1[], double w2[], double w3[],
                                  int ps1[], int ps2[], double weights[],
                                  double a1[][], double a2[][], double eps,
                                  AP_impex_methods method)
    {
        /* this procedure is internally used by IMPEX */

        double a,b,c;

        a=(t+h-t1[0])/(t1[0]-t2[0]);
        b=(t+h-t2[0])/(t1[0]-t3[0]);
        c=(t+h-t1[0])/(t2[0]-t3[0])*b;
        b *= a;
        a += 1.0+b;
        b=a+c-1.0;
        Basic.mulvec(1,n,0,z,s1,a);
        Basic.elmvec(1,n,0,z,s2,-b);
        Basic.elmvec(1,n,0,z,s3,c);
        if (impexiterate(z,s1,a1,h,t+h/2.0,weights,ps1,n,eps,method)) return true;
        Basic.dupvec(1,n,0,y,z);
        a=(t+h2-t1[0])/(t1[0]-t2[0]);
        b=(t+h2-t2[0])/(t1[0]-t3[0]);
        c=(t+h2-t1[0])/(t2[0]-t3[0])*b;
```

```
    b *= a;
    a += 1.0+b;
    b=a+c-1.0;
    Basic.mulvec(1,n,0,z,s1,a);
    Basic.elmvec(1,n,0,z,s2,-b);
    Basic.elmvec(1,n,0,z,s3,c);
    if (impexiterate(z,y,a1,h,t+3.0*h/2.0,weights,ps1,n,eps,method)) return true;
    Basic.dupvec(1,n,0,u3,u1);
    Basic.dupvec(1,n,0,u1,y);
    Basic.dupvec(1,n,0,s3,s2);
    Basic.dupvec(1,n,0,s2,s1);
    Basic.dupvec(1,n,0,s1,z);
    Basic.elmvec(1,n,0,z,w1,1.0);
    Basic.elmvec(1,n,0,z,s2,-1.0);
    if (impexiterate(z,w1,a2,h2,t+h,weights,ps2,n,eps,method)) return true;
    t3[0]=t2[0];
    t2[0]=t1[0];
    t1[0]=t+h2;
    Basic.dupvec(1,n,0,w3,w2);
    Basic.dupvec(1,n,0,w2,w1);
    Basic.dupvec(1,n,0,w1,z);
    return false;
}


static private boolean impexiterate(double z[], double y[], double a[][], double h,
                                    double t, double weights[], int ps[], int n,
                                    double eps, AP_impex_methods method)
{
    /* this procedure is internally used by IMPEXLARGESTEP (IMPEX) */

    int i,it,lit;
    double max,max1,conv,temp;
    double f1[] = new double[n+1];
    double dz[] = new double[n+1];

    max1=0.0;
    for (i=1; i<=n; i++) z[i]=(z[i]+y[i])/2.0;
    it=lit=1;
    conv=1.0;
    while (true) {
        method.deriv(t,z,f1,n);
        for (i=1; i<=n; i++) f1[i]=dz[i]=z[i]-h*f1[i]/2.0-y[i];
        Linear_algebra.sol(a,n,ps,dz);
        Basic.elmvec(1,n,0,z,dz,-1.0);
        max=0.0;
        for (i=1; i<=n; i++) {
            temp=weights[i]*dz[i];
            max += temp*temp;
        }
```

```
        max=Math.sqrt(max);
        if (max*conv < eps/10.0) break;
        it++;
        if (it != 2) {
            conv=max/max1;
            if (conv > 0.2) {
                if (lit == 0) {
                    return true;
                }
                lit=0;
                conv=1.0;
                it=1;
                if (impexrecomp(a,h,t,z,ps,n,method)) {
                    return true;
                }
            }
        }
        max1=max;
    }
    for (i=1; i<=n; i++) z[i]=2.0*z[i]-y[i];
    return false;
}


static private void impexbackdiff(int n, double u1[], double u3[], double w1[],
                                  double w2[], double w3[], double s1[],
                                  double s2[], double s3[], double r[][],
                                  double rf[][])
{
    /* this procedure is internally used by IMPEX */

    int i,k;
    double b0,b1,b2,b3;

    for (i=1; i<=n; i++) {
        b1=(u1[i]+2.0*s2[i]+u3[i])/4.0;
        b2=(w1[i]+2.0*w2[i]+w3[i])/4.0;
        b3=(s3[i]+2.0*u3[i]+s2[i])/4.0;
        b2=(b2-b1)/3.0;
        b0=b1-b2;
        b2 -= (s1[i]-2.0*s2[i]+s3[i])/16.0;
        b1=2.0*b3-(b2+rf[1][i])-(b0+r[1][i])/2.0;
        b3=0.0;
        for (k=1; k<=4; k++) {
            b1 -= b3;
            b3=r[k][i];
            r[k][i]=b0;
            b0 -= b1;
        }
        r[5][i]=b0;
```

```
for (k=1; k<=4; k++) {
    b3=rf[k][i];
    rf[k][i]=b2;
    b2 -= b3;
}
rf[5][i]=b2;
}
}
```

5.4.3 First order - Several derivatives available

A. modifiedtaylor

Solves an initial (boundary) value problem, given as a system of first order differential equations

$$Du^{(j)}(t) = h^{(j)}(u,t) \qquad (j=m0,m0+1,...,m) \tag{1}$$

where $D=d/dt$, with $u(t_0)$ prescribed, over the interval $t_0 \le t \le t_e$, by means of a one-step Taylor method [Vh70a, VhBDS71, VhK71]. This procedure is suitable for the integration of large systems arising from partial differential equations, provided that higher order derivatives can be easily obtained.

The method used has the form

$$u_{k+1} = \sum_{i=0}^{n} \beta_i c_k^{(i)} \tau_k^i \tag{2}$$

where, with u_k being an approximation to $u(t_k)$, $c_k^{(i)} = D^i u(t)$ at $t=t_k$, and τ_k is the stepsize, so that $t_{k+1} = t_k + \tau_k$ $(k=0,1,...)$. The method is p-th order consistent if $\beta_j = 1/j!$ $(j=0,1,...,p)$.

In the following $\|.\|$ denotes either the maximum or Euclidean norm. Which of these is used during the execution of *modifiedtaylor* depends upon the input parameter *norm* upon call: 1 in the first case, $\neq 1$ in the second.

The stepsize is determined by considerations of accuracy and stability. With regard to the former, it is assumed that the local error $\rho(\tau) = \|u(t_{k+1}) - u_{k+1}\|$ is approximated by the discrepancy $\|v(t_{k+1}) - u_{k+1}\|$, where $v(t)$ is the analytic solution to equation (1) with the initial values $v(t_k) = u_k$. For the p-th order exact method (2)

$$\rho_k'(\tau_k) = \sum_{i=p+1}^{n} \left| \frac{1}{i!} - \beta_i \right| \left\| c_k^{(i)} \right\| \tau_k^i \qquad (p < n) \tag{3}$$

$$\rho_k'(\tau_k) = \frac{1}{n!} \left\| c_k^{(n)} \right\| \tau_k^n \qquad (p = n)$$

are taken to be approximations to the discrepancy. It is further assumed that these estimates of the discrepancy may themselves be approximated by expressions of the form

$$\rho_k'(\tau_k) = C\tau_k^q \quad (k=1), \qquad \rho_k'(\tau_k) = (B\tau_k + C)\tau_k^q \quad (k=2)$$

$$\rho_k'(\tau_k) = (A\tau_k^2 + B\tau_k + C)\tau_k^q \quad (k \geq 3) \tag{4}$$

where A, B, and C are constants determined from fitting the given estimates at previous values of t_k, q being n if $p = n$ and $p+1$ otherwise. With the estimates (4) of

$$\rho'_k(\tau_k) = \frac{1}{n!}\left\|c_k^{(n)}\right\|\tau_k^n \qquad (p = n)$$

the discrepancy in hand, the stepsize τacc_k obtained from considerations of accuracy is to be determined by use of a relationship of the form

$$\rho_k'(\tau_k) = \eta_k \qquad (5)$$

where $\eta_k = \eta_{abs} + \eta_{re}\|u_k\|$, η_{abs} and η_{re} being absolute and relative tolerance provided by the user. *modifiedtaylor* takes

$$\tau acc_0 = \eta_0 / \left\| c_0^{(1)} \right\|$$

at the first step,

$$\tau acc_1 = \min(10\,\tau acc_0, \, (\eta_0/\| c_0^{(1)} \|)(\eta_1/\rho_0'(\tau_0))^{1/q})$$

at the second, and

$$\tau acc_k = \min(10\,\tau acc_{k-1}, \, (\eta_{k-1}/\| c_0^{(1)} \|)(\eta_k/\rho_{k-1}'(\tau_{k-1}))^{1/q})$$

at subsequent step until the value of the left hand expression in equation (5) exceeds that of the expression upon the right; then equation (5) is used.

The stepsize bound $\tau stab_k$ determined from considerations of stability is given by

$$tstab_k = \beta(n) \, / \, \sigma(J_k)$$

where $\beta(n)$ is the range constant of the stability polynomial

$$\sum_{i=0}^{n} \beta_i z^i$$

associated with the method (2), and $\sigma(J_k)$ is an upper bound for the spectral radius with respect to the eigenvalues in the left half-plane of the Jacobian matrix J_k whose elements are

$$J_k^{(i,j)}=\partial h^{(i-m0+1)}(u,t)/\partial u^{(j-m0+1)}(t) \qquad (i,j=1,2,...,m-m0+1)$$

as t ranges over $[t_k, t_k+\tau bd_k]$, τbd_k being a bound for τ_k.

The user should provide values of τmin, a lower bound for the stepsize, and $\alpha > 1$, a stepsize amplification factor. τmin is overridden by considerations of stability, but not of accuracy (thus $\max(\tau min, \tau acc_k)$ is one of the lower bounds for the stepsize). The stepsize is taken to be $\min(\max(\tau acc_0, \tau min), \tau stab_0)$ when $k=0$ and

$$\tau_k = \max(\tfrac{1}{2}\tau_{k-1}, \min(\max(\tau acc_k, \tau min), \tau stab_k, \alpha\tau_{k-1}))$$

thereafter. The upper bound τbd_k upon τ_k required in the determination of $\sigma(J_k)$ above, is τacc_0 when $k=0$ and $\alpha\tau_{k-1}$ otherwise.

The successive derivatives $D^i u^{(j)}(t)$ $(i=1,...,n;\ j=m0,...,m)$ are computed by use of a procedure *derivative* provided by the user.

Procedure parameters:

> void modifiedtaylor $(t,te,m0,m,u,method,taumin,k,order,accuracy,data,$
> $alfa,norm,eta,rho,xtmp)$

t: double $t[0:0]$;
      the independent variable;
      entry: initial value $t0$;
      exit: the final value te;

te: double;
      entry: the final value of t $(te \geq t)$;

m0,m: int;

               entry: indices of the first and last equation of the system to be solved;

u: double *u[m0:m]*;

               the dependent variable;

               entry: the initial values of the solution of the system of differential equations at $t=t0$;

               exit: the values of the solution at $t=te$;

method: a class that defines five procedures *sigma*, *derivative*, *aeta*, *reta*, and *out*, this class must implement the `AP_modifiedtaylor_methods` interface;

               double *sigma(double t[ ], int m0, int m)*

                   this procedure should deliver the spectral radius of the Jacobian matrix with respect to those eigenvalues which are located in the left half-plane;

                   if *sigma* tends to infinity, procedure *modifiedtaylor* terminates;

               void *derivative(double t[ ], int m0, int m, int i, double a[ ])*

                   when this procedure is called, array *a[m0,m]* contains the components of the $(i\text{-}1)$-th derivative of *u* at the point *t*; upon completion of *derivative*, array *a* should contain the components of the *i*-th derivative of *u* at the point *t*;

               double *aeta(double t[ ], int m0, int m)*

                   this procedure should return the absolute accuracy (value of η_{abs} above);

               double *reta(double t[ ], int m0, int m)*

                   this procedure should return the relative accuracy; (value of η_{re} above);

                   if both *aeta* and *reta* are negative, accuracy conditions will be ignored;

               void *out(double t[ ], double te, int m0, int m, double u[ ],int k[ ],*
                   double eta[ ], double rho[ ])

                   after each integration step the user can print some parameters such as *t, u, k, eta,* and *rho*;

taumin: double;

               entry: minimal step length by which the integration is performed (value of τmin above); however, actual stepsizes will always be within the interval $[\min(hmin,hstab),hstab]$, where *hstab* $(= data[0]/sigma)$ is the steplength prescribed by stability considerations;

k: int *k[0:0]*;

               exit: indicates the number of integration steps performed; on entry $k=0$;

order: int;

               the order of the highest derivative upon which the Taylor method is based (value of *n* above);

accuracy: int;

               order of accuracy of the method (value of *p* above);

data: double *data[0:order]*;

               entry: *data[0]*:   stability parameter (value of $\beta(n)$ above);

                         data[1],...,*data[order]*: polynomial coefficients (values of β_j above, $j=1,..,n$);

alfa: double;
                entry: growth factor for the integration step length (value of α above);

norm: int;
                entry: if *norm*=1 then discrepancy and tolerance are estimated in the maximum norm; otherwise, in the Euclidean norm;

eta: double *eta[0:0]*;
                exit: computed tolerance;

rho: double *rho[0:0]*;
                exit: computed discrepancy;

xtmp: double *xtmp[0:7]*;
                working storage for internal static data.

Procedure used: vecvec.

```
package numal;

public interface AP_modifiedtaylor_methods {

    double sigma(double t[], int m0, int m);
    void derivative(double t[], int m0, int m, int i, double a[]);
    double aeta(double t[], int m0, int m);
    double reta(double t[], int m0, int m);
    void out(double t[], double te, int m0, int m, double u[], int k[],
            double eta[], double rho[]);
}



public static void modifiedtaylor(double t[], double te, int m0, int m, double u[],
                                AP_modifiedtaylor_methods method, double taumin,
                                int k[], int order, int accuracy, double data[],
                                double alfa, int norm, double eta[], double rho[],
                                double xtmp[])
{
    boolean last,start,step1;
    int i,n,p,q,j;
    double ec0,ec1,ec2,tau0,tau1,tau2,taus,t2,t0,tau,taui,tauec,ecl,betan,gamma,
            ifac,tauacc,taustab,aa,bb,cc,ec,s,x,b;

    tauec=ecl=0.0;
    ec0=xtmp[0];
    ec1=xtmp[1];
    ec2=xtmp[2];
    tau0=xtmp[3];
    tau1=xtmp[4];
    tau2=xtmp[5];
    taus=xtmp[6];
    t2=xtmp[7];
    step1=true;
    n=order;
    double beta[] = new double[n+1];
```

```
double betha[] = new double[n+1];
double c[] = new double[m+1];
i=0;
start=(k[0] == 0);
t0=t[0];
/* coefficient */
ifac=1.0;
gamma=0.5;
p=accuracy;
betan=data[0];
q = (p < n) ? p+1 : n;
for (j=1; j<=n; j++) {
    beta[j]=data[j];
    ifac /= j;
    betha[j]=ifac-beta[j];
}
if (p == n) betha[n]=ifac;
last=false;
do {
    /* step size */
    s=0.0;
    if (norm == 1)
        for (j=m0; j<=m; j++) {
            x=Math.abs(u[j]);
            if (x > s) s=x;
        }
    else
        s=Math.sqrt(Basic.vecvec(m0,m,0,u,u));
    /* local error bound */
    eta[0] = method.aeta(t,m0,m)+method.reta(t,m0,m)*s;
    if (eta[0] > 0.0) {
        if (start) {
            if (k[0] == 0) {
                for (j=m0; j<=m; j++) c[j]=u[j];
                i=1;
                method.derivative(t,m0,m,i,c);
                s=0.0;
                if (norm == 1)
                    for (j=m0; j<=m; j++) {
                        x=Math.abs(c[j]);
                        if (x > s) s=x;
                    }
                else
                    s=Math.sqrt(Basic.vecvec(m0,m,0,c,c));
                tauacc=eta[0]/s;
                step1=true;
            } else if (step1) {
                tauacc=Math.pow(eta[0]/rho[0],1.0/q)*tau2;
                if (tauacc > 10.0*tau2)
                    tauacc=10.0*tau2;
```

```
                else
                    step1=false;
            } else {
                bb=(ec2-ec1)/tau1;
                cc=ec2-bb*t2;
                ec=bb*t[0]+cc;
                tauacc = (ec < 0.0) ? tau2 : Math.pow(eta[0]/ec,1.0/q);
                start=false;
            }
        } else {
            aa=((ec0-ec1)/tau0+(ec2-ec1)/tau1)/(tau1+tau0);
            bb=(ec2-ec1)/tau1-aa*(2.0*t2-tau1);
            cc=ec2-t2*(bb+aa*t2);
            ec=cc+t[0]*(bb+t[0]*aa);
            tauacc = (ec < 0.0) ? taus : Math.pow(eta[0]/ec,1.0/q);
            if (tauacc > alfa*taus) tauacc=alfa*taus;
            if (tauacc < gamma*taus) tauacc=gamma*taus;
        }
    } else
        tauacc=te-t[0];
    if (tauacc < taumin) tauacc=taumin;
    taustab=betan/method.sigma(t,m0,m);
    if (taustab < 1.0e-12*(t[0]-t0)) {
        method.out(t,te,m0,m,u,k,eta,rho);
        break;
    }
    tau = (tauacc > taustab) ? taustab : tauacc;
    taus=tau;
    if (tau >= te-t[0]) {
        tau=te-t[0];
        last=true;
    }
    tau0=tau1;
    tau1=tau2;
    tau2=tau;
    k[0]++;
    i=0;
    /* difference scheme */
    for (j=m0; j<=m; j++) c[j]=u[j];
    taui=1.0;
    do {
        i++;
        method.derivative(t,m0,m,i,c);
        taui *= tau;
        b=beta[i]*taui;
        if (eta[0] > 0.0 && i >= p) {
            /* local error construction */
            if (i == p) {
                ec1=0.0;
                tauec=1.0;
```

```
            }
            if (i > p+1) tauec *= tau;
            s=0.0;
            if (norm == 1)
                for (j=m0; j<=m; j++) {
                    x=Math.abs(c[j]);
                    if (x > s) s=x;
                }
            else
                s=Math.sqrt(Basic.vecvec(m0,m,0,c,c));
            ecl += Math.abs(betha[i])*tauec*s;
            if (i == n) {
                ec0=ecl;
                ec1=ec2;
                ec2=ecl;
                rho[0] = ecl*Math.pow(tau,q);
            }
        }
        for (j=m0; j<=m; j++) u[j] += b*c[j];
    } while (i < n);
    t2=t[0];
    if (last) {
        last=false;
        t[0]=te;
    } else
        t[0] += tau;
    method.out(t,te,m0,m,u,k,eta,rho);
  } while (t[0] != te);

xtmp[0]=ec0;
xtmp[1]=ec1;
xtmp[2]=ec2;
xtmp[3]=tau0;
xtmp[4]=tau1;
xtmp[5]=tau2;
xtmp[6]=taus;
xtmp[7]=t2;
}
```

B. eft

Solves an initial value problem, given as a system of first order differential equations

$$Du^{(j)}(t) = g^{(j)}(u,t) \qquad (j{=}m0,m0{+}1,...,m) \tag{1}$$

where $D{=}d/dt$, with $u(t_0)$ prescribed, over the interval $t0 \le t \le te$, by means of a third order, first order consistent, exponentially fitted Taylor series method [see Vh70b, VhBDS71]. Automatic stepsize control is provided. This procedure is suitable for the integration of stiff differential equations, provided that higher order derivatives can be easily obtained.

The method used has the form

$$u_{k+1} = u_k + c_k^{(1)}h_k + \beta_2(h_k)c_k^{(2)}h_k^2 + \beta_3(h_k)c_k^{(3)}h_k^3 \tag{2}$$

where, with u_k being an approximation to $u(t_k)$, $c_k^{(i)} \approx D^i u(t)$ at $t=t_k$, and h_k is the stepsize, so that $t_{k+1} = t_k + h_k$ ($k=0,1,...$).

It is supposed that over the range $t_k \le t \le t_k + hbd_k$, where hbd_k is an upper bound for h_k, those eigenvalues of the Jacobian matrix J_k, whose elements are

$$J_k^{(i,j)} = \partial g^{(i-m0+1)}(u,t) / \partial u^{(j-m0+1)}(t) \quad (i,j=1,2,...,m-m0+1),$$

lying in the left half-plane and not in the neighborhood of the origin are clustered within two circles with centers at

$$\delta_k = \sigma_{ke}^{i\phi_k} \quad \text{and} \quad \bar{\delta}_k,$$

($0 < \sigma_k < \infty$, $\pi/2 < \phi_k < \pi$) of equal radii $d\delta_k$. The coefficients $\beta_2(h_k)$ and $\beta_3(h_k)$ are determined by imposing the condition that, with

$$P_3(z) = 1 + z + \beta_2(h_k)z^2 + \beta_3(h_k)z^3$$

$P_3(z) = e^z$ when $z = h_k\delta_k$, $z = h_k\text{conj}(\delta_k)$; it then follows that

$$\beta_2(h_k) = -\frac{1}{b^2}\left(2b\cos\phi_k + 4\cos^2\phi_k - 1 + e^{b\cos\phi_k}\frac{\sin(b\sin\phi_k - 3\phi_k)}{\sin\phi_k}\right)$$

$$\beta_3(h_k) = \frac{1}{b^3}\left(b + 2\cos\phi_k + be^{b\cos\phi_k}\frac{\sin(b\sin\phi_k - 2\phi_k)}{\sin\phi_k}\right)$$

where $b = h_k\delta_k$. When b is small, $\beta_2(h_k) \approx 1/2 - b^2/24$ and $\beta_3(h_k) \approx 1/6 + b\cos\phi_k/12$.

In the following, $\|.\|$ denotes either the maximum or Euclidean norm. Which of these is used during the execution of *eft* depends upon the value of the parameter *norm* upon call: 1 in the first case, $\ne 1$ in the second.

The coefficients $\beta_2(h_k)$, $\beta_3(h_k)$ in the scheme (2) depend upon h_k: the discrepancy $\rho'(u_k)$ between the local analytic solution of equation (1) and its approximation yielded by use of the scheme (2) is not accurately estimated by substitution of the local solution into formula (2) (this process yields useful results only when the coefficients of the stability polynomial of the numerical integration procedure being used are constant, as is the case with, for example, *modifiedtaylor*). The discrepancy is now to be estimated in terms of a residual function $\zeta(h)$ (i.e., a measure of the difference between the values of the expressions upon the two sides of equation (1)) resulting from use of the scheme (2) in which the steplength is taken to be an independent variable. Taking $u'(h)$ to be an approximation to $u(t_k+h)$ produced by use of formula (2),

$$u'(h) = u_k + c_k^{(1)} + \beta_2(h)c_k^{(2)}h_2 + \beta_3(h)c_k^{(3)}h^3$$

from which

$$Du'(h) = c_k^{(1)} + \beta'_2(h)c_k^{(2)} + \beta'_3(h)c_k^{(3)}h^2$$

where, with $b = \sigma_k h$,

$$\beta'_2(h) = -\frac{1}{b}\left(2\cos\phi_k + e^{b\cos\phi_k}\frac{\sin(b\sin\phi_k - 2\phi_k)}{\sin\phi_k}\right)$$

$$\tag{3}$$

$$\beta'_3(h) = \frac{1}{b^2}\left(1 + e^{b\cos\phi_k}\frac{\sin(b\sin\phi_k - \phi_k)}{\sin\phi_k}\right).$$

The residual function $\zeta_k(h)$ is defined by the formula
$$\zeta_k(h) = Du'(h) - g(u'(h), t_k + h) \tag{4}$$
so that
$$h\|\zeta_k(h)\| = \|g(u'(h), t_k + h)h - c_k^{(1)}h - \beta'_2(h)c_k^{(2)}h^2 - \beta'_3(h)c_k^{(3)}h^3\|. \tag{5}$$
If the discrepancy $\rho'_k(h)$ is defined to be

$$\rho'_k(h) = \int_0^h \|\zeta_k(h')\| dh'$$

then approximately $\rho'_k(h) = h|\zeta_k(h)|$.
An upper bound $hacc_k$ upon h_k derived from considerations of accuracy is thus given by
$$\rho'_k(hacc_k) = \eta_k = \eta_a + \eta_{re}\|u_k\| \tag{6}$$
where, η_a and η_{re} being prescribed absolute and relative tolerances, ρ'_k is as defined by formulae (3,4,5). The value of $\rho'_k(h)$ is predicted from previous determinations by use of the formula
$$\rho'(h) = (Ah + Bt + C)h^{Q(b)}$$
where with $b = \sigma h$ (σ being the spectral radius of the Jacobian matrix derived from equation (1) at the point t) and
$Q(b) = 4 - 2b/3$ if $0 < b < 3/2$, $(30-2b)/9$ if $3/2 \le b < 6$ and 2 otherwise,
the coefficients A, B, and C are determined by three substitutions t_{k-3}, h_{k-3}; t_{k-2}, h_{k-2}; t_{k-1}, h_{k-1}. If h_{k-1} has the critical value $h_{cr} = h_{k-2}^2/h_{k-3}$, the value of A determined as above becomes infinite. For this reason h_k is, at each stage, excluded from the interval $[(1 - \sqrt{\varepsilon})h_{cr}, (1 + \sqrt{\varepsilon})h_{cr}]$, where ε is the machine precision. Also it may occur that A is negative; in this case the extrapolation formula $\rho'(h) = Ch^{Q(b)}$ is used. Except when A is negative, $hacc_k$ is determined from the equation
$$\left(A\, hacc_k + B\, t_k + C\right)hacc^{q(\sigma h_{k-1})} = \eta_k.$$

eft takes

$$hacc_0 = \frac{\eta_0}{\|c_0^{(1)}\|}$$

at the first step

$$hacc_1 = \min\left(10\, hacc_0, \ \frac{\eta_0}{\|c_0^{(1)}\|}\left(\frac{\eta_1}{\rho'_0(h_0)}\right)^{1/Q(\sigma_1 h_0)}\right)$$

at the second, and

$$hacc_k = \min\left(10\, hacc_{k-1}, \ \frac{\eta_{k-1}}{\|c_{k-1}^{(1)}\|}\left(\frac{\eta_{k-1}}{\rho'_{k-1}(h_{k-1})}\right)^{1/Q(\sigma_k h_{k-1})}\right)$$

at subsequent steps until the value of the left hand side in equation (6) (predicted as above) exceeds that of the expression upon the right; then equation (6) is used to determine $hacc_k$.

When $\sin\varphi_k$ is not small, $|P_3(z)| \le 1$ over the disc $|z - h\delta_k| \le 1 / 2\sin\varphi_k$; when $\sin\varphi_k$ is small, $|P_3(z)| \le 1$ when $|z + h\delta_k| \le (h\sigma_k)^{1/2}$. Thus if the eigenvalues of the Jacobian matrix J_k (defined above) clustered about two points δ_k, conjugate(δ_k) lie within discs of radius $d\delta_k$, the method (2) is stable if $h_k < hstab_k$ where

$hstab_k = \min [\sigma_k/(d\delta_k\sin\varphi_k), 4\sigma_k^2/d\delta_k^2] / \sigma_k$.

The user should provide values of $hmin$, a lower bound for the stepsize, and $\alpha > 1$, a stepsize amplification factor. $hmin$ is overridden by considerations of stability but not of accuracy (thus $\max(hmin,hacc_k)$ is one of the lower bounds for the stepsize). The stepsize is taken to be $\min(\max(hacc_0,hmin),hstab_0)$ when $k=0$, and $h_k = \max(h_{k-1}/2,\min(\max(hacc_k,hmin),hstab_k,\alpha_k h_{k-1}))$ thereafter, where until formula (6) is used to determine $hacc_k$ (see above) α_k is the user's amplification factor α, and thereafter

$$\alpha_k = \min\left(\alpha, \left(\frac{\eta_k}{\rho_k'(h_k)}\right)^{1/Q(\sigma_k h_{k-1})}\right)$$

The upper bound hbd_k upon h_k required in the determination of $\sigma(J_k)$ above, is $hacc_0$ when $k=0$ and $\alpha_k h_{k-1}$ otherwise.

It may occur that the user wishes to integrate equation (1) over a succession of intervals $[t_0^{(l)},te^{(l)}]$ $(l=0,1,...)$ where $t_0^{(l+1)}=te^{(l)}$. In this case he should set $k=0$ prior to the first call of eft and leave k unchanged at subsequent calls. During these subsequent calls, eft continues normal operation (without special determinations of $hacc_0$, $hacc_1$, etc.). The current value of h_k at return from a call of eft is allocated by this procedure to $hstart$, for use at the start of the next call.

The successive derivatives $D^i u^{(j)}(t)$ $(i=1,2,3; j=m0,...,m)$ are computed by use of a procedure $derivative$ which the user must provide.

Procedure parameters:

        void eft $(t,te,m0,m,u,method,phi,k,alfa,norm,eta,rho,hmin,hstart)$

t: double $t[0:0]$;
        the independent variable;
        entry: initial value $t0$;
        exit: the final value te;

te: double;
        entry: the final value of t $(te \geq t)$;

$m0,m$: int;
        entry: indices of the first and last equation of the system to be solved;

u: double $u[m0:m]$;
        the dependent variable;
        entry: the initial values of the solution of the system of differential
                  equations at $t=t0$;
        exit: the values of the solution at $t=te$;

$method$: a class that defines six procedures $sigma$, $diameter$, $derivative$, $aeta$, $reta$, and out, this class must implement the `AP_eft_methods` interface;
        double sigma(double t[ ], int m0, int m)
            this procedure should deliver the modulus of the (complex) point at which exponential fitting is desired (value of σ_k above); for example, an approximation to the modulus of the center of the left hand cluster;
        double diameter(double t[ ], int m0, int m)
            this procedure should deliver the diameter of the left hand cluster (value of $d\delta_k$ above);

void derivative(double t[ ], int m0, int m, int i, double a[ ])
> *i* assumes the values 1,2,3, and *a* is a one-dimensional array *a[m0:m]*;
> when this procedure is called, array *a* contains the components of the (*i*-1)-th derivative of *u* at the point *t*; upon completion of *derivative*, array *a* should contain the components of the *i*-th derivative of *u* at the point *t*;

double aeta(double t[ ], int m0, int m)
> this procedure should return the absolute local accuracy (value of η_a above); *aeta* should be positive;

double reta(double t[ ], int m0, int m)
> this procedure should return the relative local accuracy (value of η_{re} above); *reta* should be positive;

void out(double t[ ], double te, int m0, int m, double u[ ],int k[ ],
>   *double eta[ ], double rho[ ])*
> after each integration step the user can print some parameters such as *t, u, k, eta,* and *rho*;

phi: double;
> entry: the argument of the (complex) point at which exponential fitting is desired (value of φ_k above); *phi* should have a value in the range $[\pi/2, \pi]$;

k: int *k[0:0]*;
> exit: indicates the number of integration steps performed; on entry *k*=0;

alfa: double;
> entry: maximal growth factor for the integration step length (value of α above);

norm: int;
> entry: if *norm*=1 then discrepancy and tolerance are estimated in the maximum norm, otherwise in the Euclidean norm;

eta: double *eta[0:0]*;
> exit: computed tolerance;

rho: double *rho[0:0]*;
> exit: computed discrepancy;

hmin: double;
> entry: minimal stepsize by which the integration is performed; however, a smaller step will be taken if *hmin* exceeds the stepsize *hstab*, prescribed by the stability conditions; if *hstab* becomes zero, the procedure terminates;

hstart: double *hstart[0:0]*;
> entry: the initial stepsize;
> however, if *k* = 0 on entry then the value of *hstart* is not taken into consideration;
>
> exit: a suggestion for the stepsize (current value of h_k above), if the integration should be continued for *t* > *te*;
> *hstart* may be used in successive calls of the procedure, in order to obtain the solution in several points *te1, te2,* etc.

Procedures used: inivec, dupvec, vecvec, elmvec.

```
package numal;

public interface AP_eft_methods {

    double sigma(double t[], int m0, int m);
    double diameter(double t[], int m0, int m);
    void derivative(double t[], int m0, int m, int i, double a[]);
    double aeta(double t[], int m0, int m);
    double reta(double t[], int m0, int m);
    void out(double t[], double te, int m0, int m, double u[], int k[],
            double eta[], double rho[]);
}


public static void eft(double t[], double te, int m0, int m, double u[],
                    AP_eft_methods method, double phi, int k[], double alfa,
                    int norm, double eta[], double rho[], double hmin,
                    double hstart[])
{
    boolean extrapolate,last,start;
    int kl,j,i,ext;
    double q,ec0,ec1,ec2,h,hi,h0,h1,h2,betan,t2,sigmal,phil,s,x,hacc,hstab,hcr,
            hmax,a,b,cc,b1,b2,bb,e,beta2,beta3,c0,fc,b0,fb,a0,fa,d0,fd,fdb,fda,
            w,mb,tol,mm,p0,q0;
    double beta[] = new double[4];
    double betha[] = new double[4];
    double c[] = new double[m+1];
    double ro[] = new double[m+1];

    kl=0;
    q=h2=h0=ec2=ec1=ec0=h1=t2=fd=d0=0.0;
    start=true;
    last=false;
    Basic.dupvec(m0,m,0,c,u);
    method.derivative(t,m0,m,1,c);
    if (k[0] == 0) {
        /* local error bound */
        s=0.0;
        if (norm == 1)
            for (j=m0; j<=m; j++) {
                x=Math.abs(u[j]);
                if (x > s) s=x;
            }
        else
            s=Math.sqrt(Basic.vecvec(m0,m,0,u,u));
        eta[0] = method.aeta(t,m0,m)+method.reta(t,m0,m)*s;
        s=0.0;
        if (norm == 1)
            for (j=m0; j<=m; j++) {
                x=Math.abs(c[j]);
```

```
                  if (x > s) s=x;
            }
        else
            s=Math.sqrt(Basic.vecvec(m0,m,0,c,c));
        hstart[0] = eta[0]/s;
    }
    do {
        /* difference scheme */
        hi=1.0;
        sigmal=method.sigma(t,m0,m);
        phil=phi;
        /* step size */
        if (!start) {
            /* local error bound */
            s=0.0;
            if (norm == 1)
                for (j=m0; j<=m; j++) {
                    x=Math.abs(u[j]);
                    if (x > s) s=x;
                }
            else
                s=Math.sqrt(Basic.vecvec(m0,m,0,u,u));
            eta[0] = method.aeta(t,m0,m)+method.reta(t,m0,m)*s;
        }
        if (start) {
            h1=h2=hacc=hstart[0];
            ec2=ec1=1.0;
            kl=1;
            start=false;
        } else if (kl < 3) {
            hacc=Math.pow(eta[0]/rho[0],1.0/q)*h2;
            if (hacc > 10.0*h2)
                hacc=10.0*h2;
            else
                kl++;
        } else {
            a=(h0*(ec2-ec1)-h1*(ec1-ec0))/(h2*h0-h1*h1);
            h=h2*((eta[0] < rho[0]) ? Math.pow(eta[0]/rho[0],1.0/q) : alfa);
            if (a > 0.0) {
                b=(ec2-ec1-a*(h2-h1))/h1;
                cc=ec2-a*h2-b*t2;
                hacc=0.0;
                hmax=h;
                /* find zero */
                b0=hacc;
                fb=Math.pow(hacc,q)*(a*hacc+b*t[0]+cc)-eta[0];
                a0=hacc=h;
                fa=Math.pow(hacc,q)*(a*hacc+b*t[0]+cc)-eta[0];
                c0=a0;
                fc=fa;
```

```
ext=0;
extrapolate=true;
while (extrapolate) {
    if (Math.abs(fc) < Math.abs(fb)) {
        if (c0 != a0) {
            d0=a0;
            fd=fa;
        }
        a0=b0;
        fa=fb;
        b0=hacc=c0;
        fb=fc;
        c0=a0;
        fc=fa;
    }
    tol=1.0e-3*h2;
    mm=(c0+b0)*0.5;
    mb=mm-b0;
    if (Math.abs(mb) > tol) {
        if (ext > 2)
            w=mb;
        else {
            if (mb == 0.0)
                tol=0.0;
            else
                if (mb < 0.0) tol = -tol;
            p0=(b0-a0)*fb;
            if (ext <= 1)
                q0=fa-fb;
            else {
                fdb=(fd-fb)/(d0-b0);
                fda=(fd-fa)/(d0-a0);
                p0 *= fda;
                q0=fdb*fa-fda*fb;
            }
            if (p0 < 0.0) {
                p0 = -p0;
                q0 = -q0;
            }
            w=(p0<Double.MIN_VALUE || p0<=q0*tol) ? tol :
                    ((p0<mb*q0) ? p0/q0 : mb);
        }
        d0=a0;
        fd=fa;
        a0=b0;
        fa=fb;
        hacc = b0 += w;
        fb=Math.pow(hacc,q)*(a*hacc+b*t[0]+cc)-eta[0];
        if ((fc >= 0.0) ? (fb >= 0.0) : (fb <= 0.0)) {
            c0=a0;
```

```
                        fc=fa;
                        ext=0;
                    } else
                        ext = (w == mb) ? 0 : ext+1;
                } else
                    break;
            }
            h=c0;
            if (!((fc >= 0.0) ? (fb <= 0.0) : (fb >= 0.0)))
                hacc=hmax;
        } else
            hacc=h;
        if (hacc < 0.5*h2) hacc=0.5*h2;
    }
    if (hacc < hmin) hacc=hmin;
    h=hacc;
    if (h*sigmal > 1.0) {
        a=Math.abs(method.diameter(t,m0,m)/sigmal+Double.MIN_VALUE)/2.0;
        b=2.0*Math.abs(Math.sin(phil));
        betan=((a > b) ? 1.0/a : 1.0/b)/a;
        hstab=Math.abs(betan/sigmal);
        if (hstab < 1.0e-14*t[0]) break;
        if (h > hstab) h=hstab;
    }
    hcr=h2*h2/h1;
    if (kl > 2 && Math.abs(h-hcr) < Double.MIN_VALUE*hcr)
        h = (h < hcr) ? hcr*(1.0-Double.MIN_VALUE) : hcr*(1.0+Double.MIN_VALUE);
    if (t[0]+h > te) {
        last=true;
        hstart[0] = h;
        h=te-t[0];
    }
    h0=h1;
    h1=h2;
    h2=h;
    /* coefficient */
    b=h*sigmal;
    b1=b*Math.cos(phil);
    bb=b*b;
    if (Math.abs(b) < 1.0e-3) {
        beta2=0.5-bb/24.0;
        beta3=1.0/6.0+b1/12.0;
        betha[3]=0.5+b1/3.0;
    } else if (b1 < -40.0) {
        beta2=(-2.0*b1-4.0*b1*b1/bb+1.0)/bb;
        beta3=(1.0+2.0*b1/bb)/bb;
        betha[3]=1.0/bb;
    } else {
        e=Math.exp(b1)/bb;
        b2=b*Math.sin(phil);
```

```
        beta2=(-2.0*b1-4.0*b1*b1/bb+1.0)/bb;
        beta3=(1.0+2.0*b1/bb)/bb;
        if (Math.abs(b2/b) < 1.0e-5) {
            beta2 -= e*(b1-3.0);
            beta3 += e*(b1-2.0)/b1;
            betha[3]=1.0/bb+e*(b1-1.0);
        } else {
            beta2 -= e*Math.sin(b2-3.0*phil)/b2*b;
            beta3 += e*Math.sin(b2-2.0*phil)/b2;
            betha[3]=1.0/bb+e*Math.sin(b2-phil)/b2*b;
        }
    }
    beta[1]=betha[1]=1.0;
    beta[2]=beta2;
    beta[3]=beta3;
    betha[2]=1.0-bb*beta3;
    b=Math.abs(b);
    q = (b < 1.5) ? 4.0-2.0*b/3.0 : ((b < 6.0) ? (30.0-2.0*b)/9.0 : 2.0);
    for (i=1; i<=3; i++) {
        hi *= h;
        if (i > 1) method.derivative(t,m0,m,i,c);
        /* local error construction */
        if (i == 1) Basic.inivec(m0,m,ro,0.0);
        if (i < 4) Basic.elmvec(m0,m,0,ro,c,betha[i]*hi);
        if (i == 4) {
            Basic.elmvec(m0,m,0,ro,c,-h);
            s=0.0;
            if (norm == 1)
                for (j=m0; j<=m; j++) {
                    x=Math.abs(ro[j]);
                    if (x > s) s=x;
                }
            else
                s=Math.sqrt(Basic.vecvec(m0,m,0,ro,ro));
            rho[0]=s;
            ec0=ec1;
            ec1=ec2;
            ec2=rho[0]/Math.pow(h,q);
        }
        Basic.elmvec(m0,m,0,u,c,beta[i]*hi);
    }
    t2=t[0];
    k[0]++;
    if (last) {
        last=false;
        t[0]=te;
        start=true;
    } else
        t[0] += h;
    Basic.dupvec(m0,m,0,c,u);
```

```
        method.derivative(t,m0,m,1,c);
        /* local error construction */
        Basic.elmvec(m0,m,0,ro,c,-h);
        s=0.0;
        if (norm == 1)
            for (j=m0; j<=m; j++) {
                x=Math.abs(ro[j]);
                if (x > s) s=x;
            }
        else
            s=Math.sqrt(Basic.vecvec(m0,m,0,ro,ro));
        rho[0]=s;
        ec0=ec1;
        ec1=ec2;
        ec2=rho[0]/Math.pow(h,q);
        method.out(t,te,m0,m,u,k,eta,rho);
    } while (t[0] != te);
}
```

5.4.4 Second order - No derivatives right hand side

A. rk2

Solves an initial value problem for a single second order ordinary differential equation $d^2y/dx^2 = f(x,y,dy/dx)$, from $x=a$ to $x=b$, $y(a)$ and $dy(a)/da$ being given, by means of a 5-th order Runge-Kutta method with steplength and error control [Z64].

Procedure parameters:

$$\text{void rk2 } (x,a,b,y,ya,z,za,method,e,d,fi)$$

x: double $x[0:0]$;
       the independent variable;
a: double;
       entry: the initial value of x;
b: double;
       entry: the end value of x, ($b \leq a$ is allowed);
y: double $y[0:0]$;
       the dependent variable;
       exit: the value of $y(x)$ at $x = b$;
ya: double;
       entry: the initial value of y at $x=a$;
z: double $z[0:0]$;
       the derivative dy/dx;
       exit: the value of $z(x)$ at $x = b$;
za: double;
       entry: the initial value of dy/dx at $x=a$;

method: a class that defines a procedure *fxyz*, this class must implement the
AP_rk2_method interface;

    double fxyz(double x[ ], double y[ ], double z[ ])

        the right-hand side of the differential equation;

        fxyz depends on *x, y, z*, giving the value of d^2y/dx^2;

e: double *e[1:4]*;

    entry: *e[1]* and *e[3]* are used as relative, *e[2]* and *e[4]* are used as
absolute tolerances for *y* and *dy/dx*, respectively;

d: double *d[1:5]*;

    exit: *d[1]*: the number of steps skipped;

        d[2]: the last step length used;

        d[3]: equal to *b*;

        d[4]: equal to *y(b)*;

        d[5]: equal to *dy/dx*, for *x = b*;

fi: boolean;

    entry: if *fi* is true then the integration starts at *x=a* with a trial step *b-a*;
if *fi* is false then the integration is continued with, as initial
conditions, *x=d[3]*, *y=d[4]*, *z=d[5]*, and *a*, *ya*, and *za* are ignored.

```
package numal;

public interface AP_rk2_method {

    double fxyz(double x[], double y[], double z[]);
}


public static void rk2(double x[], double a, double b, double y[], double ya,
                       double z[], double za, AP_rk2_method method, double e[],
                       double d[], boolean fi)
{
    boolean last,first,reject,test,ta,tb;
    double e1,e2,e3,e4,xl,yl,zl,h,ind,hmin,hl,absh,k0,k1,k2,k3,k4,k5,discry,discrz,
           toly,tolz,mu,mul,fhy,fhz;

    last=true;
    mul=0.0;
    if (fi) {
        d[3]=a;
        d[4]=ya;
        d[5]=za;
    }
    d[1]=0.0;
    xl=d[3];
    yl=d[4];
    zl=d[5];
    if (fi) d[2]=b-d[3];
    absh=h=Math.abs(d[2]);
    if (b-xl < 0.0) h = -h;
    ind=Math.abs(b-xl);
```

```
hmin=ind*e[1]+e[2];
hl=ind*e[3]+e[4];
if (hl < hmin) hmin=hl;
e1=e[1]/ind;
e2=e[2]/ind;
e3=e[3]/ind;
e4=e[4]/ind;
first=true;
test=true;
if (fi) {
    last=true;
    test=false;
}
while (true) {
    if (test) {
        absh=Math.abs(h);
        if (absh < hmin) {
            h = (h > 0.0) ? hmin : -hmin;
            absh=hmin;
        }
        ta=(h >= b-xl);
        tb=(h >= 0.0);
        if ((ta && tb) || (!(ta || tb))) {
            d[2]=h;
            last=true;
            h=b-xl;
            absh=Math.abs(h);
        } else
            last=false;
    }
    test=true;
    x[0]=xl;
    y[0]=yl;
    z[0]=zl;
    k0=method.fxyz(x,y,z)*h;
    x[0]=xl+h/4.5;
    y[0]=yl+(zl*18.0+k0*2.0)/81.0*h;
    z[0]=zl+k0/4.5;
    k1=method.fxyz(x,y,z)*h;
    x[0]=xl+h/3.0;
    y[0]=yl+(zl*6.0+k0)/18.0*h;
    z[0]=zl+(k0+k1*3.0)/12.0;
    k2=method.fxyz(x,y,z)*h;
    x[0]=xl+h*0.5;
    y[0]=yl+(zl*8.0+k0+k2)/16.0*h;
    z[0]=zl+(k0+k2*3.0)/8.0;
    k3=method.fxyz(x,y,z)*h;
    x[0]=xl+h*0.8;
    y[0]=yl+(zl*100.0+k0*12.0+k3*28.0)/125.0*h;
    z[0]=zl+(k0*53.0-k1*135.0+k2*126.0+k3*56.0)/125.0;
```

```
        k4=method.fxyz(x,y,z)*h;
        x[0] = (last ? b : xl+h);
        y[0]=yl+(zl*336.0+k0*21.0+k2*92.0+k4*55.0)/336.0*h;
        z[0]=zl+(k0*133.0-k1*378.0+k2*276.0+k3*112.0+k4*25.0)/168.0;
        k5=method.fxyz(x,y,z)*h;
        discry=Math.abs((-k0*21.0+k2*108.0-k3*112.0+k4*25.0)/56.0*h);
        discrz=Math.abs(k0*21.0-k2*162.0+k3*224.0-k4*125.0+k5*42.0)/14.0;
        toly=absh*(Math.abs(zl)*e1+e2);
        tolz=Math.abs(k0)*e3+absh*e4;
        reject=(discry > toly || discrz > tolz);
        fhy=discry/toly;
        fhz=discrz/tolz;
        if (fhz > fhy) fhy=fhz;
        mu=1.0/(1.0+fhy)+0.45;
        if (reject) {
            if (absh <= hmin) {
                d[1] += 1.0;
                y[0]=yl;
                z[0]=zl;
                first=true;
                if (b == x[0]) break;
                xl = x[0];
                yl = y[0];
                zl = z[0];
            } else
                h *= mu;
        } else {
            if (first) {
                first=false;
                hl=h;
                h *= mu;
            } else {
                fhy=mu*h/hl+mu-mul;
                hl=h;
                h *= fhy;
            }
            mul=mu;
            y[0]=yl+(zl*56.0+k0*7.0+k2*36.0-k4*15.0)/56.0*hl;
            z[0]=zl+(-k0*63.0+k1*189.0-k2*36.0-k3*112.0+k4*50.0)/28.0;
            k5=method.fxyz(x,y,z)*hl;
            y[0]=yl+(zl*336.0+k0*35.0+k2*108.0+k4*25.0)/336.0*hl;
            z[0]=zl+(k0*35.0+k2*162.0+k4*125.0+k5*14.0)/336.0;
            if (b == x[0]) break;
            xl = x[0];
            yl = y[0];
            zl = z[0];
        }
    }
    if (!last) d[2]=h;
    d[3] = x[0];
```

```
    d[4] = y[0];
    d[5] = z[0];
}
```

B. rk2n

Solves an initial value problem for a system of second order ordinary differential equations $d^2y_i(x)/dx^2 = f_i(x,y,dy(x)/dx)$, $(j=1,2,...,n)$, from $x=a$ to $x=b$, $y(a)$ and $dy(a)/da$ being given, by means of a 5-th order Runge-Kutta method [Z64].
 Upon completion of a call of *rk2n* we have: $x=d[3]=b$, $y[j]=d[j+3]$ the value of the dependent variables for $x=b$, $z[j]=d[n+j+3]$, the value of the derivatives of $y[j]$ at $x=b$, $j=1,...,n$. *rk2n* uses as its minimal absolute step length
$$hmin = \min(e[2*j-1]*k+e[2*j])$$
with $1 \le j \le 2*n$ and $k = | b - (\text{if } fi \text{ is true then } a \text{ else } d[3]) |$.
 If a step of length $|h| \le hmin$ is rejected then a step $sign(h)*hmin$ is skipped. A step is rejected if the absolute value of the computed discretization error is greater than
$$(|z[j]|*e[2*j-1]+e[2*j])*|h|/k$$
or if that term is greater than
$$(|fxyzj|*e[2*(j+n)-1]+e[2*(j+n)])*|h|/k,$$
for any value of j, $1 \le j \le n$, $(k=|b-a|)$.

Procedure parameters:
    void rk2n (x,a,b,y,ya,z,za,method,e,d,fi,n)
x: double x[0:0];
    the independent variable;
    exit: upon completion of a call of *rk2n*, it is equal to *b*;
a: double;
    entry: the starting value of *x*;
b: double;
    entry: the end value of *x*;
y: double y[1:n];
    the vector of dependent variables;
    exit: the values of $y[j]$ at $x = b$, $j=1,...,n$;
ya: double ya[1:n];
    entry: the initial values of $y[j]$, i.e., values at $x=a$;
z: double z[1:n];
    the first derivatives of the dependent variables;
    exit: the value of $(d/dx)y[j]$ at $x = b$, $j=1,...,n$;
za: double za[1:n];
    entry: the initial values of $z[j]$, i.e., the values at $x=a$;
method: a class that defines a procedure *fxyzj*, this class must implement the AP_rk2n_method interface;
    double fxyzj(int n, int j, double x[ ], double y[ ], double z[ ])
     fxyzj depends on $x,j,y[j],z[j]$, $(j=1,...,n)$, giving the value of $d^2y[j]/dx^2$;
e: double e[1:4*n];
    the element $e[2*j-1]$ is a relative and $e[2*j]$ an absolute tolerance

associated with $y[j]$; $e[2*(n+j)-1]$ is a relative and $e[2*(n+j)]$ an absolute tolerance associated with $z[j]$;

d: double $d[1:2*n+3]$;

        exit: $d[1]$:   the number of steps skipped;

               $d[2]$:   the last step length used;

               $d[3]$:   equal to b;

               $d[4],...,d[n+3]$:   equal to $y[1],...,y[n]$ for $x = b$;

               $d[n+4],...,d[2*n+3]$: equal to the derivatives $z[1],...,z[n]$ for $x = b$;

fi: boolean;

        entry: if fi is true then the integration starts at $x=a$ with a trial step b-a; if fi is false then the integration is continued with, as initial conditions, $x=d[3]$, $y[j]=d[j+3]$, $z[j]=d[n+3+j]$, and step length $h=d[2]*\mathrm{sign}(b-d[3])$, and a, ya, and za are ignored;

n: int;

        entry: the number of equations.

```
package numal;

public interface AP_rk2n_method {

    double fxyzj(int n, int j, double x[], double y[], double z[]);
}



public static void rk2n(double x[], double a, double b, double y[], double ya[],
                        double z[], double za[], AP_rk2n_method method, double e[],
                        double d[], boolean fi, int n)
{
    boolean last,first,reject,test,ta,tb;
    int j,jj;
    double xl,h,ind,hmin,hl,absh,fhm,discry,discrz,toly,tolz,mu,mu1,fhy,fhz;
    double yl[]  = new double[n+1];
    double zl[]  = new double[n+1];
    double k0[]  = new double[n+1];
    double k1[]  = new double[n+1];
    double k2[]  = new double[n+1];
    double k3[]  = new double[n+1];
    double k4[]  = new double[n+1];
    double k5[]  = new double[n+1];
    double ee[]  = new double[4*n+1];

    last=true;
    hl=mu1=0.0;
    if (fi) {
        d[3]=a;
        for (jj=1; jj<=n; jj++) {
            d[jj+3]=ya[jj];
            d[n+jj+3]=za[jj];
        }
    }
```

```
d[1]=0.0;
xl=d[3];
for (jj=1; jj<=n; jj++) {
    yl[jj]=d[jj+3];
    zl[jj]=d[n+jj+3];
}
if (fi) d[2]=b-d[3];
absh=h=Math.abs(d[2]);
if (b-xl < 0.0) h = -h;
ind=Math.abs(b-xl);
hmin=ind*e[1]+e[2];
for (jj=2; jj<=2*n; jj++) {
    hl=ind*e[2*jj-1]+e[2*jj];
    if (hl < hmin) hmin=hl;
}
for (jj=1; jj<=4*n; jj++) ee[jj]=e[jj]/ind;
first=true;
test=true;
if (fi) {
    last=true;
    test=false;
}
while (true) {
    if (test) {
        absh=Math.abs(h);
        if (absh < hmin) {
            h = (h > 0.0) ? hmin : -hmin;
            absh=Math.abs(h);
        }
        ta=(h >= b-xl);
        tb=(h >= 0.0);
        if ((ta && tb) || (!(ta || tb))) {
            d[2]=h;
            last=true;
            h=b-xl;
            absh=Math.abs(h);
        } else
            last=false;
    }
    test=true;
    x[0]=xl;
    for (jj=1; jj<=n; jj++) {
        y[jj]=yl[jj];
        z[jj]=zl[jj];
    }
    for (j=1; j<=n; j++)    k0[j]=method.fxyzj(n,j,x,y,z)*h;
    x[0]=xl+h/4.5;
    for (jj=1; jj<=n; jj++) {
        y[jj]=yl[jj]+(zl[jj]*18.0+k0[jj]*2.0)/81.0*h;
        z[jj]=zl[jj]+k0[jj]/4.5;
```

```
    }
    for (j=1; j<=n; j++) k1[j]=method.fxyzj(n,j,x,y,z)*h;
    x[0]=xl+h/3.0;
    for (jj=1; jj<=n; jj++) {
        y[jj]=yl[jj]+(zl[jj]*6.0+k0[jj])/18.0*h;
        z[jj]=zl[jj]+(k0[jj]+k1[jj]*3.0)/12.0;
    }
    for (j=1; j<=n; j++) k2[j]=method.fxyzj(n,j,x,y,z)*h;
    x[0]=xl+h*0.5;
    for (jj=1; jj<=n; jj++) {
        y[jj]=yl[jj]+(zl[jj]*8.0+k0[jj]+k2[jj])/16.0*h;
        z[jj]=zl[jj]+(k0[jj]+k2[jj]*3.0)/8.0;
    }
    for (j=1; j<=n; j++) k3[j]=method.fxyzj(n,j,x,y,z)*h;
    x[0]=xl+h*0.8;
    for (jj=1; jj<=n; jj++) {
        y[jj]=yl[jj]+(zl[jj]*100.0+k0[jj]*12.0+k3[jj]*28.0)/125.0*h;
        z[jj]=zl[jj]+(k0[jj]*53.0-k1[jj]*135.0+k2[jj]*126.0+k3[jj]*56.0)/125.0;
    }
    for (j=1; j<=n; j++) k4[j]=method.fxyzj(n,j,x,y,z)*h;
    x[0] = (last ? b : xl+h);
    for (jj=1; jj<=n; jj++) {
        y[jj]=yl[jj]+(zl[jj]*336.0+k0[jj]*21.0+k2[jj]*92.0+k4[jj]*55.0)/336.0*h;
        z[jj]=zl[jj]+(k0[jj]*133.0-k1[jj]*378.0+k2[jj]*276.0+k3[jj]*112.0+
                k4[jj]*25.0)/168.0;
    }
    for (j=1; j<=n; j++) k5[j]=method.fxyzj(n,j,x,y,z)*h;
    reject=false;
    fhm=0.0;
    for (jj=1; jj<=n; jj++) {
        discry=Math.abs((-k0[jj]*21.0+k2[jj]*108.0-k3[jj]*112.0+
                    k4[jj]*25.0)/56.0*h);
        discrz=Math.abs(k0[jj]*21.0-k2[jj]*162.0+k3[jj]*224.0k4[jj]*125.0+
                    k5[jj]*42.0)/14.0;
        toly=absh*(Math.abs(zl[jj])*ee[2*jj-1]+ee[2*jj]);
        tolz=Math.abs(k0[jj])*ee[2*(jj+n)-1]+absh*ee[2*(jj+n)];
        reject=((discry > toly) || (discrz > tolz) || reject);
        fhy=discry/toly;
        fhz=discrz/tolz;
        if (fhz > fhy) fhy=fhz;
        if (fhy > fhm) fhm=fhy;
    }
    mu=1.0/(1.0+fhm)+0.45;
    if (reject) {
        if (absh <= hmin) {
            d[1] += 1.0;
            for (jj=1; jj<=n; jj++) {
                y[jj]=yl[jj];
                z[jj]=zl[jj];
            }
```

```
                 first=true;
                 if (b == x[0]) break;
                 xl = x[0];
                 for (jj=1; jj<=n; jj++) {
                     yl[jj] = y[jj];
                     zl[jj] = z[jj];
                 }
             } else
                 h *= mu;
         } else {
             if (first) {
                 first=false;
                 hl=h;
                 h *= mu;
             } else {
                 fhm=mu*h/hl+mu-mu1;
                 hl=h;
                 h *= fhm;
             }
             mu1=mu;
             for (jj=1; jj<=n; jj++) {
                 y[jj]=yl[jj]+(zl[jj]*56.0+k0[jj]*7.0+k2[jj]*36.0-
                     k4[jj]*15.0)/56.0*hl;
                 z[jj]=zl[jj]+(-k0[jj]*63.0+k1[jj]*189.0-k2[jj]*36.0-
                     k3[jj]*112.0+k4[jj]*50.0)/28.0;
             }
             for (j=1; j<=n; j++) k5[j]=method.fxyzj(n,j,x,y,z)*hl;
             for (jj=1; jj<=n; jj++) {
                 y[jj]=yl[jj]+(zl[jj]*336.0+k0[jj]*35.0+k2[jj]*108.0+
                     k4[jj]*25.0)/336.0*hl;
                 z[jj]=zl[jj]+(k0[jj]*35.0+k2[jj]*162.0+k4[jj]*125.0+
                     k5[jj]*14.0)/336.0;
             }
             if (b == x[0]) break;
             xl = x[0];
             for (jj=1; jj<=n; jj++) {
                 yl[jj] = y[jj];
                 zl[jj] = z[jj];
             }
         }
     }
 }
 if (!last) d[2]=h;
 d[3] = x[0];
 for (jj=1; jj<=n; jj++) {
     d[jj+3]=y[jj];
     d[n+jj+3]=z[jj];
 }
}
```

C. rk3

Solves an initial value problem for a single second order ordinary differential equation without first derivative

$$d^2y/dx^2 = f(x,y),$$

from $x=a$ to $x=b$, $y(a)$ and $dy(a)/da$ being given, by means of a 5-th order Runge-Kutta method [Z64].

Upon completion of a call of $rk3$ we have $x=d[3]=b$, $y=d[4]=y[b]$, and $z=d[5]$, i.e., the value of dy/dx for $x=b$. $rk3$ uses as its minimal absolute step length

$$hmin = \min \ (e[2*j-1]*k+e[2*j]) \text{ with } 1 \leq j \leq 2$$

and $k=| b -$ (if fi is true then a else $d[3])|$. If a step of length $|h| \leq hmin$ is rejected then a step $\text{sign}(h)*hmin$ is skipped. A step is rejected if the absolute value of the last term taken into account is greater than $(|dy/dx|*e[1]+e[2])*|h|/k$ or if that term is greater than $(|fxy|*e[3]+e[4])*|h|/k$ where $k=|b-a|$.

Procedure parameters:

$$\text{void rk3 } (x,a,b,y,ya,z,za,method,e,d,fi)$$

x: double $x[0:0]$;
           the independent variable;
           exit: upon completion of a call of $rk3$, it is equal to b;
a: double;
           entry: the initial value of x;
b: double;
           entry: the end value of x, ($b \leq a$ is allowed);
y: double $y[0:0]$;
           the dependent variable;
           exit: the value of $y(x)$ at $x = b$;
ya: double;
           entry: the initial value of y at $x=a$;
z: double $z[0:0]$;
           the derivative dy/dx;
           exit: the value of dy/dx at $x = b$;
za: double;
           entry: the initial value of dy/dx at $x=a$;
$method$:   a class that defines a procedure fxy, this class must implement the
           AP_rk3_method interface;
           double $fxy(double\ x[\ ], double\ y[\ ])$
               fxy depends on x and y, giving the value of d^2y/dx^2;
e: double $e[1:4]$;
           entry: $e[1]$ and $e[3]$ are used as relative tolerances, $e[2]$ and $e[4]$ are used
                    as absolute tolerances for y and dy/dx, respectively;
d: double $d[1:5]$;
           exit: $d[1]$:   the number of steps skipped;
                   $d[2]$:   the last step length used;
                   $d[3]$:   equal to b;
                   $d[4]$:   equal to $y(b)$;
                   $d[5]$:   equal to dy/dx, for $x = b$;
fi: boolean;
           entry: if fi is true then the integration starts at $x=a$ with a trial step $b-a$;

if *fi* is false then the integration is continued with, as initial conditions, *x=d[3]*, *y=d[4]*, *z=d[5]*, and steplength *h=d[2]*sign(*b-d[3]*), *a*, *ya*, and *za* are ignored.

```
package numal;

public interface AP_rk3_method {

    double fxy(double x[], double y[]);
}


public static void rk3(double x[], double a, double b, double y[], double ya,
                       double z[], double za, AP_rk3_method method, double e[],
                       double d[], boolean fi)
{
    boolean last,first,reject,test,ta,tb;
    double e1,e2,e3,e4,x1,y1,z1,h,ind,hmin,hl,absh,k0,k1,k2,k3,k4,k5,discry,discrz,
           toly,tolz,mu,mu1,fhy,fhz;

    k5=mu1=0.0;
    last=true;
    if (fi) {
        d[3]=a;
        d[4]=ya;
        d[5]=za;
    }
    d[1]=0.0;
    x1=d[3];
    y1=d[4];
    z1=d[5];
    if (fi) d[2]=b-d[3];
    absh=h=Math.abs(d[2]);
    if (b-x1 < 0.0) h = -h;
    ind=Math.abs(b-x1);
    hmin=ind*e[1]+e[2];
    hl=ind*e[3]+e[4];
    if (hl < hmin) hmin=hl;
    e1=e[1]/ind;
    e2=e[2]/ind;
    e3=e[3]/ind;
    e4=e[4]/ind;
    first=reject=true;
    test=true;
    if (fi) {
        last=true;
        test=false;
    }
    while (true) {
        if (test) {
```

```
        absh=Math.abs(h);
        if (absh < hmin) {
            h = (h > 0.0) ? hmin : -hmin;
            absh=hmin;
        }
        ta=(h >= b-xl);
        tb=(h >= 0.0);
        if ((ta && tb) || (!(ta || tb))) {
            d[2]=h;
            last=true;
            h=b-xl;
            absh=Math.abs(h);
        } else
            last=false;
    }
    test=true;
    if (reject) {
        x[0]=xl;
        y[0]=yl;
        k0=method.fxy(x,y)*h;
    } else
        k0=k5*h/hl;
    x[0]=xl+0.276393202250021*h;
    y[0]=yl+(zl*0.276393202250021+k0*0.038196601125011)*h;
    k1=method.fxy(x,y)*h;
    x[0]=xl+0.723606797749979*h;
    y[0]=yl+(zl*0.723606797749979+k1*0.261803398874989)*h;
    k2=method.fxy(x,y)*h;
    x[0]=xl+h*0.5;
    y[0]=yl+(zl*0.5+k0*0.046875+k1*0.079824155839840-k2*0.001699155839840)*h;
    k4=method.fxy(x,y)*h;
    x[0] = (last ? b : xl+h);
    y[0]=yl+(zl+k0*0.309016994374947+k2*0.190983005625053)*h;
    k3=method.fxy(x,y)*h;
    y[0]=yl+(zl+k0*0.083333333333333+k1*0.301502832395825+
        k2*0.115163834270842)*h;
    k5=method.fxy(x,y)*h;
    discry=Math.abs((-k0*0.5+k1*1.809016994374947+
                k2*0.690983005625053-k4*2.0)*h);
    discrz=Math.abs((k0-k3)*2.0-(k1+k2)*10.0+k4*16.0+k5*4.0);
    toly=absh*(Math.abs(zl)*e1+e2);
    tolz=Math.abs(k0)*e3+absh*e4;
    reject=(discry > toly || discrz > tolz);
    fhy=discry/toly;
    fhz=discrz/tolz;
    if (fhz > fhy) fhy=fhz;
    mu=1.0/(1.0+fhy)+0.45;
    if (reject) {
        if (absh <= hmin) {
            d[1] += 1.0;
```

```
                    y[0]=yl;
                    z[0]=zl;
                    first=true;
                    if (b == x[0]) break;
                    xl = x[0];
                    yl = y[0];
                    zl = z[0];
                } else
                    h *= mu;
            } else {
                if (first) {
                    first=false;
                    hl=h;
                    h *= mu;
                } else {
                    fhy=mu*h/hl+mu-mul;
                    hl=h;
                    h *= fhy;
                }
                mul=mu;
                z[0]=zl+(k0+k3)*0.083333333333333+(k1+k2)*0.416666666666667;
                if (b == x[0]) break;
                xl = x[0];
                yl = y[0];
                zl = z[0];
            }
        }
    }
    if (!last) d[2]=h;
    d[3] = x[0];
    d[4] = y[0];
    d[5] = z[0];
}
```

D. rk3n

Solves an initial value problem for a system of second order ordinary differential equations without first derivative
$$d^2y_j(x)/dx^2 = f_j(x,y), \qquad (j=1,2,...,n)$$
from $x=a$ to $x=b$, $y(a)$ and $dy(a)/da$ being given, by means of a 5-th order Runge-Kutta method [Z64].

Upon completion of a call of $rk3n$ we have: $x=d[3]=b$, $y[j]=d[j+3]$ the value of the dependent variables for $x=b$, $z[j]=d[n+j+3]$, the value of the derivatives of $y[j]$ at $x=b$. $rk3n$ uses as its minimal absolute step length
$$hmin = \min (e[2*j-1]*k+e[2*j]) \text{ with } 1 \le j \le 2n$$
and
$$k = |b - (\text{if } fi \text{ is true then } a \text{ else } d[3])|.$$
If a step of length $|h| \le hmin$ is rejected then a step $sign(h)*hmin$ is

skipped. A step is rejected if the absolute value of the last term taken into account is greater than

$$(|z[j]|*e[2*j-1]+e[2*j])*|h|/k$$

or if that term is greater than

$$(|fxyj|*e[2*(j+n)-1]+e[2*(j+n)])*|h|/k,$$

for any value of j, $1 \leq j \leq n$, $(k=|b-a|)$.

Procedure parameters:

                  void rk3n $(x,a,b,y,ya,z,za,method,e,d,fi,n)$

x: double $x[0:0]$;
             the independent variable;
             exit: upon completion of a call of *rk3n*, it is equal to b;
a: double;
             entry: the starting value of x;
b: double;
             entry: the end value of x ($b \leq a$ is allowed);
y: double $y[1:n]$;
             the vector of dependent variables;
             exit: the values of $y[j]$ at $x = b$, $j=1,\ldots,n$;
ya: double $ya[1:n]$;
             entry: the initial values of $y[j]$, i.e., values at $x=a$;
z: double $z[1:n]$;
             the first derivatives of the dependent variables, $z[j] = dy[j]/dx$;
             exit: the value of $z[j](x)$ at $x = b$, $j=1,\ldots,n$;
za: double $za[1:n]$;
             entry: the initial values of $z[j]$, i.e., the values at $x=a$;
method: a class that defines a procedure *fxyj*, this class must implement the AP_rk3n_method interface;
             double *fxyj*(int n, int j, double x[ ], double y[ ])
                fxyj depends on $x,y[1],y[2],\ldots,y[n],j$, giving the value of $d^2y[j]/dx^2$;
e: double $e[1:4*n]$;
             the element $e[2*j-1]$ is a relative and $e[2*j]$ is an absolute tolerance associated with $y[j]$; $e[2*(n+j)-1]$ is a relative and $e[2*(n+j)]$ is an absolute tolerance associated with $z[j]$;
d: double $d[1:2*n+3]$;
             exit: $d[1]$:   the number of steps skipped;
                  $d[2]$:   the last step length used;
                  $d[3]$:   equals to b;
                  $d[4],\ldots,d[n+3]$:   equal to $y[1],\ldots,y[n]$ for $x = b$;
                  $d[n+4],\ldots,d[2*n+3]$:   equal to the derivatives $z[1],\ldots,z[n]$ for $x = b$;
fi: boolean;
             entry: if *fi* is true then the integration starts at $x=a$ with a trial step $b-a$; if *fi* is false then the integration is continued with, as initial conditions, $x=d[3]$, $y[j]=d[j+3]$, $z[j]=d[n+3+j]$, and step length $h=d[2]*sign(b-d[3])$, and a, ya, and za are ignored;
n: int;
             entry: the number of equations.

```
package numal;

public interface AP_rk3n_method {

    double fxyj(int n, int j, double x[], double y[]);
}


public static void rk3n(double x[], double a, double b, double y[], double ya[],
                        double z[], double za[], AP_rk3n_method method,
                        double e[], double d[], boolean fi, int n)
{
    boolean last,first,reject,test,ta,tb;
    int j,jj;
    double xl,h,hmin,ind,hl,absh,fhm,discry,discrz,toly,tolz,mu,mu1,fhy,fhz;
    double yl[] = new double[n+1];
    double zl[] = new double[n+1];
    double k0[] = new double[n+1];
    double k1[] = new double[n+1];
    double k2[] = new double[n+1];
    double k3[] = new double[n+1];
    double k4[] = new double[n+1];
    double k5[] = new double[n+1];
    double ee[] = new double[4*n+1];

    hl=mu1=0.0;
    last=true;
    if (fi) {
        d[3]=a;
        for (jj=1; jj<=n; jj++) {
            d[jj+3]=ya[jj];
            d[n+jj+3]=za[jj];
        }
    }
    d[1]=0.0;
    xl=d[3];
    for (jj=1; jj<=n; jj++) {
        yl[jj]=d[jj+3];
        zl[jj]=d[n+jj+3];
    }
    if (fi) d[2]=b-d[3];
    absh=h=Math.abs(d[2]);
    if (b-xl < 0.0) h = -h;
    ind=Math.abs(b-xl);
    hmin=ind*e[1]+e[2];
    for (jj=2; jj<=2*n; jj++) {
        hl=ind*e[2*jj-1]+e[2*jj];
        if (hl < hmin) hmin=hl;
    }
    for (jj=1; jj<=4*n; jj++) ee[jj]=e[jj]/ind;
```

```
first=reject=true;
test=true;
if (fi) {
    last=true;
    test=false;
}
while (true) {
    if (test) {
        absh=Math.abs(h);
        if (absh < hmin) {
            h = (h > 0.0) ? hmin : -hmin;
            absh=hmin;
        }
        ta=(h >= b-xl);
        tb=(h >= 0.0);
        if ((ta && tb) || (!(ta || tb))) {
            d[2]=h;
            last=true;
            h=b-xl;
            absh=Math.abs(h);
        } else
            last=false;
    }
    test=true;
    if (reject) {
        x[0]=xl;
        for (jj=1; jj<=n; jj++) y[jj]=yl[jj];
        for (j=1; j<=n; j++) k0[j]=method.fxyj(n,j,x,y)*h;
    } else {
        fhy=h/hl;
        for (jj=1; jj<=n; jj++) k0[jj]=k5[jj]*fhy;
    }
    x[0]=xl+0.276393202250021*h;
    for (jj=1; jj<=n; jj++)
        y[jj]=yl[jj]+(zl[jj]*0.276393202250021+k0[jj]*0.038196601125011)*h;
    for (j=1; j<=n; j++) k1[j]=method.fxyj(n,j,x,y)*h;
    x[0]=xl+0.723606797749979*h;
    for (jj=1; jj<=n; jj++)
        y[jj]=yl[jj]+(zl[jj]*0.723606797749979+k1[jj]*0.261803398874989)*h;
    for (j=1; j<=n; j++) k2[j]=method.fxyj(n,j,x,y)*h;
    x[0]=xl+h*0.5;
    for (jj=1; jj<=n; jj++)
        y[jj]=yl[jj]+(zl[jj]*0.5+k0[jj]*0.046875+k1[jj]*0.079824155839840-
                k2[jj]*0.001699155839840)*h;
    for (j=1; j<=n; j++) k4[j]=method.fxyj(n,j,x,y)*h;
    x[0] = (last ? b : xl+h);
    for (jj=1; jj<=n; jj++)
        y[jj]=yl[jj]+(zl[jj]+k0[jj]*0.309016994374947+
                k2[jj]*0.190983005625053)*h;
    for (j=1; j<=n; j++) k3[j]=method.fxyj(n,j,x,y)*h;
```

```
for (jj=1; jj<=n; jj++)
    y[jj]=yl[jj]+(zl[jj]+k0[jj]*0.083333333333333+k1[jj]*
            0.301502832395825+k2[jj]*0.115163834270842)*h;
for (j=1; j<=n; j++) k5[j]=method.fxyj(n,j,x,y)*h;
reject=false;
fhm=0.0;
for (jj=1; jj<=n; jj++) {
    discry=Math.abs((-k0[jj]*0.5+k1[jj]*1.809016994374947+
                    k2[jj]*0.690983005625053-k4[jj]*2.0)*h);
    discrz=Math.abs((k0[jj]-k3[jj])*2.0-(k1[jj]+k2[jj])*10.0+
                    k4[jj]*16.0+k5[jj]*4.0);
    toly=absh*(Math.abs(zl[jj])*ee[2*jj-1]+ee[2*jj]);
    tolz=Math.abs(k0[jj])*ee[2*(jj+n)-1]+absh*ee[2*(jj+n)];
    reject=((discry > toly) || (discrz > tolz) || reject);
    fhy=discry/toly;
    fhz=discrz/tolz;
    if (fhz > fhy) fhy=fhz;
    if (fhy > fhm) fhm=fhy;
}
mu=1.0/(1.0+fhm)+0.45;
if (reject) {
    if (absh <= hmin) {
        d[1]  += 1.0;
        for (jj=1; jj<=n; jj++) {
            y[jj]=yl[jj];
            z[jj]=zl[jj];
        }
        first=true;
        if (b == x[0]) break;
        xl = x[0];
        for (jj=1; jj<=n; jj++) {
            yl[jj]=y[jj];
            zl[jj]=z[jj];
        }
    } else
        h *= mu;
} else {
    if (first) {
        first=false;
        hl=h;
        h *= mu;
    } else {
        fhy=mu*h/hl+mu-mu1;
        hl=h;
        h *= fhy;
    }
    mu1=mu;
    for (jj=1; jj<=n; jj++)
        z[jj]=zl[jj]+(k0[jj]+k3[jj])*0.083333333333333+
                    (k1[jj]+k2[jj])*0.416666666666667;
```

```
            if (b == x[0]) break;
            x1 = x[0];
            for (jj=1; jj<=n; jj++) {
                y1[jj]=y[jj];
                z1[jj]=z[jj];
            }
        }
    }
}
if (!last) d[2]=h;
d[3] = x[0];
for (jj=1; jj<=n; jj++) {
    d[jj+3]=y[jj];
    d[n+jj+3]=z[jj];
}
}
```

5.4.5 Initial boundary value problem

arkmat

Solves an initial value problem, given as a system of first order (nonlinear) differential equations

$$DU(t) = F(t, U(t)) \tag{1}$$

where $D=d/dt$ and U is an $n \times m$ matrix, with $U(t_0)$ prescribed, over the range $t_0 \leq t \leq t_e$, by means of a stabilized Runge-Kutta method [Vh71]. This procedure is suitable for the integration of systems where the dependent variable and the right-hand side are stored in a rectangular array instead of a vector. The integration stepsize used will depend on (a) the type of system to be solved (i.e., hyperbolic or parabolic); (b) the spectral radius of the Jacobian matrix of the system; and (c) the indicated order of the particular Runge-Kutta method. *arkmat* is especially intended for systems of differential equations arising from initial boundary value problems in two dimensions, e.g., when the method of lines is applied to this kind of problem, the right-hand side of the resulting system is much easier to describe in matrix than in vector form.

The method used is associated with one of four prescribed ninth degree stability polynomials (the choice being open to the user): the first has a degree of polynomial exactness equal to 9; the second and third have absolute values less than unity when their arguments lie in the range $[-\beta(9), 0]$ for the maximal values of $\beta(9)$, and have degrees of polynomial exactness equal to 1 and 2, respectively (such polynomials are suitable for the numerical solution of parabolic partial differential equations, the eigenvalues of the Jacobian matrix derived from the right-hand side of equation (1) (F and U being expanded to vector form) which lie in the closed left half-plane being real); the fourth polynomial has absolute value less than unity when its argument lies in the range $i[-8,8]$ (it is suitable for the numerical solution of hyperbolic partial differential equations, the eigenvalues of the derived Jacobian matrix lying in the closed left half-plane being confined to the imaginary axis). The Runge-Kutta scheme employed has the form (with N=9 below)

$$F_{k,0} = F(t_k, U_k)$$
$$F_{k,i} = F(t + \lambda_i \tau, U_k + \lambda_k \tau F_{k,i-1}) \qquad (i=1,\dots,N-1) \qquad (2)$$
$$U_{k+1} = U_k + F_{k,n-1}$$

where $U_k \approx U(t_k)$ and $t_{k+1} = t_k + \tau$ $(k=0,1,\dots)$, the stepsize τ being prescribed. The coefficients λ_i are derived from the coefficients β_i of the stability polynomial

$$P(z) = \sum_{i=0}^{N} \beta_i z^i$$

chosen, by use of the relationships

$$\lambda_i = \beta_{N-i+1} / \beta_{N-i} \qquad (i=1,\dots,N-1).$$

The choice (among the four described above) of stability polynomial is determined by the values allocated to the input parameters *type* and *order* upon call of *arkmat*. If *type*=1 the first polynomial is chosen, if *type*=2 and *order*=1 the second, if *type*=2 and *order*=2 the third, and if *type*=3, the fourth (if *type*≠2, the value of *order* is ignored). The fixed stepsize τ is determined partly by the range constant $\beta(n)$ of the stability polynomial associated with the method being used, and partly by the value of the input parameter *spr* (this will in general be a multiple of the spectral radius with respect to the eigenvalues in the closed left half-plane of the Jacobian matrix derived from equation (1) upon call of *arkmat*. If *type*=1, τ=4.3/*spr*, if *type*=2 and *order*=1, τ=156/*spr*, if *type*=2 and *order*=2, τ=64/*spr*, and if *type*=3, τ=8/*spr*.

Procedure parameters:

        void arkmat (*t,te,m,n,u,method,type,order,spr*)

t: double *t[0:0]*;
    the independent variable;
    entry: initial value *t0*;
    exit: the final value *te*;
te: double;
    entry: the final value of *t*;
m: int;
    entry: the number of columns of *u*;
n: int;
    entry: the number of rows of *u*;
u: double *u[1:n, 1:m]*;
    entry: the initial values of the solution of the system of differential equations at *t=t0*;
    exit: the values of the solution at *t=te*;
method: a class that defines two procedures *der* and *out*, this class must implement the AP_arkmat_methods interface;
    double der(int m, int n, double t, double v[ ][ ], double ftv[ ][ ])
        this procedure must be given by the user and performs an evaluation of the right-hand side $F(t,v)$ of the system; upon completion of *der*, the right-hand side should be stored in the array *ftv[1:n, 1:m]*;
    double out(double t[ ], double te, int m, int n, double u[ ][ ],int type,
        int order[ ], double spr[ ])
        after each integration step the user can print some parameters such as *t, u, type, order,* or possibly update *spr*;

type: int;
         entry: the type of the system of differential equations to be solved; the
              user should supply one of the following values:
              type=1:   if no specification of the type can be made;
              type=2:   if the eigenvalues of the Jacobian matrix of the right-
                     hand side are negative real;
              type=3:   if the eigenvalues of the Jacobian matrix of the right-
                     hand side are purely imaginary;
order: int *order[0:0]*;
         the order of the Runge-Kutta method used;
         entry: for *type*=2 the user may choose *order*=1 or *order*=2; *order* should
              be 2 for the other types;
         exit: if *order* is set to another value, it is assumed to be (if *type*=2 then
              1 else 2);
spr: double *spr[0:0]*;
         entry: the spectral radius of the Jacobian matrix of the right-hand side,
              when the system is written in one dimensional form (i.e., vector
              form);
              the integration step will equal constant/ *spr* (see above);
              if necessary *spr* can be updated (after each step) by means of the
              procedure *out*.

Procedures used: elmcol, dupmat.

```
package numal;

public interface AP_arkmat_methods {

    void der(int m, int n, double t, double v[][], double ftv[][]);
    void out(double t[], double te, int m, int n, double u[][], int type,
             int order[], double spr[]);
}


public static void arkmat(double t[], double te, int m, int n, double u[][],
                          AP_arkmat_methods method, int type, int order[],
                          double spr[])
{
    boolean last,ta,tb;
    int sig,l,i;
    double tau,mlt;
    double lbd1[]={1.0/9.0, 1.0/8.0, 1.0/7.0, 1.0/6.0, 1.0/5.0, 1.0/4.0, 1.0/3.0,
                  1.0/2.0, 4.3};
    double lbd2[]={0.1418519249e-2, 0.3404154076e-2, 0.0063118569, 0.01082794375,
                  0.01842733851, 0.03278507942, 0.0653627415, 0.1691078577, 156.0};
    double lbd3[]={0.3534355908e-2, 0.8532600867e-2, 0.015956206, 0.02772229155,
                  0.04812587964, 0.08848689452, 0.1863578961, 0.5, 64.0};
    double lbd4[]={1.0/8.0, 1.0/20.0, 5.0/32.0, 2.0/17.0, 17.0/80.0, 5.0/22.0,
                  11.0/32.0, 1.0/2.0, 8.0};
    double lambda[] = new double[10];
```

```
double uh[][] = new double[n+1][m+1];
double du[][] = new double[n+1][m+1];

/* initialize */
if (type != 2 && type != 3) type=1;
if (type != 2)
    order[0] = 2;
else
    if (order[0] != 2) order[0] = 1;
switch ((type == 1) ? 1 : type+order[0]-1) {
    case 1:  for (i=0; i<=8; i++) lambda[i+1]=lbd1[i]; break;
    case 2:  for (i=0; i<=8; i++) lambda[i+1]=lbd2[i]; break;
    case 3:  for (i=0; i<=8; i++) lambda[i+1]=lbd3[i]; break;
    case 4:  for (i=0; i<=8; i++) lambda[i+1]=lbd4[i]; break;
}
sig = ((te == t[0]) ? 0 : ((te > t[0]) ? 1 : -1));
last=false;
do {
    tau=((spr[0] == 0.0) ? Math.abs(te-t[0]) : Math.abs(lambda[9]/spr[0]))*sig;
    ta = t[0]+tau >= te;
    tb = tau >= 0.0;
    if ((ta && tb) || (!(ta || tb))) {
        tau=te-t[0];
        last=true;
    }
    /* difference scheme */
    method.der(m,n,t[0],u,du);
    for (i=1; i<=8; i++) {
        mlt=lambda[i]*tau;
        Basic.dupmat(1,n,1,m,uh,u);
        for (l=1; l<=m; l++) Basic.elmcol(1,n,l,l,uh,du,mlt);
        method.der(m,n,t[0]+mlt,uh,du);
    }
    for (l=1; l<=m; l++) Basic.elmcol(1,n,l,l,u,du,tau);
    t[0] = (last ? te : t[0]+tau);
    method.out(t,te,m,n,u,type,order,spr);
} while (!last);
}
```

5.5 Two point boundary value problems

5.5.1 Linear methods - Second order self adjoint

A. femlagsym

Solves a second order self-adjoint linear two point boundary value problem by means of Galerkin's method with continuous piecewise polynomials [BakH76, He75, StF73]. *femlagsym* computes approximations y_n ($n=0,...,N$) to the values at the points $x = x_n$, where

$$-\infty < a = x_0 < x_1 < ... < x_N = b < \infty$$

of the real valued function $y(x)$ which satisfies the equation

$$-D(p(x)Dy(x)) + r(x)y(x) = f(x) \tag{1}$$

where $D=d/dx$, with boundary conditions

$$e_1y(a) + e_2Dy(a) = e_3, \quad e_4y(b) + e_5Dy(b) = e_6 \quad (e_1, e_4 \neq 0). \tag{2}$$

It is assumed that

(a) $p(x) > 0$ on the interval $[x_{n-1}, x_n]$, $n=1,...,N$;

(b) $p(x)$, $r(x)$, and $f(x)$ are required to be sufficiently smooth on $[x_0, x_N]$ except at the grid points where $p(x)$ should be at least continuous; in that case the order of accuracy (2, 4, or 6) is preserved;

(c) $r(x) \geq 0$ on $[x_0, x_N]$;

if, however, the problem has pure Dirichlet boundary conditions (i.e., $e[2]=e[5]=0$) this condition can be weakened to the requirement that $r(x) > -p0*(\pi/(x_N-x_0))^2$, where $p0$ is the minimum of $p(x)$ on $[x_0, x_N]$; one should note that the problem may be ill-conditioned when $r(x)$ is quite near that lower bound; for other negative values of $r(x)$ the existence of a solution remains an open question.

The theory upon which the method is based is explained with reference to the more general equation

$$-D(p(x)Dy(x)) + q(x)Dy(x) + r(x)y(x) = f(x) \tag{3}$$

of which equation (1) and the equations treated by other procedures (*femlag*, *femlagskew*) are special cases. $y(x)$ is approximated by a sum involving $Nk+1$ functions $\varphi_{1,0}$ and $\varphi_{m,j}$:

$$y(x) = a_0\phi_{1,0}(x) + \sum_{m=1}^{N}\sum_{j=1}^{k} a_{m,j}\phi_{m,j}(x) \tag{4}$$

Inserting this expansion into equation (3), multiplying throughout by $\varphi_{n,l}(x)$ and integrating by parts, there follows

$$\left[p(x)a_0\phi_{n,l}(x)D\phi_{1,0}(x) + p(x)\sum_{m=1}^{N}\sum_{j=1}^{k} a_{m,j}\phi_{n,l}(x)D\phi_{m,j}(x) \right]_a^b$$

$$+ a_0 \int_a^b \left\{ p(x)D\phi_{n,l}(x)D\phi_{1,0}(x) + q(x)\phi_{n,l}(x)D\phi_{1,0}(x) + r(x)\phi_{n,l}(x)\phi_{1,0}(x) \right\} dx$$

$$+ \sum_{m=1}^{N}\sum_{j=1}^{k} a_{m,j} \int_a^b \left\{ p(x)D\phi_{n,l}(x)D\phi_{m,j}(x) + q(x)\phi_{n,l}(x)D\phi_{m,j}(x) + r(x)\phi_{n,l}(x)\phi_{m,j}(x) \right\} dx$$

$$= \int_a^b f(x)\phi_{n,l}(x)dx \qquad (5)$$

Let $W_v^{(b)}$ and $\xi_v^{(k)}$ be the weights and arguments of the $k+1$ Radau-Lobatto quadrature formula over $[0,1]$, so that for suitable g

$$\int_0^1 g(x)dx = \sum_{v=0}^{k} W_v^{(k)} g(\xi_v^{(k)})$$

where $\xi_0^{(k)} = 0$, $\xi_k^{(k)} = 1$, and $W_0^{(k)} = W_k^{(k)}$. Denote the interval $[x_{n-1},x_n]$ by I_n and its length $x_n - x_{n-1}$ by Δ_n $(n=1,..,N)$, and let $\Xi_{n,v}^{(k)} = x_{n-1} + \xi_v^{(k)}\Delta_n$ $(n=1,...,N; v=0,...,k)$, these displaced Radau-Lobatto points lie in I_n, and

$$\Xi_{n,0}^{(k)} = x_{n-1}, \ \Xi_{n,k}^{(k)} = x_n = \Xi_{n+1,0}^{(k)}, \ \Xi_{n,v}^{(k)} - \Xi_{n,r}^{(k)} = (\xi_{n,v}^{(k)} - \xi_{n,r}^{(k)})\Delta_n.$$

The functions $\varphi_{1,0}$ and $\varphi_{n,l}$ $(n=1,...,N; l=1,...,k-1; n=N, l=1,...,k)$ above are taken to be polynomials of degree k over the interval I_1 and I_n $(n=1,...,N)$ respectively and zero elsewhere in $[a,b]$; $\varphi_{n,k}$ $(n=1,...,N-1)$ is a polynomial of degree k over I_n, another such polynomial over I_{n+1}, and is zero elsewhere in $[a,b]$. These polynomials are fixed by the conditions

$$\varphi_0(\Xi_{1,0}^{(k)}) = \varphi_0(a) = 1, \quad \varphi_0(\Xi_{1,v}^{(k)}) = 0 \quad (v=1,...,k)$$
$$\varphi_{n,l}(\Xi_{n,l}^{(k)}) = 1, \quad \varphi_{n,l}(\Xi_{n,v}^{(k)}) = 0 \quad (0 \le v \le k \ (v \ne l))$$

for $n=1,...,N, l=1,...,k-1; n=N, l=1,...,k$, and

$$\varphi_{n,k}(\Xi_{n,k}^{(k)}) = 1$$
$$\varphi_{n,k}(\Xi_{n,v}^{(k)}) = 0 \quad (v=0,...,k-1), \quad \varphi_{n,k}(\Xi_{n+1,v}^{(k)}) = 0 \quad (v=1,...,k)$$

for $n=1,...,N-1$. For suitable g

$$\int_a^b \phi_{n,l}(x)g(x)dx = \int_{I_n} \phi_{n,l}(x)g(x)dx = W_l^{(k)}g(\Xi_{n,l}^{(k)})\Delta_n$$

for $n=1, l=0; n=1,...,N, l=1,...,k-1; n=N, l=1,...,k$, and

$$\int_a^b \phi_{n,k}(x)g(x)dx = \int_{I_n+I_{n+1}} \phi_{n,k}(x)g(x)dx = W_0^{(k)}g(\Xi_{n,k}^{(k)})[\Delta_n + \Delta_{n+1}]$$

for $n=1,...,N-1$. Setting

$$\lambda(v) = \prod_{r=0}^{k} {}^{(v)}\left(\xi_v^{(k)} - \xi_r^{(k)}\right) \qquad (v = 0,\cdots,k)$$

$$\lambda'(v,\tau) = \left(\prod_{r=0}^{k} {}^{(\tau,v)}\left(\xi_v^{(k)} - \xi_r^{(k)}\right) \right) \Big/ \lambda(\tau) \qquad (v,\tau = 0,...,k)$$

where the superscripts (v) and (τ,v) imply that the terms corresponding to $r=v$ and $r=\tau$, $r=v$ are to be omitted, the polynomials $\varphi_{1,0}$, $\varphi_{n,l}$ are given explicitly by the formulae

$$\phi_{n,l}(x) = \left(\prod_{r=0}^{k}{}^{(l)}\left(x - \Xi_{n,r}^{(k)}\right)\right)\bigg/\lambda(l)\Delta_n^k \qquad (x \in I_n ; \quad l < k)$$

$$\phi_{n,k}(x) = \left(\prod_{r=0}^{k}{}^{(k)}\left(x - \Xi_{n,r}^{(k)}\right)\right)\bigg/\lambda(k)\Delta_n^k \qquad (x \in I_n)$$

$$\phi_{n,k}(x) = \left(\prod_{r=0}^{k}{}^{(0)}\left(x - \Xi_{n+1,r}^{(k)}\right)\right)\bigg/\lambda(0)\Delta_{n+1}^k \qquad (x \in I_{n+1}).$$

Their derivatives at the displaced Radau-Lobatto points lying in the intervals over which they are not identically zero are given by

$$D\varphi_{n,l}(\Xi_{n,v}^{(k)}) = \lambda'(v,l) / \Delta_n \qquad (l < k)$$
$$D\varphi_{n,k}(\Xi_{n,v}^{(k)}) = \lambda'(v,k) / \Delta_n, \qquad D\varphi_{n,k}(\Xi_{n+1,v}^{(k)}) = \lambda'(v,0) / \Delta_{n+1}.$$

Over intervals in which the $\varphi_{n,l}$ are zero, the derivative of $\varphi_{n,l}$ is, of course, also zero. The numbers $\lambda(v)$, $\lambda'(v,l)$ are problem independent; they may be, and have been, computed in advance.

The first of the boundary conditions (2) implies that

$$e_1 a_{1,0} + \sum_{j=1}^{k} e_2 \frac{\lambda'(o, j)}{\Delta_1} a_{i,j} = e_3$$

When $n=1,\ldots,N-1$, $l=1,\ldots,k$; $n=N$, $l=1,\ldots,k-1$, the expression denoted by square brackets in formula (5) vanishes, since $\varphi_{n,l}(a) = \varphi_{n,l}(b) = 0$. With $n=1$, the following term is equal to

$$\left[\frac{1}{\Delta_1}\sum_{v=0}^{k} W_v^{(k)} p\left(\Xi_{1,v}^{(k)}\right)\lambda'(v,l)\lambda'(v,0) + q\left(\Xi_{1,l}^{(k)}\right)\lambda'(v,0)\right] a_{1,0} \qquad (6)$$

for $l=1,\ldots,k$; when $n > 1$, this term vanishes. For $n=1,\ldots,N$, $l=1,\ldots,k-1$ the next term is equal to

$$\sum_{j=1}^{k} \left\{\frac{1}{\Delta_n}\sum_{v=0}^{k} W_v^{(k)} p\left(\Xi_{n,v}^{(k)}\right)\lambda'(v,l)\lambda'(v,j) + \right.$$
$$\left. W_l^{(k)} q\left(\Xi_{n,l}^{(k)}\right)\lambda'(l,j) + W_l^{(k)}\delta_{l,j} r\left(\Xi_{n,l}^{(k)}\right)\Delta_n\right\} a_{n,j} \qquad (7)$$

and when $n=1,\ldots,N-1$, $l=k$ it is equal to the sum of the above expression with $l=k$ and

$$\sum_{j=1}^{k} \left\{\frac{1}{\Delta_{n+1}}\sum_{v=0}^{k} W_v^{(k)} p\left(\Xi_{n+1,v}^{(k)}\right)\lambda'(v,l)\lambda'(v,j) + \right.$$
$$\left. W_l^{(k)} q\left(\Xi_{n+1,l}^{(k)}\right)\lambda'(l,j) + W_l^{(k)}\delta_{l,j} r\left(\Xi_{n+1,l}^{(k)}\right)\Delta_{n+1}\right\} a_{n+1,j}$$

The value of the expression upon the right-hand side of equation (5) is $W_l^{(k)}f(\Xi_{n,l}^{(k)}) \Delta_n$ for $n=1$, $l=0,\ldots,k$, and $n=1,\ldots,N$, $l=1,\ldots,k-1$, and is equal to $W_0^{(k)}f(\Xi_{n,k}^{(k)})(\Delta_n + \Delta_{n+1})$ for $n=1,\ldots,N-1$, $l=k$.

The second of the boundary conditions (2) implies that

$$e_4 a_{N,k} + \sum_{j=1}^{k} e_5 \frac{\lambda'(k, j)}{\Delta_n} a_{N,j} = e_6$$

In summary, the two boundary conditions (2) and $Nk-1$ equations derived from equation (3) with $n=1,\ldots,N-1$, $l=1,\ldots,k$, and $n=N$, $l=1,\ldots,k-1$ yield $Nk+1$ linear equations for the $Nk+1$ coefficients a_0 and a_{mj} ($m=1,\ldots,N$, $j=1,\ldots,k$). The first $k-1$ of these equations involve a_0 and $a_{i,j}$ ($j=1,...,k$) alone. By transferring the terms involving a_0 and $a_{1,k}$ to the right-hand side, and premultiplying by the inverse of the matrix multiplying the vector $(a_{1,2},...,a_{1,k-1})$, this vector may be isolated. Its components may be eliminated from the next equation which involves the two sets $a_{1,j}$ and $a_{2,j}$ ($j=1,...,k$). The following $k-1$ equations involve $a_{2,j}$ ($j=1,...,k$) alone. The vector $(a_{2,2},...,a_{2,k-1})$ may again be isolated and its components eliminated from the equation from which $a_{1,2},...,a_{1,k-1}$ have been eliminated, and also from the next equation which involves the two sets $a_{2,j}$, $a_{3,j}$ ($j=1,...,k$). The process may be continued, and a tridiagonal system of equations for a_0 and $a_{m,k}$ ($m=1,...,N$) is obtained. The process of elimination may be carried out as soon as each complete set of equations involving $a_{m,1},...,a_{m,k-1}$ has been constructed (computer storage space is thereby saved). When $k=1$, no elimination takes place; when $k=2$ and 3 (the other two values of k permitted by *femlagsym*) the matrix elimination process is particularly simple and is programmed independently (rather than by calls of matrix operation procedures).

The tridiagonal system of equations is solved by a method of Babuska [see Bab72]. Stripping some redundant sequences from the original exposition, the set of equations $Ay=f$, where

$$A_{i+1,i} = c_i,$$
$$A_{i,i+1} = b_i \ (i=1,...,n-1),$$
$$A_{1,1} = \tau_1 - b_1,$$
$$A_{i,i} = \tau_i - c_{i-1} - b_i \ (i=2,...,n-1),$$

$A_{n,n} = \tau_n - c_n$ may be solved by means of the construction of four sequences g_1, χ_i, g_i^*, and χ_i^* as follows: with $\chi_1=\tau_1$, $g_1=f_1$

$$g_{i+1} = f_{i+1} - g_i c_i/(\chi_i - b_i), \qquad \chi_{i+1} = \tau_{i+1} - \chi_i c_i/(\chi_i - b_i) \qquad (i=1,...,n-1)$$

and with χ_n^*, $g_n^* = f_n$

$$g_i^* = f_i - b_i g_{i+1}^*/(\chi_{i+1}^* - c_i), \qquad \chi_i^* = \tau_i - b_i \chi_{i+1}^*/(\chi_{i+1}^* - c_i) \qquad (i=n-1,...,1)$$

and then

$$y_i = (g_i + g_i^* - f_i) / (\chi_i + \chi_i^* - \tau_i) \qquad (i=n,...,1).$$

The above process can be further economized (three one-dimensional arrays, with another to contain the y_i are required). The coefficients $a_{1,0}$, $a_{n,k}$ ($n=1,...,N$) obtained by means of the above process are, in order, the required approximations y_n ($n=0,...,N$) (since $\varphi_{n,k}(x_m) = \delta_{m,n}$).

Procedure parameters:

void femlagsym $(x,y,n,method,order,e)$

x: double $x[0:n]$;

     entry: $a = x_0 < x_1 < ... < x_n = b$ is a partition of the interval $[a,b]$;

y: double $y[0:n]$;

     exit: $y[i]$, $i=0,1,...,n$, is the approximate solution at $x[i]$ of the differential equation (1) with boundary conditions (2);

n: int;

     entry: the upper bound of the arrays x and y (value of N above), $n > 1$;

method: a class that defines three procedures p, r, and f, this class must implement the AP_femlagsym_methods interface;

     double p(double x)

        the procedure to compute $p(x)$, the coefficient of $Dy(x)$ in equation (1);

     double r(double x)

        the procedure to compute $r(x)$, the coefficient of $y(x)$ in equation (1);

double f(double x)
the procedure to compute *f(x)*, the right-hand side of equation (1);

order: int;
  entry: *order* denotes the order of accuracy required for the approximate
          solution of the differential equation; let $h = \max(x[i]-x[i-1])$, then
          $|y[i] - y(x[i])| \leq c*h^{order}$, $i=0,...,n$; *order* can be chosen equal to 2, 4,
          or 6 only;

e: double *e[1:6]*;
  entry: *e[1],...,e[6]* describe the boundary conditions (values of e_i, $i=1,...,6$,
          in (2));
          e[1] and *e[4]* are not allowed to vanish both.

```
package numal;

public interface AP_femlagsym_methods {

    double p(double x);
    double r(double x);
    double f(double x);
}



public static void femlagsym(double x[], double y[], int n,
                             AP_femlagsym_methods method, int order, double e[])
{
    int l,ll;
    double xll,xl,h,a12,b1,b2,tau1,tau2,ch,tl,g,yl,pp,p1,p2,p3,p4,r1,r2,r3,r4,f1,f2,
           f3,f4,e1,e2,e3,e4,e5,e6,h2,x2,h6,h15,b3,tau3,c12,c32,a13,a22,a23,x3,h12,
           h24,det,c13,c42,c43,a14,a24,a33,a34,b4,tau4,aux;
    double t[] = new double[n];
    double sub[] = new double[n];
    double chi[] = new double[n];
    double gi[] = new double[n];

    p2=p3=p4=r2=r3=r4=f2=f3=f4=ch=g=tl=yl=0.0;
    l=1;
    xl=x[0];
    e1=e[1];
    e2=e[2];
    e3=e[3];
    e4=e[4];
    e5=e[5];
    e6=e[6];
    while (l <= n) {
        ll=l-1;
        xll=xl;
        xl=x[l];
        h=xl-xll;
        if (order == 2) {
            /* element mat vec evaluation 1 */
```

```
        if (l == 1) {
            p2=method.p(xl1);
            r2=method.r(xl1);
            f2=method.f(xl1);
        }
        p1=p2;
        p2=method.p(xl);
        r1=r2;
        r2=method.r(xl);
        f1=f2;
        f2=method.f(xl);
        h2=h/2.0;
        b1=h2*f1;
        b2=h2*f2;
        tau1=h2*r1;
        tau2=h2*r2;
        a12 = -0.5*(p1+p2)/h;
    } else if (order == 4) {
        /* element mat vec evaluation 2 */
        if (l == 1) {
            p3=method.p(xl1);
            r3=method.r(xl1);
            f3=method.f(xl1);
        }
        x2=(xl1+xl)/2.0;
        h6=h/6.0;
        h15=h/1.5;
        p1=p3;
        p2=method.p(x2);
        p3=method.p(xl);
        r1=r3;
        r2=method.r(x2);
        r3=method.r(xl);
        f1=f3;
        f2=method.f(x2);
        f3=method.f(xl);
        b1=h6*f1;
        b2=h15*f2;
        b3=h6*f3;
        tau1=h6*r1;
        tau2=h15*r2;
        tau3=h6*r3;
        a12 = -(2.0*p1+p3/1.5)/h;
        a13=(0.5*(p1+p3)-p2/1.5)/h;
        a22=(p1+p3)/h/0.375+tau2;
        a23 = -(p1/3.0+p3)*2.0/h;
        c12 = -a12/a22;
        c32 = -a23/a22;
        a12=a13+c32*a12;
        b1 += c12*b2;
```

```
        b2=b3+c32*b2;
        tau1 += c12*tau2;
        tau2=tau3+c32*tau2;
    } else {
        /* element mat vec evaluation 3 */
        if (l == 1) {
            p4=method.p(xl1);
            r4=method.r(xl1);
            f4=method.f(xl1);
        }
        x2=xl1+0.27639320225*h;
        x3=xl-x2+xl1;
        h12=h/12.0;
        h24=h/2.4;
        p1=p4;
        p2=method.p(x2);
        p3=method.p(x3);
        p4=method.p(xl);
        r1=r4;
        r2=method.r(x2);
        r3=method.r(x3);
        r4=method.r(xl);
        f1=f4;
        f2=method.f(x2);
        f3=method.f(x3);
        f4=method.f(xl);
        b1=h12*f1;
        b2=h24*f2;
        b3=h24*f3;
        b4=h12*f4;
        tau1=h12*r1;
        tau2=h24*r2;
        tau3=h24*r3;
        tau4=h12*r4;
        a12 = -(4.04508497187450*p1+0.57581917135425*p3+0.25751416197911*p4)/h;
        a13=(1.5450849718747*p1-1.5075141619791*p2+0.6741808286458*p4)/h;
        a14=((p2+p3)/2.4-(p1+p4)/2.0)/h;
        a22=(5.454237476562*p1+p3/0.48+0.79576252343762*p4)/h+tau2;
        a23 = -(p1+p4)/(h*0.48);
        a24=(0.67418082864575*p1-1.50751416197910*p3+1.54508497187470*p4)/h;
        a33=(0.7957625234376*p1+p2/0.48+5.454237476562*p4)/h+tau3;
        a34 = -(0.25751416197911*p1+0.57581917135418*p2+4.0450849718747*p4)/h;
        det=a22*a33-a23*a23;
        c12=(a13*a23-a12*a33)/det;
        c13=(a12*a23-a13*a22)/det;
        c42=(a23*a34-a24*a33)/det;
        c43=(a24*a23-a34*a22)/det;
        tau1 += c12*tau2+c13*tau3;
        tau2=tau4+c42*tau2+c43*tau3;
        a12=a14+c42*a12+c43*a13;
```

```
        b1 += c12*b2+c13*b3;
        b2=b4+c42*b2+c43*b3;
    }
    if (l == 1 || l == n) {
        /* boundary conditions */
        if (l == 1 && e2 == 0.0) {
            tau1=1.0;
            b1=e3/e1;
            b2 -= a12*b1;
            tau2 -= a12;
            a12=0.0;
        } else if (l == 1 && e2 != 0.0) {
            aux=p1/e2;
            tau1 -= aux*e1;
            b1 -= e3*aux;
        } else if (l == n && e5 == 0.0) {
            tau2=1.0;
            b2=e6/e4;
            b1 -= a12*b2;
            tau1 -= a12;
            a12=0.0;
        } else if (l == n && e5 != 0.0) {
            aux=p2/e5;
            tau2 += aux*e4;
            b2 += aux*e6;
        }
    }
    /* forward babushka */
    if (l == 1) {
        chi[0]=ch=tl=tau1;
        t[0]=tl;
        gi[0]=g=yl=b1;
        y[0]=yl;
        sub[0]=a12;
        pp=a12/(ch-a12);
        ch=tau2-ch*pp;
        g=b2-g*pp;
        tl=tau2;
        yl=b2;
    } else {
        chi[ll] = ch += tau1;
        gi[ll] = g += b1;
        sub[ll]=a12;
        pp=a12/(ch-a12);
        ch=tau2-ch*pp;
        g=b2-g*pp;
        t[ll]=tl+tau1;
        tl=tau2;
        y[ll]=yl+b1;
        yl=b2;
```

```
            }
        l++;
    }
    }
    /* backward babushka */
    pp=yl;
    y[n]=g/ch;
    g=pp;
    ch=tl;
    l=n-1;
    while (l >= 0) {
        pp=sub[l];
        pp /= (ch-pp);
        tl=t[l];
        ch=tl-ch*pp;
        yl=y[l];
        g=yl-g*pp;
        y[l]=(gi[l]+g-yl)/(chi[l]+ch-tl);
        l--;
    }
}
```

B. femlag

Solves a second order self-adjoint linear two point boundary value problem by means of Galerkin's method with continuous piecewise polynomials [Bab72, BakH76, He75, StF73]. *femlag* computes approximations y_n $(n=0,...,N)$ to the values at the points $x = x_n$, where
$$-\infty < a = x_0 < x_1 < ... < x_N = b < \infty$$
of the real valued function $y(x)$ which satisfies the equation
$$-D^2y(x) + r(x)y(x) = f(x) \tag{1}$$
where $D=d/dx$, with boundary conditions
$$e_1y(a) + e_2Dy(a) = e_3, \quad e_4y(b) + e_5Dy(b) = e_6 \quad (e_1, e_4 \neq 0). \tag{2}$$
It is assumed that $r(x) \geq 0$ $(a \leq x \leq b)$; if $e_2 = e_5 = 0$, this condition may be relaxed to
$$r(x) > -(\pi/(b-a))^2 \ (a \leq x \leq b).$$
The general theory underlying the method used is that described in *femlagsym*. Now, however, (since $p(x)$ in equation (1) of *femlagsym* is unity) the term
$$\frac{1}{\Delta_n} \sum_{\upsilon=0}^{k} W_\upsilon^{(k)} p\left(\Xi_{n,\upsilon}^{(k)}\right) \lambda'(\upsilon,l)\lambda'(\upsilon,j)$$
in expression (7) of *femlagsym* becomes
$$\frac{1}{\Delta_n} \sum_{\upsilon=0}^{k} W_\upsilon^{(k)} \lambda'(\upsilon,l)\lambda'(\upsilon,j).$$
The value of the sum in this expression is problem independent and may be determined in advance (a similar remark holds with respect to expression (6) in *femlagsym*); the computations may be simplified.

Procedure parameters:

$$\text{void femlag } (x,y,n,method,order,e)$$

x: double $x[0:n]$;
    entry: $a = x_0 < x_1 < ... < x_n = b$ is a partition of the segment $[a,b]$;

y: double $y[0:n]$;
    exit: $y[i]$, $i=0,1,...,n$, is the approximate solution at $x[i]$ of the differential
      equation (1) with boundary conditions (2);

n: int;
    entry: the upper bound of the arrays x and y (value of N above), $n > 1$;

method: a class that defines two procedures r and f, this class must implement
    the `AP_femlag_methods` interface;
    double r(double x)
      the procedure to compute $r(x)$, the coefficient of $y(x)$ in equation (1);
    double f(double x)
      the procedure to compute $f(x)$, the right-hand side of equation (1);

order: int;
    entry: *order* denotes the order of accuracy required for the approximate
      solution of the differential equation; let $h = max(x[i]-x[i-1])$, then
      $$\left| y[i] - y(x[i]) \right| \le c*h^{order}, \ i=0,...,n;$$
      order can be chosen equal to 2, 4, or 6 only;

e: double $e[1:6]$;
    entry: $e[1],...,e[6]$ describe the boundary conditions (values of e_i, $i=1,...,6$,
      in (2));
      neither $e[1]$ nor $e[4]$ is allowed to vanish.

```
package numal;

public interface AP_femlag_methods {

    double r(double x);
    double f(double x);
}


public static void femlag(double x[], double y[], int n, AP_femlag_methods method,
                          int order, double e[])
{
    int l,ll;
    double xll,xl,h,a12,b1,b2,tau1,tau2,ch,tl,g,yl,pp,e1,e2,e3,e4,e5,e6,f2,r2,r1,f1,
           h2,r3,f3,x2,h6,h15,b3,tau3,c12,a13,a22,a23,r4,f4,x3,h12,h24,det,c13,c42,
           c43,a14,a24,a33,a34,b4,tau4;

    double t[]   = new double[n];
    double sub[] = new double[n];
    double chi[] = new double[n];
    double gi[]  = new double[n];

    r2=r3=r4=f2=f3=f4=ch=g=tl=yl=0.0;
    l=1;
```

```
xl=x[0];
e1=e[1];
e2=e[2];
e3=e[3];
e4=e[4];
e5=e[5];
e6=e[6];
while (l <= n) {
    l1=l-1;
    xl1=xl;
    xl=x[l];
    h=xl-xl1;
    if (order == 2) {
        /* element mat vec evaluation 1 */
        if (l == 1) {
            f2=method.f(xl1);
            r2=method.r(xl1);
        }
        a12 = -1.0/h;
        h2=h/2.0;
        r1=r2;
        r2=method.r(xl);
        f1=f2;
        f2=method.f(xl);
        b1=h2*f1;
        b2=h2*f2;
        tau1=h2*r1;
        tau2=h2*r2;
    } else if (order == 4) {
        /* element mat vec evaluation 2 */
        if (l == 1) {
            r3=method.r(xl1);
            f3=method.f(xl1);
        }
        x2=(xl1+xl)/2.0;
        h6=h/6.0;
        h15=h/1.5;
        r1=r3;
        r2=method.r(x2);
        r3=method.r(xl);
        f1=f3;
        f2=method.f(x2);
        f3=method.f(xl);
        b1=h6*f1;
        b2=h15*f2;
        b3=h6*f3;
        tau1=h6*r1;
        tau2=h15*r2;
        tau3=h6*r3;
        a12 = a23 = -8.0/h/3.0;
```

```
        a13 = -a12/8.0;
        a22 = -2.0*a12+tau2;
        c12 = -a12/a22;
        a12=a13+c12*a12;
        b2 *= c12;
        b1 += b2;
        b2 += b3;
        tau2 *= c12;
        tau1 += tau2;
        tau2=tau3+tau2;
    } else {
        /* element mat vec evaluation 3 */
        if (l == 1) {
            r4=method.r(xl1);
            f4=method.f(xl1);
        }
        x2=xl1+0.27639320225*h;
        x3=xl-x2+xl1;
        r1=r4;
        r2=method.r(x2);
        r3=method.r(x3);
        r4=method.r(xl);
        f1=f4;
        f2=method.f(x2);
        f3=method.f(x3);
        f4=method.f(xl);
        h12=h/12.0;
        h24=h/2.4;
        b1=h12*f1;
        b2=h24*f2;
        b3=h24*f3;
        b4=h12*f4;
        tau1=h12*r1;
        tau2=h24*r2;
        tau3=h24*r3;
        tau4=h12*r4;
        a12 = a34 = -4.8784183052078/h;
        a13=a24=0.7117516385412/h;
        a14 = -0.16666666666667/h;
        a23=25.0*a14;
        a22 = -2.0*a23+tau2;
        a33 = -2.0*a23+tau3;
        det=a22*a33-a23*a23;
        c12=(a13*a23-a12*a33)/det;
        c13=(a12*a23-a13*a22)/det;
        c42=(a23*a34-a24*a33)/det;
        c43=(a24*a23-a34*a22)/det;
        tau1 += c12*tau2+c13*tau3;
        tau2=tau4+c42*tau2+c43*tau3;
        a12=a14+c42*a12+c43*a13;
```

```
                b1 += c12*b2+c13*b3;
                b2=b4+c42*b2+c43*b3;
            }
            if (l == 1 || l == n) {
                /* boundary conditions */
                if (l == 1 && e2 == 0.0) {
                    tau1=1.0;
                    b1=e3/e1;
                    b2 -= a12*b1;
                    tau2 -= a12;
                    a12=0.0;
                } else if (l == 1 && e2 != 0.0) {
                    tau1 -= e1/e2;
                    b1 -= e3/e2;
                } else if (l == n && e5 == 0.0) {
                    tau2=1.0;
                    b2=e6/e4;
                    b1 -= a12*b2;
                    tau1 -= a12;
                    a12=0.0;
                } else if (l == n && e5 != 0.0) {
                    tau2 += e4/e5;
                    b2 += e6/e5;
                }
            }
            /* forward babushka */
            if (l == 1) {
                chi[0]=ch=tl=tau1;
                t[0]=tl;
                gi[0]=g=yl=b1;
                y[0]=yl;
                sub[0]=a12;
                pp=a12/(ch-a12);
                ch=tau2-ch*pp;
                g=b2-g*pp;
                tl=tau2;
                yl=b2;
            } else {
                chi[l1] = ch += tau1;
                gi[l1] = g += b1;
                sub[l1]=a12;
                pp=a12/(ch-a12);
                ch=tau2-ch*pp;
                g=b2-g*pp;
                t[l1]=tl+tau1;
                tl=tau2;
                y[l1]=yl+b1;
                yl=b2;
            }
            l++;
```

```
    }
    /* backward babushka */
    pp=yl;
    y[n]=g/ch;
    g=pp;
    ch=tl;
    l=n-1;
    while (l >= 0) {
        pp=sub[l];
        pp /= (ch-pp);
        tl=t[l];
        ch=tl-ch*pp;
        yl=y[l];
        g=yl-g*pp;
        y[l]=(gi[l]+g-yl)/(chi[l]+ch-tl);
        l--;
    }
}
```

C. femlagspher

Solves a second order self-adjoint linear two point boundary value problem with spherical coordinates by means of Galerkin's method with continuous piecewise polynomials [Bab72, BakH76, He75, StF73]. *femlagspher* computes approximations y_n ($n=0,...,N$) to the values at the points $x = x_n$, where

$$-\infty < a = x_0 < x_1 < ... < x_N = b < \infty$$

of the real valued function $y(x)$ which satisfies the equation

$$-D(x^{nc}Dy(x))/x^{nc} + r(x)y(x) = f(x) \qquad (1)$$

where $D=d/dx$, with boundary conditions

$$e_1y(a) + e_2Dy(a) = e_3, \quad e_4y(b) + e_5Dy(b) = e_6 \quad (e_1, e_4 \neq 0). \qquad (2)$$

It is assumed that $r(x)$ and $f(x)$ are sufficiently smooth on $[x_0,x_N]$ except at the grid points; furthermore, $r(x)$ should be nonnegative.

The solution is approximated by a function which is continuous on the closed interval $[x_0,x_N]$ and a polynomial of degree less than or equal to k on each segment $[x_{j-1},x_j]$ ($j=1,...,N$). This piecewise polynomial is entirely determined by the values it has at the knots x_j and on $k-1$ interior knots on each segment $[x_{j-1},x_j]$. These values are obtained by the solution of an *order*+1 diagonal linear system with a specially structured matrix. The entries of the matrix and the vector are inner products which are approximated by some piecewise k-point Gaussian quadrature. The evaluation of the matrix and the vector is done segment by segment: on each segment the contributions to the entries of the matrix and the vector are computed and embedded in the global matrix and vector. Since the function values on the interior points of each segment are not coupled with the function values outside that segment, the resulting linear system can be reduced to a tridiagonal system by means of static condensation. The final tridiagonal system, since it is of finite difference type, is solved by means of Babuska's method. For further details, see the documentation of *femlagsym*.

Procedure parameters:

$$\text{void femlagspher } (x,y,n,nc,method,order,e)$$

x: double $x[0:n]$;
            entry: $a = x_0 < x_1 < ... < x_n = b$ is a partition of the interval $[a,b]$;
y: double $y[0:n]$;
            exit: $y[i]$, $i=0,1,...,n$, is the approximate solution at $x[i]$ of the differential
                     equation (1) with boundary conditions (2);
n: int;
            entry: the upper bound of the arrays x and y (value of N above), $n > 1$;
nc: int;
            entry: if $nc = 0$, Cartesian coordinates are used;
                     if $nc = 1$, polar coordinates are used;
                     if $nc = 2$, spherical coordinates are used;
method: a class that defines two procedures r and f, this class must implement
            the AP_femlagspher_methods interface;
            double r(double x)
                the procedure to compute $r(x)$, the coefficient of $y(x)$ in equation (1);
            double f(double x)
                the procedure to compute $f(x)$, the right-hand side of equation (1);
order: int;
            entry: order denotes the order of accuracy required for the approximate
                     solution of the differential equation; let $h = \max(x[i]\text{-}x[i\text{-}1])$, then
                     $$|y[i] - y(x[i])| \le c^* h^{order}, \ i=0,...,n;$$
                     order can be chosen equal to 2 or 4 only;
e: double $e[1:6]$;
            entry: $e[1],...,e[6]$ describe the boundary conditions (values of e_i, $i=1,...,6$,
                     in (2));
                     $e[1]$ and $e[4]$ are not allowed to vanish both.

```
package numal;

public interface AP_femlagspher_methods {

  double r(double x);
  double f(double x);
}



public static void femlagspher(double x[], double y[], int n, int nc,
                        AP_femlagspher_methods method, int order, double e[])
{
    int l,ll;
    double xll,xl,h,a12,b1,b2,tau1,tau2,ch,tl,g,yl,pp,tau3,b3,a13,a22,a23,c32,c12,
           e1,e2,e3,e4,e5,e6,xm,vl,vr,wl,wr,pr,rm,fm,x12,xlxr,xr2,xlm,xrm,vlm,vrm,
           wlm,wrm,flm,frm,rlm,rrm,pl1,pl2,pl3,pr1,pr2,pr3,ql1,ql2,ql3,rlmpl1,
           rlmpl2,rrmpr1,rrmpr2,vlmql1,vlmql2,vrmqr1,vrmqr2,qr1,qr2,qr3,a,a2,a3,a4,
           b,b4,p4h,p2,p3,p4,aux1,aux2,a5,a6,a7,a8,b5,b6,b7,b8,ab4,a2b3,a3b2,a4b,
           p5,p8,p8h,aux,plm,prm;
    double t[] = new double[n];
```

```
double sub[] = new double[n];
double chi[] = new double[n];
double gi[] = new double[n];

pl1=pl2=pl3=pr1=pr2=pr3=ql1=ql2=ql3=qr1=qr2=qr3=ch=g=tl=yl=0.0;
l=1;
xl=x[0];
e1=e[1];
e2=e[2];
e3=e[3];
e4=e[4];
e5=e[5];
e6=e[6];
while (l <= n) {
    l1=l-1;
    xl1=xl;
    xl=x[l];
    h=xl-xl1;
    if (order == 2) {
        /* element mat vec evaluation 1 */
        if (nc == 0)
            vl=vr=0.5;
        else if (nc == 1) {
            vl=(xl1*2.0+xl)/6.0;
            vr=(xl1+xl*2.0)/6.0;
        } else {
            xl2=xl1*xl1/12.0;
            xlxr=xl1*xl/6.0;
            xr2=xl*xl/12.0;
            vl=3.0*xl2+xlxr+xr2;
            vr=3.0*xr2+xlxr+xl2;
        }
        wl=h*vl;
        wr=h*vr;
        pr=vr/(vl+vr);
        xm=xl1+h*pr;
        fm=method.f(xm);
        rm=method.r(xm);
        tau1=wl*rm;
        tau2=wr*rm;
        b1=wl*fm;
        b2=wr*fm;
        a12 = -(vl+vr)/h+h*(1.0-pr)*pr*rm;
    } else {
        /* element mat vec evaluation 2 */
        if (nc == 0) {
            xlm=xl1+h*0.2113248654052;
            xrm=xl1+xl-xlm;
            vlm=vrm=0.5;
            pl1=pr3=0.45534180126148;
        }
```

```
            pl3=pr1 = -0.12200846792815;
            pl2=pr2=1.0-pl1-pl3;
            ql1 = -2.15470053837925;
            ql3 = -0.15470053837925;
            ql2 = -ql1-ql3;
            qr1 = -ql3;
            qr3 = -ql1;
            qr2 = -ql2;
        } else if (nc == 1) {
            a=xl1;
            a2=a*a;
            a3=a*a2;
            a4=a*a3;
            b=xl;
            b2=b*b;
            b3=b*b2;
            b4=b*b3;
            p2=10.0*(a2+4.0*a*b+b2);
            p3=6.0*(a3+4.0*(a2*b+a*b2)+b3);
            p4=Math.sqrt(6.0*(a4+10.0*(a*b3+a3*b)+28.0*a2*b2+b4));
            p4h=p4*h;
            xlm=(p3-p4h)/p2;
            xrm=(p3+p4h)/p2;
            aux1=(a+b)/4.0;
            aux2=h*(a2+7.0*a*b+b2)/6.0/p4;
            vlm=aux1-aux2;
            vrm=aux1+aux2;
        } else {
            a=xl1;
            a2=a*a;
            a3=a*a2;
            a4=a*a3;
            a5=a*a4;
            a6=a*a5;
            a7=a*a6;
            a8=a*a7;
            b=xl;
            b2=b*b;
            b3=b*b2;
            b4=b*b3;
            b5=b*b4;
            b6=b*b5;
            b7=b*b6;
            b8=b*b7;
            ab4=a*b4;
            a2b3=a2*b3;
            a3b2=a3*b2;
            a4b=a4*b;
            p4=15.0*(a4+4.0*(a3*b+a*b3)+10.0*a2*b2+b4);
            p5=10.0*(a5+4.0*(a4b+ab4)+10.0*(a3b2+a2b3)+b5);
```

```
    p8=Math.sqrt(10.0*(a8+10.0*(a7*b+a*b7)+55.0*(a2*b6+a6*b2)+
            164.0*(a5*b3+a3*b5)+290.0*a4*b4+b8));
    aux1=(a2+a*b+b2)/6.0;
    p8h=p8*h;
    aux2=(h*(a5+7.0*(a4b+ab4)+28.0*(a3b2+a2b3)+b5))/4.8/p8;
    xlm=(p5-p8h)/p4;
    xrm=(p5+p8h)/p4;
    vlm=aux1-aux2;
    vrm=aux1+aux2;
}
if (nc > 0) {
    plm=(xlm-xl1)/h;
    prm=(xrm-xl1)/h;
    aux=2.0*plm-1.0;
    pl1=aux*(plm-1.0);
    pl3=aux*plm;
    pl2=1.0-pl1-pl3;
    aux=2.0*prm-1.0;
    pr1=aux*(prm-1.0);
    pr3=aux*prm;
    pr2=1.0-pr1-pr3;
    aux=4.0*plm;
    ql1=aux-3.0;
    ql3=aux-1.0;
    ql2 = -ql1-ql3;
    aux=4.0*prm;
    qr1=aux-3.0;
    qr3=aux-1.0;
    qr2 = -qr1-qr3;
}
wlm=h*vlm;
wrm=h*vrm;
vlm /= h;
vrm /= h;
flm=method.f(xlm)*wlm;
frm=wrm*method.f(xrm);
rlm=method.r(xlm)*wlm;
rrm=wrm*method.r(xrm);
tau1=pl1*rlm+pr1*rrm;
tau2=pl2*rlm+pr2*rrm;
tau3=pl3*rlm+pr3*rrm;
b1=pl1*flm+pr1*frm;
b2=pl2*flm+pr2*frm;
b3=pl3*flm+pr3*frm;
vlmql1=ql1*vlm;
vrmqr1=qr1*vrm;
vlmql2=ql2*vlm;
vrmqr2=qr2*vrm;
rlmpl1=rlm*pl1;
rrmpr1=rrm*pr1;
```

```
            rlmpl2=rlm*pl2;
            rrmpr2=rrm*pr2;
            a12=vlmql1*ql2+vrmqr1*qr2+rlmpl1*pl2+rrmpr1*pr2;
            a13=vlmql1*ql3+vrmqr1*qr3+rlmpl1*pl3+rrmpr1*pr3;
            a22=vlmql2*ql2+vrmqr2*qr2+rlmpl2*pl2+rrmpr2*pr2;
            a23=vlmql2*ql3+vrmqr2*qr3+rlmpl2*pl3+rrmpr2*pr3;
            c12 = -a12/a22;
            c32 = -a23/a22;
            a12=a13+c32*a12;
            b1 += c12*b2;
            b2=b3+c32*b2;
            tau1 += c12*tau2;
            tau2=tau3+c32*tau2;
        }
        if (l == 1 || l == n) {
            /* boundary conditions */
            if (l == 1 && e2 == 0.0) {
                tau1=1.0;
                b1=e3/e1;
                b2 -= a12*b1;
                tau2 -= a12;
                a12=0.0;
            } else if (l == 1 && e2 != 0.0) {
                aux=((nc == 0) ? 1.0 : Math.pow(x[0],nc))/e2;
                b1 -= e3*aux;
                tau1 -= e1*aux;
            } else if (l == n && e5 == 0.0) {
                tau2=1.0;
                b2=e6/e4;
                b1 -= a12*b2;
                tau1 -= a12;
                a12=0.0;
            } else if (l == n && e5 != 0.0) {
                aux=((nc == 0) ? 1.0 : Math.pow(x[n],nc))/e5;
                tau2 += aux*e4;
                b2 += aux*e6;
            }
        }
        /* forward babushka */
        if (l == 1) {
            chi[0]=ch=tl=tau1;
            t[0]=tl;
            gi[0]=g=yl=b1;
            y[0]=yl;
            sub[0]=a12;
            pp=a12/(ch-a12);
            ch=tau2-ch*pp;
            g=b2-g*pp;
            tl=tau2;
            yl=b2;
```

```
    } else {
        chi[l1] = ch += tau1;
        gi[l1] = g += b1;
        sub[l1]=a12;
        pp=a12/(ch-a12);
        ch=tau2-ch*pp;
        g=b2-g*pp;
        t[l1]=tl+tau1;
        tl=tau2;
        y[l1]=yl+b1;
        yl=b2;
    }
    l++;
}
/* backward babushka */
pp=yl;
y[n]=g/ch;
g=pp;
ch=tl;
l=n-1;
while (l >= 0) {
    pp=sub[l];
    pp /= (ch-pp);
    tl=t[l];
    ch=tl-ch*pp;
    yl=y[l];
    g=yl-g*pp;
    y[l]=(gi[l]+g-yl)/(chi[l]+ch-tl);
    l--;
}
}
```

5.5.2 Linear methods - Second order skew adjoint

femlagskew

Solves a second order skew-adjoint linear two point boundary value problem by means of Galerkin's method with continuous piecewise polynomials [Bab72, BakH76, He75, StF73]. *femlagskew* computes approximations y_n $(n=0,...,N)$ to the values at the points $x = x_n$, where

$$-\infty < a = x_0 < x_1 < ... < x_N = b < \infty$$

of the real valued function $y(x)$ which satisfies the equation

$$-D^2y(x) + q(x)Dy(x) + r(x)y(x) = f(x) \tag{1}$$

where $D=d/dx$, with boundary conditions

$$e_1y(a) + e_2Dy(a) = e_3, \quad e_4y(b) + e_5Dy(b) = e_6 \quad (e_1, e_4 \neq 0). \tag{2}$$

It is assumed that
(a) $r(x) \geq 0$ $(a \leq x \leq b)$, if $e_2 = e_5 = 0$ then this condition may be relaxed to
       $r(x) > -(\pi/(b-a))^2$ $(a \leq x \leq b)$;
(b) $q(x)$ is not allowed to have very large values in some sense: the product
       $q(x)^*(x_j - x_{j-1})$ should not be too large on the interval $[x_{j-1}, x_j]$; otherwise, the
       boundary value problem may degenerate to a singular perturbation or
       boundary layer problem, for which either special methods or a suitably
       chosen grid are needed;
(c) $q(x)$, $r(x)$, and $f(x)$ are required to be sufficiently differentiable on the domain
       of the boundary value problem; however, the derivatives are allowed to have
       discontinuities at the grid points, in which case the order of accuracy
       (2, 4, or 6) is preserved; and
(d)   if $q(x)$ and $r(x)$ satisfy the inequality $r(x) \geq Dq(x)/2$, the existence of a unique
       solution is guaranteed; otherwise, this remains an open question.
The general theory underlying the method used is that described in *femlagsym*.
Now a term $q(x)Dy(x)$ is being treated (it is absent from equation (1) in *femlagsym*),
however, the simplification described in *femlag* is possible.

Procedure parameters:
                              void femlagskew $(x,y,n,method,order,e)$
x: double $x[0:n]$;
              entry: $a = x_0 < x_1 < ... < x_n = b$ is a partition of the segment $[a,b]$;
y: double $y[0:n]$;
              exit: $y[i]$, $i=0,1,...,n$, is the approximate solution at $x[i]$ of the differential
                       equation (1) with boundary conditions (2);
n: int;
              entry: the upper bound of the arrays x and y (value of N above), $n > 1$;
method: a class that defines three procedures q, r, and f, this class must
              implement the AP_femlagskew_methods interface;
              double q(double x)
                  the procedure to compute $q(x)$, the coefficient of $Dy(x)$ in equation (1);
              double r(double x)
                  the procedure to compute $r(x)$, the coefficient of $y(x)$ in equation (1);
              double f(double x)
                  the procedure to compute $f(x)$, the right-hand side of equation (1);
order: int;
              entry: *order* denotes the order of accuracy required for the approximate
                       solution of the differential equation; let $h = \max(x[i]-x[i-1])$, then
                       $|y[i] - y(x[i])| \leq c^* h^{order}$, $i=0,...,n$;
                       order can be chosen equal to 2, 4, or 6 only;
e: double $e[1:6]$;
              entry: $e[1],...,e[6]$ describe the boundary conditions (values of e_i, $i=1,...,6$,
                       in (2));
                       neither $e[1]$ nor $e[4]$ is allowed to vanish.

```
package numal;

public interface AP_femlagskew_methods {

    double q(double x);
    double r(double x);
    double f(double x);
}


public static void femlagskew(double x[], double y[], int n,
                              AP_femlagskew_methods method, int order, double e[])
{
    int l,l1;
    double xl1,xl,h,a12,a21,b1,b2,tau1,tau2,ch,tl,g,yl,pp,e1,e2,e3,e4,e5,e6,q2,r2,
           f2,q1,r1,f1,h2,s12,q3,r3,f3,s13,s22,x2,h6,h15,c12,c32,a13,a31,a22,a23,
           a32,b3,tau3,q4,r4,f4,s14,s23,x3,h12,h24,det,c13,c42,c43,a14,a24,a33,a34,
           a41,a42,a43,b4,tau4;
    double t[] = new double[n];
    double ssuper[] = new double[n];
    double sub[] = new double[n];
    double chi[] = new double[n];
    double gi[] = new double[n];

    q2=q3=q4=r2=r3=r4=f2=f3=f4=ch=g=tl=yl=0.0;
    l=1;
    xl=x[0];
    e1=e[1];
    e2=e[2];
    e3=e[3];
    e4=e[4];
    e5=e[5];
    e6=e[6];
    while (l <= n) {
        xl1=xl;
        l1=l-1;
        xl=x[l];
        h=xl-xl1;
        if (order == 2) {
            /* element mat vec evaluation 1 */
            if (l == 1) {
                q2=method.q(xl1);
                r2=method.r(xl1);
                f2=method.f(xl1);
            }
            h2=h/2.0;
            s12 = -1.0/h;
            q1=q2;
            q2=method.q(xl);
            r1=r2;
```

```
            r2=method.r(xl);
            f1=f2;
            f2=method.f(xl);
            b1=h2*f1;
            b2=h2*f2;
            tau1=h2*r1;
            tau2=h2*r2;
            a12=s12+q1/2.0;
            a21=s12-q2/2.0;
        } else if (order == 4) {
            /* element mat vec evaluation 2 */
            if (l == 1) {
                q3=method.q(xl1);
                r3=method.r(xl1);
                f3=method.f(xl1);
            }
            x2=(xl1+xl)/2.0;
            h6=h/6.0;
            h15=h/1.5;
            q1=q3;
            q2=method.q(x2);
            q3=method.q(xl);
            r1=r3;
            r2=method.r(x2);
            r3=method.r(xl);
            f1=f3;
            f2=method.f(x2);
            f3=method.f(xl);
            b1=h6*f1;
            b2=h15*f2;
            b3=h6*f3;
            tau1=h6*r1;
            tau2=h15*r2;
            tau3=h6*r3;
            s12 = -1.0/h/0.375;
            s13 = -s12/8.0;
            s22 = -2.0*s12;
            a12=s12+q1/1.5;
            a13=s13-q1/6.0;
            a21=s12-q2/1.5;
            a23=s12+q2/1.5;
            a22=s22+tau2;
            a31=s13+q3/6.0;
            a32=s12-q3/1.5;
            c12 = -a12/a22;
            c32 = -a32/a22;
            a12=a13+c12*a23;
            a21=a31+c32*a21;
            b1 += c12*b2;
            b2=b3+c32*b2;
```

```
        tau1 += c12*tau2;
        tau2=tau3+c32*tau2;
    } else {
        /* element mat vec evaluation 3 */
        if (l == 1) {
            q4=method.q(xl1);
            r4=method.r(xl1);
            f4=method.f(xl1);
        }
        x2=xl1+0.27639320225*h;
        x3=xl-x2+xl1;
        h12=h/12.0;
        h24=h/2.4;
        q1=q4;
        q2=method.q(x2);
        q3=method.q(x3);
        q4=method.q(xl);
        r1=r4;
        r2=method.r(x2);
        r3=method.r(x3);
        r4=method.r(xl);
        f1=f4;
        f2=method.f(x2);
        f3=method.f(x3);
        f4=method.f(xl);
        s12 = -4.8784183052080/h;
        s13=0.7117516385414/h;
        s14 = -0.16666666666667/h;
        s23=25.0*s14;
        s22 = -2.0*s23;
        b1=h12*f1;
        b2=h24*f2;
        b3=h24*f3;
        b4=h12*f4;
        tau1=h12*r1;
        tau2=h24*r2;
        tau3=h24*r3;
        tau4=h12*r4;
        a12=s12+0.67418082864578*q1;
        a13=s13-0.25751416197912*q1;
        a14=s14+q1/12.0;
        a21=s12-0.67418082864578*q2;
        a22=s22+tau2;
        a23=s23+0.93169499062490*q2;
        a24=s13-0.25751416197912*q2;
        a31=s13+0.25751416197912*q3;
        a32=s23-0.93169499062490*q3;
        a33=s22+tau3;
        a34=s12+0.67418082864578*q3;
        a41=s14-q4/12.0;
```

```
        a42=s13+0.25751416197912*q4;
        a43=s12-0.67418082864578*q4;
        det=a22*a33-a23*a32;
        c12=(a13*a32-a12*a33)/det;
        c13=(a12*a23-a13*a22)/det;
        c42=(a32*a43-a42*a33)/det;
        c43=(a42*a23-a43*a22)/det;
        tau1 += c12*tau2+c13*tau3;
        tau2=tau4+c42*tau2+c43*tau3;
        a12=a14+c12*a24+c13*a34;
        a21=a41+c42*a21+c43*a31;
        b1 += c12*b2+c13*b3;
        b2=b4+c42*b2+c43*b3;
    }
    if (l == 1 || l == n) {
        /* boundary conditions */
        if (l == 1 && e2 == 0.0) {
            tau1=1.0;
            b1=e3/e1;
            a12=0.0;
        } else if (l == 1 && e2 != 0.0) {
            tau1 -= e1/e2;
            b1 -= e3/e2;
        } else if (l == n && e5 == 0.0) {
            tau2=1.0;
            a21=0.0;
            b2=e6/e4;
        } else if (l == n && e5 != 0.0) {
            tau2 += e4/e5;
            b2 += e6/e5;
        }
    }
    /* forward babushka */
    if (l == 1) {
        chi[0]=ch=tl=tau1;
        t[0]=tl;
        gi[0]=g=yl=b1;
        y[0]=yl;
        sub[0]=a21;
        ssuper[0]=a12;
        pp=a21/(ch-a12);
        ch=tau2-ch*pp;
        g=b2-g*pp;
        tl=tau2;
        yl=b2;
    } else {
        chi[l1] = ch += tau1;
        gi[l1] = g += b1;
        sub[l1]=a21;
        ssuper[l1]=a12;
```

```
            pp=a21/(ch-a12);
            ch=tau2-ch*pp;
            g=b2-g*pp;
            t[l1]=tl+tau1;
            tl=tau2;
            y[l1]=yl+b1;
            yl=b2;
        }
        l++;
    }
    /* backward babushka */
    pp=yl;
    y[n]=g/ch;
    g=pp;
    ch=tl;
    l=n-1;
    while (l >= 0) {
        pp=ssuper[l]/(ch-sub[l]);
        tl=t[l];
        ch=tl-ch*pp;
        yl=y[l];
        g=yl-g*pp;
        y[l]=(gi[l]+g-yl)/(chi[l]+ch-tl);
        l--;
    }
}
```

5.5.3 Linear methods - Fourth order self adjoint

femhermsym

Solves a fourth order self-adjoint linear two point boundary value problem by means of Galerkin's method with continuous differentiable piecewise polynomial functions [BakH76, He75, StF73]. *femhermsym* computes approximations y_n and dy_n $(n=0,...,N)$ to the values at the points $x=x_n$, where

$$-\infty < a = x_0 < x_1 < ... < x_N = b < \infty$$

of the real valued function $y(x)$ and its derivative $Dy(x)$, where y satisfies the equation

$$-D^2(p(x)D^2y(x)) - D(q(x)Dy(x)) + r(x)y(x) = f(x) \qquad (1)$$

where $D=d/dx$, with boundary conditions

$$y(a) = e_1, \quad Dy(a) = e_2, \quad y(b) = e_3, \quad Dy(b) = e_4. \qquad (2)$$

It is assumed that

(a) $p(x)$ should be positive on the interval $[x_0,x_N]$, and $q(x)$ and $r(x)$ should be nonnegative there;

(b) $p(x)$, $q(x)$, $r(x)$, and $f(x)$ are required to be sufficiently smooth on the interval

$[x_0,x_N]$ except at the knots, where discontinuities of the derivatives are allowed; in that case the order of accuracy is preserved.

The solution is approximated by a function which is continuously differentiable on the closed interval $[x_0,x_N]$ and a polynomial of degree less than or equal to k ($k = 1 + order/2$) on each closed segment $[x_{j-1},x_j]$ ($j=1,...,N$). This function is entirely determined by the values of the zeroth and first derivative at the knots x_j and by the value it has at k-3 interior knots on each closed segment $[x_{j-1},x_j]$. The values of the function and its derivative at the knots are obtained by the solution of an $order+1$ diagonal linear system of $(k-1)N-2$ unknowns. The entries of the matrix and the vector are inner products which are approximated by piecewise k-point Lobatto quadrature. The evaluation of the matrix and the vector is performed segment by segment. If $k>3$ then the resulting linear system can be reduced to a pentadiagonal system by means of static condensation. This is possible because the function values at the interior knots on each segment $[x_{j-1},x_j]$ do not depend on function values outside that segment. The final pentadiagonal system, since the matrix is symmetric positive definite, is solved by means of Cholesky's decomposition method.

The theory upon which the method of solving the above boundary value problem is based on an extension of that described in *femlagsym*. The function $y(x)$ is now approximated not, as was the case for the method of *femlagsym* (see formula (4) there) by a function which is simply continuous and equal to a polynomial of degree k over the interval $[x_{n-1},x_n]$, but by a (necessarily continuous) function whose first derivative possesses these properties. Radau-Lobatto quadrature is used, systems of linear equations are derived, and coefficients concerning internal points of $[x_{n-1},x_n]$ are eliminated just as is described in *femlagsym*. Now, however, the resulting system of linear equations involving coefficients equal to the y_n and dy_n is pentadiagonal and is solved by Cholesky's decomposition method.

Procedure parameters:
$$\text{void femhermsym } (x,y,n,method,order,e)$$

x: double $x[0:n]$;
        entry: $a = x_0 < x_1 < ... < x_n = b$ is a partition of the segment $[a,b]$;

y: double $y[1:2n-2]$;
        exit: $y[2i-1]$ is an approximation to $y(x[i])$, $y[2i]$ is an approximation to $dy(x[i])$, where $y(x)$ is the solution of the equation (1) with boundary conditions (2);

n: int;
        entry: the upper bound of the arrays x (value of N above), $n > 1$;

$method$:   a class that defines four procedures p, q, r, and f, this class must implement the `AP_femhermsym_methods` interface;
        double p(double x)
           the procedure to compute $p(x)$, the coefficient of $D^2y(x)$ in equation (1); $p(x)$ should be strictly positive;
        double q(double x)
           the procedure to compute $q(x)$, the coefficient of $Dy(x)$ in equation (1); $q(x)$ should be nonnegative;
        double r(double x)
           the procedure to compute $r(x)$, the coefficient of $y(x)$ in equation (1); $r(x)$ should be nonnegative;

        double f(double x)
           the procedure to compute *f(x)*, the right-hand side of equation (1);

order: int;
        entry: *order* denotes the order of accuracy required for the approximate
                solution of the differential equation; let $h = \max(x[i]-x[i-1])$, then
$$\left| y[2i\text{-}1] - y(x[i]) \right| \le c1 * h^{order},$$
$$\left| y[2i] - Dy(x[i]) \right| \le c2 * h^{order}, \quad i=1,\dots,n\text{-}1;$$
                order can be chosen equal to 4, 6, or 8 only;

e: double *e[1:4]*;
        entry: *e[1],...,e[4]* describe the boundary conditions (values of e_i, $i=1,\dots,4$,
                in (2)).

Procedure used: chldecsolbnd.

```
package numal;

public interface AP_femhermsym_methods {

    double p(double x);
    double q(double x);
    double r(double x);
    double f(double x);
}


public static void femhermsym(double x[], double y[], int n,
                        AP_femhermsym_methods method, int order, double e[])
{
    int l,n2,v,w;
    double ya,yb,za,zb,d1,d2,e1,r1,r2,xl1;
    double em[] = new double[4];
    double a[] = new double[8*(n-1)+1];
    double a11[] = new double[1];  double a12[] = new double[1];
    double a13[] = new double[1];  double a14[] = new double[1];
    double a22[] = new double[1];  double a23[] = new double[1];
    double a24[] = new double[1];  double a33[] = new double[1];
    double a34[] = new double[1];  double a44[] = new double[1];
    double b1[] = new double[1];   double b2[] = new double[1];
    double b3[] = new double[1];   double b4[] = new double[1];
    double xl[] = new double[1];
    double p3[] = new double[1];   double p4[] = new double[1];
    double p5[] = new double[1];   double q3[] = new double[1];
    double q4[] = new double[1];   double q5[] = new double[1];
    double r3[] = new double[1];   double r4[] = new double[1];
    double r5[] = new double[1];   double f3[] = new double[1];
    double f4[] = new double[1];   double f5[] = new double[1];

    l=1;
    w=v=0;
    n2=n+n-2;
```

```
        xl1=x[0];
        xl[0]=x[1];
        ya=e[1];
        za=e[2];
        yb=e[3];
        zb=e[4];
        /* element matvec evaluation */
        femhermsymeval(order,l,method,a11,a12,a13,a14,a22,a23,a24,a33,a34,a44,b1,b2,b3,
                        b4,xl,xl1,p3,p4,p5,q3,q4,q5,r3,r4,r5,f3,f4,f5);
        em[2]=Double.MIN_VALUE;
        r1=b3[0]-a13[0]*ya-a23[0]*za;
        d1=a33[0];
        d2=a44[0];
        r2=b4[0]-a14[0]*ya-a24[0]*za;
        e1=a34[0];
        l++;
        while (l < n) {
            xl1=xl[0];
            xl[0]=x[1];
            /* element matvec evaluation */
            femhermsymeval(order,l,method,a11,a12,a13,a14,a22,a23,a24,a33,a34,a44,b1,
                            b2,b3,b4,xl,xl1,p3,p4,p5,q3,q4,q5,r3,r4,r5,f3,f4,f5);
            a[w+1]=d1+a11[0];
            a[w+4]=e1+a12[0];
            a[w+7]=a13[0];
            a[w+10]=a14[0];
            a[w+5]=d2+a22[0];
            a[w+8]=a23[0];
            a[w+11]=a24[0];
            a[w+14]=0.0;
            y[v+1]=r1+b1[0];
            y[v+2]=r2+b2[0];
            r1=b3[0];
            r2=b4[0];
            v += 2;
            w += 8;
            d1=a33[0];
            d2=a44[0];
            e1=a34[0];
            l++;
        }
        l=n;
        xl1=xl[0];
        xl[0]=x[1];
        /* element matvec evaluation */
        femhermsymeval(order,l,method,a11,a12,a13,a14,a22,a23,a24,a33,a34,a44,b1,b2,b3,
                        b4,xl,xl1,p3,p4,p5,q3,q4,q5,r3,r4,r5,f3,f4,f5);
        y[n2-1]=r1+b1[0]-a13[0]*yb-a14[0]*zb;
        y[n2]=r2+b2[0]-a23[0]*yb-a24[0]*zb;
        a[w+1]=d1+a11[0];
```

```
    a[w+4]=e1+a12[0];
    a[w+5]=d2+a22[0];
    Linear_algebra.chldecsolbnd(a,n2,3,em,y);
}



static private void femhermsymeval(int order, int l, AP_femhermsym_methods method,
                                   double a11[], double a12[], double a13[],
                                   double a14[], double a22[], double a23[],
                                   double a24[], double a33[], double a34[],
                                   double a44[], double b1[], double b2[],
                                   double b3[], double b4[], double xl[],
                                   double xl1, double pp3[], double pp4[],
                                   double pp5[], double qq3[], double qq4[],
                                   double qq5[], double rr3[], double rr4[],
                                   double rr5[], double ff3[], double ff4[],
                                   double ff5[])
{
    /* this procedure is internally used by FEMHERMSYM */

    if (order == 4) {
        double x2,h,h2,h3,p1,p2,q1,q2,r1,r2,f1,f2,b11,b12,b13,b14,b22,b23,b24,b33,
               b34,b44,s11,s12,s13,s14,s22,s23,s24,s33,s34,s44,m11,m12,m13,m14,m22,
               m23,m24,m33,m34,m44;
        h=xl[0]-xl1;
        h2=h*h;
        h3=h*h2;
        x2=(xl1+xl[0])/2.0;
        if (l == 1) {
            pp3[0]=method.p(xl1);
            qq3[0]=method.q(xl1);
            rr3[0]=method.r(xl1);
            ff3[0]=method.f(xl1);
        }
        /* element bending matrix */
        p1=pp3[0];
        p2=method.p(x2);
        pp3[0]=method.p(xl[0]);
        b11=6.0*(p1+pp3[0]);
        b12=4.0*p1+2.0*pp3[0];
        b13 = -b11;
        b14=b11-b12;
        b22=(4.0*p1+p2+pp3[0])/1.5;
        b23 = -b12;
        b24=b12-b22;
        b33=b11;
        b34 = -b14;
        b44=b14-b24;
        /* element stiffness matrix */
        q1=qq3[0];
```

```
        q2=method.q(x2);
        qq3[0]=method.q(xl[0]);
        s11=1.5*q2;
        s12=q2/4.0;
        s13 = -s11;
        s14=s12;
        s24=q2/24.0;
        s22=q1/6.0+s24;
        s23 = -s12;
        s33=s11;
        s34 = -s12;
        s44=s24+qq3[0]/6.0;
        /* element mass matrix */
        r1=rr3[0];
        r2=method.r(x2);
        rr3[0]=method.r(xl[0]);
        m11=(r1+r2)/6.0;
        m12=r2/24.0;
        m13=r2/6.0;
        m14 = -m12;
        m22=r2/96.0;
        m23 = -m14;
        m24 = -m22;
        m33=(r2+rr3[0])/6.0;
        m34=m14;
        m44=m22;
        /* element load vector */
        f1=ff3[0];
        f2=method.f(x2);
        ff3[0]=method.f(xl[0]);
        b1[0]=h*(f1+2.0*f2)/6.0;
        b3[0]=h*(ff3[0]+2.0*f2)/6.0;
        b2[0]=h2*f2/12.0;
        b4[0] = -b2[0];
        a11[0]=b11/h3+s11/h+m11*h;
        a12[0]=b12/h2+s12+m12*h2;
        a13[0]=b13/h3+s13/h+m13*h;
        a14[0]=b14/h2+s14+m14*h2;
        a22[0]=b22/h+s22*h+m22*h3;
        a23[0]=b23/h2+s23+m23*h2;
        a24[0]=b24/h+s24*h+m24*h3;
        a34[0]=b34/h2+s34+m34*h2;
        a33[0]=b33/h3+s33/h+m33*h;
        a44[0]=b44/h+s44*h+m44*h3;
    } else if (order == 6) {
        double h,h2,h3,x2,x3,p1,p2,p3,q1,q2,q3,r1,r2,r3,f1,f2,f3,b11,b12,b13,b14,
               b15,b22,b23,b24,b25,b33,b34,b35,b44,b45,b55,s11,s12,s13,s14,s15,s22,
               s23,s24,s25,s33,s34,s35,s44,s45,s55,m11,m12,m13,m14,m15,m22,m23,
               m24,m25,m33,m34,m35,m44,m45,m55,a15,a25,a35,a45,a55,c1,c2,c3,c4,b5;
        if (l == 1) {
```

```
            pp4[0]=method.p(xl1);
            qq4[0]=method.q(xl1);
            rr4[0]=method.r(xl1);
            ff4[0]=method.f(xl1);
    }
    h=xl[0]-xl1;
    h2=h*h;
    h3=h*h2;
    x2=0.27639320225*h+xl1;
    x3=xl1+xl[0]-x2;
    /* element bending matrix */
    p1=pp4[0];
    p2=method.p(x2);
    p3=method.p(x3);
    pp4[0]=method.p(xl[0]);
    b11=4.0333333333333e1*p1+1.1124913866738e-1*p2+1.4422084194664e1*p3+
        8.3333333333333e0*pp4[0];
    b12=1.4666666666667e1*p1-3.3191425091659e-1*p2+2.7985809175818e0*p3+
        1.6666666666667e0*pp4[0];
    b13=1.8333333333333e1*(p1+pp4[0])+1.2666666666667e0*(p2+p3);
    b15 = -(b11+b13);
    b14 = -(b12+b13+b15/2.0);
    b22=5.3333333333333e0*p1+9.9027346441674e-1*p2+5.4305986891624e-1*p3+
        3.3333333333333e-1*pp4[0];
    b23=6.6666666666667e0*p1-3.7791278464167e0*p2+2.4579451308295e-1*p3+
        3.6666666666667e0*pp4[0];
    b25 = -(b12+b23);
    b24 = -(b22+b23+b25/2.0);
    b33=8.3333333333333e0*p1+1.4422084194666e1*p2+1.1124913866726e-1*p3+
        4.0333333333333e1*pp4[0];
    b35 = -(b13+b33);
    b34 = -(b23+b33+b35/2.0);
    b45 = -(b14+b34);
    b44 = -(b24+b34+b45/2.0);
    b55 = -(b15+b35);
    /* element stiffness matrix */
    q1=qq4[0];
    q2=method.q(x2);
    q3=method.q(x3);
    qq4[0]=method.q(xl[0]);
    s11=2.8844168389330e0*q2+2.2249827733448e-2*q3;
    s12=2.5671051872498e-1*q2+3.2894812749994e-3*q3;
    s13=2.5333333333333e-1*(q2+q3);
    s14 = -3.7453559925005e-2*q2-2.2546440074988e-2*q3;
    s15 = -(s13+s11);
    s22=8.3333333333333e-2*q1+2.2847006554164e-2*q2+4.8632677916445e-4*q3;
    s23=2.2546440075002e-2*q2+3.7453559924873e-2*q3;
    s24 = -3.3333333333333e-3*(q2+q3);
    s25 = -(s12+s23);
    s33=2.2249827733471e-2*q2+2.8844168389330e0*q3;
```

```
s34 = -3.2894812750127e-3*q2-2.5671051872496e-1*q3;
s35 = -(s13+s33);
s44=4.8632677916788e-4*q2+2.2847006554161e-2*q3+8.3333333333338e-2*qq4[0];
s45 = -(s14+s34);
s55 = -(s15+s35);
/* element mass matrix */
r1=rr4[0];
r2=method.r(x2);
r3=method.r(x3);
rr4[0]=method.r(xl[0]);
m11=8.3333333333333e-2*r1+1.0129076086083e-1*r2+7.3759058058380e-3*r3;
m12=1.3296181273333e-2*r2+1.3704853933353e-3*r3;
m13 = -2.7333333333333e-2*(r2+r3);
m14=5.0786893258335e-3*r2+3.5879773408333e-3*r3;
m15=1.3147987115999e-1*r2-3.5479871159991e-2*r3;
m22=1.7453559925000e-3*r2+2.5464400750059e-4*r3;
m23 = -3.5879773408336e-3*r2-5.0786893258385e-3*r3;
m24=6.6666666666667e-4*(r2+r3);
m25=1.7259029213333e-2*r2-6.5923625466719e-3*r3;
m33=7.3759058058380e-3*r2+1.0129076086083e-1*r3+8.3333333333333e-2*rr4[0];
m34 = -1.3704853933333e-3*r2-1.3296181273333e-2*r3;
m35 = -3.5479871159992e-2*r2+1.3147987115999e-1*r3;
m44=2.5464400750008e-4*r2+1.7453559924997e-3*r3;
m45=6.5923625466656e-3*r2-1.7259029213330e-2*r3;
m55=0.17066666666667e0*(r2+r3);
/* element load vector */
f1=ff4[0];
f2=method.f(x2);
f3=method.f(x3);
ff4[0]=method.f(xl[0]);
b1[0]=8.3333333333333e-2*f1+2.0543729868749e-1*f2-5.5437298687489e-2*f3;
b2[0]=2.6967233145832e-2*f2-1.0300566479175e-2*f3;
b3[0] = -5.5437298687489e-2*f2+2.0543729868749e-1*f3+
        8.3333333333333e-2*ff4[0];
b4[0]=1.0300566479165e-2*f2-2.6967233145830e-2*f3;
b5=2.6666666666667e-1*(f2+f3);
a11[0]=h2*(h2*m11+s11)+b11;
a12[0]=h2*(h2*m12+s12)+b12;
a13[0]=h2*(h2*m13+s13)+b13;
a14[0]=h2*(h2*m14+s14)+b14;
a15=h2*(h2*m15+s15)+b15;
a22[0]=h2*(h2*m22+s22)+b22;
a23[0]=h2*(h2*m23+s23)+b23;
a24[0]=h2*(h2*m24+s24)+b24;
a25=h2*(h2*m25+s25)+b25;
a33[0]=h2*(h2*m33+s33)+b33;
a34[0]=h2*(h2*m34+s34)+b34;
a35=h2*(h2*m35+s35)+b35;
a44[0]=h2*(h2*m44+s44)+b44;
a45=h2*(h2*m45+s45)+b45;
```

```
        a55=h2*(h2*m55+s55)+b55;
        /* static condensation */
        c1=a15/a55;
        c2=a25/a55;
        c3=a35/a55;
        c4=a45/a55;
        b1[0]=(b1[0]-c1*b5)*h;
        b2[0]=(b2[0]-c2*b5)*h2;
        b3[0]=(b3[0]-c3*b5)*h;
        b4[0]=(b4[0]-c4*b5)*h2;
        a11[0]=(a11[0]-c1*a15)/h3;
        a12[0]=(a12[0]-c1*a25)/h2;
        a13[0]=(a13[0]-c1*a35)/h3;
        a14[0]=(a14[0]-c1*a45)/h2;
        a22[0]=(a22[0]-c2*a25)/h;
        a23[0]=(a23[0]-c2*a35)/h2;
        a24[0]=(a24[0]-c2*a45)/h;
        a33[0]=(a33[0]-c3*a35)/h3;
        a34[0]=(a34[0]-c3*a45)/h2;
        a44[0]=(a44[0]-c4*a45)/h;
    } else {
        double x2,x3,x4,h,h2,h3,p1,p2,p3,p4,q1,q2,q3,q4,r1,r2,r3,r4,f1,f2,f3,f4,b11,
               b12,b13,b14,b15,b16,b22,b23,b24,b25,b26,b33,b34,b35,b36,b44,b45,b46,
               b55,b56,b66,s11,s12,s13,s14,s15,s16,s22,s23,s24,s25,s26,s33,s34,s35,
               s36,s44,s45,s46,s55,s56,s66,m11,m12,m13,m14,m15,m16,m22,m23,m24,m25,
               m26,m33,m34,m35,m36,m44,m45,m46,m55,m56,m66,c15,c16,c25,c26,c35,c36,
               c45,c46,b5,b6,a15,a16,a25,a26,a35,a36,a45,a46,a55,a56,a66,det;
        if (l == 1) {
            pp5[0]=method.p(xl1);
            qq5[0]=method.q(xl1);
            rr5[0]=method.r(xl1);
            ff5[0]=method.f(xl1);
        }
        h=xl[0]-xl1;
        h2=h*h;
        h3=h*h2;
        x2=xl1+h*0.172673164646;
        x3=xl1+h/2.0;
        x4=xl1+xl[0]-x2;
        /* element bending matrix */
        p1=pp5[0];
        p2=method.p(x2);
        p3=method.p(x3);
        p4=method.p(x4);
        pp5[0]=method.p(xl[0]);
        b11=105.8*p1+9.8*pp5[0]+7.3593121303513e-2*p2+2.2755555555556e1*p3+
            7.0565656088553e0*p4;
        b12=27.6*p1+1.4*pp5[0]-3.41554824811e-1*p2+2.8444444444444e0*p3+
            1.0113960946522e0*p4;
        b13 = -32.2*(p1+pp5[0])-7.2063492063505e-1*(p2+p4)+2.2755555555556e1*p3;
```

```
    b14=4.6*p1+8.4*pp5[0]+1.0328641222944e-1*p2-2.8444444444444e0*p3-
        3.3445562534992e0*p4;
    b15 = -(b11+b13);
    b16 = -(b12+b13+b14+b15/2.0);
    b22=7.2*p1+0.2*pp5[0]+1.5851984028581e0*p2+3.5555555555556e-1*p3+
        1.4496032730059e-1*p4;
    b23 = -8.4*p1-4.6*pp5[0]+3.3445562534992e0*p2+2.8444444444444e0*p3-
        1.0328641222944e-1*p4;
    b24=1.2*(p1+pp5[0])-4.7936507936508e-1*(p2+p4)-3.5555555555556e-1*p3;
    b25 = -(b12+b23);
    b26 = -(b22+b23+b24+b25/2.0);
    b33=7.0565656088553e0*p2+2.2755555555556e1*p3+7.3593121303513e-2*p4+
        105.8*pp5[0]+9.8*p1;
    b34 = -1.4*p1-27.6*pp5[0]-1.0113960946522e0*p2-2.8444444444444e0*p3+
        3.4155482481100e-1*p4;
    b35 = -(b13+b33);
    b36 = -(b23+b33+b34+b35/2.0);
    b44=7.2*pp5[0]+p1/5.0+1.4496032730059e-1*p2+3.5555555555556e-1*p3+
        1.5851984028581e0*p4;
    b45 = -(b14+b34);
    b46 = -(b24+b34+b44+b45/2.0);
    b55 = -(b15+b35);
    b56 = -(b16+b36);
    b66 = -(b26+b36+b46+b56/2.0);
    /* element stiffness matrix */
    q1=qq5[0];
    q2=method.q(x2);
    q3=method.q(x3);
    q4=method.q(x4);
    qq5[0]=method.q(x1[0]);
    s11=3.0242424037951e0*q2+3.1539909130065e-2*q4;
    s12=1.2575525581744e-1*q2+4.1767169716742e-3*q4;
    s13 = -3.0884353741496e-1*(q2+q4);
    s14=4.0899041243062e-2*q2+1.2842455355577e-2*q4;
    s15 = -(s13+s11);
    s16=5.9254861177068e-1*q2+6.0512612719116e-2*q4;
    s22=5.2292052865422e-3*q2+5.5310763862796e-4*q4+q1/20.0;
    s23 = -1.2842455355577e-2*q2-4.0899041243062e-2*q4;
    s24=1.7006802721088e-3*(q2+q4);
    s25 = -(s12+s23);
    s26=2.4639593097426e-2*q2+8.0134681270641e-3*q4;
    s33=3.1539909130065e-2*q2+3.0242424037951e0*q4;
    s34 = -4.1767169716742e-3*q2-1.2575525581744e-1*q4;
    s35 = -(s13+s33);
    s36 = -6.0512612719116e-2*q2-5.9254861177068e-1*q4;
    s44=5.5310763862796e-4*q2+5.2292052865422e-3*q4+qq5[0]/20.0;
    s45 = -(s14+s34);
    s46=8.0134681270641e-3*q2+2.4639593097426e-2*q4;
    s55 = -(s15+s35);
    s56 = -(s16+s36);
```

```
s66=1.1609977324263e-1*(q2+q4)+3.5555555555556e-1*q3;
/* element mass matrix */
r1=rr5[0];
r2=method.r(x2);
r3=method.r(x3);
r4=method.r(x4);
rr5[0]=method.r(x1[0]);
m11=9.7107020727310e-2*r2+1.5810259199180e-3*r4+r1/20.0;
m12=8.2354889460254e-3*r2+2.1932154960071e-4*r4;
m13=1.2390670553936e-2*(r2+r4);
m14 = -1.7188466249968e-3*r2-1.0508326752939e-3*r4;
m15=5.3089789712119e-2*r2+6.7741558661060e-3*r4;
m16 = -1.7377712856076e-2*r2+2.2173630018466e-3*r4;
m22=6.9843846173145e-4*r2+3.0424512029349e-5*r4;
m23=1.0508326752947e-3*r2+1.7188466249936e-3*r4;
m24 = -1.4577259475206e-4*(r2+r4);
m25=4.5024589679127e-3*r2+9.3971790283374e-4*r4;
m26 = -1.4737756452780e-3*r2+3.0759488725998e-4*r4;
m33=1.5810259199209e-3*r2+9.7107020727290e-2*r4+rr5[0]/20.0;
m34 = -2.1932154960131e-4*r2-8.2354889460354e-3*r4;
m35=6.7741558661123e-3*r2+5.3089789712112e-2*r4;
m36 = -2.2173630018492e-3*r2+1.7377712856071e-2*r4;
m44=3.0424512029457e-5*r2+6.9843846173158e-4*r4;
m45 = -9.3971790283542e-4*r2-4.5024589679131e-3*r4;
m46=3.0759488726060e-4*r2-1.4737756452778e-3*r4;
m55=2.9024943310657e-2*(r2+r4)+3.5555555555556e-1*r3;
m56=9.5006428402050e-3*(r4-r2);
m66=3.1098153547125e-3*(r2+r4);
/* element load vector */
f1=ff5[0];
f2=method.f(x2);
f3=method.f(x3);
f4=method.f(x4);
ff5[0]=method.f(x1[0]);
b1[0]=1.6258748099336e-1*f2+2.0745852339969e-2*f4+f1/20.0;
b2[0]=1.3788780589233e-2*f2+2.8778860774335e-3*f4;
b3[0]=2.0745852339969e-2*f2+1.6258748099336e-1*f4+ff5[0]/20.0;
b4[0] = -2.8778860774335e-3*f2-1.3788780589233e-2*f4;
b5=(f2+f4)/11.25+3.5555555555556e-1*f3;
b6=2.9095718698132e-2*(f4-f2);
a11[0]=h2*(h2*m11+s11)+b11;
a12[0]=h2*(h2*m12+s12)+b12;
a13[0]=h2*(h2*m13+s13)+b13;
a14[0]=h2*(h2*m14+s14)+b14;
a15=h2*(h2*m15+s15)+b15;
a16=h2*(h2*m16+s16)+b16;
a22[0]=h2*(h2*m22+s22)+b22;
a23[0]=h2*(h2*m23+s23)+b23;
a24[0]=h2*(h2*m24+s24)+b24;
a25=h2*(h2*m25+s25)+b25;
```

```
    a26=h2*(h2*m26+s26)+b26;
    a33[0]=h2*(h2*m33+s33)+b33;
    a34[0]=h2*(h2*m34+s34)+b34;
    a35=h2*(h2*m35+s35)+b35;
    a36=h2*(h2*m36+s36)+b36;
    a44[0]=h2*(h2*m44+s44)+b44;
    a45=h2*(h2*m45+s45)+b45;
    a46=h2*(h2*m46+s46)+b46;
    a55=h2*(h2*m55+s55)+b55;
    a56=h2*(h2*m56+s56)+b56;
    a66=h2*(h2*m66+s66)+b66;
    /* static condensation */
    det = -a55*a66+a56*a56;
    c15=(a15*a66-a16*a56)/det;
    c16=(a16*a55-a15*a56)/det;
    c25=(a25*a66-a26*a56)/det;
    c26=(a26*a55-a25*a56)/det;
    c35=(a35*a66-a36*a56)/det;
    c36=(a36*a55-a35*a56)/det;
    c45=(a45*a66-a46*a56)/det;
    c46=(a46*a55-a45*a56)/det;
    a11[0]=(a11[0]+c15*a15+c16*a16)/h3;
    a12[0]=(a12[0]+c15*a25+c16*a26)/h2;
    a13[0]=(a13[0]+c15*a35+c16*a36)/h3;
    a14[0]=(a14[0]+c15*a45+c16*a46)/h2;
    a22[0]=(a22[0]+c25*a25+c26*a26)/h;
    a23[0]=(a23[0]+c25*a35+c26*a36)/h2;
    a24[0]=(a24[0]+c25*a45+c26*a46)/h;
    a33[0]=(a33[0]+c35*a35+c36*a36)/h3;
    a34[0]=(a34[0]+c35*a45+c36*a46)/h2;
    a44[0]=(a44[0]+c45*a45+c46*a46)/h;
    b1[0]=(b1[0]+c15*b5+c16*b6)*h;
    b2[0]=(b2[0]+c25*b5+c26*b6)*h2;
    b3[0]=(b3[0]+c35*b5+c36*b6)*h;
    b4[0]=(b4[0]+c45*b5+c46*b6)*h2;
    }
}
```

5.5.4 Nonlinear methods

nonlinfemlagskew

Solves a nonlinear two point boundary value problem with spherical coordinates
[Bab72, Bak77, BakH76, StF73]. *nonlinfemlagskew* solves the differential equation
$$D(x^{nc}Dy(x))/x^{nc} = f(x,y,Dy(x)), \quad a < x < b, \tag{1}$$
where $D=d/dx$, with boundary conditions

$$e_1 y(a) + e_2 Dy(a) = e_3, \quad e_4 y(b) + e_5 Dy(b) = e_6 \quad (e_1, e_4 \neq 0). \tag{2}$$

The functions f, df/dy and df/dz are required to be sufficiently smooth in their variables on the interior of every segment $[x_i, x_{i+1}]$ ($i=0,...,n-1$).

Let $y[0](x)$ be some initial approximation of $y(x)$; then the nonlinear problem is solved by successively solving

$$-D(x^{nc} Dg[k](x))/x^{nc} + fy(x,y[k](x),Dy[k](x))*g[k](x) + fz(x,y[k](x),Dy[k](x))*Dg[k](x)$$
$$= D(x^{nc} Dy[k](x)/x^{nc} - f(x,y[k],Dy[k](x)), \quad x_0 < x < x_n,$$

with boundary conditions

$$e_1 g[k](x_0) + e_2 Dg[k](x_0) = 0, \quad e_4 g[k](x_n) + e_5 Dg[k](x_n) = 0$$

with Galerkin's method (see *femlagsym*) and putting

$$y[k+1](x) = y[k](x) + g[k](x), \quad k=0,1,....$$

This is the so-called Newton-Kantorowitch method.

Procedure parameters:

$$\text{void nonlinfemlagskew } (x,y,n,method,nc,e)$$

x: double *x[0:n]*;

    entry: $a = x_0 < x_1 < ... < x_n = b$ is a partition of the segment $[a,b]$;

y: double *y[0:n]*;

    entry: $y[i]$, $i=0,1,...,n$, is an initial approximate solution at $x[i]$ of the differential equation (1) with boundary conditions (2);

    exit: $y[i]$, $i=0,1,...,n$, is the Galerkin solution at $x[i]$ of the differential equation (1) with boundary conditions (2);

n: int;

    entry: the upper bound of the arrays x and y, $n > 1$;

method: a class that defines three procedures f, fy, and fz, this class must implement the `AP_nonlinfemlagskew_methods` interface;

    double *f(double x, double y, double z)*

        the procedure to compute the right-hand side of (1); ($f(x,y,z)$ is the right-hand side of equation (1));

    double *fy(double x, double y, double z)*

        the procedure to compute $fy(x,y,z)$, the derivative of f with respect to y;

    double *fz(double x, double y, double z)*

        the procedure to compute $fz(x,y,z)$, the derivative of f with respect to z;

nc: int;

    entry: if $nc = 0$, Cartesian coordinates are used;

        if $nc = 1$, polar coordinates are used;

        if $nc = 2$, spherical coordinates are used;

e: double *e[1:6]*;

    entry: $e[1],...,e[6]$ describe the boundary conditions (values of e_i, $i=1,...,6$, in (2));

    $e[1]$ and $e[4]$ are not allowed to vanish both.

Procedure used: dupvec.

```
package numal;

public interface AP_nonlinfemlagskew_methods {

    double f(double x, double y, double z);
    double fy(double x, double y, double z);
    double fz(double x, double y, double z);
}


public static void nonlinfemlagskew(double x[], double y[], int n,
                                    AP_nonlinfemlagskew_methods method, int nc,
                                    double e[])
{
    int l,l1,it;
    double xl1,xl,h,a12,a21,b1,b2,tau1,tau2,ch,tl,g,yl,pp,zl1,zl,e1,e2,e4,e5,eps,
           rho,xm,vl,vr,wl,wr,pr,qm,rm,fm,xl12,xl1xl,xl2,zm,zaccm;
    double t[] = new double[n];
    double ssuper[] = new double[n];
    double sub[] = new double[n];
    double chi[] = new double[n];
    double gi[] = new double[n];
    double z[] = new double[n+1];

    ch=g=tl=yl=0.0;
    Basic.dupvec(0,n,0,z,y);
    e1=e[1];
    e2=e[2];
    e4=e[4];
    e5=e[5];
    it=1;
    do {
        l=1;
        xl=x[0];
        zl=z[0];
        while (l <= n) {
            xl1=xl;
            l1=l-1;
            xl=x[l];
            h=xl-xl1;
            zl1=zl;
            zl=z[l];
            /* element mat vec evaluation 1 */
            if (nc == 0)
                vl=vr=0.5;
            else if (nc == 1) {
                vl=(xl1*2.0+xl)/6.0;
                vr=(xl1+xl*2.0)/6.0;
            } else {
                xl12=xl1*xl1/12.0;
```

```
        xl1xl=xl1*xl/6.0;
        xl2=xl*xl/12.0;
        vl=3.0*xl12+xl1xl+xl2;
        vr=3.0*xl2+xl1xl+xl12;
    }
    wl=h*vl;
    wr=h*vr;
    pr=vr/(vl+vr);
    xm=xl1+h*pr;
    zm=pr*zl+(1.0-pr)*zl1;
    zaccm=(zl-zl1)/h;
    qm=method.fz(xm,zm,zaccm);
    rm=method.fy(xm,zm,zaccm);
    fm=method.f(xm,zm,zaccm);
    tau1=wl*rm;
    tau2=wr*rm;
    b1=wl*fm-zaccm*(vl+vr);
    b2=wr*fm+zaccm*(vl+vr);
    a12 = -(vl+vr)/h+vl*qm+(1.0-pr)*pr*rm*(wl+wr);
    a21 = -(vl+vr)/h-vr*qm+(1.0-pr)*pr*rm*(wl+wr);
if (l == 1 || l == n) {
        /* boundary conditions */
        if (l == 1 && e2 == 0.0) {
            tau1=1.0;
            b1=a12=0.0;
        } else if (l == 1 && e2 != 0.0) {
            tau1 -= e1/e2;
        } else if (l == n && e5 == 0.0) {
            tau2=1.0;
            b2=a21=0.0;
        } else if (l == n && e5 != 0.0) {
            tau2 += e4/e5;
        }
    }
    /* forward babushka */
    if (l == 1) {
        chi[0]=ch=tl=tau1;
        t[0]=tl;
        gi[0]=g=yl=b1;
        y[0]=yl;
        sub[0]=a21;
        ssuper[0]=a12;
        pp=a21/(ch-a12);
        ch=tau2-ch*pp;
        g=b2-g*pp;
        tl=tau2;
        yl=b2;
    } else {
        chi[l1] = ch += tau1;
        gi[l1] = g += b1;
```

```
                sub[l1]=a21;
                ssuper[l1]=a12;
                pp=a21/(ch-a12);
                ch=tau2-ch*pp;
                g=b2-g*pp;
                t[l1]=tl+tau1;
                tl=tau2;
                y[l1]=yl+b1;
                yl=b2;
            }
            l++;
        }
        /* backward babushka */
        pp=yl;
        y[n]=g/ch;
        g=pp;
        ch=tl;
        l=n-1;
        while (l >= 0) {
            pp=ssuper[l]/(ch-sub[l]);
            tl=t[l];
            ch=tl-ch*pp;
            yl=y[l];
            g=yl-g*pp;
            y[l]=(gi[l]+g-yl)/(chi[l]+ch-tl);
            l--;
        }
        eps=0.0;
        rho=1.0;
        for (l=0; l<=n; l++) {
            rho += Math.abs(z[l]);
            eps += Math.abs(y[l]);
            z[l] -= y[l];
        }
        rho *= 1.0e-14;
        it++;
    } while (eps > rho);
    Basic.dupvec(0,n,0,y,z);
}
```

5.6 Two-dimensional boundary value problems

5.6.1 Elliptic special linear systems

A. richardson

Solves a system of linear equations with a coefficient matrix having positive real eigenvalues by means of a nonstationary second order iterative method: Richardson's method [CooHHS73, Vh68]. Since Richardson's method is particularly suitable for solving a system of linear equations that is obtained by discretizing a two-dimensional elliptic boundary value problem, the procedure *richardson* is programmed in such a way that the solution vector is given as a two-dimensional array $u[j,l]$, $lj \leq j \leq uj$, $ll \leq l \leq ul$. The coefficient matrix is not stored, but each row corresponding to a pair (j,l) is generated when needed. *richardson* can also be used to determine the eigenvalue of the coefficient matrix corresponding to the dominant eigenfunction.

richardson either (a) determines an approximation u_n to the solution of the system of equations

$$Au = f \tag{1}$$

where A is an $N{\times}N$ matrix ($N=(uj-lj+1)*(ul-ll+1)$; lj, $ll \geq 1$) whose eigenvalues λ_i are such that $0 < \lambda_1 < \lambda_2 < ... < \lambda_N = b$, and u is a vector whose components $u^{(i)}$ are stored in the locations of a two-dimensional real array U by means of the mapping $u^{((j-lj)(ul-ll+1)+l-ll+1)}$ in location (j,l) $(j=lj,lj+1,...,uj;$ $l=ll,ll+1,...,ul)$ or (b) estimates the eigenvalue λ_1 of the matrix A just described.

In both cases a sequence of vectors u_k $(k=1,...,n)$ is constructed from an initial member u_0 (if the parameter *inap* is given the value true upon call of *richardson*, the components of u_0 are those stored in the locations of U prior to call; if *inap* is given the value false, u_0 is the unit vector $(1,1,...,1)$) by means of the recursion

$$u_1 = \beta_0 u_0 - \omega_0 v_0$$
$$u_{k+1} = \beta_k u_k + (1 - \beta_k)u_{k-1} - \omega_k v_k \tag{2}$$

where

$$v_k = Au_k - f \tag{3}$$

and, with $0 < a < b$, $y_0 = (a + b)/(b - a)$

$\beta_0 = 1$, $\omega_0 = 2/(b+a)$, $\beta_k = 2y_0 T_k(y_0)/T_{k+1}(y_0)$, $\omega_k = 4T_k(y_0)/((b-a)T_{k+1}(y_0))$,

$T_k(y) = \cos(k \arccos(y))$ being a Tscheyscheff polynomial. Individually the u_k, v_k satisfy the recursions

$$u_1 = (\beta_0 - \omega_0 A)u_0 + \omega_0 f$$
$$u_{k+1} = (\beta_k - \omega_k A)u_k + (1 - \beta_k)u_{k-1} + \omega_k f$$
$$v_1 = (\beta_0 - \omega_0 A)v_0$$
$$v_{k+1} = (\beta_k - \omega_k A)v_k - (1 - \beta_k)v_{k-1}.$$

The residue vectors v_k are then given explicitly by the formula

$$v_k = C_k(a,b;A)v_0$$

where

$$C_k(a,b;\lambda) = T_k((b+a-2\lambda)/(b-a))/T_k((b+a)/(b-a)).$$

Of all k-th degree polynomials p_k with $p_k(0)=1$, $C_k(a,b;\lambda)$ is that for which

$\max |p_k(\lambda)|$ $(a \le \lambda \le b)$ is a minimum.

If, in the above, $a < \lambda_1$, $\|v_k\|$ tends to zero as k increases, and u_k tends to the solution u of equation (1). The mean rate of convergence over k steps may be defined to be

$$- \ln(\|v_k\| / \|v_0\|) / k \qquad \text{(where ln is the natural logarithm)}$$

and the measure of this quantity computed during the implementation of *richardson* is

$$r_k = -(1/(2k))[\ln\{\|v_k\|_2 / \|v_0\|_2\} + \ln\{\|v_k\|_\infty / \|v_0\|_\infty\}].$$

If, however, $\lambda_1 < a$, $\|v_k\|$ may not tend to zero, but an estimate of λ_1 may be extracted from the u_k and v_k. If, in the decomposition of v_0 in terms of the eigenvalues of e_j or A,

$$v_0 = \sum_{j=1}^{N} c_j e_j$$

$c_1 \ne 0$, then Λ_k, where

$$\mu_k = \|v_k\| / \|u_k - u_{k-1}\|, \qquad \Lambda_k = \mu_k(\sqrt{(ab)} - \mu_k)/((\sqrt{a} + \sqrt{b})^2/4 - \mu_k)$$

tends to λ_1. The estimates used in the implementation of *richardson* are $\Lambda_k = (\Lambda_k^{(2)} + \Lambda_k^{(\infty)})/2$ where with $\tau = 2, \infty$ in both cases,

$$\mu_k^{(\tau)} = \|v_k\|_\tau / \|u_k - u_{k-1}\|_\tau, \qquad \Lambda_k^{(\tau)} = \mu_k^{(\tau)}(\sqrt{(ab)} - \mu_k^{(\tau)})/((\sqrt{a} + \sqrt{b})^2/4 - \mu_k^{(\tau)}) .$$

In both cases it is required that b should be an upper bound for the eigenvalues λ_i of A: it is remarked that, denoting the elements of A by $A_{j,l}$, for all λ_i

$$\lambda_i \le \min\left(\max_j \sum_{l=1}^{N} |A_{j,l}|, \quad \max_l \sum_{j=1}^{N} |A_{j,l}| \right).$$

Procedure parameters:

void richardson $(u, lj, uj, ll, ul, inap, method, a, b, n, discr, k, rateconv, domeigval)$

u:	double $u[lj:uj, ll:ul]$;
	after each iteration the approximate solution calculated by *richardson* is stored in *u*;
	entry: if *inap* is chosen to be true then an initial approximation of the solution, otherwise arbitrary;
	exit: the final approximation of the solution;
lj,uj:	int;
	entry: lower and upper bound for the first subscript of *u*;
ll,ul:	int;
	entry: lower and upper bound for the second subscript of *u*;
inap:	boolean;
	entry: if the user wishes to introduce an initial approximation then *inap* should be chosen to be true, choosing *inap* to be false has the effect that all components of *u* are set equal to 1 before the first iteration is performed;
method:	a class that defines two procedures *residual* and *out*, this class must implement the AP_richardson_methods interface;
	void residual(int lj, int uj, int ll, int ul, double u[ ][ ])
	suppose that the system of equation at hand is $Au=f$, for any entry *u* the procedure *residual* should calculate the residual $Au-f$ in each point j,l, where $lj \le j \le uj$, $ll \le l \le ul$, and substitute these values in the array *u*;

> *void out(double u[ ][ ], int lj, int uj, int ll, int ul, int n[ ], double discr[ ],*
> *int k[ ], double rateconv[ ], double domeigval[ ])*
> by this procedure one has access to the following quantities:
> for $0 \le k \le n$ the k-th iterand in u, the Euclidean and maximum
> norm of the k-th residual in *discr[1]* and *discr[2]*, respectively;
> for $0 < k \le n$ also the average rate of convergence and the
> approximation to the dominant eigenvalue, both with respect to
> the k-th iterand in u, in *rateconv* and *domeigval*, respectively;
> moreover, *out* can be used to let n be dependent on the accuracy
> reached in approximating the dominant eigenvalue;

a,b: double;
        entry: if one wishes to find the solution of the boundary value problem
               then in a and b the user should give a lower and upper bound
               for the eigenvalues for which the corresponding eigenfunctions
               in eigenfunction expansion of the residual *Au-f*, with *u* equals
               the initial approximation, should be reduced;
        if the dominant eigenvalue is to be found then one should choose a
        greater than this eigenvalue;

n: int $n[0{:}0]$;
        entry: gives the total number of iterations to be performed;

discr: double $discr[1{:}2]$;
        exit: after each iteration *richardson* delivers in *discr[1]* the Euclidean
              norm of the residual (value of $\|v_n\|_2$ above), and in *discr[2]* the
              maximum norm of the residual (value of $\|v_n\|_\infty$ above);

k: int $k[0{:}0]$;
        exit: counts the number of iterations *richardson* is performing;

rateconv: double $rateconv[0{:}0]$;
        exit: after each iteration the average rate of convergence is assigned
              to *rateconv* (value of r_n above);

domeigval: double $domeigval[0{:}0]$;
        exit: after each iteration the dominant eigenvalue, if present, is
              assigned to *domeigval* (value of Λ_n above); if there is no
              dominant eigenvalue then the value of *domeigval* is
              meaningless; this manifests itself by showing no convergence to
              a fixed value.

```
package numal;

public interface AP_richardson_methods {

    void residual(int lj, int uj, int ll, int ul, double u[][]);
    void out(double u[][], int lj, int uj, int ll, int ul, int n[], double discr[],
            int k[], double rateconv[], double domeigval[]);
}


public static void richardson(double u[][], int lj, int uj, int ll, int ul,
                    boolean inap, AP_richardson_methods method, double a,
                    double b, int n[], double discr[], int k[],
                    double rateconv[], double domeigval[])
```

```
{
    int j,l;
    double x,y,z,y0,c,d,alfa,omega,omega0,eigmax,eigeucl,euclres,maxres,rcmax,
           rceucl,maxres0,euclres0,auxres0,auxv,auxu,auxres,eucluv,maxuv;
    double v[][] = new double[uj+1][ul+1];
    double res[][] = new double[uj+1][ul+1];

    alfa=2.0;
    omega=4.0/(b+a);
    y0=(b+a)/(b-a);
    x=0.5*(b+a);
    y=(b-a)*(b-a)/16.0;
    z=4.0*y0*y0;
    c=a*b;
    c=Math.sqrt(c);
    d=Math.sqrt(a)+Math.sqrt(b);
    d=d*d;
    if (!inap)
        for (j=lj; j<=uj; j++)
            for (l=ll; l<=ul; l++) u[j][l]=1.0;
    k[0]=0;
    for (j=lj; j<=uj; j++)
        for (l=ll; l<=ul; l++) res[j][l]=u[j][l];
    method.residual(lj,uj,ll,ul,res);
    omega0=2.0/(b+a);
    maxres0=euclres0=0.0;
    for (j=lj; j<=uj; j++)
        for (l=ll; l<=ul; l++) {
            auxres0=res[j][l];
            v[j][l]=u[j][l]-omega0*auxres0;
            auxres0=Math.abs(auxres0);
            maxres0 = (maxres0 < auxres0) ? auxres0 : maxres0;
            euclres0 += auxres0*auxres0;
        }
    euclres0=Math.sqrt(euclres0);
    discr[1]=euclres0;
    discr[2]=maxres0;
    method.out(u,lj,uj,ll,ul,n,discr,k,rateconv,domeigval);
    while (k[0] < n[0]) {
        k[0]++;
        /* calculate parameters alfa and omega for each iteration */
        alfa=z/(z-alfa);
        omega=1.0/(x-omega*y);
        /* iteration */
        eucluv=euclres=maxuv=maxres=0.0;
        for (j=lj; j<=uj; j++)
            for (l=ll; l<=ul; l++) res[j][l]=v[j][l];
        method.residual(lj,uj,ll,ul,res);
        for (j=lj; j<=uj; j++)
            for (l=ll; l<=ul; l++) {
```

```
                auxv=u[j][l];
                auxu=v[j][l];
                auxres=res[j][l];
                auxv=alfa*auxu-omega*auxres+(1.0-alfa)*auxv;
                v[j][l]=auxv;
                u[j][l]=auxu;
                auxu=Math.abs(auxu-auxv);
                auxres=Math.abs(auxres);
                maxuv = (maxuv < auxu) ? auxu : maxuv;
                maxres = (maxres < auxres) ? auxres : maxres;
                eucluv += auxu*auxu;
                euclres += auxres*auxres;
            }
        eucluv=Math.sqrt(eucluv);
        euclres=Math.sqrt(euclres);
        discr[1]=euclres;
        discr[2]=maxres;
        maxuv=maxres/maxuv;
        eucluv=euclres/eucluv;
        eigmax=maxuv*(c-maxuv)/(0.25*d-maxuv);
        eigeucl=eucluv*(c-eucluv)/(0.25*d-eucluv);
        domeigval[0]=0.5*(eigmax+eigeucl);
        rceucl = -Math.log(euclres/euclres0)/k[0];
        rcmax = -Math.log(maxres/maxres0)/k[0];
        rateconv[0]=0.5*(rceucl+rcmax);
        method.out(u,lj,uj,ll,ul,n,discr,k,rateconv,domeigval);
    }
}
```

B. elimination

Determines an approximation u'_n to the solution of the system of equations
$$Au = f \tag{1}$$
where A is an $N{\times}N$ matrix ($N=(uj-lj+1)*(ul-ll+1)$; $lj, ll \geq 1$) whose eigenvalues λ_i are such that $0 < \lambda_1 < \lambda_2 < ... < \lambda_N = b$, and u is a vector whose components $u^{(i)}$ are stored in the locations of a two-dimensional real array U by means of the mapping $u^{((j-lj)(ul-ll+1)+l-ll+1)}$ in location (j,l) $(j=lj,lj+1,...,uj; l=ll,ll+1,...,ul)$.

elimination is to be used in conjunction with richardson, one of whose possible functions is precisely the solution of the problem described above. In the implementation of richardson, a sequence of vectors u_k and residual vectors v_k, where $v_k = Au_k - f$, is produced by use of recursion (2,3) of the documentation to richardson and, with $0 < a < b$,
$$v_k = C_k(a,b;A)v_0, \qquad u_k = u + A^{-1}C_k(a,b;A)v_0$$
where of all polynomials p_k with $p_k(0) = 1$, $p_k(\lambda) = C_k(a,b;\lambda)$ is that for which $\max|p_k^{(\lambda)}|$ $(a{\leq}\lambda{\leq}b)$ is a minimum; explicitly
$$C_k(a,b;\lambda) = T_k((b+a-2\lambda)/(b-a))/T_k((b+a)/(b-a)) .$$
Decomposing v_0 in terms of the eigenvectors e_i of A,

$$v_0 = \sum_{i=1}^{N} c_i e_i$$

$$v_k = \sum_{i=1}^{N} c_i C_k(a,n;\lambda_i)e_i, \qquad u_k = u + \sum_{i=1}^{N} \lambda_i^{-1} c_i C_k(a,b;\lambda_i)e_i$$

If $a < \lambda_1$, then $u_k \to u$. An upper bound b for the λ_i may easily be given (see formula (4) of *richardson*). A lower bound for the λ_i may also be obtained from the remark that the λ_i lie in the adjunction of the discs

$$\left| \lambda - A^{(i,i)} \right| \le \sum_{\substack{j=1 \\ (j \ne i)}}^{N} \left| A^{(i,j)} \right| \qquad (i = 1,\ldots,n) \qquad (2)$$

In the above, however, it is required that the lower bound a should be positive, and this may not be true of the lower bound derived from the expressions (2).

In one of its modes of application (with $a > \lambda_1$) *richardson* yields an approximation to λ_1, and hence a lower bound a' for the λ_i. Applying *richardson* yet again, with a' in place of a and u_n used as the initial approximation u'_0 to produce a new sequence of approximations u'_k and residual vectors v'_k, the residuals

$$v'_{n'} = \sum_{j=1}^{N} C_n(a,b;\lambda_j)C_{n'}(a',b;\lambda_j)e_j$$

$$u'_{n'} = u + \sum_{j=1}^{N} C_n(a,b;\lambda_j)C_{n'}(a',b;\lambda_j)\lambda_j^{-1}e_j$$

(3)

$C_{n'}(a',b;\lambda)$ has the same properties with respect to $[a',b]$ as $C_n(a,b;\lambda)$ has with respect to $[a,b]$: $u'_{n'} \to u$.

Assuming that $a < \lambda_2$, the terms involving $j=2,3,\ldots$ in formulae (3) are negligible:

$$v'_{n'} \approx C_n(a,b,;\lambda_1)C_{n'}(a',b;\lambda_1)e_1$$

(4)

$$u'_{n'} \approx u + C_n(a,b,;\lambda_1)C_{n'}(a',b;\lambda_1)\lambda_1^{-1}e_1 .$$

By taking

$$a^* = a^*(u') = \{2\lambda_1 + b[\cos(\pi/(2n'))-1]\} / \{\cos(\pi/(2n'))+1\}$$

$C_{n'}(a^*,b;\lambda_1) = 0$, and the second term upon the right-hand side of formula (4) may be eliminated.

The above process requires a total of $n+n'$ vector constructions. The mean rate of convergence may therefore be defined as

$R(n,n') = -\ln(\|v'_{n'}\| / \|v_0\|) / (n+n')$ (where ln is the natural logarithm).

The question now arises as to what value of n' maximizes this rate of convergence. Adopting the above assumption

$$R(n,n') = -[1/(n+n')]\{\ln(C_n(a,b;\lambda_1)) + \ln(C_{n'}(a'(n'),b;\lambda_1))\}.$$

Since $T_n\{(b-a)/(b+a)\} \approx \cosh\{2n\sqrt{(a/b)}\}$ where $a << b$ and n is large,

$\max|C_n(a,b;\lambda)| \approx \cosh\{2n\sqrt{(a/b)}\}^{-1}$ $(a \le \lambda \le b)$ and $-\ln\{C_n(a,b;\lambda_1)\} \approx 2n\sqrt{(a/b)} - m(2)$.

Thus, approximately

$$R(n,n') = 2\sqrt{(a/b)} - \ln(2)/(n+n') - \{2\sqrt{(a/b)} - \ln|T_p[(b+a^*(n'))/(b-a^*(n'))]|\}/(n+n').$$

For a fixed total number of constructions (i.e., $n+n'$ constant), $R(n,n')$ is maximized when $n' \approx p$ and

$$2\sqrt{(a/b)} - (d/dp)\ln|T_p\{(b+a'(p))/(b-a'(p))\}| =$$

$$2\sqrt{(a/b)} + [\text{arc } \cos\{w(p)\} - \{b\pi\sin(\pi/(2p))\}/\{2p(b-a)\sqrt{(1-w(p)^2)}\}]\tan\{p \text{ arccos}w(p)\} \qquad (5)$$

is zero, where $w(p) = \{b\cos(\pi/(2p)) + a\} / (b - a)$. The smallest positive zero p' of the function (5) is found, and n' is taken to be p'.

In summary *elimination* is to be used as follows: firstly, *richardson* is called with $a > \lambda_1$ and the procedure *out* required by *richardson* having a form suitable for the estimation of λ_1 (e.g., increasing n by steps of unity until two successive estimates of λ_1 agree sufficiently closely); *richardson* then allocates the derived estimate of λ_1 to *domeigval*. With the parameters $u, lj, ..., domeigval$ untouched, but with the procedure *out* required by *elimination* now having a form suitable for the solution of equation (1) (e.g., inspecting *discr[1]* and *discr[2]* when $k=n+n'$, and increasing n' by unity, and repeating this operation should this be necessary) the user now calls *elimination*.

Procedure parameters:
> void elimination $(u, lj, uj, ll, ul, method, a, b, n, discr, k, rateconv, domeigval)$

u: double $u[lj:uj, ll:ul]$;

after each iteration the approximate solution calculated by *elimination* is stored in u;

entry: an initial approximation of the solution which is obtained by use of *richardson*;

exit: the final approximation to the solution;

lj, uj: int;

entry: lower and upper bound for the first subscript of u;

ll, ul: int;

entry: lower and upper bound for the second subscript of u;

$method$:   a class that defines two procedures *residual* and *out*, this class must implement the `AP_richardson_methods` interface;

void residual(int lj, int uj, int ll, int ul, double u[ ][ ])

suppose that the system of equation at hand is $Au=f$; for any entry u the procedure *residual* should calculate the residual $Au-f$ in each point j,l, where $lj \leq j \leq uj$, $ll \leq l \leq ul$, and substitute these values in the array u;

void out(double u[ ][ ], int lj, int uj, int ll, int ul, int n[ ], double discr[ ],
> *int k[ ], double rateconv[ ], double domeigval[ ])*

by this procedure one has access to the following quantities:

for $0 \leq k \leq n$ the k-th iterand in u, the Euclidean and maximum norm of the k-th residual in *discr[1]* and *discr[2]*, respectively;

for $0 < k \leq n$ also the average rate of convergence with respect to the k-th iterand in u, in *rateconv*;

for $k=n$, possibly the dominant eigenvalue of the coefficient matrix of the equation $Au=f$, in *domeigval*;

a, b: double;

a and b should have the same values as in the preceding call of *richardson* (see the description of *richardson*);

n: int $n[0:0]$;

exit: the number of iterations the procedure *elimination* needs to eliminate the eigenfunction belonging to the dominant eigenvalue is assigned to n;

discr: double *discr[1:2]*;
                exit: after each iteration *elimination* delivers in *discr[1]* the Euclidean
                        norm of the residual and in *discr[2]* the maximum norm of the
                        residual;
k: int *k[0:0]*;
                exit: counts the number of iterations *elimination* is performing;
rateconv: double *rateconv[0:0]*;
                exit: after each iteration the average rate of convergence is assigned
                        to *rateconv*;
domeigval: double *domeigval[0:0]*;
                before a call of *elimination* the value of the eigenvalue for which the
                corresponding eigenfunction has to be eliminated should be assigned
                to *domeigval*; if after application of *elimination* there is a new dominant
                eigenfunction, then *domeigval* will be equal to the corresponding
                eigenvalue; otherwise, the value of *domeigval* becomes meaningless.

Procedure used: richardson.

```
package numal;

public interface AP_richardson_methods {

    void residual(int lj, int uj, int ll, int ul, double u[][]);
    void out(double u[][], int lj, int uj, int ll, int ul, int n[], double discr[],
            int k[], double rateconv[], double domeigval[]);
}


public static void elimination(double u[][], int lj, int uj, int ll, int ul,
                            AP_richardson_methods method, double a, double b,
                            int n[], double discr[], int k[], double rateconv[],
                            double domeigval[])
{
    boolean extrapolate;
    int ext;
    double auxcos,c,d,cc,fc,bb,fb,aa,fa,dd,fd,fdb,fda,w,mb,tol,m,p,q;

    fd=dd=0.0;
    c=1.0;
    if (optpol(c,a,b,domeigval) < 0.0) {
        d=0.5*Math.PI*Math.sqrt(Math.abs(b/domeigval[0]));
        while (true) {
            d += d;
            /* finding zero */
            bb=c;
            fb=optpol(c,a,b,domeigval);
            aa=c=d;
            fa=optpol(c,a,b,domeigval);
            cc=aa;
            fc=fa;
```

```
ext=0;
extrapolate=true;
while (extrapolate) {
    if (Math.abs(fc) < Math.abs(fb)) {
        if (cc != aa) {
            dd=aa;
            fd=fa;
        }
        aa=bb;
        fa=fb;
        bb=c=cc;
        fb=fc;
        cc=aa;
        fc=fa;
    }
    tol=c*1.0e-3;
    m=(cc+bb)*0.5;
    mb=m-bb;
    if (Math.abs(mb) > tol) {
        if (ext > 2)
            w=mb;
        else {
            if (mb == 0.0)
                tol=0.0;
            else
                if (mb < 0.0) tol = -tol;
            p=(bb-aa)*fb;
            if (ext <= 1)
                q=fa-fb;
            else {
                fdb=(fd-fb)/(dd-bb);
                fda=(fd-fa)/(dd-aa);
                p *= fda;
                q=fdb*fa-fda*fb;
            }
            if (p < 0.0) {
                p = -p;
                q = -q;
            }
            w=(p<Double.MIN_VALUE || p<=q*tol) ? tol :
                                ((p<mb*q) ? p/q : mb);
        }
        dd=aa;
        fd=fa;
        aa=bb;
        fa=fb;
        c = bb += w;
        fb=optpol(c,a,b,domeigval);
        if ((fc >= 0.0) ? (fb >= 0.0) : (fb <= 0.0)) {
            cc=aa;
```

```
                        fc=fa;
                        ext=0;
                    } else
                        ext = (w == mb) ? 0 : ext+1;
                } else
                    break;
            }
            d=cc;
            if ((fc >= 0.0) ? (fb <= 0.0) : (fb >= 0.0)) {
                n[0]=(int)Math.floor(c+0.5);
                break;
            }
        }
    } else
        n[0]=1;
    auxcos=Math.cos(0.5*Math.PI/n[0]);
    richardson(u,lj,uj,ll,ul,true,method,
            (2.0*domeigval[0]+b*(auxcos-1.0))/(auxcos+1.0),b,n,discr,k,rateconv,
            domeigval);
}


static private double optpol(double x, double a, double b, double domeigval[])
{
    /* this procedure is internally used by ELIMINATION */

    double w,y;

    w=(b*Math.cos(0.5*Math.PI/x)+domeigval[0])/(b-domeigval[0]);
    if (w < -1.0) w = -1.0;
    if (Math.abs(w) <= 1.0) {
        y=Math.acos(w);
        return 2.0*Math.sqrt(a/b)+Math.tan(x*y)*(y-b*Math.PI*
                Math.sin(0.5*Math.PI/x)*0.5/(x*(b-domeigval[0])*
                Math.sqrt(Math.abs(1.0-w*w))));
    } else {
        y=Math.log(w+Math.sqrt(Math.abs(w*w-1.0)));
        return 2.0*Math.sqrt(a/b)-Special_functions.tanh(x*y)*
                (y+b*Math.PI*Math.sin(0.5*Math.PI/x)*0.5/
                (x*(b-domeigval[0])*Math.sqrt(Math.abs(w*w-1.0))));
    }
}
```

5.6 Parameter estimation in differential equations

5.6.1 Initial value problems

peide

Estimates unknown variables in a system of first order differential equations by using a given set of observed values [Vn75]. The unknown variables may appear nonlinearly both in the differential equations and its initial values.

peide determines the components p_1, p_2, ..., p_m of $p \in R^m$ occurring in the system of differential equations

$$dy_j(t,p)/dt = f_j(t,y,p) \qquad (j=1,...,n) \tag{1}$$

$(y_j(t_0,p), j=1,...,n$, being prescribed) for which the sum of squares

$$\sum_{i=1}^{nobs} \left[y^{(i)}(t^{(i)}, p) - y'^{(i)}(t^{(i)}) \right]^2 \tag{2}$$

i.e., the square of the Euclidean norm of the residual vector whose components are

$$rv_i(p) = y^{(i)}(t^{(i)},p) - y'^{(i)}(t^{(i)}) \qquad (i=1,...,nobs) \tag{3}$$

is a minimum, where the $\{y^{(i)}(t^{(i)},p)\}$ form a selection of values of components of $y(t,p)$ taken at prescribed values of $t = t^{(i)}$, and $y'^{(i)}(t^{(i)})$ are observed values of these components. The above selection is specified by an array of integers $\{cobs[i]\}$ fixing the component suffices of the observed values, and an array of real numbers $\{tobs[i]\}$ fixing the values of t at which the observations have been made. Thus if observations of all components of $y(t,p)$ at the points $t=t'_1,...,t'_N$ are available, we have $cobs[1]=1$, $cobs[2]=2,...$, $cobs[n]=n$, $cobs[n+1]=1$, $cobs[n+2]=2,...$, $cobs[nN]=n$, $tobs[1]=tobs[2]=...=tobs[n]=t'_1$, $tobs[n+1]=...=tobs[2n]=t'_2,...$, $tobs[nN]=t'_N$, with $nobs=nN$. If the first component of $y(t,p)$ alone has been observed at $t=t'_1,...,t'_N$, we have $cobs[1]=cobs[2]=...=cobs[N]=1$, $tobs[1]=t'_1,...,tobs[N]=t'_N$, and $nobs=N$. Adopting this scheme, it is possible that the distribution of components of $y(t,p)$ for which observations are available may vary from one argument t of observation to another.

The numerical integration of the system of equations (1) is carried out by Gear's method. The vector $p = p'$ yielding a minimum of the function (2) is located by use of the Levenberg-Marquardt algorithm: the user is asked to provide an initial approximation $p^{(0)}$ to p'; a sequence of vectors $\{p^{(k)}\}$ ($k=0,1...$) is produced for which a member, $p^{(k)}$, is an acceptable approximation to p'.

It may occur that it is difficult to prescribe an excellent initial approximation $p^{(0)}$ to p', and for this reason the user may instruct *peide* to carry out fitting, during the early stages of the computation, over a system of subranges of the total argument range $t = t^{(1)}$ to $t = t^{(nobs)}$. The user is then asked to prescribe a system of break-points (integers) $bp[i']$ ($i'=1,2,...,nbp$) which, in particular, specify a nondecreasing sequence of values of t, namely $T^{(i)} = t^{(bp[i'])}$ ($i'=1,...,nbp$), for which observed values of selected components of $y(t,p)$ are available. The original minimization problem is now extended so as to concern functions $y(t,T^{(i)},p'')$ defined for $T^{(i)} \leq t < T^{(i+1)}$ respectively ($i=0,1,...,nbp$) where $T^{(0)} = t^{(1)}$ and $T^{(nbp+1)} = t^{(nobs)}$, $p'' \in R^{m+nbp}$

being an extension of p. Each of these functions satisfies a system of equations of the form (1) over its range of definition. The components of the initial values $y_j(t,T^{(0)},p''^{(0)})$ at $t=T^{(0)}=t^{(1)}$ are simply the initial values $y_j(t_0,p)$ ($j=1,...,n$) for the original problem. During the first traversal of the range $[t^{(1)},t^{(nobs)}]$ the initial values $\{y_j(t,T^{(i)},p''^{(0)})\}$ at $t = T^{(i)}$ for the integration of the system corresponding to (1) over the range $[T^{(i)},T^{(i+1)}]$ are $y_j(T^{(i)},T^{(i-1)},p''^{(0)})$ (obtained from integration up to the end point of the preceding interval) for all j except that, j' say, corresponding to a component for which an observed value is available, when the initial value is simply taken to be the observed value. The components of $p''\in R^{m+nbp}$ are given by $p''_i = p_i$ ($i=1,...,n$) and $p''_{m+i'} = y(T^{(i')},T^{(i')},p)$ ($i'=1,...,nbp$). Setting $i''(i) = i'$ for all i for which $T^{(i)}\leq t^{(i)}<T^{(i'+1)}$, the function now to be minimized is

$$\sum_{i=1}^{nobs} \left[y^{(i)}(t^{(i)},T^{i''(i)},p'') - y'^{(i)}(t^{(i)})\right]^2 +$$

$$M^2 \sum_{i'=1}^{nobs} \left[y^{(bp[i'])}(T^{(i')},T^{(i'-1)},p'') - y^{(bp[i'])}(T^{(i')},T^{(i')},p'')\right]^2$$

(4)

i.e., the square of the Euclidean norm of the residual vector whose components are

$$res_i(p'') = y^{(i)}(t^{(i)},T^{i''(i)},p'') - y'^{(i)}(t^{(i)}) \qquad (i=1,...,nobs)$$

(5)

$$res_{nobs+i'}(p'') = M\{y^{(bp[i'])}(T^{(i')},T^{(i'-1)},p'') - y^{(bp[i'])}(T^{(i')},T^{(i')},p'')\} \quad (i'=1,...,nbp)$$

where M is a weight factor. After each traversal of the range $[t_0, t^{(nobs)}]$ a new vector $p''^{(k)}$ is determined by use of the Levenberg-Marquardt algorithm.

During traversals of the range subsequent to the first, the vector p used in the systems of equations corresponding to (1) is made up of the first m components of p. The initial values $y_j(t,T^{(i)},p''^{(k)})$ at $t = T^{(i)}$ for the integration of the system corresponding to (1) over the range $[T^{(i)},T^{(i+1)}]$ are again $y_j(T^{(i)},T^{(i-1)},p''^{(k)})$ for all j except j' as above, when the new initial value is taken to be $p''_{m+i'}$. The iterations of the Levenberg-Marquardt algorithm are continued until the vectors $\{p''^{(k)}\}$ have approached a limit sufficiently closely. The second sum in expression (4) corresponds to a continuity requirement. After the above $p''^{(k)}$ have attained a limit sufficiently closely, the weight factor M is increased (i.e., the continuity requirements are made more stringent). A new terminating sequence of further vectors $\{p^{(k)}\}$ is derived, again M is increased, and so on. When, at a certain break-point $bp[i'']$,

$$\left| y^{(bp[i''])}(T^{(i'')},T^{(i''-1)},p''^{(k)}) - y^{(bp[i''])}(T^{(i'')},T^{(i'')},p''^{(k)}) \right|$$

becomes sufficiently small (i.e., the discontinuity at that break-point has virtually disappeared) the term corresponding to $i' = i'''$ is removed from expression (4), and the component with suffix $m+i'''$ is discarded from p'', the computations being continued with reduced sum and vector. This process is continued until all break-points have been discarded, and the original minimization problem is solved by a final use of the Levenberg-Marquardt algorithm. Naturally cases occur for which an excellent approximation $p^{(0)}$ to p'' is available, and in such a case (by setting $nbp=0$) the user may avoid the above excursion. For the sake of completeness, it is mentioned that certain of the above break points may correspond to equal values of t, (the components of $y(t,p)$ for which observed values are available at this value of t having differing suffixes). In such a case in the above exposition certain of the intervals of the form $[T^{(i)},T^{(i+1)}]$ reduce to a single point. Numerical integration of equations of the form (1) over such an interval is dispensed with, but the process of eliminating discontinuities as described is retained.

That the initial values $y_j(t_0,p)$ ($j=1,...,n$) for the equations (1) may depend upon p is also permitted (see the procedure *ystart* below).

Procedure parameters:

void peide (*n,m,nobs,nbp,par,res,bp,jtjinv,in,out,method*)

n: int;
  entry: the number of differential equations;
m: int;
  entry: the number of unknown variables;
nobs: int;
  entry: the number of observations; *nobs* should satisfy $nobs \geq m$;
nbp: int *nbp[0:0]*;
  entry: the number of break-points; if no break-points are used then set $nbp = 0$;
  exit: with normal termination of the process $nbp = 0$;
  otherwise, if the process has been broken off (see *out[1]*), the value of *nbp* is the number of break-points used before the process broke off;
par: double *par[1:m+nbp]*;
  entry: *par[1:m]* should contain an initial approximation to the required parameter vector (values of $p_i^{(0)}$, *i=1,...,m*, above);
  exit: *par[1:m]* contains the calculated parameter vector;
  if *out[1]* > 0 and *nbp* > 0 then *par[m+1:m+nbp]* contains the values of the newly introduced parameters before the process broke off;
res: double *res[1:nobs+nbp]*;
  exit: *res[1:nobs]* contains the residual vector at the calculated minimum (*res[i]* contains $y^{(i)}(t^{(i)},p') - y^{(i)}(t^{(i)})$, *i=1,...,nobs*, in the above);
  if *out[1]* > 0 and *nbp* > 0 then *res[nobs+1:nobs+nbp]* contains the additional continuity requirements at the break-points before the process broke off (*res[nobs+i]* contains $y^{(bp[i])}(T^{(i)},T^{(i-1)},p''^{(k)}) - y^{(bp[i])}(T^{(i)},T^{(i)},p''^{(k)})$, *i=1,...,nbp*, in the above);
bp: int *bp[0:nbp]*;
  entry: *bp[i]*, *i=1,...,nbp*, should correspond to the index of that time of observation which will be used as a break-point ($1 \leq bp[i] \leq nobs$); the break-points have to be ordered such that $bp[i] \leq bp[j]$ if $i \leq j$;
  exit: with normal termination of the process *bp[1:nbp]* contains no information; otherwise, if *out[1]* > 0 and *nbp* > 0 then *bp[i]*, *i=1,...,nbp*, contains the index of that time of observation which was used as a break-point before the process broke off;
jtjinv: double *jtjinv[1:m,1:m]*;
  exit: the inverse of the matrix $J'*J$ where J denotes the matrix of partial derivatives *dres[i]/dpar[k]* (*i=1,...,nobs*; *k=1,...,m*) and J' denotes the transpose of J; this matrix can be used if additional information about the result is required; e.g., statistical data such as the covariance matrix, correlation matrix and confidence intervals can easily be calculated from *jtjinv* and *out[2]*;

in: double *in[0:6]*;
             entry: *in[0]*:   the machine precision;
                     in[1]:   the ratio: the minimal steplength for the integration of the
                                differential equations divided by the distance between two
                                neighboring observations; mostly, a suitable value is 10^{-4};
                     in[2]:   the relative local error bound for the integration process;
                                this value should satisfy *in[2]* $\leq$ *in[3]*; this parameter
                                controls the accuracy of the numerical integration;
                                mostly, a suitable value is *in[3]*/100;
                     in[3]:   the relative tolerance for the difference between the
                                Euclidean norm of the ultimate and penultimate residual
                                vector; see *in[4]* below;
                     in[4]:   the absolute tolerance for the difference between the
                                Euclidean norm of the ultimate and penultimate residual
                                vector;
                                the process is terminated if the improvement of the sum
                                of squares is less than
                                     in[3] * (sum of squares) + *in[4]* * *in[4]*;
                                in[3] and *in[4]* should be chosen in accordance with the
                                relative and absolute errors in the observations; note that
                                the Euclidean norm of the residual vector is defined as
                                the square root of the sum of squares;
                     in[5]:   the maximum number of times that the integration of the
                                differential equations is performed;
                     in[6]:   a starting value used for the relation between the gradient
                                and the Gauss-Newton direction; if the problem is well
                                conditioned then a suitable value for *in[6]* will be 0.01; if
                                the problem is ill conditioned then *in[6]* should be greater,
                                but the value of *in[6]* should satisfy: *in[0]* < *in[6]* $\leq$ 1/*in[0]*;
out: double *out[1:7]*;
             exit: *out[1]*: this value gives information about the termination of the
                                process;
                                out[1]=0:   normal termination;
                                if *out[1]* > 0 then the process has been broken off and this
                                may occur because of the following reasons:
                                out[1]=1:   the number of integrations performed
                                              exceeded the number given in *in[5]*;
                                out[1]=2:   the differential equations are very nonlinear;
                                              during an integration the value of *in[1]* was
                                              decreased by a factor of 10000, and it is
                                              advised to decrease *in[1]*, although this will
                                              increase computing time;
                                out[1]=3:   a call of *deriv* delivered the value zero;
                                out[1]=4:   a call of *jacdfdy* delivered the value zero;
                                out[1]=5:   a call of *jacdfdp* delivered the value zero;
                                out[1]=6:   the precision asked for cannot be attained;
                                              this precision is possibly chosen too high,
                                              relative to the precision in which the residual
                                              vector is calculated (see *in[3]*);

out[2]: the Euclidean norm of the residual vector calculated with values of the unknowns delivered;

out[3]: the Euclidean norm of the residual vector calculated with the initial values of the unknown variables;

out[4]: the number of integrations performed, needed to obtain the calculated result; if *out[4]* = 1 and *out[1]* > 0 then the matrix *jtjinv* cannot be used;

out[5]: the maximum number of times that the requested local error bound was exceeded in one integration; if it is a large number then it may be better to decrease the value of *in[1]*;

out[6]: the improvement of the Euclidean norm of the residual vector in the last integration step of the process of Marquardt;

out[7]: the condition number of $J'*J$, i.e., the ratio of its largest to smallest eigenvalues;

method: a class that defines six procedures *deriv*, *jacdfdy*, *jacdfdp*, *callystart*, *data*, and *monitor*, this class must implement the AP_peide_methods interface;

boolean deriv(int n, int m, double par[ ], double y[ ], double t[ ], double df[ ])

entry: *par[1:m]* contains the current values of the unknowns and should not be altered;

y[1:n] contains the solutions of the differential equations at time *t* and should not be altered;

exit: an array element *df[i]* (*i*=1,...,*n*) should contain the right-hand side of the *i*-th differential equation;

after a successful call of *deriv*, the procedure should deliver the value true; however, if *deriv* delivers the value false then the process is terminated (see *out[1]*); hence, proper programming of *deriv* makes it possible to avoid calculation of the right-hand side with values of the unknown variables which cause overflow in the computation;

boolean jacdfdy(int n, int m, double par[ ], double y[ ], double t[ ],
     double fy[ ][ ])

entry: for parameters *par*, *y*, and *t*, see *deriv* above;

exit: an array element *fy[i,j]* (*i,j*=1,...,*n*) should contain the partial derivative of the right-hand side of the *i*-th differential equation with respect to *y[j]*, i.e., *df[i]/dy[j]*;

a boolean value should be assigned to this procedure in the same way as is done for the value of *deriv*;

boolean jacdfdp(int n, int m, double par[ ], double y[ ], double t[ ],
     double fp[ ][ ])

entry: for parameters *par*, *y* and *t*, see *deriv* above;

exit: an array element *fp[i,j]* should contain the partial derivative of the right-hand side of the *i*-th differential equation with respect to *par[j]*, i.e., *df[i]/dpar[j]*;

a boolean value should be assigned to this procedure in the same way as is done for the value of *deriv*;

void callystart(int n, int m, double par[ ], double y[ ],double ymax[ ])

    entry: *par[1:m]* contains the current values of the unknown variables and should not be altered;

    exit: *y[1:n]* should contain the initial values of the corresponding differential equations; the initial values may be functions of the unknown variables *par*; in that case, the initial values of *dy/dpar* also have to be supplied; note that *dy[i]/dpar[j]* corresponds with *y[5*n+j*n+i]* (i=1,...,n, j=1,...,m); *ymax[i]*, i=1,...,n, should contain a rough estimate to the maximal absolute value of *y[i]* over the integration interval;

void data(int nobs, double tobs[ ], double obs[ ], int cobs[ ])

    this procedure takes the data to fit into the procedure *peide*;

    entry: *nobs* has the same meaning as in *peide*;

    exit: *tobs*:   double *tobs[0:nobs]*;

                the array element *tobs[0]* should contain the time, corresponding to the initial values of *y* given in the procedure *callystart*; an array element *tobs[i]*, 1≤i≤*nobs*, should contain the *i*-th time of observation; the observations have to be ordered such that *tobs[i]* ≤ *tobs[j]* if i ≤ j;

        cobs: int *cobs[1:nobs]*;

                an array element *cobs[i]* should contain the component of *y* observed at time *tobs[i]*; note that 1 ≤ *cobs[i]* ≤ *n*;

        obs: double *obs[1:nobs]*;

                an array element *obs[i]* should contain the observed value of the component *cobs[i]* of *y* at the time *tobs[i]*;

void monitor(int post, int ncol, int nrow, double par[ ],double rv[ ],
        int weight, int nis[ ])

    this procedure can be used to obtain information about the course of the iteration process; if no intermediate results are desired then a dummy procedure satisfies;

    inside *peide*, the procedure *monitor* is called at two different places, and this is denoted by the value of *post*;

    post=1:   *monitor* is called after an integration of the differential equations; at this place are available: the current values of the unknown variables *par[1:ncol]*, where *ncol=m+nbp*, the calculated residual vector *res[1:nrow]*, where *nrow=nobs+nbp*, and the value of *nis*, which is the number of integration steps performed during the solution of the last initial value problem;

    post=2:   *monitor* is called before a minimization of the Euclidean norm of the residual vector with the Levenberg-Marquardt algorithm is started; available are the current values of *par[1:ncol]* and the value of the *weight*, with which the continuity requirements at the break-points are added to the original least squares problem.

Procedures used: inivec, inimat, mulvec, mulrow, dupvec, dupmat, vecvec, matvec, elmevc, sol, dec, mulcol, tamvec, mattam, qrisngvaldec.

```
package numal;

public interface AP_peide_methods {

    boolean deriv(int n, int m, double par[], double y[], double t[], double df[]);
    boolean jacdfdy(int n, int m, double par[], double y[], double t[],
                    double fy[][]);
    boolean jacdfdp(int n, int m, double par[], double y[], double t[],
                    double fp[][]);
    void callystart(int n, int m, double par[], double y[], double ymax[]);
    void data(int nobs, double tobs[], double obs[], int cobs[]);
    void monitor(int post, int ncol, int nrow, double par[], double rv[],
                 int weight, int nis[]);
}


public static void peide(int n, int m, int nobs, int nbp[], double par[],
                         double res[], int bp[], double jtjinv[][], double in[],
                         double out[], AP_peide_methods method)
{
    boolean emergency,first,clean;
    int i,j,weight,ncol,nrow,away,nfe,nbpold,maxfe,fe,it,err;
    double eps1,res1,in3,in4,fac3,fac4,w,temp,vv,ww,w2,mu,res2,fpar,fparpres,lambda,
           lambdamin,p,pw,reltolres,abstolres;
    double save1[]={1.0, 1.0, 9.0, 4.0, 0.0, 2.0/3.0, 1.0, 1.0/3.0, 36.0, 20.25,
                    1.0, 6.0/11.0, 1.0, 6.0/11.0, 1.0/11.0, 84.028, 53.778, 0.25,
                    0.48, 1.0, 0.7, 0.2, 0.02, 156.25, 108.51, 0.027778,
                    120.0/274.0, 1.0, 225.0/274.0, 85.0/274.0, 15.0/274.0,
                    1.0/274.0, 0.0, 187.69, 0.0047361};
    double aux[] = new double[4];
    double em[] = new double[8];
    boolean sec[] = new boolean[1];
    int max[] = new int[1];
    int nis[] = new int[1];

    nbpold=nbp[0];
    int cobs[] = new int[nobs+1];
    double obs[] = new double[nobs+1];
    double tobs[] = new double[nobs+1];
    double save[] = new double[6*n+39];
    double ymax[] = new double[n+1];
    double y[] = new double[6*n*(nbpold+m+1)+1];
    double yp[][] = new double[nbpold+nobs+1][nbpold+m+1];
    double fy[][] = new double[n+1][n+1];
    double fp[][] = new double[n+1][m+nbpold+1];
    double aid[][] = new double[m+nbpold+1][m+nbpold+1];

    nfe=0;
    res1=fac3=fac4=lambda=0.0;
```

```
for (i=0; i<=34; i++) save[i]=save1[i];
method.data(nobs,tobs,obs,cobs);
weight=1;
first=sec[0]=false;
clean=(nbp[0] > 0);
aux[2]=Double.MIN_VALUE;
eps1=1.0e10;
out[1]=0.0;
bp[0]=max[0]=0;
/* smooth integration without break-points */
if (!peidefunct(nobs,m,par,res,n,m,nobs,nbp,first,sec,max,nis,eps1,weight,bp,
                save,ymax,y,yp,fy,fp,cobs,tobs,obs,in,aux,clean,method)) {
    if (save[35] != 0.0) out[1]=save[35];
    out[3]=res1;
    out[4]=nfe;
    out[5]=max[0];
    return;
}
res1=Math.sqrt(Basic.vecvec(1,nobs,0,res,res));
nfe=1;
if (in[5] == 1.0) {
    out[1]=1.0;
    if (save[35] != 0.0) out[1]=save[35];
    out[3]=res1;
    out[4]=nfe;
    out[5]=max[0];
    return;
}
if (clean) {
    first=true;
    clean=false;
    fac3=Math.sqrt(Math.sqrt(in[3]/res1));
    fac4=Math.sqrt(Math.sqrt(in[4]/res1));
    eps1=res1*fac4;
    if (!peidefunct(nobs,m,par,res,n,m,nobs,nbp,first,sec,max,nis,eps1,weight,
                    bp,save,ymax,y,yp,fy,fp,cobs,tobs,obs,in,aux,clean,method)) {
        if (out[1] == 3.0)
            out[1]=2.0;
        else
            if (out[1] == 4.0) out[1]=6.0;
        if (save[35] != 0.0) out[1]=save[35];
        out[3]=res1;
        out[4]=nfe;
        out[5]=max[0];
        return;
    }
    first=false;
} else
    nfe=0;
ncol=m+nbp[0];
```

```
nrow=nobs+nbp[0];
sec[0]=true;
in3=in[3];
in4=in[4];
in[3]=res1;
weight=away=0;
out[4]=out[5]=w=0.0;
temp=Math.sqrt(weight)+1.0;
weight=(int)(temp*temp);
while (weight != 16 && nbp[0] > 0) {
    if (away == 0 && w != 0.0) {
        /* if no break-points were omitted then one
           function evaluation is saved */
        w=weight/w;
        for (i=nobs+1; i<=nrow; i++) {
            for (j=1; j<=ncol; j++) yp[i][j] *= w;
            res[i] *= w;
        }
        sec[0]=true;
        nfe--;
    }
    in[3] *= fac3*weight;
    in[4]=eps1;
    method.monitor(2,ncol,nrow,par,res,weight,nis);
    /* marquardt's method */
    double val[] = new double[ncol+1];
    double b[] = new double[ncol+1];
    double bb[] = new double[ncol+1];
    double parpres[] = new double[ncol+1];
    double jaco[][] = new double[nrow+1][ncol+1];
    vv=10.0;
    w2=0.5;
    mu=0.01;
    ww = (in[6] < 1.0e-7) ? 1.0e-8 : 1.0e-1*in[6];
    em[0]=em[2]=em[6]=in[0];
    em[4]=10*ncol;
    reltolres=in[3];
    abstolres=in[4]*in[4];
    maxfe=(int)in[5];
    err=0;
    fe=it=1;
    p=fpar=res2=0.0;
    pw = -Math.log(ww*in[0])/2.30;
    if (!peidefunct(nrow,ncol,par,res,n,m,nobs,nbp,first,sec,max,nis,eps1,
            weight,bp,save,ymax,y,yp,fy,fp,cobs,tobs,obs,in,aux,clean,method))
        err=3;
    else {
        fpar=Basic.vecvec(1,nrow,0,res,res);
        out[3]=Math.sqrt(fpar);
        emergency=false;
```

```
        it=1;
        do {
            Basic.dupmat(1,nrow,1,ncol,jaco,yp);
            i=Linear_algebra.qrisngvaldec(jaco,nrow,ncol,val,aid,em);
            if (it == 1)
                lambda=in[6]*Basic.vecvec(1,ncol,0,val,val);
            else
                if (p == 0.0) lambda *= w2;
            for (i=1; i<=ncol; i++)
                b[i]=val[i]*Basic.tamvec(1,nrow,i,jaco,res);
            while (true) {
                for (i=1; i<=ncol; i++)
                    bb[i]=b[i]/(val[i]*val[i]+lambda);
                for (i=1; i<=ncol; i++)
                    parpres[i]=par[i]-Basic.matvec(1,ncol,i,aid,bb);
                fe++;
                if (fe >= maxfe)
                    err=1;
                else
                    if (!peidefunct(nrow,ncol,parpres,res,n,m,nobs,nbp,first,sec,
                                    max,nis,eps1,weight,bp,save,ymax,y,yp,fy,fp,
                                    cobs,tobs,obs,in,aux,clean,method))
                        err=2;
                if (err != 0) {
                    emergency=true;
                    break;
                }
                fparpres=Basic.vecvec(1,nrow,0,res,res);
                res2=fpar-fparpres;
                if (res2 < mu*Basic.vecvec(1,ncol,0,b,bb)) {
                    p += 1.0;
                    lambda *= vv;
                    if (p == 1.0) {
                        lambdamin=ww*Basic.vecvec(1,ncol,0,val,val);
                        if (lambda < lambdamin) lambda=lambdamin;
                    }
                    if (p >= pw) {
                        err=4;
                        emergency=true;
                        break;
                    }
                } else {
                    Basic.dupvec(1,ncol,0,par,parpres);
                    fpar=fparpres;
                    break;
                }
            }
            if (emergency) break;
            it++;
        } while (fpar>abstolres && res2>reltolres*fpar+abstolres);
```

```
        for (i=1; i<=ncol; i++)
            Basic.mulcol(1,ncol,i,i,jaco,aid,1.0/(val[i]+in[0]));
        for (i=1; i<=ncol; i++)
            for (j=1; j<=i; j++)
                aid[i][j]=aid[j][i]=Basic.mattam(1,ncol,i,j,jaco,jaco);
        lambda=lambdamin=val[1];
        for (i=2; i<=ncol; i++)
            if (val[i] > lambda)
                lambda=val[i];
            else
                if (val[i] < lambdamin) lambdamin=val[i];
        temp=lambda/(lambdamin+in[0]);
        out[7]=temp*temp;
        out[2]=Math.sqrt(fpar);
        out[6]=Math.sqrt(res2+fpar)-out[2];
    }
    out[4]=fe;
    out[5]=it-1;
    out[1]=err;
    if (out[1] > 0.0) {
        if (out[1] == 3.0)
            out[1]=2.0;
        else
            if (out[1] == 4.0) out[1]=6.0;
        if (save[35] != 0.0) out[1]=save[35];
        out[3]=res1;
        out[4]=nfe;
        out[5]=max[0];
        return;
    }
    /* the relative starting value of lambda is adjusted
       to the last value of lambda used */
    away=(int)(out[4]-out[5]-1.0);
    in[6] *= Math.pow(5.0,away)*Math.pow(2.0,away-out[5]);
    nfe += out[4];
    w=weight;
    temp=Math.sqrt(weight)+1.0;
    eps1=temp*temp*in[4]*fac4;
    away=0;
    /* omit useless break-points */
    for (j=1; j<=nbp[0]; j++)
        if (Math.abs(obs[bp[j]]+res[bp[j]]-par[j+m]) < eps1) {
            nbp[0]--;
            for (i=j; i<=nbp[0]; i++) bp[i]=bp[i+1];
            Basic.dupvec(j+m,nbp[0]+m,1,par,par);
            j--;
            away++;
            bp[nbp[0]+1]=0;
        }
    ncol -= away;
```

```
        nrow -= away;
        temp=Math.sqrt(weight)+1.0;
        weight=(int)(temp*temp);
    }
    in[3]=in3;
    in[4]=in4;
    nbp[0]=0;
    weight=1;
    method.monitor(2,m,nobs,par,res,weight,nis);
    /* marquardt's method */
    double val[] = new double[m+1];
    double b[] = new double[m+1];
    double bb[] = new double[m+1];
    double parpres[] = new double[m+1];
    double jaco[][] = new double[nobs+1][m+1];
    vv=10.0;
    w2=0.5;
    mu=0.01;
    ww = (in[6] < 1.0e-7) ? 1.0e-8 : 1.0e-1*in[6];
    em[0]=em[2]=em[6]=in[0];
    em[4]=10*m;
    reltolres=in[3];
    abstolres=in[4]*in[4];
    maxfe=(int)in[5];
    err=0;
    fe=it=1;
    p=fpar=res2=0.0;
    pw = -Math.log(ww*in[0])/2.30;
    if (!peidefunct(nobs,m,par,res,n,m,nobs,nbp,first,sec,max,nis,eps1,weight,
                    bp,save,ymax,y,yp,fy,fp,cobs,tobs,obs,in,aux,clean,method))
        err=3;
    else {
        fpar=Basic.vecvec(1,nobs,0,res,res);
        out[3]=Math.sqrt(fpar);
        emergency=false;
        it=1;
        do {
            Basic.dupmat(1,nobs,1,m,jaco,yp);
            i=Linear_algebra.qrisngvaldec(jaco,nobs,m,val,jtjinv,em);
            if (it == 1)
                lambda=in[6]*Basic.vecvec(1,m,0,val,val);
            else
                if (p == 0.0) lambda *= w2;
            for (i=1; i<=m; i++)
                b[i]=val[i]*Basic.tamvec(1,nobs,i,jaco,res);
            while (true) {
                for (i=1; i<=m; i++)
                    bb[i]=b[i]/(val[i]*val[i]+lambda);
                for (i=1; i<=m; i++)
                    parpres[i]=par[i]-Basic.matvec(1,m,i,jtjinv,bb);
```

```
            fe++;
            if (fe >= maxfe)
                err=1;
            else
                if (!peidefunct(nobs,m,parpres,res,n,m,nobs,nbp,first,sec,max,
                                 nis,eps1,weight,bp,save,ymax,y,yp,fy,fp,cobs,
                                 tobs,obs,in,aux,clean,method))
                    err=2;
            if (err != 0) {
                emergency=true;
                break;
            }
            fparpres=Basic.vecvec(1,nobs,0,res,res);
            res2=fpar-fparpres;
            if (res2 < mu*Basic.vecvec(1,m,0,b,bb)) {
                p += 1.0;
                lambda *= vv;
                if (p == 1.0) {
                    lambdamin=ww*Basic.vecvec(1,m,0,val,val);
                    if (lambda < lambdamin) lambda=lambdamin;
                }
                if (p >= pw) {
                    err=4;
                    emergency=true;
                    break;
                }
            } else {
                Basic.dupvec(1,m,0,par,parpres);
                fpar=fparpres;
                break;
            }
        }
        if (emergency) break;
        it++;
    } while (fpar>abstolres && res2>reltolres*fpar+abstolres);
    for (i=1; i<=m; i++)
        Basic.mulcol(1,m,i,i,jaco,jtjinv,1.0/(val[i]+in[0]));
    for (i=1; i<=m; i++)
        for (j=1; j<=i; j++)
            jtjinv[i][j]=jtjinv[j][i]=Basic.mattam(1,m,i,j,jaco,jaco);
    lambda=lambdamin=val[1];
    for (i=2; i<=m; i++)
        if (val[i] > lambda)
            lambda=val[i];
        else
            if (val[i] < lambdamin) lambdamin=val[i];
    temp=lambda/(lambdamin+in[0]);
    out[7]=temp*temp;
    out[2]=Math.sqrt(fpar);
    out[6]=Math.sqrt(res2+fpar)-out[2];
```

```
    }
    out[4]=fe;
    out[5]=it-1;
    out[1]=err;
    nfe += out[4];

    if (out[1] == 3.0)
        out[1]=2.0;
    else
        if (out[1] == 4.0) out[1]=6.0;
    if (save[35] != 0.0) out[1]=save[35];
    out[3]=res1;
    out[4]=nfe;
    out[5]=max[0];
}


static private boolean peidefunct(int nrow, int ncol, double par[], double res[],
                                  int n, int m, int nobs, int nbp[], boolean first,
                                  boolean sec[], int max[], int nis[], double eps1,
                                  int weight, int bp[], double save[],
                                  double ymax[], double y[], double yp[][],
                                  double fy[][], double fp[][], int cobs[],
                                  double tobs[], double obs[], double in[],
                                  double aux[], boolean clean,
                                  AP_peide_methods method)
{
    /* this procedure is internally used by PEIDE */

    boolean evaluate,evaluated,conv,newstart,errortestok;
    int l,k,fails,same,kpold,n6,nnpar,j5n,cobsii,extra,npar,i,j,jj,ii;
    double xold,hold,error,dfi,tobsdif,xend,hmax,hmin,eps,s,aa,t,c;
    double a[] = new double[6];
    boolean decompose[] = new boolean[1];
    int knew[] = new int[1];
    double ch[] = new double[1];
    double x[] = new double[1];
    double h[] = new double[1];
    double a0[] = new double[1];
    double tolup[] = new double[1];
    double tol[] = new double[1];
    double toldwn[] = new double[1];
    double tolconv[] = new double[1];
    double chnew[] = new double[1];

    int p[] = new int[n+1];
    double delta[] = new double[n+1];
    double lastdelta[] = new double[n+1];
    double df[] = new double[n+1];
    double y0[] = new double[n+1];
```

```
double jacob[][] = new double[n+1][n+1];

conv=true;
error=tobsdif=0.0;
if (sec[0]) {
    sec[0]=false;
    if (save[36] > max[0]) max[0]=(int)save[36];
    if (!first) method.monitor(1,ncol,nrow,par,res,weight,nis);
    return (save[37] <= 40.0 && save[35] == 0.0);
}
xend=tobs[nobs];
eps=in[2];
npar=m;
extra=nis[0]=0;
ii=1;
jj = (nbp[0] == 0) ? 0 : 1;
n6=n*6;
Basic.inivec(35,37,save,0.0);
Basic.inivec(n6+1,(6+m)*n,y,0.0);
Basic.inimat(1,nobs+nbp[0],1,m+nbp[0],yp,0.0);
t=tobs[1];
x[0]=tobs[0];
method.callystart(n,m,par,y,ymax);
hmax=tobs[1]-tobs[0];
hmin=hmax*in[1];
/* evaluate jacobian */
evaluate=false;
decompose[0]=evaluated=true;
if (!method.jacdfdy(n,m,par,y,x,fy)) {
    save[35]=4.0;
    if (save[36] > max[0]) max[0]=(int)save[36];
    if (!first) method.monitor(1,ncol,nrow,par,res,weight,nis);
    return (save[37] <= 40.0 && save[35] == 0.0);
}
nnpar=n*npar;

newstart=true;
while (newstart) {
    newstart=false;
    k=1;
    kpold=0;
    same=2;
    peideorder(n,k,eps,a,save,tol,tolup,toldwn,tolconv,a0,decompose);
    if (!method.deriv(n,m,par,y,x,df)) {
        save[35]=3.0;
        if (save[36] > max[0]) max[0]=(int)save[36];
        if (!first)
            method.monitor(1,ncol,nrow,par,res,weight,nis);
        return (save[37] <= 40.0 && save[35] == 0.0);
    }
```

```
s=Double.MIN_VALUE;
for (i=1; i<=n; i++) {
    aa=Basic.matvec(1,n,i,fy,df)/ymax[i];
    s += aa*aa;
}
h[0]=Math.sqrt(2.0*eps/Math.sqrt(s));
if (h[0] > hmax)
    h[0]=hmax;
else
    if (h[0] < hmin) h[0]=hmin;
xold=x[0];
hold=h[0];
ch[0]=1.0;
for (i=1; i<=n; i++) {
    save[i+38]=y[i];
    save[n+i+38]=y[n+i]=df[i]*h[0];
}
fails=0;
while (x[0] < xend && !newstart) {
    if (x[0]+h[0] <= xend)
        x[0] += h[0];
    else {
        h[0]=xend-x[0];
        x[0]=xend;
        ch[0]=h[0]/hold;
        c=1.0;
        for (j=n; j<=k*n; j += n) {
            c *= ch[0];
            for (i=j+1; i<=j+n; i++) y[i] *= c;
        }
        same = (same < 3) ? 3 : same+1;
    }
    /* prediction */
    for (l=1; l<=n; l++) {
        for (i=1; i<=(k-1)*n+l; i += n)
            for (j=(k-1)*n+l; j>=i; j -= n) y[j] += y[j+n];
        delta[l]=0.0;
    }
    evaluated=false;
    /* correction and estimation local error */
    for (l=1; l<=3; l++) {
        if (!method.deriv(n,m,par,y,x,df)) {
            save[35]=3;
            if (save[36] > max[0]) max[0]=(int)save[36];
            if (!first)
                method.monitor(1,ncol,nrow,par,res,weight,nis);
            return (save[37] <= 40.0 && save[35] == 0.0);
        }
        for (i=1; i<=n; i++) df[i]=df[i]*h[0]-y[n+i];
        if (evaluate) {
```

```
        /* evaluate jacobian */
        evaluate=false;
        decompose[0]=evaluated=true;
        if (!method.jacdfdy(n,m,par,y,x,fy)) {
            save[35]=4.0;
            if (save[36] > max[0]) max[0]=(int)save[36];
            if (!first)
                method.monitor(1,ncol,nrow,par,res,weight,nis);
            return (save[37] <= 40.0 && save[35] == 0.0);
        }
    }
    if (decompose[0]) {
        /* decompose jacobian */
        decompose[0]=false;
        c = -a0[0]*h[0];
        for (j=1; j<=n; j++) {
            for (i=1; i<=n; i++) jacob[i][j]=fy[i][j]*c;
            jacob[j][j] += 1.0;
        }
        Linear_algebra.dec(jacob,n,aux,p);
    }
    Linear_algebra.sol(jacob,n,p,df);
    conv=true;
    for (i=1; i<=n; i++) {
        dfi=df[i];
        y[i]   += a0[0]*dfi;
        y[n+i] += dfi;
        delta[i] += dfi;
        conv=(conv && (Math.abs(dfi) < tolconv[0]*ymax[i]));
    }
    if (conv) {
        s=Double.MIN_VALUE;
        for (i=1; i<=n; i++) {
            aa=delta[i]/ymax[i];
            s += aa*aa;
        }
        error=s;
        break;
    }
}
/* acceptance or rejection */
if (!conv) {
    if (!evaluated)
        evaluate=true;
    else {
        ch[0] /= 4.0;
        if (h[0] < 4.0*hmin) {
            save[37] += 10.0;
            hmin /= 10.0;
            if (save[37] > 40.0) {
```

```
                    if (save[36] > max[0]) max[0]=(int)save[36];
                    if (!first)
                        method.monitor(1,ncol,nrow,par,res,weight,nis);
                    return (save[37] <= 40.0 && save[35] == 0.0);
                }
            }
        }
        peidereset(n,k,hmin,hmax,hold,xold,y,save,ch,x,h,decompose);
    } else {
        errortestok=true;
        if (error > tol[0]) {
            errortestok=false;
            fails++;
            if (h[0] > 1.1*hmin) {
                if (fails > 2) {
                    peidereset(n,k,hmin,hmax,hold,xold,y,save,ch,x,h,
                            decompose);
                    newstart=true;
                } else {
                    /* calculate step and order */
                    peidestep(n,k,fails,tolup,toldwn,tol,error,delta,
                            lastdelta,y,ymax,knew,chnew);
                    if (knew[0] != k) {
                        k=knew[0];
                        peideorder(n,k,eps,a,save,tol,tolup,toldwn,tolconv,
                                a0,decompose);
                    }
                    ch[0] *= chnew[0];
                    peidereset(n,k,hmin,hmax,hold,xold,y,save,ch,x,h,
                            decompose);
                }
            } else {
                if (k == 1) {
                    /* violate eps criterion */
                    save[36] += 1.0;
                    same=4;
                    errortestok=true;
                } else {
                    k=1;
                    peidereset(n,k,hmin,hmax,hold,xold,y,save,ch,x,h,
                            decompose);
                    peideorder(n,k,eps,a,save,tol,tolup,toldwn,tolconv,a0,
                            decompose);
                    same=2;
                }
            }
        }
        if (errortestok && !newstart) {
            fails=0;
            for (i=1; i<=n; i++) {
```

```
        c=delta[i];
        for (l=2; l<=k; l++) y[l*n+i] += a[l]*c;
        if (Math.abs(y[i]) > ymax[i])
            ymax[i]=Math.abs(y[i]);
    }
    same--;
    if (same == 1)
        Basic.dupvec(1,n,0,lastdelta,delta);
    else if (same == 0) {
        /* calculate step and order */
        peidestep(n,k,fails,tolup,toldwn,tol,error,delta,lastdelta,y,
                    ymax,knew,chnew);
        if (chnew[0] > 1.1) {
            if (k != knew[0]) {
                if (knew[0] > k)
                    Basic.mulvec(knew[0]*n+1,knew[0]*n+n,-knew[0]*n,y,
                                    delta,a[k]/knew[0]);
                k=knew[0];
                peideorder(n,k,eps,a,save,tol,tolup,toldwn,tolconv,
                            a0,decompose);
            }
            same=k+1;
            if (chnew[0]*h[0] > hmax) chnew[0]=hmax/h[0];
            h[0] *= chnew[0];
            c=1.0;
            for (j=n; j<=k*n; j += n) {
                c *= chnew[0];
                Basic.mulvec(j+1,j+n,0,y,y,c);
            }
            decompose[0]=true;
        } else
            same=10;
    }
    nis[0]++;
    /* start of an integration step of yp */
    if (clean) {
        hold=h[0];
        xold=x[0];
        kpold=k;
        ch[0]=1.0;
        Basic.dupvec(39,k*n+n+38,-38,save,y);
    } else {
        if (h[0] != hold) {
            ch[0]=h[0]/hold;
            c=1.0;
            for (j=n6+nnpar; j<=kpold*nnpar+n6; j += nnpar) {
                c *= ch[0];
                for (i=j+1; i<=j+nnpar; i++) y[i] *= c;
            }
            hold=h[0];
```

```
    }
    if (k > kpold)
        Basic.inivec(n6+k*nnpar+1,n6+k*nnpar+nnpar,y,0.0);
    xold=x[0];
    kpold=k;
    ch[0]=1.0;
    Basic.dupvec(39,k*n+n+38,-38,save,y);
    /* evaluate jacobian */
    evaluate=false;
    decompose[0]=evaluated=true;
    if (!method.jacdfdy(n,m,par,y,x,fy)) {
        save[35]=4.0;
        if (save[36] > max[0]) max[0]=(int)save[36];
        if (!first)
            method.monitor(1,ncol,nrow,par,res,weight,nis);
        return (save[37] <= 40.0 && save[35] == 0.0);
    }
    /* decompose jacobian */
    decompose[0]=false;
    c = -a0[0]*h[0];
    for (j=1; j<=n; j++) {
        for (i=1; i<=n; i++) jacob[i][j]=fy[i][j]*c;
        jacob[j][j] += 1.0;
    }
    Linear_algebra.dec(jacob,n,aux,p);
    if (!method.jacdfdp(n,m,par,y,x,fp)) {
        save[35]=5.0;
        if (save[36] > max[0]) max[0]=(int)save[36];
        if (!first)
            method.monitor(1,ncol,nrow,par,res,weight,nis);
        return (save[37] <= 40.0 && save[35] == 0.0);
    }
    if (npar > m) Basic.inimat(1,n,m+1,npar,fp,0.0);
    /* prediction */
    for (l=0; l<=k-1; l++)
        for (j=k-1; j>=l; j--)
            Basic.elmvec(j*nnpar+n6+1,j*nnpar+n6+nnpar,nnpar,
                        y,y,1.0);
    /* correction */
    for (j=1; j<=npar; j++) {
        j5n=(j+5)*n;
        Basic.dupvec(1,n,j5n,y0,y);
        for (i=1; i<=n; i++)
            df[i]=h[0]*(fp[i][j]+Basic.matvec(1,n,i,fy,y0))-
                    y[nnpar+j5n+i];
        Linear_algebra.sol(jacob,n,p,df);
        for (l=0; l<=k; l++) {
            i=l*nnpar+j5n;
            Basic.elmvec(i+1,i+n,-i,y,df,a[l]);
        }
```

```
            }
        }
        while (x[0] >= t) {
            /* calculate a row of the jacobian matrix and an
                element of the residual vector */
            tobsdif=(tobs[ii]-x[0])/h[0];
            cobsii=cobs[ii];
            res[ii]=peideinterpol(cobsii,n,k,tobsdif,y)-obs[ii];
            if (!clean) {
                for (i=1; i<=npar; i++)
                    yp[ii][i]=peideinterpol(cobsii+(i+5)*n,nnpar,k,
                                                tobsdif,y);
                /* introducing break-points */
                if (bp[jj] != ii) {
                } else if (first && Math.abs(res[ii]) < eps1) {
                    nbp[0]--;
                    for (i=jj; i<=nbp[0]; i++) bp[i]=bp[i+1];
                    bp[nbp[0]+1]=0;
                } else {
                    extra++;
                    if (first) par[m+jj]=obs[ii];
                    /* introducing a jacobian row and a residual vector
                        element for continuity requirements */
                    yp[nobs+jj][m+jj] = -weight;
                    Basic.mulrow(1,npar,nobs+jj,ii,yp,yp,weight);
                    res[nobs+jj]=weight*(res[ii]+obs[ii]-par[m+jj]);
                }
            }
            if (ii == nobs) {
                if (save[36] > max[0]) max[0]=(int)save[36];
                if (!first)
                    method.monitor(1,ncol,nrow,par,res,weight,nis);
                return (save[37] <= 40.0 && save[35] == 0.0);
            }
            else {
                t=tobs[ii+1];
                if (bp[jj] == ii && jj < nbp[0]) jj++;
                hmax=t-tobs[ii];
                hmin=hmax*in[1];
                ii++;
            }
        }
        /* break-points introduce new initial values for y & yp */
        if (extra > 0) {
            for (i=1; i<=n; i++) {
                y[i]=peideinterpol(i,n,k,tobsdif,y);
                for (j=1; j<=npar; j++)
                    y[i+(j+5)*n]=peideinterpol(i+(j+5)*n,nnpar,k,
                                                tobsdif,y);
            }
```

```
                    for (l=1; l<=extra; l++) {
                        cobsii=cobs[bp[npar-m+l]];
                        y[cobsii]=par[npar+l];
                        for (i=1; i<=npar+extra; i++)
                            y[cobsii+(5+i)*n]=0.0;
                        Basic.inivec(1+nnpar+(l+5)*n,nnpar+(l+6)*n,y,0.0);
                        y[cobsii+(5+npar+l)*n]=1.0;
                    }
                    npar += extra;
                    extra=0;
                    x[0]=tobs[ii-1];
                    /* evaluate jacobian */
                    evaluate=false;
                    decompose[0]=evaluated=true;
                    if (!method.jacdfdy(n,m,par,y,x,fy)) {
                        save[35]=4.0;
                        if (save[36] > max[0]) max[0]=(int)save[36];
                        if (!first)
                            method.monitor(1,ncol,nrow,par,res,weight,nis);
                        return (save[37] <= 40.0 && save[35] == 0.0);
                    }
                    nnpar=n*npar;
                    newstart=true;
                }
            }  /*  errortestok */
        }
    }  /* while (x[0] < xend && !newstart) loop */
    }  /* newstart loop */

    if (save[36] > max[0]) max[0]=(int)save[36];
    if (!first) method.monitor(1,ncol,nrow,par,res,weight,nis);
    return (save[37] <= 40.0 && save[35] == 0.0);
}



static private void peidereset(int n, int k, double hmin, double hmax, double hold,
                               double xold, double y[], double save[], double ch[],
                               double x[], double h[], boolean decompose[])
{
    /* this procedure is internally used by PEIDEFUNCT of PEIDE */

    int i,j;
    double c;

    if (ch[0] < hmin/hold)
        ch[0] = hmin/hold;
    else
        if (ch[0] > hmax/hold) ch[0] = hmax/hold;
    x[0] = xold;
    h[0] = hold*ch[0];
```

```
    c=1.0;
    for (j=0; j<=k*n; j += n) {
        for (i=1; i<=n; i++) y[j+i]=save[j+i+38]*c;
        c *= ch[0];
    }
    decompose[0] = true;
}


static private void peideorder(int n, int k, double eps, double a[], double save[],
                               double tol[], double tolup[], double toldwn[],
                               double tolconv[], double a0[], boolean decompose[])
{
    /* this procedure is internally used by PEIDEFUNCT of PEIDE */

    int i,j;
    double c;

    c=eps*eps;
    j=((k-1)*(k+8))/2-38;
    for (i=0; i<=k; i++) a[i]=save[i+j+38];
    j += k+1;
    tolup[0] = c*save[j+38];
    tol[0] = c*save[j+39];
    toldwn[0] = c*save[j+40];
    tolconv[0] = eps/(2*n*(k+2));
    a0[0] = a[0];
    decompose[0] = true;
}


static private void peidestep(int n, int k, int fails, double tolup[],
                              double toldwn[], double tol[], double error,
                              double delta[], double lastdelta[], double y[],
                              double ymax[], int knew[], double chnew[])
{
    /* this procedure is internally used by PEIDEFUNCT of PEIDE */

    int i;
    double a1,a2,a3,aa,s;

    if (k <= 1)
        a1=0.0;
    else {
        s=Double.MIN_VALUE;
        for (i=1; i<=n; i++) {
            aa=y[k*n+i]/ymax[i];
            s += aa*aa;
        }
        a1=0.75*Math.pow(toldwn[0]/s,0.5/k);
```

```
    }
    a2=0.80*Math.pow(tol[0]/error,0.5/(k+1));
    if (k >= 5 || fails != 0)
        a3=0.0;
    else {
        s=Double.MIN_VALUE;
        for (i=1; i<=n; i++) {
            aa=(delta[i]-lastdelta[i])/ymax[i];
            s += aa*aa;
        }
        a3=0.70*Math.pow(tolup[0]/s,0.5/(k+2));
    }
    if (a1 > a2 && a1 > a3) {
        knew[0] = k-1;
        chnew[0] = a1;
    } else if (a2 > a3) {
        knew[0] = k;
        chnew[0] = a2;
    } else {
        knew[0] = k+1;
        chnew[0] = a3;
    }
}


static private double peideinterpol(int startindex, int jump, int k, double tobsdif,
                                    double y[])
{
    /* this procedure is internally used by PEIDEFUNCT of PEIDE */

    int i;
    double s,r;

    s=y[startindex];
    r=tobsdif;
    for (i=1; i<=k; i++) {
        startindex += jump;
        s += y[startindex]*r;
        r *= tobsdif;
    }
    return s;
}
```

6. Special Functions

All procedures in this chapter are class methods of the following *Special_functions* class.

```
package numal;

import numal.*;

public class Special_functions extends Object {
    // all procedures in this chapter are to be inserted here
}
```

6.1 Elementary functions

6.1.1 Hyperbolic functions

A. arcsinh

Computes the inverse hyperbolic sine of the argument x.
If $|x| \leq 10^{10}$ then we use the procedure *logoneplusx* by writing
$$arcsinh(x) = ln(x + \sqrt{(x*x+1)}) = ln(1 + x + x^2/(1+\sqrt{(1+x^2)})).$$
If $|x| > 10^{10}$ we use the formula
$$arcsinh(x) = sign(x) * (ln(2) + ln(|x|)).$$

Procedure parameters:
$$\text{double arcsinh } (x)$$
arcsinh: delivers the inverse hyperbolic sine of the argument x;
x: double;
           entry: the real argument of *arcsinh(x)*.

Procedure used: logoneplusx.

```
public static double arcsinh(double x)
{
    double y;

    if (Math.abs(x) > 1.0e10)
        return ((x > 0.0) ? 0.69314718055995+Math.log(Math.abs(x)) :
                        -0.69314718055995+Math.log(Math.abs(x)));
    else {
```

```
        y=x*x;
        return ((x == 0.0) ? 0.0 : ((x > 0.0) ?
                logoneplusx(Math.abs(x)+y/(1.0+Math.sqrt(1.0+y))) :
                -logoneplusx(Math.abs(x)+y/(1.0+Math.sqrt(1.0+y))))));
    }
}
```

B. arccosh

Computes the inverse hyperbolic cosine of the argument x.
If $x = 1$ then the value 0 is delivered.
If $1 < x \le 10^{10}$ then we use the formula
$$arccosh(x) = ln(x + \sqrt{(x*x-1)}).$$
If $x > 10^{10}$ we use the formula
$$arccosh(x) = ln(2) + ln(x).$$
If x is close to 1, say $x = 1 + y$, then it is advised to use the procedure *logoneplusx*
by writing
$$arccosh(x) = ln(1 + y + \sqrt{(y*(y+2))}).$$
For example, if $x = exp(t)$, $t > 0$, t is small, then $y = exp(t) - 1$ is available in good
relative accuracy, $y = 2 * exp(t/2) * sinh(t/2)$.

Procedure parameters:
$$double\ arccosh\ (x)$$
arccosh: delivers the inverse hyperbolic cosine of the argument x;
x: double;
              entry: the real argument of *arccosh(x)*, $x \ge 1$.

```
public static double arccosh(double x)
{
    return ((x <= 1.0) ? 0.0 : ((x > 1.0e10) ? 0.69314718055995+Math.log(x) :
                                Math.log(x+Math.sqrt((x-1.0)*(x+1.0))))));
}
```

C. arctanh

Computes the inverse hyperbolic tangent of the argument x.
If $|x| < 1$ then we use the procedure *logoneplusx* by writing
$$arctanh(x) = 0.5 * ln((1+x)/(1-x)) = 0.5 * ln(1 + 2*x/(1-x)).$$
If $|x| = 1$ then the value is $sign(x) * MAX\_VALUE$, where MAX_VALUE is a large
number.

Procedure parameters:
$$double\ arctanh\ (x)$$
arctanh: delivers the inverse hyperbolic tangent of the argument x;
x: double;
              entry: the real argument of *arctanh(x)*.

Procedure used: logoneplusx.

```
public static double arctanh(double x)
{
    double ax;

    if (Math.abs(x) >= 1.0)
        return ((x > 0.0) ? Double.MAX_VALUE : -Double.MAX_VALUE);
    else {
        ax=Math.abs(x);
        return ((x == 0.0) ? 0.0 : ((x > 0.0) ? 0.5*logoneplusx(2.0*ax/(1.0-ax)) :
                                                -0.5*logoneplusx(2.0*ax/(1.0-ax))));

    }
}
```

6.1.2 Logarithmic functions

logoneplusx

Computes the function $ln(1 + x)$ for $x > -1$. For values of x near zero, loss of relative accuracy in the computation of $ln(1+x)$ by use of the formulae $z=1+x$, $ln(1+x)=ln(z)$ occurs (since $1+x \approx 1$ for small x); the use of *logoneplusx* avoids this loss [HaCL68].

    For $x < -0.2929$ or $x > 0.4142$, $ln(1+x)$ is evaluated by use of the standard function ln directly; otherwise, a polynomial expression of the form

$$\ln(1+x) = \sum_{i=0}^{6} c_i \left(\frac{x}{x+2} \right)^{2i+1}$$

is used, and for small x loss of relative accuracy does not take place.

Procedure parameters:
$$\text{double logoneplusx } (x)$$

logoneplusx: delivers the value of $ln(1+x)$;
x: double;
            entry: the real argument of $ln(1+x)$, $x > -1$.

```
public static double logoneplusx(double x)
{
    double y,z;

    if (x == 0.0)
        return 0.0;
    else if (x < -0.2928 || x > 0.4142)
        return Math.log(1.0+x);
    else {
```

```
        z=x/(x+2.0);
        y=z*z;
        return z*(2.0+y*(0.666666666663366+y*(0.400000001206045+y*
                (0.285714091590488+y*(0.22223823332791+y*
                (0.1811136267967+y*0.16948212488))))));
    }
}
```

6.2 Exponential integral

6.2.1 Exponential integral

A. ei

Calculates the exponential integral

$$E_i(x) = \int_{-\infty}^{x} \frac{e^t}{t} dt$$

where the integral is to be interpreted as the Cauchy principal value. When $x > 0$, the related function

$$E_1(x) = \int_{x}^{\infty} \frac{e^{-t}}{t} dt$$

may be obtained from that of $E_i(x)$ by use of the relationship $E_1(x) = -E_i(-x)$. For $x=0$ the integral is undefined and the procedure will cause overflow.

The exponential integral is computed by means of the rational Chebyshev approximations given in [AbS65, CodT68, CodT69]. Only ratios of polynomials with equal degree l are considered.

Procedure parameters:
$$\text{double ei } (x)$$
ei: delivers the value of the exponential integral;
x: double;
         entry: the argument of the integral.

Procedures used: chepolsum, pol, jfrac.

```
public static double ei(double x)
{
    double p[] = new double[8];
    double q[] = new double[8];

    if (x > 24.0) {
        p[0]= 1.00000000000058;    q[1] = 1.99999999924131;
```

```
    p[1]=x-3.00000016782085;      q[2] = -2.99996432944446;
    p[2]=x-5.00140345515924;      q[3] = -7.90404992298926;
    p[3]=x-7.49289167792884;      q[4] = -4.31325836146628;
    p[4]=x-3.08336269051763e1;    q[5] =  2.95999399486831e2;
    p[5]=x-1.39381360364405;      q[6] = -6.74704580465832;
    p[6]=x+8.91263822573708;      q[7] =  1.04745362652468e3;
    p[7]=x-5.31686623494482e1;
    return Math.exp(x)*(1.0+Algebraic_eval.jfrac(7,q,p)/x)/x;
} else if (x > 12.0) {
    p[0]=  9.99994296074708e-1;   q[1] =  1.00083867402639;
    p[1]=x-1.95022321289660;      q[2] = -3.43942266899870;
    p[2]=x+1.75656315469614;      q[3] =  2.89516727925135e1;
    p[3]=x+1.79601688769252e1;    q[4] =  7.60761148007735e2;
    p[4]=x-3.23467330305403e1;    q[5] =  2.57776384238440e1;
    p[5]=x-8.28561994140641;      q[6] =  5.72837193837324e1;
    p[6]=x-1.86545454883399e1;    q[7] =  6.95000655887434e1;
    p[7]=x-3.48334653602853;
    return Math.exp(x)*Algebraic_eval.jfrac(7,q,p)/x;
} else if (x > 6.0) {
    p[0]=  1.00443109228078;      q[1] =  5.27468851962908e-1;
    p[1]=x-4.32531132878135e1;    q[2] =  2.73624119889328e3;
    p[2]=x+6.01217990830080e1;    q[3] =  1.43256738121938e1;
    p[3]=x-3.31842531997221e1;    q[4] =  1.00367439516726e3;
    p[4]=x+2.50762811293561e1;    q[5] = -6.25041161671876;
    p[5]=x+9.30816385662165;      q[6] =  3.00892648372915e2;
    p[6]=x-2.19010233854880e1;    q[7] =  3.93707701852715;
    p[7]=x-2.18086381520724;
    return Math.exp(x)*Algebraic_eval.jfrac(7,q,p)/x;
} else if (x > 0.0) {
    double t,r,x0,xmx0;
    p[0]= -1.95773036904548e8;    q[0] = -8.26271498626055e7;
    p[1]=  3.89280421311201e6;    q[1] =  8.91925767575612e7;
    p[2]= -2.21744627758845e7;    q[2] = -2.49033375740540e7;
    p[3]= -1.19623669349247e5;    q[3] =  4.28559624611749e6;
    p[4]= -2.49301393458648e5;    q[4] = -4.83547436162164e5;
    p[5]= -4.21001615357070e3;    q[5] =  3.57300298058508e4;
    p[6]= -5.49142265521085e2;    q[6] = -1.60708926587221e3;
    p[7]= -8.66937339951070;      q[7] =  3.41718750000000e1;
    x0=0.372507410781367;
    t=x/3.0-1.0;
    r=Algebraic_eval.chepolsum(7,t,p)/Algebraic_eval.chepolsum(7,t,q);
    xmx0=(x-409576229586.0/1099511627776.0)-0.767177250199394e-12;
    if (Math.abs(xmx0) > 0.037)
        t=Math.log(x/x0);
    else {
        double z,z2;
        p[0] =  0.837207933976075e1;
        p[1] = -0.652268740837103e1;
        p[2] =  0.569955700306720;
        q[0] =  0.418603966988037e1;
```

```
            q[1] = -0.465669026080814e1;
            q[2] =  0.1e1;
            z=xmx0/(x+x0);
            z2=z*z;
            t=z*Algebraic_eval.pol(2,z2,p)/Algebraic_eval.pol(2,z2,q);
        }
        return t+xmx0*r;
    } else if (x > -1.0) {
        double y;
        p[0] = -4.41785471728217e4;  q[0]=7.65373323337614e4;
        p[1] =  5.77217247139444e4;  q[1]=3.25971881290275e4;
        p[2] =  9.93831388962037e3;  q[2]=6.10610794245759e3;
        p[3] =  1.84211088668000e3;  q[3]=6.35419418378382e2;
        p[4] =  1.01093806161906e2;  q[4]=3.72298352833327e1;
        p[5] =  5.03416184097568;    q[5]=1.0;
        y = -x;
        return Math.log(y)-Algebraic_eval.pol(5,y,p)/Algebraic_eval.pol(5,y,q);
    } else if (x > -4.0) {
        double y;
        p[0]=8.67745954838444e-8;  q[0]=1.0;
        p[1]=9.99995519301390e-1;  q[1]=1.28481935379157e1;
        p[2]=1.18483105554946e1;   q[2]=5.64433569561803e1;
        p[3]=4.55930644253390e1;   q[3]=1.06645183769914e2;
        p[4]=6.99279451291003e1;   q[4]=8.97311097125290e1;
        p[5]=4.25202034768841e1;   q[5]=3.14971849170441e1;
        p[6]=8.83671808803844;     q[6]=3.79559003762122;
        p[7]=4.01377664940665e-1;  q[7]=9.08804569188869e-2;
        y = -1.0/x;
        return -Math.exp(x)*Algebraic_eval.pol(7,y,p)/Algebraic_eval.pol(5,y,q);
    } else {
        double y;
        p[0] = -9.99999999998447e-1;  q[0]=1.0;
        p[1] = -2.66271060431811e1;   q[1]=2.86271060422192e1;
        p[2] = -2.41055827097015e2;   q[2]=2.92310039388533e2;
        p[3] = -8.95927957772937e2;   q[3]=1.33278537748257e3;
        p[4] = -1.29885688756484e3;   q[4]=2.77761949509163e3;
        p[5] = -5.45374158883133e2;   q[5]=2.40401713225909e3;
        p[6] = -5.66575206533869;     q[6]=6.31657483280800e2;
        y = -1.0/x;
        return -Math.exp(x)*y*(1.0+y*Algebraic_eval.pol(6,y,p)/
                               Algebraic_eval.pol(5,y,q));
    }
}
}
```

B. eialpha

Calculates a sequence of integrals [AbS65] of the form

$$\alpha_i(x) = \int_1^\infty e^{-xt} t^i\, dt \qquad\qquad (x > 0; \quad i = 0,\ldots,n)$$

by use of the recursion

$$\alpha_0(x) = x^{-1}, \quad \alpha_i(x) = \alpha_0(x) + (i/x)\alpha_{i-1}(x) \qquad (i=1,\ldots,n).$$

Procedure parameters:

$$\text{void eialpha } (x,n,alpha)$$

x: double;
          entry: the real x in the integrand;
n: int;
          entry: the integer n in the integrand;
alpha: double alpha[0:n];
          exit: the value of the integral is stored in alpha[i], i=0,...,n.

```
public static void eialpha(double x, int n, double alpha[])
{
    int k;
    double a,b,c;

    c=1.0/x;
    a=Math.exp(-x);
    b=alpha[0]=a*c;
    for (k=1; k<=n; k++) alpha[k]=b=(a+k*b)*c;
}
```

C. enx

Calculates a sequence of integrals of the form

$$\alpha_n(x) = \int_1^\infty \frac{e^{-xt}}{t^n}\, dt \qquad (x > 0; \quad n = n1,\ldots,n2) \qquad (1)$$

The value of $\alpha_{n0}(x)$ where $n0 = \lfloor x \rfloor$ is first computed using (a) a call of *ei* if $n0 = 1$, (b) a Taylor series expansion in powers of $x - n0$ if $n0 \le 10$, and (c) (calling *nonexpenx* for this purpose) the relationship $\alpha_n(x) = e^x \alpha'_n(x)$, where

$$\alpha'_n(x) = \frac{1}{x+} \frac{n}{1+} \frac{1}{x+} \frac{n+1}{1+} \cdots \qquad (2)$$

with $n=n0$ if $n0 > 10$. Thereafter, the recursion

$$\alpha_{n+1}(x) = n^{-1}\{e^x - x\alpha_n(x)\}$$

is used in a forward direction if $n0 < n2$, and in a backward direction if $n0 > n1$ to compute the required values of the functions (1). The successive convergence $C_r(x)$ of expansion (2) are computed until $|1 - C_r(x)/C_{r+1}(x)| \le \delta$, where δ is the value of the machine precision.

Procedure parameters:
$$\text{void enx } (x,n1,n2,a)$$

x: double;
           entry: the real positive *x* in the integrand;
n1,n2: int;
           entry: lower and upper bound, respectively, of the integer *n* in the
                   integrand;
a: double *a[n1:n2]*;
           exit: the value of the integral is stored in *a[i]*, *i=n1,...,n2*.

Procedures used: ei, nonexpenx.

```
public static void enx(double x, int n1, int n2, double a[])
{
    if (x <= 1.5) {
        int i;
        double w,e;
        e=0.0;
        w = -ei(-x);
        if (n1 == 1) a[1]=w;
        if (n2 > 1) e=Math.exp(-x);
        for (i=2; i<=n2; i++) {
            w=(e-x*w)/(i-1);
            if (i >= n1) a[i]=w;
        }
    } else {
        int i,n;
        double w,e,an;
        n=(int)Math.ceil(x);
        if (n <= 10) {
            double f,w1,t,h;
            double p[] = new double[20];
            p[2] =0.37534261820491e-1;  p[11]=0.135335283236613;
            p[3] =0.89306465560228e-2;  p[12]=0.497870683678639e-1;
            p[4] =0.24233983686581e-2;  p[13]=0.183156388887342e-1;
            p[5] =0.70576069342458e-3;  p[14]=0.673794699908547e-2;
            p[6] =0.21480277819013e-3;  p[15]=0.247875217666636e-2;
            p[7] =0.67375807781018e-4;  p[16]=0.911881965554516e-3;
            p[8] =0.21600730159975e-4;  p[17]=0.335462627902512e-3;
            p[9] =0.70411579854292e-5;  p[18]=0.123409804086680e-3;
            p[10]=0.23253026570282e-5;  p[19]=0.453999297624848e-4;
            f=w=p[n];
            e=p[n+9];
            w1=t=1.0;
            h=x-n;
            i=n-1;
            do {
                f=(e-i*f)/n;
                t = -h*t/(n-i);
                w1=t*f;
```

```
            w += w1;
            i--;
        } while (Math.abs(w1) > 1.0e-15*w);
    } else {
        double b[] = new double[n+1];
        nonexpenx(x,n,n,b);
        w=b[n]*Math.exp(-x);
    }
    if (n1 == n2 && n1 == n)
        a[n]=w;
    else {
        e=Math.exp(-x);
        an=w;
        if (n <= n2 && n >= n1) a[n]=w;
        for (i=n-1; i>=n1; i--) {
            w=(e-i*w)/x;
            if (i <= n2) a[i]=w;
        }
        w=an;
        for (i=n+1; i<=n2; i++) {
            w=(e-x*w)/(i-1);
            if (i >= n1) a[i]=w;
        }
    }
  }
}
```

D. nonexpenx

Calculates a sequence of integrals of the form

$$\alpha_n(x) = e^x \int_1^\infty \frac{e^{-xt}}{t^n} dt \qquad (x > 0; \quad n = n1,\ldots,n2) \tag{1}$$

The value of $\alpha_{n0}(x)$ where $n0 = \lfloor x \rfloor$ is first computed using the methods (a) and (b) described in the documentation to *enx* if $n0 \le 10$ (calling *enx* for this purpose) and the continued fraction expansion (2) of that documentation if $n0 > 10$. Thereafter, the recursion

$$\alpha_{n+1}(x) = n^{-1}\{1 - x\alpha_n(x)\}$$

is used as described in the documentation to *enx* to compute the required values of the functions (1). See [AbS65, CodT68, CodT69, G61, G73].

Procedure parameters:
$$\text{void nonexpenx } (x,n1,n2,a)$$

x: double;
             entry: the real positive x in the integrand;
n1,n2:   int;
             entry: lower and upper bound, respectively, of the integer n in the integrand;

a: double *a[n1:n2]*;
            exit: the value of the integral is stored in *a[i]*, *i=n1,...,n2*.

Procedure used: enx.

```
public static void nonexpenx(double x, int n1, int n2, double a[])
{
    int i,n;
    double w,an;

    n = (x <= 1.5) ? 1 : (int)Math.ceil(x);
    if (n <= 10) {
        double b[] = new double[n+1];
        enx(x,n,n,b);
        w=b[n]*Math.exp(x);
    } else {
        int k,k1;
        double ue,ve,we,we1,uo,vo,wo,wo1,r,s;
        ue=1.0;
        ve=we=1.0/(x+n);
        we1=0.0;
        uo=1.0;
        vo = -n/(x*(x+n+1.0));
        wo1=1.0/x;
        wo=vo+wo1;
        w=(we+wo)/2.0;
        k1=1;
        k=k1;
        while (wo-we > 1.0e-15*w && we > we1 && wo < wo1) {
            we1=we;
            wo1=wo;
            r=n+k;
            s=r+x+k;
            ue=1.0/(1.0-k*(r-1.0)*ue/((s-2.0)*s));
            uo=1.0/(1.0-k*r*uo/(s*s-1.0));
            ve *= (ue-1.0);
            vo *= (uo-1.0);
            we += ve;
            wo += vo;
            w=(we+wo)/2.0;
            k1++;
            k=k1;
        }
    }
    an=w;
    if (n <= n2 && n >= n1) a[n]=w;
    for (i=n-1; i>=n1; i--) {
        w=(1.0-i*w)/x;
        if (i <= n2) a[i]=w;
    }
```

```
    w=an;
    for (i=n+1; i<=n2; i++) {
        w=(1.0-x*w)/(i-1);
        if (i >= n1) a[i]=w;
    }
}
```

6.2.2 Sine and cosine integral

A. sincosint

Calculates the sine integral $si(x)$ and cosine integral $ci(x)$ defined by

$$si(x) = \int_0^x \frac{\sin(t)}{t}\,dt, \qquad ci(x) = \gamma + \ln|x| + \int_0^x \frac{\cos(t)-1}{t}\,dt.$$

If $|x| \leq 4$ then $si(x)$ is approximated by means of a Chebyshev series of the form

$$x \sum_{j=0}^{10} a_j T_{2j}(x),$$

and $ci(x)$ by a similar series of the form

$$\gamma + \ln|x| - x^2 \sum_{j=0}^{10} a_j T_{2j}(x).$$

For values of x outside this range, the functions $f(x)$ and $g(x)$ occurring in the formulae

$$si(x) = \tfrac{1}{2}\pi\,\mathrm{sign}(x) - f(x)\cos(x) - g(x)\sin(x)$$
$$ci(x) = f(x)\sin(x) - g(x)\cos(x)$$

are evaluated by means of a call of $sincosfg$, and the values of $si(x)$ and $ci(x)$ are obtained by use of these relationships [AbS65, Bul67].

When using the procedure $sincosint$ for large values of x, the relative accuracy mainly depends on the accuracy of the functions $sin(x)$ and $cos(x)$.

Procedure parameters:
$$\text{void sincosint } (x,si,ci)$$
x: double;
        entry: the real argument of $si(x)$ and $ci(x)$;
si: double $si[0:0]$;
        exit: the value of $si(x)$;
ci: double $ci[0:0]$;
        exit: the value of $ci(x)$.

Procedures used: sincosfg, chepolsum.

```
public static void sincosint(double x, double si[], double ci[])
{
    double absx,z,gg;
```

```
    double f[] = new double[1];
    double g[] = new double[1];

    absx=Math.abs(x);
    if (absx <= 4.0) {
        double z2;
    double a[] = new double[11];
        a[0] =2.7368706803630e0;    a[1] = -1.1106314107894e0;
        a[2] =1.4176562194666e-1;   a[3] = -1.0252652579174e-2;
        a[4] =4.6494615619880e-4;   a[5] = -1.4361730896642e-5;
        a[6] =3.2093684948229e-7;   a[7] = -5.4251990770162e-9;
        a[8] =7.1776288639895e-11;  a[9] = -7.6335493723482e-13;
        a[10]=6.6679958346983e-15;
        z=x/4.0;
        z2=z*z;
        gg=z2+z2-1.0;
        si[0] = z*Algebraic_eval.chepolsum(10,gg,a);
        a[0] =2.9659610400727e0;    a[1] = -9.4297198341830e-1;
        a[2] =8.6110342738169e-2;   a[3] = -4.7776084547139e-3;
        a[4] =1.7529161205146e-4;   a[5] = -4.5448727803752e-6;
        a[6] =8.7515839180060e-8;   a[7] = -1.2998699938109e-9;
        a[8] =1.5338974898831e-11;  a[9] = -1.4724256070277e-13;
        a[10]=1.1721420798429e-15;
        ci[0]=0.577215664901533+Math.log(absx)-z2*Algebraic_eval.chepolsum(10,gg,a);
    } else {
        double cx,sx;
        sincosfg(x,f,g);
        cx=Math.cos(x);
        sx=Math.sin(x);
        si[0] = 1.570796326794897;
        if (x < 0.0) si[0] = -si[0];
        si[0] -= f[0]*cx+g[0]*sx;
        ci[0] = f[0]*sx-g[0]*cx;
    }
}
```

B. sincosfg

Evaluates the functions $f(x)$, $g(x)$ related to the sine and cosine integrals [AbS65, Bul67] by means of the formulae

$$f(x) = ci(x)\sin(x) - \{si(x) - \pi/2\}\cos(x)$$
$$g(x) = -ci(x)\cos(x) - \{si(x) - \pi/2\}\sin(x).$$

When $|x| \leq 4$ then the functions $si(x)$ and $ci(x)$ are evaluated by means of a call of *sincosint*, and the values of $f(x)$ and $g(x)$ are obtained by use of the above relationships. When $|x| > 4$ then $f(x)$ is approximated by use of a Chebyshev series of the form

$$\frac{1}{x} \sum_{k=0}^{21} a_i T_i \left(\frac{8}{|x|} - 1 \right),$$

and *g(x)* by a use of a similar series of the form

$$\frac{1}{x^2} \sum_{k=0}^{23} a_i T_i \left(\frac{8}{|x|} - 1 \right).$$

Procedure parameters:

void sincosfg (*x,f,g*)

x: double;
           entry: the real argument of *f(x)* and *g(x)*;
f: double *f[0:0]*;
           exit: the value of *f(x)*;
g: double *g[0:0]*;
           exit: the value of *g(x)*.

Procedures used: sincosint, chepolsum.

```
public static void sincosfg(double x, double f[], double g[])
{
    double absx;
    double si[] = new double[1];
    double ci[] = new double[1];

    absx=Math.abs(x);
    if (absx <= 4.0) {
        double cx,sx;
        sincosint(x,si,ci);
        cx=Math.cos(x);
        sx=Math.sin(x);
        si[0] -= 1.570796326794897;
        f[0] = ci[0]*sx-si[0]*cx;
        g[0] = -ci[0]*cx-si[0]*sx;
    } else {
        double a[] = new double[24];
        a[0]  =  9.6578828035185e-1;   a[1]  = -4.3060837778597e-2;
        a[2]  = -7.3143711748104e-3;   a[3]  =  1.4705235789868e-3;
        a[4]  = -9.8657685732702e-5;   a[5]  = -2.2743202204655e-5;
        a[6]  =  9.8240257322526e-6;   a[7]  = -1.8973430148713e-6;
        a[8]  =  1.0063435941558e-7;   a[9]  =  8.0819364822241e-8;
        a[10] = -3.8976282875288e-8;   a[11] =  1.0335650325497e-8;
        a[12] = -1.4104344875897e-9;   a[13] = -2.5232078399683e-10;
        a[14] =  2.5699831325961e-10;  a[15] = -1.0597889253948e-10;
        a[16] =  2.8970031570214e-11;  a[17] = -4.1023142563083e-12;
        a[18] = -1.0437693730018e-12;  a[19] =  1.0994184520547e-12;
        a[20] = -5.2214239401679e-13;  a[21] =  1.7469920787829e-13;
        a[22] = -3.8470012979279e-14;
        f[0] = Algebraic_eval.chepolsum(22,8.0/absx-1.0,a)/x;
```

```
            a[0]  =   2.2801220638241e-1;  a[1]   =  -2.6869727411097e-2;
            a[2]  =  -3.5107157280958e-3;  a[3]   =   1.2398008635186e-3;
            a[4]  =  -1.5672945116862e-4;  a[5]   =  -1.0664141798094e-5;
            a[6]  =   1.1170629343574e-5;  a[7]   =  -3.1754011655614e-6;
            a[8]  =   4.4317473520398e-7;  a[9]   =   5.5108696874463e-8;
            a[10] =  -5.9243078711743e-8;  a[11]  =   2.2102573381555e-8;
            a[12] =  -5.0256827540623e-9;  a[13]  =   3.1519168259424e-10;
            a[14] =   3.6306990848979e-10; a[15]  =  -2.2974764234591e-10;
            a[16] =   8.5530309424048e-11; a[17]  =  -2.1183067724443e-11;
            a[18] =   1.7133662645092e-12; a[19]  =   1.7238877517248e-12;
            a[20] =  -1.2930281366811e-12; a[21]  =   5.7472339223731e-13;
            a[22] =  -1.8415468268314e-13; a[23]  =   3.5937256571434e-14;
            g[0]  = 4.0*Algebraic_eval.chepolsum(23,8.0/absx-1.0,a)/absx/absx;

      }

}
```

6.3 Gamma function

A. recipgamma

Calculates the reciprocal of the gamma function for arguments in the range [0.5,1.5]; moreover, odd and even parts are delivered.

recipgamma computes the values of the functions

$$\Gamma'(x) = \{\Gamma(1 - x)\}^{-1}$$

where

$$\Gamma(x) = \int_0^\infty t^{x-1} e^{-t} dt$$

and

$$\Gamma'_o(x) = (2+x)^{-1}\{\Gamma'(x) - \Gamma'(-x)\}$$
$$\Gamma'_e(x) = \tfrac{1}{2}\{\Gamma'(x) + \Gamma'(-x)\}$$

for $-0.5 \leq x \leq 0.5$.

The functions $\Gamma'_o(x)$, $\Gamma'_e(x)$ are evaluated by use of truncated Chebyshev series, and the function $\Gamma'(x)$ recovered from them by use of the formula

$$\Gamma'(x) = \Gamma'_e(x) + (1 + x/2)\Gamma'_o(x).$$

Procedure parameters:

$$\text{double recipgamma } (x, odd, even)$$

recipgamma: delivers $1/\Gamma(1-x)$;

x: double;

                entry: this argument should satisfy $-0.5 \leq x \leq 0.5$;
                         (actually the gamma function Γ is calculated for $1-x$, i.e., if one wants to calculate $1/\Gamma(1)$, one has to set x to 0);

odd: double *odd[0:0]*;

                exit: the odd part of $1/\Gamma(1-x)$ divided by $(2x)$;
                         i.e., $(1/\Gamma(1-x) - 1/\Gamma(1+x)) / (2x)$;

even: double *even[0:0]*;
                   exit: the even part of $1/\Gamma(1-x)$ divided by 2;
                              i.e., $(1/\Gamma(1-x) + 1/\Gamma(1+x)) / 2$.

```
public static double recipgamma(double x, double odd[], double even[])
{
    int i;
    double alfa,beta,x2;
    double b[] = new double[13];

    b[1]  = -0.283876542276024;    b[2]  = -0.076852840844786;
    b[3]  =  0.001706305071096;    b[4]  =  0.001271927136655;
    b[5]  =  0.000076309597586;    b[6]  = -0.000004971736704;
    b[7]  = -0.000000865920800;    b[8]  = -0.000000033126120;
    b[9]  =  0.000000001745136;    b[10] =  0.000000000242310;
    b[11] =  0.000000000009161;    b[12] = -0.000000000000170;
    x2=x*x*8.0;
    alfa = -0.000000000000001;
    beta=0.0;
    for (i=12; i>=2; i -= 2) {
        beta = -(alfa*2.0+beta);
        alfa = -beta*x2-alfa+b[i];
    }
    even[0]=(beta/2.0+alfa)*x2-alfa+0.921870293650453;
    alfa = -0.000000000000034;
    beta=0.0;
    for (i=11; i>=1; i -= 2) {
        beta = -(alfa*2.0+beta);
        alfa = -beta*x2-alfa+b[i];
    }
    odd[0]=(alfa+beta)*2.0;
    return (odd[0])*x+(even[0]);
}
```

B. gamma

Computes the value of the gamma function at x

$$\Gamma(x) = \int_0^\infty t^{x-1} e^{-t} dt .$$

We distinguish between the following cases for the argument x:
$x < 0.5$:

In this case the formula $\Gamma(x) * \Gamma(1-x) = \pi/\sin(\pi*x)$ is used. However the sine function is not calculated directly on the argument $\pi*x$ but on the argument $\pi*(x \bmod 0.5)$, in this way a big decrease of precision is avoided. The precision here depends strongly on the precision of the sine function.
$0.5 \le x \le 1.5$:

Here the procedure *recipgamma* is called; moreover, $\Gamma(1) = 1$.

$1.5 < x \leq 22$:

The recursion formula $\Gamma(1+x) = x * \Gamma(x)$ is used. The precision depends on the number of recursions needed. The upper bound of 22 has been chosen because now it is assured that for all integer arguments for which the value of the gamma function is representable (and this is the case for all integer arguments in the range [1,22]), this value is obtained, i.e., $\Gamma(i) = 1 * 2 * \ldots * (i\text{-}1)$.

$x > 22$:

Now the procedures *loggamma* and *exp* are used. The precision strongly depends on the precision of the exponential function, and no bound for the error can be given.

Procedure parameters:

$$\text{double gamma } (x)$$

gamma: delivers the value of the gamma function at x;

x: double;

entry: the argument; if one of the following three conditions is fulfilled then overflow will occur:
1. the argument is too large;
2. the argument is a nonpositive integer;
3. the argument is too close to a large (in absolute value) nonpositive integer.

Procedures used: recipgamma, loggamma.

```
public static double gamma(double x)
{
    boolean inv;
    double y,s,f,g;
    double odd[] = new double[1];
    double even[] = new double[1];

    f=0.0;
    if (x < 0.5) {
        y=x-Math.floor(x/2.0)*2;
        s = Math.PI;
        if (y >= 1.0) {
            s = -s;
            y=2.0-y;
        }
        if (y >= 0.5) y=1.0-y;
        inv=true;
        x=1.0-x;
        f=s/Math.sin(Math.PI*y);
    } else
        inv=false;
    if (x > 22.0)
        g=Math.exp(loggamma(x));
    else {
        s=1.0;
```

```
        while (x > 1.5) {
            x=x-1.0;
            s *= x;
        }
        g=s/recipgamma(1.0-x,odd,even);
    }
    return (inv ? f/g : g);
}
```

C. loggamma

Computes the natural logarithm of the gamma function at x: $ln(\Gamma(x))$.   We distinguish between the following cases for the argument x (in most cases nothing is said about precision, as this highly depends on the precision of the natural logarithm):

$0 < x < 1$:

> Here the recursion formula $loggamma(x) = loggamma(1+x) - ln(x)$ is used.

$1 \leq x \leq 2$:

> On the interval the truncated Chebyshev series for the function $loggamma(x) / ((x-1)*(x-2))$ is used.

$2 < x \leq 13$:

> The recursion formula $loggamma(x) = loggamma(1-x) + ln(x)$ is used.

$13 < x \leq 22$:

> As for $x < 1$ the formula $loggamma(x) = loggamma(1+x) - ln(x)$ is used.

$x > 22$:

> In this case $loggamma$ is calculated by use of the asymptotic expansion for $loggamma(x) - (x-0.5) * ln(x)$.

Procedure parameters:
$$double\ loggamma\ (x)$$
loggamma: delivers the value of the natural logarithm of the gamma function at
               x;
x: double;
               entry: this argument must be positive.

```
public static double loggamma(double x)
{
    int i;
    double r,x2,y,f,u0,u1,u,z;
    double b[] = new double[19];

    if (x > 13.0) {
        r=1.0;
        while (x <= 22.0) {
            r /= x;
            x += 1.0;
        }
```

```
    x2 = -1.0/(x*x);
    r=Math.log(r);
    return Math.log(x)*(x-0.5)-x+r+0.918938533204672+
        ((((0.595238095238095e-3*x2+0.793650793650794e-3)*x2+
        0.277777777777778e-2)*x2+0.833333333333333e-1)/x;
    } else {
        f=1.0;
        u0=u1=0.0;
        b[1]  = -0.0761141616704358;  b[2]  = 0.0084323249659328;
        b[3]  = -0.0010794937263286;  b[4]  = 0.0001490074800369;
        b[5]  = -0.0000215123998886;  b[6]  = 0.0000031979329861;
        b[7]  = -0.0000004851693012;  b[8]  = 0.0000000747148782;
        b[9]  = -0.0000000116382967;  b[10] = 0.0000000018294004;
        b[11] = -0.0000000002896918;  b[12] = 0.0000000000461570;
        b[13] = -0.0000000000073928;  b[14] = 0.0000000000011894;
        b[15] = -0.0000000000001921;  b[16] = 0.0000000000000311;
        b[17] = -0.0000000000000051;  b[18] = 0.0000000000000008;
        if (x < 1.0) {
            f=1.0/x;
            x += 1.0;
        } else
            while (x > 2.0) {
                x -= 1.0;
                f *= x;
            }
        f=Math.log(f);
        y=x+x-3.0;
        z=y+y;
        for (i=18; i>=1; i--) {
            u=u0;
            u0=z*u0+b[i]-u1;
            u1=u;
        }
        return (u0*y+0.491415393029387-u1)*(x-1.0)*(x-2.0)+f;
    }
}
```

D. incomgam

Computes the incomplete gamma functions based on Padé approximations [AbS65, Lu70]. *incomgam* evaluates the functions

$$\gamma(a,x) = \int_0^x t^{a-1}e^{-t}\,dt, \qquad \Gamma(a,x) = \int_x^\infty t^{a-1}e^{-t}\,dt$$

to a prescribed relative accuracy ε.

If (a) $a,x < 3$ or (b) $x < a$ and $a \geq 3$, $\gamma(a,x)$ is computed by use of a continued fraction derived from a power series in ascending powers of x for this function, and $\Gamma(a,x)$ is determined from the relationship $\Gamma(a,x) = \Gamma(a) - \gamma(a,x)$. If neither of the above conditions holds, $\Gamma(a,x)$ is computed by use of a continued fraction derived from an asymptotic series in descending powers of x for this function, and the relationship $\gamma(a,x) = \Gamma(a) - \Gamma(a,x)$ is used.

The relative accuracy of the results depends not only on the quantity ε, but also on the accuracy of the functions *exp* and *gamma*. Especially for large values of x and a, the desired accuracy cannot be guaranteed.

Procedure parameters:
$$\text{void incomgam } (x,a,klgam,grgam,gam,eps)$$

x: double;
      entry: the independent argument x, $x \geq 0$;
a: double;
      entry: the independent parameter a, $a > 0$;
klgam: double *klgam[0:0]*;
      exit: the integral $\gamma(a,x)$ is delivered in *klgam*;
grgam: double *grgam[0:0]*;
      exit: the integral $\Gamma(a,x)$ is delivered in *grgam*;
gam: double;
      entry: the value of $\Gamma(a)$;
                for this expression, the procedure *gamma* may be used;
eps: double;
      entry: the desired relative accuracy (value of ε above);
                the value of *eps* should not be smaller than the machine accuracy.

```
public static void incomgam(double x, double a, double klgam[], double grgam[],
                            double gam, double eps)
{
    int n;
    double c0,c1,c2,d0,d1,d2,x2,ax,p,q,r,s,r1,r2,scf;

    s=Math.exp(-x+a*Math.log(x));
    scf=Double.MAX_VALUE;
    if (x <= ((a < 3.0) ? 1.0 : a)) {
        x2=x*x;
        ax=a*x;
        d0=1.0;
        p=a;
        c0=s;
        d1=(a+1.0)*(a+2.0-x);
        c1=((a+1.0)*(a+2.0)+x)*s;
        r2=c1/d1;
        n=1;
        do {
            p += 2.0;
            q=(p+1.0)*(p*(p+2.0)-ax);
            r=n*(n+a)*(p+2.0)*x2;
            c2=(q*c1+r*c0)/p;
            d2=(q*d1+r*d0)/p;
            r1=r2;
```

```
                r2=c2/d2;
                c0=c1;
                c1=c2;
                d0=d1;
                d1=d2;
                if (Math.abs(c1) > scf || Math.abs(d1) > scf) {
                    c0 /= scf;
                    c1 /= scf;
                    d0 /= scf;
                    d1 /= scf;
                }
                n++;
            } while (Math.abs((r2-r1)/r2) > eps);
            klgam[0] = r2/a;
            grgam[0] = gam-klgam[0];
        } else {
            c0=a*s;
            c1=(1.0+x)*c0;
            q=x+2.0-a;
            d0=x;
            d1=x*q;
            r2=c1/d1;
            n=1;
            do {
                q += 2.0;
                r=n*(n+1-a);
                c2=q*c1-r*c0;
                d2=q*d1-r*d0;
                r1=r2;
                r2=c2/d2;
                c0=c1;
                c1=c2;
                d0=d1;
                d1=d2;
                if (Math.abs(c1) > scf || Math.abs(d1) > scf) {
                    c0 /= scf;
                    c1 /= scf;
                    d0 /= scf;
                    d1 /= scf;
                }
                n++;
            } while (Math.abs((r2-r1)/r2) > eps);
            grgam[0] = r2/a;
            klgam[0] = gam-grgam[0];
        }
    }
}
```

E. incbeta

The incomplete beta function is defined as

$$B_x(p,q) = \int_0^x t^{p-1}(1-t)^{q-1}\,dt$$

$p > 0$, $q > 0$, $0 \le x \le 1$, and the incomplete beta function ratio is
$$I_x(p,q) = B_x(p,q) / B_1(p,q).$$
incbeta computes $I_x(p,q)$ for $0 \le x \le 1$, $p > 0$, $q > 0$ by use of the continued fraction corresponding to formula 26.5.8 in [AbS65], see also [G64, G67]. If $x > 0.5$ then the relation $I_x(p,q)=1-I_{1-x}(q,p)$ is used. The value of the continued fraction is approximated by the convergent $C_r(x)$, where r is the smallest integer for which $\left|1 - C_{r-1}(x)/C_r(x)\right| \le \varepsilon$, ε being a small positive real number supplied by the user.

It is advised to use in *incbeta* only small values of p and q, say $0<p\le5$, $0<q\le5$. For other ranges of the parameters p and q, the procedures *ibpplusn* and *ibqplusn* can be used. *incbeta* satisfies *incbeta=x* if $x=0$ or $x=1$, whatever p and q. There is no control on the parameters x, p, and q for their intended ranges.

Procedure parameters:
$$\text{double incbeta }(x,p,q,eps)$$
incbeta: delivers the value of $I_x(p,q)$;
x: double;
              entry: this argument should satisfy: $0 \le x \le 1$;
p: double;
              entry: the parameter p, $p > 0$;
q: double;
              entry: the parameter q, $q > 0$;
eps: double;
              entry: the desired relative accuracy (value of ε above);
                        the value of *eps* should not be smaller than the machine accuracy.

Procedure used: gamma.

```
public static double incbeta(double x, double p, double q, double eps)
{
    int m,n;
    boolean neven,recur;
    double g,f,fn,fn1,fn2,gn,gn1,gn2,dn,pq;

    if (x == 0.0 || x == 1.0)
        return x;
    else {
        if (x > 0.5) {
            f=p;
            p=q;
            q=f;
            x=1.0-x;
            recur=true;
        } else
```

```
            recur=false;
        g=fn2=0.0;
        m=0;
        pq=p+q;
        f=fn1=gn1=gn2=1.0;
        neven=false;
        n=1;
        do {
            if (neven) {
                m++;
                dn=m*x*(q-m)/(p+n-1.0)/(p+n);
            } else
                dn = -x*(p+m)*(pq+m)/(p+n-1.0)/(p+n);
            g=f;
            fn=fn1+dn*fn2;
            gn=gn1+dn*gn2;
            neven=(!neven);
            f=fn/gn;
            fn2=fn1;
            fn1=fn;
            gn2=gn1;
            gn1=gn;
            n++;
        } while (Math.abs((f-g)/f) > eps);
        f=f*Math.pow(x,p)*Math.pow(1.0-x,q)*gamma(p+q)/gamma(p+1.0)/gamma(q);
        if (recur) f=1.0-f;
        return f;
    }
}
```

F. ibpplusn

The incomplete beta function is defined as

$$B_x(p,q) = \int_0^x t^{p-1}(1-t)^{q-1}\,dt$$

$p > 0$, $q > 0$, $0 \le x \le 1$, and the incomplete beta function ratio is
$$I_x(p,q) = B_x(p,q) / B_1(p,q).$$
ibpplusn computes $I_x(p+n,q)$ for $n=0,1,...,nmax$, $0 \le x \le 1$, $p > 0$, $q > 0$ (see [G64, G67]). In [G64] the procedure *ibpplusn* is called "incomplete beta q fixed". There is no control on the parameters x, p, q, and *nmax* for their intended ranges.

Procedure parameters:
$$\text{void ibpplusn } (x,p,q,nmax,eps,i)$$
x: double;
          entry: this argument should satisfy: $0 \le x \le 1$;
p: double;
          entry: the parameter p, $p > 0$; it is advised to take $0 < p \le 1$;

q: double;
            entry: the parameter *q*, *q* > 0;
nmax: int;
            entry: *nmax* indicates the maximum number of function values $I_x(p+n,q)$
                    to be generated;
eps: double;
            entry: the desired relative accuracy (value of ε above);
                    the value of *eps* should not be smaller than the machine accuracy;
i: double *i[0:nmax]*;
            exit: *i[n]* = $I_x(p+n,q)$ for *n*=0,1,...,*nmax*.

Procedures used: ixqfix, ixpfix.

```
public static void ibpplusn(double x, double p, double q, int nmax, double eps,
                            double i[])
{
    int n;

    if (x == 0.0 || x == 1.0)
        for (n=0; n<=nmax; n++) i[n]=x;
    else {
        if (x <= 0.5)
            ixqfix(x,p,q,nmax,eps,i);
        else {
            ixpfix(1.0-x,q,p,nmax,eps,i);
            for (n=0; n<=nmax; n++) i[n]=1.0-i[n];
        }
    }
}
```

G. ibqplusn

The incomplete beta function is defined as

$$B_x(p,q) = \int_0^x t^{p-1}(1-t)^{q-1} dt$$

p > 0, *q* > 0, 0 ≤ *x* ≤ 1, and the incomplete beta function ratio is
$$I_x(p,q) = B_x(p,q) / B_1(p,q).$$
ibqplusn computes $I_x(p,q+n)$ for *n*=0,1,...,*nmax*, 0 ≤ *x* ≤ 1, *p* > 0, *q* > 0 (see [G64, G67]). In [G64] the procedure *ibqplusn* is called "incomplete beta p fixed". There is no control on the parameters *x*, *p*, *q*, and *nmax* for their intended ranges.

Procedure parameters:
                            void ibqplusn (*x,p,q,nmax,eps,i*)
x: double;
            entry: this argument should satisfy: 0 ≤ *x* ≤ 1;
p: double;
            entry: the parameter *p*, *p* > 0;

q: double;
      entry: the parameter q, $q > 0$; it is advised to take $0 < q \leq 1$;
nmax: int;
      entry: *nmax* indicates the maximum number of function values $I_x(p,q+n)$
             to be generated;
eps: double;
      entry: the desired relative accuracy (value of ε above);
             the value of *eps* should not be smaller than the machine accuracy;
i: double *i[0:nmax]*;
      exit: $i[n] = I_x(p,q+n)$ for $n=0,1,\dots,nmax$.

Procedures used: ixqfix, ixpfix.

```
public static void ibqplusn(double x, double p, double q, int nmax, double eps,
                            double i[])
{
    int n;

    if (x == 0.0 || x == 1.0)
        for (n=0; n<=nmax; n++) i[n]=x;
    else {
        if (x <= 0.5)
            ixpfix(x,p,q,nmax,eps,i);
        else {
            ixqfix(1.0-x,q,p,nmax,eps,i);
            for (n=0; n<=nmax; n++) i[n]=1.0-i[n];
        }
    }
}
```

H. ixqfix

The four auxiliary procedures *ixqfix, ixpfix, forward,* and *backward* are for procedures *incbeta, ibpplusn,* and *ibqplusn.*

    These auxiliary procedures are not described here. More information can be found in [G64] of *ibqplusn,* where the procedures *forward* and *backward* have the same name, while *ixqfix* and *ixpfix* are called "Isubx q fixed" and "Isubx p fixed", respectively. In the procedure *backward* we changed the starting value *nu* for the backward recurrence algorithm. The new value of *nu* is more realistic. Its computation is based on some asymptotic estimations. Also the initial value *r*=0 is changed into *r*=*x*.

Procedures used: incbeta, forward, backward.

```
public static void ixqfix(double x, double p, double q, int nmax, double eps,
                          double i[])
{
    int m,mmax;
    double s,iq0,iq1,q0;

    iq1=0.0;
    m=(int)Math.floor(q);
    s=q-m;
    q0 = (s > 0.0) ? s : s+1.0;
    mmax = (s > 0.0) ? m : m-1;
    iq0=incbeta(x,p,q0,eps);
    if (mmax > 0) iq1=incbeta(x,p,q0+1.0,eps);
    double iq[] = new double[mmax+1];
    forward(x,p,q0,iq0,iq1,mmax,iq);
    backward(x,p,q,iq[mmax],nmax,eps,i);
}
```

I. ixpfix

See the documentation of *ixqfix*.

Procedures used: incbeta, forward, backward.

```
public static void ixpfix(double x, double p, double q, int nmax, double eps,
                          double i[])
{
    int m,mmax;
    double s,p0,i0,i1,iq0,iq1;

    m=(int)Math.floor(p);
    s=p-m;
    p0 = (s > 0.0) ? s : s+1.0;
    mmax = (s > 0.0) ? m : m-1;
    i0=incbeta(x,p0,q,eps);
    i1=incbeta(x,p0,q+1.0,eps);
    double ip[] = new double[mmax+1];
    backward(x,p0,q,i0,mmax,eps,ip);
    iq0=ip[mmax];
    backward(x,p0,q+1.0,i1,mmax,eps,ip);
    iq1=ip[mmax];
    forward(x,p,q,iq0,iq1,nmax,i);
}
```

J. forward

See the documentation of *ixqfix*.

```
public static void forward(double x, double p, double q, double i0, double i1,
                           int nmax, double i[])
{
    int m,n;
    double y,r,s;

    i[0]=i0;
    if (nmax > 0) i[1]=i1;
    m=nmax-1;
    r=p+q-1.0;
    y=1.0-x;
    for (n=1; n<=m; n++) {
        s=(n+r)*y;
        i[n+1]=((n+q+s)*i[n]-s*i[n-1])/(n+q);
    }
}
```

K. backward

See the documentation of *ixqfix*.

```
public static void backward(double x, double p, double q, double i0, int nmax,
                            double eps, double i[])
{
    int m,n,nu;
    boolean finish;
    double r,pq,y,logx;
    double iapprox[] = new double[nmax+1];

    i[0]=i0;
    if (nmax > 0) {
        for (n=1; n<=nmax; n++) iapprox[n]=0.0;
        pq=p+q-1.0;
        logx=Math.log(x);
        r=nmax+(Math.log(eps)+q*Math.log(nmax))/logx;
        nu=(int)Math.floor(r-q*Math.log(r)/logx);
        while (true) {
            n=nu;
            r=x;
            while (true) {
                y=(n+pq)*x;
                r=y/(y+(n+p)*(1.0-r));
                if (n <= nmax) i[n]=r;
```

```
        n--;
        if (n < 1) break;
    }
    r=i0;
    for (n=1; n<=nmax; n++) r = i[n] *= r;
    finish=true;
    for (n=1; n<=nmax; n++)
        if (Math.abs((i[n]-iapprox[n])/i[n]) > eps) {
            for (m=1; m<=nmax; m++) iapprox[m]=i[m];
            nu += 5;
            finish=false;
            break;
        }
    if (finish) break;
    }
  }
}
```

6.4 Error function

A. errorfunction

Computes the error function *erf(x)* and complementary error function *erfc(x)* for a real argument, i.e.,

$$erf(x) = \frac{2}{\sqrt{\pi}} \int_0^x e^{-t^2} dt \quad \text{and} \quad erfc(x) = \frac{2}{\sqrt{\pi}} \int_x^\infty e^{-t^2} dt .$$

When $x > 26$ then *erf(x)* = 1 and *erfc(x)* = 0; when $x < -5.5$ then *erf(x)* = -1 and *erfc(x)* = 2. Over the range $|x| \le 0.5$, *erf(x)* is approximated by a Chebyshev rational function of the form

$$x \sum_{k=0}^{3} n_k x^{2k} \Big/ \sum_{k=0}^{3} d_k x^{2k}$$

and the relationship *erfc(x)* = *1 - erf(x)* is used. If $0.5 < x$ or $-5.5 \le x < 0.5$ then the function *nonexperfc(x)* = *exp(x²)* * *erfc(|x|)* is evaluated by means of a call of *nonexperfc*, the value of *erfc(|x|)* is recovered by multiplication by *exp(x²)*, and the relationship *erf(|x|)* = *1 - erfc(|x|)* is used.; if x lies in the second of the above ranges the further relationships

$$erf(x) = -erfc(|x|), \quad erfc(x) = 2 - erfc(|x|)$$

are used.

Procedure parameters:

                    void errorfunction (*x,erf,erfc*)

x: double;
         entry: the real argument of *erf(x)* and *erfc(x)*;
erf: double erf[0:0];
         exit: the value of *erf(x)*;

erfc: double *erfc[0:0]*;
      exit: the value of *erfc(x)*.

Procedure used: nonexperfc.

```
public static void errorfunction(double x, double erf[], double erfc[])
{
    if (x > 26.0) {
        erf[0] = 1.0;
        erfc[0] = 0.0;
        return;
    } else if (x < -5.5) {
        erf[0] = -1.0;
        erfc[0] = 2.0;
        return;
    } else {
        double absx,c,p,q;
        absx=Math.abs(x);
        if (absx <= 0.5) {
            c=x*x;
            p=((-0.356098437018154e-1*c+0.699638348861914e1)*c+
                0.219792616182942e2)*c+0.242667955230532e3;
            q=((c+0.150827976304078e2)*c+0.911649054045149e2)*c+0.215058875869861e3;
            erf[0] = x*p/q;
            erfc[0] = 1.0-erf[0];
        } else {
            erfc[0] = Math.exp(-x*x)*nonexperfc(absx);
            erf[0] = 1.0-erfc[0];
            if (x < 0.0) {
                erf[0] = -erf[0];
                erfc[0] = 2.0-erfc[0];
            }
        }
    }
}
```

B. nonexperfc

The error function *erf(x)* and complementary error function *erfc(x)* for a real argument are given by

$$erf(x) = \frac{2}{\sqrt{\pi}} \int_0^x e^{-t^2} dt \quad \text{and} \quad erfc(x) = \frac{2}{\sqrt{\pi}} \int_x^\infty e^{-t^2} dt \ .$$

nonexperfc computes *exp(x²)*erfc(x)*, i.e.,

$$non\exp erfc(x) = \frac{2e^{x^2}}{\sqrt{\pi}} \int_x^\infty e^{-t^2} dt \ .$$

When $|x| \leq 0.5$ then the function *erfc(x)* is evaluated by means of a call of *errorfunction*, and the value of *nonexperfc(x)* is recovered by multiplication by $exp(x^2)$. If $0.5 < |x| < 4$ then *nonexperfc($|x|$)* is approximated by a Chebyshev rational approximation (see [Cod69]) of the form

$$\sum_{k=0}^{4} n_k |x|^k \Big/ \sum_{k=0}^{4} d_k |x|^k$$

if, in addition, x is negative, the relationship

$$nonexperfc(x) = 2 * exp(x^2) - nonexperfc(|x|) \tag{1}$$

is used. If $|x| \geq 4$ then *nonexperfc(x)* is approximated by a Chebyshev rational approximation (see [Cod69]) of the form

$$\frac{1}{|x|}\left\{ \frac{1}{\sqrt{\pi}} + \frac{1}{x^2}\sum_{k=0}^{3} n_k x^{-2k} \Big/ \sum_{k=0}^{3} d_k x^{-2k} \right\}$$

and again if x is negative then relationship (1) is used.

Procedure parameters:
$$\text{double nonexperfc } (x)$$
nonexperfc: delivers the value of $exp(x^2)*erfc(x)$;
x: double;
              entry: the real argument of *nonexperfc*.

Procedure used: errorfunction.

```
public static double nonexperfc(double x)
{
    double absx,c,p,q;
    double erf[]  = new double[1];
    double erfc[] = new double[1];

    absx=Math.abs(x);
    if (absx <= 0.5) {
        errorfunction(x,erf,erfc);
        return Math.exp(x*x)*erfc[0];
    } else if (absx < 4.0) {
        c=absx;
        p=((((((-0.136864857382717e-6*c+0.564195517478974e0)*c+
            0.721175825088309e1)*c+0.431622272220567e2)*c+
            0.152989285046940e3)*c+0.339320816734344e3)*c+
            0.451918953711873e3)*c+0.300459261020162e3;
        q=((((((c+0.127827273196294e2)*c+0.770001529352295e2)*c+
            0.277585444743988e3)*c+0.638980264465631e3)*c+
            0.931354094850610e3)*c+0.790950925327898e3)*c+0.300459260956983e3;
        return ((x > 0.0) ? p/q : Math.exp(x*x)*2.0-p/q);
    } else {
        c=1.0/x/x;
        p=(((0.223192459734185e-1*c+0.278661308609648e0)*c+
            0.226956593539687e0)*c+0.494730910623251e-1)*c+0.299610707703542e-2;
        q=(((c+0.198733201817135e1)*c+0.105167510706793e1)*c+
            0.191308926107830e0)*c+0.106209230528468e-1;
```

```
        c=(c*(-p)/q+0.564189583547756)/absx;
        return ((x > 0.0) ? c : Math.exp(x*x)*2.0-c);
    }
}
```

C. inverseerrorfunction

Evaluates the inverse error function $y(x)$, where

$$x = \frac{2}{\sqrt{\pi}} \int_0^{y(x)} e^{-t^2} dt$$

the values of x and $1-x$ being supplied (for values of x near 1, loss of accuracy results from the formation of $1-x$; this loss is avoided by defining y as a function of $1-x$ for such values of x, provided the supplied value of $1-x$ is sufficiently accurate).

When $|x| \le 0.8$, a telescoped power series is used to evaluate $y(x)$: y satisfies the equation

$$\left(\frac{dy}{dx}\right)^{-1} = 2 \int_0^x y(t)dt - \frac{2}{\sqrt{\pi}}$$

so that if

$$y(x) = \sum_{n=0}^{\infty} C_n x^{2n-1}$$

then the $\{C_n\}$ may be extracted from the equation

$$1 + \left(\sum_{m=1}^{\infty} (2m-1)C_m x^{2m-2}\right)\left(\sum_{n=1}^{\infty} \frac{C_n x^{2n}}{n} - \frac{2}{\sqrt{\pi}}\right) = 0$$

and from the $\{C_n\}$ a Chebyshev series (see [Str68]) of the form

$$y(x) \approx x\left[\xi_0 + \sum_{n=1}^{22} \xi_n T_n\left(\frac{x^2}{0.32} - 1\right)\right]$$

valid for $|x| \le 0.8$ may be obtained. Over the range $0.8 < |x| \le 0.9975$, an expansion of the form $y(x) = \beta(x)R(x)$, where

$$\beta(x) = \sqrt{-\ln(1-x^2)}, \qquad R(x) = sign(x)\sum_{n=0}^{j} \lambda_n T_n\left(D_1\beta(x) + D_2\right)$$

with $j=15$ is used; over the range $2.5*10^{-3} > 1-x \ge 0.5*10^{-15}$ a similar expansion with $j=23$ is used, and over the range $0.5*10^{-15} > 1-x$, yet another similar expansion with $j=12$ is used.

Procedure parameters:

                void inverseerrorfunction (x,oneminx,inverf)

x: double;
                entry: the argument of *inverseerrorfunction(x)*; it is necessary that
                -1 < x < 1;
                if $|x| > 0.8$ then the value of x is not used in the procedure;

oneminx: double;
            entry: if $|x| \leq 0.8$ then the value of *oneminx* is not used in the
                    procedure;
                    if $|x| > 0.8$ then *oneminx* has to contain the value of $1-|x|$;
                    in the case that $|x|$ is in the neighborhood of 1, cancellation of
                    digits take place in the calculation of $1-|x|$; if the value $1-|x|$
                    is known exactly from another source then *oneminx* has to
                    contain this value, which will give better results;
inverf: double *inverf[0:0]*;
            exit: the result of the procedure.

Procedure used: chepolsum.

```
public static void inverseerrorfunction(double x, double oneminx, double inverf[])
{
    double absx,p,betax;
    double a[] = new double[24];

    absx=Math.abs(x);
    if (absx > 0.8 && oneminx > 0.2) oneminx=0.0;
    if (absx <= 0.8) {
        a[0]  = 0.992885376618941;  a[1]  = 0.120467516143104;
        a[2]  = 0.016078199342100;  a[3]  = 0.002686704437162;
        a[4]  = 0.000499634730236;  a[5]  = 0.000098898218599;
        a[6]  = 0.000020391812764;  a[7]  = 0.000004327271618;
        a[8]  = 0.000000938081413;  a[9]  = 0.000000206734720;
        a[10] = 0.000000046159699;  a[11] = 0.000000010416680;
        a[12] = 0.000000002371501;  a[13] = 0.000000000543928;
        a[14] = 0.000000000125549;  a[15] = 0.000000000029138;
        a[16] = 0.000000000006795;  a[17] = 0.000000000001591;
        a[18] = 0.000000000000374;  a[19] = 0.000000000000088;
        a[20] = 0.000000000000021;  a[21] = 0.000000000000005;
        inverf[0] = Algebraic_eval.chepolsum(21,x*x/0.32-1.0,a)*x;
    } else if (oneminx >= 25.0e-4) {
        a[0]  =  0.912158803417554;  a[1]  = -0.016266281867664;
        a[2]  =  0.000433556472949;  a[3]  =  0.000214438570074;
        a[4]  =  0.000002625751076;  a[5]  = -0.000003021091050;
        a[6]  = -0.000000012406062;  a[7]  =  0.000000062406609;
        a[8]  = -0.000000000540125;  a[9]  = -0.000000001423208;
        a[10] =  0.000000000034384;  a[11] =  0.000000000033584;
        a[12] = -0.000000000001458;  a[13] = -0.000000000000810;
        a[14] =  0.000000000000053;  a[15] =  0.000000000000020;
        betax=Math.sqrt(-Math.log((1.0+absx)*oneminx));
        p = -1.54881304237326*betax+2.56549012314782;
        p=Algebraic_eval.chepolsum(15,p,a);
        inverf[0] = (x < 0.0) ? -betax*p : betax*p;
    } else if (oneminx >= 5.0e-16) {
        a[0]  =  0.956679709020493;  a[1]  = -0.023107004309065;
        a[2]  = -0.004374236097508;  a[3]  = -0.000576503422651;
        a[4]  = -0.000010961022307;  a[5]  =  0.000025108547025;
```

```
    a[6]  =  0.000010562336068;    a[7]  =  0.000002754412330;
    a[8]  =  0.000000432484498;    a[9]  = -0.000000020530337;
    a[10] = -0.000000043891537;    a[11] = -0.000000017684010;
    a[12] = -0.000000003991289;    a[13] = -0.000000000186932;
    a[14] =  0.000000000272923;    a[15] =  0.000000000132817;
    a[16] =  0.000000000031834;    a[17] =  0.000000000001670;
    a[18] = -0.000000000002036;    a[19] = -0.000000000000965;
    a[20] = -0.000000000000220;    a[21] = -0.000000000000010;
    a[22] =  0.000000000000014;    a[23] =  0.000000000000006;
    betax=Math.sqrt(-Math.log((1.0+absx)*oneminx));
    p = -0.559457631329832*betax+2.28791571626336;
    p=Algebraic_eval.chepolsum(23,p,a);
    inverf[0] = (x < 0.0) ? -betax*p : betax*p;
} else if (oneminx >= Double.MIN_VALUE) {
    a[0]  =  0.988575064066189;    a[1]  =  0.010857705184599;
    a[2]  = -0.001751165102763;    a[3]  =  0.000021196993207;
    a[4]  =  0.000015664871404;    a[5]  = -0.000000519041687;
    a[6]  = -0.000000037135790;    a[7]  =  0.000000001217431;
    a[8]  = -0.000000000176812;    a[9]  = -0.000000000011937;
    a[10] =  0.000000000000380;    a[11] = -0.000000000000066;
    a[12] = -0.000000000000009;
    betax=Math.sqrt(-Math.log((1.0+absx)*oneminx));
    p = -9.19999235883015/Math.sqrt(betax)+2.79499082012460;
    p=Algebraic_eval.chepolsum(12,p,a);
    inverf[0] = (x < 0.0) ? -betax*p : betax*p;
} else
    inverf[0] = (x > 0.0) ? 26.0 : -26.0;
}
```

D. fresnel

Evaluates the Fresnel integrals

$$S(x) = \int_0^x \sin\left(\frac{\pi}{2}t^2\right)dt, \qquad C(x) = \int_0^x \cos\left(\frac{\pi}{2}t^2\right)dt .$$

If $|x| \le 1.2$ then $S(x)$ is approximated by means of a Chebyshev rational function (see [Cod68]) of the form

$$x^3 \sum_{k=0}^{j} n_k x^{4k} \Bigg/ \sum_{k=0}^{j} d_k x^{4k}$$

and $C(x)$ by a similar function of the form

$$x \sum_{k=0}^{j} n_k x^{4k} \Bigg/ \sum_{k=0}^{j} d_k x^{4k}$$

where $j=4$. When $1.2 < |x| \le 1.6$, similar approximations with $j=5$ are used. Over the range $1.6 < |x|$, the functions $f(x)$ and $g(x)$ occurring in the formulae

$$S(x) = \frac{1}{2} - f(x)\cos(\pi x^2/2) - g(x)\sin(\pi x^2/2)$$
$$C(x) = \frac{1}{2} + f(x)\sin(\pi x^2/2) - g(x)\cos(\pi x^2/2)$$

are evaluated by means of a call of *fg*, and the values of *S(x)* and *C(x)* are obtained by use of these relationships.

Procedure parameters:
$$\text{void fresnel } (x,c,s)$$

x: double;
           entry: the real argument of *C(x)* and *S(x)*;
c: double *c[0:0]*;
           exit: the value of *C(x)*;
s: double *s[0:0]*;
           exit: the value of *S(x)*.

Procedure used: fg.

```
public static void fresnel(double x, double c[], double s[])
{
    double absx,x3,x4,a,p,q,c1,s1;
    double f[] = new double[1];
    double g[] = new double[1];

    absx=Math.abs(x);
    if (absx <= 1.2) {
        a=x*x;
        x3=a*x;
        x4=a*a;
        p=(((5.47711385682687e-6*x4-5.28079651372623e-4)*x4+
                1.76193952543491e-2)*x4-1.99460898826184e-1)*x4+1.0;
        q=(((1.18938901422876e-7*x4+1.55237885276994e-5)*x4+
                1.09957215025642e-3)*x4+4.72792112010453e-2)*x4+1.0;
        c[0] = x*p/q;
        p=(((6.71748466625141e-7*x4-8.45557284352777e-5)*x4+
                3.87782123463683e-3)*x4-7.07489915144523e-2)*x4+5.23598775598299e-1;
        q=(((5.95281227678410e-8*x4+9.62690875939034e-6)*x4+
                8.17091942152134e-4)*x4+4.11223151142384e-2)*x4+1.0;
        s[0] = x3*p/q;
    } else if (absx <= 1.6) {
        a=x*x;
        x3=a*x;
        x4=a*a;
        p=((((-5.68293310121871e-8*x4+1.02365435056106e-5)*x4-
                6.71376034694922e-4)*x4+1.91870279431747e-2)*x4-
                2.07073360335324e-1)*x4+1.00000000000111e0;
        q=((((4.41701374065010e-10*x4+8.77945377892369e-8)*x4+
                1.01344630866749e-5)*x4+7.88905245052360e-4)*x4+
                3.96667496952323e-2)*x4+1.0;
        c[0] = x*p/q;
        p=((((-5.76765815593089e-9*x4+1.28531043742725e-6)*x4-
                1.09540023911435e-4)*x4+4.30730526504367e-3)*x4-
                7.37766914010191e-2)*x4+5.23598775598344e-1;
        q=((((2.05539124458580e-10*x4+5.03090581246612e-8)*x4+
```

```
                6.87086265718620e-6)*x4+6.18224620195473e-4)*x4+
                3.53398342767472e-2)*x4+1.0;
        s[0] = x3*p/q;
    } else if (absx < 1.0e15) {
        fg(x,f,g);
        a=x*x;
        a=(a-Math.floor(a/4.0)*4.0)*1.57079632679490;
        c1=Math.cos(a);
        s1=Math.sin(a);
        a = (x < 0.0) ? -0.5 : 0.5;
        c[0] = f[0]*s1-g[0]*c1+a;
        s[0] = -f[0]*c1-g[0]*s1+a;
    } else
        c[0] = s[0] = ((x > 0.0) ? 0.5 : -0.5);
}
```

E. fg

Evaluates the functions $f(x)$, $g(x)$ related to the Fresnel integrals by means of the formulae

$$f(x) = \{½ - S(x)\}\cos(\pi x^2/2) - \{½ - C(x)\}\sin(\pi x^2/2)$$
$$g(x) = \{½ - C(x)\}\cos(\pi x^2/2) + \{½ - S(x)\}\sin(\pi x^2/2).$$

When $|x| \leq 1.6$ the functions $S(x)$ and $C(x)$ are evaluated by means of a call of *fresnel*, and the values of $f(x)$ and $g(x)$ are obtained by use of the above relationships. When $1.6 < |x| \leq 1.9$, $f(x)$ is approximated by use of a Chebyshev rational function (see [Cod68]) of the form

$$\frac{1}{x}\sum_{k=0}^{i} n_k x^{-4k} \bigg/ \sum_{k=0}^{i} d_k x^{-4k}$$

and $g(x)$ by use of a similar function of the form

$$\frac{1}{x^3}\sum_{k=0}^{j} n_k x^{-4k} \bigg/ \sum_{k=0}^{j} d_k x^{-4k}$$

with $i=4$ and $j=5$. When $|x| \geq 2.4$, similar expansions with $i=j=5$ are used. When $|x| > 2.4$, the approximating functions are

$$\frac{1}{x}\left(\frac{1}{\pi} + \frac{1}{x^4}\sum_{k=0}^{5} n_k x^{-4k} \bigg/ \sum_{k=0}^{5} d_k x^{-4k}\right)$$

for $f(x)$ and

$$\frac{1}{x^3}\left(\frac{1}{\pi^2} + \frac{1}{x^4}\sum_{k=0}^{6} n_k x^{-4k} \bigg/ \sum_{k=0}^{6} d_k x^{-4k}\right)$$

for $g(x)$.

Procedure parameters:

$$\text{void fg }(x,f,g)$$

x: double;

entry: the real argument of *f(x)* and *g(x)*;
f: double *f[0:0]*;
exit: the value of *f(x)*;
g: double *g[0:0]*;
exit: the value of *g(x)*.

Procedure used: fresnel.

```
public static void fg(double x, double f[], double g[])
{
    double absx,c1,s1,a,xinv,x3inv,c4,p,q;
    double c[] = new double[1];
    double s[] = new double[1];

    absx=Math.abs(x);
    if (absx <= 1.6) {
        fresnel(x,c,s);
        a=x*x*1.57079632679490;
        c1=Math.cos(a);
        s1=Math.sin(a);
        a = (x < 0.0) ? -0.5 : 0.5;
        p=a-c[0];
        q=a-s[0];
        f[0] = q*c1-p*s1;
        g[0] = p*c1+q*s1;
    } else if (absx <= 1.9) {
        xinv=1.0/x;
        a=xinv*xinv;
        x3inv=a*xinv;
        c4=a*a;
        p=(((1.35304235540388e1*c4+6.98534261601021e1)*c4+
                4.80340655577925e1)*c4+8.03588122803942e0)*c4+3.18309268504906e-1;
        q=(((6.55630640083916e1*c4+2.49561993805172e2)*c4+
                1.57611005580123e2)*c4+2.55491618435795e1)*c4+1.0;
        f[0] = xinv*p/q;
        p=((((2.05421432498501e1*c4+1.96232037971663e2)*c4+
                1.99182818678903e2)*c4+5.31122813480989e1)*c4+
                4.44533827550512e0)*c4+1.01320618810275e-1;
        q=((((1.01379483396003e3*c4+3.48112147856545e3)*c4+
                2.54473133181822e3)*c4+5.83590575716429e2)*c4+
                4.53925019673689e1)*c4+1.0;
        g[0] = x3inv*p/q;
    } else if (absx <= 2.4) {
        xinv=1.0/x;
        a=xinv*xinv;
        x3inv=a*xinv;
        c4=a*a;
        p=(((((7.17703249365140e2*c4+3.09145161574430e3)*c4+
                1.93007640786716e3)*c4+3.39837134926984e2)*c4+
                1.95883941021969e1)*c4+3.18309881822017e-1;
```

```
        q=((((3.36121699180551e3*c4+1.09334248988809e4)*c4+
            6.33747155851144e3)*c4+1.08535067500650e3)*c4+
            6.18427138172887e1)*c4+1.0;
        f[0] = xinv*p/q;
        p=((((3.13330163068756e2*c4+1.59268006085354e3)*c4+
            9.08311749529594e2)*c4+1.40959617911316e2)*c4+
            7.11205001789783e0)*c4+1.01321161761805e-1;
        q=((((1.15149832376261e4*c4+2.41315567213370e4)*c4+
            1.06729678030581e4)*c4+1.49051922797329e3)*c4+
            7.17128596939302e1)*c4+1.0;
        g[0] = x3inv*p/q;
    } else {
        xinv=1.0/x;
        a=xinv*xinv;
        x3inv=a*xinv;
        c4=a*a;
        p=((((2.61294753225142e4*c4+6.13547113614700e4)*c4+
            1.34922028171857e4)*c4+8.16343401784375e2)*c4+
            1.64797712841246e1)*c4+9.67546032967090e-2;
        q=((((1.37012364817226e6*c4+1.00105478900791e6)*c4+
            1.65946462621853e5)*c4+9.01827596231524e3)*c4+
            1.73871690673649e2)*c4+1.0;
        f[0] = (c4*(-p)/q+0.318309886183791)*xinv;
        p=(((((1.72590224654837e6*c4+6.66907061668636e6)*c4+
            1.77758950838030e6)*c4+1.35678867813756e5)*c4+
            3.87754141746378e3)*c4+4.31710157823358e1)*c4+1.53989733819769e-1;
        q=(((((1.40622441123580e8*c4+9.38695862531635e7)*c4+
            1.62095600500232e7)*c4+1.02878693056688e6)*c4+
            2.69183180396243e4)*c4+2.86733194975899e2)*c4+1.0;
        g[0] = (c4*(-p)/q+0.101321183642338)*x3inv;
    }
}
```

6.5 Bessel functions of integer order

6.5.1 Bessel functions J and Y

A. bessj0

Computes the value of the ordinary Bessel function of the first kind of order zero $J_0(x)$. When $|x| < 8$, $J_0(x)$ is evaluated by use of a Chebyshev polynomial approximation [Cle62] in the variable $2(x^2/16 - 1)$, and $|x| \geq 8$ by use of the formula

$$J_0(x) = \sqrt{\frac{2}{\pi |x|}} \left\{ P_0(x) \cos\left(x - \frac{9\pi}{4}\right) - Q_0(x) \sin\left(x - \frac{9\pi}{4}\right) \right\}$$

(*besspq0* being called to evaluate the functions P_0 and Q_0).

Procedure parameters:
$$\text{double bessj0 } (x)$$

bessj0:   delivers the ordinary Bessel function of the first kind of order zero with argument x;

x: double;

             entry: the argument of the Bessel function.

Procedure used: besspq0.

```
public static double bessj0(double x)
{
    if (x == 0.0) return 1.0;
    if (Math.abs(x) < 8.0) {
        int i;
        double z,z2,b0,b1,b2;
        double ar[]={-0.75885e-15, 0.4125321e-13,
                    -0.194383469e-11, 0.7848696314e-10, -0.267925353056e-8,
                    0.7608163592419e-7, -0.176194690776215e-5,
                    0.324603288210051e-4, -0.46062616620628e-3,
                    0.48191800694676e-2, -0.34893769411409e-1,
                    0.158067102332097, -0.37009499387265, 0.265178613203337,
                    -0.872344235285222e-2};
        x /= 8.0;
        z=2.0*x*x-1.0;
        z2=z+z;
        b1=b2=0.0;
        for (i=0; i<=14; i++) {
            b0=z2*b1-b2+ar[i];
            b2=b1;
            b1=b0;
        }
        return z*b1-b2+0.15772797147489;
    } else {
        double c,cosx,sinx;
        double p0[] = new double[1];
        double q0[] = new double[1];
        x=Math.abs(x);
        c=0.797884560802865/Math.sqrt(x);
        cosx=Math.cos(x-0.706858347057703e1);
        sinx=Math.sin(x-0.706858347057703e1);
        besspq0(x,p0,q0);
        return c*(p0[0]*cosx-q0[0]*sinx);
    }
}
```

B. bessj1

Computes the value of the ordinary Bessel function of the first kind of order one $J_1(x)$. When $|x| < 8$, $J_1(x)$ is evaluated by use of a Chebyshev polynomial approximation [Cle62] in the variable $2(x^2/16 - 1)$, and $|x| \geq 8$ by use of the formula

$$J_1(x) = sign(x) \sqrt{\frac{2}{\pi |x|}} \left\{ P_1(x) \sin\left(x - \frac{9\pi}{4} \right) + Q_1(x) \cos\left(x - \frac{9\pi}{4} \right) \right\}$$

(*besspq1* being called to evaluate the functions P_1 and Q_1).

Procedure parameters:
$$\text{double bessj1 } (x)$$
bessj1:   delivers the ordinary Bessel function of the first kind of order one with
             argument x;
x: double;
             entry: the argument of the Bessel function.

Procedure used: besspq1.

```
public static double bessj1(double x)
{
    if (x == 0.0) return 1.0;
    if (Math.abs(x) < 8.0) {
        int i;
        double z,z2,b0,b1,b2;
        double ar[]={-0.19554e-15, 0.1138572e-13,
                    -0.57774042e-12, 0.2528123664e-10, -0.94242129816e-9,
                    0.2949707007278e-7, -0.76175878054003e-6,
                    0.158870192399321e-4, -0.260444389348581e-3,
                    0.324027018268386e-2, -0.291755248061542e-1,
                    0.177709117239728e0, -0.661443934134543e0,
                    0.128799409885768e1, -0.119180116054122e1};
        x /= 8.0;
        z=2.0*x*x-1.0;
        z2=z+z;
        b1=b2=0.0;
        for (i=0; i<=14; i++) {
            b0=z2*b1-b2+ar[i];
            b2=b1;
            b1=b0;
        }
        return x*(z*b1-b2+0.648358770605265);
    } else {
        int sgnx;
        double c,cosx,sinx;
        double p1[] = new double[1];
        double q1[] = new double[1];
        sgnx = (x > 0.0) ? 1 : -1;
```

```
    x=Math.abs(x);
    c=0.797884560802865/Math.sqrt(x);
    cosx=Math.cos(x-0.706858347057703e1);
    sinx=Math.sin(x-0.706858347057703e1);
    besspq1(x,p1,q1);
    return sgnx*c*(p1[0]*sinx+q1[0]*cosx);
  }
}
```

C. bessj

Generates an array of ordinary Bessel functions of the first kind of order k, $J_k(x)$, $k=0,...,n$. The method used [G67] is suitable for the determination of a sequence of numbers $f_0, f_1, ...$ which satisfy a recursion of the form

$$f_{k+1} + a_k f_k + b_k f_{k-1} = 0 \tag{1}$$

and decrease in magnitude with k more rapidly than any other solution of this difference equation, and in addition satisfy a relationship of the form

$$\sum_{k=0}^{\infty} \lambda_k f_k = s .$$

Setting $r_k = f_{k+1} / f_k$, formally

$$r_{k-1} = \frac{-b_k}{a_k -} \frac{b_{k+1}}{a_{k+1} -} \cdots$$

and with

$$s_{k-1} = \left(\sum_{m=k}^{\infty} \lambda_m f_m \right) \Big/ f_{k-1}$$

$$s_{k-1} = r_{k-1} \left(\lambda_k + s_k \right) .$$

By definition $f_0 = s/(\lambda_0+s_0)$. For suitably large v, the numbers

$$r_{k-1}^{(v)} = \frac{-b_k}{a_k -} \frac{b_{k+1}}{a_{k+1} -} \cdots \frac{b_v}{a_v}$$

$$s_{k-1}^{(v)} = \sum_{m=k+1}^{v} \lambda_m r_k^{(v)} r_{k+1}^{(v)} r_{m-1}^{(v)}$$

are computed by use of the recursions

$$r_v^{(v)} = 0, \quad r_{k-1}^{(v)} = -b_k/(a_k+r_k^{(v)})$$
$$s_v^{(v)} = 0, \quad s_{k-1}^{(v)} = r_{k-1}^{(v)}(\lambda_k+s_k^{(v)})$$

for $k=v,v-1,...,1$. Thereafter the required numbers $f_0,f_1,...,f_n$ are recovered by use of the relationships

$$f_0^{(v)} = s / (\lambda_0+s_0^{(v)}), \quad f_k^{(v)} = r_{k-1}^{(v)} f_{k-1}^{(v)}, \quad (k=1,...,n).$$

The functions $J_{a+k}(x)$ satisfy the recursion

$$J_{a+k+1}(x) - \{2(a+k)/x\}J_{a+k}(x) + J_{a+k-1}(x) = 0$$

and also the relationship

$$J_a(x) + \sum_{m=1}^{\infty} \frac{(a+2m)\,\Gamma(a+m)}{m!\,\Gamma(a+1)} J_{a+2m}(x) = \frac{(x/2)^2}{\Gamma(a+1)}.$$

Thus when $a = 0$, $a_k = 2k/x$, $b_k = -1$ and $\lambda_0 = 1$, $\lambda_{2m-1} = 0$, $\lambda_{2m} = 2$ $(m=1,2,...)$, $s = 1$ may be taken in the above.

Procedure parameters:
$$\text{void bessj } (x,n,j)$$

x: double;
         entry: the argument of the Bessel functions;
n: int;
         entry: the upper bound of the indices of array j; $n \geq 0$;
j: double $j[0:n]$;
         exit: $j[k]$ is the ordinary Bessel function of the first kind of order k and
                  argument x.

Procedure used: start.

```java
public static void bessj(double x, int n, double j[])
{
    if (x == 0.0) {
        j[0]=1.0;
        for (; n>=1; n--) j[n]=0.0;
    } else {
        int l,m,nu,signx;
        double x2,r,s;
        signx = (x > 0.0) ? 1 : -1;
        x=Math.abs(x);
        r=s=0.0;
        x2=2.0/x;
        l=0;
        nu=start(x,n,0);
        for (m=nu; m>=1; m--) {
            r=1.0/(x2*m-r);
            l=2-l;
            s=r*(l+s);
            if (m <= n) j[m]=r;
        }
        j[0]=r=1.0/(1.0+s);
        for (m=1; m<=n; m++) r = j[m] *= r;
        if (signx < 0.0)
            for (m=1; m<=n; m += 2) j[m] = -j[m];
    }
}
```

D. bessy01

Computes the ordinary Bessel functions of the second kind of orders zero and one: $Y_0(x)$ and $Y_1(x)$ for $x > 0$. When $x < 8$, $Y_0(x)$ and $Y_1(x)$ are determined from relationships of the form

$$Y_0(x) = (2/\pi)\ln(x)J_0(x) + y_0(x)$$
$$Y_1(x) = (2/\pi)\ln(x)J_1(x) - 2/(\pi x) + xy_1(x)$$

where $y_0(x)$ and $y_1(x)$ are approximated by use of Chebyshev polynomials [Cle62]. When $x \geq 8$, the relationships

$$Y_0(x) = (2/(\pi x))^{1/2}\{P_0(x)\sin(x - 9\pi/4) + Q_0(x)\cos(x - 9\pi/4)\}$$
$$Y_1(x) = (2/(\pi x))^{1/2}\{Q_1(x)\sin(x - 9\pi/4) - P_1(x)\cos(x - 9\pi/4)\}$$

are used.

Procedure parameters:

$$\text{void bessy01 } (x, y0, y1)$$

x: double;

         entry: the argument of the Bessel function; $x > 0$;

$y0$:   double $y0[0:0]$;

         exit: $y0$ has the value of the ordinary Bessel function of the second kind of order zero with argument x;

$y1$:   double $y1[0:0]$;

         exit: $y1$ has the value of the ordinary Bessel function of the second kind of order one with argument x.

Procedures used: bessj0, bessj1, besspq0, besspq1.

```
public static void bessy01(double x, double y0[], double y1[])
{
    if (x < 8.0) {
        int i;
        double z,z2,c,lnx,b0,b1,b2;
        double ar1[]={0.164349e-14, -0.8747341e-13,
                    0.402633082e-11, -0.15837552542e-9, 0.524879478733e-8,
                    -0.14407233274019e-6, 0.32065325376548e-5,
                    -0.563207914105699e-4, 0.753113593257774e-3,
                    -0.72879624795521e-2, 0.471966895957634e-1,
                    -0.177302012781143, 0.261567346255047,
                    0.179034314077182, -0.274474305529745};
        double ar2[]={0.42773e-15, -0.2440949e-13,
                    0.121143321e-11, -0.5172121473e-10, 0.187547032473e-8,
                    -0.5688440039919e-7, 0.141662436449235e-5,
                    -0.283046401495148e-4, 0.440478629867099e-3,
                    -0.51316411610611e-2, 0.423191803533369e-1,
                    -0.226624991556755, 0.675615780772188,
                    -0.767296362886646, -0.128697384381350};
        c=0.636619772367581;
        lnx=c*Math.log(x);
        c /= x;
        x /= 8.0;
```

```
        z=2.0*x*x-1.0;
        z2=z+z;
        b1=b2=0.0;
        for (i=0; i<=14; i++) {
            b0=z2*b1-b2+ar1[i];
            b2=b1;
            b1=b0;
        }
        y0[0] = lnx*bessj0(8.0*x)+z*b1-b2-0.33146113203285e-1;
        b1=b2=0.0;
        for (i=0; i<=14; i++) {
            b0=z2*b1-b2+ar2[i];
            b2=b1;
            b1=b0;
        }
        y1[0] = lnx*bessj1(8.0*x)-c+x*(z*b1-b2+0.2030410588593425e-1);
    } else {
        double c,cosx,sinx;
        double p0[] = new double[1];
        double q0[] = new double[1];
        double p1[] = new double[1];
        double q1[] = new double[1];
        c=0.797884560802865/Math.sqrt(x);
        besspq0(x,p0,q0);
        besspq1(x,p1,q1);
        x -= 0.706858347057703e1;
        cosx=Math.cos(x);
        sinx=Math.sin(x);
        y0[0] = c*(p0[0]*sinx+q0[0]*cosx);
        y1[0] = c*(q1[0]*sinx-p1[0]*cosx);
    }
}
```

E. bessy

Generates an array of ordinary Bessel functions of the second kind of order k, $Y_k(x)$, $k=0,...,n$, for $x > 0$.

The functions $Y_0(x)$ and (if $n > 0$) $Y_1(x)$ are evaluated by means of a call of *bessy01*, and the recursion
$$Y_{k+1}(x) = (2k/x)Y_k(x) - Y_{k-1}(x)$$
is used.

Procedure parameters:
$$\text{void bessy } (x,n,y)$$

x: double;
            entry: this argument should satisfy $x > 0$;
n: int;
            entry: the upper bound of the indices of array y; $n \geq 0$;

y: double $y[0:n]$;
         exit: $y[k]$ is the value of the ordinary Bessel function of the second
                   kind of order k and argument x, $k=0,...,n$.

Procedure used: bessy01.

```
public static void bessy(double x, int n, double y[])
{
    int i;
    double y0,y1,y2;
    double tmp1[] = new double[1];
    double tmp2[] = new double[1];

    bessy01(x,tmp1,tmp2);
    y0 = tmp1[0];
    y1 = tmp2[0];
    y[0]=y0;
    if (n > 0) y[1]=y1;
    x=2.0/x;
    for (i=2; i<=n; i++) {
        y[i]=y2=(i-1)*x*y1-y0;
        y0=y1;
        y1=y2;
    }
}
```

F. besspq0

This procedure is an auxiliary procedure for the computation of the ordinary
Bessel functions of order zero for large values of their argument.

$besspq0$ computes the values of the functions $P_0(x)$, $Q_0(x)$ occurring in the
formulae

$$J_0(x) = (2/(\pi x))^{1/2}\{P_0(x)\cos(x - \pi/4) - Q_0(x)\sin(x - \pi/4)\}$$
$$Y_0(x) = (2/(\pi x))^{1/2}\{P_0(x)\sin(x - \pi/4) + Q_0(x)\cos(x - \pi/4)\}.$$

When $|x| < 8$, the above equations are used in the form

$$P_0(x) = (\pi x/2)^{1/2}\{Y_0(x)\sin(x - \pi/4) + J_0(x)\cos(x - \pi/4)\}$$
$$Q_0(x) = (\pi x/2)^{1/2}\{Y_0(x)\cos(x - \pi/4) - J_0(x)\sin(x - \pi/4)\}$$

($bessj0$ and $bessy0$ being used to evaluate the functions J_0 and Y_0) and when
$|x| \geq 8$, $P_0(x)$ and $Q_0(x)$ are evaluated by use of Chebyshev polynomial
approximations [Cle62] in the variable $2(128/x^2 - 1)$.

Procedure parameters:

$$\text{void besspq0 } (x,p0,q0)$$

x: double;
          entry: this argument should satisfy $x > 0$;
$p0$:   double $p0[0:0]$;
          exit: the value of $P_0(x)$;

q0:   double *q0[0:0]*;
          exit: the value of $Q_0(x)$.

Procedures used: bessj0, bessy01.

```
public static void besspq0(double x, double p0[], double q0[])
{
    if (x < 8.0) {
        double b,cosx,sinx;
        double j0x[] = new double[1];
        double y0[] = new double[1];
        b=Math.sqrt(x)*1.25331413731550;
        bessy01(x,y0,j0x);
        j0x[0]=bessj0(x);
        x -= 0.785398163397448;
        cosx=Math.cos(x);
        sinx=Math.sin(x);
        p0[0] = b*(y0[0]*sinx+j0x[0]*cosx);
        q0[0] = b*(y0[0]*cosx-j0x[0]*sinx);
    } else {
        int i;
        double x2,b0,b1,b2,y;
        double ar1[]={-0.10012e-15, 0.67481e-15, -0.506903e-14,
                      0.4326596e-13, -0.43045789e-12, 0.516826239e-11,
                      -0.7864091377e-10, 0.163064646352e-8, -0.5170594537606e-7,
                      0.30751847875195e-5, -0.536522046813212e-3};
        double ar2[]={-0.60999e-15, 0.425523e-14,
                      -0.3336328e-13, 0.30061451e-12, -0.320674742e-11,
                      0.4220121905e-10, -0.72719159369e-9, 0.1797245724797e-7,
                      -0.74144984110606e-6, 0.683851994261165e-4};
        y=8.0/x;
        x=2.0*y*y-1.0;
        x2=x+x;
        b1=b2=0.0;
        for (i=0; i<=10; i++) {
            b0=x2*b1-b2+ar1[i];
            b2=b1;
            b1=b0;
        }
        p0[0] = x*b1-b2+0.99946034934752;
        b1=b2=0.0;
        for (i=0; i<=9; i++) {
            b0=x2*b1-b2+ar2[i];
            b2=b1;
            b1=b0;
        }
        q0[0] = (x*b1-b2-0.015555854605337)*y;
    }
}
```

G. besspq1

This procedure is an auxiliary procedure for the computation of the ordinary Bessel functions of order one for large values of their argument.

besspq1 computes the values of the functions $P_1(x)$, $Q_1(x)$ occurring in the formulae

$$J_1(x) = (2/(\pi x))^{1/2}\{P_1(x)\cos(x - 3\pi/4) - Q_1(x)\sin(x - 3\pi/4)\}$$
$$Y_1(x) = (2/(\pi x))^{1/2}\{P_1(x)\sin(x - 3\pi/4) + Q_1(x)\cos(x - 3\pi/4)\}.$$

When $|x| < 8$, the above equations are used in the form

$$P_1(x) = (\pi x/2)^{1/2}\{Y_1(x)\sin(x - 3\pi/4) + J_1(x)\cos(x - 3\pi/4)\}$$
$$Q_1(x) = (\pi x/2)^{1/2}\{Y_1(x)\cos(x - 3\pi/4) - J_1(x)\sin(x - 3\pi/4)\}$$

(*bessj1* and *bessy1* being used to evaluate the functions J_1 and Y_1) and when $|x| \geq 8$, $P_1(x)$ and $Q_1(x)$ are evaluated by use of Chebyshev polynomial approximations [Cle62] in the variable $2(128/x^2 - 1)$.

Procedure parameters:
$$\text{void besspq1 } (x, p1, q1)$$

x: double;
          entry: this argument should satisfy $x > 0$;
p1:   double $p1[0:0]$;
          exit: the value of $P_1(x)$;
q1:   double $q1[0:0]$;
          exit: the value of $Q_1(x)$.

Procedures used: bessj1, bessy01.

```
public static void besspq1(double x, double p1[], double q1[])
{
    if (x < 8.0) {
        double b,cosx,sinx;
        double j1x[] = new double[1];
        double y1[] = new double[1];
        b=Math.sqrt(x)*1.25331413731550;
        bessy01(x,j1x,y1);
        j1x[0]=bessj1(x);
        x -= 0.785398163397448;
        cosx=Math.cos(x);
        sinx=Math.sin(x);
        p1[0] = b*(j1x[0]*sinx-y1[0]*cosx);
        q1[0] = b*(j1x[0]*cosx+y1[0]*sinx);
    } else {
        int i;
        double x2,b0,b1,b2,y;
        double ar1[]={0.10668e-15, -0.72212e-15, 0.545267e-14,
                -0.4684224e-13, 0.46991955e-12, -0.570486364e-11,
                0.881689866e-10, -0.187189074911e-8, 0.6177633960644e-7,
                -0.39872843004889e-5, 0.89898983308594e-3};
        double ar2[]={-0.10269e-15, 0.65083e-15, -0.456125e-14,
                0.3596777e-13, -0.32643157e-12, 0.351521879e-11,
```

```
                    -0.4686363688e-10, 0.82291933277e-9, -0.2095978138408e-7,
                    0.91386152579555e-6, -0.96277235491571e-4};
    y=8.0/x;
    x=2.0*y*y-1.0;
    x2=x+x;
    b1=b2=0.0;
    for (i=0; i<=10; i++) {
        b0=x2*b1-b2+ar1[i];
        b2=b1;
        b1=b0;
    }
    p1[0]  = x*b1-b2+1.0009030408600137;
    b1=b2=0.0;
    for (i=0; i<=10; i++) {
        b0=x2*b1-b2+ar2[i];
        b2=b1;
        b1=b0;
    }
    q1[0]  = (x*b1-b2+0.46777787069535e-1)*y;
  }
}
```

6.5.2 Bessel functions I and K

A. bessi0

Computes the value of the modified Bessel function of the first kind of order zero $I_0(x)$. For $|x| \leq 15$, a Chebyshev rational function approximation [Bla74] of the form

$$\sum_{k=0}^{14} n_k x^{2k} \bigg/ \sum_{k=0}^{3} d_k x^{2k}$$

is used. When $|x| > 15$, the function $I_0(x) = e^{-x}I_0(x)$ is evaluated by means of a call of *nonexpbessi0*, and the relationship $I_0(x) = e^x I_0'(x)$ is used.

Procedure parameters:
$$\text{double bessi0 } (x)$$
bessi0:   delivers the modified Bessel function of the first kind of order zero with
              argument x;
x: double;
              entry: the argument of the Bessel function.

Procedure used: nonexpbessi0.

```
public static double bessi0(double x)
{
    if (x == 0.0) return 1.0;
    if (Math.abs(x) <= 15.0) {
        double z,denominator,numerator;
        z=x*x;
        numerator=(z*(z*(z*(z*(z*(z*(z*(z*(z*(z*(z*(z*(z*
            0.210580722890567e-22+0.380715242345326e-19)+
            0.479440257548300e-16)+0.435125971262668e-13)+
            0.300931127112960e-10)+0.160224679395361e-7)+
            0.654858370096785e-5)+0.202591084143397e-2)+
            0.463076284721000e0)+0.754337328948189e2)+
            0.830792541809429e4)+0.571661130563785e6)+
            0.216415572361227e8)+0.356644482244025e9)+
            0.144048298227235e10);
        denominator=(z*(z*(z-0.307646912682801e4)+
            0.347626332405882e7)-0.144048298227235e10);
        return -numerator/denominator;
    } else {
        return Math.exp(Math.abs(x))*nonexpbessi0(x);
    }
}
```

B. bessi1

Computes the value of the modified Bessel function of the first kind of order one $I_1(x)$. For $|x| \leq 15$, a Chebyshev rational function approximation [Bla74] of the form

$$\sum_{k=0}^{14} n_k x^{2k} \Big/ \sum_{k=0}^{3} d_k x^{2k}$$

for the function $x^{-1}I_1(x)$ is used, and the value of $I_1(x)$ is recovered. When $|x| > 15$, the function $I_1'(x) = e^{-x}I_1(x)$ is evaluated by means of a call of *nonexpbessi1*, and the relationship $I_1(x) = e^x I_1'(x)$ is used.

Procedure parameters:
$$\text{double bessi1 } (x)$$
bessi1:   delivers the modified Bessel function of the first kind of order one with argument x;

x: double;

entry: the argument of the Bessel function.

Procedure used: nonexpbessi1.

```
public static double bessi1(double x)
{
    if (x == 0.0) return 0.0;
    if (Math.abs(x) <= 15.0) {
        double z,denominator,numerator;
        z=x*x;
        denominator=z*(z-0.222583674000860e4)+0.136293593052499e7;
        numerator=(z*(z*(z*(z*(z*(z*(z*(z*(z*(z*(z*(z*(z*
            0.207175767232792e-26+0.257091905584414e-23)+
            0.306279283656135e-20)+0.261372772158124e-17)+
            0.178469361410091e-14)+0.963628891518450e-12)+
            0.410068906847159e-9)+0.135455228841096e-6)+
            0.339472890308516e-4)+0.624726195127003e-2)+
            0.806144878821295e0)+0.682100567980207e2)+
            0.341069752284422e4)+0.840705772877836e5)+0.681467965262502e6);
        return x*(numerator/denominator);
    } else {
        return Math.exp(Math.abs(x))*nonexpbessi1(x);
    }
}
```

C. bessi

Generates an array of modified Bessel functions of the first kind of order one $I_j(x)$, $j=0,...,n$.

The functions $I_j'(x) = e^{-|x|}I_j(x)$, $j=0,...,n$, are first evaluated by means of a call of *nonexpbessi*, and the required values of $I_j(x)$ are recovered by multiplication by e^x.

Procedure parameters:
$$\text{void bessi } (x,n,i)$$

x: double;
        entry: the argument of the Bessel functions;
n: int;
        entry: the upper bound of the indices of the array i;
i: double i[0:n];
        exit: i[j] contains the value of the modified Bessel function of the first
                  kind of order j, j=0,...,n.

Procedure used: nonexpbessi.

```
public static void bessi(double x, int n, double i[])
{
    if (x == 0.0) {
        i[0]=1.0;
        for (; n>=1; n--) i[n]=0.0;
    } else {
```

```
        double expx;
        expx=Math.exp(Math.abs(x));
        nonexpbessi(x,n,i);
        for (; n>=0; n--) i[n] *= expx;
    }
}
```

D. bessk01

Computes the modified Bessel functions of the third kind of orders zero and one: $K_0(x)$ and $K_1(x)$ for $x>0$.

For $0 < x < 1.5$, $K_0(x)$ and $K_1(x)$ are computed by use of truncated versions of the Taylor series expansions [AbS65]

$$I_0(x) = \sum_{\upsilon=0}^{\infty} \left(\frac{x^2}{4}\right)^{\upsilon} \Big/ (\upsilon!)^2$$

$$K_0(x) = -\left[\ln\left(\frac{x}{2}\right) + \gamma\right] I_0(x) + \sum_{\upsilon=1}^{\infty}\left(\sum_{\tau=1}^{\upsilon}\frac{1}{\tau}\right)\left(\frac{x^2}{4}\right)^{\upsilon} \Big/ (\upsilon!)^2$$

$$I_1(x) = \frac{x}{2}\sum_{\upsilon=0}^{\infty}\left(\frac{x^2}{4}\right)^{\upsilon} \Big/ (\upsilon!(\upsilon+1)!)$$

$$K_1(x) = \frac{1}{x} + \frac{\gamma x}{2} - \frac{x}{4} + \ln\left(\frac{x}{2}\right)I_1(x) - \frac{x}{4}\sum_{\upsilon=1}^{\infty}\left(\frac{1}{\upsilon+1} - 2\gamma + 2\sum_{\tau=1}^{\upsilon}\frac{1}{\tau}\right)\left(\frac{x^2}{4}\right)^{\upsilon} \Big/ (\upsilon!(\upsilon+1)!).$$

For $x \geq 1.5$, the functions $K_j'(x) = e^x K_j(x)$, $j=0,1$, are evaluated by a call of *nonexpbessk01*, and the relationship $K_j(x)=e^{-x}K_j'(x)$ is used.

Procedure parameters:

$$\text{void bessk01 } (x,k0,k1)$$

x: double;
          entry: the argument of the Bessel functions; $x > 0$;
k0:   double $k0[0:0]$;
          exit: $k0$ has the value of the modified Bessel function of the third kind of order zero with argument x;
k1:   double $k1[0:0]$;
          exit: $k1$ has the value of the modified Bessel function of the third kind of order one with argument x.

Procedure used: nonexpbessk01.

```
public static void bessk01(double x, double k0[], double k1[])
{
    if (x <= 1.5) {
        int k;
        double c,d,r,sum0,sum1,t,term,t0,t1;
```

```
        sum0=d=Math.log(2.0/x)-0.5772156649015328606;
        sum1 = c = -1.0-2.0*d;
        r=term=1.0;
        t=x*x/4.0;
        k=1;
        do {
            term *= t*r*r;
            d += r;
            c -= r;
            r=1.0/(k+1);
            c -= r;
            t0=term*d;
            t1=term*c*r;
            sum0 += t0;
            sum1 += t1;
            k++;
        } while (Math.abs(t0/sum0)+Math.abs(t1/sum1) > 1.0e-15);
        k0[0] = sum0;
        k1[0] = (1.0+t*sum1)/x;
    } else {
        double expx;
        expx=Math.exp(-x);
        nonexpbessk01(x,k0,k1);
        k1[0] *= expx;
        k0[0] *= expx;
    }
}
```

E. bessk

Generates an array of modified Bessel functions of the third kind of order j, $K_j(x)$, $j=0,...,n$, for $x > 0$.

The functions $K_0(x)$ and $K_1(x)$ are first evaluated by means of a call of *bessk01*, and the recursion [AbS65]

$$K_{j+1}(x) = K_{j-1}(x) + (2_j/x)K_j(x)$$

is then used.

Procedure parameters:

$$\text{void bessk } (x,n,k)$$

x: double;

          entry: the argument of the Bessel functions; $x > 0$;

n: int;

          entry: the upper bound of the indices of array k; $n \geq 0$;

k: double $k[0:n]$;

          exit: $k[j]$ is the value of the modified Bessel function of the third kind of
                   order j with argument x, $j=0,...,n$.

Procedure used: bessk01.

```
public static void bessk(double x, int n, double k[])
{
    int i;
    double k0,k1,k2;
    double tmp1[] = new double[1];
    double tmp2[] = new double[1];

    bessk01(x,tmp1,tmp2);
    k0 = tmp1[0];
    k1 = tmp2[0];
    k[0]=k0;
    if (n > 0) k[1]=k1;
    x=2.0/x;
    for (i=2; i<=n; i++) {
        k[i]=k2=k0+x*(i-1)*k1;
        k0=k1;
        k1=k2;
    }
}
```

F. nonexpbessi0

Computes the value of the modified Bessel function of the first kind of order zero multiplied by $e^{-|x|}$.

nonexpbessi0 evaluates the function $I_0'(x) = e^{-|x|}I_0(x)$. When $|x| \le 15$, the function $I_0(x)$ is evaluated by means of a call of bessi0, and the function I_0' is computed by use of its defining relationship. When $|x| > 15$, a Chebyshev rational function approximation [Bla74] of the form

$$\sum_{k=0}^{4} n_k T_k\left(\frac{30}{x}-1\right) \Big/ \sum_{k=0}^{3} d_k T_k\left(\frac{30}{x}-1\right)$$

for the function $x^{1/2}I_0'(x)$ is used, and the value of $I_0'(x)$ is recovered.

Procedure parameters:
$$\text{double nonexpbessi0 } (x)$$
nonexpbessi0:   delivers the modified Bessel function of the first kind of order
                 zero with argument x, multiplied by $e^{-|x|}$;
x: double;
                 entry: the argument of the Bessel function.

Procedure used: bessi0.

```
public static double nonexpbessi0(double x)
{
    if (x == 0.0) return 1.0;
    if (Math.abs(x) <= 15.0) {
        return Math.exp(-Math.abs(x))*bessi0(x);
    } else {
        int i;
        double sqrtx,br,br1,br2,z,z2,numerator,denominator;
        double ar1[]={0.2439260769778, -0.115591978104435e3, 0.784034249005088e4,
                      -0.143464631313583e6};
        double ar2[]={1.0, -0.325197333369824e3, 0.203128436100794e5,
                      -0.361847779219653e6};
        x=Math.abs(x);
        sqrtx=Math.sqrt(x);
        br1=br2=0.0;
        z=30.0/x-1.0;
        z2=z+z;
        for (i=0; i<=3; i++) {
            br=z2*br1-br2+ar1[i];
            br2=br1;
            br1=br;
        }
        numerator=z*br1-br2+0.346519833357379e6;
        br1=br2=0.0;
        for (i=0; i<=3; i++) {
            br=z2*br1-br2+ar2[i];
            br2=br1;
            br1=br;
        }
        denominator=z*br1-br2+0.865665274832055e6;
        return (numerator/denominator)/sqrtx;
    }
}
```

G. nonexpbessi1

Computes the value of the modified Bessel function of the first kind of order one multiplied by $e^{-|x|}$.

nonexpbessi1 evaluates the function $I_1'(x)$ = $\text{sign}(x)e^{-|x|}I_1(x)$. When $|x| \leq 15$, the function $I_1(x)$ is evaluated by means of a call of bessi1, and the function I_1' is computed by use of its defining relationship. When $|x| > 15$, a Chebyshev rational function approximation [Bla74] of the form

$$\sum_{k=0}^{4} n_k T_k\left(\frac{30}{x}-1\right) \Big/ \sum_{k=0}^{3} d_k T_k\left(\frac{30}{x}-1\right)$$

for the function $x^{1/2}I_1'(x)$ is used, and the value of $I_1'(x)$ is recovered.

Procedure parameters:
$$\text{double nonexpbessi1 } (x)$$
nonexpbessi1:   delivers the modified Bessel function of the first kind of order one
with argument x, multiplied by $e^{-|x|}$;

x: double;
entry: the argument of the Bessel function.

Procedure used: bessi1.

```
public static double nonexpbessi1(double x)
{
    if (x == 0.0) return 0.0;
    if (Math.abs(x) > 15.0) {
        int i,signx;
        double br,br1,br2,z,z2,sqrtx,numerator,denominator;
        double ar1[]={0.1494052814740e1, -0.362026420242263e3, 0.220549722260336e5,
                      -0.408928084944275e6};
        double ar2[]={1.0, -0.631003200551590e3, 0.496811949533398e5,
                      -0.100425428133695e7};
        signx = (x > 0.0) ? 1 : -1;
        x=Math.abs(x);
        sqrtx=Math.sqrt(x);
        z=30.0/x-1.0;
        z2=z+z;
        br1=br2=0.0;
        for (i=0; i<=3; i++) {
            br=z2*br1-br2+ar1[i];
            br2=br1;
            br1=br;
        }
        numerator=z*br1-br2+0.102776692371524e7;
        br1=br2=0.0;
        for (i=0; i<=3; i++) {
            br=z2*br1-br2+ar2[i];
            br2=br1;
            br1=br;
        }
        denominator=z*br1-br2+0.26028876789105e7;
        return ((numerator/denominator)/sqrtx)*signx;
    } else {
        return Math.exp(-Math.abs(x))*bessi1(x);
    }
}
```

H. nonexpbessi

Generates an array of modified Bessel functions of the first kind of order one $I_j(x)$, multiplied by $e^{-|x|}$, $j=0,...,n$.

nonexpbessi evaluates the functions $I_j'(x) = e^{-|x|}I_j(x)$, $j=0,...,n$. The method used [G67] is that described in the documentation to *bessj*. The functions $I_{a+k}'(x)$ satisfy the recursion

$$I_{a+k+1}'(x) + \{2(a+k)/x\}I_{a+k}'(x) - I_{a+k-1}'(x) = 0$$

and the relationship

$$I_a(x) + 2\sum_{k=1}^{\infty} \frac{(a+k)\,\Gamma(2a+k)}{k!\,\Gamma(2a+1)}\,I_{a+k}'(x) = \frac{(x/2)^a}{\Gamma(a+1)}$$

when $x=0$, a_k and b_k in formula (1) of the documentation to *bessj* are $a_k = 2k/x$, $b_k = -1$, and $\lambda_k=1$, $s=1$ may be taken in the further formulae of that documentation.

Procedure parameters:

$$\text{void nonexpbessi } (x,n,i)$$

x: double;
        entry: the argument of the Bessel functions;
n: int;
        entry: the upper bound of the indices of the array i, $n \geq 0$;
i: double $i[0:n]$;
        exit: $i[j]$ contains the value of the modified Bessel function of the first
                kind of order j, multiplied by $e^{-|x|}$, $j=0,...,n$.

Procedure used: start.

```
public static void nonexpbessi(double x, int n, double i[])
{
    if (x == 0.0) {
        i[0]=1.0;
        for (; n>=1; n--) i[n]=0.0;
    } else {
        int k;
        boolean negative;
        double x2,r,s;
        negative = (x < 0.0);
        x=Math.abs(x);
        r=s=0.0;
        x2=2.0/x;
        k=start(x,n,1);
        for (; k>=1; k--) {
            r=1.0/(r+x2*k);
            s=r*(2.0+s);
            if (k <= n) i[k]=r;
        }
        i[0]=r=1.0/(1.0+s);
        if (negative)
```

```
            for (k=1; k<=n; k++) r = i[k] *= (-r);
        else
            for (k=1; k<=n; k++) r = i[k] *= r;
    }
}
```

I. nonexpbessk01

Computes the modified Bessel functions of the third kind of orders zero and one, $K_0(x)$ and $K_1(x)$ for $x>0$, multiplied by e^x.

   nonexpbessk01 evaluates the functions $K_j'(x) = e^x K_j(x)$ for $x > 0$, $j=0,1$. For $0<x\leq1.5$, the functions $K_0(x)$ and $K_1(x)$ are computed by a call of *bessk01* and the K_j' are evaluated by use of their defining relationship. For $1.5<x\leq5$, the trapezoidal rule (see [Hu64] of *bessi0*) is used to evaluate the integrals (see [AbS65] of *bessi0*)

$$K_j'(x) = \sqrt{\frac{\pi}{2x}} \frac{\mu^{j+\frac{1}{2}}}{\Gamma(j+\frac{1}{2})} \int_{-\infty}^{\infty} e^{-\mu t^2} t^{2j} \left(1 + \frac{t^2\mu}{2x}\right) dt \qquad (j=0,1)$$

with $\mu = 2/5$. For $x > 5$, truncated Chebyshev expansions (see [Cle62, Lu69]) of the form

$$K_j'(x) = \sqrt{\frac{\pi}{2x}} \sum_{n=0} d_n T_n\left(\tfrac{10}{x} - 1\right)$$

are used.

Procedure parameters:
$$\text{void nonexpbessk01 } (x,k0,k1)$$
x: double;
          entry: the argument of the Bessel functions; $x > 0$;
k0:   double k0[0:0];
          exit: k0 has the value of the modified Bessel function of the third kind of order zero with argument x, multiplied by e^x;
k1:   double k1[0:0];
          exit: k1 has the value of the modified Bessel function of the third kind of order one with argument x, multiplied by e^x.

Procedure used: bessk01.

```
public static void nonexpbessk01(double x, double k0[], double k1[])
{
    if (x <= 1.5) {
        double expx;
        expx=Math.exp(x);
        bessk01(x,k0,k1);
        k0[0] *= expx;
        k1[0] *= expx;
    } else if (x <= 5.0) {
        int i,r;
```

```
    double t2,s1,s2,term1,term2,sqrtexpr,exph2,x2;
    double fac[]={0.90483741803596, 0.67032004603564,
        0.40556965974060, 0.20189651799466, 0.82084998623899e-1,
        0.27323722447293e-1, 0.74465830709243e-2,
        0.16615572731739e-2, 0.30353913807887e-3,
        0.45399929762485e-4, 0.55595132416500e-5,
        0.55739036926944e-6, 0.45753387694459e-7,
        0.30748798795865e-8, 0.16918979226151e-9,
        0.76218651945127e-11, 0.28111852987891e-12,
        0.84890440338729e-14, 0.2098791048793e-15,
        0.42483542552916e-17};
    s1=0.5;
    s2=0.0;
    r=0;
    x2=x+x;
    exph2=1.0/Math.sqrt(5.0*x);
    for (i=0; i<=19; i++) {
        r += 1.0;
        t2=r*r/10.0;
        sqrtexpr=Math.sqrt(t2/x2+1.0);
        term1=fac[i]/sqrtexpr;
        term2=fac[i]*sqrtexpr*t2;
        s1 += term1;
        s2 += term2;
    }
    k0[0] = exph2*s1;
    k1[0] = exph2*s2*2.0;
} else {
    int r,i;
    double br,br1,br2,cr,cr1,cr2,ermin1,erplus1,er,f0,f1,expx,y,y2;
    double dr[]={0.27545e-15, -0.172697e-14,
            0.1136042e-13, -0.7883236e-13, 0.58081063e-12,
            -0.457993633e-11, 0.3904375576e-10, -0.36454717921e-9,
            0.379299645568e-8, -0.450473376411e-7,
            0.63257510850049e-6, -0.11106685196665e-4,
            0.26953261276272e-3, -0.11310504646928e-1};
    y=10.0/x-1.0;
    y2=y+y;
    r=30;
    br1=br2=cr1=cr2=erplus1=er=0.0;
    for (i=0; i<=13; i++) {
        r -= 2;
        br=y2*br1-br2+dr[i];
        cr=cr1*y2-cr2+er;
        ermin1=r*dr[i]+erplus1;
        erplus1=er;
        er=ermin1;
        br2=br1;
        br1=br;
        cr2=cr1;
```

```
            cr1=cr;
        }
        f0=y*br1-br2+0.9884081742308258;
        f1=y*cr1-cr2+er/2.0;
        expx=Math.sqrt(1.5707963267949/x);
        k0[0] = f0 *=expx;
        k1[0] = (1.0+0.5/x)*f0+(10.0/x/x)*expx*f1;
    }
}
```

J. nonexpbessk

Generates an array of modified Bessel functions of the third kind of order j, $K_j(x)$, multiplied by e^x, $j=0,...,n$, for $x > 0$.

nonexpbessk computes the values of the functions $K_j'(x) = e^x K_j(x)$, $j=0,...,n$, where $K_j(x)$ is a modified Bessel function of the third kind of order j, for $x > 0$.

The functions $K_0'(x)$ and $K_1'(x)$ are first evaluated by means of a call of nonexpbessk01, and the recursion [AbS65]
$$K_{j+1}'(x) = K_{j-1}'(x) + (2_j/x)K_j'(x)$$
is then used.

Procedure parameters:
$$\text{void nonexpbessk } (x,n,k)$$

x: double;
        entry: the argument of the Bessel functions; $x > 0$;
n: int;
        entry: the upper bound of the indices of array k; $n \geq 0$;
k: double $k[0:n]$;
        exit: $k[j]$ is value of the modified Bessel function of the third kind of order j multiplied by e^x, $j=0,...,n$.

Procedure used: nonexpbessk01.

```
public static void nonexpbessk(double x, int n, double k[])
{
    int i;
    double k0,k1,k2;
    double tmp1[] = new double[1];
    double tmp2[] = new double[1];

    nonexpbessk01(x,tmp1,tmp2);
    k0 = tmp1[0];
    k1 = tmp2[0];
    k[0]=k0;
    if (n > 0) k[1]=k1;
    x=2.0/x;
    for (i=2; i<=n; i++) {
```

```
        k[i]=k2=k0+x*(i-1)*k1;
        k0=k1;
        k1=k2;
    }
}
```

6.6 Bessel functions of real order

6.6.1 Bessel functions J and Y

A. bessjaplusn

Calculates the Bessel functions of the first kind of order $a+k$, $J_{a+k}(x)$, ($0 \le k \le n$, $0 \le a < 1$).

If $x=0$ then $J_a(x)$ equals 0 if $a>0$ and equals 1 if $a=0$, and $J_{a+k}(x)=0$ for $k=1,...,n$. If $a=\frac{1}{2}$, the functions $J_{a+k}(x)$ are evaluated by means of a call of *spherbessj*; otherwise, the method used is that described in the documentation to *bessj*. Now a_k and b_k in formula (1) of that documentation are

$$a_k = 2(a+k)/x, \; b_k = -1,$$

and

$$\lambda_0 = 1, \; \lambda_{2m-1} = 0, \; \lambda_{2m} = (a+2m)\Gamma(a+m)/\{m!\Gamma(a+m)\} \; (m=1,2,...),$$

$s=(x/2)^a/\Gamma(a+1)$ may be taken in the further formulae of that documentation, see [G67].

Procedure parameters:

$$\text{void bessjaplusn } (a,x,n,ja)$$

a: double;
    entry: the noninteger part of the order; $0 \le a < 1$;
x: double;
    entry: the argument value, $x \ge 0$;
n: int;
    entry: the upper bound of the indices of the array *ja*;
ja: double *ja[0:n]*;
    exit: *ja[k]* is assigned the value of the Bessel function of the first kind $J_{a+k}(x)$, $0 \le k \le n$.

Procedures used: bessj, spherbessj, gamma, start.

```
public static void bessjaplusn(double a, double x, int n, double ja[])
{
    if (x == 0.0) {
        ja[0] = (a == 0.0) ? 1.0 : 0.0;
        for (; n>=1; n--) ja[n]=0.0;
    } else if (a == 0.0) {
```

```
            bessj(x,n,ja);
    } else if (a == 0.5) {
            double s;
            s=Math.sqrt(x)*0.797884560802865;
            spherbessj(x,n,ja);
            for (; n>=0; n--) ja[n] *= s;
    } else {
            int k,m,nu;
            double a2,x2,r,s,l,labda;
            l=1.0;
            nu=start(x,n,0);
            for (m=1; m<=nu; m++) l=l*(m+a)/(m+1);
            r=s=0.0;
            x2=2.0/x;
            k = -1;
            a2=a+a;
            for (m=nu+nu; m>=1; m--) {
                    r=1.0/(x2*(a+m)-r);
                    if (k == 1)
                            labda=0.0;
                    else {
                            l=l*(m+2)/(m+a2);
                            labda=l*(m+a);
                    }
                    s=r*(labda+s);
                    k = -k;
                    if (m <= n) ja[m]=r;
            }
            ja[0]=r=1.0/gamma(1.0+a)/(1.0+s)/Math.pow(x2,a);
            for (m=1; m<=n; m++) r = ja[m] *= r;
    }
}
```

B. bessya01

Computes the Bessel functions of the second kind (also called Neumann's functions) of order a and $a+1$: $Y_a(x)$ and $Y_{a+1}(x)$ for $x > 0$, $a \geq 0$.

    For $x < 3$, the above functions are evaluated by use of truncated Taylor series (see [T76a]). For $x \geq 3$, the functions $P_\alpha(x)$, $Q_\alpha(x)$ occurring in the formula

$$Y_\alpha(x) = (2/(\pi x))^{1/2}\{P_\alpha(x)\sin(x - (\alpha+\tfrac{1}{2})\pi/2) + Q_\alpha(x)\cos(x - (\alpha+\tfrac{1}{2})\pi/2)\}$$

are evaluated for $\alpha = a$, $a+1$ by means of a call of *besspqa01*; the values of $Y_a(x)$, $Y_{a+1}(x)$ are then recovered.

    Note that the hyperbolic sine, cosine, and tangent functions are included here.

Procedure parameters:

$$\text{void bessya01 } (a,x,ya,ya1)$$

a: double;

    entry: the order;

x: double;

    entry: this argument should satisfy $x > 0$;

ya: double *ya[0:0]*;

    exit: the Neumann function of order *a* and argument *x*;

ya1: double *ya1[0:0]*;

    exit: the Neumann function of order *a*+1.

Procedures used: bessy01, recipgamma, besspqa01.

```
public static void bessya01(double a, double x, double ya[], double ya1[])
{
    if (a == 0.0) {
        bessy01(x,ya,ya1);
    } else {
        int n,na;
        boolean rec,rev;
        double b,c,d,e,f,g,h,p,q,r,s;
        double tmp1[] = new double[1];
        double tmp2[] = new double[1];
        double tmp3[] = new double[1];
        double tmp4[] = new double[1];
        na=(int)Math.floor(a+0.5);
        rec = (a >= 0.5);
        rev = (a < -0.5);
        if (rev || rec) a -= na;
        if (a == -0.5) {
            p=Math.sqrt(2.0/Math.PI/x);
            f=p*Math.sin(x);
            g = -p*Math.cos(x);
        } else if (x < 3.0) {
            b=x/2.0;
            d = -Math.log(b);
            e=a*d;
            c = (Math.abs(a) < 1.0e-8) ? 1.0/Math.PI : a/Math.sin(a*Math.PI);
            s = (Math.abs(e) < 1.0e-8) ? 1.0 : sinh(e)/e;
            e=Math.exp(e);
            g=recipgamma(a,tmp1,tmp2)*e;
            p=tmp1[0];
            q=tmp2[0];
            e=(e+1.0/e)/2.0;
            f=2.0*c*(p*e+q*s*d);
            e=a*a;
            p=g*c;
            q=1.0/g/Math.PI;
            c=a*Math.PI/2.0;
            r = (Math.abs(c) < 1.0e-8) ? 1.0 : Math.sin(c)/c;
```

```
        r *= Math.PI*c*r;
        c=1.0;
        d = -b*b;
        ya[0] = f+r*q;
        ya1[0] = p;
        n=1;
        do {
            f=(f*n+p+q)/(n*n-e);
            c=c*d/n;
            p /= (n-a);
            q /= (n+a);
            g=c*(f+r*q);
            h=c*p-n*g;
            ya[0] += g;
            ya1[0] += h;
            n++;
        } while (Math.abs(g/(1.0+Math.abs(ya[0])))+
                Math.abs(h/(1.0+Math.abs(ya1[0]))) > 1.0e-15);
        f = -ya[0];
        g = -ya1[0]/b;
    } else {
        b=x-Math.PI*(a+0.5)/2.0;
        c=Math.cos(b);
        s=Math.sin(b);
        d=Math.sqrt(2.0/x/Math.PI);
        besspqa01(a,x,tmp1,tmp2,tmp3,tmp4);
        p=tmp1[0];
        q=tmp2[0];
        b=tmp3[0];
        h=tmp4[0];
        f=d*(p*s+q*c);
        g=d*(h*s-b*c);
    }
    if (rev) {
        x=2.0/x;
        na = -na-1;
        for (n=0; n<=na; n++) {
            h=x*(a-n)*f-g;
            g=f;
            f=h;
        }
    } else if (rec) {
        x=2.0/x;
        for (n=1; n<=na; n++) {
            h=x*(a+n)*g-f;
            f=g;
            g=h;
        }
    }
    ya[0] = f;
```

```
        ya1[0] = g;
    }
}


public static double sinh(double x)
{
    return 0.5*(Math.exp(x)-Math.exp(-x));
}


public static double cosh(double x)
{
    return 0.5*(Math.exp(x)+Math.exp(-x));
}


public static double tanh(double x)
{
    double p = Math.exp(x);
    double q = Math.exp(-x);
    return (p-q)/(p+q);
}
```

C. bessyaplusn

Generates an array of Bessel functions of the second kind of order $a+n$, $Y_{a+n}(x)$, $n=0,...,nmax$, for $x > 0$, $a \geq 0$.

The values of the functions $Y_a(x)$, $Y_{a+1}(x)$ are first obtained by means of a call of *bessya01*, and the recursion
$$Y_{a+n+1}(x) = -Y_{a+n-1}(x) + \{2(n+a)/x\}Y_{a+n}(x)$$
is then used.

Procedure parameters:
$$\text{void bessyaplusn } (a,x,nmax,yan)$$
a: double;
         entry: the order;
x: double;
         entry: the argument value, $x > 0$;
nmax: int;
         entry: the upper bound of the indices of the array *yan*; $nmax \geq 0$;
yan: double *yan[0:nmax]*;
         exit: the values of the Bessel functions of the second kind of order $a+n$,
                 for argument x are assigned to *yan[n]*, $0 \leq n \leq nmax$.

Procedure used: bessya01.

```
public static void bessyaplusn(double a, double x, int nmax, double yan[])
{
    int n;
    double y1;
    double tmp1[] = new double[1];
    double tmp2[] = new double[1];

    bessya01(a,x,tmp1,tmp2);
    yan[0] = tmp1[0];
    y1 = tmp2[0];
    a -= 1.0;
    x=2.0/x;
    if (nmax > 0) yan[1]=y1;
    for (n=2; n<=nmax; n++) yan[n] = -yan[n-2]+(a+n)*x*yan[n-1];
}
```

D. besspqa01

This procedure is an auxiliary procedure for the computation of the Bessel functions for large values of their argument.

$besspqa01$ evaluates the functions $P_\alpha(x)$, $Q_\alpha(x)$ occurring in the formulae

$$J_\alpha(x) = (2/(\pi x))^{1/2}\{P_\alpha(x)\cos(x - (\alpha+\tfrac{1}{2})\pi/2) - Q_\alpha(x)\sin(x - (\alpha+\tfrac{1}{2})\pi/2)\}$$
$$Y_\alpha(x) = (2/(\pi x))^{1/2}\{P_\alpha(x)\sin(x - (\alpha+\tfrac{1}{2})\pi/2) + Q_\alpha(x)\cos(x - (\alpha+\tfrac{1}{2})\pi/2)\}$$

for $\alpha = a$, $a+1$ ($a \geq 0$ and $x > 0$).
If $x < 3$ then the formulae

$$P_a(x) = (\pi x/2)^{1/2}\{Y_a(x)\sin(x - (a+\tfrac{1}{2})\pi/2) + J_a(x)\cos(x - (a+\tfrac{1}{2})\pi/2)\}$$
$$Q_a(x) = (\pi x/2)^{1/2}\{Y_a(x)\cos(x - (a+\tfrac{1}{2})\pi/2) - J_a(x)\sin(x - (a+\tfrac{1}{2})\pi/2)\}$$
$$P_{a+1}(x) = (\pi x/2)^{1/2}\{J_{a+1}(x)\sin(x - (a+\tfrac{1}{2})\pi/2) - Y_{a+1}(x)\cos(x - (a+\tfrac{1}{2})\pi/2)\}$$
$$Q_{a+1}(x) = (\pi x/2)^{1/2}\{J_{a+1}(x)\cos(x - (a+\tfrac{1}{2})\pi/2) + Y_{a+1}(x)\sin(x - (a+\tfrac{1}{2})\pi/2)\}$$

are used, $bessjaplusn$ and $bessya01$ being called to evaluate the functions $J_a,...,Y_{a+1}$. When $x \geq 3$, Chebyshev expansions are used (see [T76a, W45]).

Procedure parameters:

$$\text{void besspqa01 } (a,x,pa,qa,pa1,qa1)$$

a: double;
         entry: the order;
x: double;
         entry: this argument should satisfy $x > 0$;
pa: double $pa[0:0]$;
         exit: the value of $P_a(x)$;
qa: double $qa[0:0]$;
         exit: the value of $Q_a(x)$;
$pa1$:   double $pa1[0:0]$;
         exit: the value of $P_{a+1}(x)$;
$qa1$:   double $qa1[0:0]$;
         exit: the value of $Q_{a+1}(x)$.

Procedures used: besspq0, besspq1, bessjaplusn, bessya01.

```
public static void besspqa01(double a, double x, double pa[], double qa[],
                             double pa1[], double qa1[])
{
    if (a == 0.0) {
        besspq0(x,pa,qa);
        besspq1(x,pa1,qa1);
    } else {
        int n,na;
        boolean rec,rev;
        double b,p0,q0;
        na=0;
        rev = (a < -0.5);
        if (rev) a = -a-1.0;
        rec = (a >= 0.5);
        if (rec) {
            na=(int)Math.floor(a+0.5);
            a -= na;
        }
        if (a == -0.5) {
            pa[0] = pa1[0] = 1.0;
            qa[0] = qa1[0] = 0.0;
        } else if (x >= 3.0) {
            double c,d,e,f,g,p,q,r,s,temp;
            c=0.25-a*a;
            b=x+x;
            f=r=1.0;
            g = -x;
            s=0.0;
            temp=x*Math.cos(a*Math.PI)/Math.PI*1.0e15;
            e=temp*temp;
            n=2;
            do {
                d=(n-1+c/n);
                p=(2*n*f+b*g-d*r)/(n+1);
                q=(2*n*g-b*f-d*s)/(n+1);
                r=f;
                f=p;
                s=g;
                g=q;
                n++;
            } while ((p*p+q*q)*n*n < e);
            e=f*f+g*g;
            p=(r*f+s*g)/e;
            q=(s*f-r*g)/e;
            f=p;
            g=q;
            n--;
            while (n > 0) {
```

```
            r=(n+1)*(2.0-p)-2.0;
            s=b+(n+1)*q;
            d=(n-1+c/n)/(r*r+s*s);
            p=d*r;
            q=d*s;
            e=f;
            f=p*(e+1.0)-g*q;
            g=q*(e+1.0)+p*g;
            n--;
        }
        f += 1.0;
        d=f*f+g*g;
        pa[0] = f/d;
        qa[0] = -g/d;
        d=a+0.5-p;
        q += x;
        pa1[0] = (pa[0]*q-qa[0]*d)/x;
        qa1[0] = (qa[0]*q+pa[0]*d)/x;
    } else {
        double c,s,chi;
        double ya[] = new double[1];
        double ya1[] = new double[1];
        double ja[] = new double[2];
        b=Math.sqrt(Math.PI*x/2.0);
        chi=x-Math.PI*(a/2.0+0.25);
        c=Math.cos(chi);
        s=Math.sin(chi);
        bessya01(a,x,ya,ya1);
        bessjaplusn(a,x,1,ja);
        pa[0] = b*(ya[0]*s+c*ja[0]);
        qa[0] = b*(c*ya[0]-s*ja[0]);
        pa1[0] = b*(s*ja[1]-c*ya1[0]);
        qa1[0] = b*(c*ja[1]+s*ya1[0]);
    }
    if (rec) {
        x=2.0/x;
        b=(a+1.0)*x;
        for (n=1; n<=na; n++) {
            p0=pa[0]-qa1[0]*b;
            q0=qa[0]+pa1[0]*b;
            pa[0] = pa1[0];
            pa1[0] = p0;
            qa[0] = qa1[0];
            qa1[0] = q0;
            b += x;
        }
    }
    if (rev) {
        p0 = pa1[0];
        pa1[0] = pa[0];
```

```
        pa[0] = p0;
        q0 = qa1[0];
        qa1[0] = qa[0];
        qa[0] = q0;
    }
  }
}
```

E. besszeros

Calculates the first n zeros of a Bessel function of the first or the second kind or its derivative.

    besszero computes the first n zeros of either 1) $J_a(z)$, 2) $Y_a(z)$, 3) $dJ_a(z)/dz$, or 4) $dY_a(z)/dz$, $(a \geq 0)$. Which of the above set of zeros is derived is determined by the value of the parameter d upon call: thus 1 for the zeros of $J_a(z)$, and so on.

    Each zero is obtained by use of an initial approximation derived from an asymptotic expansion [AbS65], and subsequent improvement by use of a Newton-Raphson process [T76b, T78].

    If $a < 3$, then with $\mu = 4a^2$, the s-th zero of $J_a(z)$ or $Y_a(z)$ is given by

$$z_{a,s} \approx \beta - (\mu-1)/(8\beta) - 4(\mu-1)(7\mu-31)/\{3(8\beta)^2\} - 32(\mu-1)(83\mu^2-982\mu+3779)/\{15(8\beta)^5\} \ldots$$

where $\beta = (s + a/2 - 1/4)\pi$ for $J_a(z)$ and $\beta = (s + a/2 - 3/4)\pi$ for $Y_a(z)$.
Similarly the s-th zero of $dJ_a(z)/dz$ or $dY_a(z)/dz$ is given by

$$z'_{a,s} \approx \beta' - (\mu+3)/(8\beta') - 4(7\mu^2+82\mu-9)/\{3(8\beta')^3\} -$$
$$32(83\mu^3+2075\mu^2-3039\mu+3537)/\{15(8\beta')^5\} \ldots$$

where $\beta' = (s + a/2 - 3/4)\pi$ for $dJ_a(z)/dz$ and $\beta' = (s + a/2 - 1/4)\pi$ for $dY_a(z)/dz$.

    If $a \geq 3$, then with $w(u)$ defined by

$$(2/3)(-u)^{3/2} = \{w(u)^2 - 1\}^{1/2} - \arccos(w(u)^{-1})$$

i.e., $w(u) = 1/\cos\varphi(u)$ where

$$\{\tan\varphi(u)\} - \varphi(u) = (2/3)(-u)^{3/2}$$

the s-th zero of $J_a(z)$ or $Y_a(z)$ is given by

$$z_{a,s} \approx aw[-a^{2/3}f\{3\pi(4s-2k-1)/8\}]$$

where

$$f\{x\} \approx x^{2/3}(1 + (5/48)x^2 - (5/36)x^4 + \ldots)$$

where $k=0$ for $J_a(z)$ and $k=1$ for $Y_a(z)$. Similarly, the s-th zero of $dJ_a(z)/dz$ or $dY_a(z)/dz$ is given by

$$z'_{a,s} \approx aw[-a^{2/3}g\{3\pi(4s-2k-1)/8\}]$$

where

$$g\{x\} \approx x^{2/3}(1 - (7/48)x^2 + (35/288)x^4 + \ldots)$$

where now $k=1$ for $dJ_a(z)/dz$ and $k=0$ for $dY_a(z)/dz$.

Procedure parameters:
$$\text{void besszeros } (a,n,z,d)$$

a: double;
           entry: the order of the Bessel function, $a \geq 0$;

n: int;
           entry: the number of zeros to be evaluated, $n \geq 1$;

z: double *z[1:n]*;
            exit: *z[j]* is the *j*-th zero of the selected Bessel function;
d: int;
            entry: the choice of *d* determines the type of the Bessel function of which
                     the zeros are computed:
                     if *d*=1 then $J_a(z)$;
                     if *d*=2 then $Y_a(z)$;
                     if *d*=3 then $dJ_a(z)/dz$;
                     if *d*=4 then $dY_a(z)/dz$.

Procedure used: besspqa01.

```
public static void besszeros(double a, int n, double z[], int d)
{
    int j,s;
    double aa,a2,b,bb,c,chi,co,mu,mu2,mu3,mu4,p,p0,p1,pp1,q,q1,qq1,ro,si,t,tt,u,v,w,
            x,xx,x4,y,yy,fi;
    double pa[] = new double[1];
    double pa1[] = new double[1];
    double qa[] = new double[1];
    double qa1[] = new double[1];

    aa=a*a;
    mu=4.0*aa;
    mu2=mu*mu;
    mu3=mu*mu2;
    mu4=mu2*mu2;
    if (d < 3) {
        p=7.0*mu-31.0;
        p0=mu-1.0;
        p1=4.0*(253.0*mu2-3722.0*mu+17869.0)/15.0/p*p0;
        q1=8.0*(83.0*mu2-982.0*mu+3779.0)/5.0/p;
    } else {
        p=7.0*mu2+82.0*mu-9.0;
        p0=mu+3.0;
        p1=(4048.0*mu4+131264.0*mu3-221984.0*mu2-417600.0*mu+1012176.0)/60.0/p;
        q1=1.6*(83.0*mu3+2075.0*mu2-3039.0*mu+3537.0)/p;
    }
    t = (d == 1 || d == 4) ? 0.25 : 0.75;
    tt=4.0*t;
    if (d < 3) {
        pp1=5.0/48.0;
        qq1 = -5.0/36.0;
    } else {
        pp1 = -7.0/48.0;
        qq1=35.0/288.0;
    }
    y=3.0*Math.PI/8.0;
    bb = (a >= 3.0) ? Math.pow(a,-2.0/3.0) : 0.0;
    for (s=1; s<=n; s++) {
```

```
if (a == 0.0 && s == 1 && d == 3) {
    x=0.0;
    j=0;
} else {
    if (s >= 3.0*a-8.0) {
        b=(s+a/2.0-t)*Math.PI;
        c=1.0/b/b/64.0;
        x=b-1.0/b/8.0*(p0-p1*c)/(1.0-q1*c);
    } else {
        if (s == 1)
            x = ((d == 1) ? -2.33811 : ((d == 2) ? -1.17371 :
                    ((d == 3) ? -1.01879 : -2.29444)));
        else {
            x=y*(4.0*s-tt);
            v=1.0/x/x;
            x = -Math.pow(x,2.0/3.0)*(1.0+v*(pp1+qq1*v));
        }
        u=x*bb;
        yy=2.0/3.0*Math.pow(-u,1.5);
        if (yy == 0.0)
            fi=0.0;
        else if (yy > 1.0e5)
            fi=1.570796;
        else {
            double r,pp;
            if (yy <1.0) {
                p=Math.pow(3.0*yy,1.0/3.0);
                pp=p*p;
                p *= (1.0+pp*(-210.0+pp*(27.0-2.0*pp)))/1575.0);
            } else {
                p=1.0/(yy+1.570796);
                pp=p*p;
                p=1.570796-p*(1.0+pp*(2310.0+pp*(3003.0+pp*
                    (4818.0+pp*(8591.0+pp*16328.0)))))/3465.0);
            }
            pp=(yy+p)*(yy+p);
            r=(p-Math.atan(p+yy))/pp;
            fi=p-(1.0+pp)*r*(1.0+r/(p+yy));
        }
        v=fi;
        w=1.0/Math.cos(v);
        xx=1.0-w*w;
        c=Math.sqrt(u/xx);
        x=w*(a+c/a/u*((d < 3) ? -5.0/48.0/u-c*(-5.0/24.0/xx+1.0/8.0) :
                            7.0/48.0/u+c*(-7.0/24.0/xx+3.0/8.0)));
    }
    j=0;
    do {
        xx=x*x;
        x4=xx*xx;
```

```
       a2=aa-xx;
       besspqa01(a,x,pa,qa,pa1,qa1);
       chi=x-Math.PI*(a/2.0+0.25);
       si=Math.sin(chi);
       co=Math.cos(chi);
       ro = ((d == 1) ? (pa[0]*co-qa[0]*si)/(pa1[0]*si+qa1[0]*co) :
              ((d == 2) ? (pa[0]*si+qa[0]*co)/(qa1[0]*si-pa1[0]*co) :
               ((d == 3) ? a/x-(pa1[0]*si+qa1[0]*co)/(pa[0]*co-qa[0]*si) :
                a/x-(qa1[0]*si-pa1[0]*co)/(pa[0]*si+qa[0]*co))));
       j++;
       if (d < 3) {
           u=ro;
           p=(1.0-4.0*a2)/6.0/x/(2.0*a+1.0);
           q=(2.0*(xx-mu)-1.0-6.0*a)/3.0/x/(2.0*a+1.0);
       } else {
           u = -xx*ro/a2;
           v=2.0*x*a2/(aa+xx)/3.0;
           w=a2*a2*a2;
           q=v*(1.0+(mu2+32.0*mu*xx+48.0*x4)/32.0/w);
           p=v*(1.0+(-mu2+40.0*mu*xx+48.0*x4)/64.0/w);
       }
       w=u*(1.0+p*ro)/(1.0+q*ro);
       x += w;
    } while (Math.abs(w/x) > 1.0e-13 && j < 5);
  }
  z[s]=x;
 }
}
```

F. start

This is an auxiliary procedure which computes a starting value of an algorithm used in several Bessel function procedures.

Certain stable methods for evaluating functions $f_0, f_1, ..., f_n$ which satisfy a three term recurrence relationship of the form $f_{k+1} + a_k f_k + b_k f_{k-1} = 0$ require a value of v to be determined such that the value of the continued fraction

$$r_{k-1} = \frac{-b_k}{a_k -} \frac{b_{k+1}}{a_{k+1} -} \cdots \qquad (1)$$

and that of its convergent

$$r_{k-1}^{(v)} = \frac{-b_k}{a_k -} \frac{b_{k+1}}{a_{k+1} -} \cdots \frac{b_v}{a_v}$$

should agree. (The theory of such methods [see T76b], which involve backward recursion, is described in outline in the documentation to *bessj* which concerns the case in which $f_k = J_k(x)$; the same methods are also implemented by *bessjaplusn* (for which $f_k = J_{a+k}(x)$), by *nonexpbessi* (for which $f_k = I_k(x)$), by *nonexpbessiaplusn* (for which $f_k = e^x I_{a+k}(x)$), by *spherbessj* (for which $f_k = (\pi/(2x))^{1/2} J_{k+\frac{1}{2}}(x)$), and by *nonexpspherbessi* (for which $f_k = e^{-x}(\pi/(2x))^{1/2} I_{k+\frac{1}{2}}(x)$).)

The above requirement is equivalent to the condition that the tail

$$t(\upsilon) = \frac{-b_{\upsilon+1}}{a_{\upsilon+1} -} \frac{b_{\upsilon+2}}{a_{\upsilon+2} -} \cdots$$

of expansion (1) should be negligible. For the special cases considered, $t(v)$ represents the ratio of two Bessel functions of contiguous orders. Estimates of $t(v)$ may be obtained by the use of such formulae as

$$J_r(x) \approx \{2\pi r \tanh(\alpha)\}^{-1/2} \exp[r\{\tanh(\alpha)-\alpha\}]$$

where $r=x \cosh(\alpha)$, and

$$I_r(x) \approx (2\pi r)^{-1/2}(1+z^2)^{-1/4} e^{r\eta(z)}$$

where $z=x/r$ and

$$\eta(z) = (1+z^2)^{1/2} + \ln[z/\{1+(1+z^2)^{1/2}\}].$$

Procedure parameters:

$$\text{int start } (x,n,t)$$

start: a starting value for the Miller algorithm for computing an array of Bessel functions;

x: double;
           entry: the argument of the Bessel functions, $x > 0$;

n: int;
           entry: the number of Bessel functions to be computed, $n \geq 0$;

t: int;
           entry: the type of Bessel function in question;
                   $t=0$ corresponds to ordinary Bessel functions;
                   $t=1$ corresponds to modified Bessel functions.

```
public static int start(double x, int n, int t)
{
    int s;
    double p,q,r,y;

    s=2*t-1;
    p=36.0/x-t;
    r=n/x;
    if (r > 1.0 || t == 1) {
        q=Math.sqrt(r*r+s);
        r=r*Math.log(q+r)-q;
    } else
        r=0.0;
    q=18.0/x+r;
    r = (p > q) ? p : q;
    p=Math.sqrt(2.0*(t+r));
    p=x*((1.0+r)+p)/(1.0+p);
    y=0.0;
    q=y;
    do {
        y=p;
        p /= x;
```

```
        q=Math.sqrt(p*p+s);
        p=x*(r+q)/Math.log(p+q);
        q=y;
    } while (p > q || p < q-1.0);
    return ((t == 1) ? (int)Math.floor(p+1.0) : -(int)Math.floor(-p/2.0)*2);
}
```

6.6.2 Bessel functions I and K

A. bessiaplusn

Generates an array of modified Bessel functions of the first kind of order $a+j$, $I_{a+j}(x)$, $(0 \le j \le n, 0 \le a < 1)$.

When $x=0$ the above functions are evaluated directly; when $a=0$ or $a=0.5$ the procedures *bessi* or *nonexpspherbessi*, as is appropriate, is used. Otherwise the functions

$$e^{-|x|} I_{a+j}(x) \ (j=0,...,n)$$

are evaluated by means of a call of *nonexpbessiaplusn*, and the values of the functions $I_{a+j}(x)$ $(j=0,...,n)$ are recovered by multiplication by $e^{|x|}$.

Procedure parameters:
$$\text{void bessiaplusn } (a,x,n,ia)$$

a: double;
           entry: the noninteger part of the order of the Bessel functions; $0 \le a < 1$;
x: double;
           entry: the argument value of the Bessel functions, $x \ge 0$;
n: int;
           entry: the upper bound of the indices of the array *ia*; $n \ge 0$;
ia: double *ia[0:n]*;
           exit: *ia[j]* is assigned the value of the modified Bessel function of the first
                   kind of order $a+j$ and argument x, $I_{a+j}(x)$, $0 \le j \le n$.

Procedures used: nonexpbessiaplusn, bessi, nonexpspherbessi.

```
public static void bessiaplusn(double a, double x, int n, double ia[])
{
    if (x == 0.0) {
        ia[0] = (a == 0.0) ? 1.0 : 0.0;
        for (; n>=1; n--) ia[n]=0.0;
    } else if (a == 0.0) {
        bessi(x,n,ia);
    } else if (a == 0.5) {
        double c;
        c=0.797884560802865*Math.sqrt(Math.abs(x))*Math.exp(Math.abs(x));
        nonexpspherbessi(x,n,ia);
```

```
        for (; n>=0; n--) ia[n] *= c;
    } else {
        double expx;
        expx=Math.exp(Math.abs(x));
        nonexpbessiaplusn(a,x,n,ia);
        for (; n>=0; n--) ia[n] *= expx;
    }
}
```

B. besska01

Computes the modified Bessel functions of the third kind of order a and $a+1$: $K_a(x)$ and $K_{a+1}(x)$ for $x>0$, $a \geq 0$.

For $0 < x < 1$, $K_a(x)$ and $K_{a+1}(x)$ are computed by using Taylor series (see [T75]). For $x \geq 1$ the procedure calls for *nonexpbesska01*.

Procedure parameters:
$$\text{void besska01 } (a, x, ka, ka1)$$

a: double;
      entry: the order;
x: double;
      entry: this argument should satisfy $x > 0$;
ka: double *ka[0:0]*;
      exit: the value of the modified Bessel function of the third kind of order
            a and argument x;
ka1:   double *ka1[0:0]*;
      exit: the value of the modified Bessel function of the third kind of order
            $a+1$ and argument x.

Procedures used: bessk01, recipgamma, nonexpbesska01.

```
public static void besska01(double a, double x, double ka[], double ka1[])
{
    if (a == 0.0) {
        bessk01(x,ka,ka1);
    } else {
        int n,na;
        boolean rec,rev;
        double f,g,h;
        na=0;
        rev = (a < -0.5);
        if (rev) a = -a-1.0;
        rec = (a >= 0.5);
        if (rec) {
            na=(int)Math.floor(a+0.5);
            a -= na;
        }
```

```
if (a == 0.5)
    f=g=Math.sqrt(Math.PI/x/2.0)*Math.exp(-x);
else if (x < 1.0) {
    double a1,b,c,d,e,p,q,s;
    double tmp1[] = new double[1];
    double tmp2[] = new double[1];
    b=x/2.0;
    d = -Math.log(b);
    e=a*d;
    c=a*Math.PI;
    c = (Math.abs(c) < 1.0e-15) ? 1.0 : c/Math.sin(c);
    s = (Math.abs(e) < 1.0e-15) ? 1.0 : sinh(e)/e;
    e=Math.exp(e);
    a1=(e+1.0/e)/2.0;
    g=recipgamma(a,tmp1,tmp2)*e;
    p=tmp1[0];
    q=tmp2[0];
    ka[0] = f = c*(p*a1+q*s*d);
    e=a*a;
    p=0.5*g*c;
    q=0.5/g;
    c=1.0;
    d=b*b;
    ka1[0] = p;
    n=1;
    do {
        f=(f*n+p+q)/(n*n-e);
        c=c*d/n;
        p /= (n-a);
        q /= (n+a);
        g=c*(p-n*f);
        h=c*f;
        ka[0] += h;
        ka1[0] += g;
        n++;
    } while (h/ka[0]+Math.abs(g)/ka1[0] > 1.0e-15);
    f=ka[0];
    g=ka1[0]/b;
} else {
    double expon;
    expon=Math.exp(-x);
    nonexpbesska01(a,x,ka,ka1);
    f=expon*ka[0];
    g=expon*ka1[0];
}
if (rec) {
    x=2.0/x;
    for (n=1; n<=na; n++) {
        h=f+(a+n)*x*g;
        f=g;
```

```
                g=h;
            }
        }
        if (rev) {
            ka1[0] = f;
            ka[0] = g;
        } else {
            ka[0] = f;
            ka1[0] = g;
        }
    }
}
```

C. besskaplusn

Generates an array of modified Bessel functions of the third kind of order $a+n$, $K_{a+n}(x)$, $n=0,...,nmax$, for $x > 0$, $a \geq 0$.

The values of the functions $K_a(x)$, $K_{a+1}(x)$ are first obtained by means of a call of *besska01*, and the recursion

$$K_{a+n+1}(x) = K_{a+n-1}(x) + \{2(n+a)/x\}K_{a+n}(x)$$

is then used.

Procedure parameters:
$$\text{void besskaplusn } (a,x,nmax,kan)$$

a: double;
            entry: the order; $a \geq 0$;
x: double;
            entry: the argument value, $x > 0$;
nmax: int;
            entry: the upper bound of the indices of the array *kan*; $nmax \geq 0$;
kan: double *kan[0:nmax]*;
            exit: the values of the modified Bessel functions of the third kind of
                     order $a+n$, for argument x are assigned to *kan[n]*, $0 \leq n \leq nmax$.

Procedure used: besska01.

```
public static void besskaplusn(double a, double x, int nmax, double kan[])
{
    int n;
    double k1;
    double tmp1[] = new double[1];
    double tmp2[] = new double[1];

    besska01(a,x,tmp1,tmp2);
    kan[0]=tmp1[0];
    k1=tmp2[0];
    a -= 1.0;
```

```
    x=2.0/x;
    if (nmax > 0) kan[1]=k1;
    for (n=2; n<=nmax; n++) kan[n]=kan[n-2]+(a+n)*x*kan[n-1];
}
```

D. nonexpbessiaplusn

Generates an array of modified Bessel functions of the first kind of order $a+j$, $I_{a+j}(x)$, ($0 \le j \le n$, $0 \le a < 1$), multiplied by the factor e^x. Thus, apart from the exponential factor the array entries are the same as those computed by *bessiaplusn*.

   nonexpbessiaplusn computes the values of the functions $e^{-x}I_{a+j}(x)$ ($j=0,...,n$). When $x=0$, the above functions are evaluated directly; when $a=0$ or $a=0.5$ the procedure *nonexpbessi* or *nonexpspherbessi*, as is appropriate, is used. Otherwise the above functions are evaluated by the methods described in the documentation to *nonexpbessi*, where now $a_k=2(a+k)/x$, $b_k=-1$, $\lambda_0=1$, $\lambda_k=2(a+k)\Gamma(a+2k)/\{k!\Gamma(2a+1)\}$, and $s=(x/2)^a/\Gamma(a+1)$. See [G67].

Procedure parameters:
$$\text{void nonexpbessiaplusn } (a,x,n,ia)$$
a: double;
          entry: the noninteger part of the order $a+n$; $0 \le a < 1$;
x: double;
          entry: the argument of the Bessel functions, $x \ge 0$;
n: int;
          entry: the upper bound of the indices of the array *ia*; $n \ge 0$;
ia: double *ia[0:n]*;
          exit: *ia[j]* is assigned the value of the modified Bessel function of the first kind of order $a+j$ and argument x, multiplied by e^{-x}, $0 \le j \le n$.

Procedures used: nonexpbessi, nonexpspherbessi, gamma, start.

```
public static void nonexpbessiaplusn(double a, double x, int n, double ia[])
{
    if (x == 0.0) {
        ia[0] = (a == 0.0) ? 1.0 : 0.0;
        for (; n>=1; n--) ia[n]=0.0;
    } else if (a == 0.0) {
        nonexpbessi(x,n,ia);
    } else if (a == 0.5) {
        double c;
        c=0.797884560802865*Math.sqrt(x);
        nonexpspherbessi(x,n,ia);
        for (; n>=0; n--) ia[n] *= c;
    } else {
        int m,nu;
        double r,s,labda,l,a2,x2;
        a2=a+a;
```

```
        x2=2.0/x;
        l=1.0;
        nu=start(x,n,1);
        r=s=0.0;
        for (m=1; m<=nu; m++) l=l*(m+a2)/(m+1);
        for (m=nu; m>=1; m--) {
            r=1.0/(x2*(a+m)+r);
            l=l*(m+1)/(m+a2);
            labda=l*(m+a)*2.0;
            s=r*(labda+s);
            if (m <= n) ia[m]=r;
        }
        ia[0]=r=1.0/(1.0+s)/gamma(1.0+a)/Math.pow(x2,a);
        for (m=1; m<=n; m++) r = ia[m] *= r;
    }
}
```

E. nonexpbesska01

Computes the modified Bessel functions of the third kind of order a and $a+1$, $K_a(x)$ and $K_{a+1}(x)$, multiplied by the factor e^x, for $x>0$, $a \geq 0$. Thus, apart from the exponential factor, the functions are the same as those computed by *besska01*.

nonexpbesska01 evaluates the functions $K_\alpha'(x) = e^x K_\alpha(x)$ $(\alpha=a,a+1)$. For $0<x<1$, the procedure *besska01* is called. For $x \geq 1$ the Bessel functions are computed by a Miller algorithm for confluent hypergeometric functions (see [T75]).

Procedure parameters:
$$\text{void nonexpbesska01 } (a,x,ka,ka1)$$
a: double;
          entry: the order;
x: double;
          entry: this argument should satisfy $x > 0$;
ka: double *ka[0:0]*;
          exit: the value of the modified Bessel function of the third kind of order
                  a and argument x;
ka1:   double *ka1[0:0]*;
          exit: the value of the modified Bessel function of the third kind of order
                  $a+1$ and argument x.

Procedures used: besska01, nonexpbessk01.

```
public static void nonexpbesska01(double a, double x, double ka[], double ka1[])
{
    if (a == 0.0) {
        nonexpbessk01(x,ka,ka1);
    } else {
        int n,na;
```

```
boolean rec,rev;
double f,g,h;
na=0;
rev = (a < -0.5);
if (rev) a = -a-1.0;
rec = (a >= 0.5);
if (rec) {
    na=(int)Math.floor(a+0.5);
    a -= na;
}
if (a == -0.5)
    f=g=Math.sqrt(Math.PI/x/2.0);
else if (x < 1.0) {
    double expon;
    expon=Math.exp(x);
    besska01(a,x,ka,ka1);
    f=expon*ka[0];
    g=expon*ka1[0];
} else {
    double b,c,e,p,q;
    c=0.25-a*a;
    b=x+x;
    g=1.0;
    f=0.0;
    e=Math.cos(a*Math.PI)/Math.PI*x*1.0e15;
    n=1;
    do {
        h=(2.0*(n+x)*g-(n-1+c/n)*f)/(n+1);
        f=g;
        g=h;
        n++;
    } while (h*n < e);
    p=q=f/g;
    e=b-2.0;
    do {
        p=(n-1+c/n)/(e+(n+1)*(2.0-p));
        q=p*(1.0+q);
        n--;
    } while (n > 0);
    f=Math.sqrt(Math.PI/b)/(1.0+q);
    g=f*(a+x+0.5-p)/x;
}
if (rec) {
    x=2.0/x;
    for (n=1; n<=na; n++) {
        h=f+(a+n)*x*g;
        f=g;
        g=h;
    }
}
```

```
    if (rev) {
        ka1[0] = f;
        ka[0] = g;
    } else {
        ka[0] = f;
        ka1[0] = g;
    }
  }
}
```

F. nonexpbesskaplusn

Generates an array of modified Bessel functions of the third kind of order $a+n$, $K_{a+n}(x)$, $n=0,...,nmax$, multiplied by the factor e^x, for $x > 0$, $a \geq 0$. Thus, apart from the exponential factor, the functions are the same as those computed by *besskaplusn*.

      nonexpbesskaplusn computes the values of the functions $K_{a+n}'(x) = e^x K_{a+n}(x)$ $(n=0,...,nmax)$. The values of the functions $K_a'(x)$, $K_{a+1}'(x)$ are first obtained by means of a call of *nonexpbesska01*, and the recursion
$$K_{a+n+1}'(x) = K_{a+n-1}'(x) + \{2(n+a)/x\}K_{a+n}'(x)$$
is then used.

Procedure parameters:
$$\text{void nonexpbesskaplusn } (a,x,nmax,kan)$$
a: double;
           entry: the order; $a \geq 0$;
x: double;
           entry: the argument value, $x > 0$;
nmax: int;
           entry: the upper bound of the indices of the array *kan*; $nmax \geq 0$;
kan: double *kan[0:nmax]*;
           exit: the values of the modified Bessel functions of the third kind of order $a+n$, for argument x multiplied by e^x are assigned to *kan[n]*, $0 \leq n \leq nmax$.

Procedure used: nonexpbesska01.

```
public static void nonexpbesskaplusn(double a, double x, int nmax, double kan[])
{
    int n;
    double k1;
    double tmp1[] = new double[1];
    double tmp2[] = new double[1];

    nonexpbesska01(a,x,tmp1,tmp2);
    kan[0]=tmp1[0];
    k1=tmp2[0];
```

```
    a -= 1.0;
    x=2.0/x;
    if (nmax > 0) kan[1]=k1;
    for (n=2; n<=nmax; n++) kan[n]=kan[n-2]+(a+n)*x*kan[n-1];
}
```

6.6.3 Spherical Bessel functions

A. spherbessj

Calculates the spherical Bessel functions $J_{k+0.5}'(x) = (\pi/(2x))^{1/2}J_{k+0.5}(x)$, $k=0,...,n$, where $J_{k+0.5}(x)$ denotes the Bessel function of the first kind of order $k+0.5$, for $x \geq 0$.
The method used [G67] is that described in the documentation to *bessj*, with $a = 0.5$ in formula (1). Since
$$J_{0.5}'(x) = x^{-1}\sin(x),\ \lambda_0 = 1,\ \lambda_k = 0\ (k=1,2,...),$$
$s = x^{-1}\sin(x)$ may be taken in the further formulae of that documentation, which in consequence become a little simpler.

Procedure parameters:
$$\text{void spherbessj } (x,n,j)$$
x: double;
           entry: the value of the argument; $x \geq 0$;
n: int;
           entry: the upper bound of the indices of array j; $n \geq 0$;
j: double $j[0:n]$;
           exit: $j[k]$ is the value of the spherical Bessel function $J_{k+0.5}'(x)$, $k=0,...,n$.

Procedure used: start.

```
public static void spherbessj(double x, int n, double j[])
{
    if (x == 0.0) {
        j[0]=1.0;
        for (; n>=1; n--) j[n]=0.0;
    } else if (n == 0) {
        double x2;
        if (Math.abs(x) < 0.015) {
            x2=x*x/6.0;
            j[0]=1.0+x2*(x2*0.3-1.0);
        } else
            j[0]=Math.sin(x)/x;
    } else {
        int m;
        double r,s;
        r=0.0;
```

```
        m=start(x,n,0);
        for (; m>=1; m--) {
            r=1.0/((m+m+1)/x-r);
            if (m <= n) j[m]=r;
        }
        if (x < 0.015) {
            s=x*x/6.0;
            j[0]=r=s*(s*0.3-1.0)+1.0;
        } else
            j[0]=r=Math.sin(x)/x;
        for (m=1; m<=n; m++) r = j[m] *= r;
    }
}
```

B. spherbessy

Calculates the spherical Bessel functions $Y_{j+0.5}'(x) = (\pi/(2x))^{1/2}Y_{j+0.5}(x)$, $j=0,...,n$, where $Y_{j+0.5}(x)$ denotes the Bessel function of the third kind of order $j+0.5$, for $x > 0$. The recursion [AbS65]

$$Y_{0.5}'(x) = -x^{-1}\cos(x),$$
$$Y_{1.5}'(x) = -x^{-2}\cos(x) - x^{-1}\sin(x)$$
$$Y_{j+0.5}'(x) = ((2j-1)/x)Y_{j-0.5}'(x) + Y_{j-1.5}'(x) \qquad (j=2,...,n)$$

is used.

Procedure parameters:
$$\text{void spherbessy }(x,n,y)$$

x: double;
          entry: the argument of the Bessel functions; $x > 0$;
n: int;
          entry: the upper bound of the indices of array y; $n \geq 0$;
y: double $y[0:n]$;
          exit: $y[j]$ is the value of the spherical Bessel function $Y_{j+0.5}'(x)$, $j=0,...,n$.

```
public static void spherbessy(double x, int n, double y[])
{
    if (n == 0)
        y[0] = -Math.cos(x)/x;
    else {
        int i;
        double yi,yi1,yi2;
        yi2 = y[0] = -Math.cos(x)/x;
        yi1=y[1]=(yi2-Math.sin(x))/x;
        for (i=2; i<=n; i++) {
            y[i] = yi = -yi2+(i+i-1)*yi1/x;
            yi2=yi1;
            yi1=yi;
        }
}
```

```
        }
}
```

C. spherbessi

Calculates the modified spherical Bessel functions $I_{j+0.5}'(x) = (\pi/(2x))^{1/2}I_{j+0.5}(x)$, $j=0,...,n$, where $I_{j+0.5}(x)$ denotes the modified Bessel function of the first kind of order $j+0.5$, for $x \geq 0$.

The functions $I_{j+0.5}''(x) = e^{-x}I_{j+0.5}'(x)$, $j=0,...,n$, are first evaluated by means of a call of *nonexpspherbessi*; the values of the functions $I_{j+0.5}'(x)$ are then recovered by multiplication by e^x.

Procedure parameters:
$$\text{void spherbessi } (x,n,i)$$
x:	double;
	entry: the argument of the Bessel functions; $x \geq 0$;
n:	int;
	entry: the upper bound of the indices of array i; $n \geq 0$;
i:	double $i[0:n]$;
	exit: $i[j]$ is the value of the modified spherical Bessel function $I_{j+0.5}'(x)$, $j=0,...,n$.

Procedure used: nonexpspherbessi.

```
public static void spherbessi(double x, int n, double i[])
{
    if (x == 0.0) {
        i[0]=1.0;
        for (; n>=1; n--) i[n]=0.0;
    } else {
        double expx;
        expx=Math.exp(x);
        nonexpspherbessi(x,n,i);
        for (; n>=0; n--) i[n] *= expx;
    }
}
```

D. spherbessk

Calculates the modified spherical Bessel functions $K_{j+0.5}'(x) = (\pi/(2x))^{1/2}K_{j+0.5}(x)$, $j=0,...,n$, where $K_{j+0.5}(x)$ denotes the modified Bessel function of the third kind of order $j+0.5$, for $x > 0$.

The functions $K_{j+0.5}''(x) = e^xK_{j+0.5}'(x)$, $j=0,...,n$, are first evaluated by means of a call of *nonexpspherbessk*; the functions $K_{j+0.5}'(x)$ are then recovered by multiplication by e^{-x}.

Procedure parameters:

$$\text{void spherbessk } (x,n,k)$$

x: double;

        entry: the argument value; $x > 0$;

n: int;

        entry: the upper bound of the indices of array k; $n \geq 0$;

k: double $k[0{:}n]$;

        exit: $k[j]$ is the value of the modified spherical Bessel function $K_{j+0.5}'(x)$,
                 $j=0,\ldots,n$.

Procedure used: nonexpspherbessk.

```
public static void spherbessk(double x, int n, double k[])
{
    double expx;
    expx=Math.exp(-x);
    nonexpspherbessk(x,n,k);
    for (; n>=0; n--) k[n] *= expx;
}
```

E. nonexpspherbessi

Calculates the modified spherical Bessel functions multiplied by e^x. *nonexpspherbessi* evaluates $I_{j+0.5}''(x) = e^{-x}(\pi/(2x))^{1/2}I_{j+0.5}(x)$, $j=0,\ldots,n$, where $I_{j+0.5}(x)$ denotes the modified Bessel function of the first kind of order $j+0.5$, for $x \geq 0$.

The ratio of two subsequent elements is computed using a backward recurrence formula according to Miller's method (see [G67]). Since the zeroth element is known to be $(1-e^{-2x})/(2x)$, the other elements follow immediately. The starting value is computed by *start*.

Procedure parameters:

$$\text{void nonexpspherbessi } (x,n,i)$$

x: double;

        entry: the argument of the Bessel functions; $x \geq 0$;

n: int;

        entry: the upper bound of the indices of array i; $n \geq 0$;

i: double $i[0{:}n]$;

        exit: $i[j]$ is the value of the modified spherical Bessel function $I_{j+0.5}''(x)$,
                 $j=0,\ldots,n$.

Procedure used: start.

```
public static void nonexpspherbessi(double x, int n, double i[])
{
    if (x == 0.0) {
        i[0]=1.0;
```

```
    for (; n>=1; n--) i[n]=0.0;
} else {
    int m;
    double x2,r;
    x2=x+x;
    i[0] = x2 = ((x == 0.0) ? 1.0 : ((x2 < 0.7) ?
        sinh(x)/(x*Math.exp(x)) : (1.0-Math.exp(-x2))/x2));
    if (n != 0) {
        r=0.0;
        m=start(x,n,1);
        for (; m>=1; m--) {
            r=1.0/((m+m+1)/x+r);
            if (m <= n) i[m]=r;
        }
        for (m=1; m<=n; m++) x2 = i[m] *= x2;
    }
}
}
```

F. nonexpspherbessk

Calculates the modified spherical Bessel functions multiplied by e^x. *nonexpspherbessk* evaluates the functions $K_{j+0.5}''(x) = e^x(\pi/(2x))^{1/2}K_{j+0.5}(x)$, $j=0,...,n$, where $K_{j+0.5}(x)$ denotes the modified Bessel function of the third kind of order $j+0.5$, for $x > 0$.

The recursion

$$K_{0.5}''(x) = \pi/(2x),$$
$$K_{1.5}''(x) = (1 + 1/x)K_{0.5}''(x)$$
$$K_{j+0.5}''(x) = ((2j-1)/x)K_{j-0.5}''(x) + K_{j-1.5}''(x) \qquad (j=2,...,n)$$

is used.

Procedure parameters:

$$\text{void nonexpspherbessk } (x,n,k)$$

x: double;

       entry: the argument of the Bessel functions; $x > 0$;

n: int;

       entry: the upper bound of the indices of array k; $n \geq 0$;

k: double $k[0:n]$;

       exit: $k[j]$ is the value of the modified spherical Bessel function $K_{j+0.5}''(x)$, $j=0,...,n$.

```
public static void nonexpspherbessk(double x, int n, double k[])
{
    int i;
    double ki,ki1,ki2;
    x=1.0/x;
    k[0]=ki2=x*1.5707963267949;
```

```
if (n != 0) {
    k[1]=ki1=ki2*(1.0+x);
    for (i=2; i<=n; i++) {
        k[i]=ki=ki2+(i+i-1)*x*ki1;
        ki2=ki1;
        ki1=ki;
    }
}
}
```

6.6.4 Airy functions

A. airy

Evaluates the functions $e^{-expon(z)}Ai(z)$ and $e^{expon(z)}Bi(z)$, where the Airy functions Ai and Bi (see [AbS65, Gor69]) may be defined by the formulae
$$Ai(z) = c_1f(z) - c_2g(z), \quad Bi(z) = 3^{1/2}\{c_1f(z) + c_2g(z)\}$$
where

$$f(z) = \sum_{k=0}^{\infty} \frac{f_k z^{3k}}{(3k)!}, \qquad g(z) = \sum_{k=0}^{\infty} \frac{g_k z^{3k+1}}{(3k+1)!}$$

$$f_k = \prod_{\tau=0}^{k-1}(3\tau+1), \qquad g_k = \prod_{\tau=0}^{k-1}(3\tau+2)$$

$$c_1 = 3^{-\frac{2}{3}} \Gamma(\tfrac{2}{3}), \qquad c_2 = 3^{-\frac{1}{3}} \Gamma(\tfrac{1}{3})$$

and $expon(z) = 0$ for $z < 9$, and $expon(z) = (2/3)z^{3/2}$ for $z \geq 9$. $airy$ also evaluates the derivatives of these functions:
$$e^{-expon(z)}dAi(z)/dz \text{ and } e^{expon(z)}dBi(z)/dz.$$

When $-5 \leq z \leq 8$, the above defining formulae, and those obtained by differentiation are used to compute the Airy functions and their derivatives.

When $z \geq 8$, approximate versions of the formulae

$$Ai(z) = \frac{1}{2\sqrt{\pi}\, z^{\frac{1}{4}}} e^{-\frac{3}{2}z^{\frac{3}{2}}} \int_0^{\infty} \frac{\rho(x)}{1+\left(3x/2z^{3/2}\right)} dx$$

$$Bi(z) = \frac{1}{\sqrt{\pi}\, z^{\frac{1}{4}}} e^{\frac{3}{2}z^{\frac{3}{2}}} \int_0^{\infty} \frac{\rho(x)}{1-\left(3x/2z^{3/2}\right)} dx$$

where

$$\rho(x) = \frac{1}{\sqrt{\pi}\, 2^{11/6}\, 3^{2/3}} e^{-x} Ai\left((3x/2)^{2/3}\right)$$

are used. These versions have the form

$$Ai(z) = \frac{1}{2\sqrt{\pi}\ z^{\frac{1}{4}}}\ e^{-\frac{3}{2}z^{\frac{3}{2}}} \sum_{i=1}^{n} \frac{w_i}{1 + \left(3x_i / 2z^{3/2}\right)}$$

$$Bi(z) = \frac{1}{\sqrt{\pi}\ z^{\frac{1}{4}}}\ e^{\frac{3}{2}z^{\frac{3}{2}}} \sum_{i=1}^{n} \frac{w_i}{1 - \left(3x_i / 2z^{3/2}\right)}$$

where

$$\int_0^{\infty} x^k \rho(x)dx = \frac{\Gamma(3k + \frac{1}{2})}{54^k\ k!\ \Gamma(k + \frac{1}{2})} = \sum_{i=1}^{n} x_i^k\ w_i \qquad \left(k = 0,\dots,2n-1\right).$$

When $z < 0$, similar use is made of the formulae

$$Ai(-z) = \frac{1}{\sqrt{\pi}\ z^{\frac{1}{4}}} \int_0^{\infty} \frac{\cos\left(\beta - \frac{\pi}{4}\right) + \left(\frac{x}{\beta}\right)\sin\left(\beta - \frac{\pi}{4}\right)}{1 + \left(\frac{x}{\beta}\right)^2} \rho(x)dx$$

$$Bi(-z) = \frac{1}{\sqrt{\pi}\ z^{\frac{1}{4}}} \int_0^{\infty} \frac{\left(\frac{x}{\beta}\right)\cos\left(\beta - \frac{\pi}{4}\right) - \sin\left(\beta - \frac{\pi}{4}\right)}{1 + \left(\frac{x}{\beta}\right)^2} \rho(x)dx$$

where $\beta = 2z^{3/2}/3$. In both of the above cases the derivatives of Ai and Bi are obtained by differentiation of their approximation formulae.

Procedure parameters:
$$\text{void airy } (z,ai,aid,bi,bid,expon,first,xtmp)$$

z: double;

  entry: the real argument of the Airy functions;

ai: double *ai[0:0]*;

  exit: the value of the Airy function given by $e^{-expon(z)}Ai(z)$;

aid: double *aid[0:0]*;

  exit: the value of the derivative of the Airy function given by $e^{-expon(z)}Ai(z)$;

bi: double *bi[0:0]*;

  exit: the value of the Airy function given by $e^{expon(z)}Bi(z)$;

bid: double *bid[0:0]*;

  exit: the value of the derivative of the Airy function given by $e^{expon(z)}Bi(z)$;

expon: double *expon[0:0]*;

  exit: if $z < 9$ then 0 else $(2/3)z^{3/2}$;

first: boolean;

  entry: the value of *first* should be false unless the procedure is called for the first time; if *first* is true then two internal static arrays are built up;

xtmp: double *xtmp[1:25]*;

  working storage for internal static data.

```
public static void airy(double z, double ai[], double aid[], double bi[],
                double bid[], double expon[], boolean first, double xtmp[])
{
    int n,l;
    double s,t,u,v,sc,tc,uc,vc,x,k1,k2,k3,k4,c,zt,si,co,expzt,sqrtz,wwl,pl,pl1,pl2,
        pl3;
```

```
if (first) {
    xtmp[1] =1.4083081072180964e1;
    xtmp[2] =1.0214885479197331e1;
    xtmp[3] =7.4416018450450930;
    xtmp[4] =5.3070943061781927;
    xtmp[5] =3.6340135029132462;
    xtmp[6] =2.3310652303052450;
    xtmp[7] =1.3447970842609268;
    xtmp[8] =6.4188858369567296e-1;
    xtmp[9] =2.0100345998121046e-1;
    xtmp[10]=8.0594359172052833e-3;
    xtmp[11] =3.1542515762964787e-14;
    xtmp[12] =6.6394210819584921e-11;
    xtmp[13] =1.7583889061345669e-8;
    xtmp[14] =1.3712392370435815e-6;
    xtmp[15] =4.4350966639284350e-5;
    xtmp[16] =7.1555010917718255e-4;
    xtmp[17] =6.4889566103335381e-3;
    xtmp[18] =3.6440415875773282e-2;
    xtmp[19] =1.4399792418590999e-1;
    xtmp[20]=8.1231141336261486e-1;
    xtmp[21]=0.355028053887817;
    xtmp[22]=0.258819403792807;
    xtmp[23]=1.73205080756887729;
    xtmp[24]=0.78539816339744831;
    xtmp[25]=0.56418958354775629;
}
expon[0]=0.0;
if (z >= -5.0 && z <= 8.0) {
    u=v=t=uc=vc=tc=1.0;
    s=sc=0.5;
    n=3;
    x=z*z*z;
    while (Math.abs(u)+Math.abs(v)+Math.abs(s)+Math.abs(t) > 1.0e-18) {
        u=u*x/(n*(n-1));
        v=v*x/(n*(n+1));
        s=s*x/(n*(n+2));
        t=t*x/(n*(n-2));
        uc += u;
        vc += v;
        sc += s;
        tc += t;
        n += 3;
    }
    bi[0]=xtmp[23]*(xtmp[21]*uc+xtmp[22]*z*vc);
    bid[0]=xtmp[23]*(xtmp[21]*z*z*sc+xtmp[22]*tc);
    if (z < 2.5) {
        ai[0]=xtmp[21]*uc-xtmp[22]*z*vc;
        aid[0]=xtmp[21]*sc*z*z-xtmp[22]*tc;
        return;
```

```
        }
    }
    k1=k2=k3=k4=0.0;
    sqrtz=Math.sqrt(Math.abs(z));
    zt=0.666666666666667*Math.abs(z)*sqrtz;
    c=xtmp[25]/Math.sqrt(sqrtz);
    if (z < 0.0) {
        z = -z;
        co=Math.cos(zt-xtmp[24]);
        si=Math.sin(zt-xtmp[24]);
        for (l=1; l<=10; l++) {
            wwl=xtmp[l+10];
            pl=xtmp[l]/zt;
            pl2=pl*pl;
            pl1=1.0+pl2;
            pl3=pl1*pl1;
            k1 += wwl/pl1;
            k2 += wwl*pl/pl1;
            k3 += wwl*pl*(1.0+pl*(2.0/zt+pl))/pl3;
            k4 += wwl*(-1.0-pl*(1.0+pl*(zt-pl))/zt)/pl3;
        }
        ai[0]=c*(co*k1+si*k2);
        aid[0]=0.25*ai[0]/z-c*sqrtz*(co*k3+si*k4);
        bi[0]=c*(co*k2-si*k1);
        bid[0]=0.25*bi[0]/z-c*sqrtz*(co*k4-si*k3);
    } else {
        if (z < 9.0)
            expzt=Math.exp(zt);
        else {
            expzt=1.0;
            expon[0]=zt;
        }
        for (l=1; l<=10; l++) {
            wwl=xtmp[l+10];
            pl=xtmp[l]/zt;
            pl1=1.0+pl;
            pl2=1.0-pl;
            k1 += wwl/pl1;
            k2 += wwl*pl/(zt*pl1*pl1);
            k3 += wwl/pl2;
            k4 += wwl*pl/(zt*pl2*pl2);
        }
        ai[0]=0.5*c*k1/expzt;
        aid[0]=ai[0]*(-0.25/z-sqrtz)+0.5*c*sqrtz*k2/expzt;
        if (z >= 8.0) {
            bi[0]=c*k3*expzt;
            bid[0]=bi[0]*(sqrtz-0.25/z)-c*k4*sqrtz*expzt;
        }
    }
}
```

B. airyzeros

Computes the zeros and associated values of the Airy functions $Ai(x)$ and $Bi(x)$, and their derivatives.

Denoting by a_s, a_s', b_s, b_s' the s-th zero of $Ai(x)$, $dAi(x)/dx$, $Bi(x)$, $dBi(x)/dx$ respectively (see [AbS65]):

$$a_s \approx -f\{3\pi(4s-1)/8\}, \quad a_s' \approx -g\{3\pi(4s-3)/8\}$$
$$b_s \approx -f\{3\pi(4s-3)/8\}, \quad b_s' \approx -g\{3\pi(4s-1)/8\}$$

where the functions f and g are defined in the documentation of *airy*. The appropriate member of the above set of approximations is used as the first iterate in a quadratic interpolation process for determining the zeros in question. The values of the Airy functions (and the associated values delivered in the array *vai*) are calculated by means of the procedure *airy*.

Procedure parameters:

$$\text{double airyzeros } (n,d,zai,vai)$$

airyzeros: delivers the n-th zero of the selected Airy function;
n: int;
    entry: the number of zeros to be calculated;
d: int;
    entry: an integer which selects the required Airy function, $d = 0$, 1, 2, or 3;
zai: double *zai[1:n]*;
    exit: *zai[j]* contains the *j*-th zero of the selected Airy function:
      if $d=0$ then $Ai(x)$,
      if $d=1$ then $(d/dx)Ai(x)$,
      if $d=2$ then $Bi(x)$,
      if $d=3$ then $(d/dx)Bi(x)$;
vai: double *vai[1:n]*;
    exit: *vai[j]* contains the value at $x = zai[j]$ of the following function:
      if $d=0$ then $(d/dx)Ai(x)$,
      if $d=1$ then $Ai(x)$,
      if $d=2$ then $(d/dx)Bi(x)$,
      if $d=3$ then $Bi(x)$.

Procedure used: airy.

```
public static double airyzeros(int n, int d, double zai[], double vai[])
{
    boolean a,found;
    int i;
    double c,e,r,zaj,zak,vaj,daj,kaj,zz;
    double statictemp[] = new double[26];
    double tmp1[] = new double[1];
    double tmp2[] = new double[1];
    double tmp3[] = new double[1];
    double tmp4[] = new double[1];
    double tmp5[] = new double[1];
```

```
a=((d == 0) || (d == 2));
r = (d == 0 || d == 3) ? -1.17809724509617 : -3.53429173528852;
airy(0.0,tmp1,tmp2,tmp3,tmp4,tmp5,true,statictemp);
zaj=tmp1[0];
vaj=tmp2[0];
daj=tmp3[0];
kaj=tmp4[0];
zz=tmp5[0];
for (i=1; i<=n; i++) {
    r += 4.71238898038469;
    zz=r*r;
    zaj = (i == 1 && d == 1) ? -1.01879297 : ((i == 1 && d == 2) ? -1.17371322 :
            Math.pow(r,0.666666666666667)*
            (a ? -(1.0+(5.0/48.0-(5.0/36.0-(77125.0/82944.0-
            (108056875.0/6967296.0-(162375596875.0/334430208.0)/
            zz)/zz)/zz)/zz)/zz) : -(1.0-(7.0/48.0-(35.0/288.0-
            (181223.0/207360.0-(18683371.0/1244160.0-
            (91145884361.0/191102976.0)/zz)/zz)/zz)/zz)/zz)));
    if (d <= 1.0) {
        airy(zaj,tmp1,tmp2,tmp3,tmp4,tmp5,false,statictemp);
        vaj=tmp1[0];
        daj=tmp2[0];
        c=tmp3[0];
        e=tmp4[0];
        zz=tmp5[0];
    }
    else {
        airy(zaj,tmp1,tmp2,tmp3,tmp4,tmp5,false,statictemp);
        c=tmp1[0];
        e=tmp2[0];
        vaj=tmp3[0];
        daj=tmp4[0];
        zz=tmp5[0];
    }
    found=(Math.abs(a ? vaj : daj) < 1.0e-12);
    while (!found) {
        if (a) {
            kaj=vaj/daj;
            zak=zaj-kaj*(1.0+zaj*kaj*kaj);
        } else {
            kaj=daj/(zaj*vaj);
            zak=zaj-kaj*(1.0+kaj*(kaj*zaj+1.0/zaj));
        }
        if (d <= 1) {
            airy(zak,tmp1,tmp2,tmp3,tmp4,tmp5,false,statictemp);
            vaj=tmp1[0];
            daj=tmp2[0];
            c=tmp3[0];
            e=tmp4[0];
            zz=tmp5[0];
```

```
            }
            else {
                airy(zak,tmp1,tmp2,tmp3,tmp4,tmp5,false,statictemp);
                c=tmp1[0];
                e=tmp2[0];
                vaj=tmp3[0];
                daj=tmp4[0];
                zz=tmp5[0];
            }
            found=(Math.abs(zak-zaj) < 1.0e-14*Math.abs(zak) ||
                    Math.abs(a ? vaj : daj) < 1.0e-12);
            zaj=zak;
        }
        vai[i]=(a ? daj : vaj);
        zai[i]=zaj;
    }
    return zai[n];
}
```

7. Interpolation and Approximation

All procedures in this chapter are class methods of the following *Approximation* class.

```
package numal;

import numal.*;

public class Approximation extends Object {
    // all procedures in this chapter are to be inserted here
}
```

7.1 Real data in one dimension

7.1.1 Interpolation with general polynomials

newton

Calculates the coefficients of the Newton polynomial through given interpolation points and corresponding function values.

newton computes the divided differences $\delta_x^j f(x_0)$, $j=0,...,n$, derived from the function values f_i and their associated arguments x_i, $i=0,...,n$, (the x_i being assumed distinct). The $\{\delta_x^j f(x_0)\}$ are computed by use of the defining recursion

$$\delta_i^0 = f_i \qquad (i=0,...,n)$$
$$\delta_i^{j+1} = (\delta_{i+1}^j - \delta_i^j) / (x_{i+j+1} - x_i) \qquad (j=0,...,n-1; \; i=0,...,n-j-1)$$

when $\delta_x^j f(x_0) = \delta_0^j$.

Procedure parameters:

$$\text{void newton } (n,x,f)$$

n: int;
     entry: the degree of the polynomial;
x: double $x[0:n]$;
     entry: the interpolation points;
f: double $f[0:n]$;
     entry: the function values at the interpolation points;
     exit: the coefficients of the Newton polynomial.

```
public static void newton(int n, double x[], double f[])
{
    int k,i,im1;
    double xim1,fim1;
```

```
    im1=0;
    for (i=1; i<=n; i++) {
        fim1=f[im1];
        xim1=x[im1];
        for (k=i; k<=n; k++) f[k]=(f[k]-fim1)/(x[k]-xim1);
        im1=i;
    }
}
```

7.1.2 Approximation in infinity norm

A. ini

Selects a set of integers out of a given set of integers. With integers m,n $(m > n)$, *ini* determines a sequence of integer approximations $s(j)$, $j=0,...,n$; $s(j) > s(j-1)$, to the points in the range $[0,m]$ at which $T_n((2x-m)/m)$, where
$$T_n(y) = \cos\{n \text{ arc } \cos(y)\}$$
being a Chebyshev polynomial and $(2x-m)/m$ maps $x\in[-1,1]$ onto $[0,m]$, assumes its maximal absolute values.

    ini is an auxiliary procedure used in *minmaxpol*.

Procedure parameters:
$$\text{void ini } (n,m,s)$$
n,m: int;
        entry: the number of points to be selected equals $n+1$; the reference set
                contains the numbers $0,1,...,m$, $(m \geq n)$;
s: int $s[0:n]$;
        exit: the selected integers are delivered in *s*.

```
public static void ini(int n, int m, int s[])
{
    int i,j,k,l;
    double pin2,temp;

    pin2=Math.atan(1.0)*2.0/n;
    k=1;
    l=n-1;
    j=s[0]=0;
    s[n]=m;
    while (k < l) {
        temp=Math.sin(k*pin2);
        i=(int)(temp*temp*m);
        j = s[k] = ((i <= j) ? j+1 : i);
        s[l]=m-j;
```

```
        l--;
        k++;
    }
    if (l*2 == n) s[l]=m/2;
}
```

B. sndremez

Exchanges at most $n+1$ numbers with numbers out of a reference set   (see [Me64]). Given a sequence of real numbers g_i, $i=0,...,m$, which contains a subsequence $g_{s(j)}$, $j=0,1,...,n<m$; $s(j) > s(j-1)$, for which a)
$$|g_{s(j)}| = G, j=0,...,n,$$
and b)
$$\text{sign } g_{s(j)} = -\text{sign } g_{s(j-1)}, j=1,...,n,$$
a sequence of integers $S(j)$, $j=0,1,...,n$;
$$S(j) > S(j-1); S(0) \geq 0, S(n) \leq m,$$
is determined such that A)
$$|g_{S(j)}| \geq G, j=0,...,n,$$
B)
$$\text{sign } g_{S(j)} = -\text{sign } g_{S(j-1)}, j=1,...,n,$$
and C) $|g_{S(j)}| > G$ for at least one j in the range $0 \leq j \leq n$ if $|g_i| > G$ for at least one i in the range $0 \leq i \leq n$.

The process used is as follows: let
$$H(1) = \max|g_i| \text{ for } s(0) < i < s(1)$$
and $|g_{I(1)}| = H(1)$ ($H(1)$ is zero if $s(0) = s(1)$ and $I(1)$ is then undefined; a similar convention is adopted below). If $H(1) \leq G$ then
$$S(0) = s(0), S(1) = s(1);$$
otherwise if
$$\text{sign } g_{I(1)} = \text{sign } g_{s(0)}$$
then
$$S(0) = I(1), S(1)=s(1),$$
whereas if
$$\text{sign } g_{I(1)} \neq \text{sign } g_{s(0)},$$
then
$$S(0) = s(0), S(1) = I(1).$$
Let
$$H(2) = \max|g_i| \text{ for } s(1) < i < s(2) \text{ and } |g_{I(2)}| = H(2).$$
If $H(2) \leq G$ then $S(2) = s(2)$; otherwise if sign $g_{I(2)} = $ sign $g_{S(1)}$ then $S(1)$ is replaced by $I(2)$ and $S(2) = s(2)$, whereas if sign $g_{I(2)} \neq$ sign $g_{S(1)}$ then $S(2) = I(2)$. The numbers g_i, $s(2)<i<s(3)$, are inspected, $S(2)$ is possibly replaced, and $S(3)$ is determined as before. In this way $S(0),...,S(n)$ are determined. Let
$$H' = \max|g_i| \text{ for } 0 \leq i < s(0) \text{ with } |g_{I'}|=H',$$
and
$$H'' = \max|g_i| \text{ for } s(n) < i \leq m, \text{ with } |g_{I''}| = H''.$$
If H', $H'' < G$, the numbers $S(j)$ are left as they stand. If not, then if $H' > H''$ then a) if sign $g_{I'} = $ sign $g_{S(0)}$ then a') if $H'>|g_{S(0)}|$, $S(0)$ is replaced by I' and a'') if sign $g_{I'} = g_{S(n)}$ and $H'' > |g_{S(n)}|$, $S(n)$ is replaced by I''; b) if sign $g_{I'} \neq$ sign $g_{S(0)}$ then b') if $H' > |g_{S(n)}|$, $S(k)$ is replaced by $S(k-1)$, $k=n,...,1$, and

$S(0)$ is replaced by I' and b") (with the new value of $S(n)$) if sign $g_{I'}$ = sign $g_{S(n)}$ and $H'' > |g_{S(n)}|$, $S(n)$ is replaced by I''. If $H'' \geq H'$ then similar modifications take place.

    sndremez is an auxiliary procedure used in *minmaxpol*.

Procedure parameters:
$$\text{void sndremez } (n,m,s,g,em)$$

n,m: int;

        entry: the number of points to be exchanged is smaller than or equal to $n+1$;

             the reference set contains the numbers $0,1,...,m$, $(m \geq n)$;

s: int $s[0:n]$;

        entry: in s one must give $n+1$ (strictly) monotone increasing numbers out of $0,1,...,m$;

        exit: $n+1$ (strictly) monotone increasing numbers out of the numbers $0,1,...,m$;

g: double $g[0:m]$;

        entry: in array $g[0:m]$ one must give the function values;

em: double $em[0:1]$;

        entry: $0 < em[0] \leq g[i]$, $i=0,...,m$;

        exit: $em[1]$ = infinity norm of array $g[0:m]$.

Procedure used: infnrmvec.

```
public static void sndremez(int n, int m, int s[], double g[], double em[])
{
    int s0,sn,sjp1,i,j,k,up,low,nml;
    double max,msjp1,hi,hj,he,abse,h,temp1,temp2;
    int itmp[] = new int[1];
    int jtmp[] = new int[1];

    s0=sjp1=s[0];
    he=em[0];
    low=s0+1;
    max=msjp1=abse=Math.abs(he);
    nml=n-1;
    for (j=0; j<=nml; j++) {
        up=s[j+1]-1;
        h=Basic.infnrmvec(low,up,itmp,g);
        i=itmp[0];
        if (h > max) max=h;
        if (h > abse)
            if (he*g[i] > 0.0) {
                s[j] = (msjp1 < h) ? i : sjp1;
                sjp1=s[j+1];
                msjp1=abse;
            } else {
                s[j]=sjp1;
                sjp1=i;
                msjp1=h;
            }
```

```
        else {
            s[j]=sjp1;
            sjp1=s[j+1];
            msjp1=abse;
        }
        he = -he;
        low=up+2;
    }
    sn=s[n];
    s[n]=sjp1;
    hi=Basic.infnrmvec(0,s0-1,itmp,g);
    i=itmp[0];
    hj=Basic.infnrmvec(sn+1,m,jtmp,g);
    j=jtmp[0];
    if (j > m) j=m;
    if (hi > hj) {
        if (hi > max) max=hi;
        temp1 = (g[i] == 0.0) ? 0.0 : ((g[i] > 0.0) ? 1.0 : -1.0);
        temp2 = (g[s[0]]==0.0) ? 0.0 : ((g[s[0]]>0.0) ? 1.0 : -1.0);
        if (temp1 == temp2) {
            if (hi > Math.abs(g[s[0]])) {
                s[0]=i;
                if (g[j]/g[s[n]] > 1.0) s[n]=j;
            }
        }
        else {
            if (hi > Math.abs(g[s[n]])) {
                s[n] = (g[j]/g[s[nm1]] > 1.0) ? j : s[nm1];
                for (k=nm1; k>=1; k--) s[k]=s[k-1];
                s[0]=i;
            }
        }
    } else {
        if (hj > max) max=hj;
        temp1 = (g[j] == 0.0) ? 0.0 : ((g[j] > 0.0) ? 1.0 : -1.0);
        temp2 = (g[s[n]]==0.0) ? 0.0 : ((g[s[n]]>0.0) ? 1.0 : -1.0);
        if (temp1 == temp2) {
            if (hj > Math.abs(g[s[n]])) {
                s[n]=j;
                if (g[i]/g[s[0]] > 1.0) s[0]=i;
            }
        } else
            if (hj > Math.abs(g[s[0]])) {
                s[0] = (g[i]/g[s[1]] > 1.0) ? i : s[1];
                for (k=1; k<=nm1; k++) s[k]=s[k+1];
                s[n]=j;
            }
    }
    em[1]=max;
}
```

C. minmaxpol

Calculates the coefficients of the polynomial (as a sum of powers) which approximates a function, given for discrete arguments, in such a way that the infinity norm of the error vector is minimized.

With y_j $(j=0,...,m)$ being a sequence of real arguments $(y_{j-1} < y_j; j=1,...,m)$ and $f(y_j)$ corresponding function values, *minmaxpol* determines the coefficients $\{c_i\}$ of that polynomial

$$\sum_{i=0}^{n-1} c_i y^i$$

for which

$$\max \left| f(y_j) - \sum_{i=0}^{n-1} c_i y_j^i \right| \tag{1}$$

is a minimum as j ranges over the values $0,...,m$ $(m > n)$.

The method used [Me64] involves the iterative construction of polynomials

$$\sum_{i=0}^{n-1} c_{k,i} y^i$$

for which

$$f(y_{s(k,j)}) - \sum_{i=0}^{n-1} c_{k,i} y_{s(k,j)}^i = (-1)^j \lambda_k \qquad (j = 0,...,n) \tag{2}$$

where $s(k,j)$, $j=0,...,n$, is a strictly increasing sequence selected from $0,1,...,m$. The discrepancies

$$g'_{k,j} = f(y_j) - \sum_{i=0}^{n-1} c_{k,i} y_j^i$$

at all points $j=0,...,m$ are then inspected, and from these a sequence $g_{k+1,j} = g_{k,s(k+1,j)}'$, $j=0,...,n$, is constructed which a) possesses the alternating property sign $g_{k+1,j}$ = -sign $g_{k+1,j-1}$, $j=1,...,n$, and b) for which $|g_{k+1,j}| \geq |g_{k,j}|$, $j=0,...,n$, with (unless the process is to be terminated) $|g_{k+1,j}| > |g_{k,j}|$ for at least one j (for the details of this construction, see the documentation to *sndremez*). The coefficients $\{c_{k+1,i}\}$ in the succeeding polynomial are determined (by use of Newton's interpolation formula, and subsequent reduction to the given polynomial form) from condition (2) with k replaced by $k+1$.

Initially $s(0,j)$, $j=0,...,n$, is a sequence of integer approximations to the points at which $T_n((2x-m)/m)$, where $T_n(y)$ = cos{n arc cos(y)} being a Chebyshev polynomial (see the documentation of *ini*) assumes its maximal absolute values in the range $0 \leq x \leq m$. The procedure *ini* is used to set the $s(0,j)$.

Procedure parameters:

$$\text{void minmaxpol } (n,m,y,fy,co,em)$$

n: int;

        entry: the degree of the approximating polynomial; $n \geq 0$;

m: int;

entry: the number of reference function values is $m+1$;

y,fy: double $y[0:m]$, $fy[0:m]$;

entry: $fy[i]$ is the function values at $y[i]$, $i=0,...,m$;

co: double $co[0:n]$;

exit: the coefficients of the approximating polynomial ($co[i]$ is the coefficient of y^i);

em: double $em[0:3]$;

entry: $em[2]$: maximum allowed number of iterations, say $10*n+5$;

exit: $em[0]$: the difference of the given function and the polynomial in the first approximation point;

$em[1]$: the infinity norm of the error of approximation over the discrete interval;

$em[3]$: the number of iterations performed.

Procedures used: elmvec, dupvec, newton, pol, newgrn, ini, sndremez.

```
public static void minmaxpol(int n, int m, double y[], double fy[], double co[],
                            double em[])
{
    int np1,k,pomk,count,cnt,j,mi,sjm1,sj,s0,up;
    double e,abse,abseh;
    int s[] = new int[n+2];
    double x[] = new double[n+2];
    double b[] = new double[n+2];
    double coef[] = new double[n+2];
    double g[] = new double[m+1];

    sj=0;
    np1=n+1;
    ini(np1,m,s);
    mi=(int)em[2];
    abse=0.0;
    count=1;
    do {
        pomk=1;
        for (k=0; k<=np1; k++) {
            x[k]=y[s[k]];
            coef[k]=fy[s[k]];
            b[k]=pomk;
            pomk = -pomk;
        }
        newton(np1,x,coef);
        newton(np1,x,b);
        em[0]=e=coef[np1]/b[np1];
        Basic.elmvec(0,n,0,coef,b,-e);
        Algebraic_eval.newgrn(n,x,coef);
        s0=sjm1=s[0];
        g[s0]=e;
        for (j=1; j<=np1; j++) {
            sj=s[j];
```

```
            up=sj-1;
            for (k=sjm1+1; k<=up; k++)
                g[k]=fy[k]-Algebraic_eval.pol(n,y[k],coef);
            g[sj] = e = -e;
            sjm1=sj;
        }
        for (k=s0-1; k>=0; k--)
            g[k]=fy[k]-Algebraic_eval.pol(n,y[k],coef);
        for (k=sj+1; k<=m; k++)
            g[k]=fy[k]-Algebraic_eval.pol(n,y[k],coef);
        sndremez(np1,m,s,g,em);
        abseh=abse;
        abse=Math.abs(e);
        cnt=count;
        count++;
    } while (count <= mi && abse > abseh);
    em[2]=mi;
    em[3]=cnt;
    Basic.dupvec(0,n,0,co,coef);
}
```

Addenda

All procedures in this chapter are class methods of the following *FFT* class.

```
package numal;

import numal.*;

public class FFT extends Object {
  // all procedures in this chapter are to be inserted here
}
```

I. Fast Fourier transforms

A. cfftp

Computes the values of

$$a_{k+1} = \sum_{j=0}^{n-1} a'_{j+1} e^{2\pi ijk/n} \qquad (k = 0, \ldots, n-1)$$

using a fast Fourier transform, where the $\{a'_{j+1}\}$ are complex numbers.

Procedure parameters:

$$\text{void cfftp } (a,n)$$

a: double *a[1:2][1:n]*;
       entry: Re(a'_{j+1}) and Im(a'_{j+1}) in locations (1,j+1) and (2,j+1) respectively
               (j=0,..,n-1);
       exit: Re(a_{k+1}) and Im(a_{k+1}) in locations (1,k+1) and (2,k+1) respectively
               (k=0,..,n-1);
n: integer; the value of *n* above.

```
public static void cfftp(double a[][], int n)
{
  int mp,m,ic,id,ill,ird,icc,iss,ick,isk,isf,iap,kd2,ibp,k,iam,ibm,mm1,i,ja,kt,ka,
      ita,itb,idm1,ikt,im,jj,j,kb,kn,jf,ktp,icf,l,mm,kf,isp,k0,k1,k2,k3,ikb,ija,ii,
      jk,kh;
  int iwk[] = new int[6*n+150];
  boolean l1,more,outloop,inloop;
  boolean ll[] = new boolean[6*n+150];
  double wk[] = new double[6*n+150];
  double ak2[] = new double[3];
  double cm,sm,c1,c2,c3,s1,s2,s3,c30,rad,a0,a1,a4,b4,a2,a3,b0,b1,b2,b3,zero,half,
         one,two;
  rad=2.0*Math.PI;
```

```
c30=0.8660254037844386;
zero=0.0;
half=0.5;
one=1.0;
two=2.0;
cm=sm=c2=s2=c3=s3=0.0;
k2=0;
if (n == 1) return;
k = n;
m = 0;
j = 2;
jj = 4;
jf = 0;
iwk[1] = 1;
while (true) {
  i = k/jj;
  if (i*jj == k) {
    m++;
    iwk[m+1] = j;
    k = i;
  } else {
    j += 2;
    if (j == 4) j = 3;
    jj = j * j;
    if (jj > k) break;
  }
}
kt = m;
j = 2;
while (true) {
  i = k / j;
  if (i*j == k) {
    m++;
    iwk[m+1] = j;
    k = i;
  } else {
    j = j + 1;
    if (j != 3) {
      j++;
      if (j > k) break;
    }
  }
}
k = iwk[m+1];
if (iwk[kt+1] > iwk[m+1]) k = iwk[kt+1];
if (kt > 0) {
  ktp = kt + 2;
  for (i=1; i<=kt; i++) {
    j = ktp - i;
    m++;
```

```
      iwk[m+1] = iwk[j];
   }
}
mp = m+1;
ic = mp+1;
id = ic+mp;
ill = id+mp;
ird = ill+mp+1;
icc = ird+mp;
iss = icc+mp;
ick = iss+mp;
isk = ick+k;
icf = isk+k;
isf = icf+k;
iap = isf+k;
kd2 = (k-1) / 2 + 1;
ibp = iap + kd2;
iam = ibp + kd2;
ibm = iam + kd2;
mm1 = m-1;
i = 1;
do {
   l = mp - i;
   j = ic - i;
   ll[ill+l] = (iwk[j-1] + iwk[j]) == 4;
   if (ll[ill+l]) {
      i++;
      l--;
      ll[ill+l] = false;
   }
   i++;
} while (i <= mm1);
ll[ill+1] = false;
ll[ill+mp] = false;
iwk[ic] = 1;
iwk[id] = n;
for (j=1; j<=m; j++) {
   k = iwk[j+1];
   iwk[ic+j] = iwk[ic+j-1] * k;
   iwk[id+j] = iwk[id+j-1] / k;
   wk[ird+j] = rad/iwk[ic+j];
   c1 = rad/k;
   if (k > 2) {
      wk[icc+j] = Math.cos(c1);
      wk[iss+j] = Math.sin(c1);
   }
}
mm = m;
if (ll[ill+m]) mm = m - 1;
if (mm > 1) {
```

```
      sm = iwk[ic+mm-2] * wk[ird+m];
      cm = Math.cos(sm);
      sm = Math.sin(sm);
    }
    kb = 0;
    kn = n;
    jj = 0;
    i = 1;
    c1 = one;
    s1 = zero;
    ll = true;
    outloop: for (;;) {
      if (ll[ill+i+1]) {
        kf = 4;
        i++;
      } else
        kf = iwk[i+1];
      isp = iwk[id+i];
      if (!ll) {
        s1 = jj * wk[ird+i];
        c1 = Math.cos(s1);
        s1 = Math.sin(s1);
      }
      inloop: for (;;) {
        if (kf <= 4) {
          switch (kf) {
            case 1:case 2:
              k0 = kb + isp;
              k2 = k0 + isp;
              if (!ll) {
                while (true) {
                  k0--;
                  if (k0 < kb) {
                    if ( i < mm ) {
                      i++;
                      continue outloop;
                    }
                    i = mm;
                    ll = false;
                    kb = iwk[id+i-1] + kb;
                    if (kb >= kn) break outloop;
                    more=true;
                    while (more) {
                      jj = iwk[ic+i-2] + jj;
                      if (jj < iwk[ic+i-1]) {
                        more=false;
                      } else {
                        i--;
                        jj -= iwk[ic+i];
                      }
```

```
      }
      if (i == mm) {
        c2 = c1;
        c1 = cm * c1 - sm * s1;
        s1 = sm * c2 + cm * s1;
        continue inloop;
      } else {
        if (ll[ill+i]) i++;
        continue outloop;
      }
    }
    k2--;
    a4 = a[1][k2+1];
    b4 = a[2][k2+1];
    a0 = a4*c1-b4*s1;
    b0 = a4*s1+b4*c1;
    a[1][k2+1] = a[1][k0+1]-a0;
    a[2][k2+1] = a[2][k0+1]-b0;
    a[1][k0+1] = a[1][k0+1]+a0;
    a[2][k0+1] = a[2][k0+1]+b0;
  }
}
while (true) {
  k0--;
  if (k0 < kb) {
    if ( i < mm ) {
      i++;
      continue outloop;
    }
    i = mm;
    ll = false;
    kb = iwk[id+i-1] + kb;
    if (kb >= kn) break outloop;
    more=true;
    while (more) {
      jj = iwk[ic+i-2] + jj;
      if (jj < iwk[ic+i-1]) {
        more=false;
      } else {
        i--;
        jj -= iwk[ic+i];
      }
    }
    if (i == mm) {
      c2 = c1;
      c1 = cm * c1 - sm * s1;
      s1 = sm * c2 + cm * s1;
      continue inloop;
    } else {
      if (ll[ill+i]) i++;
```

```
       continue outloop;
     }
   }
   k2--;
   ak2[1] = a[1][k2+1];
   ak2[2] = a[2][k2+1];
   a[1][k2+1] = a[1][k0+1]-ak2[1];
   a[2][k2+1] = a[2][k0+1]-ak2[2];
   a[1][k0+1] = a[1][k0+1]+ak2[1];
   a[2][k0+1] = a[2][k0+1]+ak2[2];
 }

case 3:
 if (!l1) {
   c2 = c1 * c1 - s1 * s1;
   s2 = two * c1 * s1;
 }
 ja = kb + isp - 1;
 ka = ja + kb;
 ikb = kb+1;
 ija = ja+1;
 for (ii=ikb; ii<=ija; ii++) {
   k0 = ka - ii + 1;
   k1 = k0 + isp;
   k2 = k1 + isp;
   a0 = a[1][k0+1];
   b0 = a[2][k0+1];
   if (l1) {
     a1 = a[1][k1+1];
     b1 = a[2][k1+1];
     a2 = a[1][k2+1];
     b2 = a[2][k2+1];
   } else {
     a4 = a[1][k1+1];
     b4 = a[2][k1+1];
     a1 = a4*c1-b4*s1;
     b1 = a4*s1+b4*c1;
     a4 = a[1][k2+1];
     b4 = a[2][k2+1];
     a2 = a4*c2-b4*s2;
     b2 = a4*s2+b4*c2;
   }
   a[1][k0+1] = a0+a1+a2;
   a[2][k0+1] = b0+b1+b2;
   a0 = -half * (a1+a2) + a0;
   a1 = (a1-a2) * c30;
   b0 = -half * (b1+b2) + b0;
   b1 = (b1-b2) * c30;
   a[1][k1+1] = a0-b1;
   a[2][k1+1] = b0+a1;
```

```
    a[1][k2+1] = a0+b1;
    a[2][k2+1] = b0-a1;
  }
  if ( i < mm ) {
    i++;
    continue outloop;
  }
  i = mm;
  l1 = false;
  kb = iwk[id+i-1] + kb;
  if (kb >= kn) break outloop;
  more=true;
  while (more) {
    jj = iwk[ic+i-2] + jj;
    if (jj < iwk[ic+i-1]) {
      more=false;
    } else {
      i--;
      jj -= iwk[ic+i];
    }
  }
  if (i == mm) {
    c2 = c1;
    c1 = cm * c1 - sm * s1;
    s1 = sm * c2 + cm * s1;
    continue inloop;
  } else {
    if (l1[ill+i]) i++;
    continue outloop;
  }

case 4:
  if (!l1) {
    c2 = c1 * c1 - s1 * s1;
    s2 = two * c1 * s1;
    c3 = c1 * c2 - s1 * s2;
    s3 = s1 * c2 + c1 * s2;
  }
  ja = kb + isp - 1;
  ka = ja + kb;
  ikb = kb+1;
  ija = ja+1;
  for (ii = ikb; ii<=ija; ii++) {
    k0 = ka - ii + 1;
    k1 = k0 + isp;
    k2 = k1 + isp;
    k3 = k2 + isp;
    a0 = a[1][k0+1];
    b0 = a[2][k0+1];
    if (l1) {
```

```
        a1 = a[1][k1+1];
        b1 = a[2][k1+1];
        a2 = a[1][k2+1];
        b2 = a[2][k2+1];
        a3 = a[1][k3+1];
        b3 = a[2][k3+1];
      } else {
        a4 = a[1][k1+1];
        b4 = a[2][k1+1];
        a1 = a4*c1-b4*s1;
        b1 = a4*s1+b4*c1;
        a4 = a[1][k2+1];
        b4 = a[2][k2+1];
        a2 = a4*c2-b4*s2;
        b2 = a4*s2+b4*c2;
        a4 = a[1][k3+1];
        b4 = a[2][k3+1];
        a3 = a4*c3-b4*s3;
        b3 = a4*s3+b4*c3;
      }
      a[1][k0+1] = a0+a2+a1+a3;
      a[2][k0+1] = b0+b1+b2+b3;
      a[1][k1+1] = a0+a2-a1-a3;
      a[2][k1+1] = b0+b2-b1-b3;
      a[1][k2+1] = a0-a2-b1+b3;
      a[2][k2+1] = b0-b2+a1-a3;
      a[1][k3+1] = a0-a2+b1-b3;
      a[2][k3+1] = b0-b2-a1+a3;
    }

    if ( i < mm ) {
      i++;
      continue outloop;
    }
    i = mm;
    l1 = false;
    kb += iwk[id+i-1];
    if (kb >= kn) break outloop;
    more=true;
    while (more) {
      jj = iwk[ic+i-2] + jj;
      if (jj < iwk[ic+i-1]) {
        more=false;
      } else {
        i--;
        jj -= iwk[ic+i];
      }
    }
    if (i == mm) {
      c2 = c1;
```

```
          c1 = cm * c1 - sm * s1;
          s1 = sm * c2 + cm * s1;
          continue inloop;
        } else {
          if (ll[ill+i]) i++;
          continue outloop;
        }

    }  // end switch
  }
  jk = kf - 1;
  kh = jk/2;
  k3 = iwk[id+i-1];
  k0 = kb + isp;
  if (!ll) {
    k = jk - 1;
    wk[icf+1] = c1;
    wk[isf+1] = s1;
    for (j=1; j<=k; j++) {
      wk[icf+j+1] = wk[icf+j] * c1 - wk[isf+j] * s1;
      wk[isf+j+1] = wk[icf+j] * s1 + wk[isf+j] * c1;
    }
  }
  if (kf != jf) {
    c2 = wk[icc+i];
    wk[ick+1] = c2;
    wk[ick+jk] = c2;
    s2 = wk[iss+i];
    wk[isk+1] = s2;
    wk[isk+jk] = -s2;
    for (j = 1; j<=kh; j++) {
      k = jk - j;
      wk[ick+k] = wk[ick+j] * c2 - wk[isk+j] * s2;
      wk[ick+j+1] = wk[ick+k];
      wk[isk+j+1] = wk[ick+j] * s2 + wk[isk+j] * c2;
      wk[isk+k] = -wk[isk+j+1];
    }
  }
  do {
    k0--;
    k1 = k0;
    k2 = k0 + k3;
    a0 = a[1][k0+1];
    b0 = a[2][k0+1];
    a3 = a0;
    b3 = b0;
    for (j = 1; j<=kh; j++) {
      k1 += isp;
      k2 -= isp;
      if (ll) {
```

```
            a1 = a[1][k1+1];
            b1 = a[2][k1+1];
            a2 = a[1][k2+1];
            b2 = a[2][k2+1];
          } else {
            k = kf - j;
            a4 = a[1][k1+1];
            b4 = a[2][k1+1];
            a1 = a4*wk[icf+j]-b4*wk[isf+j];
            b1 = a4*wk[isf+j]+b4*wk[icf+j];
            a4 = a[1][k2+1];
            b4 = a[2][k2+1];
            a2 = a4*wk[icf+k]-b4*wk[isf+k];
            b2 = a4*wk[isf+k]+b4*wk[icf+k];
          }
          wk[iap+j] = a1 + a2;
          wk[iam+j] = a1 - a2;
          wk[ibp+j] = b1 + b2;
          wk[ibm+j] = b1 - b2;
          a3 += a1 + a2;
          b3 += b1 + b2;
        }
        a[1][k0+1] = a3;
        a[2][k0+1] = b3;
        k1 = k0;
        k2 = k0 + k3;
        for (j=1; j<=kh; j++) {
          k1 += isp;
          k2 -= isp;
          jk = j;
          a1 = a0;
          b1 = b0;
          a2 = zero;
          b2 = zero;
          for (k=1; k<=kh; k++) {
            a1 += wk[iap+k] * wk[ick+jk];
            a2 += wk[iam+k] * wk[isk+jk];
            b1 += wk[ibp+k] * wk[ick+jk];
            b2 += wk[ibm+k] * wk[isk+jk];
            jk += j;
            if (jk >= kf) jk -= kf;
          }
          a[1][k1+1] = a1-b2;
          a[2][k1+1] = b1+a2;
          a[1][k2+1] = a1+b2;
          a[2][k2+1] = b1-a2;
        }
      } while (k0 > kb);

      jf = kf;
```

```
      if ( i < mm ) {
        i++;
        continue outloop;
      }
      i = mm;
      l1 = false;
      kb += iwk[id+i-1];
      if (kb >= kn) break outloop;
      more=true;
      while (more) {
        jj = iwk[ic+i-2] + jj;
        if (jj < iwk[ic+i-1]) {
          more=false;
        } else {
          i--;
          jj -= iwk[ic+i];
        }
      }
      if (i == mm) {
        c2 = c1;
        c1 = cm * c1 - sm * s1;
        s1 = sm * c2 + cm * s1;
        continue inloop;
      } else {
        if (ll[ill+i]) i++;
        continue outloop;
      }
    }  //  inloop
  }    //  outloop

i = 1;
ja = kt - 1;
ka = ja + 1;
if (ja >= 1) {
  for (ii=1; ii<=ja; ii++) {
    j = ka - ii;
    iwk[j+1]--;
    i = iwk[j+1] + i;
  }
}
if (kt > 0) {
  j = 1;
  i = 0;
  kb = 0;
  while (true) {
    k2 = iwk[id+j] + kb;
    k3 = k2;
    jj = iwk[ic+j-1];
    jk = jj;
    k0 = kb + jj;
```

```java
    isp = iwk[ic+j] - jj;
    while (true) {
      k = k0 + jj;
      do {
        a4 = a[1][k0+1];
        b4 = a[2][k0+1];
        a[1][k0+1] = a[1][k2+1];
        a[2][k0+1] = a[2][k2+1];
        a[1][k2+1] = a4;
        a[2][k2+1] = b4;
        k0++;
        k2++;
      } while (k0 < k);
      k0 += isp;
      k2 += isp;
      if (k0 >= k3) {
        if (k0 >= k3 + isp) {
          k3 += iwk[id+j];
          if (k3 - kb >= iwk[id+j-1]) break;
          k2 = k3 + jk;
          jk += jj;
          k0 = k3 - iwk[id+j] + jk;
        } else
          k0 = k0 - iwk[id+j] + jj;
      }
    }
    if (j < kt) {
      k = iwk[j+1] + i;
      j++;
      do {
        i = i + 1;
        iwk[ill+i] = j;
      } while (i < k);
    } else {
      kb = k3;
      if (i > 0) {
        j = iwk[ill+i];
        i--;
      } else {
        if (kb >= n) break;
        j = 1;
      }
    }
  }
}
jk = iwk[ic+kt];
isp = iwk[id+kt];
m -= kt;
kb = isp/jk-2;
if (kt >= m-1 ) return;
```

```
ita = ill+kb+1;
itb = ita+jk;
idm1 = id-1;
ikt = kt+1;
im = m+1;
for (j=ikt; j<=im; j++)
  iwk[idm1+j] = iwk[idm1+j]/jk;
jj = 0;
for (j = 1; j<=kb; j++) {
  k = kt;
  while (true) {
    jj += iwk[id+k+1];
    if (jj < iwk[id+k]) break;
    jj -= iwk[id+k];
    k++;
  }
  iwk[ill+j] = jj;
  if (jj == j) iwk[ill+j] = -j;
}
for (j=1; j<=kb; j++) {
  if (iwk[ill+j] > 0) {
    k2 = j;
    while (true) {
      k2 = Math.abs(iwk[ill+k2]);
      if (k2 == j) break;
      iwk[ill+k2] = -iwk[ill+k2];
    }
  }
}
i = 0;
j = 0;
kb = 0;
kn = n;
while (true) {
  j++;
  if (iwk[ill+j] >= 0) {
    k = iwk[ill+j];
    k0 = jk * k + kb;
    do {
      a4 = a[1][k0+i+1];
      b4 = a[2][k0+i+1];
      wk[ita+i] = a4;
      wk[itb+i] = b4;
      i = i + 1;
    } while (i < jk);
    i = 0;
    do {
      k = -iwk[ill+k];
      jj = k0;
      k0 = jk * k + kb;
```

```
        do {
          a[1][jj+i+1] = a[1][k0+i+1];
          a[2][jj+i+1] = a[2][k0+i+1];
          i++;
        } while (i < jk);
        i = 0;
      } while (k != j);
      do {
        a[1][k0+i+1] = wk[ita+i];
        a[2][k0+i+1] = wk[itb+i];
        i++;
      } while (i < jk);
      i = 0;
      if (j >= k2) {
        j = 0;
        kb += isp;
        if (kb >= kn) break;
      }
    }
  }
}
}
```

B. orderf

Replaces the numbers in locations (i,j+1) (j=0,1,...,2^m-1; i=1,2) of the real array A
by those in the locations (i,r(j)+1) (j=0,1,...,2^m-1; i=1,2) where when

$$j = \sum_{\upsilon=0}^{m-1} d_\upsilon 2^\upsilon \quad (d_\upsilon = 0 \quad \text{or} \quad 1), \qquad r(j) = \sum_{\upsilon=0}^{m-1} d_{m-\upsilon-1} 2^\upsilon .$$

Procedure parameters:

$$\text{void orderf } (a,m)$$

a: double $a[1:2][1:2^m]$; the array A above;
m: integer; the value of m above.

```
public static void orderf(double a[][], int m)
{
  int i,mp,k,j,jj,jk,n2,n4,n8,lm,nn;
  int iwk[] = new int[m+2];
  double temp[] = new double[3];

  if (m <= 1) return;
  n8=0;
  mp = m+1;
  jj = 1;
  iwk[1] = 1;
  for (i = 2; i<=mp; i++)
```

```
   iwk[i] = iwk[i-1] * 2;
n4 = iwk[mp-2];
if (m > 2) n8 = iwk[mp-3];
n2 = iwk[mp-1];
lm = n2;
nn = iwk[mp]+1;
mp -= 4;
j = 2;
do {
  jk = jj + n2;
  temp[1] = a[1][j];
  temp[2] = a[2][j];
  a[1][j] = a[1][jk];
  a[2][j] = a[2][jk];
  a[1][jk] = temp[1];
  a[2][jk] = temp[2];
  j++;
  if (jj <= n4) {
      jj += n4;
  } else {
    jj -= n4;
    if (jj <= n8) {
      jj += n8;
    } else {
      jj -= n8;
      k = mp;
      while (iwk[k] < jj) {
        jj -= iwk[k];
        k--;
      }
      jj += iwk[k];
    }
  }
  if (jj > j) {
    k = nn - j;
    jk = nn - jj;
    temp[1] = a[1][j];
    temp[2] = a[2][j];
    a[1][j] = a[1][jj];
    a[2][j] = a[2][jj];
    a[1][jj] = temp[1];
    a[2][jj] = temp[2];
    temp[1] = a[1][k];
    temp[2] = a[2][k];
    a[1][k] = a[1][jk];
    a[2][k] = a[2][jk];
    a[1][jk] = temp[1];
    a[2][jk] = temp[2];
  }
  j++;
```

```
  } while (j <= lm);
}
```

C. cfft2p

Computes the values of

$$a_{r(k)+1} = \sum_{j=0}^{n-1} a'_{j+1} e^{2\pi ijk/n} \qquad (k = 0,\dots,n-1)$$

using a fast Fourier transform, where the $\{a'_{j+1}\}$ are complex numbers, $n=2^m$, and with

$$k = \sum_{\upsilon=0}^{m-1} d_\upsilon 2^\upsilon \quad (d_\upsilon = 0 \quad \text{or} \quad 1), \qquad r(k) = \sum_{\upsilon=0}^{m-1} d_{m-\upsilon-1} 2^\upsilon \qquad (k = 0,\dots,n-1).$$

Procedure parameters:
$$\text{void cfft2p }(a,m)$$

a: double $a[1:2][1:n]$;
         entry: the values of $Re(a'_{j+1})$ and $Im(a'_{j+1})$ in locations $(1,j+1)$ and $(2,j+1)$ respectively $(j=0,..,n-1)$;
         exit: the values of $Re(a_{r(k)+1})$ and $Im(a_{r(k)+1})$ in locations $(1,k+1)$ and $(2,k+1)$ respectively $(k=0,..,n-1)$;
m: integer; the value of m above.

```
public static void cfft2p(double a[][], int m)
{
  int i,mp,k,j,jj,mm,n,kb,mk,kn,k0,k1,k2,k3,isp,jsp;
  int iwk[] = new int[m+2];
  double ak2[] = new double[3];
  double zero,one,rad,c1,c2,c3,s1,s2,s3,ck,sk,sq,a0,a1,a2,a3,b0,b1,b2,b3,temp;

  sq=0.7071067811865475;
  sk=0.3826834323650898;
  ck=0.9238795325112868;
  zero=0.0;
  one=1.0;
  c2=c3=s2=s3=0.0;
  k3=0;
  mp = m+1;
  n = (int)Math.pow(2,m);
  iwk[1] = 1;
  mm = (m/2)*2;
  kn = n+1;
  for (i=2;i<=mp;i++)
    iwk[i] = iwk[i-1]+iwk[i-1];
  rad = (2.0*Math.PI)/n;
  mk = m - 4;
```

```
kb = 1;
if (mm != m) {
  k2 = kn;
  k0 = iwk[mm+1] + kb;
  do {
    k2--;
    k0--;
    ak2[1] = a[1][k2];
    ak2[2] = a[2][k2];
    a[1][k2] = a[1][k0] - ak2[1];
    a[2][k2] = a[2][k0] - ak2[2];
    a[1][k0] = a[1][k0] + ak2[1];
    a[2][k0] = a[2][k0] + ak2[2];
  } while (k0 > kb);
}
c1 = one;
s1 = zero;
jj = 0;
k = mm - 1;
j = 4;
if (k < 1) return;
isp = iwk[k];
if (jj != 0) {
  c2 = jj * isp * rad;
  c1 = Math.cos(c2);
  s1 = Math.sin(c2);
  c2 = c1 * c1 - s1 * s1;
  s2 = c1 * (s1 + s1);
  c3 = c2 * c1 - s2 * s1;
  s3 = c2 * s1 + s2 * c1;
}
while (true) {
  jsp = isp + kb;
  for (i=1;i<=isp;i++) {
    k0 = jsp - i;
    k1 = k0 + isp;
    k2 = k1 + isp;
    k3 = k2 + isp;
    a0 = a[1][k0];
    b0 = a[2][k0];
    a1 = a[1][k1];
    b1 = a[2][k1];
    a2 = a[1][k2];
    b2 = a[2][k2];
    a3 = a[1][k3];
    b3 = a[2][k3];
    if (s1 != zero) {
      temp = a1;
      a1 = a1 * c1 - b1 * s1;
      b1 = temp * s1 + b1 * c1;
```

```
          temp = a2;
          a2 = a2 * c2 - b2 * s2;
          b2 = temp * s2 + b2 * c2;
          temp = a3;
          a3 = a3 * c3 - b3 * s3;
          b3 = temp * s3 + b3 * c3;
        }
        temp = a0 + a2;
        a2 = a0 - a2;
        a0 = temp;
        temp = a1 + a3;
        a3 = a1 - a3;
        a1 = temp;
        temp = b0 + b2;
        b2 = b0 - b2;
        b0 = temp;
        temp = b1 + b3;
        b3 = b1 - b3;
        b1 = temp;
        a[1][k0] = a0+a1;
        a[2][k0] = b0+b1;
        a[1][k1] = a0-a1;
        a[2][k1] = b0-b1;
        a[1][k2] = a2-b3;
        a[2][k2] = b2+a3;
        a[1][k3] = a2+b3;
        a[2][k3] = b2-a3;
      }
      if (k > 1) {
        k -= 2;
        isp = iwk[k];
        if (jj != 0) {
          c2 = jj * isp * rad;
          c1 = Math.cos(c2);
          s1 = Math.sin(c2);
          c2 = c1 * c1 - s1 * s1;
          s2 = c1 * (s1 + s1);
          c3 = c2 * c1 - s2 * s1;
          s3 = c2 * s1 + s2 * c1;
        }
      } else {
        kb = k3 + isp;
        if (kn <= kb) return;
        if (j == 1) {
          k = 3;
          j = mk;
          while (true) {
            if (iwk[j] > jj) break;
            jj -= iwk[j];
            j--;
```

```
      if (iwk[j] > jj) break;
      jj -= iwk[j];
      j--;
      k += 2;
    }
    jj += iwk[j];
    j = 4;
    isp = iwk[k];
    if (jj != 0) {
      c2 = jj * isp * rad;
      c1 = Math.cos(c2);
      s1 = Math.sin(c2);
      c2 = c1 * c1 - s1 * s1;
      s2 = c1 * (s1 + s1);
      c3 = c2 * c1 - s2 * s1;
      s3 = c2 * s1 + s2 * c1;
    }
  } else {
    j--;
    c2 = c1;
    if (j == 2) {
      c1 = c1 * ck + s1 * sk;
      s1 = s1 * ck - c2 * sk;
    } else {
      c1 = (c1 - s1) * sq;
      s1 = (c2 + s1) * sq;
    }
    c2 = c1 * c1 - s1 * s1;
    s2 = c1 * (s1 + s1);
    c3 = c2 * c1 - s2 * s1;
    s3 = c2 * s1 + s2 * c1;
  }
}
}
}
}
```

D. cfft2r

Computes the values of

$$a_{k+1} = \sum_{j=0}^{n-1} a'_{r(j)+1} e^{2\pi ijk/n} \qquad (k = 0,\ldots,n-1)$$

where the $\{a'_{r(j)+1}\}$ are complex numbers, $n=2^m$, and with

$$j = \sum_{\upsilon=0}^{m-1} d_{\upsilon} 2^{\upsilon} \quad (d_{\upsilon} = 0 \quad \text{or} \quad 1), \qquad r(j) = \sum_{\upsilon=0}^{m-1} d_{m-\upsilon-1} 2^{\upsilon} \qquad (j = 0,\ldots,n-1)$$

r(j) being the binary permutation of the numbers 0,...,n-1 of the procedure *orderf*.

Procedure parameters:

$$\text{void cfft2r } (a,m)$$

a: double *a[1:2][1:n]*;

entry: the values of $\text{Re}(a'_{r(j)+1})$ and $\text{Im}(a'_{r(j)+1})$ in locations $(1,j+1)$ and $(2,j+1)$
respectively ($j=0,..,n-1$);

exit: the values of $\text{Re}(a_{k+1})$ and $\text{Im}(a_{k+1})$ in locations $(1,k+1)$ and $(2,k+1)$
respectively ($k=0,..,n-1$);

m: integer; the value of *m* above.

```
public static void cfft2r(double a[][], int m)
{
  int i,mp1,k,j,jj,n,kb,kn,ks,mk,k0,k1,k2,k3,k4,nt,isp;
  int iwk[] = new int[m+2];
  double ak2[] = new double[3];
  double rad,c1,c2,c3,s1,s2,s3,ck,sk,sq,a0,a1,a2,a3,b0,b1,b2,b3,zero,one,pie,tr,ti,
      temp;

  sq=0.7071067811865475;
  sk=0.3826834323650898;
  ck=0.9238795325112867;
  zero=0.0;
  one=1.0;
  if (m == 0) return;
  n = (int)Math.pow(2,m);
  c2=c3=s2=s3=0.0;
  mp1 = m + 1;
  iwk[1] = 1;
  kn = 0;
  k4 = 4;
  mk = m - 3;
  ks = 1;
  for (i=2; i<=mp1; i++) {
    ks = ks + ks;
    iwk[i] = ks;
  }
  rad = Math.PI / ks;
  do {
    kb = kn + k4;
    kn += ks;
    if (m != 1) {
      jj = 0;
      k = 1;
      j = mk;
      nt = 3;
      c1 = one;
      isp = iwk[k];
      s1 = zero;
      if (jj != 0) {
        c2 = jj * isp * rad;
```

```
    c1 = Math.cos(c2);
    s1 = Math.sin(c2);
    c2 = c1 * c1 - s1 * s1;
    s2 = 2 * c1 * s1;
    c3 = c1 * c2 - s1 * s2;
    s3 = s1 * c2 + c1 * s2;
}
k3 = kb - isp + 1;
while (true) {
  k2 = k3 - isp;
  k1 = k2 - isp;
  k0 = k1 - isp;
  a0 = a[1][k0];
  b0 = a[2][k0];
  a1 = a[1][k1];
  b1 = a[2][k1];
  a2 = a[1][k2];
  b2 = a[2][k2];
  a3 = a[1][k3];
  b3 = a[2][k3];
  temp = a0 + a1;
  a1 = a0 - a1;
  a0 = temp;
  temp = a2 + a3;
  a3 = a2 - a3;
  a2 = temp;
  temp = b0 + b1;
  b1 = b0 - b1;
  b0 = temp;
  temp = b2 + b3;
  b3 = b2 - b3;
  b2 = temp;
  a[1][k0] = a0+a2;
  a[2][k0] = b0+b2;
  if (s1 == zero) {
    a[1][k1] = a1-b3;
    a[2][k1] = b1+a3;
    a[1][k2] = a0-a2;
    a[2][k2] = b0-b2;
    a[1][k3] = a1+b3;
    a[2][k3] = b1-a3;
  } else {
    tr = a1 - b3;
    ti = b1 + a3;
    a[1][k1] = tr*c1-ti*s1;
    a[2][k1] = tr*s1+ti*c1;
    tr = a0 - a2;
    ti = b0 - b2;
    a[1][k2] = tr*c2-ti*s2;
    a[2][k2] = tr*s2+ti*c2;
```

```
    tr = a1 + b3;
    ti = b1 - a3;
    a[1][k3] = tr*c3-ti*s3;
    a[2][k3] = tr*s3+ti*c3;
}
k3++;
if (k3 > kb) {
  nt--;
  if (nt >= 0) {
    c2 = c1;
    if (nt != 1) {
      c1 = (c1 - s1) * sq;
      s1 = (c2 + s1) * sq;
    } else {
      c1 = c1 * ck + s1 * sk;
      s1 = s1 * ck - c2 * sk;
    }
    kb += k4;
    if (kb <= kn) {
      c2 = c1 * c1 - s1 * s1;
      s2 = 2 * c1 * s1;
      c3 = c1 * c2 - s1 * s2;
      s3 = s1 * c2 + c1 * s2;
      k3 = kb - isp + 1;
    } else
      break;
  } else {
    if (nt == -1) {
      k = 3;
      isp = iwk[k];
      s1 = zero;
      if (jj != 0) {
        c2 = jj * isp * rad;
        c1 = Math.cos(c2);
        s1 = Math.sin(c2);
        c2 = c1 * c1 - s1 * s1;
        s2 = 2 * c1 * s1;
        c3 = c1 * c2 - s1 * s2;
        s3 = s1 * c2 + c1 * s2;
      }
      k3 = kb - isp + 1;
    } else {
      if (iwk[j] <= jj) {
        jj -= iwk[j];
        j--;
        if (iwk[j] <= jj) {
          jj -= iwk[j];
          j--;
          k += 2;
        } else {
```

```
                       jj += iwk[j];
                       j = mk;
                   }
               } else {
                   jj += iwk[j];
                   j = mk;
               }
               if (j < mk) {
                   isp = iwk[k];
                   s1 = zero;
                   if (jj != 0) {
                       c2 = jj * isp * rad;
                       c1 = Math.cos(c2);
                       s1 = Math.sin(c2);
                       c2 = c1 * c1 - s1 * s1;
                       s2 = 2 * c1 * s1;
                       c3 = c1 * c2 - s1 * s2;
                       s3 = s1 * c2 + c1 * s2;
                   }
                   k3 = kb - isp + 1;
               } else {
                   k = 1;
                   nt = 3;
                   kb += k4;
                   if (kb <= kn) {
                       isp = iwk[k];
                       s1 = zero;
                       if (jj != 0) {
                           c2 = jj * isp * rad;
                           c1 = Math.cos(c2);
                           s1 = Math.sin(c2);
                           c2 = c1 * c1 - s1 * s1;
                           s2 = 2 * c1 * s1;
                           c3 = c1 * c2 - s1 * s2;
                           s3 = s1 * c2 + c1 * s2;
                       }
                       k3 = kb - isp + 1;
                   } else
                       break;
               }
           }
       }
     }
   }
}
k = (m/2) * 2;
if (k != m) {
  k2 = kn + 1;
  j = kn - iwk[k+1] + 1;
  k0 = j;
```

```
        do {
          k2--;
          k0--;
          ak2[1]  =  a[1][k2];
          ak2[2]  =  a[2][k2];
          a[1][k2]  =  a[1][k0]  -  ak2[1];
          a[2][k2]  =  a[2][k0]  -  ak2[2];
          a[1][k0]  =  a[1][k0]  +  ak2[1];
          a[2][k0]  =  a[2][k0]  +  ak2[2];
        } while (k2 > j);
      }
    } while (kn < n);
}
```

E. Test_cfftp

Procedures tested: cfftp, cfft2p, cfft2r, orderf.

Example:
Computes

$$(1) \quad a_{k+1} = \sum_{j=0}^{n-1} e^{2ij\left\{\beta + \frac{\pi k}{n}\right\}}$$

$$(2) \qquad = \frac{\sin(n\beta)}{\sin\left(\beta + \frac{\pi k}{n}\right)}\left[\cos\left\{(n-1)\beta - \frac{\pi k}{n}\right\} + i\,\sin\left\{(n-1)\beta - \frac{\pi k}{n}\right\}\right] \qquad (k = 0,\dots,n-1)$$

with $\beta = 0.001 - \pi$
    a)   when n=25 by a call of *cfftp*,
    b)   when n=32 by a call of *cfft2p*, and
    c)   when n=16 by a call of *cfft2r*.

```
package numal;

import java.text.DecimalFormat;
import numal.*;

public class Test_cfftp extends Object {

    public static void main(String args[]) {

        double beta;
        double a[][] = new double[3][301];

        Test_cfftp testcfftp = new Test_cfftp();
        beta=0.001-Math.PI;
        testcfftp.setfft(a,25,beta);
        FFT.cfftp(a,25);
```

```
    System.out.println("results produced by fast fourier transform program:\n");
    testcfftp.out(a,25);
    testcfftp.fftans(a,25,beta);
    System.out.println("correct answers:\n");
    testcfftp.out(a,25);
    testcfftp.setfft(a,32,beta);
    FFT.cfft2p(a,5);
    FFT.orderf(a,5);
    System.out.println("results produced by fast fourier transform program:\n");
    testcfftp.out(a,32);
    testcfftp.fftans(a,32,beta);
    System.out.println("correct answers:\n");
    testcfftp.out(a,32);
    testcfftp.setfft(a,16,beta);
    FFT.orderf(a,4);
    FFT.cfft2r(a,4);
    System.out.println("results produced by fast fourier transform program:\n");
    testcfftp.out(a,16);
    testcfftp.fftans(a,16,beta);
    System.out.println("correct answers:\n");
    testcfftp.out(a,16);
}


public void setfft(double a[][], int n, double beta)
{
  int k,i;
  double theta,tbeta;
  tbeta=2.0*beta;
  theta=-tbeta;
  for (i=1; i<=n; i++) {
    theta=theta+tbeta;
    a[1][i]=Math.cos(theta);
    a[2][i]=Math.sin(theta);
  }
}


public void fftans(double a[][], int n, double beta)
{
  int k;
  double incr,pkdn,snbet,nm1bet,numarg,fact;

  incr=Math.PI/n;
  pkdn=-incr;
  snbet=Math.sin(n*beta);
  nm1bet=(n-1)*beta;
  for (k=1; k<=n; k++) {
    pkdn=pkdn+incr;
    numarg=nm1bet-pkdn;
```

```
        fact=snbet/Math.sin(beta+pkdn);
        a[1][k]=fact*Math.cos(numarg);
        a[2][k]=fact*Math.sin(numarg);
      }
  }


  public void out(double a[][], int n)
  {
    int i,j,k,p,rows;
    DecimalFormat fourDigit = new DecimalFormat("0.0000E0");
    rows=5;
    for (i=1; i<=2; i++) {
      j=0;
      for (p=1; p<=(n/rows); p++) {
        for (k=1; k<=rows; k++) {
          j++;
          System.out.print(fourDigit.format(a[i][j]) + "\t");
        }
        System.out.println();
      }
      for (p=1; p<=n-rows*(n/rows); p++) {
        j++;
        System.out.print(fourDigit.format(a[i][j]) + "\t");
      }
      if (n > rows*(n/rows))
        System.out.println();
      System.out.println();
    }
  }

}
```

Output:

```
results produced by fast fourier transform program:

2.4990E1        1.9686E-1       9.7550E-2       6.3557E-2       4.5973E-2
3.4948E-2       2.7183E-2       2.1256E-2       1.6449E-2       1.2354E-2
8.7169E-3       5.3660E-3       2.1720E-3       -9.7243E-4      -4.1681E-3
-7.5224E-3      -1.1165E-2      -1.5269E-2      -2.0090E-2      -2.6040E-2
-3.3843E-2      -4.4939E-2      -6.2676E-2      -9.7109E-2      -1.9879E-1

5.9988E-1       -2.0083E-2      -2.2566E-2      -2.3416E-2      -2.3856E-2
-2.4131E-2      -2.4325E-2      -2.4474E-2      -2.4594E-2      -2.4696E-2
-2.4787E-2      -2.4871E-2      -2.4951E-2      -2.5030E-2      -2.5109E-2
-2.5193E-2      -2.5284E-2      -2.5387E-2      -2.5508E-2      -2.5656E-2
-2.5851E-2      -2.6129E-2      -2.6572E-2      -2.7433E-2      -2.9976E-2

correct answers:

2.4990E1        1.9686E-1       9.7550E-2       6.3557E-2       4.5973E-2
3.4948E-2       2.7183E-2       2.1256E-2       1.6449E-2       1.2354E-2
8.7169E-3       5.3660E-3       2.1720E-3       -9.7243E-4      -4.1681E-3
-7.5224E-3      -1.1165E-2      -1.5269E-2      -2.0090E-2      -2.6040E-2
-3.3843E-2      -4.4939E-2      -6.2676E-2      -9.7109E-2      -1.9879E-1
```

5.9988E-1	-2.0083E-2	-2.2566E-2	-2.3416E-2	-2.3856E-2
-2.4131E-2	-2.4325E-2	-2.4474E-2	-2.4594E-2	-2.4696E-2
-2.4787E-2	-2.4871E-2	-2.4951E-2	-2.5030E-2	-2.5109E-2
-2.5193E-2	-2.5284E-2	-2.5387E-2	-2.5508E-2	-2.5656E-2
-2.5851E-2	-2.6129E-2	-2.6572E-2	-2.7433E-2	-2.9976E-2

results produced by fast fourier transform program:

3.1979E1	3.2241E-1	1.6095E-1	1.0606E-1	7.8008E-2
6.0707E-2	4.8779E-2	3.9910E-2	3.2938E-2	2.7214E-2
2.2345E-2	1.8075E-2	1.4232E-2	1.0689E-2	7.3513E-3
4.1409E-3	9.9167E-4	-2.1582E-3	-5.3705E-3	-8.7117E-3
-1.2260E-2	-1.6110E-2	-2.0390E-2	-2.5274E-2	-3.1019E-2
-3.8021E-2	-4.6939E-2	-5.8947E-2	-7.6397E-2	-1.0477E-1
-1.6059E-1	-3.2702E-1			

9.9167E-1	-2.1690E-2	-2.6859E-2	-2.8616E-2	-2.9514E-2
-3.0068E-2	-3.0449E-2	-3.0733E-2	-3.0957E-2	-3.1140E-2
-3.1296E-2	-3.1432E-2	-3.1555E-2	-3.1669E-2	-3.1776E-2
-3.1878E-2	-3.1979E-2	-3.2080E-2	-3.2183E-2	-3.2290E-2
-3.2403E-2	-3.2527E-2	-3.2664E-2	-3.2820E-2	-3.3004E-2
-3.3228E-2	-3.3513E-2	-3.3898E-2	-3.4456E-2	-3.5365E-2
-3.7151E-2	-4.2479E-2			

correct answers:

3.1979E1	3.2241E-1	1.6095E-1	1.0606E-1	7.8008E-2
6.0707E-2	4.8779E-2	3.9910E-2	3.2938E-2	2.7214E-2
2.2345E-2	1.8075E-2	1.4232E-2	1.0689E-2	7.3513E-3
4.1409E-3	9.9167E-4	-2.1582E-3	-5.3705E-3	-8.7117E-3
-1.2260E-2	-1.6110E-2	-2.0390E-2	-2.5274E-2	-3.1019E-2
-3.8021E-2	-4.6939E-2	-5.8947E-2	-7.6397E-2	-1.0477E-1
-1.6059E-1	-3.2702E-1			

9.9167E-1	-2.1690E-2	-2.6859E-2	-2.8616E-2	-2.9514E-2
-3.0068E-2	-3.0449E-2	-3.0733E-2	-3.0957E-2	-3.1140E-2
-3.1296E-2	-3.1432E-2	-3.1555E-2	-3.1669E-2	-3.1776E-2
-3.1878E-2	-3.1979E-2	-3.2080E-2	-3.2183E-2	-3.2290E-2
-3.2403E-2	-3.2527E-2	-3.2664E-2	-3.2820E-2	-3.3004E-2
-3.3228E-2	-3.3513E-2	-3.3898E-2	-3.4456E-2	-3.5365E-2
-3.7151E-2	-4.2479E-2			

results produced by fast fourier transform program:

1.5998E1	8.0261E-2	3.8768E-2	2.4146E-2	1.6221E-2
1.0922E-2	6.8635E-3	3.4214E-3	2.3998E-4	-2.9427E-3
-6.3891E-3	-1.0456E-2	-1.5773E-2	-2.3738E-2	-3.8474E-2
-8.0590E-2				

2.3998E-1	-1.4717E-2	-1.5381E-2	-1.5615E-2	-1.5742E-2
-1.5827E-2	-1.5892E-2	-1.5947E-2	-1.5998E-2	-1.6048E-2
-1.6104E-2	-1.6169E-2	-1.6254E-2	-1.6381E-2	-1.6617E-2
-1.7291E-2				

correct answers:

1.5998E1	8.0261E-2	3.8768E-2	2.4146E-2	1.6221E-2
1.0922E-2	6.8635E-3	3.4214E-3	2.3998E-4	-2.9427E-3
-6.3891E-3	-1.0456E-2	-1.5773E-2	-2.3738E-2	-3.8474E-2
-8.0590E-2				

2.3998E-1	-1.4717E-2	-1.5381E-2	-1.5615E-2	-1.5742E-2
-1.5827E-2	-1.5892E-2	-1.5947E-2	-1.5998E-2	-1.6048E-2
-1.6104E-2	-1.6169E-2	-1.6254E-2	-1.6381E-2	-1.6617E-2
-1.7291E-2				

F. rfftr

Computes the values of

$$a_{k+1} = \sum_{j=0}^{n-1} a'_{j+1} e^{2\pi i j k / n} \qquad (k = 0, \ldots, n-1)$$

using a fast Fourier transform, where the $\{a'_{j+1}\}$ are real numbers, and n is assumed to be an even positive integer.

Procedure parameters:

$$\text{void rfftr } (a, gamn, n)$$

a: double $a[1:n]$;
         entry: the values of a'_{j+1} in location (j+1), (j=0,..,n-1);
         exit: $Re(a_{k+1})$ and $Im(a_{k+1})$ in locations (2k+1) and (2k+2) respectively
              (k=0,..,½n);
$gamn$: double $gamn[1:2]$;
         $Re(a_{k+1})$ and $Im(a_{k+1})$ in locations (1) and (2) respectively when k=½n+1
         (for the remaining $\{a_{k+1}\}$, $a_{n+2-j}=\bar{a}_j$ (j=2,3,..., ½n));
n: integer; the value of n above.

```
public static void rfftr(double a[], double gamn[], int n)
{
  int i,k,nd2,nd4,mtwo,imax,m,np2,nmk;
  double zero,one,theta,tp,half,ai,ar;
  double s1[] = new double[3];
  double ximag[] = new double[3];
  double beta[] = new double[3];
  double gam[] = new double[3];
  double alph[] = new double[3];
  double a1[][] = new double[3][(n/2)+1];

  zero=0.0;
  half=0.5;
  one=1.0;
  imax=24;
  nd2 = n/2;
  for (i=1; i<=nd2; i++) {
    a1[1][i]=a[2*i-1];
    a1[2][i]=a[2*i];
  }
  if (n == 2) {
    ar = a1[1][1];
    ai = a1[2][1];
    theta = ar;
    tp = ai;
    gamn[1] = theta-tp;
    gamn[2] = zero;
    a[1] = theta+tp;
    a[2] = zero;
```

```
    return;
  }
  gam[1] = zero;
  gam[2] = zero;
  for (i=1; i<=nd2; i++) {
    gam[1] += a1[1][i];
    gam[2] += a1[2][i];
  }
  tp = gam[1]-gam[2];
  gam[1] = tp;
  gam[2] = zero;
  mtwo = 2;
  m = 1;
  for (i=1; i<=imax; i++) {
    if (nd2 <= mtwo) break;
    mtwo = mtwo+mtwo;
    m++;
  }
  if (nd2 != mtwo) {
    cfftp(a1,nd2);
  } else {
    cfft2p(a1,m);
    orderf(a1,m);
  }
  alph[1] = a1[1][1];
  alph[2] = a1[2][1];
  a1[1][1] = alph[1] + alph[2];
  a1[2][1] = zero;
  nd4 = (nd2+1)/2;
  if (nd4 >= 2) {
    np2 = nd2 + 2;
    theta = Math.PI/nd2;
    tp = theta;
    ximag[1] = zero;
    ximag[2] = one;
    for (k=2; k<=nd4; k++) {
      nmk = np2 - k;
      s1[1] = a1[1][nmk];
      s1[2] = -a1[2][nmk];
      alph[1] = a1[1][k] + s1[1];
      alph[2] = a1[2][k] + s1[2];
      beta[1] = ximag[1]*(s1[1]-a1[1][k]) - ximag[2]*(s1[2]-a1[2][k]);
      beta[2] = ximag[1]*(s1[2]-a1[2][k]) + ximag[2]*(s1[1]-a1[1][k]);
      s1[1] = Math.cos(theta);
      s1[2] = Math.sin(theta);
      a1[1][k] = (alph[1] + (beta[1]*s1[1] - beta[2]*s1[2])) * half;
      a1[2][k] = (alph[2] + (beta[1]*s1[2] + beta[2]*s1[1])) * half;
      a1[1][nmk] = (alph[1] - (beta[1]*s1[1] - beta[2]*s1[2])) * half;
      a1[2][nmk] = -(alph[2] - (beta[1]*s1[2] + beta[2]*s1[1])) * half;
      theta += tp;
```

```
      }
    }
  gamn[1] = gam[1];
  gamn[2] = gam[2];
  for (i=1; i<=n; i+=2) {
    k = i/2 + 1;
    a[i] = a1[1][k];
    a[i+1] = a1[2][k];
  }
}
```

G. Test_rfftr

Procedures tested: rfftr, cfftp, cfft2p, orderf.

Example:
Computes the values of

$$(1) \qquad a_{k+1} = \sum_{j=0}^{n-1} x^j e^{2\pi i j k / n}$$

$$(2) \qquad = \frac{\left(1 - x^n\right)\left\{1 - x\cos\frac{2\pi k}{n} + i\,x\sin\frac{2\pi k}{n}\right\}}{1 - 2x\cos\frac{2\pi k}{n} + x^2} \qquad (k = 0,\ldots,\tfrac{1}{2}n+1)$$

for x = 0.9 and n = 26, by use of *rfftr*, printing out first a list of values of the sums (1), followed by a list of values of expressions (2) for comparison.

```
package numal;

import java.text.DecimalFormat;
import numal.*;

public class Test_rfftr extends Object {

    public static void main(String args[]) {

        int i,k,n,n2,i1,i2;
        double xi,x,fact1,fact2,term,tx,tpn,den,sint,r1,r2,cost;
        double a[] = new double[101];
        double gamn[] = new double[3];

        x=0.9;
        n=26;
        n2=n/2;
        xi=1.0;
        for (i=1; i<=n; i++) {
            a[i]=xi;
            xi=x*xi;
```

```
    }
    FFT.rfftr(a,gamn,n);
    DecimalFormat sevenDigit = new DecimalFormat("0.0000000E0");
    System.out.println("results produced by fast fourier " +
      "transform program:\n");
    for (i=1; i<=n2; i++) {
      i2=2*i;
      i1=i2-1;
      System.out.println("  " + sevenDigit.format(a[i1]) + "     " +
                    sevenDigit.format(a[i2]));
    }
    System.out.println("\n  gamn = " + sevenDigit.format(gamn[1]) + "     " +
                    sevenDigit.format(gamn[2]));
    fact1=1.0-Math.pow(x,n);
    fact2=x*fact1;
    term=1.0+x*x;
    tpn=Math.PI/n2;
    tx=2.0*x;
    n2=n2+1;
    System.out.println("\ncorrect answers :\n");
    for (i=1; i<=n2; i++) {
      k=i-1;
      cost=Math.cos(tpn*k);
      sint=Math.sin(tpn*k);
      den=term-tx*cost;
      r1=(fact1-fact2*cost)/den;
      r2=fact2*sint/den;
      if (i == n2)
        System.out.println();
      System.out.println("  " + sevenDigit.format(r1) + "     " +
                    sevenDigit.format(r2));
    }
  }
}
}
```

Output:

```
results produced by fast fourier transform program:

  9.3538918E0     0.0000000E0
  1.8939422E0     3.2335901E0
  8.7875167E-1    1.8097366E0
  6.5975357E-1    1.2065557E0
  5.8053756E-1    8.7980157E-1
  5.4353407E-1    6.7178983E-1
  5.2347618E-1    5.2460415E-1
  5.1153448E-1    4.1229711E-1
  5.0399013E-1    3.2150769E-1
  4.9906668E-1    2.4459846E-1
  4.9583934E-1    1.7681138E-1
```

```
4.9380113E-1      1.1493770E-1
4.9267199E-1      5.6628783E-2


gamn = 4.9231010E-1      0.0000000E0


correct answers :


9.3538918E0      0.0000000E0
1.8939422E0      3.2335901E0
8.7875167E-1     1.8097366E0
6.5975357E-1     1.2065557E0
5.8053756E-1     8.7980157E-1
5.4353407E-1     6.7178983E-1
5.2347618E-1     5.2460415E-1
5.1153448E-1     4.1229711E-1
5.0399013E-1     3.2150769E-1
4.9906668E-1     2.4459846E-1
4.9583934E-1     1.7681138E-1
4.9380113E-1     1.1493770E-1
4.9267199E-1     5.6628783E-2


4.9231010E-1     2.8558704E-17
```

II. Time series analysis

A. powsp

Computes a) the Fourier transform of the power-spectrum of a time series and, if
so requested, b) that of a second time series together with the Fourier transform of
the cross-spectrum of the two time series.

The values of two positive integers n and l must be supplied at call and it
is assumed that n is divisible by l and that l is a power of 2; if n is not divisible by
l, powsp is given that value false and no computations are performed.

With the real numbers x_i, $(i=0,...,n-1)$ provided, and with the value
allocated to crossp false upon call, the numbers

$$\hat{x}_j^{(h)} = w_j x_{hl+j} \qquad (h = 0,\ldots,m-1; \; j = 0,\ldots,l-1)$$

where $n=ml$ and

$$w_j = 1 - \frac{\left| j - \frac{1}{2}(l-1) \right|}{\frac{1}{2}(l+1)} \qquad (j = 0,\ldots,l-1)$$

and

$$f_k^{(h)} = \sum_{j=0}^{l-1} \hat{x}_j^{(h)} e^{2\pi ijk/n} \qquad (k = 0,\ldots,l-1)$$

$$\psi_x(k) = \frac{3}{n}\sum_{h=0}^{m-1}\left|f_k^{(h)}\right|^2 \qquad (k = 0,\ldots,\tfrac{1}{2}l)$$

are computed.

With the real numbers x_i, y_i ($i=0,\ldots,n$-1) provided, and with the value allocated to *crossp* true upon call, the above calculations are performed, and in addition the numbers

$$\hat{y}_j^{(h)} = w_j w_{hl+j} \qquad (h = 0,\ldots,m-1; \; j = 0,\ldots,l-1)$$

$$g_k^{(h)} = \sum_{j=0}^{l-1} \hat{y}_j^{(h)} e^{2\pi ijk/n} \qquad (k = 0,\ldots,l-1)$$

$$\psi_y(k) = \frac{3}{n}\sum_{h=0}^{m-1}\left|g_k^{(h)}\right|^2 \qquad (k = 0,\ldots,l-1)$$

$$\phi_{xy}(k) = \frac{3}{n}\sum_{h=0}^{m-1} f_k^{(h)} \overline{g}_k^{(h)} \qquad (k = 0,\ldots,l-1)$$

$$\rho^{(k)} = \left|\phi_{xy}(k)\right|^2 \qquad (k = 0,\ldots,l-1)$$

$$\theta(k) = [\arg\{\phi_{xy}(k)\}]/2\pi \qquad (k = 0,\ldots,l-1)$$

are computed (the argument in the last formula is taken to lie in the range $[0,2\pi)$).

The above computations are only meaningful if the x_i have (or each set x_i and y_i has) zero mean. It is assumed that, if necessary, the user has, before call of *powsp*, centered the input sequences by subtracting the mean of a nonzero mean sequence from each member.

Procedure parameters:

boolean powsp (*x,y,n,l,crossp,psx,psy,xps*)

powsp: given the value true if n is divisible by l, false otherwise;

x: double x*[1:n]*; entry: the value of x_i in location (i+1) (i=0,...,n-1);

y: double y*[1:n]*;
entry: (if *crossp* is true upon call, unused otherwise) the value of y_i in location (i+1) (i=0,...,n-1);

n,l: integers; the value of n and l respectively;

crossp: boolean;
entry: false if computations a) above are requested, and true if computations a) and b) are required;

psx: double *psx[1:½l+1]*;
exit: the value of $\psi_x(k)$ in location (k+1) (k=0,...,½l);

psy: double *psy[1:½l+1]*;
exit: (if the value of *crossp* is true upon call) the value of $\psi_y(k)$ in location (k+1) (k=0,...,½l);

xps: double *xps[1:l+2]*;
exit: (if the value of *crossp* is true upon call) the value of $\rho(k)$ in location (k+1) and the value of $\theta(k)$ in location (k + ½l + 2) (k=0,...,½l).

```java
public static boolean powsp(double x[], double y[], int n, int l, boolean crossp,
                            double psx[], double psy[], double xps[])
{
  int m,lp1,ld2,nf,k,i,j,km1sl,lpk,ipnf,jm1;
  double pi2,pii,xm,xsave,ysave,phase,xim1,c1,c2;
  double wk[] = new double[(l/2)+1];
  double gamn[] = new double[3];
  double hamn[] = new double[3];
  double tmp[] = new double[l+1];

  if ((n/l)*l != n)
    return (false);
  m = n / l;
  lp1 = l + 1;
  ld2 = l / 2;
  nf = ld2 + 1;
  for (i=1; i<=nf; i++)
    psx[i] = 0.0;
  if (crossp)
    for (i=1; i<=nf; i++) {
      psy[i] = 0.0;
      xps[i] = 0.0;
      xps[i+nf] = 0.0;
    }
  c1 = 0.5 * (l-1);
  c2 = 2.0 / (l+1);
  for (i=1; i<=ld2; i++) {
    xim1 = i - 1;
    wk[i] = 1.0 + c2*(xim1-c1);
  }
  for (k=1; k<=m; k++) {
    km1sl = (k-1) * l;
    lpk = lp1 + km1sl;
    for (i=1; i<=ld2; i++) {
      j = i + km1sl;
      x[j] = x[j] * wk[i];
      j = lpk - i;
      x[j] = x[j] * wk[i];
    }
    for (i=1; i<=l; i++)
      tmp[i] = x[km1sl+i];
    rfftr(tmp,gamn,l);
    for (i=1; i<=l; i++)
      x[km1sl+i] = tmp[i];
    for (i=1; i<=ld2; i++) {
      j = km1sl + i + i;
      psx[i] = psx[i] + Math.pow(x[j-1],2) + Math.pow(x[j],2);
    }
    psx[nf] = psx[nf] + Math.pow(gamn[1],2);
    if (crossp) {
```

```
    for (i=1; i<=ld2; i++) {
      j = i + km1sl;
      y[j] *= wk[i];
      j = lpk - i;
      y[j] *= wk[i];
    }
    for (i=1; i<=l; i++)
      tmp[i] = y[km1sl+i];
    rfftr (tmp,hamn,l);
    for (i=1; i<=l; i++)
      y[km1sl+i] = tmp[i];
    for (i=1; i<=ld2; i++) {
      ipnf = i + nf;
      j = km1sl + i + i;
      jm1 = j - 1;
      psy[i] = psy[i] + Math.pow(y[jm1],2) + Math.pow(y[j],2);
      xps[i] = xps[i] + x[jm1]*y[jm1] + x[j]*y[j];
      xps[ipnf] = xps[ipnf] + x[j]*y[jm1] - x[jm1]*y[j];
    }
    psy[nf] += Math.pow(hamn[1],2);
    xps[nf] += gamn[1]*hamn[1];
  }
}
xm = 3.0 / n;
for (i=1; i<=nf; i++)
  psx[i] *= xm;
if (!crossp) return (true);
xps[nf+nf] = 0.0;
pi2 = Math.PI + Math.PI;
pii = 1.0 / pi2;
for (i=1; i<=nf; i++) {
  ipnf = i + nf;
  psy[i] *= xm;
  xsave = xps[i] * xm;
  ysave = xps[ipnf] * xm;
  xps[i] = xsave*xsave + ysave*ysave;
  if ((xsave == 0.0) && (ysave == 0.0)) {
    xps[ipnf] = 0.0;
  } else {
    phase = Math.atan2(ysave,xsave);
    if (phase < 0.0) phase += pi2;
    xps[ipnf] = phase * pii;
  }
}
return (true);
}
```

B. Test_powsp

Procedures tested: powsp, rfftr, cfftp, cfft2p, orderf.

Example:

Computes firstly the Fourier transform of the power spectrum of the time series with 16 elements:

   1.4696, -0.085540, 0.50465, 0.24784, 0.95925, -1.4027, 1.3512, 0.75327,
   1.4586, 0.24690, -2.2006, -1.0595, -0.73718, -1.5057, 0.0, 0.0

segmented into sections of four elements, printing the three results, and secondly the Fourier transforms of the power-spectra and cross-spectrum of the two time series whose elements are $x_j = \sin(2\pi j/n)$, $y_j = \cos(2\pi j/n)$ $(j=0,...,n-1)$ where $n=120$, segmented into sections of eight elements, printing the 5+5+10 results.

```
package numal;

import java.text.DecimalFormat;
import numal.*;

public class Test_powsp extends Object {

    public static void main(String args[]) {

        int n,l,i,i1,i2;
        double fact,argu;
        double x[] = new double[121];
        double y[] = new double[121];
        double psx[] = new double[6];
        double psy[] = new double[6];
        double xps[] = new double[11];

        x[1]=1.4696;   x[2]=-0.08554;  x[3]=0.50465;  x[4]=0.24784;
        x[5]=0.95925; x[6]=-1.4027;    x[7]=1.3512;    x[8]=0.75327;
        x[9]=1.4586;  x[10]=0.2469;    x[11]=-2.2006; x[12]=-1.0595;
        x[13]=-0.73718; x[14]=-1.5057; x[15]=0.0;      x[16]=0.0;
        if (!FFT.powsp(x,y,16,4,false,psx,psy,xps))   return;
        DecimalFormat sevenDigit = new DecimalFormat("0.0000000E0");
        System.out.println("first part of the example produces:\n\n" + "  psx :  " +
                            sevenDigit.format(psx[1]) + "\n          " +
                            sevenDigit.format(psx[2]) + "\n          " +
                            sevenDigit.format(psx[3]) + "\n\n" +
                            "second part of the example produces:\n");
        n=120;
        l=8;
        fact=6.283185307179586;
        for (i=1; i<=n; i++) {
          argu=(i-1)*fact;
          x[i]=Math.sin(argu);
          y[i]=Math.cos(argu);
```

```
    }
    if (!FFT.powsp(x,y,n,l,true,psx,psy,xps))  return;
    i1=l/2+1;
    for (i=1; i<=i1; i++)
      System.out.println("  psx(" + i + ") = " + sevenDigit.format(psx[i]) +
                         "    " + "psy(" + i + ") = " + sevenDigit.format(psy[i]));
    i2=l+2;
    System.out.println("\nvalues of xps:\n");
    for (i=1; i<=i2; i++) {
      System.out.println("  " + sevenDigit.format(xps[i]));
      if (i == i1)
        System.out.println();
    }
  }
 }
}
```

Output:

first part of the example produces:

```
  psx :  1.0644592E0
         1.8735611E0
         1.4771787E0
```

second part of the example produces:

```
  psx(1) = 2.1796035E-27   psy(1) = 7.4074074E0
  psx(2) = 8.0117701E-28   psy(2) = 7.3701833E-1
  psx(3) = 6.7099525E-28   psy(3) = 0.0000000E0
  psx(4) = 5.4502660E-28   psy(4) = 3.7224098E-3
  psx(5) = 2.3930780E-28   psy(5) = 0.0000000E0
```

values of xps:

```
  1.1881655E-26
  6.7233224E-29
  0.0000000E0
  3.5374838E-32
  0.0000000E0

  5.0000000E-1
  5.3189042E-1
  0.0000000E0
  6.3102790E-1
  0.0000000E0
```

C. timser

Computes upon request some or all of: the mean, the variance, the autocovariances, the autocorrelations, and the partial autocorrelations of the elements of a given time series.

With the elements w_i ($i=1,...,n$) supplied, the numbers which may be computed are as follows:

The mean:
$$\mu = \frac{1}{n}\sum_{i=1}^{n} w_i$$

the variance:
$$v = \frac{1}{n}\sum_{i=1}^{n} (w_i - \mu)^2$$

the autocovariances:
$$\alpha_j = \frac{1}{n}\sum_{i=1}^{n-j} (w_i - \mu)(w_{i+j} - \mu) \qquad (j=1,...,k)$$

the autocorrelations:
$$\beta_j = \alpha_j/v \qquad (j=1,...,k)$$

the partial autocorrelations:
$$\gamma_j = \frac{\beta_j - \sum_{i=1}^{j-1}\beta_{j-i}\phi_{j-1,i}}{1 - \sum_{i=1}^{j-1}\beta_i\phi_{j-1,i}} \qquad (j=1,...,l \le k)$$

where
$$\phi_j = \phi_{j-1} - \gamma_j\phi_{j-1,j-k} \qquad (k=1,...,j-1),$$
$$\phi_{j,j} = \gamma_j \qquad\qquad (j=1,...,l \le k).$$

(If, in the above, all elements w_i ($i=1,...,n$) are equal, α_j, β_j, and γ_j are defined to be zero.)

Which of the above quantities are to be computed is decided by the value given to the integer mode upon call of *timser*. With *mode* = 1, μ and v are computed; with *mode* = 2, α_j ($j=1,...,k$); with *mode* = 3, μ, v, and α_j ($j=1,...,k$); with *mode* = 4, α_j and β_j ($j=1,...,k$); with *mode* = 5, μ, v, and α_j, β_j ($j=1,...,k$); with *mode* = 6, α_j, β_j ($j=1,...,k$), and γ_j ($j=1,...,l$); with *mode* = 7, μ, v, and α_j, β_j ($j=1,...,k$), and γ_j ($j=1,...,l$). If *timser* is called with an even value of *mode*, it is assumed that the values of μ and v have been allocated to *mean* and *var* before call.

Procedure parameters:
> void timser ($w,n,k,l,mode,mean,var,alpha,beta, gamma$)

w:	double $w[1:n]$; the value of w_i in location (i) ($i=1,...,n$);
n:	int; the value of n above;
k:	int; (required if *mode* > 1) the value of k;
l:	int; (required if *mode* > 5) the value of l;
mode:	int; the value of *mode* above;
mean, var:	double $mean[0:0]$, $var[0:0]$;
	entry: (required if *mode* is even) the values of μ and v, respectively;
	exit: the value of μ and v;

alpha: double *alpha[1:k]*;
               exit (if *mode* > 1): the value of α_j in location j (j=1,...,k);
beta: double *beta[1:k]*;
               exit (if *mode* > 3): the value of β_j in location j (j=1,...,k);
gamma: double *gamma[1:l]*;
               exit (if *mode* > 5): the value of γ_j in location j (j=1,...,l).

```
public static void timser(double w[], int n, int k, int l, int mode, double mean[],
                          double var[], double alpha[], double beta[],
                          double gamma[])
{
  boolean contd;
  int i,j,im,iflag,j1,j2,kend,k0,j1mk;
  double temp2,zero,temp,temp1;
  double wkarea[] = new double[l+1];

  zero=0.0;
  if (mode != 1) {
    iflag = 0;
    contd=true;
    if (k > 0) {
      for (i=1; i<=k; i++)
        alpha[i] = zero;
      if (mode < 4)
        contd=false;
      else
        for (i=1; i<=k; i++)
          beta[i] = zero;
    }
    if (contd) {
      if (l > 0)
        for (i=1; i<=l; i++)
          gamma[i] = zero;
    }
    if ((mode/2)*2 != mode)  var[0] = zero;
    contd=true;
    for (i=2; i<=n; i++)
      if (w[i] != w[i-1]) {
        contd=false;
        break;
      }
    if (contd) {
      iflag = 1;
      mean[0] = w[1];
    }
    if (iflag == 1) return;
  }
  im = (mode/2)*2-mode;
  if (im != 0) {
    temp = 0.0;
```

```
      for (i=1; i<=n; i++)
        temp += w[i];
      mean[0] = temp/n;
      temp = 0.0;
      for (i=1; i<=n; i++)
        temp += (w[i]-mean[0])*(w[i]-mean[0]);
      var[0] = temp/n;
      if (mode == 1) return;
    }
    for (j=1; j<=k; j++) {
      kend = n-j;
      temp=0.0;
      for (i=1; i<=kend; i++)
        temp += (w[i]-mean[0])*(w[i+j]-mean[0]);
      alpha[j] = temp/n;
    }
    if (mode < 4) return;
    for (j=1; j<=k; j++)
      beta[j] = alpha[j]/var[0];
    if (mode < 6) return;
    gamma[1] = beta[1];
    for (j=2; j<=l; j++) {
      j1 = j-1;
      wkarea[j1] = gamma[j1];
      j2 = j1/2;
      if (j != 2) {
        for (k0=1; k0<=j2; k0++) {
          j1mk = j1-k0;
          temp2 = wkarea[k0]-gamma[j1]*wkarea[j1mk];
          wkarea[j1mk] -= gamma[j1]*wkarea[k0];
          wkarea[k0] = temp2;
        }
      }
      temp = 0.0;
      temp1 = 0.0;
      for (i=1; i<=j1; i++) {
        temp += beta[j-i]*wkarea[i];
        temp1 += beta[i]*wkarea[i];
      }
      gamma[j] = (beta[j]-temp)/(1.0-temp1);
    }
  }
}
```

D. Test_timser

Procedure tested: timser

Example:
Computes the mean, variance, the first five autocovariances, autocorrelations, and
partial autocorrelations of the elements of the series $w_i = i$ ($i=1,\ldots,25$).

```
package numal;

import java.text.DecimalFormat;
import numal.*;

public class Test_timser extends Object {

    public static void main(String args[]) {

    int cn,n,i;
    double m[] = new double[1];
    double v[] = new double[1];
    double w[] = new double[26];
    double a[] = new double[6];
    double b[] = new double[6];
    double g[] = new double[6];

    cn=12;
    n=2*cn+1;
    for (i=1; i<=n; i++)
    w[i]=i;
    FFT.timser(w,n,5,5,7,m,v,a,b,g);
    DecimalFormat fiveDigit = new DecimalFormat("0.00000");
    System.out.println("mean and variance:\n     " + m[0] + "      " + v[0] +
                       "\n\nautocovariances:");
    for (i=1; i<=5; i++)
      System.out.print("    " + a[i]);
    System.out.println("\n\nautocorrelations:");
    for (i=1; i<=5; i++)
      System.out.print("    " + fiveDigit.format(b[i]));
    System.out.println("\n\npartial autocorrelations:");
    for (i=1; i<=5; i++)
      System.out.print("    " + fiveDigit.format(g[i]));
  }
}
```

Output:

```
mean and variance:
      13.0       52.0

autocovariances:
      45.76      39.56      33.44      27.44      21.6

autocorrelations:
      0.88000    0.76077    0.64308    0.52769    0.41538

partial autocorrelations:
      0.88000    -0.06042   -0.06086   -0.06129   -0.06168
```

E. timspc

Computes upon request some of all of: the means, variances, the autocovariances, the frequencies at which spectral estimates are to be determined, and the power spectral estimates of two time series, together with the cross covariances, the real part (the cospectrum), the imaginary part (the quadrature spectrum) of the cross spectrum, the amplitudes and phases of the transfer functions, and the squared coherences derived from the two time series.

In the following exposition it is assumed that n (the number of elements in each time series), x_i, y_i ($i=1,...,n$) (the elements of the two time series), m (the number of lags), δt (a time interval), and τ (a whitening factor) are available.

- Whitening
  Elements $\{x_i'\}$ are produced. If no whitening is carried out then in the sequel the $\{x_i'\}$ are the same as the $\{x_i\}$. If whitening is performed
  (*) $\qquad x_i' = x_{i+1} - \tau x_i$ ($i=1,...,n-1$), $x_n' = x_n - x_1$

- Calculation of the mean ($\mu^{(x)}$)
  $$\mu^{(x)} = \frac{1}{n}\sum_{i=1}^{n} x_i$$

- Detrending
  Elements $\{x_i''\}$ are produced. If detrending is requested then in the sequal the $\{x_i''\}$ are the same as the $\{x_i'\}$. If detrending is not requested,
  (*) $\qquad x_i'' = x_i' - \mu^{(x)}$ $\qquad$ ($i=1,...,n-1$)

- Calculation of the variances ($v^{(x)}$)
  If no detrending has been requested
  $$v^{(x)} = \frac{1}{n}\sum_{i=1}^{n}(x_i'' - \mu^{(x)})^2$$

If detrending has taken place, then with $k = (n+2) \div 3$ and

$$\Xi^{(x)} = \frac{1}{k(n-k)} \sum_{i=1}^{k} (x''_{i+n-k} - x''_i)$$

$$v^{(x)} = v^{(x)} - \left(n\Xi^{(x)}\right)^2$$

- Calculation of the autocovariances $(\alpha_i^{(x)})$
  If no detrending has been requested

$$\alpha_i^{(x)} = \frac{1}{n-i} \sum_{j=1}^{n-i} (x''_j - \mu^{(x)})(x''_{i+j} - \mu^{(x)}) \qquad (i = 1, \ldots, m)$$

If detrending has taken place

$$\alpha_i^{(x)} = \frac{1}{n-i} \sum_{j=1}^{n-i} (x''_j - \mu^{(x)})(x''_{i+j} - \mu^{(x)}) - \mu^{(x)2} - \frac{\{n(n-2i) - 2i^2\} (x)^2}{12}$$

$$(i = 1, \ldots, m)$$

- Calculation of frequencies at which spectral quantities are calculated in cycles per unit of time $\{f_i\}$

$$f_i = \frac{i-1}{2m\delta t} \qquad (i = 1, \ldots, m+1)$$

- Calculation of power spectral estimates
  Firstly, the following numbers are produced by use of the relationship

$$\hat{\psi}_1^{(x)} = \left\{ \frac{1}{2} v^{(x)} + \sum_{j=1}^{m-1} \alpha_j^{(x)} \cos\left(\frac{(i-1)j\pi}{m}\right) + \frac{1}{2} \alpha_m^{(x)} \cos((i-1)\pi) \right\} \frac{2\delta t}{\pi}$$

$$(i = 1, \ldots, m+1)$$

This sequence of numbers is then smoothed:

$$\tilde{\psi}_1^{(x)} = 0.54\hat{\psi}_1^{(x)} + 0.46\hat{\psi}_2^{(x)}$$

$$\tilde{\psi}_i^{(x)} = 0.23\left(\hat{\psi}_{i-1}^{(x)} + \hat{\psi}_{i+1}^{(x)}\right) + 0.54\hat{\psi}_i^{(x)} \qquad (i = 2, \ldots, m)$$

$$\tilde{\psi}_{m+1}^{(x)} = 0.54\hat{\psi}_{m+1}^{(x)} + 0.46\hat{\psi}_m^{(x)}$$

If the $\{x_i\}$ have not been whitened, then

$$\psi_i^{(x)} = \tilde{\psi}_i^{(x)} \qquad (i = 1, \ldots, m+1)$$

otherwise

$$\psi_i^{(x)} = \frac{\tilde{\psi}_i^{(x)}}{1 + \tau\left(\tau - 2\cos\left\{\frac{(i-1)\pi}{m}\right\}\right)} \qquad (i = 1, \ldots, m+1)$$

If any of the $\psi_i^{(x)} < 0$, then $\psi_i^{(x)}$ is replaced by zero.

A whitened sequence $\{ y_i' \}$, mean $\mu^{(y)}$, detrended sequence $\{ y_i'' \}$, variance $v^{(y)}$, autocovariances $\{ \alpha_i^{(y)} \}$, power spectral estimates $\{ \psi_i^{(y)} \}$, and the sum $\Xi^{(y)}$ are obtained from $\{ y_i \}$ in a similar fashion.

In the following formulae concerning cross spectral estimates the $\{ x_i'' \}$ have the detrended form (**) (they are used in the computations whether the user has requested detrending in the calculation of single time series estimates or not); if the user has requested whitening, the $\{\,x_i''\,\}$ are produced from $\{\,x_i'\,\}$ having the form (*), otherwise directly from the original time series elements $\{\,x_i\,\}$.

- Calculation of cross covariances
  Firstly, the following numbers are produced by use of the relationships

$$\hat{\phi}_i = \frac{1}{n+i-m-1} \sum_{j=1}^{n+i-m-1} x''_{j+m+1-i} y''_j \qquad (i=1,\ldots,m+1)$$

$$\hat{\phi}_{i+m+1} = \frac{1}{n-i} \sum_{j=1}^{n-i} x''_j y''_{j+i} \qquad (i=1,\ldots,m)$$

If the variances $v^{(x)}$ and $v^{(y)}$ have the detrended form (**), numbers $\{\,\phi_i\,\}$ are produced by use of the relationships

$$\phi_{m+1-i} = \hat{\phi}_{m+1-i} - \mu^{(x)}\mu^{(y)} - \frac{(\delta t)^2}{12}\Big\{n(n-2i)-2i^2\Big\}\Xi^{(x)}\Xi^{(y)} + \frac{i}{2}\Big\{\mu^{(x)}\Xi^{(y)} - v^{(x)}\Xi^{(x)}\Big\}$$

$$(i=1,\ldots,m)$$

$$\phi_{m+1} = \hat{\phi}_{m+1} - \mu^{(x)}\mu^{(y)} - \frac{(n\delta t)^2}{12}\Xi^{(x)}\Xi^{(y)}$$

$$\phi_{m+1-i} = \hat{\phi}_{m+1-i} - \mu^{(x)}\mu^{(y)} - \frac{(\delta t)^2}{12}\Big\{n(n-2i)-2i^2\Big\}\Xi^{(x)}\Xi^{(y)} - \frac{i}{2}\Big\{\mu^{(x)}\Xi^{(y)} - v^{(x)}\Xi^{(x)}\Big\}$$

$$(i=1,\ldots,m)$$

otherwise

$$\phi_i = \hat{\phi}_i \qquad (i=1,\ldots,2m+1)$$

- Calculation of the real part (the cospectrum) and the imaginary part (the quadrature spectrum) of the cross spectrum
  Firstly, the following numbers are produced by use of the relationships

$$\hat{\sigma}_i = \frac{\delta t}{\pi}\left[\phi_{m+1} + \sum_{j=1}^{m-1}\Big\{\phi_{m+j+1} + \phi_{m+1-j}\Big\}\cos\Big\{\tfrac{j(i-1)\pi}{m}\Big\} + \tfrac{1}{2}\big(\phi_1 + \phi_{2m+1}\big)\cos\{(i-1)\pi\}\right]$$

$$(i=1,\ldots,m+1)$$

$$\hat{\theta}_i = \frac{\delta t}{\pi}\left[\sum_{j=1}^{m-1}\Big\{\phi_{m+j+1} - \phi_{m+1-j}\Big\}\sin\Big\{\tfrac{j(i-1)\pi}{m}\Big\} + \tfrac{1}{2}\big(\phi_{2m+1} - \phi_1\big)\sin\{(i-1)\pi\}\right]$$

$$(i=1,\ldots,m+1)$$

These sequences are then smoothed:

$$\sigma_1 = 0.54\hat{\sigma}_1 + 0.46\hat{\sigma}_2\,, \qquad\qquad \theta_1 = 0.54\hat{\theta}_1 + 0.46\hat{\theta}_2$$

$$\sigma_i = 0.23\big(\hat{\sigma}_{i-1} + \hat{\sigma}_{i+1}\big) + 0.54\hat{\sigma}_i \qquad (i=2,\ldots,m)$$

$$\theta_1 = 0.23\big(\hat{\theta}_{i-1} + \hat{\theta}_{i+1}\big) + 0.54\hat{\theta}_i \qquad (i=2,\ldots,m)$$

$$\sigma_{m+1} = 0.54\hat{\sigma}_{m+1} + 0.46\hat{\sigma}_m\,, \qquad \theta_{m+1} = 0.54\hat{\theta}_{m+1} + 0.46\hat{\theta}_m$$

- Calculation of amplitudes and phases of cross spectrum ($\{\,\lambda_i\,\}$ and $\{\,\omega_i\,\}$)
  Firstly,

$$\lambda_i = \sqrt{\sigma_i^2 + \theta_i^2} \qquad (i = 1, \ldots, m+1)$$

Then, for $i=1,\ldots,m+1$, if $\sigma_i = 0$ and $\theta_i \geq 0$, $\omega_i = 0.25$, and if $\sigma_i = 0$ and $\theta_i < 0$, $\omega_i = 0.75$; otherwise, $\omega_i = \{$ arc tan (θ_i/σ_i) (in the range $[0, \pi)) \}/2\pi$

- Calculation of amplitudes of transfer functions from x to y $\{\beta_i^{(xy)}\}$ and from y to x $\{\beta_i^{(yx)}\}$ (undefined if any $\psi_i^{(x)}$ or $\psi_i^{(y)}$ is nonpositive)

$$\beta_i^{(xy)} = \lambda_i \big/ \psi_i^{(x)} \qquad \beta_i^{(yx)} = \lambda_i \big/ \psi_i^{(y)} \qquad (i = 1, \ldots, m+1)$$

- Calculation of squared coherences $\{\gamma_i\}$ (undefined if any $\psi_i^{(x)}$ or $\psi_i^{(y)}$ is nonpositive)

$$\gamma_i = \lambda_i^2 \big/ \big\{\psi_i^{(x)}\psi_i^{(y)}\big\} \qquad (i = 1, \ldots, m+1)$$

The choice of computations to be performed is determined by two Boolean variables (*dtrend* for detrending and *whiten* for whitening), a further Boolean variable *cross*, and an integer variable *mode*. If *mode* is given the value 1 upon call and *cross* the value false, then $\{x_i^{"}\}$, $\mu^{(x)}$, $v^{(x)}$ and $\{\alpha_i^{(x)}\}$ are computed; if *mode*=1 and *cross*=true, (and *timspc* has previously been called with *mode*=1 and *cross*=false) then $\{y_i^{"}\}$, $\mu^{(y)}$, $v^{(y)}$, and $\{\alpha_i^{(y)}\}$ and $\{\phi_i\}$ are computed; if *mode*=2 and *cross*=false, then $\{x_i^{"}\}$, $\mu^{(x)}$, $v^{(x)}$, $\{\alpha_i^{(x)}\}$, $\{f_i\}$, and $\{\psi_i^{(x)}\}$ are computed; if *mode*=2 and *cross*=true (and *timspc* has previously been called with *mode*=2 and *cross*=false) then $\{y_i^{"}\}$, $\mu^{(y)}$, $v^{(y)}$, $\{\alpha_i^{(y)}\}$, $\{\psi_i^{(y)}\}$, $\{\phi_i\}$, $\{\sigma_i\}$, and $\{\theta_i\}$ are computed; if *mode*=3 and *cross*=false, then $\{x_i^{"}\}$, $\mu^{(x)}$, $v^{(x)}$, $\{\alpha_i^{(x)}\}$, $\{f_i\}$, and $\{\psi_i^{(x)}\}$ are computed (i.e., as for *mode*=2); if *mode*=3 and *cross*=true (and *timspc* has previously been called with *mode*=2 or 3 and *cross*=false) then $\{y_i^{"}\}$, $\mu^{(y)}$, $v^{(y)}$, $\{\alpha_i^{(y)}\}$, $\{\psi_i^{(y)}\}$, $\{\phi_i\}$, $\{\sigma_i\}$, $\{\theta_i\}$, $\{\lambda_i\}$, $\{\omega_i\}$ and, if defined (see above), $\{\beta_i^{(xy)}\}$, $\{\beta_i^{(yx)}\}$, and $\{\gamma_i\}$.

If *timspc* is called with *whiten* and *cross* true, then *mode* must be given the value 1.

The value of *timspc* upon return from call functions is a failure indicator. If all computations were performed as intended, *timspc* is given the value 0; if $n<4$, the value 1, if $m \geq n$, the value 2, if $m<3$, the value 3; if $\delta t \leq 0$, the value 4; if *whiten*=true and $|whfact| \geq 1$, the value 5; if *whiten*=*cross*=true and *mode*=3, the value 6. In all of the preceding nonzero cases, no computations are performed. If one of the $\psi_i^{(x)}$ is nonpositive, *timspc* is given 7, and if one of the $\psi_i^{(y)}$ is nonpositive, the value 8.

timspc may be used in the following way: it is first called with *cross*=false, values allocated to the $\{x_i\}$, and with, for example, *mode*=1. Values of $\{x_i^{"}\}$, $\mu^{(x)}$, $v^{(x)}$, and $\{\alpha_i^{(x)}\}$ are then computed. The values of the $\{x_i\}$ (they are now in detrended form), and those of $\mu^{(x)}$, $v^{(x)}$, and $\{\alpha_i^{(x)}\}$ are left undisturbed, values are allocated to the $\{y_i\}$, with *cross*=true and *mode*=1, and *timspc* is called a second time. The second call of *timspc*, with different values of $\{y_i\}$ may be repeated, if desired. More extensive computations result from double calls of *timspc* with *mode*=2 or 3 (see above).

Procedure parameters:
        int timspc (*x,y,n,m,deltat,dtrend,whiten,whfact,mode,cross,meanx,varx,*
                meany,vary,aucvx,aucvy,freq,powspx,powspy,croscv,recrsp,imcrsp,
                amcrsp,phcrsp,amtfxy,amtfyx,coher)

timspc: int; given a value in the range 0,...,7 as described above;
x,y: double *x[1:n]*, *y[1:n]*;
                entry: the value of x_i, y_i in locations (i) respectively;
                exit: the values of x_i'', y_i'' in locations (i) respectively;
n,m: int; the values of *n* and *m* above;
deltat: double; the value of δt;
dtrend: boolean; given the value true if detrending is desired, false otherwise;
whiten: boolean; given the value true if whitening is desired, false otherwise;
whfact: double; the value of τ above (not required if no whitening is desired);
mode: int; the value of *mode* above;
cross: boolean; the value of *cross* above;
meanx, varx, meany, vary: double; $\mu^{(x)}$, $v^{(x)}$, $\mu^{(y)}$ and $v^{(y)}$ above;
aucvx,aucvy: double *aucvx[1:m]*, *aucvy[1:m]*;
                              exit: the values of $\alpha_i^{(x)}$ and $\alpha_i^{(y)}$ respectively in
                                     position (i), (i=1,..,m);
freq: double *freq[1:m+1]*;
                              exit: the value of f_i in location (i) (i=1,...,m+1);
powspx,powspy: double *powspx[1:m+1]*, *powspy[1:m+1]*;
                              exit: the values of $\psi_i^{(x)}$ and $\psi_i^{(y)}$ respectively in
                                     position (i), (i=1,..,m+1);
croscv: double *croscv[1:m+1]*;
                              exit: the value of ϕ_i in location (i),
                                     (i=1,...,2m+1);
recrsp,imcrsp,amcrsp,phcrsp: double *recrsp[1:m+1],imcrsp[1:m+1],amcrsp[1:m+1]*,
                                phcrsp[1:m+1];
                              exit: the values of σ_i, θ_i, λ_i, and ω_i respectively
                                     in position (i), (i=1,..,m+1);
amtfxy,amtfyx,coher: double *amtfxy[1:m+1]*, *amtfyx[1:m+1]*, *coher[1:m+1]*;
                              exit: the values of $\beta_i^{(xy)}$, $\beta_i^{(yx)}$ and γ_i respectively
                                     in position (i), (i=1,..,m+1).

```
public static int timspc(double x[], double y[], int n, int m, double deltat,
                    boolean dtrend, boolean whiten, double whfact, int mode,
                    boolean cross, double meanx[], double varx[],
                    double meany[], double vary[], double aucvx[],
                    double aucvy[], double freq[], double powspx[],
                    double powspy[], double croscv[], double recrsp[],
                    double imcrsp[], double amcrsp[], double phcrsp[],
                    double amtfxy[], double amtfyx[], double coher[])
{
  int i,j,nm1,nsq,mp1,mm1,k,mp1mi,nim1,nmi,mp1pi,mp1pj,mp1mj,im1,result;
  double piom,dtopi,dtsq,ndtsq,st,fact,sum,ksixd,ksiyd,meansq,fact1,term,st1,sst,c1,
       c2,c3,f,g,hfpg,hfmg,im1pi,sum1,e,sst1,tpi,st2,ps,c;
  boolean ncross,pslto;
  double duma[] = new double[2];
  double tmp[] = new double[1];

  result=-1;
  if (!cross) result=0;
  if (n < 4) return (1);
```

```
if (m >= n) return (2);
if (m < 3) return (3);
if (deltat <= 0.0) return (4);
if (whiten) {
  if (whfact*whfact >= 1.0) return (5);
  if (cross && (mode == 3)) return (6);
}
st=meansq=st1=st2=term=ps=ksiyd=ksixd=0.0;
pslto=true;
nm1=n-1;
nsq=n*n;
mp1=m+1;
mm1=m-1;
ncross = !cross;
piom=Math.PI/m;
dtopi=deltat/Math.PI;
dtsq=deltat*deltat;
ndtsq=nsq*dtsq;
for (i=1; i<=mp1; i++) {
  if (powspx[i] <= 0.0) result=7;
  if (powspy[i] <= 0.0) result=8;
}
if (!cross) pslto=false;
if (whiten) {
  if (ncross) st=whfact*x[n]-x[1];
  if (cross)  st=whfact*y[n]-y[1];
  for (i=1; i<=nm1; i++) {
    if (ncross) x[i]=x[i+1]-whfact*x[i];
    if (cross)  y[i]=y[i+1]-whfact*y[i];
  }
  if (ncross) x[n]=st;
  if (cross)  y[n]=st;
}
if (ncross) timser(x,n,0,0,1,meanx,varx,duma,duma,duma);
if (cross)  timser(y,n,0,0,1,meany,vary,duma,duma,duma);
if (dtrend && ncross) varx[0] += meanx[0]*meanx[0];
if (dtrend && cross)  vary[0] += meany[0]*meany[0];
if (!dtrend)
  for (i=1; i<=n; i++) {
    if (ncross) x[i] -= meanx[0];
    if (cross)  y[i] -= meany[0];
  }
st=0.0;
tmp[0]=0.0;
if (ncross) timser(x,n,m,0,2,tmp,tmp,aucvx,duma,duma);
if (cross)  timser(y,n,m,0,2,tmp,tmp,aucvy,duma,duma);
st=tmp[0];
for (i=1; i<=m; i++) {
  fact=(double)(n)/(double)(n-i);
  if (ncross) aucvx[i] *= fact;
```

```
      if (cross)  aucvy[i] *= fact;
  }
  if (dtrend) {
    k=(n+2)/3;
    sum=0.0;
    for (i=1; i<=k; i++) {
      if (ncross) sum += x[i+n-k]-x[i];
      if (cross)  sum += y[i+n-k]-y[i];
    }
    sum /= (k*(n-k)*deltat);
    if (ncross) ksixd=sum;
    if (cross)  ksiyd=sum;
    sum=sum*sum/12.0;
    if (ncross) meansq=meanx[0]*meanx[0];
    if (cross)  meansq=meany[0]*meany[0];
    if (ncross) varx[0]=varx[0]-meansq-ndtsq*sum;
    if (cross)  vary[0]=vary[0]-meansq-ndtsq*sum;
    fact=dtsq*sum;
    for (i=1; i<=m; i++) {
      fact1=n*(n-2*i)-2*i*i;
      term=meansq+fact*fact1;
      if (ncross) aucvx[i] -= term;
      if (cross)  aucvy[i] -= term;
    }
  }
  if (mode != 1) {
    if (!cross) {
      fact=0.5/(m*deltat);
      for (i=1; i<=mp1; i++)
        freq[i]=(i-1)*fact;
    }
    c=2.0*dtopi;
    if (ncross) st1=0.5*varx[0];
    if (cross)  st1=0.5*vary[0];
    if (ncross) st2=0.5*aucvx[m];
    if (cross)  st2=0.5*aucvy[m];
    for (i=1; i<=mp1; i++) {
      sum=st1;
      fact=(i-1)*piom;
      for (j=1; j<=mm1; j++) {
        if (ncross) term=aucvx[j];
        if (cross)  term=aucvy[j];
        sum += term*Math.cos(j*fact);
      }
      sum += st2*Math.cos((i-1)*Math.PI);
      if (ncross) powspx[i]=c*sum;
      if (cross)  powspy[i]=c*sum;
    }
    if (ncross) st=powspx[1];
    if (cross)  st=powspy[1];
```

```
      if (ncross) powspx[1]=0.54*st+0.46*powspx[2];
      if (cross)  powspy[1]=0.54*st+0.46*powspy[2];
      for (i=2; i<=m; i++) {
        if (ncross) st1=powspx[i];
        if (cross)  st1=powspy[i];
        if (ncross) powspx[i]=0.23*(st+powspx[i+1])+0.54*st1;
        if (cross)  powspy[i]=0.23*(st+powspy[i+1])+0.54*st1;
        st=st1;
      }
      if (ncross) powspx[mp1]=0.54*powspx[mp1]+0.46*st;
      if (cross)  powspy[mp1]=0.54*powspy[mp1]+0.46*st;
      if (whiten) {
        st=1.0+whfact*whfact;
        st1=2.0*whfact;
        for (i=1; i<=mp1; i++) {
          sst=st-st1*Math.cos((i-1)*piom);
          if (ncross) powspx[i] /= sst;
          if (cross)  powspy[i] /= sst;
        }
      }
      for (i=1; i<=mp1; i++) {
        if (ncross) ps=powspx[i];
        if (cross)  ps=powspy[i];
        if (ps <= 0.0) {
          if (ncross) result=7;
          if (cross)  result=8;
          if (ncross) powspx[i]=0.0;
          if (cross)  powspy[i]=0.0;
          pslto=true;
        }
      }
    }
    if (ncross) return (result);
    for (i=1; i<=mp1; i++) {
      mp1mi=mp1-i;
      sum=0.0;
      nim1=n-mp1mi;
      for (j=1; j<=nim1; j++)
        sum += x[j+mp1mi]*y[j];
      croscv[i]=sum/nim1;
    }
    for (i=1; i<=m; i++) {
      nmi=n-i;
      sum=0.0;
      for (j=1; j<=nmi; j++)
        sum += x[j]*y[j+i];
      croscv[i+mp1]=sum/nmi;
    }
    if (dtrend) {
      c1=meanx[0]*meany[0];
```

```
    c2=0.5*(meanx[0]*ksiyd-varx[0]*ksixd);
    c3=ksixd*ksiyd/12.0;
    for (i=1; i<=m; i++) {
      fact=dtsq*(n*(n-2*i)-2*i*i);
      mp1pi=mp1+i;
      mp1mi=mp1-i;
      st=-c1-fact*c3;
      st1=i*c2;
      croscv[mp1pi] += st-st1;
      croscv[mp1mi] += st+st1;
    }
    croscv[mp1]=croscv[mp1]-c1-c3*ndtsq;
  }
  if (mode == 1) return (result);
  f=croscv[2*m+1];
  g=croscv[1];
  hfpg=0.5*(f+g);
  hfmg=0.5*(f-g);
  for (i=1; i<=mp1; i++) {
    im1=i-1;
    fact=im1*piom;
    im1pi=m*fact;
    sum=croscv[mp1];
    sum1=0.0;
    for (j=1; j<=mm1; j++) {
      mp1pj=mp1+j;
      mp1mj=mp1-j;
      e=j*fact;
      f=croscv[mp1pj];
      g=croscv[mp1mj];
      sum += (f+g)*Math.cos(e);
      sum1 += (f-g)*Math.sin(e);
    }
    recrsp[i]=dtopi*(sum+hfpg*Math.cos(im1pi));
    imcrsp[i]=dtopi*(sum1+hfmg*Math.sin(im1pi));
  }
  st=recrsp[1];
  st1=imcrsp[1];
  recrsp[1]=0.54*st+0.46*recrsp[2];
  imcrsp[1]=0.54*st1+0.46*imcrsp[2];
  for (i=2; i<=m; i++) {
    sst=recrsp[i];
    sst1=imcrsp[i];
    recrsp[i]=0.54*sst+0.23*(st+recrsp[i+1]);
    imcrsp[i]=0.54*sst1+0.23*(st1+imcrsp[i+1]);
    st=sst;
    st1=sst1;
  }
  recrsp[mp1]=0.54*recrsp[mp1]+0.46*st;
  imcrsp[mp1]=0.54*imcrsp[mp1]+0.46*st1;
```

```
    if (mode == 2) return (result);
    tpi=2.0*Math.PI;
    for (i=1; i<=mp1; i++) {
       st=recrsp[i];
       st1=imcrsp[i];
       amcrsp[i]=Math.sqrt(st*st+st1*st1);
       if (st == 0.0) {
          if (st1 >= 0.0) phcrsp[i]=0.25;
          if (st1 < 0.0) phcrsp[i]=0.75;
       } else {
          st1=Math.atan(st1/st);
          if (st < 0.0) st1 += Math.PI;
          st1 /= tpi;
          if (st1 < 0.0) st1 += 1.0;
          phcrsp[i]=st1;
       }
    }
    if (pslto) return (result);
    for (i=1; i<=mp1; i++) {
       st1=powspx[i];
       sst=powspy[i];
       st=amcrsp[i];
       amtfxy[i]=st/st1;
       amtfyx[i]=st/sst;
       st=st*st/(st1*sst);
       if (st > 1.0) {
          st=1.0;
          result=9;
       }
       coher[i]=st;
    }
    return (result);
}
```

F. Test_timspc

Procedures tested: timspc, timser.

Example:
Computes the means, variances, autocovariances, power spectral estimates, cross covariances, and the real parts, imaginary parts, amplitudes, and phases of the cross spectrum, together with the frequencies derived from the two time series with elements

$$\text{x: } 1.1, \quad 1.3, \quad 1.2, \quad 1.2$$
$$\text{y: } 2.1, \quad 2.3, \quad 2.5, \quad 2.5$$

with m=3 lags and time interval=1.0.

```
package numal;

import java.text.DecimalFormat;
import numal.*;

public class Test_timspc extends Object {

    public static void main(String args[]) {

    int i,j;
    boolean cross;
    double meanx[] = new double[1];
    double varx[] = new double[1];
    double meany[] = new double[1];
    double vary[] = new double[1];
    double aucvx[] = new double[4];
    double aucvy[] = new double[4];
    double freq[] = new double[5];
    double powspx[] = new double[5];
    double powspy[] = new double[5];
    double croscv[] = new double[8];
    double recrsp[] = new double[5];
    double imcrsp[] = new double[5];
    double y[] = new double[5];
    double amcrsp[] = new double[5];
    double phcrsp[] = new double[5];
    double amtfxy[] = new double[5];
    double amtfyx[] = new double[5];
    double coher[] = new double[5];
    double x[] = new double[5];

    x[1]=1.1;   x[2]=1.3;    x[3]=1.2;    x[4]=1.2;
    y[1]=2.1;   y[2]=2.3;    y[3]=2.5;    y[4]=2.5;
    DecimalFormat fiveDigit = new DecimalFormat("0.00000E0");
    System.out.println("elements of first time series:");
    for (i=1; i<=4; i++)
      System.out.print("    " + x[i]);
    System.out.println("\n\nelements of second time series:");
    for (i=1; i<=4; i++)
      System.out.print("    " + y[i]);
    cross=false;
    while (true) {
      j=FFT.timspc(x,y,4,3,1.0,false,false,0.0,3,cross,meanx,varx,meany,vary,aucvx,
                   aucvy,freq,powspx,powspy,croscv,recrsp,imcrsp,amcrsp,phcrsp,
                   amtfxy,amtfyx,coher);
      if (cross) break;
      if ((j == 0) || (j == 7))
        cross=true;
      else {
        System.out.println("\npreliminary call of timspc " +
```

```
                              "not satisfactorily executed");
      return;
    }
  }
  System.out.println("\n\nelements of first processed " + "time series x");
  for (i=1; i<=4; i++)
    System.out.print("   " + fiveDigit.format(x[i]));
  System.out.println("\n\nelements of second processed " + "time series y");
  for (i=1; i<=4; i++)
    System.out.print("   " + fiveDigit.format(y[i]));
  System.out.println("\n\nmean and variance of x\n   " +
    fiveDigit.format(meanx[0]) + "   " + fiveDigit.format(varx[0]) +
    "\n\nmean and variance of y\n   " + fiveDigit.format(meany[0]) + "   " +
    fiveDigit.format(vary[0]));
  System.out.println("\nautocovariances of x");
  for (i=1; i<=3; i++)
    System.out.print("   " + fiveDigit.format(aucvx[i]));
  System.out.println("\n\nautocovariances of y");
  for (i=1; i<=3; i++)
    System.out.print("   " + fiveDigit.format(aucvy[i]));
  System.out.println("\n\nfrequencies");
  for (i=1; i<=4; i++)
    System.out.print("   " + fiveDigit.format(freq[i]));
  System.out.println("\n\npower spectral estimates of x");
  for (i=1; i<=4; i++)
    System.out.print("   " + fiveDigit.format(powspx[i]));
  System.out.println("\n\npower spectral estimates of y");
  for (i=1; i<=4; i++)
    System.out.print("   " + fiveDigit.format(powspy[i]));
  System.out.println("\n\ncross covariances");
  for (i=1; i<=7; i++) {
    if (i == 5)
      System.out.println();
    System.out.print("   " + fiveDigit.format(croscv[i]));
  }
  System.out.println("\n\ncospectrum");
  for (i=1; i<=4; i++)
    System.out.print("   " + fiveDigit.format(recrsp[i]));
  System.out.println("\n\nquadrature spectrum");
  for (i=1; i<=4; i++)
    System.out.print("   " + fiveDigit.format(imcrsp[i]));
  System.out.println("\n\namplitudes of the cross spectrum");
  for (i=1; i<=4; i++)
    System.out.print("   " + fiveDigit.format(amcrsp[i]));
  System.out.println("\n\nphases of the cross spectrum");
  for (i=1; i<=4; i++)
    System.out.print("   " + fiveDigit.format(phcrsp[i]));
  if (j != 0) {
    if ((j-7)*(j-8) == 0) {
      System.out.println("\n\nNonpositive power spectral" +
```

```
          " estimate:\n  amplitudes of transfer functions" +
          " from x to y and y to x,\n  and the coherences are all undefined");
        return;
      }
    }
    System.out.println("\n\namplitudes of transfer function from x to y");
    for (i=1; i<=4; i++)
      System.out.print("    " + amtfxy[i]);
    System.out.println("\n\namplitudes of transfer function from y to x");
    for (i=1; i<=4; i++)
      System.out.print("    " + amtfyx[i]);
    System.out.println("\n\nsquared coherences");
    for (i=1; i<=4; i++)
      System.out.print("    " + coher[i]);
  }
}
```

Output:

```
elements of first time series:
    1.1     1.3     1.2     1.2

elements of second time series:
    2.1     2.3     2.5     2.5

elements of first processed time series x
   -1.00000E-1    1.00000E-1    -2.22045E-16    -2.22045E-16

elements of second processed time series y
   -2.50000E-1    -5.00000E-2    1.50000E-1    1.50000E-1

mean and variance of x
    1.20000E0    5.00000E-3

mean and variance of y
    2.35000E0    2.75000E-2

autocovariances of x
   -3.33333E-3    2.46519E-32    2.22045E-17

autocovariances of y
    9.16667E-3    -2.25000E-2    -3.75000E-2

frequencies
    0.00000E0    1.66667E-1    3.33333E-1    5.00000E-1

power spectral estimates of x
    0.00000E0    7.74554E-4    2.40854E-3    3.22554E-3

power spectral estimates of y
```

```
    7.85164E-3    1.41754E-2    7.77207E-3    7.74554E-4

cross covariances
    5.55112E-17    3.33067E-17    -8.33333E-3    5.00000E-3
    6.66667E-3    -1.64799E-17    -1.50000E-2

cospectrum
    9.92066E-4    1.57829E-3    1.60481E-3    2.19103E-3

quadrature spectrum
    1.90208E-3    3.18392E-3    3.18392E-3    1.90208E-3

amplitudes of the cross spectrum
    2.14526E-3    3.55364E-3    3.56550E-3    2.90147E-3

phases of the cross spectrum
    1.73486E-1    1.76756E-1    1.75695E-1    1.13783E-1

Nonpositive power spectral estimate:
    amplitudes of transfer functions from x to y and y to x,
    and the coherences are all undefined
```

Worked Examples

Examples for chapter 1 procedures

Procedures tested:

hshcomcol, hshcomprd

Example:
The matrix

$$\begin{pmatrix} 3 & 4i \\ 4i & 5 \end{pmatrix}$$

is reduced to upper triangular form by successive calls of *hshcomcol* and *hshcomprd*.

```java
package numal;

import numal.*;

public class Test_hshcomcol extends Object {

  public static void main(String args[]) {

    double ar[][] = new double[3][3];
    double ai[][] = new double[3][3];
    double k[] = new double[1];
    double c[] = new double[1];
    double s[] = new double[1];
    double t[] = new double[1];
    double tt;

    ar[1][1]=3.0;
    ar[1][2]=ar[2][1]=0.0;
    ar[2][2]=5.0;
    ai[1][1]=0.0;
    ai[1][2]=ai[2][1]=4.0;
    ai[2][2]=0.0;
    if (Basic.hshcomcol(1,2,1,ar,ai,25.0e-28,k,c,s,t)) {
      tt = t[0];
      Basic.hshcomprd(1,2,2,2,1,ar,ai,ar,ai,tt);
    }
    System.out.println("After using hshcomcol and hshcomprd:\n  " +
      ar[1][1] + "+" + ai[1][1] + "*I  " + ar[1][2] + "+" + ai[1][2] + "*I\n  " +
      ar[2][1] + "+" + ai[2][1] + "*I  " + ar[2][2] + "+" + ai[2][2] + "*I\n" +
      "k, c, s, t\n  " + k[0] + "   " + c[0] + "   " + s[0] + "   " + t[0]);
```

```
    }
}
```

Output:

```
After using hshcomcol and hshcomprd:
  8.0+0.0*I   0.0+1.6*I
  0.0+4.0*I   6.2+0.0*I
k, c, s, t
  5.0    -1.0    -0.0    40.0
```

Procedure tested:

elmcomcol

Example:

With the matrix

$$A = \begin{pmatrix} 1 + 2\mathrm{i} & -9 + 2\mathrm{i} \\ -1 + 2\mathrm{i} & -1 - 2\mathrm{i} \end{pmatrix}$$

this example sets $A_{k,2} = A_{k,2} + (1-4i)A_{k,1}$, for $k=1,2$.

```
package numal;

import numal.*;

public class Test_elmcomcol extends Object {

    public static void main(String args[]) {

        double ar[][] = new double[3][3];
        double ai[][] = new double[3][3];
        ar[1][1]=1.0;   ar[1][2] = -9.0;    ar[2][1]=ar[2][2] = -1.0;
        ai[1][1]=ai[1][2]=ai[2][1]=2.0;     ai[2][2] = -2.0;
        Basic.elmcomcol(1,2,2,1,ar,ai,ar,ai,1,-4);
        System.out.println("Matrix after elimination:");
        System.out.print("    " + ar[1][1] + "+" + ai[1][1] + "*I");
        System.out.println("   " + ar[1][2] + "+" + ai[1][2] + "*I");
        System.out.print("   " + ar[2][1] + "+" + ai[2][1] + "*I");
        System.out.println("   " + ar[2][2] + "+" + ai[2][2] + "*I");
    }
}
```

Output:

```
Matrix after elimination:
   1.0+2.0*I   0.0+0.0*I
  -1.0+2.0*I   6.0+4.0*I
```

Procedure tested:
rotcomcol

Example:
The matrix

$$\begin{pmatrix} 4+3i & 5 \\ -5 & 4-3i \end{pmatrix}$$

is post multiplied by the rotation matrix

$$\begin{pmatrix} 0.08-0.06i & -0.1 \\ 0.1 & 0.08+0.06i \end{pmatrix}.$$

```
package numal;

import java.text.DecimalFormat;
import numal.*;

public class Test_rotcomcol extends Object {

  public static void main(String args[]) {

    double ar[][] = new double[3][3];
    double ai[][] = new double[3][3];

    ar[1][1]=4.0; ar[1][2]=5.0; ar[2][1] = -5.0;  ar[2][2]=4.0;
    ai[1][1]=3.0; ai[1][2]=ai[2][1]=0.0;  ai[2][2] = -3.0;
    Basic.rotcomcol(1,2,1,2,ar,ai,0.08,0.06,-0.1);
    DecimalFormat oneDigit = new DecimalFormat("0.0");
    System.out.println("After postmultiplication:\n  " +
                ar[1][1] + "+" + ai[1][1] + "*I   " +
                ar[1][2] + oneDigit.format(ai[1][2]) + "*I\n  " +
                ar[2][1] + oneDigit.format(ai[2][1]) + "*I   " +
                ar[2][2] + "+" + ai[2][2] + "*I");
  }
}
```

Output:

```
After postmultiplication:
   1.0+0.0*I   0.0-0.0*I
   0.0-0.0*I   1.0+0.0*I
```

Procedure tested:

comabs

Example:
Compute $\left|0.3+0.4i\right|$.

```
package numal;

import numal.*;

public class Test_comabs extends Object {

    public static void main(String args[]) {

        System.out.println("The modulus of .3+.4*i equals " + Basic.comabs(0.3,0.4));
    }
}
```

Output:

```
The modulus of .3+.4*i equals 0.5
```

Procedure tested:

comsqrt

Example:
Compute the value of $(-3+4i)^{1/2}$ with positive real part.

```
package numal;

import numal.*;

public class Test_comsqrt extends Object {

  public static void main(String args[]) {

    double r[] = new double[1];
    double i[] = new double[1];

    Basic.comsqrt(-3.0,4.0,r,i);
    System.out.println("The square root of -3+4*i is " + r[0] + "+" + i[0] + "*i");
  }
}
```

Output:

```
The square root of -3+4*i is 1.0+2.0*i
```

Procedure tested:
carpol

Example:
Compute the value of r, $\cos\varphi$ and $\sin\varphi$, where $re^{i\varphi} = 0.3+0.4$i.

```
package numal;

import numal.*;

public class Test_carpol extends Object {

    public static void main(String args[]) {

        double r[] = new double[1];
        double c[] = new double[1];
        double s[] = new double[1];

        Basic.carpol(0.3,0.4,r,c,s);
        System.out.println("The polar coordinates of 0.3+0.4*i are \n" +
                        "  modulus:  " + r[0] + "\n  cosine of argument: " +
                        c[0] + "\n  sine of argument:   " + s[0]);
    }
}
```

Output:

```
The polar coordinates of 0.3+0.4*i are
  modulus:   0.5
  cosine of argument: 0.6
  sine of argument:   0.8
```

Procedure tested:
commul

Example:
Compute the value of (0.1+0.2i)(0.3+0.4i).

```
package numal;

import java.text.DecimalFormat;
import numal.*;

public class Test_commul extends Object {

    public static void main(String args[]) {
```

```
        double r[] = new double[1];
        double i[] = new double[1];

        Basic.commul(0.1,0.2,0.3,0.4,r,i);
        DecimalFormat twoDigit = new DecimalFormat("0.00");
        System.out.println("(.1+.2i)*(.3+.4*i) = " + twoDigit.format(r[0]) +
                        "+" + twoDigit.format(i[0]) + "*i");
    }
}
```

Output:

```
(.1+.2i)*(.3+.4*i) = -0.05+0.10*i
```

Procedure tested:

comdiv

Example:
Compute the value of (-0.05+0.1i)/(0.1+0.2i).

```
package numal;

import java.text.DecimalFormat;
import numal.*;

public class Test_comdiv extends Object {

    public static void main(String args[]) {

        double r[] = new double[1];
        double i[] = new double[1];

        Basic.comdiv(-0.05,0.1,0.1,0.2,r,i);
        DecimalFormat twoDigit = new DecimalFormat("0.00");
        System.out.println("(-.05+.1*i)/(.1+.2*i) = " + twoDigit.format(r[0]) +
                        "+" + twoDigit.format(i[0]) + "*i");
    }
}
```

Output:

```
(-.05+.1*i)/(.1+.2*i) = 0.30+0.40*i
```

Procedures tested:
Ingintadd, Ingintsbtract, Ingintmult, Ingintdivide, Ingintpower

Example:
Compute the sum, difference, product, quotient, and remainder of $u = 3370707070$ and $v = 444$, and also compute v^4.

```
package numal;

import numal.*;

public class Test_lngint extends Object {

  public static void main(String args[]) {
    int i;
    int u[] = new int[100];
    int v[] = new int[100];
    int r1[] = new int[100];
    int r2[] = new int[100];

    u[0]=5;
    u[1]=33;
    u[2]=u[3]=u[4]=u[5]=70;
    v[0]=2;
    v[1]=4;
    v[2]=44;
    System.out.println("\nInput numbers:");
    for (i=1; i<=u[0]; i++) System.out.print("  " + u[i]);
    System.out.print("\n   ");
    for (i=1; i<=v[0]; i++) System.out.print(v[i] + "  ");
    System.out.print("\n\nadd:       ");
    Basic.lngintadd(u,v,r1);
    for (i=1; i<=r1[0]; i++) System.out.print("  " + r1[i]);
    System.out.print("\nsubtract:  ");
    Basic.lngintsubtract(u,v,r1);
    for (i=1; i<=r1[0]; i++) System.out.print("  " + r1[i]);
    System.out.print("\nmultiple:  ");
    Basic.lngintmult(u,v,r1);
    for (i=1; i<=r1[0]; i++) System.out.print("  " + r1[i]);
    System.out.print("\ndivide:\n  Quotient:  ");
    Basic.lngintdivide(u,v,r1,r2);
    for (i=1; i<=r1[0]; i++) System.out.print("  " + r1[i]);
    System.out.print("\n  Remainder: ");
    for (i=1; i<=r2[0]; i++) System.out.print("  " + r2[i]);
    System.out.print("\npower:     ");
    Basic.lngintpower(v,4,r1);
    for (i=1; i<=r1[0]; i++) System.out.print("  " + r1[i]);
  }
}
```

Output:

```
Input numbers:
  33  70  70  70  70
   4  44

add:          33  70  70  75  14
subtract:     33  70  70  66  26
multiple:      1  49  65  93  93  90  80
divide:
  Quotient:    7  59  16  82
  Remainder:   2  62
power:         3  88  62  60  24  96
```

Examples for chapter 2 procedures

Procedure tested:

derpol

Example:
Compute the derivatives $d^jp(x)/dx^j$, for $j=0,...,3$, of
$$p(x) = -1 + x - 2x^2 + 3x^3$$
at the point $x = 1$.

```
package numal;

import numal.*;

public class Test_derpol extends Object {

    public static void main(String args[]) {

        double[] a = {-1.0, 1.0, -2.0, 3.0};

        Algebraic_eval.derpol(3,3,1.0,a);
        System.out.println("The 0-th until and including the 3rd derivatives:\n  "
                      + "    " + a[0] + "    " + a[1] + "    " + a[2] + "    " + a[3]);
    }
}
```

Output:

```
The 0-th until and including the 3rd derivatives:
   1.0  6.0  14.0  18.0
```

Procedure tested:

allortpol

Example:

Compute the Laguerre polynomials $L_i(x)$, for $i=0,...,5$, which satisfy the recursion

$$L_0(x) = 1, \qquad L_1(x) = x - 1$$
$$L_{k+1}(x) = (x-2k-1)L_k(x) - k^2 L_{k-1}(x) \qquad (k=1,2,...)$$

when $x = 0$.

```
package numal;

import numal.*;

public class Test_allortpol extends Object {

    public static void main(String args[]) {

        int i;
        double b[] = new double[5];
        double c[] = new double[5];
        double p[] = new double[6];

        b[0]=1.0;
        for (i=1; i<=4; i++) {
            b[i]=2*i+1;
            c[i]=i*i;
        }
        Algebraic_eval.allortpol(5,0.0,b,c,p);
        System.out.println("ALLORTPOL delivers: " + "   " + p[0] + "   " + p[1] +
                        "   " + p[2] + "   " + p[3] +"   " + p[4] + "   " + p[5]);
    }
}
```

Output:

```
ALLORTPOL delivers:    1.0   -1.0   2.0   -6.0   24.0   -120.0
```

Procedure tested:

chepolsum

Example:

Compute the polynomial

$$1 + (1/2)T_1(x) + (1/4)T_2(x)$$

for $i=-1,0,1$, where $T_1(x)$ and $T_2(x)$ are the Chebyshev polynomials of first and second degree, respectively.

```
package numal;

import numal.*;

public class Test_chepolsum extends Object {

    public static void main(String args[]) {

        double a[] = {1.0, 0.5, 0.25};

        System.out.println("CHEPOLSUM delivers: " +
                    Algebraic_eval.chepolsum(2,-1.0,a) + "   " +
                    Algebraic_eval.chepolsum(2,0.0,a) + "   " +
                    Algebraic_eval.chepolsum(2,1.0,a));
    }
}
```

Output:

```
CHEPOLSUM delivers: 0.75   0.75   1.75
```

Procedure tested:

oddchepolsum

Example:
Compute the polynomial
$$1 + (1/2)T_1(x) + (1/5)T_3(x)$$
for i=-1,0,1, where $T_1(x)$ and $T_3(x)$ are the Chebyshev polynomials of first and third degree, respectively.

```
package numal;

import numal.*;

public class Test_oddchepolsum extends Object {

    public static void main(String args[]) {

        double a[] = {0.5, 0.2};

        System.out.println("ODDCHEPOLSUM delivers: " +
                    Algebraic_eval.oddchepolsum(1,-1.0,a) + "   " +
                    Algebraic_eval.oddchepolsum(1,0.0,a) + "   " +
                    Algebraic_eval.oddchepolsum(1,1.0,a));
    }
}
```

Output:

```
ODDCHEPOLSUM delivers:  -0.7   -0.0   0.7
```

Procedures tested:
chepol, allchepol

Example:
The Chebyshev polynomials of the first kind of degrees 0, 1, 2 are evaluated at -1, 0, 1.

```java
package numal;

import numal.*;

public class Test_chepol extends Object {

  public static void main(String args[]) {

    double t[] = new double[3];

    Algebraic_eval.allchepol(2,-1.0,t);
    System.out.println("Delivers:\n" + "   " + t[0] + "   " + t[1] + "    " + t[2]);
    Algebraic_eval.allchepol(2,0.0,t);
    System.out.println("   " + t[0] + "    " + t[1] + "   " + t[2]);
    Algebraic_eval.allchepol(2,1.0,t);
    System.out.println("   " + t[0] + "    " + t[1] + "    " + t[2]);
    System.out.println("\n  " + Algebraic_eval.chepol(2,-1.0) + "   " +
                    Algebraic_eval.chepol(2,0.0) + "    " +
                    Algebraic_eval.chepol(2,1.0));
  }
}
```

Output:

```
Delivers:
  1.0  -1.0   1.0
  1.0   0.0  -1.0
  1.0   1.0   1.0

  1.0  -1.0   1.0
```

Procedure tested:

fouser

Example:
Evaluate $0.5 + \cos\theta + \sin\theta$ for $\theta = 0$, $\pi/2$ and π.

```
package numal;

import java.text.DecimalFormat;
import numal.*;

public class Test_fouser extends Object {

    public static void main(String args[]) {

        double a[] = {0.5, 1.0};

        DecimalFormat twoDigit = new DecimalFormat("0.00");
        System.out.println("FOUSER delivers: " +
          twoDigit.format(Algebraic_eval.fouser(1,0.0,a)) + "  " +
          twoDigit.format(Algebraic_eval.fouser(1,Math.PI/2.0,a)) + "  " +
          twoDigit.format(Algebraic_eval.fouser(1,Math.PI,a)));
    }
}
```

Output:

```
FOUSER delivers: 1.50   1.50   -0.50
```

Procedure tested:

jfrac

Example:
Compute the value of the n-th convergent of the continued fraction
$$1 + \frac{1}{2+} \frac{1}{2+} \cdots$$
for $n = 7,...,10$.

```
package numal;

import java.text.DecimalFormat;
import numal.*;

public class Test_jfrac extends Object {

    public static void main(String args[]) {

        int i;
```

```
        double p[] = new double[11];
        double q[] = new double[11];

        for (i=1; i<=10; i++) {
            p[i]=1.0;
            q[i]=2.0;
        }
        q[0]=1.0;
        DecimalFormat sevenDigit = new DecimalFormat("0.0000000");
        System.out.println("JFRAC delivers: " +
            sevenDigit.format(Algebraic_eval.jfrac(7,p,q)) + "   " +
            sevenDigit.format(Algebraic_eval.jfrac(8,p,q)) + "   " +
            sevenDigit.format(Algebraic_eval.jfrac(9,p,q)) + "   " +
            sevenDigit.format(Algebraic_eval.jfrac(10,p,q)));
    }
}
```

Output:

```
JFRAC delivers: 1.4142157   1.4142132   1.4142136   1.4142136
```

Procedures tested:

chspol, polchs

Example:

Convert the power sum $1+2x+3x^2$ into a sum of Chebyshev polynomials $b_0+b_1T_1(x)+b_2T_2(x)$ (by means of a call of *polchs*) and, as a check, transform the latter representation back into the original power sum (by means of a call of *chspol*).

```
package numal;

import numal.*;

public class Test_chspol extends Object {

    public static void main(String args[]) {

        double a[] = {1.0, 2.0, 3.0};

        System.out.println("           a[0]   a[1]   a[2]\ninput:" + "   " + a[0] +
                        "   " + a[1] + "   " + a[2]);
        Algebraic_eval.polchs(2,a);
        System.out.println("polchs:  " + a[0] + "   " + a[1] + "   " + a[2]);
        Algebraic_eval.chspol(2,a);
        System.out.println("chspol:  " + a[0] + "   " + a[1] + "   " + a[2]);
    }
}
```

Output:

```
          a[0]   a[1]   a[2]
input:    1.0    2.0    3.0
polchs:   2.5    2.0    1.5
chspol:   1.0    2.0    3.0
```

Procedures tested:

polshtchs, shtchspol

Example:

Convert the power sum $1+2x+3x^2$ into a sum of shifted Chebyshev polynomials $b_0+b_1T_1'(x)+b_2T_2'(x)$ (by means of a call of *polshtchs*) and, as a check, transform the latter representation back into the original power sum (by means of a call of *shtchspol*).

```java
package numal;

import java.text.DecimalFormat;
import numal.*;

public class Test_polshtchs extends Object {

    public static void main(String args[]) {

        double a[] = new double[3];

        a[0]=1.0; a[1]=2.0; a[2]=3.0;
        DecimalFormat twoDigit = new DecimalFormat("0.00");
        System.out.println("            a[0]   a[1]   a[2]\ninput: " + "       " +
                    twoDigit.format(a[0]) + "   " + twoDigit.format(a[1]) +
                    "   " + twoDigit.format(a[2]));
        Algebraic_eval.polshtchs(2,a);
        System.out.println("polshtchs:  " + twoDigit.format(a[0]) + "   " +
                    twoDigit.format(a[1]) + "   " + twoDigit.format(a[2]));
        Algebraic_eval.shtchspol(2,a);
        System.out.println("shtchspol:  " + twoDigit.format(a[0]) + "   " +
                    twoDigit.format(a[1]) + "   " + twoDigit.format(a[2]));
    }
}
```

Output:

```
             a[0]   a[1]   a[2]
input:       1.00   2.00   3.00
polshtchs:   3.12   2.50   0.38
shtchspol:   1.00   2.00   3.00
```

Procedures tested:

newgrn, grnnew

Example:
Convert the power sum $1+2x+3x^2$ into its Newton sum of the form
$$\delta_0^0+(x\text{-}x_0)\delta_0^1+(x\text{-}x_0)(x\text{-}x_1)\delta_0^2$$
(by means of a call of *newgrn*) and, as a check, transform the latter representation
back into the original power sum (by means of a call of *grnnew*).

```
package numal;

import numal.*;

public class Test_grnnew extends Object {

  public static void main(String args[]) {

    double x[] = { 1.0, 2.0 };
    double a[] = { 1.0, 2.0, 3.0 };

    System.out.println("             a[0]    a[1]    a[2]\ninput:" + "       " + a[0] +
                  "    " + a[1] + "    " + a[2]);
    Algebraic_eval.grnnew(2,x,a);
    System.out.println("grnnew: " + "     " + a[0] + "     " + a[1] + "     " + a[2]);
    Algebraic_eval.newgrn(2,x,a);
    System.out.println("newgrn:    " + a[0] + "        " + a[1] + "     " + a[2]);
  }
}
```

Output:

	a[0]	a[1]	a[2]
input:	1.0	2.0	3.0
grnnew:	6.0	11.0	3.0
newgrn:	1.0	2.0	3.0

Procedure tested:

lintfmpol

Example:
Convert the power sum $1+2x+3x^2$ into its power sum in y with $x = 2y + 3$, and, as
a check, transform the latter representation back into the original power sum
(both transformations are carried out by means of a call of *lintfmpol*).

```
package numal;

import numal.*;

public class Test_lintfmpol extends Object {

    public static void main(String args[]) {

        double a[] = {1.0, 2.0, 3.0};

        System.out.println("                    a[0]     a[1]      a[2]\ninput:" +
                    "             " + a[0] + "       " + a[1] + "       " + a[2]);
        Algebraic_eval.lintfmpol(2.0,3.0,2,a);
        System.out.println("lintfmpol: " + "   " + a[0] + "       " + a[1] + "      " +
                    a[2] + "    (power sum in y)");
        Algebraic_eval.lintfmpol(0.5,-1.5,2,a);
        System.out.println("lintfmpol:     " + a[0] + "       " + a[1] + "        " +
                    a[2] + "    (power sum in x)");
    }
}
```

Output:

```
              a[0]      a[1]      a[2]
input:        1.0       2.0       3.0
lintfmpol:    34.0      40.0      12.0    (power sum in y)
lintfmpol:    1.0       2.0       3.0     (power sum in x)
```

Procedure tested:

intchs

Example:
Determine the coefficients $\{b_j\}$ in the expansion

$$\int_0^x \left(1 + \frac{1}{2}T_1(x) + \frac{1}{5}T_2(x) + \frac{1}{10}T_3(x)\right) dx = \sum_{j=1}^{4} b_j T_j(x)$$

where $\{T_j(x)\}$ are Chebyshev polynomials.

```
package numal;

import java.text.DecimalFormat;
import numal.*;

public class Test_intchs extends Object {

    public static void main(String args[]) {

        double a[] = {1.0, 0.5, 0.2, 0.1};
        double b[] = new double[5];
```

```
        DecimalFormat fourDigit = new DecimalFormat("0.0000");
        Algebraic_eval.intchs(3,a,b);
        System.out.println("INTCHS delivers:   " + fourDigit.format(b[1]) +
            "   " + fourDigit.format(b[2]) + "   " + fourDigit.format(b[3]) +
            "   " + fourDigit.format(b[4]));
    }
}
```

Output:

```
INTCHS delivers:   0.9000   0.1000   0.0333   0.0125
```

Examples for chapter 3 procedures

Procedures tested:
determ, gsselm

Example:
Compute the determinant of the 4×4 Hilbert matrix A for which
$A_{i,j} = 1/(i+j-1)$, $i,j=1,...,4$.

```
package numal;

import java.text.DecimalFormat;
import numal.*;

public class Test_determ extends Object {

    public static void main(String args[]) {
        int i,j;
        int ri[] = new int[5];
        int ci[] = new int[5];
        double d;
        double aux[] = new double[8];
        double a[][] = new double[5][5];

        for (i=1; i<=4; i++)
            for (j=1; j<=4; j++) a[i][j]=1.0/(i+j-1);
        aux[2]=1.0e-5;
        aux[4]=8;
        Linear_algebra.gsselm(a,4,aux,ri,ci);
        d = (aux[3] == 4) ? Linear_algebra.determ(a,4,(int)aux[1]) : 0.0;
        DecimalFormat fiveDigit = new DecimalFormat("0.00000E0");
        System.out.println("Determinant = " + fiveDigit.format(d));
    }
}
```

Output:

```
Determinant = 1.65344E-7
```

Procedure tested:

decsol

Example:
Solve the 4×4 linear system of equations $Ax=b$, where A is the Hilbert matrix, for which $A_{i,j}=1/(i+j-1)$, $i,j=1,...,4$, and $b_i = 1/(2+i)$, $i=1,...,4$, (so that the solution is the third unit vector: $(0,0,1,0)^T$).

```java
package numal;

import java.text.DecimalFormat;
import numal.*;

public class Test_decsol extends Object {

  public static void main(String args[]) {

    int i,j;
    double b[] = new double[5];
    double aux[] = new double[4];
    double a[][] = new double[5][5];

    for (i=1; i<=4; i++) {
      for (j=1; j<=4; j++) a[i][j]=1.0/(i+j-1);
      b[i]=a[i][3];
    }
    aux[2]=1.0e-5;
    Linear_algebra.decsol(a,4,aux,b);
    DecimalFormat fiveDigit = new DecimalFormat("0.00000E0");
    System.out.println("Solution:  " + fiveDigit.format(b[1]) + "   " +
                    fiveDigit.format(b[2]) + "   " + fiveDigit.format(b[3]) +
                    "   " + fiveDigit.format(b[4]));
    System.out.println("Sign(Det) = " + (int)aux[1] +
                    "\nNumber of eliminations = " + (int)aux[3]);
  }
}
```

Output:

```
Solution:  0.00000E0  0.00000E0  1.00000E0  -0.00000E0
Sign(Det) = 1
Number of eliminations = 4
```

Procedure tested:

gsssol

Example:

Solve the 4×4 linear system of equations $Ax=b$, where A is the Hilbert matrix, for which $A_{i,j}=1/(i+j-1)$, $i,j=1,...,4$, and $b_i = 1/(2+i)$, $i=1,...,4$, (so that the solution is the third unit vector: $(0,0,1,0)^T$).

```java
package numal;

import java.text.DecimalFormat;
import numal.*;

public class Test_gsssol extends Object {

  public static void main(String args[]) {

    int i,j;
    double b[] = new double[5];
    double aux[] = new double[8];
    double a[][] = new double[5][5];

    for (i=1; i<=4; i++) {
      for (j=1; j<=4; j++) a[i][j]=1.0/(i+j-1);
      b[i]=a[i][3];
    }
    aux[2]=1.0e-5;
    aux[4]=8;
    Linear_algebra.gsssol(a,4,aux,b);
    DecimalFormat fiveDigit = new DecimalFormat("0.00000E0");
    System.out.println("Solution:  " + fiveDigit.format(b[1]) + "  " +
                    fiveDigit.format(b[2]) + "  " + fiveDigit.format(b[3]) +
                    "  " + fiveDigit.format(b[4]));
    System.out.println("Sign(Det) = " + (int)aux[1] +
                    "\nNumber of eliminations = " + (int)aux[3] +
                    "\nMax(abs(a[i,j])) = " + fiveDigit.format(aux[5]) +
                    "\nUpper bound growth = " + fiveDigit.format(aux[7]));
  }
}
```

Output:

```
Solution:  4.99600E-16  -2.66454E-15  1.00000E0  0.00000E0
Sign(Det) = 1
Number of eliminations = 4
Max(abs(a[i,j])) = 1.00000E0
Upper bound growth = 1.59722E0
```

Procedure tested:

gsssolerb

Example:
Solve the 4×4 linear system of equations $Ax=b$, where A is the Hilbert matrix, for which $A_{i,j}=1/(i+j-1)$, $i,j=1,...,4$, and $b_i = 1/(2+i)$, $i=1,...,4$, (so that the solution is the third unit vector: $(0,0,1,0)^T$), and provide an upper bound for the relative error in x.

```java
package numal;

import java.text.DecimalFormat;
import numal.*;

public class Test_gsssolerb extends Object {

  public static void main(String args[]) {

    int i,j;
    double b[] = new double[5];
    double aux[] = new double[12];
    double a[][] = new double[5][5];

    for (i=1; i<=4; i++) {
      for (j=1; j<=4; j++) a[i][j]=1.0/(i+j-1);
      b[i]=a[i][3];
    }
    aux[0]=aux[2]=aux[6]=1.0e-14;
    aux[4]=8;
    Linear_algebra.gsssolerb(a,4,aux,b);
    DecimalFormat fiveDigit = new DecimalFormat("0.00000E0");
    System.out.println("Solution:  " + fiveDigit.format(b[1]) + "  " +
                    fiveDigit.format(b[2]) + "  " + fiveDigit.format(b[3]) +
                    "  " +fiveDigit.format(b[4]));
    System.out.println("Sign(Det) = " + (int)aux[1] +
                    "\nNumber of eliminations = " + (int)aux[3] +
                    "\nMax(abs(a[i,j])) = " + fiveDigit.format(aux[5]) +
                    "\nUpper bound growth = " + fiveDigit.format(aux[7]) +
                    "\n1-norm of the inverse matrix = " +
                    fiveDigit.format(aux[9]) +
                    "\nUpper bound rel. error in the calc. solution = " +
                    fiveDigit.format(aux[11]));
  }
}
```

Output:

```
Solution:   4.99600E-16   -2.66454E-15  1.00000E0   0.00000E0
Sign(Det) = 1
Number of eliminations = 4
Max(abs(a[i,j])) = 1.00000E0
Upper bound growth = 1.59722E0
1-norm of the inverse matrix = 1.36200E4
Upper bound rel. error in the calc. solution = 2.78075E-8
```

Procedure tested:

decinv

Example:
Compute the inverse of the 4×4 matrix

$$\begin{pmatrix} 4 & 2 & 4 & 1 \\ 30 & 20 & 45 & 12 \\ 20 & 15 & 36 & 10 \\ 35 & 28 & 70 & 20 \end{pmatrix}.$$

```
package numal;

import numal.*;

public class Test_decinv extends Object {

  public static void main(String args[]) {

    int i;
    double aux[] = new double[4];
    double a[][] = new double[5][5];

    a[1][1]=4.0;  a[1][2]=2.0;  a[1][3]=4.0;  a[1][4]=1.0;
    a[2][1]=30.0; a[2][2]=20.0; a[2][3]=45.0; a[2][4]=12.0;
    a[3][1]=20.0; a[3][2]=15.0; a[3][3]=36.0; a[3][4]=10.0;
    a[4][1]=35.0; a[4][2]=28.0; a[4][3]=70.0; a[4][4]=20.0;
    aux[2]=1.0e-5;
    Linear_algebra.decinv(a,4,aux);
    System.out.println("Calculated inverse:");
    for (i=1; i<=4; i++)
      System.out.println("\t" + (int)a[i][1] + "\t" + (int)a[i][2] + "\t" +
                        (int)a[i][3] + "\t" + (int)a[i][4]);
    System.out.println("\naux[1] = " + (int)aux[1] + "\naux[3] = " + (int)aux[3]);
  }
}
```

Output:

```
Calculated inverse:
        4        -2        4        -1
      -30        20      -45        12
       20       -15       36       -10
      -35        28      -70        20

aux[1] = 1
aux[3] = 4
```

Procedure tested:

gssinv

Example:

Compute the inverse of the 4×4 matrix

$$\begin{pmatrix} 4 & 2 & 4 & 1 \\ 30 & 20 & 45 & 12 \\ 20 & 15 & 36 & 10 \\ 35 & 28 & 70 & 20 \end{pmatrix}.$$

```java
package numal;

import java.text.DecimalFormat;
import numal.*;

public class Test_gssinv extends Object {

  public static void main(String args[]) {
    int i;
    double aux[] = new double[10];
    double a[][] = new double[5][5];

    a[1][1]=4.0;  a[1][2]=2.0;  a[1][3]=4.0;  a[1][4]=1.0;
    a[2][1]=30.0; a[2][2]=20.0; a[2][3]=45.0; a[2][4]=12.0;
    a[3][1]=20.0; a[3][2]=15.0; a[3][3]=36.0; a[3][4]=10.0;
    a[4][1]=35.0; a[4][2]=28.0; a[4][3]=70.0; a[4][4]=20.0;
    aux[2]=1.0e-5;
    aux[4]=8;
    Linear_algebra.gssinv(a,4,aux);
    DecimalFormat fiveDigit = new DecimalFormat("0.00000E0");
    System.out.println("Calculated inverse:");
    for (i=1; i<=4; i++)
      System.out.println("\t" + (int)a[i][1] + "\t" + (int)a[i][2] + "\t" +
                         (int)a[i][3] + "\t" + (int)a[i][4]);
    System.out.println("\nAUX elements:\n  " + fiveDigit.format(aux[1]) + "    " +
                      fiveDigit.format(aux[3]) + "    " + fiveDigit.format(aux[5]) +
                      "    " + fiveDigit.format(aux[7]) + "    " +
                      fiveDigit.format(aux[9]));
```

```
  }
}
```

Output:

```
Calculated inverse:
        4         -2        4         -1
       -30        20       -45        12
        20       -15        36        -10
       -35        28       -70        20

AUX elements:
  1.00000E0   4.00000E0   7.00000E1   1.12529E2   1.55000E2
```


Procedure tested:

gssinverb

Example:
Compute the inverse of the 4×4 matrix

$$\begin{pmatrix} 4 & 2 & 4 & 1 \\ 30 & 20 & 45 & 12 \\ 20 & 15 & 36 & 10 \\ 35 & 28 & 70 & 20 \end{pmatrix},$$

and obtain an error bound for the computed inverse.

```
package numal;

import java.text.DecimalFormat;
import numal.*;

public class Test_gssinverb extends Object {

  public static void main(String args[]) {

    int i;
    double aux[] = new double[12];
    double a[][] = new double[5][5];

    a[1][1]=4.0;  a[1][2]=2.0;  a[1][3]=4.0;  a[1][4]=1.0;
    a[2][1]=30.0; a[2][2]=20.0; a[2][3]=45.0; a[2][4]=12.0;
    a[3][1]=20.0; a[3][2]=15.0; a[3][3]=36.0; a[3][4]=10.0;
    a[4][1]=35.0; a[4][2]=28.0; a[4][3]=70.0; a[4][4]=20.0;
    aux[0]=aux[2]=aux[6]=1.0e-14;
    aux[4]=8;
    Linear_algebra.gssinverb(a,4,aux);
    DecimalFormat fiveDigit = new DecimalFormat("0.00000E0");
    System.out.println("Calculated inverse:");
    for (i=1; i<=4; i++)
```

```
      System.out.println("\t" + (int)a[i][1] + "\t" + (int)a[i][2] + "\t" +
                       (int)a[i][3] + "\t" + (int)a[i][4]);
    System.out.println("\nAUX elements:\n" + fiveDigit.format(aux[1]) + "   " +
                     fiveDigit.format(aux[3]) + "   " + fiveDigit.format(aux[5]) +
                     "   " + fiveDigit.format(aux[7]) + "   " +
                     fiveDigit.format(aux[9]) + "   " + fiveDigit.format(aux[11]));
  }
}
```

Output:

```
Calculated inverse:
      4        -2       4       -1
     -30       20      -45      12
      20      -15       36      -10
     -35       28      -70      20

AUX elements:
1.00000E0   4.00000E0   7.00000E1   1.12529E2   1.55000E2   2.22946E-8
```

Procedure tested:

gssitisol

Example:
Solve the 4×4 linear system of equations $Ax=b$, where $A_{i,j}=840/(i+j-1)$, $i,j=1,...,4$, and b is the third column of A (so that the solution is the third unit vector: $(0,0,1,0)^T$).

```
package numal;

import java.text.DecimalFormat;
import numal.*;

public class Test_gssitisol extends Object {

  public static void main(String args[]) {

    int i,j;
    double b[]    = new double[5];
    double aux[]  = new double[14];
    double a[][]  = new double[5][5];

    for (i=1; i<=4; i++) {
      for (j=1; j<=4; j++) a[i][j]=840/(i+j-1);
      b[i]=a[i][3];
    }
    aux[2]=aux[10]=1.0e-5;
    aux[4]=8;
    aux[12]=5.0;
    Linear_algebra.gssitisol(a,4,aux,b);
    DecimalFormat fiveDigit = new DecimalFormat("0.00000E0");
```

```
    System.out.println("Solution:  " + fiveDigit.format(b[1]) + "   " +
                       fiveDigit.format(b[2]) + "   " + fiveDigit.format(b[3]) +
                       "   " + fiveDigit.format(b[4]));
    System.out.println("Sign(Det) = " + (int)aux[1] + "\nNumber of eliminations = "
                       + (int)aux[3] + "\nMax(abs(a[i,j])) = " +
                       fiveDigit.format(aux[5]) + "\nUpper bound growth = " +
                       fiveDigit.format(aux[7]) + "\nNorm last correction vector = "
                       + fiveDigit.format(aux[11]) + "\nNorm residual vector = " +
                       fiveDigit.format(aux[13]));
  }
}
```

Output:

```
Solution:  -4.26484E-15  5.11654E-14  1.00000E0  8.52651E-14
Sign(Det) = 1
Number of eliminations = 4
Max(abs(a[i,j])) = 8.40000E2
Upper bound growth = 1.34080E3
Norm last correction vector = 2.63510E-13
Norm residual vector = 3.25218E-24
```

Procedure tested:

gssitisolerb

Example:

Solve the 4×4 linear system of equations $Ax=b$, where $A_{i,j}=840/(i+j-1)$, $i,j=1,...,4$, and b is the third column of A (so that the solution is the third unit vector: $(0,0,1,0)^T$), and provide a realistic upper bound for the relative error in it.

```
package numal;

import java.text.DecimalFormat;
import numal.*;

public class Test_gssitisolerb extends Object {

  public static void main(String args[]) {

    int i,j;
    double b[]   = new double[5];
    double aux[] = new double[14];
    double a[][] = new double[5][5];

    for (i=1; i<=4; i++) {
      for (j=1; j<=4; j++) a[i][j]=840/(i+j-1);
      b[i]=a[i][3];
    }
```

```
    aux[0]=aux[2]=aux[10]=1.0e-14;
    aux[4]=8;
    aux[6]=aux[8]=0.0;
    aux[12]=5.0;
    Linear_algebra.gssitisolerb(a,4,aux,b);
    DecimalFormat fiveDigit = new DecimalFormat("0.00000E0");
    System.out.println("Solution:  " + fiveDigit.format(b[1]) +
        "  " + fiveDigit.format(b[2]) + "  " +
        fiveDigit.format(b[3]) + "  " + fiveDigit.format(b[4]));
    System.out.println("Sign(Det) = " + (int)aux[1] +
        "\nNumber of eliminations = " + (int)aux[3] +
        "\nMax(abs(a[i,j])) = " + fiveDigit.format(aux[5]) +
        "\nUpper bound growth = " + fiveDigit.format(aux[7]) +
        "\nNorm calculated inverse matrix = " + fiveDigit.format(aux[9]) +
        "\nUpper bound for the relative error = " + fiveDigit.format(aux[11]) +
        "\nNorm residual vector = " + fiveDigit.format(aux[13]));
  }
}
```

Output:

```
Solution:  -4.26484E-15  5.11654E-14  1.00000E0  8.52651E-14
Sign(Det) = 1
Number of eliminations = 4
Max(abs(a[i,j])) = 8.40000E2
Upper bound growth = 1.34080E3
Norm calculated inverse matrix = 1.62143E1
Upper bound for the relative error = 2.59336E-24
Norm residual vector = 1.59943E-25
```

Procedures tested:

chldec2, chlsol2, chlinv2

Example:

Solve the 4×4 linear system of equations $Ax=b$, where A is the symmetric positive definite coefficient matrix (the Pascal matrix of order 4) for which

$$A_{i,j} = \binom{i+j-2}{j-1}, \quad i,j = 1,\dots,4$$

and $b_i = 2^i$, $i=1,\dots,4$, and obtain the inverse of A.

```java
package numal;
import java.text.DecimalFormat;
import numal.*;

public class Test_chldec2 extends Object {

  public static void main(String args[]) {
    int i,j;
    double b[] = new double[5];
    double aux[] = new double[4];
    double pascal2[][] = new double[5][5];

    for (j=1; j<=4; j++) {
      pascal2[1][j]=1.0;
      for (i=2; i<=j; i++)
        pascal2[i][j] = (i == j) ? pascal2[i-1][j]*2.0 :
                                   pascal2[i][j-1]+pascal2[i-1][j];
      b[j]=Math.pow(2.0,j);
    }
    aux[2]=1.0e-11;
    Linear_algebra.chldec2(pascal2,4,aux);
    if (aux[3] == 4) {
      Linear_algebra.chlsol2(pascal2,4,b);
      Linear_algebra.chlinv2(pascal2,4);
    } else
      System.out.println("Matrix not positive definite");
    DecimalFormat fiveDigit = new DecimalFormat("0.00000E0");
    System.out.println("Solution with CHLDEC2 and CHLSOL2:\n  " +
      fiveDigit.format(b[1]) + "   " + fiveDigit.format(b[2]) + "   " +
      fiveDigit.format(b[3]) + "   " + fiveDigit.format(b[4]));
    System.out.println("\nInverse matrix with CHLINV2:");
    for (i=1; i<=4; i++) {
      for (j=1; j<=4; j++)
        if (j < i)
          System.out.print("\t");
        else
          System.out.print("\t" + pascal2[i][j]);
      System.out.print("\n");
    }
  }
}
```

Output:

```
Solution with CHLDEC2 and CHLSOL2:
  0.00000E0   4.00000E0   -4.00000E0   2.00000E0

Inverse matrix with CHLINV2:
        4.0      -6.0      4.0        -1.0
                 14.0      -11.0      3.0
                           10.0       -3.0
                                      1.0
```

Procedures tested:

chldec1, chlsol1, chlinv1

Example:

Solve the 4×4 linear system of equations $Ax=b$, where A is the symmetric positive definite coefficient matrix (the Pascal matrix of order 4) for which

$$A_{i,j} = \binom{i+j-2}{j-1}, \quad i,j = 1,...,4$$

and $b_i = 2^i$, $i=1,...,4$, and obtain the inverse of A.

```
package numal;

import java.text.DecimalFormat;
import numal.*;

public class Test_chldec1 extends Object {

  public static void main(String args[]) {

    int i,j,jj;
    double b[] = new double[5];
    double aux[] = new double[4];
    double pascal1[] = new double[(((4+1)*4)/2)+1];

    jj=1;
    for (j=1; j<=4; j++) {
      pascal1[jj]=1.0;
      for (i=2; i<=j; i++)
        pascal1[jj+i-1] = (i == j) ? pascal1[jj+i-2]*2.0 :
                                     pascal1[jj+i-2]+pascal1[jj+i-j];
      b[j]=Math.pow(2.0,j);
      jj += j;
    }
    aux[2]=1.0e-11;
    Linear_algebra.chldec1(pascal1,4,aux);
    if (aux[3] == 4) {
      Linear_algebra.chlsol1(pascal1,4,b);
      Linear_algebra.chlinv1(pascal1,4);
    } else
      System.out.println("Matrix not positive definite");
    DecimalFormat fiveDigit = new DecimalFormat("0.00000E0");
    System.out.println("Solution with CHLDEC1 and CHLSOL1:\n  " +
      fiveDigit.format(b[1]) + "  " + fiveDigit.format(b[2]) + "  " +
      fiveDigit.format(b[3]) + "  " + fiveDigit.format(b[4]));
    System.out.println("\nInverse matrix with CHLINV1:");
    for (i=1; i<=4; i++) {
      for (j=1; j<=4; j++)
        if (j < i)
```

```
        System.out.print("\t");
      else
        System.out.print("\t" + pascal1[((j-1)*j)/2+i]);
      System.out.print("\n");
    }
  }
}
```

Output:

```
Solution with CHLDEC1 and CHLSOL1:
  0.00000E0   4.00000E0   -4.00000E0   2.00000E0

Inverse matrix with CHLINV1:
        4.0      -6.0       4.0         -1.0
                 14.0      -11.0         3.0
                           10.0         -3.0
                                         1.0
```

Procedures tested:

chldecsol2, chldeterm2, chldecinv2

Example:
Solve the 4×4 linear system of equations $Ax=b$, where A is the symmetric positive definite coefficient matrix (the Pascal matrix of order 4) for which

$$A_{i,j} = \binom{i+j-2}{j-1}, \quad i,j = 1,...,4$$

and $b_i = 2^i$, $i=1,...,4$, and obtain the determinant and the inverse of A.

```
package numal;

import java.text.DecimalFormat;
import numal.*;

public class Test_chldecsol2 extends Object {

  public static void main(String args[]) {

    int i,j;
    double determinant;
    double b[] = new double[5];
    double aux[] = new double[4];
    double pascal2[][] = new double[5][5];

    for (j=1; j<=4; j++) {
      pascal2[1][j]=1.0;
      for (i=2; i<=j; i++)
        pascal2[i][j] = (i == j) ? pascal2[i-1][j]*2.0 :
                          pascal2[i][j-1]+pascal2[i-1][j];
      b[j]=Math.pow(2.0,j);
```

```
    }
    aux[2]=1.0e-11;
    Linear_algebra.chldecsol2(pascal2,4,aux,b);
    determinant=0.0;
    if (aux[3] == 4)
      determinant=Linear_algebra.chldeterm2(pascal2,4);
    else
      System.out.println("Matrix not positive definite");
    DecimalFormat fiveDigit = new DecimalFormat("0.00000E0");
    System.out.println("Solution with CHLDECSOL2:\n  " +
      fiveDigit.format(b[1]) + "  " + fiveDigit.format(b[2]) + "   " +
      fiveDigit.format(b[3]) + "   " + fiveDigit.format(b[4]));
    System.out.println("\nDeterminant with CHLDETERM2: " + determinant);
    for (j=1; j<=4; j++) {
      pascal2[1][j]=1.0;
      for (i=2; i<=j; i++)
        pascal2[i][j] = (i == j) ? pascal2[i-1][j]*2.0 :
                                   pascal2[i][j-1]+pascal2[i-1][j];
    }
    Linear_algebra.chldecinv2(pascal2,4,aux);
    System.out.println("\nInverse matrix with CHLDECINV2:");
    for (i=1; i<=4; i++) {
      for (j=1; j<=4; j++)
        if (j < i)
          System.out.print("\t");
        else
          System.out.print("\t" + pascal2[i][j]);
      System.out.print("\n");
    }
  }
}
```

Output:

```
Solution with CHLDECSOL2:
  0.00000E0   4.00000E0  -4.00000E0   2.00000E0

Determinant with CHLDETERM2: 1.0

Inverse matrix with CHLDECINV2:
      4.0       -6.0       4.0       -1.0
                14.0      -11.0       3.0
                          10.0      -3.0
                                     1.0
```

Procedures tested:

chldecsol1, chldeterm1, chldecinv1

Example:

Solve the 4×4 linear system of equations $Ax=b$, where A is the symmetric positive definite coefficient matrix (the Pascal matrix of order 4) for which

$$A_{i,j} = \binom{i + j - 2}{j - 1}, \quad i, j = 1, \dots, 4$$

and $b_i = 2^i$, $i = 1, \dots, 4$, and obtain the determinant and the inverse of A.

```
package numal;

import java.text.DecimalFormat;
import numal.*;

public class Test_chldecsol1 extends Object {

  public static void main(String args[]) {

    int i,j,jj;
    double determinant;
    double b[] = new double[5];
    double aux[] = new double[4];
    double pascal1[] = new double[(((4+1)*4)/2)+1];

    jj=1;
    for (j=1; j<=4; j++) {
      pascal1[jj]=1.0;
      for (i=2; i<=j; i++)
        pascal1[jj+i-1] = (i == j) ? pascal1[jj+i-2]*2.0 :
                                     pascal1[jj+i-2]+pascal1[jj+i-j];
      b[j]=Math.pow(2.0,j);
      jj += j;
    }
    aux[2]=1.0e-11;
    Linear_algebra.chldecsol1(pascal1,4,aux,b);
    determinant=0.0;
    if (aux[3] == 4)
      determinant=Linear_algebra.chldeterm1(pascal1,4);
    else
      System.out.println("Matrix not positive definite");
    DecimalFormat fiveDigit = new DecimalFormat("0.00000E0");
    System.out.println("Solution with CHLDECSOL1:\n  " +
      fiveDigit.format(b[1]) + "  " + fiveDigit.format(b[2]) + "  " +
      fiveDigit.format(b[3]) + "  " + fiveDigit.format(b[4]));
    System.out.println("\nDeterminant with CHLDETERM1: " + determinant);
    jj=1;
    for (j=1; j<=4; j++) {
      pascal1[jj]=1.0;
      for (i=2; i<=j; i++)
        pascal1[jj+i-1] = (i == j) ? pascal1[jj+i-2]*2.0 :
                                     pascal1[jj+i-2]+pascal1[jj+i-j];
      jj += j;
    }
```

```
    Linear_algebra.chldecinv1(pascal1,4,aux);
    System.out.println("\nInverse matrix with CHLDECINV1:");
    for (i=1; i<=4; i++) {
      for (j=1; j<=4; j++)
        if (j < i)
          System.out.print("\t");
        else
          System.out.print("\t" + pascal1[((j-1)*j)/2+i]);
      System.out.print("\n");
    }
  }
}
```

Output:

```
Solution with CHLDECSOL1:
  0.00000E0   4.00000E0   -4.00000E0   2.00000E0

Determinant with CHLDETERM1: 1.0

Inverse matrix with CHLDECINV1:
      4.0      -6.0       4.0       -1.0
               14.0      -11.0       3.0
                         10.0       -3.0
                                     1.0
```

Procedure tested:

determsym2

Example:
Evaluate the determinant of the matrix

$$\begin{pmatrix} -3 & -3 & -18 & -30 & 18 \\ -3 & -1 & -4 & -48 & 8 \\ -18 & -4 & -6 & -274 & 6 \\ -30 & -48 & -274 & 119 & 19 \\ 18 & 8 & 6 & 19 & 216 \end{pmatrix}.$$

```
package numal;

import java.text.DecimalFormat;
import numal.*;

public class Test_determsym2 extends Object {

  public static void main(String args[]) {

    int i,j;
    int aux[] = new int[6];
    int p[] = new int[6];
```

```
    double tol,determinant;
    double detaux[]   = new double[6];
    double a[][] = new double[6][6];

    a[1][1]=a[1][2]   = -3.0; a[1][3]  = -18.0;
    a[1][4] = -30.0;   a[1][5]=18.0;
    a[2][2]  = -1.0; a[2][3]  = -4.0; a[2][4]  = -48.0; a[2][5]=8.0;
    a[3][3]  = -6.0; a[3][4]  = -274.0;   a[3][5]=6.0;
    a[4][4]=119.0;   a[4][5]=19.0;       a[5][5]=216.0;
    for (i=1; i<=5; i++)
      for (j=i+1; j<=5; j++) a[j][i]=a[i][j];
    tol=1.0e-6;
    Linear_algebra.decsym2(a,5,tol,aux,p,detaux);
    if (aux[2] == 1)
      System.out.println("\nThe matrix is symmetric.");
    else
      System.out.println("The matrix is asymmetric, results are meaningless.");
    determinant=Linear_algebra.determsym2(detaux,5,aux);
    System.out.println("\nThe determinant of the matrix : " + determinant);
  }
}
```

Output:

```
The matrix is symmetric.

The determinant of the matrix : 168.0
```

Procedure tested:

decsolsym2

Example:
Solve the system of equations

$$
\begin{pmatrix}
-3 & -3 & -18 & -30 & 18 \\
-3 & -1 & -4 & -48 & 8 \\
-18 & -4 & -6 & -274 & 6 \\
-30 & -48 & -274 & 119 & 19 \\
18 & 8 & 6 & 19 & 216
\end{pmatrix}
\begin{pmatrix}
x_1 \\ x_2 \\ x_3 \\ x_4 \\ x_5
\end{pmatrix}
=
\begin{pmatrix}
327 \\ 291 \\ 1290 \\ 275 \\ 1720
\end{pmatrix}.
$$

```
package numal;

import java.text.DecimalFormat;
import numal.*;
```

```
public class Test_decsolsym2 extends Object {

  public static void main(String args[]) {

    int i,j;
    int aux[] = new int[6];
    double tol;
    double b[] = new double[6];
    double a[][] = new double[6][6];

    a[1][1]=a[1][2] = -3.0;    a[1][3] = -18.0;
    a[1][4] = -30.0; a[1][5]=18.0;
    a[2][2] = -1.0; a[2][3] = -4.0; a[2][4] = -48.0;   a[2][5]=8.0;
    a[3][3] = -6.0; a[3][4] = -274.0; a[3][5]=6.0;
    a[4][4]=119.0;   a[4][5]=19.0; a[5][5]=216.0;
    b[1]=327.0; b[2]=291.0; b[3]=1290.0; b[4]=275.0;   b[5]=1720.0;
    for (i=1; i<=5; i++)
      for (j=i+1; j<=5; j++) a[j][i]=a[i][j];
    tol=1.0e-6;
    Linear_algebra.decsolsym2(a,5,b,tol,aux);
    if (aux[2] == 1)
      System.out.println("The matrix is symmetric.");
    else
      System.out.println("The matrix is asymmetric, results are meaningless.");
    DecimalFormat fiveDigit = new DecimalFormat("0.00000E0");
    System.out.println("Inertia :   " + (int)(aux[3]) + ",  " + (int)(aux[4]) +
                      ",  " + (int)(aux[5]));
    System.out.println("The computed solution:\n  " + fiveDigit.format(b[1]) +
                      "  " + fiveDigit.format(b[2]) + "  " +
                      fiveDigit.format(b[3]) + "  " + fiveDigit.format(b[4]) +
                      "  " + fiveDigit.format(b[5]));
  }
}
```

Output:

```
The matrix is symmetric.
Inertia :  3,  2,  0
The computed solution:
  -7.00000E0  -2.00000E0  -1.00000E0  -4.00000E0  9.00000E0
```

Procedures tested:

Isqortdec, Isqsol, Isqdglinv

Example:

Derive the least squares solution of the system of equations

$$-2x_1 + x_2 = 0$$
$$- x_1 + x_2 = 1$$
$$x_1 + x_2 = 2$$
$$2x_1 + x_2 = 2$$
$$x_1 + 2x_2 = 3$$

```java
package numal;

import java.text.DecimalFormat;
import numal.*;

public class Test_lsqortdec extends Object {

  public static void main(String args[]) {

    int i;
    int piv[] = new int[3];
    double sum,temp;
    double b[] = new double[6];
    double x[] = new double[6];
    double diag[] = new double[3];
    double aid[] = new double[3];
    double aux[] = new double[6];
    double a[][] = new double[6][3];
    double c[][] = new double[6][3];

    aux[2]=1.0e-6;
    a[1][1]=c[1][1] = -2.0;
    a[2][1]=c[2][1] = -1.0;
    a[3][1]=c[3][1]=1.0;
    a[4][1]=c[4][1]=2.0;
    a[5][1]=c[5][1]=1.0;
    a[1][2]=a[2][2]=a[3][2]=a[4][2]=c[1][2]=c[2][2]=c[3][2]=c[4][2]=1.0;
    a[5][2]=c[5][2]=2.0;
    b[1]=x[1]=0.0;
    b[2]=x[2]=1.0;
    b[3]=x[3]=b[4]=x[4]=2.0;
    b[5]=x[5]=3.0;
    Linear_algebra.lsqortdec(a,5,2,aux,aid,piv);
    if (aux[3] == 2) {
      Linear_algebra.lsqsol(a,5,2,aid,piv,x);
      Linear_algebra.lsqdglinv(a,2,aid,piv,diag);
      sum=0.0;
      for (i=1; i<=5; i++) {
```

```
        temp=b[i]-c[i][1]*x[1]-c[i][2]*x[2];
        sum += temp*temp;
    }
    DecimalFormat fiveDigit = new DecimalFormat("0.00000E0");
    System.out.println("Aux[2, 3, 5] = " + fiveDigit.format(aux[2]) +
        " " + fiveDigit.format(aux[3]) + " " + fiveDigit.format(aux[5]) +
        "\nLSQ solution : " + fiveDigit.format(x[1]) + " " +
        fiveDigit.format(x[2]) + "\nResidue (delivered) : " +
        fiveDigit.format(Math.sqrt(Basic.vecvec(3,5,0,x,x))) +
        "\nResidue (checked)   : " + fiveDigit.format(Math.sqrt(sum)) +
        "\nDiagonal of inverse M'M : " + fiveDigit.format(diag[1]) + " " +
        fiveDigit.format(diag[2]));
  }
 }
}
```

Output:

```
Aux[2, 3, 5] = 1.00000E-6  2.00000E0  3.31662E0
LSQ solution :  5.00000E-1  1.25000E0
Residue (delivered) :  5.00000E-1
Residue (checked)   :  5.00000E-1
Diagonal of inverse M'M :  9.52381E-2  1.30952E-1
```

Procedure tested:

lsqortdecsol

Example:
Derive the least squares solution of the system of equations
$$-2x_1 + x_2 = 0$$
$$-x_1 + x_2 = 1$$
$$x_1 + x_2 = 2$$
$$2x_1 + x_2 = 2$$
$$x_1 + 2x_2 = 3$$

```
package numal;

import java.text.DecimalFormat;
import numal.*;

public class Test_lsqortdecsol extends Object {

  public static void main(String args[]) {

    int i;
    double sum,temp;
    double b[] = new double[6];
    double x[] = new double[6];
```

```
        double diag[]  = new double[3];
        double aux[]  = new double[6];
        double a[][]  = new double[6][3];
        double c[][]  = new double[6][3];

        aux[2]=1.0e-6;
        a[1][1]=c[1][1]  = -2.0;
        a[2][1]=c[2][1]  = -1.0;
        a[3][1]=c[3][1]=1.0;
        a[4][1]=c[4][1]=2.0;
        a[5][1]=c[5][1]=1.0;
        a[1][2]=a[2][2]=a[3][2]=a[4][2]=c[1][2]=c[2][2]=c[3][2]=c[4][2]=1.0;
        a[5][2]=c[5][2]=2.0;
        b[1]=x[1]=0.0;
        b[2]=x[2]=1.0;
        b[3]=x[3]=b[4]=x[4]=2.0;
        b[5]=x[5]=3.0;
        Linear_algebra.lsqortdecsol(a,5,2,aux,diag,x);
        if (aux[3] == 2) {
          sum=0.0;
          for (i=1; i<=5; i++) {
            temp=b[i]-c[i][1]*x[1]-c[i][2]*x[2];
            sum += temp*temp;
          }
          DecimalFormat fiveDigit = new DecimalFormat("0.00000E0");
          System.out.println("Aux[2, 3, 5] =  " + fiveDigit.format(aux[2]) +
            "  " + fiveDigit.format(aux[3]) + "  " + fiveDigit.format(aux[5]) +
            "\nLSQ solution :  " + fiveDigit.format(x[1]) + "  " +
            fiveDigit.format(x[2]) + "\nResidue (delivered) :  " +
            fiveDigit.format(Math.sqrt(Basic.vecvec(3,5,0,x,x))) +
            "\nResidue (checked)   :  " + fiveDigit.format(Math.sqrt(sum)) +
            "\nDiagonal of inverse M'M :  " + fiveDigit.format(diag[1]) +
            "  " + fiveDigit.format(diag[2]));
        }
      }
    }
  }
```

Output:

```
Aux[2, 3, 5] =  1.00000E-6  2.00000E0  3.31662E0
LSQ solution :  5.00000E-1  1.25000E0
Residue (delivered) :  5.00000E-1
Residue (checked)   :  5.00000E-1
Diagonal of inverse M'M :  9.52381E-2  1.30952E-1
```

Procedure tested:

lsqinv

Example:

Computes the inverse T of $S'S$, where S' is the transpose of S given by

$$S' = \begin{pmatrix} -2 & -1 & 1 & 2 & 1 \\ 1 & 1 & 1 & 1 & 2 \end{pmatrix}.$$

```
package numal;

import java.text.DecimalFormat;
import numal.*;

public class Test_lsqinv extends Object {

  public static void main(String args[]) {

    int i,j;
    int piv[] = new int[3];
    double aid[] = new double[3];
    double aux[] = new double[6];
    double a[][] = new double[6][3];
    double c[][] = new double[6][3];
    double t[][] = new double[3][3];

    aux[2]=1.0e-6;
    a[1][1]=c[1][1] = -2.0;
    a[2][1]=c[2][1] = -1.0;
    a[3][1]=c[3][1]=1.0;
    a[4][1]=c[4][1]=2.0;
    a[5][1]=c[5][1]=1.0;
    a[1][2]=a[2][2]=a[3][2]=a[4][2]=c[1][2]=c[2][2]=c[3][2]=c[4][2]=1.0;
    a[5][2]=c[5][2]=2.0;
    Linear_algebra.lsqortdec(a,5,2,aux,aid,piv);
    if (aux[3] == 2) {
      Linear_algebra.lsqinv(a,2,aid,piv);
      t[1][1]=a[1][1];
      t[2][2]=a[2][2];
      t[2][1]=t[1][2]=a[1][2];
      for (j=1; j<=2; j++)
        for (i=1; i<=5; i++) a[i][j]=Basic.matmat(1,2,i,j,c,t);
      DecimalFormat fiveDigit = new DecimalFormat("0.00000E0");
      System.out.println("Aux[2, 3, 5] = " + fiveDigit.format(aux[2]) +
        "  " + fiveDigit.format(aux[3]) + "  " + fiveDigit.format(aux[5]) +
        "\n\nInverse:\n   " + fiveDigit.format(t[1][1]) + "  " +
        fiveDigit.format(t[1][2]) + "\n  " + fiveDigit.format(t[2][1]) + "   " +
        fiveDigit.format(t[2][2]) + "\n\nCheck:  S' * (S * T) :\n  " +
        fiveDigit.format(Basic.tammat(1,5,1,1,c,a)) + "   " +
```

```
        fiveDigit.format(Basic.tammat(1,5,1,2,c,a)) + "\n  " +
        fiveDigit.format(Basic.tammat(1,5,2,1,c,a)) + "  " +
        fiveDigit.format(Basic.tammat(1,5,2,2,c,a)));
    }
  }
}
```

Output:

```
Aux[2, 3, 5] =  1.00000E-6  2.00000E0  3.31662E0

Inverse:
   9.52381E-2  -2.38095E-2
  -2.38095E-2   1.30952E-1

Check:  S' * (S * T) :
  1.00000E0     8.32667E-17
  1.38778E-17  1.00000E0
```

Procedures tested:

lsqdecomp, lsqrefsol

Example:
Minimize $|b\text{-}Ax|_E$, where $x \in R^3$ and

$$A = \begin{pmatrix} 1 & 0 & 8 \\ 0 & 3 & 2 \\ 1 & 2 & 10^{-5} \\ 0 & 0 & 0 \end{pmatrix}, \quad b = \begin{pmatrix} 25 \\ 12 \\ 5.00003 \\ 1 \end{pmatrix}$$

subject to the constraint
$$x_1 + 1000x_2 + 5x_3 = 2016 \,.$$

```
package numal;

import java.text.DecimalFormat;
import numal.*;

public class Test_lsqdecomp extends Object {

  public static void main(String args[]) {

    int n,m,n1,i,j;
    int ci[] = new int[4];
    double ldx[] = new double[1];
    double aux[] = new double[8];
    double b[] = new double[6];
    double res[] = new double[6];
```

```
    double aid[] = new double[4];
    double x[] = new double[4];
    double a[][] = new double[6][4];
    double qr[][] = new double[6][4];

    n=5; m=3; n1=1;
    a[1][1]=1.0; a[1][2]=1000.0; a[1][3]=5.0;
    a[2][1]=1.0; a[2][2]=0.0; a[2][3]=8.0;
    a[3][1]=0.0; a[3][2]=3.0; a[3][3]=2.0;
    a[4][1]=1.0; a[4][2]=2.0; a[4][3]=1.0e-5;
    a[5][1]=a[5][2]=a[5][3]=0.0;
    b[1]=2016.0; b[2]=25.0; b[3]=12.0; b[4]=5.00003; b[5]=1.0;
    aux[2]=1.0e-6;
    aux[6]=5.0;
    for (i=1; i<=5; i++)
      for (j=1; j<=3; j++) qr[i][j]=a[i][j];
    Linear_algebra.lsqdecomp(qr,n,m,n1,aux,aid,ci);
    Linear_algebra.lsqrefsol(a,qr,n,m,n1,aux,aid,ci,b,ldx,x,res);
    DecimalFormat fiveDigit = new DecimalFormat("0.00000E0");
    System.out.println("The solution vector:\n  " + fiveDigit.format(x[1]) + "  " +
      fiveDigit.format(x[2]) + "  " + fiveDigit.format(x[3]) +
      "\n\nThe residual vector:\n   " + fiveDigit.format(res[2]) + "\n  " +
      fiveDigit.format(res[3]) + "\n  " + fiveDigit.format(res[4]) + "\n   " +
      fiveDigit.format(res[5]) + "\nNumber of iterations:  " + (int)aux[7] +
      "\nNorm last correction of x:  " + fiveDigit.format(ldx[0]));
  }
}
```

Output:

```
The solution vector:
  1.00000E0  2.00000E0  3.00000E0

The residual vector:
   5.18881E-17
  -2.09386E-16
  -5.26215E-17
   1.00000E0
Number of iterations:  2
Norm last correction of x:  1.72985E-34
```

Procedure tested:

solovr

Example:

Determine that $x \in R^5$ with minimum $|x|_2$ which minimizes $|Ax\text{-}b|_2$, where

$$A = \begin{pmatrix} 22 & 10 & 2 & 3 & 7 \\ 14 & 7 & 10 & 0 & 8 \\ -1 & 13 & -1 & -11 & 3 \\ -3 & -2 & 13 & -2 & 4 \\ 9 & 8 & 1 & -2 & 4 \\ 9 & 1 & -7 & 5 & -1 \\ 2 & -6 & 6 & 5 & 1 \\ 4 & 5 & 0 & -2 & 2 \end{pmatrix}, \quad b = \begin{pmatrix} -1 \\ 2 \\ 1 \\ 4 \\ 0 \\ -3 \\ 1 \\ 0 \end{pmatrix}.$$

```
package numal;

import java.text.DecimalFormat;
import numal.*;

public class Test_solovr extends Object {

  public static void main(String args[]) {

    int i;
    double b[]  = new double[9];
    double em[] = new double[8];
    double a[][] = new double[9][6];

    a[1][1]=22; a[1][2]=a[2][3]=10.0; a[1][3]=a[7][1]=a[8][5]=2.0;
    a[1][4]=a[3][5]=3.0; a[1][5]=a[2][2]=7.0; a[2][1]=14.0;
    a[2][5]=8.0;  a[2][4]=a[8][3]=0.0;
    a[3][1]=a[3][3]=a[6][5] = -1.0;  a[3][2]=13.0;
    a[3][4]  = -11.0; a[4][1] = -3.0;
    a[4][2]=a[4][4]=a[5][4]=a[8][4] = -2.0;
    a[4][3]=13.0;  a[4][5]=a[5][5]=a[8][1]=4.0;
    a[5][1]=a[6][1]=9.0;
    a[5][2]=8.0; a[5][3]=a[6][2]=a[7][5]=1.0; a[6][3] = -7.0;
    a[6][4]=a[7][4]=a[8][2]=5.0; a[7][2] = -6.0; a[7][3]=6.0;
    b[1] = -1.0; b[2]=2.0; b[3]=b[7]=1.0; b[4]=4.0; b[5]=b[8]=0.0;
    b[6] = -3.0;
    em[0]=1.0e-14; em[2]=1.0e-12; em[4]=80.0; em[6]=1.0e-10;
    i=Linear_algebra.solovr(a,8,5,b,em);
    DecimalFormat fiveDigit = new DecimalFormat("0.00000E0");
    System.out.println("Number of singular values not found : " + i +
```

```
        "\nNorm :   " + fiveDigit.format(em[1]) +
        "\nMaximal neglected subdiagonal element : " + fiveDigit.format(em[3]) +
        "\nNumber of iterations : " + (int)em[5] + "\nRank :   " + (int)em[7] +
        "\n\nSolution vector\n  " + fiveDigit.format(b[1]) + "\n   " +
        fiveDigit.format(b[2]) + "\n   " + fiveDigit.format(b[3]) + "\n   " +
        fiveDigit.format(b[4]) + "\n   " + fiveDigit.format(b[5]));
  }
}
```

Output:

```
Number of singular values not found : 0
Norm :   4.40000E1
Maximal neglected subdiagonal element :   6.03352E-16
Number of iterations :   6
Rank :   3

Solution vector
  -8.33333E-2
   3.92364E-17
   2.50000E-1
  -8.33333E-2
   8.33333E-2
```

Procedure tested:

solund

Example:

Determine that $x \in R^8$ with minimum $|x|_2$ which minimizes $|A^Tx-b|_2$, where

$$A = \begin{pmatrix} 22 & 10 & 2 & 3 & 7 \\ 14 & 7 & 10 & 0 & 8 \\ -1 & 13 & -1 & -11 & 3 \\ -3 & -2 & 13 & -2 & 4 \\ 9 & 8 & 1 & -2 & 4 \\ 9 & 1 & -7 & 5 & -1 \\ 2 & -6 & 6 & 5 & 1 \\ 4 & 5 & 0 & -2 & 2 \end{pmatrix}, \quad b = \begin{pmatrix} -1 \\ 2 \\ 1 \\ 4 \\ 0 \end{pmatrix}.$$

```
package numal;

import java.text.DecimalFormat;
import numal.*;
```

```java
public class Test_solund extends Object {
  public static void main(String args[]) {
    int i;
    double b[] = new double[9];
    double em[] = new double[8];
    double a[][] = new double[9][6];

    a[1][1]=22; a[1][2]=a[2][3]=10.0; a[1][3]=a[7][1]=a[8][5]=2.0;
    a[1][4]=a[3][5]=3.0;  a[1][5]=a[2][2]=7.0;
    a[2][1]=14.0; a[2][5]=8.0;
    a[2][4]=a[8][3]=0.0;  a[3][1]=a[3][3]=a[6][5] = -1.0;
    a[3][2]=13.0;  a[3][4] = -11.0;  a[4][1] = -3.0;
    a[4][2]=a[4][4]=a[5][4]=a[8][4] = -2.0;
    a[4][3]=13.0; a[4][5]=a[5][5]=a[8][1]=4.0;
    a[5][1]=a[6][1]=9.0;
    a[5][2]=8.0; a[5][3]=a[6][2]=a[7][5]=1.0; a[6][3] = -7.0;
    a[6][4]=a[7][4]=a[8][2]=5.0; a[7][2] = -6.0; a[7][3]=6.0;
    b[1] = -1.0; b[2]=2.0; b[3]=1.0; b[4]=4.0; b[5]=0.0;
    em[0]=1.0e-14; em[2]=1.0e-12; em[4]=80.0; em[6]=1.0e-10;
    i=Linear_algebra.solund(a,8,5,b,em);
    DecimalFormat fiveDigit = new DecimalFormat("0.00000E0");
    System.out.println("Number of singular values not found : " + i +
      "\nNorm :  " + fiveDigit.format(em[1]) +
      "\nMaximal neglected subdiagonal element :  " + fiveDigit.format(em[3]) +
      "\nNumber of iterations :  " + (int)em[5] + "\nRank :  " + (int)em[7] +
      "\n\nSolution vector\n  " + fiveDigit.format(b[1]) + "\n  " +
      fiveDigit.format(b[2]) + "\n  " + fiveDigit.format(b[3]) + "\n  " +
      fiveDigit.format(b[4]) + "\n  " + fiveDigit.format(b[5]) + "\n  " +
      fiveDigit.format(b[6]) + "\n  " + fiveDigit.format(b[7]) + "\n  " +
      fiveDigit.format(b[8]));
  }
}
```

Output:

```
Number of singular values not found : 0
Norm :  4.40000E1
Maximal neglected subdiagonal element :  6.03352E-16
Number of iterations :  6
Rank :  3

Solution vector
   1.64103E-2
   1.48077E-2
  -4.83974E-2
   1.00000E-2
  -6.79487E-3
   1.16026E-2
   3.00000E-2
  -8.39744E-3
```

Procedure tested:
homsol

Example:
With the 8×5 matrix A of rank 3 given by

$$A = \begin{pmatrix} 22 & 10 & 2 & 3 & 7 \\ 14 & 7 & 10 & 0 & 8 \\ -1 & 13 & -1 & -11 & 3 \\ -3 & -2 & 13 & -2 & 4 \\ 9 & 8 & 1 & -2 & 4 \\ 9 & 1 & -7 & 5 & -1 \\ 2 & -6 & 6 & 5 & 1 \\ 4 & 5 & 0 & -2 & 2 \end{pmatrix}$$

the vectors $v_j \in R^5$ (j=4,5) defining the complementary column subspace in which $x \in R^5$ must lie for the condition $Ax=0$ to hold, and the vectors u_j^T (j=4,5) defining the complementary row subspace in which $u^T \in R^8$ must lie for the condition $x^T A=0$ to hold are determined.

```
package numal;

import java.text.DecimalFormat;
import numal.*;

public class Test_homsol extends Object {

  public static void main(String args[]) {

    int i,j;
    double em[]  = new double[8];
    double a[][] = new double[9][6];
    double v[][] = new double[6][6];

    a[1][1]=22; a[1][2]=a[2][3]=10.0; a[1][3]=a[7][1]=a[8][5]=2.0;
    a[1][4]=a[3][5]=3.0; a[1][5]=a[2][2]=7.0; a[2][1]=14.0;
    a[2][5]=8.0;  a[2][4]=a[8][3]=0.0;
    a[3][1]=a[3][3]=a[6][5]  = -1.0;
    a[3][2]=13.0;  a[3][4]  = -11.0;  a[4][1]  = -3.0;
    a[4][2]=a[4][4]=a[5][4]=a[8][4]  = -2.0;
    a[4][3]=13.0;  a[4][5]=a[5][5]=a[8][1]=4.0;
    a[5][1]=a[6][1]=9.0;
    a[5][2]=8.0; a[5][3]=a[6][2]=a[7][5]=1.0; a[6][3]  = -7.0;
    a[6][4]=a[7][4]=a[8][2]=5.0; a[7][2]  = -6.0; a[7][3]=6.0;
```

```
      em[0]=1.0e-14; em[2]=1.0e-12; em[4]=80.0; em[6]=1.0e-10;
      i=Linear_algebra.homsol(a,8,5,v,em);
      DecimalFormat fiveDigit = new DecimalFormat("0.00000E0");
      System.out.println("Number of singular values not found : " + i + "\nNorm :  " +
        fiveDigit.format(em[1]) + "\nMaximal neglected subdiagonal element :  " +
        fiveDigit.format(em[3]) + "\nNumber of iterations :  " + (int)em[5] +
        "\nRank :  " + (int)em[7]);
      for (j=(int)(em[7])+1; j<=5; j++) {
        System.out.println("\nColumn number :  " + j);
        for (i=1; i<=5; i++)
          System.out.println("\t" + fiveDigit.format(a[i][j]) + "\t" +
                             fiveDigit.format(v[i][j]));
        System.out.println("\t" + fiveDigit.format(a[6][j]) + "\n\t" +
                           fiveDigit.format(a[7][j]) + "\n\t" +
                           fiveDigit.format(a[8][j]));
      }
    }
  }
}
```

Output:

```
Number of singular values not found : 0
Norm :  4.40000E1
Maximal neglected subdiagonal element :  6.03352E-16
Number of iterations :  6
Rank :  3

Column number :  4
        -3.47086E-1        -4.19095E-1
        6.07234E-1         4.40509E-1
        -1.22075E-1        -5.20045E-2
        -6.18826E-1        6.76059E-1
        4.63444E-3         4.12977E-1
        -3.34099E-1
        3.35284E-2
        1.35472E-2

Column number :  5
        -2.55331E-1        -0.00000E0
        -1.73598E-1        4.18548E-1
        -2.20812E-1        3.48790E-1
        4.11655E-2         2.44153E-1
        9.20442E-1         -8.02217E-1
        -2.88960E-2
        6.13276E-2
        -4.90581E-2
```

Procedure tested:

psdinv

Example:
Compute the generalized inverse of the 8×5 matrix given by

$$\begin{pmatrix} 22 & 10 & 2 & 3 & 7 \\ 14 & 7 & 10 & 0 & 8 \\ -1 & 13 & -1 & -11 & 3 \\ -3 & -2 & 13 & -2 & 4 \\ 9 & 8 & 1 & -2 & 4 \\ 9 & 1 & -7 & 5 & -1 \\ 2 & -6 & 6 & 5 & 1 \\ 4 & 5 & 0 & -2 & 2 \end{pmatrix}.$$

```
package numal;

import java.text.DecimalFormat;
import numal.*;

public class Test_psdinv extends Object {

  public static void main(String args[]) {

    int i,j;
    double em[] = new double[8];
    double a[][] = new double[9][6];

    a[1][1]=22; a[1][2]=a[2][3]=10.0; a[1][3]=a[7][1]=a[8][5]=2.0;
    a[1][4]=a[3][5]=3.0; a[1][5]=a[2][2]=7.0; a[2][1]=14.0;
    a[2][5]=8.0;   a[2][4]=a[8][3]=0.0;
    a[3][1]=a[3][3]=a[6][5]  = -1.0;
    a[3][2]=13.0;   a[3][4]  = -11.0;   a[4][1]  = -3.0;
    a[4][2]=a[4][4]=a[5][4]=a[8][4]  = -2.0;
    a[4][3]=13.0; a[4][5]=a[5][5]=a[8][1]=4.0;
    a[5][1]=a[6][1]=9.0;
    a[5][2]=8.0; a[5][3]=a[6][2]=a[7][5]=1.0; a[6][3]  = -7.0;
    a[6][4]=a[7][4]=a[8][2]=5.0; a[7][2]  = -6.0; a[7][3]=6.0;
    em[0]=1.0e-14; em[2]=1.0e-12; em[4]=80.0; em[6]=1.0e-10;
    i=Linear_algebra.psdinv(a,8,5,em);
    DecimalFormat fiveDigit = new DecimalFormat("0.00000E0");
    System.out.println("Number of singular values not found : " + i + "\nNorm :   " +
      fiveDigit.format(em[1]) + "\nMaximal neglected subdiagonal element :   " +
      fiveDigit.format(em[3]) + "\nNumber of iterations :   " + (int)em[5] +
      "\nRank :   " + (int)em[7]);
```

```
    System.out.println("\nTranspose of pseudo-inverse, first three columns:");
    for (i=1; i<=8; i++)
      System.out.println("\t" + fiveDigit.format(a[i][1]) + "\t" +
                         fiveDigit.format(a[i][2]) + "\t" +
                         fiveDigit.format(a[i][3]));
    System.out.println("\nLast two columns:");
    for (i=1; i<=8; i++)
      System.out.println("\t\t" + fiveDigit.format(a[i][4]) + "\t" +
                         fiveDigit.format(a[i][5]));
  }
}
```

Output:

```
Number of singular values not found : 0
Norm :   4.40000E1
Maximal neglected subdiagonal element :   6.03352E-16
Number of iterations :   6
Rank :   3

Transpose of pseudo-inverse,first three columns:
        2.11298E-2        4.61538E-3        -2.10737E-3
        9.31090E-3        2.21154E-3        2.05288E-2
        -1.10978E-2       2.74038E-2        -3.88622E-3
        -7.91667E-3       -5.00000E-3       3.37500E-2
        5.51282E-3        9.80769E-3        -8.97436E-4
        1.43189E-2        -2.59615E-3       -2.01362E-2
        4.89583E-3        -1.50000E-2       1.53125E-2
        1.50641E-3        7.40385E-3        -1.69872E-3

Last two columns:
        7.60417E-3        3.80609E-3
        -2.08333E-4       1.00160E-2
        -2.76042E-2       4.20673E-3
        -5.41667E-3       1.04167E-2
        -5.00000E-3       3.20513E-3
        1.28125E-2        -6.20994E-3
        1.23958E-2        2.60417E-3
        -5.00000E-3       1.60256E-3
```

Procedures tested:

solbnd, decbnd, determbnd

Example:
Solve the system of equations
$$
\begin{aligned}
2x_1 - x_2 \quad\quad\quad\quad\quad &= 1 \\
-x_1 + 2x_2 - x_3 \quad\quad\quad &= 0 \\
- x_2 + 2x_3 - x_4 \quad\quad &= 0 \\
- x_3 + 2x_4 - x_5 &= 0 \\
- x_4 + 2x_5 &= 1
\end{aligned}
$$
and evaluate the determinant of the coefficient matrix.

```
package numal;

import java.text.DecimalFormat;
import numal.*;
```

```
public class Test_solbnd extends Object {

  public static void main(String args[]) {

    int i;
    int rowind[] = new int[6];
    double band[] = new double[14];
    double mult[] = new double[5];
    double right[] = new double[6];
    double aux[] = new double[6];

    for (i=1; i<=13; i++)
      band[i] = (((i+1)/3)*3 < i) ? 2.0 : -1.0;
    right[1]=right[5]=1.0;
    right[2]=right[3]=right[4]=0.0;
    aux[2]=1.0e-12;
    Linear_algebra.decbnd(band,5,1,1,aux,mult,rowind);
    if (aux[3] == 5) {
      Linear_algebra.solbnd(band,5,1,1,mult,rowind,right);
      DecimalFormat fiveDigit = new DecimalFormat("0.00000E0");
      System.out.println("Delivers: " + fiveDigit.format(right[1]) + "  " +
        fiveDigit.format(right[2]) + "  " + fiveDigit.format(right[3]) + "  " +
        fiveDigit.format(right[4]) + "  " + fiveDigit.format(right[5]) +
        "\nDeterminant is  " +
        fiveDigit.format(Linear_algebra.determbnd(band,5,1,1,(int)aux[1])));
    }
  }
}
```

Output:

```
Delivers: 1.00000E0  1.00000E0  1.00000E0  1.00000E0  1.00000E0
Determinant is  6.00000E0
```

Procedure tested:

decsolbnd

Example:
Solve the system of equations
$$
\begin{aligned}
2x_1 - x_2 \qquad\qquad &= 1 \\
-x_1 + 2x_2 - x_3 \qquad &= 0 \\
- x_2 + 2x_3 - x_4 \qquad &= 0 \\
- x_3 + 2x_4 - x_5 &= 0 \\
- x_4 + 2x_5 &= 1
\end{aligned}
$$
and evaluate the determinant of the coefficient matrix.

```
package numal;

import java.text.DecimalFormat;
import numal.*;

public class Test_decsolbnd extends Object {

  public static void main(String args[]) {

    int i;
    double band[] = new double[14];
    double right[] = new double[6];
    double aux[] = new double[6];

    for (i=1; i<=13; i++)
      band[i] = (((i+1)/3)*3 < i) ? 2.0 : -1.0;
    right[1]=right[5]=1.0;
    right[2]=right[3]=right[4]=0.0;
    aux[2]=1.0e-12;
    Linear_algebra.decsolbnd(band,5,1,1,aux,right);
    DecimalFormat fiveDigit = new DecimalFormat("0.00000E0");
    if (aux[3] == 5)
      System.out.println("Delivers: " + fiveDigit.format(right[1]) + "   " +
        fiveDigit.format(right[2]) + "   " + fiveDigit.format(right[3]) + "   " +
        fiveDigit.format(right[4]) + "   " + fiveDigit.format(right[5]) +
        "\nDeterminant is  " +
        fiveDigit.format(Linear_algebra.determbnd(band,5,1,1,(int)aux[1])));
  }
}
```

Output:

```
Delivers: 1.00000E0  1.00000E0   1.00000E0   1.00000E0   1.00000E0
Determinant is  6.00000E0
```

Procedure tested:

decsoltri

Example:
Solve the system of equations $Ax=b$, where A is a 30×30 tridiagonal matrix for which $A_{i+1,i} = 2*i$, $A_{i,i} = i+10$, $A_{i,i+1} = i$, and b is the second column of A (the solution is given by the second unit vector).

```
package numal;

import java.text.DecimalFormat;
import numal.*;
```

```
public class Test_decsoltri extends Object {

  public static void main(String args[]) {

    int i;
    double d[] = new double[31];
    double sub[] = new double[31];
    double supre[] = new double[31];
    double b[] = new double[31];
    double aux[] = new double[6];

    for (i=1; i<=30; i++) {
      sub[i]=i*2;
      supre[i]=i;
      d[i]=i+10;
      b[i]=0.0;
    }
    b[1]=1.0;  b[2]=12.0;  b[3]=4.0;
    aux[2]=1.0e-6;
    Linear_algebra.decsoltri(sub,d,supre,30,aux,b);
    b[2]--;
    DecimalFormat fiveDigit = new DecimalFormat("0.00000E0");
    System.out.println("AUX[3] and AUX[5]:  " + fiveDigit.format(aux[3]) + "  " +
      fiveDigit.format(aux[5]) + "\nError in the solution:  " +
      fiveDigit.format(Math.sqrt(Basic.vecvec(1,30,0,b,b))));
  }
}
```

Output:

```
AUX[3] and AUX[5]:  3.00000E1  1.24000E2
Error in the solution:  0.00000E0
```

Procedure tested:

soltripiv

Example:
Solve the two systems of equations $Ax=b1$, $Ax=b2$, where A is a 30×30 tridiagonal matrix for which $A_{i+1,i} = 2*i$, $A_{i,i} = i+10$, $A_{i,i+1} = i$, and $b1$, $b2$ are the second and third column of A, respectively, (the solutions of the systems are given by the second and third unit vector, respectively).

```
package numal;

import java.text.DecimalFormat;
import numal.*;
```

```
public class Test_soltripiv extends Object {

  public static void main(String args[]) {

    int i;
    boolean piv[] = new boolean[30];
    double d[] = new double[31];
    double sub[] = new double[31];
    double supre[] = new double[31];
    double aid[] = new double[31];
    double b1[] = new double[31];
    double b2[] = new double[31];
    double aux[] = new double[6];

    for (i=1; i<=30; i++) {
      sub[i]=i*2;
      supre[i]=i;
      d[i]=i+10;
      b1[i]=b2[i]=0.0;
    }
    b1[1]=1.0;  b1[2]=12.0;  b1[3]=4.0;
    b2[2]=2.0;  b2[3]=13.0;  b2[4]=6.0;
    aux[2]=1.0e-6;
    Linear_algebra.dectripiv(sub,d,supre,30,aid,aux,piv);
    Linear_algebra.soltripiv(sub,d,supre,30,aid,piv,b1);
    Linear_algebra.soltripiv(sub,d,supre,30,aid,piv,b2);
    b1[2]--;  b2[3]--;
    DecimalFormat fiveDigit = new DecimalFormat("0.00000E0");
    System.out.println("AUX[3] and AUX[5]:  " + fiveDigit.format(aux[3]) + "  " +
      fiveDigit.format(aux[5]) + "\nError in b1:  " +
      fiveDigit.format(Math.sqrt(Basic.vecvec(1,30,0,b1,b1))) + "\nError in b2:  " +
      fiveDigit.format(Math.sqrt(Basic.vecvec(1,30,0,b2,b2))));
  }
}
```

Output:

```
AUX[3] and AUX[5]:  3.00000E1  1.24000E2
Error in b1:  0.00000E0
Error in b2:  0.00000E0
```

Procedure tested:

decsoltripiv

Example:
Solve the system of equations $Ax=b$, where A is a 30×30 tridiagonal matrix for which $A_{i+1,i} = 2{*}i$, $A_{i,i} = i+10$, $A_{i,i+1} = i$, and b is the second column of A (the solution is given by the second unit vector).

```
package numal;
import java.text.DecimalFormat;
import numal.*;

public class Test_decsoltripiv extends Object {

  public static void main(String args[]) {
    int i;
    double d[] = new double[31];
    double sub[] = new double[31];
    double supre[] = new double[31];
    double b[] = new double[31];
    double aux[] = new double[6];

    for (i=1; i<=30; i++) {
      sub[i]=i*2;
      supre[i]=i;
      d[i]=i+10;
      b[i]=0.0;
    }
    b[1]=1.0;  b[2]=12.0;  b[3]=4.0;
    aux[2]=1.0e-6;
    Linear_algebra.decsoltripiv(sub,d,supre,30,aux,b);
    b[2]--;
    DecimalFormat fiveDigit = new DecimalFormat("0.00000E0");
    System.out.println("AUX[3] and AUX[5]:  " + fiveDigit.format(aux[3]) + "  " +
      fiveDigit.format(aux[5]) + "\nError in the solution:  " +
      fiveDigit.format(Math.sqrt(Basic.vecvec(1,30,0,b,b))));
  }
}
```

Output:

```
AUX[3] and AUX[5]:  3.00000E1  1.24000E2
Error in the solution:  0.00000E0
```

Procedures tested:

chlsolbnd, chldecbnd, chldetermbnd

Example:
Solve the system of equations

$$
\begin{aligned}
2x_1 - x_2 &= 1 \\
-x_1 + 2x_2 - x_3 &= 0 \\
-x_2 + 2x_3 - x_4 &= 0 \\
-x_3 + 2x_4 - x_5 &= 0 \\
-x_4 + 2x_5 &= 1
\end{aligned}
$$

and evaluate the determinant of the coefficient matrix.

```
package numal;

import java.text.DecimalFormat;
import numal.*;

public class Test_chlsolbnd extends Object {

  public static void main(String args[]) {

    int i;
    double symband[] = new double[10];
    double right[] = new double[6];
    double aux[] = new double[4];

    for (i=1; i<=9; i++)
      symband[i] = ((i/2)*2 < i) ? 2.0 : -1.0;
    right[1]=right[5]=1.0;
    right[2]=right[3]=right[4]=0.0;
    aux[2]=1.0e-12;
    Linear_algebra.chldecsolbnd(symband,5,1,aux,right);
    if (aux[3] == 5) {
      DecimalFormat fiveDigit = new DecimalFormat("0.00000E0");
      System.out.println("Delivers: " + fiveDigit.format(right[1]) + "   " +
        fiveDigit.format(right[2]) + "   " + fiveDigit.format(right[3]) + "   " +
        fiveDigit.format(right[4]) + "   " + fiveDigit.format(right[5]) +
        "\nDeterminant is  " +
        fiveDigit.format(Linear_algebra.chldetermbnd(symband,5,1)));
    }
  }
}
```

Output:

```
Delivers: 1.00000E0  1.00000E0  1.00000E0  1.00000E0  1.00000E0
Determinant is  6.00000E0
```

Procedures tested:

chldecsolbnd, chldetermbnd

Example:
Solve the system of equations

$$
\begin{aligned}
2x_1 - x_2 & = 1 \\
-x_1 + 2x_2 - x_3 & = 0 \\
- x_2 + 2x_3 - x_4 & = 0 \\
- x_3 + 2x_4 - x_5 &= 0 \\
- x_4 + 2x_5 &= 1
\end{aligned}
$$

and evaluate the determinant of the coefficient matrix.

```
package numal;

import java.text.DecimalFormat;
import numal.*;

public class Test_chldecsolbnd extends Object {

  public static void main(String args[]) {

    int i;
    double symband[] = new double[10];
    double right[] = new double[6];
    double aux[] = new double[4];

    for (i=1; i<=9; i++)
      symband[i] = ((i/2)*2 < i) ? 2.0 : -1.0;
    right[1]=right[5]=1.0;
    right[2]=right[3]=right[4]=0.0;
    aux[2]=1.0e-12;
    Linear_algebra.chldecsolbnd(symband,5,1,aux,right);
    if (aux[3] == 5) {
      DecimalFormat fiveDigit = new DecimalFormat("0.00000E0");
      System.out.println("Delivers: " + fiveDigit.format(right[1]) + "   " +
        fiveDigit.format(right[2]) + "   " + fiveDigit.format(right[3]) + "    " +
        fiveDigit.format(right[4]) + "   " + fiveDigit.format(right[5]) +
        "\nDeterminant is   " +
        fiveDigit.format(Linear_algebra.chldetermbnd(symband,5,1)));
    }
  }
}
```

Output:

```
Delivers: 1.00000E0  1.00000E0  1.00000E0  1.00000E0  1.00000E0
Determinant is  6.00000E0
```

Procedure tested:

decsolsymtri

Example:
Solve the system of equations $Ax=b$, where A is a 100×100 symmetric tridiagonal matrix for which $A_{i+1,i} = A_{i,i+1} = 2*i$, $A_{i,i} = i$, and b is the second column of A (the solution is given by the second unit vector).

```
package numal;

import java.text.DecimalFormat;
import numal.*;
```

```
public class Test_decsolsymtri extends Object {

  public static void main(String args[]) {

    int i;
    double d[]  = new double[101];
    double co[] = new double[101];
    double b[]  = new double[101];
    double aux[] = new double[6];

    for (i=1; i<=100; i++) {
      d[i]=i;
      co[i]=i*2;
      b[i]=0.0;
    }
    b[1]=b[2]=2.0;  b[3]=4.0;
    aux[2]=1.0e-6;
    Linear_algebra.decsolsymtri(d,co,100,aux,b);
    b[2]--;
    DecimalFormat fiveDigit = new DecimalFormat("0.00000E0");
    System.out.println("AUX[3] and AUX[5]:  " + fiveDigit.format(aux[3]) + "   " +
      fiveDigit.format(aux[5]) + "\nError in the solution:  " +
      fiveDigit.format(Math.sqrt(Basic.vecvec(1,100,0,b,b))));
  }
}
```

Output:

```
AUX[3] and AUX[5]:  1.00000E2  4.93000E2
Error in the solution:  0.00000E0
```

Procedure tested:

conjgrad

Example:
Solve the system of equations $Ax=b$, $b \in R^{13}$, where
$$A_{i+1,i} = A_{i,i+1} = -1, A_{i,i} = 2, A_{i,j} = 0 \text{ for } |i-j| > 1,$$
and $b_1 = 1$, $b_i = 0$ $(i=2,...,12)$, $b_{13} = 4$, using $x=(0,0,...,0)$ as an initial approximation to the solution.

```
package numal;

import java.text.DecimalFormat;
import numal.*;
```

```
public class Test_conjgrad extends Object implements LA_conjgrad_methods {

  public static void main(String args[]) {

    int i;
    int it[] = new int[1];
    double no[] = new double[1];
    double x[] = new double[13];
    double b[] = new double[13];

    DecimalFormat fiveDigit = new DecimalFormat("0.00000E0");
    for (i=0; i<=12; i++) x[i]=b[i]=0.0;
    b[0]=1.0;  b[12]=4.0;
    Test_conjgrad testconjgrad = new Test_conjgrad();;
    Linear_algebra.conjgrad(testconjgrad,x,b,0,12,it,no);
    System.out.println("Delivers:\n  IT = " + it[0] + ",    NO = " +
                       fiveDigit.format(no[0]) + "\n\n      X           R\n");
    for (i=0; i<=12; i++)
      System.out.println("  " + fiveDigit.format(x[i]) +
                         "    " + fiveDigit.format(b[i]));
  }

  public void matvec(double x[], double b[]) {

    int i;

    b[0]=2.0*x[0]-x[1];
    for (i=1; i<=11; i++) b[i] = -x[i-1]+2.0*x[i]-x[i+1];
    b[12]=2.0*x[12]-x[11];
  }

  public boolean goon(int iter[], double norm[]) {
    return (iter[0]<20 && norm[0]>1.0e-10);
  }
}
```

Output:

```
Delivers:
  IT = 13,    NO = 1.54161E-31

      X           R

  1.21429E0   0.00000E0
  1.42857E0   -1.49040E-16
  1.64286E0   3.66494E-17
  1.85714E0   -6.41016E-17
  2.07143E0   3.88843E-17
  2.28571E0   -1.81011E-16
  2.50000E0   1.68777E-16
```

```
2.71429E0    -1.68116E-16
2.92857E0     9.34072E-17
3.14286E0    -9.61890E-17
3.35714E0     9.52193E-17
3.57143E0    -8.18090E-17
3.78571E0    -4.16334E-17
```

Procedure tested:

eqilbrcom

Example:

Equilibrate the 3×3 matrix

$$\begin{pmatrix} 1 & 0 & 1024\mathrm{i} \\ 0 & 1 & 0 \\ \mathrm{i}/1024 & 0 & 2 \end{pmatrix}.$$

```
package numal;

import numal.*;

public class Test_eqilbrcom extends Object {

  public static void main(String args[]) {

    int inter[] = new int[4];
    double em[] = new double[8];
    double d[] = new double[4];
    double a1[][] = new double[4][4];
    double a2[][] = new double[4][4];

    em[0]=5.0e-7;
    em[6]=10.0;
    Basic.inimat(1,3,1,3,a1,0.0);
    Basic.inimat(1,3,1,3,a2,0.0);
    a1[1][1]=a1[2][2]=1.0;
    a1[3][3]=2.0;
    a2[1][3]=Math.pow(2.0,10.0);
    a2[3][1]=Math.pow(2.0,-10.0);
    Linear_algebra.eqilbrcom(a1,a2,3,em,d,inter);
    System.out.println("Equilibrated matrix:\n  " + a1[1][1] + "  " + a1[1][2] +
      "  " + a1[1][3] + "\n  " + a1[2][1] + "  " + a1[2][2] + "  I" + "\n  " +
      a1[3][1] + "  I    " + a1[3][3] + "\n\nEM[7]:       " + (int)em[7] +
      "\nD[1:3]:      " + d[1] + "  " + d[2] + "  " + d[3] + "\nINTER[1:3]:  " +
      inter[1] + "    " + inter[2] + "      " + inter[3]);
  }
}
```

Output:

```
Equilibrated matrix:
   1.0   0.0   0.0
   0.0   1.0   I
   0.0   I     2.0

EM[7]:          4
D[1:3]:         1.0  1024.0  1.0
INTER[1:3]:  2     0       0
```

Procedure tested:

hshhrmtri

Example:

Reduce the 4×4 complex matrix

$$A = \begin{pmatrix} 3 & 1 & 0 & 2\mathrm{i} \\ 1 & 3 & -2\mathrm{i} & 0 \\ 0 & 2\mathrm{i} & 1 & 1 \\ -2\mathrm{i} & 0 & 1 & 1 \end{pmatrix}$$

to real tridiagonal form, and estimate the value of $|A|$.

```java
package numal;

import java.text.DecimalFormat;
import numal.*;

public class Test_hshhrmtri extends Object {

  public static void main(String args[]) {

    double d[]  = new double[5];
    double b[]  = new double[5];
    double bb[] = new double[5];
    double tr[] = new double[4];
    double ti[] = new double[4];
    double em[] = new double[2];
    double a[][] = new double[5][5];

    Basic.inimat(1,4,1,4,a,0.0);
    a[1][1]=a[2][2]=3.0;
    a[1][2]=a[3][3]=a[3][4]=a[4][4]=1.0;
    a[3][2]=2.0;
    a[4][1]  = -2.0;
    em[0]=1.0e-6;
```

```
      Linear_algebra.hshhrmtri(a,4,d,b,bb,em,tr,ti);
      DecimalFormat threeeDigit = new DecimalFormat("0.000");
      System.out.println("HSHHRMTRI delivers\n\nD[1:4]:   " + threeeDigit.format(d[1])
        + "  " + threeeDigit.format(d[2]) + "  " + threeeDigit.format(d[3]) + "  " +
        threeeDigit.format(d[4]) + "\nB[1:3]:   " + threeeDigit.format(b[1]) + "  " +
        threeeDigit.format(b[2]) + "  " + threeeDigit.format(b[3]) + "\nBB[1:3]:  " +
        threeeDigit.format(bb[1]) + "  " + threeeDigit.format(bb[2]) + "  " +
        threeeDigit.format(bb[3]) + "\nEM[1]:    " + threeeDigit.format(em[1]));
  }
}
```

Output:

```
HSHHRMTRI delivers

D[1:4]:    3.000  1.400  2.600  1.000
B[1:3]:    2.236  0.800  2.236
BB[1:3]:   5.000  0.640  5.000
EM[1]:     6.000
```

Procedures tested:
valsymtri, vecsymtri

Example:
Compute the largest and second largest eigenvalues, and their corresponding
eigenvectors, of the 4×4 matrix A for which
$$A_{i,i}=2 \ (i=1,2,3,4), \ A_{i,i-1}=A_{i-1,i}=-1 \ (i=2,3,4), \ A_{i,j}=0 \ (|i-j|\geq 2).$$

```
package numal;

import java.text.DecimalFormat;
import numal.*;

public class Test_valsymtri extends Object {

  public static void main(String args[]) {

    double b[] = new double[5];
    double d[] = new double[5];
    double bb[] = new double[4];
    double val[] = new double[3];
    double em[] = new double[10];
    double vec[][] = new double[5][3];

    em[0]=1.0e-6; em[1]=4.0;   em[2]=1.0e-5;
    em[4]=1.0e-3;   em[6]=1.0e-6; em[8]=5.0;
    d[1]=d[2]=d[3]=d[4]=2.0;
    b[4]=0.0; b[1]=b[2]=b[3] = -1.0;
```

```
    bb[1]=bb[2]=bb[3]=1.0;
    Linear_algebra.valsymtri(d,bb,4,1,2,val,em);
    Linear_algebra.vecsymtri(d,b,4,1,2,val,vec,em);
    DecimalFormat fiveDigit = new DecimalFormat("0.00000E0");
    System.out.println("The eigenvalues:\n   " + fiveDigit.format(val[1]) + "    " +
      fiveDigit.format(val[2]) + "\n\nThe eigenvectors:\n  " +
      fiveDigit.format(vec[1][1]) + "   " + fiveDigit.format(vec[1][2]) + "\n   " +
      fiveDigit.format(vec[2][1]) + "  " + fiveDigit.format(vec[2][2]) + "\n  " +
      fiveDigit.format(vec[3][1]) + "  " + fiveDigit.format(vec[3][2]) + "\n   " +
      fiveDigit.format(vec[4][1]) + "   " + fiveDigit.format(vec[4][2]) +
      "\n\nEM[7] =  " + fiveDigit.format(em[7]) + "\nEM[3] =  " + (int)em[3] +
      "\nEM[5] =  " + (int)em[5] + "\nEM[9] =  " + (int)em[9]);
  }
}
```

Output:

```
The eigenvalues:
   3.61802E0    2.61804E0

The eigenvectors:
  -3.71748E-1   6.01501E-1
   6.01501E-1  -3.71748E-1
  -6.01501E-1  -3.71748E-1
   3.71748E-1   6.01501E-1


EM[7] =  1.47554E-5
EM[3] =  21
EM[5] =  1
EM[9] =  6
```

Procedure tested:

eigsym1

Example:
Compute the two largest eigenvalues and corresponding eigenvectors of the 4×4 matrix A for which $A_{i,j} = 1/(i+j-1)$, $i,j=1,2,3,4$.

```
package numal;

import java.text.DecimalFormat;
import numal.*;

public class Test_eigsym1 extends Object {

  public static void main(String args[]) {
    int i,j;
    double a[] = new double[11];
```

```
      double val[] = new double[3];
      double em[] = new double[10];
      double vec[][] = new double[5][3];

      em[0]=1.0e-6; em[2]=1.0e-5; em[4]=1.0e-3;
      em[6]=1.0e-5; em[8]=5.0;
      for (i=1; i<=4; i++)
        for (j=i; j<=4; j++) a[(j*j-j)/2+i]=1.0/(i+j-1);
      Linear_algebra.eigsym1(a,4,2,val,vec,em);
      DecimalFormat fiveDigit = new DecimalFormat("0.00000E0");
      System.out.println("The eigenvalues:\n   " + fiveDigit.format(val[1]) + "    " +
        fiveDigit.format(val[2]) + "\n\nThe eigenvectors:\n   " +
        fiveDigit.format(vec[1][1]) + "    " + fiveDigit.format(vec[1][2]) + "\n   " +
        fiveDigit.format(vec[2][1]) + "    " + fiveDigit.format(vec[2][2]) + "\n   " +
        fiveDigit.format(vec[3][1]) + "    " + fiveDigit.format(vec[3][2]) + "\n   " +
        fiveDigit.format(vec[4][1]) + "    " + fiveDigit.format(vec[4][2]) +
        "\n\nEM[1] = " + fiveDigit.format(em[1]) +   "\nEM[7] = " +
        fiveDigit.format(em[7]) + "\nEM[3] = " + (int)em[3] + "\nEM[5] = " +
        (int)em[5] + "\nEM[9] = " + (int)em[9]);
  }
}
```

Output:

```
The eigenvalues:
   1.50021E0    1.69142E-1

The eigenvectors:
  -7.92608E-1    5.82076E-1
  -4.51923E-1   -3.70502E-1
  -3.22416E-1   -5.09578E-1
  -2.52161E-1   -5.14049E-1

EM[1] =   2.08333E0
EM[7] =   8.28039E-6
EM[3] =   27
EM[5] =   1
EM[9] =   1
```

Procedure tested:

symeigimp

Example:

Determine approximations to the eigenvalues and eigenvectors of the matrix

$$\begin{pmatrix} 6 & 4 & 4 & 1 \\ 4 & 6 & 1 & 4 \\ 4 & 1 & 6 & 4 \\ 1 & 4 & 4 & 6 \end{pmatrix}.$$

```java
package numal;

import java.text.DecimalFormat;
import numal.*;

public class Test_symeigimp extends Object {

  public static void main(String args[]) {

    int i,j;
    double val[] = new double[5];
    double lbound[] = new double[5];
    double ubound[] = new double[5];
    double em[] = new double[6];
    double aux[] = new double[6];
    double a[][] = new double[5][5];
    double x[][] = new double[5][5];

    a[1][1]=a[2][2]=a[3][3]=a[4][4]=6.0;
    a[1][2]=a[2][1]=a[3][1]=a[1][3]=4.0;
    a[4][2]=a[2][4]=a[3][4]=a[4][3]=4.0;
    a[1][4]=a[4][1]=a[3][2]=a[2][3]=1.0;
    for (i=1; i<=4; i++)
      for (j=i; j<=4; j++) x[i][j]=x[j][i]=a[i][j];
    em[0]=1.0e-6; em[4]=100.0;  em[2]=1.0e-5;
    Linear_algebra.qrisym(x,4,val,em);
    aux[0]=0.0;    aux[4]=10.0;  aux[2]=1.0e-6;
    Linear_algebra.symeigimp(4,a,x,val,lbound,ubound,aux);
    DecimalFormat fiveDigit = new DecimalFormat("0.00000E0");
    System.out.println("\nThe exact eigenvalues are:  -1,  5,  5,  15\n\n" +
      "The computed eigenvalues:\n  " + fiveDigit.format(val[1]) + "\n  " +
      fiveDigit.format(val[2]) + "\n   " + fiveDigit.format(val[3]) + "\n   " +
      fiveDigit.format(val[4]) + "\n\n Lowerbounds   Upperbounds\n " +
      fiveDigit.format(lbound[1]) + "   " + fiveDigit.format(lbound[2]) + "   " +
      fiveDigit.format(ubound[2]) + "\n " + fiveDigit.format(lbound[3]) + "   " +
      fiveDigit.format(ubound[3]) + "\n " + fiveDigit.format(lbound[4]) + "   " +
      fiveDigit.format(ubound[4]) + "\n\nNumber of iterations = " + (int)aux[5] +
```

```
        "\nInfinity norm of A    = " + (int)aux[1] +
        "\nMaximum absolute element of residu = " + fiveDigit.format(aux[3]));
    }
}
```

Output:

```
The exact eigenvalues are:   -1,   5,   5,   15

The computed eigenvalues:
   -1.00000E0
    5.00000E0
    5.00000E0
    1.50000E1

  Lowerbounds     Upperbounds
  7.74431E-28     7.74431E-28
  3.21292E-15     3.21292E-15
  6.75590E-14     6.75590E-14
  3.55480E-30     3.55480E-30

Number of iterations = 1
Infinity norm of A    = 15
Maximum absolute element of residu = 6.16174E-15
```

Procedures tested:

comvalqri, comveches

Example:
Determine all eigenvalues and corresponding eigenvectors of the 4×4 matrix

$$\begin{pmatrix} -1 & -1 & -1 & -1 \\ 1 & 0 & 0 & 0 \\ 0 & 1 & 0 & 0 \\ 0 & 0 & 1 & 0 \end{pmatrix}.$$

```
package numal;

import java.text.DecimalFormat;
import numal.*;

public class Test_comvalqri extends Object {

  public static void main(String args[]) {

    int i,j,m,k;
    double re[] = new double[5];
```

```
        double im[] = new double[5];
        double u[] = new double[5];
        double v[] = new double[5];
        double em[] = new double[10];
        double a[][] = new double[5][5];

        em[0]=1.0e-6;  em[2]=1.0e-6;  em[1]=4.0;  em[4]=40.0;
        em[6]=1.0e-5;  em[8]=5.0;
        for (i=1; i<=4; i++)
          for (j=1; j<=4; j++)
            a[i][j] = (i == 1) ? -1.0 : ((i-j == 1) ? 1.0 : 0.0);
        m=Linear_algebra.comvalqri(a,4,em,re,im);
        DecimalFormat fiveDigit = new DecimalFormat("0.00000E0");
        System.out.println("The number of not calculated eigenvalues: " + m +
                           "\n\nThe eigenvalues and eigenvectors:");
        for (j=m+1; j<=4; j++) {
          for (i=1; i<=4; i++)
            for (k=1; k<=4; k++)
              a[i][k] = (i == 1) ? -1.0 : ((i-k == 1) ? 1.0 : 0.0);
          Linear_algebra.comveches(a,4,re[j],im[j],em,u,v);
          System.out.println("\n  " + fiveDigit.format(re[j]) + "      " +
            fiveDigit.format(im[j]) + "\n\n\t" + fiveDigit.format(u[1]) + "\t" +
            fiveDigit.format(v[1]) + "\n\t" + fiveDigit.format(u[2]) + "\t" +
            fiveDigit.format(v[2]) + "\n\t" + fiveDigit.format(u[3]) + "\t" +
            fiveDigit.format(v[3]) + "\n\t" + fiveDigit.format(u[4]) + "\t" +
            fiveDigit.format(v[4]));
        }
        System.out.println("\nEM[3] =  " + fiveDigit.format(em[3]) +
          "\nEM[7] =  " + fiveDigit.format(em[7]) + "\nEM[5] =  " + (int)em[5] +
          "\nEM[9] =  " + (int)em[9]);
    }
}
```

Output:

```
The number of not calculated eigenvalues: 0

The eigenvalues and eigenvectors:

  3.09017E-1     9.51057E-1

      -2.93894E-1                -4.04508E-1
      -4.75527E-1                1.54509E-1
      7.60850E-7                 5.00000E-1
      4.75528E-1                 1.54509E-1

  3.09017E-1    -9.51057E-1

      -2.93894E-1                4.04508E-1
      -4.75527E-1                -1.54509E-1
      7.60850E-7                 -5.00000E-1
      4.75528E-1                 -1.54509E-1

 -8.09017E-1     5.87785E-1

      3.80421E-7                 5.00000E-1
      2.93893E-1                 -4.04508E-1
```

```
      -4.75528E-1              1.54509E-1
       4.75529E-1              1.54509E-1

  -8.09017E-1     -5.87785E-1

       3.80421E-7             -5.00000E-1
       2.93893E-1              4.04508E-1
      -4.75528E-1             -1.54509E-1
       4.75529E-1             -1.54509E-1
EM[3]  =   5.08074E-11
EM[7]  =   7.60846E-7
EM[5]  =   8
EM[9]  =   1
```

Procedure tested:

reaeig3

Example:

Compute all eigenvalues and corresponding eigenvectors of the 4×4 matrix

$$\begin{pmatrix} 1 & 1 & 1 & 1 \\ 1/2 & 1/3 & 1/4 & 1/5 \\ 1/3 & 1/4 & 1/5 & 1/6 \\ 1/4 & 1/5 & 1/6 & 1/7 \end{pmatrix}.$$

```
package numal;

import java.text.DecimalFormat;
import numal.*;

public class Test_reaeig3 extends Object {

  public static void main(String args[]) {

    int i,j,m;
    double val[] = new double[5];
    double em[] = new double[6];
    double a[][] = new double[5][5];
    double vec[][] = new double[5][5];

    for (i=1; i<=4; i++)
      for (j=1; j<=4; j++)
        a[i][j] = (i == 1) ? 1.0 : 1.0/(i+j-1);
    em[0]=1.0e-6;
    em[2]=1.0e-5;
    em[4]=40.0;
    m=Linear_algebra.reaeig3(a,4,em,val,vec);
    DecimalFormat fiveDigit = new DecimalFormat("0.00000E0");
    System.out.println("The number of not calculated eigenvalues: " + m +
                  "\n\nThe eigenvalues and corresponding eigenvectors:\n ");
    for (i=m+1; i<=4; i++)
```

```
    System.out.println("\t" + fiveDigit.format(val[i]) + "\t" +
      fiveDigit.format(vec[1][i]) + "\n\t\t\t" + fiveDigit.format(vec[2][i]) +
      "\n\t\t\t" + fiveDigit.format(vec[3][i]) + "\n\t\t\t" +
      fiveDigit.format(vec[4][i]) + "\n");
  System.out.println("EM[1] =  " + fiveDigit.format(em[1]) + "\nEM[3] =  " +
                    fiveDigit.format(em[3]) + "\nEM[5] =  " + (int)em[5]);
  }
}
```

Output:

```
The number of not calculated eigenvalues: 0

The eigenvalues and corresponding eigenvectors:

        1.88663E0          1.00000E0
                           3.94224E-1
                           2.77320E-1
                           2.15088E-1

       -1.98015E-1         1.00000E0
                          -7.38848E-1
                          -3.11624E-1
                          -1.47542E-1

       -1.22829E-2        -4.63473E-1
                           1.00000E0
                          -1.54257E-1
                          -3.76577E-1

       -1.44131E-4         1.09571E-1
                          -6.20840E-1
                           1.00000E0
                          -4.88747E-1

EM[1]  =   4.00000E0
EM[3]  =   7.25659E-8
EM[5]  =   3
```

Procedure tested:

eighrm

Example:
Compute the largest eigenvalue and corresponding eigenvector of the 4×4 matrix

$$\begin{pmatrix} 3 & 1 & 0 & 2i \\ 1 & 3 & -2i & 0 \\ 0 & 2i & 1 & 1 \\ -2i & 0 & 1 & 1 \end{pmatrix}.$$

```
package numal;

import java.text.DecimalFormat;
import numal.*;
```

```java
public class Test_eighrm extends Object {
  public static void main(String args[]) {
    int i;
    double val[] = new double[2];
    double em[] = new double[10];
    double a[][] = new double[5][5];
    double vecr[][] = new double[5][2];
    double veci[][] = new double[5][2];

    Basic.inimat(1,4,1,4,a,0.0);
    a[1][1]=a[2][2]=3.0;
    a[1][2]=a[3][3]=a[3][4]=a[4][4]=1.0;
    a[3][2]=2.0;  a[4][1] = -2.0;
    em[0]=5.0e-6;
    em[2]=1.0e-5;
    em[4]=0.01;
    em[6]=1.0e-5;
    em[8]=5.0;
    Linear_algebra.eighrm(a,4,1,val,vecr,veci,em);
    Basic.sclcom(vecr,veci,4,1,1);
    DecimalFormat threeDigit = new DecimalFormat("0.000");
    DecimalFormat fiveDigit = new DecimalFormat("0.00000E0");
    System.out.println("Largest eigenvalue:  " + fiveDigit.format(val[1]) +
                       "\n\nCorresponding eigenvector:");
    for (i=1; i<=4; i++)
      System.out.println("  " + threeDigit.format(vecr[i][1]) + "  " +
                         threeDigit.format(veci[i][1]) + "*I");
    System.out.println("\nEM[1]:  " + fiveDigit.format(em[1]) +
                       "\nEM[3]:  " + fiveDigit.format(em[3]) +
                       "\nEM[5]:  " + fiveDigit.format(em[5]) +
                       "\nEM[7]:  " + fiveDigit.format(em[7]) +
                       "\nEM[9]:  " + fiveDigit.format(em[9]));
  }
}
```

Output:

```
Largest eigenvalue:  4.82843E0

Corresponding eigenvector:
  1.000   0.000*I
  1.000   0.000*I
  0.000   0.414*I
  0.000  -0.414*I

EM[1]:  6.00000E0
EM[3]:  1.60000E1
EM[5]:  1.00000E0
EM[7]:  5.30318E-6
EM[9]:  1.00000E0
```

Procedure tested:

qrihrm

Example:

Compute all eigenvalues and corresponding eigenvectors of the 4×4 matrix

$$\begin{pmatrix} 3 & 1 & 0 & 2i \\ 1 & 3 & -2i & 0 \\ 0 & 2i & 1 & 1 \\ -2i & 0 & 1 & 1 \end{pmatrix}.$$

```java
package numal;

import java.text.DecimalFormat;
import numal.*;

public class Test_qrihrm extends Object {

  public static void main(String args[]) {

    int i;
    double val[] = new double[5];
    double em[] = new double[6];
    double a[][] = new double[5][5];
    double vr[][] = new double[5][5];
    double vi[][] = new double[5][5];

    Basic.inimat(1,4,1,4,a,0.0);
    a[1][1]=a[2][2]=3.0;
    a[3][2]=2.0;  a[4][1] = -2.0;
    a[1][2]=a[3][3]=a[3][4]=a[4][4]=1.0;
    em[0]=em[2]=5.0e-5;
    em[4]=20.0;
    DecimalFormat twoDigit = new DecimalFormat("##");
    DecimalFormat threeDigit = new DecimalFormat("0.000");
    DecimalFormat fiveDigit = new DecimalFormat("0.00000E0");

    System.out.println("QRIHRM: " + Linear_algebra.qrihrm(a,4,val,vr,vi,em));
    Basic.sclcom(vr,vi,4,2,3);
    System.out.println("\nEigenvalues:" +
                       "\n  VAL[1]:   " + threeDigit.format(val[1]) +
                       "\n  VAL[2]:   " + threeDigit.format(val[2]) +
                       "\n  VAL[3]:   " + threeDigit.format(val[3]) +
                       "\n  VAL[4]:   " + threeDigit.format(val[4]) +
                       "\n\nEigenvectors corresponding to\n    VAL[2]      VAL[3]");
    for (i=1; i<=4; i++)
      System.out.println("   " + twoDigit.format(vr[i][2]) + "  " +
                         twoDigit.format(vi[i][2]) + "*I" + "       " +
```

```
                    twoDigit.format(vr[i][3]) + "   " +
                    twoDigit.format(vi[i][3]) + "*I");
      System.out.println("\nEM[1]:  " + fiveDigit.format(em[1]) +
                    "\nEM[3]:  " + fiveDigit.format(em[3]) +
                    "\nEM[5]:  " + fiveDigit.format(em[5]));
    }
}
```

Output:

```
QRIHRM: 0


Eigenvalues:
   VAL[1]:    4.828
   VAL[2]:    4.000
   VAL[3]:   -0.000
   VAL[4]:   -0.828


Eigenvectors corresponding to
   VAL[2]        VAL[3]
   0   1*I       0  -1*I
   0  -1*I       0   1*I
   1   0*I       1  -0*I
   1   0*I       1   0*I


EM[1]:   6.00000E0
EM[3]:   5.63494E-7
EM[5]:   4.00000E0
```

Procedure tested:

valqricom

Example:
Determine the roots of the polynomial
$$x^4 + (4+2i)x^3 + (5+6i)x^2 + (2+6i)x + 2i$$
by computing the eigenvalues of the companion matrix
$$\begin{pmatrix} -4-2i & -5-6i & -2-6i & -2i \\ 1 & 0 & 0 & 0 \\ 0 & 1 & 0 & 0 \\ 0 & 0 & 1 & 0 \end{pmatrix}.$$

```
package numal;


import java.text.DecimalFormat;
import numal.*;
```

```
public class Test_valqricom extends Object {

  public static void main(String args[]) {

    int i;
    double b[] = new double[4];
    double val1[] = new double[5];
    double val2[] = new double[5];
    double em[] = new double[6];
    double a1[][] = new double[5][5];
    double a2[][] = new double[5][5];

    Basic.inimat(1,4,1,4,a1,0.0);
    Basic.inimat(1,4,1,4,a2,0.0);
    a1[1][1] = -4.0;   a1[1][2] = -5.0;
    a1[1][3] = a2[1][1] = a2[1][4] = -2.0;
    a2[1][2] = a2[1][3] = -6.0;
    b[1]=b[2]=b[3]=1.0;
    em[0]=5.0e-6;  em[1]=27.0;  em[2]=1.0e-6;  em[4]=15.0;
    DecimalFormat fiveDigit = new DecimalFormat("0.00000E0");
    System.out.println("VALQRICOM:  " +
                       Linear_algebra.valqricom(a1,a2,b,4,em,val1,val2));
    System.out.println("\nEigenvalues:\n\t" + "Real part\tImaginary part");
    for (i=1; i<=4; i++)
      System.out.println("\t" + fiveDigit.format(val1[i]) +
                         "\t" + fiveDigit.format(val2[i]));
    System.out.println("\nEM[3]:  " + fiveDigit.format(em[3]) + "\nEM[5]:  " +
                       (int)em[5]);
  }
}
```

Output:

```
VALQRICOM:   0

Eigenvalues:
        Real part          Imaginary part
        -1.00002E0         -1.00004E0
        -9.99984E-1        -9.99958E-1
        -1.00473E0         -3.43588E-3
        -9.95271E-1         3.43232E-3

EM[3]:  8.84848E-6
EM[5]:  8
```

Procedure tested:

qricom

Example:
Compute the eigenvalues and eigenvectors of the matrix

$$\begin{pmatrix} -4-2i & -5-6i & -2-6i & -2i \\ 1 & 0 & 0 & 0 \\ 0 & 1 & 0 & 0 \\ 0 & 0 & 1 & 0 \end{pmatrix}.$$

```
package numal;

import java.text.DecimalFormat;
import numal.*;

public class Test_qricom extends Object {

  public static void main(String args[]) {

    int i;
    double b[] = new double[4];
    double val1[] = new double[5];
    double val2[] = new double[5];
    double em[] = new double[6];
    double a1[][] = new double[5][5];
    double a2[][] = new double[5][5];
    double vec1[][] = new double[5][5];
    double vec2[][] = new double[5][5];

    Basic.inimat(1,4,1,4,a1,0.0);
    Basic.inimat(1,4,1,4,a2,0.0);
    a1[1][1] = -4.0;   a1[1][2] = -5.0;
    a1[1][3] = a2[1][1] = a2[1][4] = -2.0;
    a2[1][2] = a2[1][3] = -6.0;
    b[1]=b[2]=b[3]=1.0;
    em[0]=5.0e-6;   em[1]=27.0;   em[2]=1.0e-6;   em[4]=15.0;
    DecimalFormat fiveDigit = new DecimalFormat("0.00000E0");
    System.out.println("QRICOM:  " +
                    Linear_algebra.qricom(a1,a2,b,4,em,val1,val2,vec1,vec2));
    System.out.println("\nEigenvalues:\n\t" + "Real part\tImaginary part");
    for (i=1; i<=4; i++)
      System.out.println("\t" + fiveDigit.format(val1[i]) +
                      "\t" + fiveDigit.format(val2[i]));
    Basic.sclcom(vec1,vec2,4,1,4);
    System.out.println("\nFirst eigenvector:\n\t" + "Real part\tImaginary part");
    for (i=1; i<=4; i++)
      System.out.println("\t" + fiveDigit.format(vec1[i][1]) +
```

```
                                "\t" + fiveDigit.format(vec2[i][1]));
       System.out.println("\nEM[3]:  " + fiveDigit.format(em[3]) + "\nEM[5]:   " +
                          (int)em[5]);
   }
}
```

Output:

```
QRICOM:   0

Eigenvalues:
        Real part          Imaginary part
        -9.99984E-1        -9.99958E-1
        -1.00002E0         -1.00004E0
        -1.00473E0         -3.43588E-3
        -9.95271E-1         3.43232E-3

First eigenvector:
        Real part          Imaginary part
        1.00000E0          0.00000E0
        -5.00021E-1        5.00008E-1
        1.27974E-5         -5.00029E-1
        2.50012E-1         2.50031E-1

EM[3]:   8.84848E-6
EM[5]:   8
```

Procedure tested:

eigcom

Example:
Compute the eigenvalues and eigenvectors of the matrix

$$\begin{pmatrix} 1+3i & 2+i & 3+2i & 1+i \\ 3+4i & 1+2i & 2+i & 4+3i \\ 2+3i & 1+5i & 3+i & 5+2i \\ 1+2i & 3+i & 1+4i & 5+3i \end{pmatrix}.$$

```
package numal;

import java.text.DecimalFormat;
import numal.*;

public class Test_eigcom extends Object {

  public static void main(String args[]) {

    int i;
    double em[]  = new double[8];
    double valr[] = new double[5];
    double vali[] = new double[5];
    double ar[][] = new double[5][5];
    double ai[][] = new double[5][5];
```

```
    double vr[][] = new double[5][5];
    double vi[][] = new double[5][5];

    ar[1][1]=ar[1][4]=ar[2][2]=ar[3][2]=ar[4][1]=ar[4][3]=
            ai[1][2]=ai[1][4]=ai[2][3]=ai[3][3]=ai[4][2]=1.0;
    ar[1][2]=ar[2][3]=ar[3][1]=ai[1][3]=ai[2][2]=ai[3][4]=
            ai[4][1]=2.0;
    ar[1][3]=ar[2][1]=ar[3][3]=ar[4][2]=
    ai[1][1]=ai[2][4]=ai[3][1]=ai[4][4]=3.0;
    ar[2][4]=ai[2][1]=ai[4][3]=4.0;
    ar[3][4]=ar[4][4]=ai[3][2]=5.0;
    em[0]=5.0e-6;  em[2]=1.0e-5;  em[4]=10.0;  em[6]=10.0;
    DecimalFormat fiveDigit = new DecimalFormat("0.00000E0");
    System.out.println("EIGCOM:  " +
                        Linear_algebra.eigcom(ar,ai,4,em,valr,vali,vr,vi));
    System.out.println("\nEigenvalues:");
    for (i=1; i<=4; i++)
      System.out.println("\t" + fiveDigit.format(valr[i]) + "\t" +
                        fiveDigit.format(vali[i]) + " * I ");
    System.out.println("\nFirst eigenvector:");
    for (i=1; i<=4; i++)
      System.out.println("\t" + fiveDigit.format(vr[i][1]) +
                        "\t" + fiveDigit.format(vi[i][1]));
    System.out.println("\nEM[1]:  " + fiveDigit.format(em[1]) +
                        "\nEM[3]:  " + fiveDigit.format(em[3]) +
                        "\nEM[5]:  " + (int)em[5] + "\nEM[7]:  " + (int)em[7]);
  }
}
```

Output:

```
EIGCOM:  0

Eigenvalues:
        -3.37101E0              -7.70453E-1 * I
        9.78366E0              9.32251E0 * I
        1.36568E0              -1.40105E0 * I
        2.22167E0              1.84899E0 * I

First eigenvector:
        -5.06096E-1              5.83452E-1
        1.00000E0              0.00000E0
        5.18319E-1              -7.14656E-1
        -5.53485E-1              1.87559E-2

EM[1]:  1.52971E1
EM[3]:  2.12413E-5
EM[5]:  5

EM[7]:  4
```

Procedure tested:

qzival

Example:
With

$$A = \begin{pmatrix} 2 & 3 & -3 & 4 \\ 1 & -1 & 5 & 1 \\ 0 & 2 & 6 & 8 \\ 1 & 1 & 0 & 4 \end{pmatrix}, \quad B = \begin{pmatrix} 1 & 5 & 9 & 0 \\ 2 & 6 & 10 & 2 \\ 3 & 7 & 11 & -1 \\ 4 & 8 & 12 & 3 \end{pmatrix}$$

determine complex numbers $\alpha^{(j)}$ and real numbers $\beta^{(j)}$ such that $\beta^{(j)}A - \alpha^{(j)}B$ is singular ($j=1,...,4$) and evaluate the quotients $\lambda^{(j)} = \alpha^{(j)}/\beta^{(j)}$.

```
package numal;

import java.text.DecimalFormat;
import numal.*;

public class Test_qzival extends Object {

  public static void main(String args[]) {

    int k;
    int iter[] = new int[5];
    double alfr[] = new double[5];
    double alfi[] = new double[5];
    double beta[] = new double[5];
    double em[] = new double[2];
    double a[][] = new double[5][5];
    double b[][] = new double[5][5];

    a[1][1]=2.0;  a[1][2]=3.0;   a[1][3] = -3.0;  a[1][4]=4.0;
    a[2][1]=1.0;  a[2][2] = -1.0; a[2][3]=5.0;     a[2][4]=1.0;
    a[3][1]=0.0;  a[3][2]=2.0;   a[3][3]=6.0;     a[3][4]=8.0;
    a[4][1]=1.0;  a[4][2]=1.0;   a[4][3]=0.0;     a[4][4]=4.0;
    b[1][1]=1.0;  b[1][2]=5.0;   b[1][3]=9.0;     b[1][4]=0.0;
    b[2][1]=2.0;  b[2][2]=6.0;   b[2][3]=10.0;    b[2][4]=2.0;
    b[3][1]=3.0;  b[3][2]=7.0;   b[3][3]=11.0;    b[3][4] = -1.0;
    b[4][1]=4.0;  b[4][2]=8.0;   b[4][3]=12.0;    b[4][4]=3.0;
    em[0]=1.0e-35;
    em[1]=1.0e-6;
    Linear_algebra.qzival(4,a,b,alfr,alfi,beta,iter,em);
    DecimalFormat fiveDigit = new DecimalFormat("0.00000E0");
    for (k=1; k<=4; k++)
      System.out.println("ITER[" + k + "] =  " + iter[k]);
    System.out.println("\nALFA(real part)\tALFA(imaginary part)\tBETA");
    for (k=1; k<=4; k++)
      System.out.println(fiveDigit.format(alfr[k]) + "\t" +
```

```
                              fiveDigit.format(alfi[k]) + "\t\t" +
                              fiveDigit.format(beta[k]));
        System.out.println("\nLAMBDA(real part)\tLAMBDA(imaginary part)");
        for (k=1; k<=4; k++)
          if (beta[k] == 0.0)
            System.out.println("  INFINITE\t\tINDEFINITE");
          else
            System.out.println(fiveDigit.format(alfr[k]/beta[k]) + "\t\t" +
                              fiveDigit.format(alfi[k]/beta[k]));
    }
}
```

Output:

```
ITER[1]  =   0
ITER[2]  =   0
ITER[3]  =   0
ITER[4]  =   8
```

```
ALFA(real part)        ALFA(imaginary part)        BETA
-4.43471E0             0.00000E0                   0.00000E0
-6.08855E-1            0.00000E0                   3.02265E-1
-1.67937E-1            5.34994E-1                  1.69765E0
-1.29564E-3            -4.12750E-3                 1.30974E-2
```

```
LAMBDA(real part)      LAMBDA(imaginary part)
   INFINITE            INDEFINITE
-2.01431E0             0.00000E0
-9.89233E-2            3.15138E-1
-9.89233E-2            -3.15138E-1
```

Procedure tested:

qzi

Example:
With

$$A = \begin{pmatrix} 2 & 3 & -3 & 4 \\ 1 & -1 & 5 & 1 \\ 0 & 2 & 6 & 8 \\ 1 & 1 & 0 & 4 \end{pmatrix}, \quad B = \begin{pmatrix} 1 & 5 & 9 & 0 \\ 2 & 6 & 10 & 2 \\ 3 & 7 & 11 & -1 \\ 4 & 8 & 12 & 3 \end{pmatrix}$$

determine complex numbers $\alpha^{(j)}$ and real numbers $\beta^{(j)}$ such that $\beta^{(j)}A - \alpha^{(j)}B$ is singular ($j=1,...,4$), and evaluate the quotients $\lambda^{(j)} = \alpha^{(j)}/\beta^{(j)}$, and also determine the components of normalized eigenvectors $x^{(j)}$ such that $\beta^{(j)}Ax^{(j)} = \alpha^{(j)}Bx^{(j)}$.

```
package numal;

import java.text.DecimalFormat;
import numal.*;
```

```java
public class Test_qzi extends Object {

  public static void main(String args[]) {

    int k,l;
    int iter[] = new int[5];
    double alfr[] = new double[5];
    double alfi[] = new double[5];
    double beta[] = new double[5];
    double em[] = new double[2];
    double a[][] = new double[5][5];
    double b[][] = new double[5][5];
    double x[][] = new double[5][5];

    a[1][1]=2.0;  a[1][2]=3.0;   a[1][3] = -3.0;  a[1][4]=4.0;
    a[2][1]=1.0;  a[2][2] = -1.0; a[2][3]=5.0;    a[2][4]=1.0;
    a[3][1]=0.0;  a[3][2]=2.0;    a[3][3]=6.0;    a[3][4]=8.0;
    a[4][1]=1.0;  a[4][2]=1.0;    a[4][3]=0.0;    a[4][4]=4.0;
    b[1][1]=1.0;  b[1][2]=5.0;    b[1][3]=9.0;    b[1][4]=0.0;
    b[2][1]=2.0;  b[2][2]=6.0;    b[2][3]=10.0;   b[2][4]=2.0;
    b[3][1]=3.0;  b[3][2]=7.0;    b[3][3]=11.0;   b[3][4] = -1.0;
    b[4][1]=4.0;  b[4][2]=8.0;    b[4][3]=12.0;   b[4][4]=3.0;
    for (k=1; k<=4; k++)
      for (l=1; l<=4; l++)  x[k][l] = (k == l) ? 1.0 : 0.0;
    em[0]=1.0e-35;
    em[1]=1.0e-6;
    Linear_algebra.qzi(4,a,b,x,alfr,alfi,beta,iter,em);
    DecimalFormat fiveDigit = new DecimalFormat("0.00000E0");
    for (k=1; k<=4; k++)
      System.out.println("ITER[" + k + "] =  " + iter[k]);
    System.out.println("\nEigenvectors:");
    for (k=1; k<=4; k++)
      System.out.println(" " + fiveDigit.format(x[k][1]) + "\t" +
        fiveDigit.format(x[k][2]) + "\t" + fiveDigit.format(x[k][3]) + "\t" +
        fiveDigit.format(x[k][4]));
    System.out.println("\nALFA(real part)\tALFA(imaginary part)\tBETA");
    for (k=1; k<=4; k++)
      System.out.println(fiveDigit.format(alfr[k]) + "\t" +
                         fiveDigit.format(alfi[k]) + "\t\t" +
                         fiveDigit.format(beta[k]));
    System.out.println("\nLAMBDA(real part)\tLAMBDA(imaginary part)");
    for (k=1; k<=4; k++)
      if (beta[k] == 0.0)
        System.out.println("  INFINITE\t\tINDEFINITE");
      else
        System.out.println(fiveDigit.format(alfr[k]/beta[k]) + "\t\t" +
                           fiveDigit.format(alfi[k]/beta[k]));
  }
}
```

Output:

```
ITER[1] = 0
ITER[2] = 0
ITER[3] = 0
ITER[4] = 8
```

Eigenvectors:

```
 -5.00000E-1     1.00000E0      -6.29122E-1     6.52070E-1
  1.00000E0     -3.82667E-2      1.00000E0     -1.23732E-17
 -5.00000E-1    -3.04662E-2      1.65891E-1     1.09321E-1
 -2.71948E-16   -7.63333E-1     -5.84842E-1     1.77454E-1
```

```
ALFA(real part)          ALFA(imaginary part)              BETA
-4.43471E0               0.00000E0                         0.00000E0
-6.08855E-1              0.00000E0                         3.02265E-1
-1.67937E-1              5.34994E-1                        1.69765E0
-1.29564E-3             -4.12750E-3                        1.30974E-2
```

```
LAMBDA(real part)        LAMBDA(imaginary part)
  INFINITE               INDEFINITE
-2.01431E0               0.00000E0
-9.89233E-2              3.15138E-1
-9.89233E-2             -3.15138E-1
```

Procedure tested:

qrisngvaldec

Example:
Compute the singular value decomposition of the 6×5 matrix A for which
$$A_{i,j} = 1/(i+j-1).$$

```java
package numal;

import java.text.DecimalFormat;
import numal.*;

public class Test_qrisngvaldec extends Object {

  public static void main(String args[]) {

    int i,j;
    double val[] = new double[6];
    double em[] = new double[8];
    double a[][] = new double[7][6];
    double v[][] = new double[6][6];

    for (i=1; i<=6; i++)
      for (j=1; j<=5; j++) a[i][j]=1.0/(i+j-1);
    em[0]=1.0e-6;  em[2]=1.0e-5;  em[4]=25.0;  em[6]=1.0e-5;
    i=Linear_algebra.qrisngvaldec(a,6,5,val,v,em);
    DecimalFormat fiveDigit = new DecimalFormat("0.00000E0");
    System.out.println("Number of singular values not found :  " + i +
      "\n\nInfinity norm :  " + fiveDigit.format(em[1]) +
      "\nMax neglected subdiagonal element :  " + fiveDigit.format(em[3]) +
```

```
    "\nNumber of iterations :   " + (int)em[5] + "\nNumerical rank :   " +
    (int)em[7] + "\n\nSingular values :");
  for (i=1; i<=5; i++)
    System.out.println("   " + fiveDigit.format(val[i]));
  System.out.println("\nMatrix U, first 3 columns :");
  for (i=1; i<=6; i++)
    System.out.println("   " + fiveDigit.format(a[i][1]) + "\t" +
                      fiveDigit.format(a[i][2]) + "\t" +
                      fiveDigit.format(a[i][3]));
  System.out.println("Last 2 columns :");
  for (i=1; i<=6; i++)
    System.out.println("   " +fiveDigit.format(a[i][4]) + "\t" +
                      fiveDigit.format(a[i][5]));
  }
}
```

Output:

```
Number of singular values not found :   0

Infinity norm :   2.28333E0
Max neglected subdiagonal element :   2.11719E-5
Number of iterations :   3
Numerical rank :   5

Singular values :
  1.59212E0
  2.24496E-1
  1.36102E-2
  1.25733E-4
  1.41703E-5

Matrix U, first 3 columns :
  -7.54979E-1    6.10111E-1      -2.33268E-1
  -4.39093E-1   -2.26021E-1      7.05899E-1
  -3.17031E-1   -3.73070E-1      2.11269E-1
  -2.49995E-1   -3.95578E-1      -1.47776E-1
  -2.07050E-1   -3.84833E-1      -3.68054E-1
  -1.76997E-1   -3.64582E-1      -4.95335E-1
Last 2 columns :
  9.87458E-3    -2.57541E-2
  -6.88740E-3    2.67620E-1
  -3.89435E-1   -5.02005E-1
  8.52031E-1    -1.34877E-1
  -2.25331E-1    7.40444E-1
  -2.67328E-1   -3.30543E-1
```

Procedures tested:

zerpol, bounds

Example:
Determine the zeros of the polynomial
$$z^7 - 3z^6 - 3z^5 + 25z^4 - 46z^3 + 38z^2 - 12z$$
by means of a call of *zerpol*, and derive the real and imaginary parts of the centers and corresponding radii of discs containing the true zeros of the above polynomial.

```java
package numal;

import java.text.DecimalFormat;
import numal.*;

public class Test_zerpol extends Object {

  public static void main(String args[]) {

    int i,j;
    double a[] = new double[8];
    double d[] = new double[8];
    double re[] = new double[8];
    double im[] = new double[8];
    double em[] = new double[5];
    double recentre[] = new double[8];
    double imcentre[] = new double[8];
    double bound[] = new double[8];

    a[7]=1.0;    a[6]=a[5] = -3.0;  a[4]=25.0;   a[3] = -46.0;
    a[2]=38.0;   a[1] = -12.0;      a[0]=0.0;
    em[0]=1.0e-6;   em[1]=40.0;
    i=Linear_algebra.zerpol(7,a,em,re,im,d);
    DecimalFormat fiveDigit = new DecimalFormat("0.00000E0");
    System.out.print("Coefficients of polynomial:\n  ");
    for (j=7; j>=0; j--) System.out.print("  " + a[j]);
    System.out.println("\n\nNumber of not found zeros:  " + i +
                    "\nFail indication:           " + (int)em[2] +
                    "\nNumber of new starts:      " + (int)em[3] +
                    "\nNumber of iterations:      " + (int)em[4] + "\n\nZeros:");
    for (j=i+1; j<=7; j++)
      if (im[j] == 0.0)
        System.out.println("  " + fiveDigit.format(re[j]));
      else
        System.out.println("  " + fiveDigit.format(re[j]) + "\t" +
                        fiveDigit.format(im[j]));
    if (i == 0) {
      Linear_algebra.bounds(7,a,re,im,0.0,0.0,recentre,imcentre,bound);
      System.out.println("\nReal and imaginary part of centre + radius");
      for (j=1; j<=7; j++)
        System.out.println("  " + fiveDigit.format(recentre[j]) + "\t" +
                        fiveDigit.format(imcentre[j]) + "\t" +
                        fiveDigit.format(bound[j]));
    }
  }
}
```

Output:

```
Coefficients of polynomial:
    1.0   -3.0   -3.0   25.0   -46.0   38.0   -12.0   0.0

Number of not found zeros:   0
Fail indication:             0
Number of new starts:        0
Number of iterations:        11

Zeros:
  1.99994E0
 -3.00000E0
  9.99974E-1    -9.99995E-1
  9.99974E-1     9.99995E-1
  9.93026E-1
  1.00708E0
  0.00000E0

Real and imaginary part of centre + radius
  1.99994E0      0.00000E0      6.84647E-5
 -3.00000E0      0.00000E0      1.03045E-6
  9.99974E-1    -9.99995E-1     3.30028E-5
  9.99974E-1     9.99995E-1     3.30028E-5
  1.00005E0      0.00000E0      1.48901E-2
  1.00005E0      0.00000E0      1.48901E-2
  0.00000E0      0.00000E0      0.00000E0
```

Procedure tested:

allzerortpol

Example:

Determine the roots of the Chebyshev polynomial $T_3(x)$ for which
$$T_0(x)=1, \quad T_1(x)=x, \quad T_{k+1}(x)=xT_k(x)-2^{-k}T_{k-1}(x), \quad k=1,2.$$

```java
package numal;

import java.text.DecimalFormat;
import numal.*;

public class Test_allzerortpol extends Object {

  public static void main(String args[]) {
    double b[] = new double[4];
    double c[] = new double[4];
    double zer[] = new double[4];
    double em[] = new double[6];

    em[0]=em[2]=1.0e-6;   em[4]=15.0;
    b[2]=b[1]=b[0]=0.0;
    c[0]=0.0;   c[1]=0.5;   c[2]=0.25;
    Linear_algebra.allzerortpol(3,b,c,zer,em);
    DecimalFormat fiveDigit = new DecimalFormat("0.00000E0");
    System.out.println("The three zeros:\n  " + fiveDigit.format(zer[1]) + "\n  " +
      fiveDigit.format(zer[2]) + "\n  " + fiveDigit.format(zer[3]) + "\n\n" +
      "EM[1]:  " + em[1] + "\nEM[3]:  " + fiveDigit.format(em[3]) + "\nEM[5]:  " +
      (int)em[5]);
```

```
  }
}
```

Output:

```
The three zeros:
  -8.66026E-1
   8.66025E-1
  -1.00000E-6

EM[1]:   1.5
EM[3]:   7.07107E-7
EM[5]:   1
```

Procedure tested:

lupzerortpol

Example:

Determine the two smaller and two larger roots of the Laguerre polynomial $L_3(x)$ for which

$$L_0(x)=1, \quad L_1(x)=x\text{-}1, \quad L_{k+1}(x)=(x\text{-}2k\text{-}1)L_k(x)\text{-}k^2L_{k\text{-}1}(x), \quad k=1,2.$$

```
package numal;

import java.text.DecimalFormat;
import numal.*;

public class Test_lupzerortpol extends Object {

  public static void main(String args[]) {

    int i;
    double b[] = new double[4];
    double c[] = new double[4];
    double zer[] = new double[3];
    double em[] = new double[7];

    em[0]=em[2]=1.0e-6;   em[4]=45.0;   em[6]=1.0;
    for (i=0; i<=2; i++) {
      b[i]=2*i+1;
      c[i]=i*i;
    }
    Linear_algebra.lupzerortpol(3,2,b,c,zer,em);
    DecimalFormat fiveDigit = new DecimalFormat("0.00000E0");
    System.out.println("The two lower zeros:\n  " + fiveDigit.format(zer[1]) +
      "\n  " + fiveDigit.format(zer[2]) + "\n\n" + "EM[1]:  " + em[1] +
      "\nEM[3]:  " + fiveDigit.format(em[3]) + "\nEM[5]:  " + (int)em[5]);
    em[6]=0.0;
```

```
    for (i=0; i<=2; i++) {
      b[i] = -2*i-1;
      c[i]=i*i;
    }
    Linear_algebra.lupzerortpol(3,2,b,c,zer,em);
    System.out.println("\nThe two upper zeros:\n  " + fiveDigit.format(-zer[1]) +
      "\n  " + fiveDigit.format(-zer[2]) + "\n\n" + "EM[1]:  " + em[1] +
      "\nEM[3]:  " + fiveDigit.format(em[3]) + "\nEM[5]:  " + (int)em[5]);
  }
}
```

Output:

```
The two lower zeros:
  4.15775E-1
  2.29428E0

EM[1]:  9.0
EM[3]:  4.79226E-8
EM[5]:  10

The two upper zeros:
  6.28995E0
  2.29428E0

EM[1]:  9.0
EM[3]:  3.68340E-10
EM[5]:  12
```

Procedure tested:

selzerortpol

Example:
Determine the third root of the Legendre polynomial $P_4(x)$ for which
$$P_0(x)=1, \quad P_1(x)=x, \quad P_{k+1}(x)=xP_k(x)-\{k^2/(4k^2-1)\}P_{k-1}(x), \; k=1,2,3.$$

```
package numal;

import java.text.DecimalFormat;
import numal.*;

public class Test_selzerortpol extends Object {

  public static void main(String args[]) {

    int i;
    double b[] = new double[5];
    double c[] = new double[5];
```

```
    double zer[] = new double[4];
    double em[] = new double[6];

    em[0]=em[2]=1.0e-6;
    for (i=0; i<=3; i++) {
      b[i]=0.0;
      c[i]=i*i/(4.0*i*i-1.0);
    }
    Linear_algebra.selzerortpol(4,3,3,b,c,zer,em);
    DecimalFormat fiveDigit = new DecimalFormat("0.00000E0");
    System.out.println("The third zero:\n  " + fiveDigit.format(zer[3]) +
      "\n\nEM[1]:  " + fiveDigit.format(em[1]) + "\nEM[5]:  " + (int)em[5]);
  }
}
```

Output:

```
The third zero:
  -3.39981E-1

EM[1]:  1.33333E0
EM[5]:  10
```

Procedure tested:

alljaczer

Example:

Compute the roots of the Jacobi polynomial $P_3^{(-1/2,-1/2)}(x)$.

```
package numal;

import java.text.DecimalFormat;
import numal.*;

public class Test_alljaczer extends Object {

  public static void main(String args[]) {

    double x[] = new double[4];

    Linear_algebra.alljaczer(3,-0.5,-0.5,x);
    DecimalFormat fiveDigit = new DecimalFormat("0.00000E0");
    System.out.println("Delivers:\n  " + fiveDigit.format(x[1]) + "   " +
                    fiveDigit.format(x[2]) + "   " + fiveDigit.format(x[3]));
  }
}
```

Output:

```
Delivers:
  -8.66025E-1    0.00000E0    8.66025E-1
```

Procedure tested:
alllagzer

Example:
Compute the roots of the Laguerre polynomial $L_3^{(-1/2)}(x)$.

```
package numal;

import java.text.DecimalFormat;
import numal.*;

public class Test_alllagzer extends Object {

  public static void main(String args[]) {

    double x[] = new double[4];

    Linear_algebra.alllagzer(3,-0.5,x);
    DecimalFormat fiveDigit = new DecimalFormat("0.00000E0");
    System.out.println("Delivers:\n  " + fiveDigit.format(x[1]) + "    " +
                    fiveDigit.format(x[2]) + "    " + fiveDigit.format(x[3]));
  }
}
```

Output:

```
Delivers:
  5.52534E0    1.78449E0    1.90164E-1
```

Procedure tested:
comkwd

Example:
Compute the roots of $x^2 - 2(-0.1+0.3i)x - (0.11+0.02i)$.

```
package numal;

import java.text.DecimalFormat;
import numal.*;
```

```
public class Test_comkwd extends Object {

  public static void main(String args[]) {

    double gr[] = new double[1];
    double gi[] = new double[1];
    double kr[] = new double[1];
    double ki[] = new double[1];

    Linear_algebra.comkwd(-0.1,0.3,0.11,0.02,gr,gi,kr,ki);
    DecimalFormat twoDigit = new DecimalFormat("0.00");
    System.out.println("x**2-2(-0.1+0.3*i)*x-(0.11+0.02*i)" + " has roots\n  " +
                    twoDigit.format(gr[0]) + "+" + twoDigit.format(gi[0]) +
                    "*i\n  " + twoDigit.format(kr[0]) + "+" +
                    twoDigit.format(ki[0]) + "*i");
  }
}
```

Output:

```
x**2-2(-0.1+0.3*i)*x-(0.11+0.02*i) has roots
  -0.30+0.40*i
   0.10+0.20*i
```

Examples for chapter 4 procedures

Procedure tested:

euler

Example:
Apply the Euler transformation to the series

$$\sum_{i=1}^{\infty} \frac{(-1)^{i-1}}{i^2}.$$

```
package numal;

import java.text.DecimalFormat;
import numal.*;

public class Test_euler extends Object implements AE_euler_method {
  public static void main(String args[]) {
    Test_euler testeuler = new Test_euler();
    DecimalFormat fiveDigit = new DecimalFormat("0.00000E0");
    System.out.println("Delivers:  " + fiveDigit.format(
                    Analytic_eval.euler(testeuler,1.0e-6,100)));
  }
```

```
  public double ai(int i) {
    return (Math.pow(-1,i)/((i+1)*(i+1)));
  }
}
```

Output:

```
Delivers:  8.22467E-1
```

Procedure tested:

sumposseries

Example:
Evaluate the sum

$$\sum_{i=1}^{\infty} \frac{1}{i^2}.$$

```
package numal;

import java.text.DecimalFormat;
import numal.*;

public class Test_sumposseries extends Object implements AE_sumposseries_method {

  public static void main(String args[]) {

    Test_sumposseries testsumposseries = new Test_sumposseries();
    DecimalFormat fiveDigit = new DecimalFormat("0.00000E0");
    System.out.println("SUMPOSSERIES delivers:  " + fiveDigit.format(
            Analytic_eval.sumposseries(testsumposseries,100,1.0e-6,8,100,10)));
  }


  public double ai(double i) {
    return 1.0/(i*i);
  }
}
```

Output:

```
SUMPOSSERIES delivers:  1.64493E0
```

Procedure tested:

qadrat

Example:
Evaluate the integral

$$\int_0^\pi \sin(x)dx .$$

```
package numal;

import java.text.DecimalFormat;
import numal.*;

public class Test_qadrat extends Object implements AE_qadrat_method {

  public static void main(String args[]) {

    double q;
    double e[] = new double[4];
    Test_qadrat testqadrat = new Test_qadrat();
    e[1]=e[2]=1.0e-6;
    q=Analytic_eval.qadrat(0.0,Math.PI,testqadrat,e);
    DecimalFormat fiveDigit = new DecimalFormat("0.00000E0");
    System.out.println("Delivers:  " + fiveDigit.format(q) + "    " + (int)e[3]);
  }


  public double fx(double x) {
    return (Math.sin(x));
  }
}
```

Output:

```
Delivers:  2.00000E0   0
```

Procedure tested:

integral

Example:
Evaluate the integral

$$\int_\alpha^\beta \frac{10}{x^2} dx$$

with, in succession, $\alpha = -1$ and $\beta = -2, -4, -20, -100$.

```
package numal;

import java.text.DecimalFormat;
import numal.*;

public class Test_integral extends Object implements AE_integral_method {

  public static void main(String args[]) {

    boolean ua,ub;
    int i;
    double a;
    double e[] = new double[7];
    double b[]={2.0, 4.0, 20.0, 100.0};

    Test_integral testintegral = new Test_integral();
    DecimalFormat fiveDigit = new DecimalFormat("0.00000E0");
    System.out.println("INTEGRAL delivers:");
    ua=true;
    e[1]=e[2]=1.0e-6;
    for (i=0; i<=3; i++) {
      ub=(b[i] < 50.0);
      a=Analytic_eval.integral(-1.0,-b[i],testintegral,e,ua,ub);
      System.out.println(" " + fiveDigit.format(a) + "\t" + (int)e[3] + "\t" +
                        fiveDigit.format(e[4]) + "\t" + e[5] + "\t" + e[6]);
      ua=false;
    }
  }


  public double fx(double x) {
    return 10.0/(x*x);
  }
}
```

Output:

```
INTEGRAL delivers:
 -5.00000E0  0  -5.00000E0  -2.0     2.5
 -7.50000E0  0   -7.50000E0  -4.0     0.625
 -9.50000E0  0   -9.50000E0  -20.0   0.025
 -9.99999E0  1   -9.99999E0  0.0      0.0
```

Procedure tested:

tricub

Example:
Evaluate the double integral

$$\iint_{\Omega} \cos(x)\cos(y)dxdy$$

over the triangle Ω in the x-y plane with vertices $(0,0)$, $(0, \pi/2)$, $(\pi/2, \pi/2)$.

```
package numal;

import java.text.DecimalFormat;
import numal.*;

public class Test_tricub extends Object implements AE_tricub_method {

  public static void main(String args[]) {

    int i;
    double acc;

    Test_tricub testtricub = new Test_tricub();
    DecimalFormat oneDigit = new DecimalFormat("0.0E0");
    DecimalFormat fiveDigit = new DecimalFormat("0.00000E0");
    System.out.println("TRICUB delivers:");
    acc=1.0;
    for (i=0; i<=5; i++) {
      acc *= 1.0e-1;
      System.out.println("  " + oneDigit.format(acc) + "    " + fiveDigit.format(
          Analytic_eval.tricub(0.0,0.0,0.0,Math.PI/2.0,Math.PI/2.0,Math.PI/2.0,
                        testtricub,acc,acc)));
    }
  }


  public double g(double x, double y) {
    return Math.cos(x)*Math.cos(y);
  }
}
```

Output:

```
TRICUB delivers:
    1.0E-1    5.00640E-1
    1.0E-2    5.00640E-1
    1.0E-3    5.00640E-1
    1.0E-4    4.99991E-1
    1.0E-5    4.99998E-1
    1.0E-6    4.99998E-1
```

Procedure tested:

reccof

Example:

Evaluate the coefficients c_1, c_2 in the recursion

$$C_{k+1}(x) = xC_k(x) - c_kC_{k-1}(x) \quad (k=1,2)$$

for the Chebyshev polynomials of the second kind which are orthogonal over $[-1,1]$ with respect to the weight function $w(x)=(1-x^2)^{1/2}$.

```
package numal;

import java.text.DecimalFormat;
import numal.*;

public class Test_reccof extends Object implements AE_reccof_method {

  public static void main(String args[]) {

    double x;
    double b[] = new double[3];
    double c[] = new double[3];
    double l[] = new double[3];

    Test_reccof testreccof = new Test_reccof();
    DecimalFormat threeDigit = new DecimalFormat("0.000");
    Analytic_eval.reccof(2,200,testreccof,b,c,l,true);
    System.out.println("Delivers:   " + threeDigit.format(c[1]) + "   " +
                       threeDigit.format(c[2]));
  }

  public double wx(double x) {
    return Math.sqrt(1.0-x*x);
  }
}
```

Output:

```
Delivers:   0.250   0.250
```

Procedure tested:

gsswtssym

Example:

Compute the weights w_i which render the formula

$$\pi \sum_{i=1}^{5} w_i f(z_i) = \int_{-1}^{1} \frac{1}{\sqrt{\left(1-x^2\right)}} f(x)dx$$

where

$$z_i = \cos\left(\frac{2(n-i)-1}{2n}\pi\right) \qquad i = 1,\dots,n/2, \quad n = 5$$

exact for all polynomials of degree < 10 by means of a call of *gsswtssym* with $c_1=1/2$, $c_k=1/4$ (k=2,3,...).

```
package numal;

import java.text.DecimalFormat;
import numal.*;

public class Test_gsswtssym extends Object {

  public static void main(String args[]) {

    double zer[]  = new double[3];
    double w[]    = new double[4];
    double c[]    = new double[5];

    c[1]=0.5; c[2]=0.25;  c[3]=0.25;  c[4]=0.25;
    zer[1]=Math.cos(0.9*Math.PI);
    zer[2]=Math.cos(0.7*Math.PI);
    Analytic_eval.gsswtssym(5,zer,c,w);
    DecimalFormat threeDigit = new DecimalFormat("0.000");
    System.out.println("Results:\n  " +
                  threeDigit.format(w[1]*Math.PI) + "   " +
                  threeDigit.format(w[2]*Math.PI) + "   " +
                  threeDigit.format(w[3]*Math.PI) + "   " +
                  threeDigit.format(w[2]*Math.PI) + "   " +
                  threeDigit.format(w[1]*Math.PI));
  }
}
```

Output:

```
Results:
  0.628   0.628   0.628   0.628   0.628
```

Procedure tested:

gssjacwghts

Example:

Evaluate the integral

$$\int_{-1}^{1} (1-x)(1+x)^2 e^x dx$$

by use of a fifth order Gauss-Jacobi quadrature formula. The exact value is $2e - 10/e$. (The error is printed out.)

```java
package numal;

import java.text.DecimalFormat;
import numal.*;

public class Test_gssjacwghts extends Object {

  public static void main(String args[]) {
    int n;
    double alfa,beta,ind;
    double x[] = new double[6];
    double w[] = new double[6];

    alfa=1.0;
    beta=2.0;
    n=5;
    ind=0.0;
    Analytic_eval.gssjacwghts(n,alfa,beta,x,w);
    for (n=1; n<=5; n++) ind += w[n]*Math.exp(x[n]);
    DecimalFormat fiveDigit = new DecimalFormat("0.00000E0");
    System.out.println("Delivers:  " +
      fiveDigit.format(ind-2.0*Math.exp(1.0)+10.0/Math.exp(1.0)));
  }
}
```

Output:

```
Delivers:  -1.59371E-10
```

Procedure tested:

gsslagwghts

Example:

Evaluate the integral

$$\int_{0}^{\infty} \sin(x)e^{-x} dx$$

by use of a ten point Gauss-Laguerre quadrature formula. The exact value is 0.5. (The error is printed out.)

```
package numal;

import java.text.DecimalFormat;
import numal.*;

public class Test_gsslagwghts extends Object {

  public static void main(String args[]) {

    int n;
    double ind;
    double x[] = new double[11];
    double w[] = new double[11];

    Analytic_eval.gsslagwghts(10,0.0,x,w);
    ind=0.0;
    for (n=10; n>=1; n--) ind += w[n]*Math.sin(x[n]);
    DecimalFormat fiveDigit = new DecimalFormat("0.00000E0");
    System.out.println("Delivers:  " + fiveDigit.format(ind-0.5));
  }
}
```

Output:

```
Delivers:   2.04965E-7
```

Procedure tested:

jacobnnf

Example:
Compute approximations to the partial derivatives $\partial f_i / \partial x_j$ $(i,j=1,2)$ of the components of the function $f(x)$ given by
$$f_1 = x_1{}^3 + x_2 \qquad\qquad f_2 = 10x_2$$
at the point $x = (2,1)$.

```
package numal;

import java.text.DecimalFormat;
import numal.*;

public class Test_jacobnnf extends Object implements AE_jacobnnf_methods {

  public static void main(String args[]) {

    int i;
    double x[]  = new double[3];
    double f[]  = new double[3];
    double jac[][] = new double[3][3];
```

```
    Test_jacobnnf testjacobnnf = new Test_jacobnnf();
    x[1]=2.0; x[2]=1.0;
    testjacobnnf.funct(2,x,f);
    Analytic_eval.jacobnnf(2,x,f,jac,testjacobnnf);
    DecimalFormat fiveDigit = new DecimalFormat("0.00000E0");
    System.out.println("The calculated jacobian is:\n  " +
                       fiveDigit.format(jac[1][1]) + "    " +
                       fiveDigit.format(jac[1][2]) + "\n  " +
                       fiveDigit.format(jac[2][1]) + "    " +
                       fiveDigit.format(jac[2][2]));
  }


  public double di(int i, int n) {
    return ((i == 1) ? 1.0e-5 : 1.0);
  }


  public void funct(int n, double x[], double f[]) {
    f[1]=x[1]*x[1]*x[1]+x[2];
    f[2]=x[2]*10.0;
  }
}
```

Output:

```
The calculated jacobian is:
  1.20001E1    1.00000E0
  0.00000E0    1.00000E1
```

Procedure tested:

jacobnmf

Example:

Compute approximations to the partial derivatives $\partial f_i / \partial x_j$ (i=1,2,3; j=1,2) of the components of the function $f(x)$ given by

$$f_1 = x_1^3 + x_2, \qquad f_2 = 10x_2 + x_1x_2, \qquad f_3 = x_1x_2$$

at the point $x = (2,1)$.

```
package numal;

import java.text.DecimalFormat;
import numal.*;
```

```
public class Test_jacobnmf extends Object implements AE_jacobnmf_methods {

  public static void main(String args[]) {

    int i;
    double x[] = new double[3];
    double f[] = new double[4];
    double jac[][] = new double[4][3];

    Test_jacobnmf testjacobnmf = new Test_jacobnmf();
    x[1]=2.0; x[2]=1.0;
    testjacobnmf.funct(3,2,x,f);
    Analytic_eval.jacobnmf(3,2,x,f,jac,testjacobnmf);
    DecimalFormat fiveDigit = new DecimalFormat("0.00000E0");
    System.out.println("The calculated jacobian is:\n  " +
                       fiveDigit.format(jac[1][1]) + "    " +
                       fiveDigit.format(jac[1][2]) + "\n  " +
                       fiveDigit.format(jac[2][1]) + "    " +
                       fiveDigit.format(jac[2][2]) + "\n  " +
                       fiveDigit.format(jac[3][1]) + "    " +
                       fiveDigit.format(jac[3][2]));
  }


  public double di(int i, int m) {
    return ((i == 2) ? 1.0 : 1.0e-5);
  }


  public void funct(int n, int m, double x[], double f[]) {
    f[1]=x[1]*x[1]*x[1]+x[2];
    f[2]=x[2]*10.0+x[2]*x[1]*x[1];
    f[3]=x[1]*x[2];
  }
}
```

Output:

```
The calculated jacobian is:
  1.20001E1    1.00000E0
  4.00001E0    1.40000E1
  1.00000E0    2.00000E0
```

Procedure tested:

jacobnbndf

Example:

Compute approximations to the partial derivatives

$$\partial f_i / \partial x_j, \qquad i=1,...,5; \; j=\max(1,i\text{-}1),...,\min(5,i\text{+}1)$$

of the components of the function $f(x)$ given by

$$f_1 = (3\text{-}2x_1)x_1 - 2x_2 + 1$$
$$f_i = (3\text{-}2x_i)x_i - 2x_{i+1} - x_{i-1} + 1 \qquad (i=2,3,4)$$
$$f_5 = 4 - x_4 - 2x_5$$

at the point $x = (\text{-}1,\text{-}1,\text{-}1,\text{-}1,\text{-}1)$.

```java
package numal;

import java.text.DecimalFormat;
import numal.*;

public class Test_jacobnbndf extends Object implements AE_jacobnbndf_methods {

  public static void main(String args[]) {

    int i;
    double x[]  = new double[6];
    double f[]  = new double[6];
    double jac[] = new double[14];

    Test_jacobnbndf testjacobnbndf = new Test_jacobnbndf();
    for (i=1; i<=5; i++) x[i] = -1.0;
    testjacobnbndf.funct(5,1,5,x,f);
    Analytic_eval.jacobnbndf(5,1,1,x,f,jac,testjacobnbndf);
    DecimalFormat oneDigit = new DecimalFormat("0.0E0");
    System.out.println("The calculated tridiagonal jacobian is:\n" +
      oneDigit.format(jac[1]) + "\t" + oneDigit.format(jac[2]) + "\n" +
      oneDigit.format(jac[3]) + "\t" + oneDigit.format(jac[4]) + "\t" +
      oneDigit.format(jac[5]) + "\n\t" +oneDigit.format(jac[6]) + "\t" +
      oneDigit.format(jac[7]) + "\t" + oneDigit.format(jac[8]) + "\n\t\t" +
      oneDigit.format(jac[9]) + "\t" + oneDigit.format(jac[10]) + "\t" +
      oneDigit.format(jac[11]) + "\n\t\t\t" + oneDigit.format(jac[12]) + "\t" +
      oneDigit.format(jac[13]));
  }


  public double di(int i, int m) {
    return ((i == 5) ? 1.0 : 1.0e-6);
  }
```

```java
public void funct(int n, int l, int u, double x[], double f[]) {

    int i;

    for (i=1; i<=((u == 5) ? 4 : u); i++) {
        f[i]=(3.0-2.0*x[i])*x[i]+1.0-2.0*x[i+1];
        if (i != 1) f[i] -= x[i-1];
    }
    if (u == 5) f[5]=4.0-x[4]-x[5]*2.0;
}
}
```

Output:

```
The calculated tridiagonal jacobian is:
7.0E0    -2.0E0
-1.0E0   7.0E0      -2.0E0
         -1.0E0    7.0E0     -2.0E0
                   -1.0E0    7.0E0     -2.0E0
                             -1.0E0    -2.0E0
```

Examples for chapter 5 procedures

Procedure tested:

zeroin

Example:
Determine a zero of $e^{-3x}(x-1)+x^3$ in the interval $[0,1]$.

```java
package numal;

import java.text.DecimalFormat;
import numal.*;

public class Test_zeroin extends Object implements AP_zeroin_methods {

    public static void main(String args[]) {
        double x[] = new double[1];
        double y[] = new double[1];

        Test_zeroin testzeroin = new Test_zeroin();
        DecimalFormat fiveDigit = new DecimalFormat("0.00000E0");
        x[0]=0.0;
        y[0]=1.0;
        if (Analytic_problems.zeroin(x,y,testzeroin))
            System.out.println("Calculated zero:  " + fiveDigit.format(x[0]));
        else
            System.out.println("No zero found.");
    }
```

```
  public double fx(double  x[])
  {
    return Math.exp(-x[0]*3.0)*(x[0]-1.0)+x[0]*x[0]*x[0];
  }


  public double tolx(double  x[])
  {
    return Math.abs(x[0])*1.0e-6+1.0e-6;
  }
}
```

Output:

```
Calculated zero:   4.89703E-1
```

Procedure tested:

zeroinrat

Example:
Determine a zero of $e^{-3x}(x-1)+x^3$ in the interval $[0,1]$.

```
package numal;

import java.text.DecimalFormat;
import numal.*;

public class Test_zeroinrat extends Object implements AP_zeroinrat_methods {

  public static void main(String args[]) {
    double x[] = new double[1];
    double y[] = new double[1];

    Test_zeroinrat testzeroinrat = new Test_zeroinrat();
    DecimalFormat fiveDigit = new DecimalFormat("0.00000E0");
    x[0]=0.0;
    y[0]=1.0;
    if (Analytic_problems.zeroinrat(x,y,testzeroinrat))
      System.out.println("Calculated zero:   " + fiveDigit.format(x[0]));
    else
      System.out.println("No zero found.");
  }


  public double fx(double  x[])
  {
    return Math.exp(-x[0]*3.0)*(x[0]-1.0)+x[0]*x[0]*x[0];
  }
```

```
  public double tolx(double  x[])
  {
    return Math.abs(x[0])*1.0e-6+1.0e-6;
  }
}
```

Output:

```
Calculated zero:  4.89703E-1
```

Procedure tested:

zeroinder

Example:
Determine a zero of $e^{-3x}(x-1)+x^3$ in the interval $[0,1]$.

```
package numal;

import java.text.DecimalFormat;
import numal.*;

public class Test_zeroinder extends Object implements AP_zeroinder_methods {

  public static void main(String args[]) {

    double x[] = new double[1];
    double y[] = new double[1];

    Test_zeroinder testzeroinder = new Test_zeroinder();
    DecimalFormat fiveDigit = new DecimalFormat("0.00000E0");
    x[0]=0.0;
    y[0]=1.0;
    if (Analytic_problems.zeroinder(x,y,testzeroinder))
      System.out.println("Calculated zero and function value:" + "\n  " +
        fiveDigit.format(x[0]) + "    " + fiveDigit.format(testzeroinder.fx(x)) +
        "\nOther straddling approximation and function value:" + "\n  " +
        fiveDigit.format(y[0]) + "    " + fiveDigit.format(testzeroinder.fx(y)));
    else
      System.out.println("No zero found.");
  }


  public double fx(double  x[])
  {
    return Math.exp(-x[0]*3.0)*(x[0]-1.0)+x[0]*x[0]*x[0];
  }
}
```

```
  public double tolx(double  x[])
  {
    return Math.abs(x[0])*1.0e-6+1.0e-6;
  }


  public double dfx(double  x[])
  {
    return Math.exp(-x[0]*3.0)*(-3.0*x[0]+4.0)+3.0*x[0]*x[0];
  }
}
```

Output:

```
Calculated zero and function value:
  4.89703E-1    -5.82126E-10
Other straddling approximation and function value:
  4.89704E-1    1.93881E-6
```

Procedure tested:

quanewbnd1

Example:

Solve the system of equations $f(x) = 0$ $(f,x \in R^n)$ where

$$f^{(1)} = (3 - 2x^{(1)})x^{(1)} - 2x^{(2)} + 1$$
$$f^{(k)} = (3 - 2x^{(k)})x^{(k)} - 2x^{(k+1)} - x^{(k-1)} + 1 \qquad (k=2,...,n-1)$$
$$f^{(n)} = (3 - 2x^{(n)})x^{(n)} - x^{(n-1)} + 1$$

with initial approximation $x_0^{(k)} = -1$ $(k=1,...,n)$, for $n = 600$.

```
package numal;

import java.text.DecimalFormat;
import numal.*;

public class Test_quanewbnd1 extends Object implements AP_quanewbnd_method {

  public static void main(String args[]) {

    int i;
    double in[]  = new double[6];
    double out[] = new double[6];
    double x[]= new double[601];
    double f[]= new double[601];

    Test_quanewbnd1 testquanewbnd1 = new Test_quanewbnd1();
    DecimalFormat fiveDigit = new DecimalFormat("0.00000E0");
    for (i=1; i<=600; i++) x[i] = -1.0;
    in[0]=1.0e-6;   in[1]=in[2]=in[3]=1.0e-5;  in[4]=20000.0;
```

```
    in[5]=0.001;
    Analytic_problems.quanewbnd1(600,1,1,x,f,testquanewbnd1,in,out);
    System.out.println("Norm Residual vector:   " + fiveDigit.format(out[2]) +
       "\nLength of last step:    " + fiveDigit.format(out[1]) +
       "\nNumber of function component evaluations:   " + (int)out[3] +
       "\nNumber of iterations:   " + (int)out[4] + "\nReport:   " + (int)out[5]);
  }


  public boolean funct(int n, int l, int u, double x[], double f[])
  {
    int i;
    double x1,x2,x3;

    x1 = (l == 1) ? 0.0 : x[l-1];
    x2=x[l];
    x3 = (l == n) ? 0.0 : x[l+1];
    for (i=1; i<=u; i++) {
      f[i]=(3.0-2.0*x2)*x2+1.0-x1-x3*2.0;
      x1=x2;
      x2=x3;
      x3 = (i <= n-2) ? x[i+2] : 0.0;
    }
    return (true);
  }
}
```

Output:

```
Norm Residual vector:   2.13360E-6
Length of last step:    2.70373E-5
Number of function component evaluations:   5998
Number of iterations:   6
Report:   0
```

Procedure tested:

minin

Example:
Determine an approximation to the point in the interval [1,4] at which the function

$$f(x) = \sum_{i=1}^{20} \left(\frac{2i-5}{x-i^2} \right)^2$$

assumes a minimum value.

```
package numal;

import java.text.DecimalFormat;
```

```
import numal.*;

public class Test_minin extends Object implements AP_minin_methods {

  public static void main(String args[]) {

    double m;
    double x[] = new double[1];
    double a[] = new double[1];
    double b[] = new double[1];
    double tmp1[] = {1.0};
    double tmp2[] = {4.0};

    Test_minin testminin = new Test_minin();
    DecimalFormat fiveDigit = new DecimalFormat("0.00000E0");
    a[0]=1.0+testminin.tolx(tmp1);
    b[0]=4.0-testminin.tolx(tmp2);
    m=Analytic_problems.minin(x,a,b,testminin);
    System.out.println("Minimum is   " + fiveDigit.format(m) + "\nFor x is   " +
      fiveDigit.format(x[0]) + "\nin the interval with endpoints  " +
      fiveDigit.format(a[0]) + "   " + fiveDigit.format(b[0]));
  }


  public double fx(double x[])
  {
    int i;
    double s,temp;

    s=0.0;
    for (i=1; i<=20; i++) {
      temp=(i*2-5)/(x[0]-i*i);
      s += temp*temp;
    }
    return s;
  }


  public double tolx(double x[])
  {
    return (Math.abs(x[0])*1.0e-6+1.0e-6);
  }
}
```

Output:

```
Minimum is   3.67670E0
For x is  3.02292E0
in the interval with endpoints  3.02291E0   3.02292E0
```

Procedure tested:

mininder

Example:

Determine an approximation to the point in the interval [1.01,3.99] at which the function

$$f(x) = \sum_{i=1}^{20} \left(\frac{2i-5}{x-i^2} \right)^2$$

assumes a minimum value. The derivative of this function is

$$\frac{df(x)}{dx} = -2 \sum_{i=1}^{20} \frac{(2i-5)^2}{(x-i^2)^3} .$$

```java
package numal;

import java.text.DecimalFormat;
import numal.*;

public class Test_mininder extends Object implements AP_mininder_methods {

  public static void main(String args[]) {

    double m;
    double x[] = new double[1];
    double y[] = new double[1];

    Test_mininder testmininder = new Test_mininder();
    DecimalFormat fiveDigit = new DecimalFormat("0.00000E0");
    x[0]=1.01;
    y[0]=3.99;
    m=Analytic_problems.mininder(x,y,testmininder);
    System.out.println("Minimum is  " + fiveDigit.format(m) +  "\nFor x is  " +
            fiveDigit.format(x[0]) + "  and y is  " + fiveDigit.format(y[0]));
  }


  public double fx(double x[])
  {
    int i;
    double s,temp;

    s=0.0;
    for (i=1; i<=20; i++) {
      temp=(i*2-5)/(x[0]-i*i);
      s += temp*temp;
    }
    return s;
  }
```

```
public double dfx(double x[])
{
   int i;
   double s,temp1,temp2;

   s=0.0;
   for (i=1; i<=20; i++) {
     temp1=i*2-5;
     temp2=x[0]-i*i;
     s += (temp1*temp1)/(temp2*temp2*temp2);
   }
   return -s*2.0;
}


public double tolx(double x[])
{
   return (Math.abs(x[0])*1.0e-6+1.0e-6);
}
}
```

Output:

```
Minimum is  3.67670E0
For x is  3.02292E0  and y is  3.02291E0
```

Procedure tested:
praxis

Example:
Calculate the minimum of the function
$$f(x) = 100(x_2-x_1^2)^2 + (1-x_1)^2$$
using (-1.2,1) as an initial estimate.

```
package numal;

import java.text.DecimalFormat;
import numal.*;

public class Test_praxis extends Object implements AP_praxis_method {

  public static void main(String args[]) {

    double x[] = new double[3];
    double in[] = new double[10];
    double out[] = new double[7];
```

```
Test_praxis testpraxis = new Test_praxis();
DecimalFormat fiveDigit = new DecimalFormat("0.00000E0");
in[0]=1.0e-6;   in[1]=in[2]=1.0e-6;   in[5]=250.0;
in[6]=in[7]=in[8]=in[9]=1.0;
x[1] = -1.2;   x[2]=1.0;
Analytic_problems.praxis(2,x,testpraxis,in,out);
if (out[1] == 0.0)
  System.out.println("Normal Termination\n");
System.out.println("Minimum is  " + fiveDigit.format(out[2]) + "\nFor x is  " +
  fiveDigit.format(x[1]) + "     " + fiveDigit.format(x[2]) +
  "\nThe initial function value was  " + fiveDigit.format(out[3]) +
  "\nThe number of function evaluations needed was  " + (int)out[4] +
  "\nThe number of line searches was  " + (int)out[5] +
  "\nThe step size in the last " +
  "iteration step was  " + fiveDigit.format(out[6]));
}


  public double funct(int n, double x[])
  {
    double temp;

    temp=x[2]-x[1]*x[1];
    return temp*temp*100.0+(1.0-x[1])*(1.0-x[1]);
  }
}
```

Output:

```
Normal Termination

Minimum is  3.52015E-26
For x is  1.00000E0     1.00000E0
The initial function value was  2.42000E1
The number of function evaluations needed was  188
The number of line searches was  72
The step size in the last iteration step was  1.92891E-10
```

Procedures tested:

rnk1min, flemin

Example:

Determine the value of $x = (x_1, x_2)$ yielding a minimum of the function
$$f(x) = 100(x_2 - x_1^2)^2 + (1 - x_1)^2$$
twice: firstly by use of *rnk1min*, and secondly by *flemin*. In both cases the initial approximation is taken to be (-1.2,1).

```
package numal;

import java.text.DecimalFormat;
import numal.*;

public class Test_rnk1min extends Object implements AP_linemin_method {

  public static void main(String args[]) {

    boolean again;
    double f;
    double x[] = new double[3];
    double g[] = new double[3];
    double h[] = new double[4];
    double in[] = new double[9];
    double out[] = new double[5];

    Test_rnk1min testrnk1min = new Test_rnk1min();
    DecimalFormat fiveDigit = new DecimalFormat("0.00000E0");
    in[0]=1.0e-6;  in[1]=in[2]=1.0e-5;  in[3]=1.0e-4;
    in[4]=1.0e-5;  in[5] = -10.0;  in[6]=1.0;  in[7]=100.0;
    in[8]=0.01;
    x[1] = -1.2;  x[2]=1.0;
    again=true;
    f=Analytic_problems.rnk1min(2,x,g,h,testrnk1min,in,out);
    while (true) {
      System.out.println("\nLeast value: " + fiveDigit.format(f) + "\nx:          "
        + fiveDigit.format(x[1]) + "     " + fiveDigit.format(x[2]) + "\nGradient:  "
        + fiveDigit.format(g[1]) + "     " + fiveDigit.format(g[2]) + "\nMetric:    "
        + fiveDigit.format(h[1]) + "     " + fiveDigit.format(h[2]) + "\n          "
        + "              " + fiveDigit.format(h[3]) + "\nOUT:   " +
        fiveDigit.format(out[0]) + "\n       " + fiveDigit.format(out[1]) +
        "\n      " + (int)out[2] + "\n       " + (int)out[3] +
        "\n      " + (int)out[4]);
      if (again) {
        x[1] = -1.2;  x[2]=1.0;
        again=false;
        f=Analytic_problems.flemin(2,x,g,h,testrnk1min,in,out);
      } else
        break;
    }
  }


  public double funct(int n, double x[], double g[])
  {
    double temp;

    temp=x[2]-x[1]*x[1];
    g[1]=(-temp*400.0+2.0)*x[1]-2.0;
```

```
    g[2]=temp*200.0;
    return temp*temp*100.0+(1.0-x[1])*(1.0-x[1]);
  }
}
```

Output:

```
Least value:  4.80041E-19
x:            1.00000E0    1.00000E0
Gradient:   -1.93268E-8    1.01358E-8
Metric:      4.99655E-1    9.99305E-1
                           2.00360E0
OUT:   1.10115E-9
       2.18234E-8
       54
       9
       5


Least value:  8.11976E-17
x:            1.00000E0    1.00000E0
Gradient:   3.59827E-7    -1.80155E-7
Metric:      5.01085E-1    1.00198E0
                           2.00862E0
OUT:   1.33803E-9
       4.02407E-7
       44
       7
       0
```

Procedure tested:

marquardt

Example:
Determine the parameters p_1, p_2, and p_3 of best fit of the function
$$g(x,p) = p_1 + p_2*exp(p_3x)$$
when $x=x_i$, to data readings y_i ($i=1,...,b$), where $x_1,...,x_6$ are -5, -3, -1, 1, 3, 5, and $y_1,...,y_6$ are 127.0, 151.0, 379.0, 421.0, 460.0, 426.0.
  The components of the residual vector are
$$p_1 + p_2*exp(p_3x_i) - y_i \qquad (i=1,...,6).$$
The elements of the Jacobian matrix are
$$\partial g_i/\partial p_1 = 1, \quad \partial g_i/\partial p_2 = exp(p_3x_i), \quad \partial g_i/\partial p_3 = p_2*x_i*exp(p_3x_i), \quad (i=1,...,6).$$

```
package numal;

import java.text.DecimalFormat;
import numal.*;
```

```
public class Test_marquardt extends Object implements AP_marquardt_methods {

  static double x[] = new double[7];
  static double y[] = new double[7];

  public static void main(String args[]) {

    double in[] = new double[7];
    double out[] = new double[8];
    double rv[] = new double[7];
    double par[] = new double[4];
    double jjinv[][] =  new double[4][4];

    Test_marquardt testmarquardt = new Test_marquardt();
    DecimalFormat fiveDigit = new DecimalFormat("0.00000E0");
    DecimalFormat oneDigit = new DecimalFormat("0.0");
    in[0]=1.0e-6;  in[3]=1.0e-4;  in[4]=1.0e-1;  in[5]=75.0;
    in[6]=1.0e-2;
    x[1] = -5.0;  x[2] = -3.0;  x[3] = -1.0;  x[4]=1.0;
    x[5]=3.0;  x[6]=5.0;
    y[1]=127.0;  y[2]=151.0;  y[3]=379.0;  y[4]=421.0;
    y[5]=460.0;  y[6]=426.0;
    par[1]=580.0;  par[2] = -180.0;  par[3] = -0.160;
    Analytic_problems.marquardt(6,3,par,rv,jjinv,testmarquardt,in,out);
    System.out.println("Parameters:\n  " + fiveDigit.format(par[1]) + "    " +
      fiveDigit.format(par[2]) + "    " + fiveDigit.format(par[3]) + "\n\nOUT:\n    "
      + fiveDigit.format(out[7]) + "\n    " + fiveDigit.format(out[2]) + "\n    " +
      fiveDigit.format(out[6]) + "\n    " + fiveDigit.format(out[3]) + "\n    " +
      fiveDigit.format(out[4]) + "\n    " + fiveDigit.format(out[5]) + "\n    " +
      fiveDigit.format(out[1]) + "\n\nLast residual vector:\n  " +
      oneDigit.format(rv[1]) + "  " + oneDigit.format(rv[2]) + "  " +
      oneDigit.format(rv[3]) + "  " + oneDigit.format(rv[4]) + "  " +
      oneDigit.format(rv[5]) + "  " + oneDigit.format(rv[6]));
  }


  public boolean funct(int m, int n,double par[], double rv[])
  {
    int i;

    for (i=1; i<=m; i++) {
      if (par[3]*x[i] > 680.0) return false;
      rv[i]=par[1]+par[2]*Math.exp(par[3]*x[i])-y[i];
    }
    return true;
  }
}
```

```
public void jacobian(int m, int n, double par[], double rv[], double jac[][])
{
  int i;
  double ex;

  for (i=1; i<=m; i++) {
    jac[i][1]=1.0;
    jac[i][2]=ex=Math.exp(par[3]*x[i]);
    jac[i][3]=x[i]*par[2]*ex;
  }
}
}
```

Output:

```
Parameters:
  5.23225E2    -1.56839E2    -1.99778E-1

OUT:
      7.22078E7
      1.15716E2
      1.72802E-3
      1.65459E2
      2.30000E1
      2.20000E1
      0.00000E0


Last residual vector:
  -29.6   86.6   -47.3   -26.2   -22.9   39.5
```

Procedure tested:

gssnewton

Example:

Determine the parameters p_1, p_2, and p_3 of best fit of the function
$$g(x,p) = p_1 + p_2*exp(p_3x)$$
when $x=x_i$, to data readings y_i ($i=1,...,b$), where $x_1,...,x_6$ are -5, -3, -1, 1, 3, 5, and $y_1,...,y_6$ are 127.0, 151.0, 379.0, 421.0, 460.0, 426.0.

The components of the residual vector are
$$p_1 + p_2*exp(p_3x_i) - y_i \qquad (i=1,...,6).$$
The elements of the Jacobian matrix are
$$\partial g_i/\partial p_1 = 1, \quad \partial g_i/\partial p_2 = exp(p_3x_i), \quad \partial g_i/\partial p_3 = p_2*x_i*exp(p_3x_i), \quad (i=1,...,6).$$

```
package numal;

import java.text.DecimalFormat;
import numal.*;
```

```
public class Test_gssnewton extends Object implements AP_gssnewton_methods {

  static double x[] = new double[7];
  static double y[] = new double[7];

  public static void main(String args[]) {

    double in[] = new double[8];
    double out[] = new double[10];
    double g[] = new double[7];
    double par[] = new double[4];
    double v[][] =  new double[4][4];

    Test_gssnewton testgssnewton = new Test_gssnewton();
    DecimalFormat fiveDigit = new DecimalFormat("0.00000E0");
    DecimalFormat oneDigit = new DecimalFormat("0.0");
    in[0]=1.0e-6;  in[1]=in[2]=1.0e-6;  in[5]=75.0;  in[4]=1.0e-6;
    in[6]=14.0;  in[7]=1.0;
    x[1] = -5.0;  x[2] = -3.0;  x[3] = -1.0;  x[4]=1.0;
    x[5]=3.0;  x[6]=5.0;
    y[1]=127.0;  y[2]=151.0;  y[3]=379.0;  y[4]=421.0;
    y[5]=460.0;  y[6]=426.0;
    par[1]=580.0;  par[2] = -180.0;  par[3] = -0.160;
    Analytic_problems.gssnewton(6,3,par,g,v,testgssnewton,in,out);
    System.out.println("Parameters:\n  " + fiveDigit.format(par[1]) + "    " +
      fiveDigit.format(par[2]) + "    " + fiveDigit.format(par[3]) + "\n\nOUT:\n      "
      + fiveDigit.format(out[6]) + "\n     " + fiveDigit.format(out[2]) + "\n      " +
      fiveDigit.format(out[3]) + "\n     " + fiveDigit.format(out[4]) + "\n      " +
      fiveDigit.format(out[5]) + "\n     " + fiveDigit.format(out[1]) + "\n      " +
      fiveDigit.format(out[7]) + "\n     " + fiveDigit.format(out[8]) + "\n      " +
      fiveDigit.format(out[9]) + "\n\nLast residual vector:\n  " +
      oneDigit.format(g[1]) + "    " + oneDigit.format(g[2]) + "    " +
      oneDigit.format(g[3]) + "    " + oneDigit.format(g[4]) + "    " +
      oneDigit.format(g[5]) + "    " + oneDigit.format(g[6]));
  }


  public boolean funct(int m, int n, double par[], double g[])
  {
    int i;

    for (i=1; i<=m; i++) {
      if (par[3]*x[i] > 680.0) return false;
      g[i]=par[1]+par[2]*Math.exp(par[3]*x[i])-y[i];
    }
    return true;
  }
```

```
public void jacobian(int m, int n, double par[], double g[], double jac[][])
{
    int i;
    double ex;

    for (i=1; i<=m; i++) {
        jac[i][1]=1.0;
        jac[i][2]=ex=Math.exp(par[3]*x[i]);
        jac[i][3]=x[i]*par[2]*ex;
    }
}
}
```

Output:

```
Parameters:
  5.23305E2   -1.56948E2   -1.99665E-1

OUT:
    5.26048E-4
    1.15716E2
    1.65459E2
    1.60000E1
    1.60000E1
    2.00000E0
    0.00000E0
    3.00000E0
    2.33953E3

Last residual vector:
  -29.6  86.6  -47.3  -26.2  -22.9  39.5
```

Procedure tested:

rk1

Example:
Compute the solution at $x = 1$ of the differential equation
$$dy/dx = -y$$
with initial condition $y(0) = 1$.

```
package numal;

import java.text.DecimalFormat;
import numal.*;
```

```
public class Test_rk1 extends Object implements AP_rk1_method {

  public static void main(String args[]) {

    boolean first;
    double x[] = new double[1];
    double y[] = new double[1];
    double d[] = new double[5];
    double e[] = new double[3];

    Test_rk1 testrk1 = new Test_rk1();
    DecimalFormat fiveDigit = new DecimalFormat("0.00000E0");
    e[1]=e[2]=1.0e-4;
    first=true;
    Analytic_problems.rk1(x,0.0,1.0,y,1.0,testrk1,e,d,first);
    System.out.println("RK1 delivers:\n  x = " + fiveDigit.format(x[0]) +
      "\n  y = " + fiveDigit.format(y[0]) + "      yexact = " +
      fiveDigit.format(Math.exp(-x[0])));
  }


  public double fxy(double x[], double y[])
  {
    return -y[0];
  }
}
}
```

Output:

```
RK1 delivers:
  x = 1.00000E0
  y = 3.67877E-1      yexact = 3.67879E-1
```

Procedure tested:

rke

Example:

Compute the solution at $t = 1$ and $t = -1$ of the system

$$dx/dt = y - z$$
$$dy/dt = x^2 + 2y + 4t$$
$$dz/dt = x^2 + 5x + z^2 + 4t,$$

with $x = y = 0$ and $z = 2$ at $t = 0$.

```
package numal;

import java.text.DecimalFormat;
import numal.*;
```

```java
public class Test_rke extends Object implements AP_rke_methods {

  public static void main(String args[]) {

    double t[] = new double[1];
    double te[] = new double[1];
    double y[] = new double[4];
    double data[] = new double[7];

    Test_rke testrke = new Test_rke();
    te[0]=1.0;
    while (true) {
      y[1]=y[2]=0.0;
      y[3]=2.0;
      t[0]=0.0;
      data[1]=data[2]=1.0e-5;
      Analytic_problems.rke(t,te,3,y,testrke,data,true);
      if (te[0] != 1.0) break;
      te[0] = -1.0;
    }
  }


  public void der(int n, double t, double y[])
  {
    double xx,yy,zz;

    xx=y[1];
    yy=y[2];
    zz=y[3];
    y[1]=yy-zz;
    y[2]=xx*xx+2.0*yy+4.0*t;
    y[3]=xx*(xx+5.0)+2.0*zz+4.0*t;
  }


  public void out(int n, double t[], double te[], double y[], double data[])
  {
    double et,t2,aex,aey,aez,rex,rey,rez;

    DecimalFormat oneDigit = new DecimalFormat("0.0E0");
    if (t[0] == te[0]) {
      et=Math.exp(t[0]);
      t2=2.0*t[0];
      rex = -et*Math.sin(t2);
      aex=rex-y[1];
      rex=Math.abs(aex/rex);
      rey=et*et*(8.0+2.0*t2-Math.sin(2.0*t2))/8.0-t2-1.0;
      rez=et*(Math.sin(t2)+2.0*Math.cos(t2))+rey;
      aey=rey-y[2];
```

```
    rey=Math.abs(aey/rey);
    aez=rez-y[3];
    rez=Math.abs(aez/rez);
    System.out.println("\nT = " + t[0] +
      "\nRelative and absolute errors in x, y and z:\n" +
      "  RE(X)    RE(Y)    RE(Z)    AE(X)    AE(Y)    AE(Z)" + "\n  " +
      oneDigit.format(rex) + "  " + oneDigit.format(rey) + "  " +
      oneDigit.format(rez) + "  " + oneDigit.format(Math.abs(aex)) + "  " +
      oneDigit.format(Math.abs(aey)) + "  " + oneDigit.format(Math.abs(aez)) +
      "\nNumber of integration steps performed : " + (int)data[4] +
      "\nNumber of integration steps skipped   : " + (int)data[6] +
      "\nNumber of integration steps rejected  : " + (int)data[5]);
  }
 }
}
```

Output:

```
T =  1.0
Relative and absolute errors in x, y and z:
  RE(X)    RE(Y)    RE(Z)    AE(X)    AE(Y)    AE(Z)
 3.7E-7   1.5E-6   1.3E-6   9.1E-7   1.3E-5   1.1E-5
Number of integration steps performed :  9
Number of integration steps skipped   :  0
Number of integration steps rejected  :  5


T =  -1.0
Relative and absolute errors in x, y and z:
  RE(X)    RE(Y)    RE(Z)    AE(X)    AE(Y)    AE(Z)
 2.2E-7   5.2E-8   1.9E-7   7.5E-8   5.5E-8   7.7E-8
Number of integration steps performed :  10
Number of integration steps skipped   :  0
Number of integration steps rejected  :  7
```

Procedure tested:

rk4a

Example:

The solution of the differential equation

$$dy/dx = 1 - 2(x^2+y), \quad x \geq 0,$$
$$y = 0 \text{ at } x = 0,$$

is represented by the parabola $y = x(1-x)$. Find the value of x for which the curve of the solution intersects the line $y+x=0$.

```
package numal;


import java.text.DecimalFormat;
import numal.*;
```

```java
public class Test_rk4a extends Object implements AP_rk4a_methods {

  public static void main(String args[]) {

    double x[] = new double[1];
    double y[] = new double[1];
    double d[] = new double[5];
    double e[] = new double[6];

    DecimalFormat fiveDigit = new DecimalFormat("0.00000E0");
    Test_rk4a testrk4a = new Test_rk4a();
    e[0]=e[1]=e[2]=e[3]=e[4]=e[5]=1.0e-4;
    Analytic_problems.rk4a(x,0.0,testrk4a,y,0.0,e,d,true,true,true);
    System.out.println("x =  " + fiveDigit.format(x[0]) +
      "          Exactly : 2.00000\ny = " + fiveDigit.format(y[0]) + "\ny-x*(1-x) = "
      + fiveDigit.format(y[0]-x[0]*(1-x[0])));
  }


  public double b(double x[], double y[])
  {
    return x[0]+y[0];
  }


  public double fxy(double x[], double y[])
  {
    return 1.0-2.0*(x[0]*x[0]+y[0]);
  }
}
```

Output:

```
x =  2.00000E0        Exactly : 2.00000
y = -2.00011E0
y-x*(1-x) = -1.16523E-4
```

Procedure tested:

rk4na

Example:
Obtain the period of the solution of the van der Pol equation
$$dx_1/dt = x_2$$
$$dx_2/dt = 10(1-x_1^2)x_2-x_1, \quad t \geq 0.$$

```
package numal;

import java.text.DecimalFormat;
import numal.*;

public class Test_rk4na extends Object implements AP_rk4na_methods {

  public static void main(String args[]) {

    boolean first;
    int j;
    double x0;
    double e[] = new double[8];
    double xa[] = new double[3];
    double x[] = new double[3];
    double d[] = new double[6];

    DecimalFormat eightDigit = new DecimalFormat("0.00000000");
    Test_rk4na testrk4na = new Test_rk4na();
    for (j=0; j<=5; j++) e[j]=0.1e-6;
    e[6]=e[7]=1.0e-8;
    System.out.println("VAN DER POL\n\nEPS = " + eightDigit.format(e[0]) +
                       "\n\nThe values of x[0],x[1],x[2],p:");
    x0=xa[0]=xa[2]=0.0;
    xa[1]=2.0;
    System.out.println("  " + eightDigit.format(xa[0]) + "\t" +
                       eightDigit.format(xa[1]) + "\t" + eightDigit.format(xa[2]) +
                       "\t" + eightDigit.format(x0));
    first=true;
    for (j=1; j<=4; j++) {
      Analytic_problems.rk4na(x,xa,testrk4na,e,d,first,2,0,true);
      x0=x[0]-x0;
      System.out.println("  " + eightDigit.format(x[0]) + "\t" +
                         eightDigit.format(x[1]) + "\t" + eightDigit.format(x[2]) +
                         "\t" + eightDigit.format(x0));
      first=false;
      x0=x[0];
    }
  }


  public double b(int n, double x[])
  {
    return x[2];
  }


  public double fxj(int n, int k, double x[])
  {
    return ((k == 1) ? x[2] : 10.0*(1.0-x[1]*x[1])*x[2]-x[1]);
```

```
  }
}
```

Output:

```
VAN DER POL

EPS = 0.00000010

The values of x[0],x[1],x[2],p:
   0.00000000    2.00000000 0.00000000   0.00000000
   9.32386574   -2.01428536 0.00000000   9.32386574
  18.86305053    2.01428536 0.00000000   9.53918478
  28.40223531   -2.01428536 0.00000000   9.53918478
  37.94142009    2.01428536 0.00000000   9.53918478
```

Procedure tested:

rk5na

Example:
The van der Pol equation in the phase plane
$$dx_1/dx_0 = (10(1-x_0^2)x_1-x_0)/x_1$$
can be integrated by *rk5na*. The starting values are $x_0 = 2$, $x_1 = 0$. The integration
proceeds until the next zero of x_1, then it continues until the next zero and so on
until the fourth zero is encountered.

```
package numal;

import java.text.DecimalFormat;
import numal.*;

public class Test_rk5na extends Object implements AP_rk5na_methods {

  public static void main(String args[]) {

    boolean first;
    int j;
    double e[] = new double[6];
    double xa[] = new double[2];
    double x[] = new double[2];
    double d[] = new double[5];

    DecimalFormat eightDigit = new DecimalFormat("0.00000000");
    Test_rk5na testrk5na = new Test_rk5na();
    for (j=0; j<=3; j++) e[j]=1.0e-6;
    e[4]=e[5]=1.0e-10;
    d[1]=0.0;
    xa[0]=2.0;   xa[1]=0.0;
```

```
    j=0;
    System.out.println("Results:\n      " + "x[0]                x[1]                s\n" +
      "  " + eightDigit.format(xa[0]) + "\t" + eightDigit.format(xa[1]) + "\t" +
      eightDigit.format(Math.abs(d[1])));
    do {
      first=(j == 0);
      Analytic_problems.rk5na(x,xa,testrk5na,e,d,first,1,1,false);
      j++;
      System.out.println("  " + eightDigit.format(x[0]) + "\t" +
        eightDigit.format(x[1]) + "\t" + eightDigit.format(Math.abs(d[1])));
    } while (j < 4);
  }


  public double b(int n, double x[])
  {
    return x[1];
  }


  public double fxj(int n, int k, double x[])
  {
    return ((k == 0) ? x[1] : 10.0*(1.0-x[0]*x[0])*x[1]-x[0]);
  }
}
```

Output:

```
Results:
    x[0]              x[1]                s
  2.00000000    0.00000000    0.00000000
 -2.01428537   -0.00000000   29.38738341
  2.01428537    0.00000000   58.78843319
 -2.01428537   -0.00000000   88.18948298
  2.01428537    0.00000000  117.59053276
```

Procedure tested:

multistep

Example:

Compute the solution at $x = 1$ and at $x = 10$ of the differential equations
$$dy_1/dx = 0.04(1 - y_1 - y_2) - y_1(10^4 y_2 + 3*10^7 y_1)$$
$$dy_2/dx = 3*10^7 y_1^2$$
with the initial conditions $y_1 = 0$ and $y_2 = 0$ at $x = 0$.

```
package numal;

import java.text.DecimalFormat;
```

```
import numal.*;

public class Test_multistep extends Object implements AP_multistep_methods {

  public static void main(String args[]) {

    int i;
    double xend,hmin,eps;
    boolean first[] = new boolean[1];
    boolean btmp[] = new boolean[2];
    int itmp[] = new int[3];
    double xtmp[] = new double[7];
    double x[] = new double[1];
    double y[] = new double[13];
    double ymax[] = new double[3];
    double d[] = new double[53];
    double jac[][] = new double[3][3];

    Test_multistep testmultistep = new Test_multistep();
    hmin=1.0e-10;
    eps=1.0e-9;
    first[0]=true;
    x[0]=0.0;
    y[1]=y[2]=0.0;
    ymax[1]=0.0001;
    ymax[2]=1.0;
    DecimalFormat fiveDigit = new DecimalFormat("0.00000E0");
    System.out.println("Delivers with hmin = " + fiveDigit.format(hmin) +
                       " and eps = " + fiveDigit.format(eps));
    for (i=1; i<=10; i+=9) {
      xend=i;
      Analytic_problems.multistep(x,xend,y,hmin,5,ymax,eps,first,d,testmultistep,
                                  jac,true,2,btmp,itmp,xtmp);
      System.out.println("   " + fiveDigit.format(y[1]) +
                         "   " + fiveDigit.format(y[2]));
    }
  }


  public void deriv(double f[], int n, double x[], double y[])
  {
    double r;

    f[2]=r=3.0e7*y[1]*y[1];
    f[1]=0.04*(1-y[1]-y[2])-1.0e4*y[1]*y[2]-r;
  }
```

```
public boolean available(int n, double x[], double y[], double jac[][])
{
  double r;

  jac[2][1]=r=6.0e7*y[1];
  jac[1][1] = -0.04-1.0e4*y[2]-r;
  jac[1][2] = -0.04-1.0e4*y[1];
  jac[2][2]=0.0;
  return true;
}


public void out(double h, int k, int n, double x[], double y[])
{
  return;
}
}
```

Output:

```
Delivers with hmin = 1.00000E-10  and eps = 1.00000E-9
   3.07463E-5    3.35095E-2
   1.62339E-5    1.58614E-1
```

Procedure tested:

diffsys

Example:
Solve the system of differential equations

$$y_1' = y_2$$
$$y_2' = y_1 + 2y_4 - \mu_1(y_1+\mu)/s_1 - \mu(y_1-\mu_1)/s_2$$
$$y_3' = y_4$$
$$y_4' = y_3 + 2y_2 - \mu_1 y_3/s_1 - \mu y_3/s_2$$

where

$$s_1 = \{(y_1+\mu)^2 + y_3^2\}, \quad s_2 = \{(y_1-\mu_1)^2 + y_3^2\}, \quad \mu_1 = 1-\mu.$$

This problem arises from the restricted problem of three bodies. When $\mu = 1/82.45$, the solution is a closed orbit with period $t = 6.192169331396$.

```
package numal;

import java.text.DecimalFormat;
import numal.*;


public class Test_diffsys extends Object implements AP_diffsys_methods {

  static int passes,k;
```

```
public static void main(String args[]) {

    int i;
    double xe,tol,h0;
    double y[] = new double[5];
    double s[] = new double[5];
    double x[] = new double[1];

    Test_diffsys testdiffsys = new Test_diffsys();
    System.out.println("Results with DIFFSYS are :\n" +
                       "   K    DER.EV.    Y[1]          Y[3]");
    tol=1.0e-2;
    for (i=1; i<=5; i++) {
      tol *= 1.0e-2;
      passes=k=0;
      x[0]=0.0;
      xe=6.192169331396;
      y[1]=1.2;
      y[2]=y[3]=0.0;
      y[4] = -1.04935750983;
      s[1]=s[2]=s[3]=s[4]=0.0;
      h0=0.2;
      Analytic_problems.diffsys(x,xe,4,y,testdiffsys,tol,tol,s,h0);
    }
}


public void derivative(int n, double x, double y[], double dy[])
{
    double mu,mu1,y1,y2,y3,y4,s1,s2;

    mu=1.0/82.45;
    mu1=1.0-mu;
    passes++;
    y1=y[1];
    y2=dy[1]=y[2];
    y3=y[3];
    y4=dy[3]=y[4];
    s1=(y1+mu)*(y1+mu)+y3*y3;
    s2=(y1-mu1)*(y1-mu1)+y3*y3;
    s1 *= Math.sqrt(s1);
    s2 *= Math.sqrt(s2);
    dy[2]=y1+2.0*y4-mu1*(y1+mu)/s1-mu*(y1-mu1)/s2;
    dy[4]=y3-2.0*y2-mu1*y3/s1-mu*y3/s2;
}


public void output(int n, double x[], double xe, double y[], double s[])
{
    k++;
```

```
    DecimalFormat fiveDigit = new DecimalFormat("0.00000E0");
    if (x[0] >= xe)
      System.out.println("   " + k + "    " + passes + "    " +
                          fiveDigit.format(y[1]) + "    " + fiveDigit.format(y[3]));
  }
}
```

Output:

```
Results with DIFFSYS are :
  K   DER.EV.    Y[1]         Y[3]
  30    2591    1.32036E0    -3.26455E-2
  33    3414    1.20008E0    -5.39061E-5
  37    4213    1.20000E0    -2.36375E-6
  44    4618    1.20000E0    -8.97779E-11
  56    6299    1.20000E0    -8.24296E-11
```

Procedure tested:

ark

Example:
Compute the values of:

(1) $y(1)$ and $y(2)$ of the initial value problem
$$dy/dx = y - 2x/y, \quad y(0) = 1$$
and

(2) $u(0.6,0)$ of the Cauchy problem
$$du/dt = 0.5*du/dx, \quad u(0,x) = exp(-x^2).$$

```
package numal;

import java.text.DecimalFormat;
import numal.*;

public class Test_ark_1 extends Object implements AP_ark_methods {

  public static void main(String args[]) {

    int i;
    double dat1[]={3.0, 3.0, 1.0, 1.0, 1.0e-3, 1.0e-6,
                   1.0e-6, 0.0, 0.0, 0.0, 1.0, 0.5, 1.0/6.0};
    double data[] = new double[15];
    int m[] = new int[1];
    int m0[] = new int[1];
    double t[] = new double[1];
    double te[] = new double[1];
    double y[] = new double[2];

    Test_ark_1 testark = new Test_ark_1();
```

```
    for (i=1; i<=13; i++) data[i]=dat1[i-1];
    t[0]=0.0;
    y[1]=1.0;
    te[0]=1.0;
    m0[0]=m[0]=1;
    Analytic_problems.ark(t,te,m0,m,y,testark,data);
  }


  public void derivative(int m0[], int m[], double t[], double v[])
  {
    v[1] -= 2.0*(t[0])/v[1];
  }


  public void out(int m0[], int m[], double t[], double te[], double y[],
                  double data[])
  {
    DecimalFormat fiveDigit = new DecimalFormat("0.00000E0");
    if (t[0] == te[0]) {
      if (t[0] == 1.0)
        System.out.println("\nProblem 1\n\n" +
                            " x   integration steps   y(computed)    y(exact)");
      System.out.println(" " + (int)t[0] + "              " + (int)data[8] +
                         "               " + fiveDigit.format(y[1]) + "       " +
                         fiveDigit.format(Math.sqrt(2.0*(t[0])+1)));
      te[0] = 2.0;
    }
  }
}
```

Output:

```
Problem 1

x    integration steps    y(computed)    y(exact)
1              38              1.73205E0      1.73205E0
2              56              2.23609E0      2.23607E0


package numal;

import java.text.DecimalFormat;
import numal.*;

public class Test_ark_2 extends Object implements AP_ark_methods {

  public static void main(String args[]) {

    int i;
```

```
   double dat2[]={4.0, 3.0, 0.0, 500.0/3.0, 0.0, -1.0, -1.0,
                  0.0, 0.0, 0.0, 1.0, 0.5, 1.0/6.0, 1.0/24.0};
   double data[] = new double[15];
   int m[] = new int[1];
   int m0[] = new int[1];
   double t[] = new double[1];
   double te[] = new double[1];
   double y[] = new double[2];
   double u[] = new double[301];

   Test_ark_2 testark = new Test_ark_2();
   for (i=1; i<=14; i++) data[i]=dat2[i-1];
   data[3]=Math.sqrt(8.0);
   data[5]=data[3]/data[4];
   m0[0] = 0;
   m[0]=300;
   t[0]=0.0;
   u[150]=1.0;
   for (i=1; i<=150; i++)
     u[i+150]=u[150-i]=Math.exp(-(0.003*i)*(0.003*i));
   te[0]=0.6;
   Analytic_problems.ark(t,te,m0,m,u,testark,data);
 }


public void derivative(int m0[], int m[], double t[], double v[])
{
  int j;
  double v1,v2,v3;

  v2=v[m0[0]];
  m0[0]++;
  m[0]--;
  v3=v[m0[0]];
  for (j=m0[0]; j<=m[0]; j++) {
    v1=v2;
    v2=v3;
    v3=v[j+1];
    v[j]=250.0*(v3-v1)/3.0;
  }
}


public void out(int m0[], int m[], double t[], double te[], double u[],
                double data[])
{
  DecimalFormat fiveDigit = new DecimalFormat("0.00000E0");
  if (Math.abs((t[0])-0.6) < 1.0e-5)
    System.out.println("\n\nProblem 2\n\nderivative " +
      "calls   u(.6,0)(computed)   u(.6,0)exact\n " + data[1]*data[8] +
```

```
            "                  " + fiveDigit.format(u[150]) + "                    " +
         fiveDigit.format(Math.exp(-0.09)));
   }
}
```

Output:

```
Problem 2

derivative calls    u(.6,0)(computed)    u(.6,0)exact
   144.0                9.13933E-1           9.13931E-1
```

Procedure tested:

efrk

Example:

Consider the system of differential equations:

$$dy_1/dx = -y_1 + y_1y_2 + 0.99y_2$$
$$dy_2/dx = -1000*(-y_1 + y_1y_2 + y_2)$$

with the initial conditions $y_1 = 1$ and $y_2 = 0$ at $x = 0$. The solution at $x = 50$ is approximately $y_1 = 0.7658783202487$ and $y_2 = 0.4337103535768$. The following program shows some different calls of the procedure *efrk*.

From these results it appears that calls with *thirdorder* equals true in *efrk* are less advisable.

```
package numal;

import java.text.DecimalFormat;
import numal.*;

public class Test_efrk extends Object implements AP_efrk_methods {

   static int passes;

   public static void main(String args[]) {

      int r,l,i,kk;
      int k[] = new int[1];
      double xe,step;
      double y[] = new double[3];
      double beta[] = new double[7];
      double x[] = new double[1];
      double sigma[] = new double[1];
      double phi[] = new double[1];
      double diameter[] = new double[1];

      Test_efrk testefrk = new Test_efrk();
      System.out.println("The results with EFRK are:\n\n" +
```

```
                            "  R   L   K   DER.EV.       Y[1]            Y[2]");
   phi[0]=Math.PI;
   xe=0.0;
   beta[0]=beta[1]=1.0;
   for (r=1; r<=3; r++)
     for (l=1; l<=3; l++) {
        for (kk=2; kk<=r; kk++) beta[kk]=beta[kk-1]/kk;
        for (i=1; i<=2; i++) {
          step = ((i == 1) ? 1.0 : 0.1);
          passes=k[0]=0;
          x[0]=y[2]=0.0;
          y[1]=1.0;
          testefrk.output(1,2,x,xe,y,sigma,phi,diameter, k,step,r,l);
          Analytic_problems.efrk(x,50.0,1,2,y,sigma,phi,diameter,testefrk,k,step,
                                  r,l,beta,r>=3,1.0e-3);
        }
        System.out.println();
      }
 }


public void derivative(int m0, int m, double x, double y[])
{
   double y1,y2;

   y1=y[1];
   y2=y[2];
   y[1]=(y1+0.99)*(y2-1.0)+0.99;
   y[2]=1000.0*((1.0+y1)*(1.0-y2)-1.0);
   passes++;
}


public void output(int m0, int m, double x[], double xe, double y[],
                   double sigma[], double phi[], double diameter[], int k[],
                   double step, int r, int l)
{
   double s;

   DecimalFormat fiveDigit = new DecimalFormat("0.00000E0");
   s=(-1000.0*y[1]-1001.0+y[2])/2.0;
   sigma[0]=Math.abs(s-Math.sqrt(s*s+10.0*(y[2]-1.0)));
   diameter[0]=2.0*step*Math.abs(1000.0*
              (1.99*y[2]-2.0*y[1]*(1.0-y[2])));
   if (x[0] == 50.0)
     System.out.println("  " + r + "   " + l + "   " + k[0] + "\t" + passes + "\t"
                        + fiveDigit.format(y[1]) + "   " + fiveDigit.format(y[2]));
 }
}
```

Output:

```
The results with EFRK are:
```

R	L	K	DER.EV.	Y[1]	Y[2]
1	1	237	474	7.65813E-1	4.33689E-1
1	1	501	1002	7.65848E-1	4.33701E-1
1	2	52	156	7.65571E-1	4.33615E-1
1	2	501	1503	7.65848E-1	4.33701E-1
1	3	52	208	7.65571E-1	4.33615E-1
1	3	500	2000	7.65849E-1	4.33701E-1
2	1	3317	9951	7.65878E-1	4.33710E-1
2	1	1050	3150	7.65878E-1	4.33710E-1
2	2	174	696	7.65878E-1	4.33710E-1
2	2	501	2004	7.65878E-1	4.33709E-1
2	3	57	285	7.65881E-1	4.30817E-1
2	3	501	2505	7.65878E-1	4.33710E-1
3	1	7010	28040	7.65878E-1	4.33710E-1
3	1	3255	13020	7.65878E-1	4.33710E-1
3	2	949	4745	7.65878E-1	4.33712E-1
3	2	1384	6920	7.65860E-1	4.52099E-1
3	3	918	5508	7.65877E-1	4.34709E-1
3	3	1147	6882	7.65859E-1	4.47259E-1

Procedure tested:

efsirk

Example:

Solve the differential equation:

$$dy/dx = -e^x\{y - ln(x)\} + 1/x$$

with the initial value $y(0.01) = ln(0.01)$ over the two ranges [0.01,0.4] and [0.4,8]. The analytic solution is $y(x) = ln(x)$.

The above equation is written in autonomous form (taking $y^{(2)}(x) = x$) as

$$dy^{(1)}/dx = - exp(y^{(2)})\{y^{(1)} - ln(y^{(2)})\} + 1/y^{(2)}$$
$$dy^{(2)}/dx = 1.$$

The Jacobian matrix J_n for this set of equations has elements

$$J_n^{(1,1)} = -exp(y^{(2)}), \quad J_n^{(1,2)} = -\{y^{(1)} - ln(y^{(2)}) - 1/y^{(2)}\}exp(y^{(2)}) - 1/(y^{(2)})^2$$

$J_n^{(2,1)} = J_n^{(2,2)} = 0$, and its eigenvalue with largest absolute value is $\delta_n = -exp(y^{(2)})$.

```java
package numal;

import java.text.DecimalFormat;
import numal.*;

public class Test_efsirk extends Object implements AP_efsirk_methods {

  static double lnx;

  public static void main(String args[]) {

    int n[] = new int[1];
    double xe;
    double x[] = new double[1];
    double delta[] = new double[1];
    double y[] = new double[3];
    double j[][] = new double[3][3];

    Test_efsirk testefsirk = new Test_efsirk();
    System.out.println("EFSIRK delivers:");
    xe=0.4;
    x[0]=0.01;
    y[1]=Math.log(0.01);
    y[2]=x[0];
    Analytic_problems.efsirk(x,xe,2,y,delta,testefsirk,j,n,1.0e-2,1.0e-2,0.005,
                            1.5,false);
    xe=8.0;
    x[0]=0.01;
    y[1]=Math.log(0.01);
    y[2]=x[0];
    Analytic_problems.efsirk(x,xe,2,y,delta,testefsirk,j,n,1.0e-2,1.0e-2,0.005,
                            1.5,false);
  }


  public void derivative(int m, double y[], double delta[])
  {
    double y2;

    y2=y[2];
    delta[0] = -Math.exp(y2);
    lnx=Math.log(y2);
    y[1]=(y[1]-lnx)*delta[0]+1.0/y2;
    y[2]=1.0;
  }
```

```
public void jacobian(int m, double j[][], double y[], double delta[])
{
  double y2;

  y2=y[2];
  j[1][1]=delta[0];
  j[1][2]=(y[1]-lnx-1.0/y2)*delta[0]-1.0/(y2*y2);
  j[2][1]=j[2][2]=0.0;
}


public void output(double x[], double xe, int m, double y[], double delta[],
                   double j[][], int n[])
{
  double y1;

  DecimalFormat oneDigit = new DecimalFormat("0.0E0");
  DecimalFormat twoDigit = new DecimalFormat("0.00");
  DecimalFormat fiveDigit = new DecimalFormat("0.00000");
  if (x[0] == xe) {
    y1=y[1];
    lnx=Math.log(x[0]);
    System.out.println("\n N = " + n[0] + "    X = " + x[0] + "     Y(X) = " +
      fiveDigit.format(y1) + "   DELTA = " + twoDigit.format(delta[0]) +
      "\n ABS.ERR. = " + oneDigit.format(Math.abs(y1-lnx)) + "    REL.ERR. = " +
      oneDigit.format(Math.abs((y1-lnx)/lnx)));
  }
 }
}
```

Output:

```
EFSIRK delivers:

N = 10    X = 0.4    Y(X) = -0.91093    DELTA = -1.44
ABS.ERR. = 5.4E-3    REL.ERR. = 5.8E-3


N = 96    X = 8.0    Y(X) = 2.07797    DELTA = -2967.79
ABS.ERR. = 1.5E-3    REL.ERR. = 7.1E-4
```

Procedure tested:

eferk

Example:
Consider the system of differential equations:
$$dy_1/dx = -y_1 + y_1 y_2 + 0.99y_2$$
$$dy_2/dx = -1000*(-y_1 + y_1 y_2 + y_2)$$
with the initial conditions $y_1 = 1$ and $y_2 = 0$ at $x = 0$. The solution at $x = 50$ is

approximately $y_1 = 0.7658783202487$ and $y_2 = 0.4337103535768$. The following program shows some different calls of the procedure *eferk*.

```java
package numal;

import java.text.DecimalFormat;
import numal.*;

public class Test_eferk extends Object implements AP_eferk_methods {

  static int passes,pasjac;

  public static void main(String args[]) {

    int i;
    int k[] = new int[1];
    double xe,phi,tol;
    double y[] = new double[3];
    double j[][] = new double[3][3];
    double x[] = new double[1];
    double sigma[] = new double[1];

    Test_eferk testeferk = new Test_eferk();
    System.out.println("The results with EFERK are:\n\n" +
                    " K     DER.EV.   JAC.EV.    Y[1]           Y[2]");
    phi=4.0*Math.atan(1.0);
    tol=1.0;
    for (i=1; i<=4; i++) {
        passes=pasjac=0;
        x[0]=y[2]=0.0;
        y[1]=1.0;
        xe=50.0;
        Analytic_problems.eferk(x,xe,2,y,sigma,phi,testeferk,j,k,1,true,tol,tol,
                          1.0e-6,50.0,false);
        tol *= 1.0e-1;
    }
  }


  public void derivative(int m, double y[])
  {
    double y1,y2;

    y1=y[1];
    y2=y[2];
    y[1]=(y1+0.99)*(y2-1.0)+0.99;
    y[2]=1000.0*((1.0+y1)*(1.0-y2)-1.0);
    passes++;
  }
```

```java
public void jacobian(int m, double j[][], double y[], double sigma[])
{
  j[1][1]=y[2]-1.0;
  j[1][2]=0.99+y[1];
  j[2][1]=1000.0*(1.0-y[2]);
  j[2][2] = -1000.0*(1.0+y[1]);
  sigma[0]=Math.abs(j[2][2]+j[1][1]-Math.sqrt((j[2][2]-j[1][1])*
          (j[2][2]-j[1][1])+4.0*j[2][1]*j[1][2]))/2.0;
  pasjac++;
}


public void output(double x[], double xe, int m, double y[], double j[][],
                  int k[])
{
  DecimalFormat fiveDigit = new DecimalFormat("0.00000E0");
   if (x[0] == 50.0)
      System.out.println(" " + k[0] + "\t" + passes + "\t" + pasjac + "\t" +
                          fiveDigit.format(y[1]) + "\t" + fiveDigit.format(y[2]));
  }
}
```

Output:

```
The results with EFERK are:
```

K	DER.EV.	JAC.EV.	Y[1]	Y[2]
93	186	93	7.65883E-1	4.28753E-1
105	210	105	7.65878E-1	4.33570E-1
147	294	147	7.65878E-1	4.33709E-1
266	532	266	7.65878E-1	4.33710E-1

Procedure tested:

liniger1vs

Example:

Consider the system of differential equations:
$$dy_1/dx = -y_1 + y_1y_2 + 0.99y_2$$
$$dy_2/dx = -1000*(-y_1 + y_1y_2 + y_2)$$
with the initial conditions $y_1 = 1$ and $y_2 = 0$ at $x = 0$. The solution at $x = 50$ is approximately $y_1 = 0.7658783202487$ and $y_2 = 0.4337103535768$. The following program shows some integration of this problem with variable and constant stepsizes.

```java
package numal;

import java.text.DecimalFormat;
```

```
import numal.*;

public class Test_liniger1vs extends Object implements AP_liniger1vs_methods {

  public static void main(String args[]) {

    int i,itmax;
    double reta;
    double y[] = new double[3];
    double info[] = new double[10];
    double x[] = new double[1];
    double sigma[] = new double[1];
    double j[][] = new double[3][3];

    Test_liniger1vs testliniger1vs = new Test_liniger1vs();
    System.out.println("The results with LINIGER1VS are:\n");
    reta=1.0;
    for (i=1; i<=3; i++) {
        reta *= 1.0e-2;
        x[0]=y[2]=0.0;
        y[1]=1.0;
        Analytic_problems.liniger1vs(x,50.0,2,y,sigma,testliniger1vs,j,10,0.1,
                            50.0,reta,reta,info);
    }
    System.out.println();
    reta = -1.0;
    for (i=1; i<=3; i++) {
        reta *= 1.0e-2;
        x[0]=y[2]=0.0;
        y[1]=1.0;
        Analytic_problems.liniger1vs(x,50.0,2,y,sigma,testliniger1vs,j,10,0.1,
                            1.0,reta,reta,info);
    }
  }


  public void derivative(int m, double a[], double sigma[])
  {
    double a1,a2;

    a1=a[1];
    a2=a[2];
    a[1]=(a1+0.99)*(a2-1.0)+0.99;
    a[2]=1000.0*((1.0+a1)*(1.0-a2)-1.0);
  }
```

```
public void jacobian(int m, double j[][], double y[], double sigma[])
{
  j[1][1]=y[2]-1.0;
  j[1][2]=0.99+y[1];
  j[2][1]=1000.0*(1.0-y[2]);
  j[2][2] = -1000.0*(1.0+y[1]);
  sigma[0]=Math.abs(j[2][2]+j[1][1]-Math.sqrt((j[2][2]-j[1][1])*
          (j[2][2]-j[1][1])+4.0*j[2][1]*j[1][2]))/2.0;
}


public void output(double x[], double xe, int m, double y[], double sigma[],
                   double j[][], double info[])
{
  DecimalFormat fiveDigit = new DecimalFormat("0.00000E0");
  DecimalFormat oneDigit = new DecimalFormat("0.E0");
   if (x[0] == 50.0)
      System.out.println(" " + (int)info[1] + "\t" + (int)info[2] + "\t" +
                        (int)info[3] + "\t" + (int)info[4] + "\t" + (int)info[5] +
                        "\t" + (int)info[6] + "\t" + oneDigit.format(info[9]) +
                        "\t" + fiveDigit.format(y[1])+ "\t" +
                        fiveDigit.format(y[2]));
  }
}
```

Output:

```
The results with LINIGER1VS are:

17      21      8       2   0       2   2E-2    7.72017E-1  4.35672E-1
13      25      23      2   0       2   2E-2    7.67414E-1  4.34202E-1
105     210     105     2   0       2   2E-2    7.66027E-1  4.33758E-1

50      52      1       0   50      2   0E0     7.66670E-1  4.33081E-1
50      104     3       0   50      3   0E0     7.66183E-1  4.33811E-1
50      152     12      0   50      4   0E0     7.66185E-1  4.33809E-1
```

Procedure tested:

liniger2

Example:

Consider the system of differential equations:
$$dy_1/dx = -y_1 + y_1y_2 + 0.99y_2$$
$$dy_2/dx = -1000*(-y_1 + y_1y_2 + y_2)$$
with the initial conditions $y_1 = 1$ and $y_2 = 0$ at $x = 0$. The solution at $x = 50$ is approximately $y_1 = 0.7658783202487$ and $y_2 = 0.4337103535768$. The following program shows some different calls of *liniger2*.

```
package numal;

import java.text.DecimalFormat;
import numal.*;

public class Test_liniger2_1 extends Object implements AP_liniger2_methods {

  static int passes,pasjac;

  public static void main(String args[]) {

    int i,itmax;
    int k[] = new int[1];
    double step;
    double y[] = new double[3];
    double j[][] = new double[3][3];
    double x[] = new double[1];
    double sigma1[] = new double[1];
    double sigma2[] = new double[1];

    Test_liniger2_1 testliniger2 = new Test_liniger2_1();
    System.out.println("The results with LINIGER2 (second order)" +
      " are:\n K    DER.EV.  JAC.EV.    Y[1]           Y[2]");
    for (i=1; i<=2; i++) {
      step = (i == 1) ? 10.0 : 1.0;
      for (itmax=1; itmax<=3; itmax += 2) {
        passes=pasjac=0;
        x[0]=y[2]=0.0;
        y[1]=1.0;
        sigma2[0]=0.0;
        Analytic_problems.liniger2(x,50.0,2,y,sigma1,sigma2,testliniger2,j,k,itmax,
                         step,1.0e-4,1.0e-4);
      }
    }
  }

  public double f(int m, double y[], int i, double sigma1[], double sigma2[])
  {
    if (i == 1)
      return (y[1]+0.99)*(y[2]-1.0)+0.99;
    else {
      passes++;
      return 1000.0*((1.0+y[1])*(1.0-y[2])-1.0);
    }
  }
}
```

```
  public void jacobian(int m, double j[][], double y[], double sigma1[],
                       double sigma2[])
  {
    j[1][1]=y[2]-1.0;
    j[1][2]=0.99+y[1];
    j[2][1]=1000.0*(1.0-y[2]);
    j[2][2] = -1000.0*(1.0+y[1]);
    sigma1[0]=Math.abs(j[2][2]+j[1][1]-Math.sqrt((j[2][2]-j[1][1])*
              (j[2][2]-j[1][1])+4.0*j[2][1]*j[1][2]))/2.0;
    pasjac++;
  }



  public boolean evaluate(int i)
  {
    return (i == 1);
  }



  public void output(double x[], double xe, int m, double y[],double sigma1[],
                     double sigma2[], double j[][], int k[])
  {
    DecimalFormat fiveDigit = new DecimalFormat("0.00000E0");
    if (x[0] == 50.0)
      System.out.println(" " + k[0] + "\t" + passes + "\t" + pasjac + "\t" +
                         fiveDigit.format(y[1]) + "\t" + fiveDigit.format(y[2]));
  }
}
```

Output:

```
The results with LINIGER2 (second order) are:
  K     DER.EV.   JAC.EV.     Y[1]           Y[2]
  5       5         5      7.66392E-1     4.34219E-1
  5      15         5      7.65756E-1     4.33671E-1
 50      50        50      7.65884E-1     4.33716E-1
 50     101        50      7.65877E-1     4.33710E-1


package numal;

import java.text.DecimalFormat;
import numal.*;

public class Test_liniger2_2 extends Object implements AP_liniger2_methods {

  static int passes,pasjac;

  public static void main(String args[]) {
```

```
      int i,itmax;
      int k[] = new int[1];
      double step;
      double y[] = new double[3];
      double j[][] = new double[3][3];
      double x[] = new double[1];
      double sigma1[] = new double[1];
      double sigma2[] = new double[1];

      Test_liniger2_2 testliniger2 = new Test_liniger2_2();
      System.out.println("The results with LINIGER2 (third order)" +
        " are:\n K   DER.EV.  JAC.EV.    Y[1]            Y[2]");
      for (i=1; i<=2; i++) {
        step = (i == 1) ? 10.0 : 1.0;
        for (itmax=1; itmax<=3; itmax += 2) {
          passes=pasjac=0;
          x[0]=y[2]=0.0;
          y[1]=1.0;
          sigma2[0]=0.0;
          Analytic_problems.liniger2(x,50.0,2,y,sigma1,sigma2,testliniger2,j,k,itmax,
                              step,1.0e-4,1.0e-4);
        }
      }
}


public double f(int m, double y[], int i, double sigma1[], double sigma2[])
{
  if (i == 1)
    return (y[1]+0.99)*(y[2]-1.0)+0.99;
  else {
    passes++;
    return 1000.0*((1.0+y[1])*(1.0-y[2])-1.0);
  }
}


public void jacobian(int m, double j[][], double y[], double sigma1[],
                  double sigma2[])
{
  j[1][1]=y[2]-1.0;
  j[1][2]=0.99+y[1];
  j[2][1]=1000.0*(1.0-y[2]);
  j[2][2] = -1000.0*(1.0+y[1]);
  sigma1[0]=Math.abs(j[2][2]+j[1][1]-Math.sqrt((j[2][2]-j[1][1])*
          (j[2][2]-j[1][1])+4.0*j[2][1]*j[1][2]))/2.0;
  pasjac++;
}
```

```java
public boolean evaluate(int i)
{
  return true;
}


public void output(double x[], double xe, int m, double y[],double sigma1[],
                   double sigma2[], double j[][], int k[])
{
  DecimalFormat fiveDigit = new DecimalFormat("0.00000E0");
  if (x[0] == 50.0)
    System.out.println(" " + k[0] + "\t" + passes + "\t" + pasjac + "\t" +
                       fiveDigit.format(y[1]) + "\t" + fiveDigit.format(y[2]));
}
}
```

Output:

```
The results with LINIGER2 (third order) are:
K      DER.EV.    JAC.EV.      Y[1]            Y[2]
5       5          5        7.66392E-1      4.34219E-1
5       15         15       7.65882E-1      4.33712E-1
50      50         50       7.65884E-1      4.33716E-1
50      101        101      7.65879E-1      4.33711E-1
```

Procedure tested:

gms

Example:
Solve the system of differential equations:
$$dy^{(1)}/dx = -1000y^{(1)}(y^{(1)}+y^{(2)}-1.999987)$$
$$dy^{(2)}/dx = -2500y^{(2)}(y^{(1)}+y^{(2)}-2)$$
with the initial value $y^{(1)}(0) = y^{(2)}(0) = 1$ over the range [0,50]. The Jacobian matrix
J_n for this set of equations has elements
$$J_n^{(1,1)} = 1999.987 - 1000(2y^{(1)}+y^{(2)}), \quad J_n^{(1,2)} = -1000y^{(2)}$$
$$J_n^{(2,1)} = -2500y^{(2)}, \quad J_n^{(2,2)} = 2500(2-y^{(1)}-y^{(2)}).$$

```java
package numal;

import java.text.DecimalFormat;
import numal.*;

public class Test_gms extends Object implements AP_gms_methods {

  public static void main(String args[]) {

    int n[] = new int[1];
    int jev[] = new int[1];
```

```
    int lu[] = new int[1];
    double x[] = new double[1];
    double y[] = new double[3];
    double delta[] = new double[1];

    Test_gms testgms = new Test_gms();
    System.out.println("The results with GMS are:\n");
    y[1]=y[2]=1.0;
    x[0]=0.0;
    delta[0]=0.0;
    Analytic_problems.gms(x,50.0,2,y,0.01,0.001,0.5,delta,testgms,1.0e-5,1.0e-5,
                          n,jev,lu,0,false);
}


public void derivative(int r, double y[], double delta[])
{
  double y1,y2;

  y1=y[1];
  y2=y[2];
  y[1] = -1000.0*y1*(y1+y2-1.999987);
  y[2] = -2500.0*y2*(y1+y2-2.0);
}


public void jacobian(int r, double j[][], double y[], double delta[])
{
  double y1,y2;

  y1=y[1];
  y2=y[2];
  j[1][1] = 1999.987-1000.0*(2.0*y1+y2);
  j[1][2] = -1000.0*y1;
  j[2][1] = -2500.0*y2;
  j[2][2] = 2500.0*(2.0-y1-2.0*y2);
}


public void out(double x[], double xe, int r, double y[], double delta[], int n[],
                int jev[], int lu[])
{
  double ye1,ye2;

  DecimalFormat fiveDigit = new DecimalFormat("0.00000E0");
  DecimalFormat twoDigit = new DecimalFormat("0.00E0");
  if (x[0] == 50.0) {
    ye1=0.5976546988;
    ye2=1.4023434075;
    System.out.println(" X = " + (int)x[0] + "     N = " + n[0] + "     JEV = " +
```

```
        jev[0] + "    LU = " + lu[0] + "\n\n Y1 = " + fiveDigit.format(y[1]) +
        "    REL.ERR. = " + twoDigit.format(Math.abs((y[1]-ye1)/ye1)) +
        "\n Y2 = " + fiveDigit.format(y[2]) + "       REL.ERR. = " +
        twoDigit.format(Math.abs((y[2]-ye2)/ye2)));
    }
  }
}
```

Output:

```
The results with GMS are:

 X = 50      N = 109     JEV = 3     LU = 12

Y1 = 5.97655E-1        REL.ERR. = 6.02E-8
Y2 = 1.40234E0         REL.ERR. = 2.54E-8
```

Procedure tested:

impex

Example:

Consider the autonomous system of differential equations:

$$dy_1/dx = 0.2(y_2 - y_1)$$
$$dy_2/dx = 10y_1 - (60 - y_3/8)y_2 + y_3/8$$
$$dy_3/dx = 1$$

with initial conditions $y_1=y_2=y_3=0$ at $x=0$. The solution at several points in the interval $[0,400]$ may be obtained by the following program. (The solution at $x=400$ is $y_1=22.24222011$ and $y_2=27.11071335$)

```
package numal;

import java.text.DecimalFormat;
import numal.*;

public class Test_impex extends Object implements AP_impex_methods {

  static int nfe,nje,point;
  static double print[] = new double[6];

  public static void main(String args[]) {

    boolean fail[] = new boolean[1];
    int n,i,it;
    double t,tend,eps,hmax,l,h2,n1,n2;
    double y[] = new double[4];
    double sw[] = new double[4];
    double f1[] = new double[4];
    double f2[] = new double[4];
```

```
   double z[] = new double[4];
   double x[] = new double[4];

   Test_impex testimpex = new Test_impex();
   DecimalFormat fiveDigit = new DecimalFormat("0.00000E0");
   System.out.println("The results with IMPEX are:\n");
   n=3;
   nje=nfe=0;
   t=0.0;
   tend=400.0;
   eps=1.0e-5;
   hmax=400.0;
   y[1]=y[2]=y[3]=0.0;
   sw[1]=sw[2]=sw[3]=1.0;
   print[1]=0.1;  print[2]=1.0;  print[3]=10.0;
   print[4]=100.0; print[5]=400.0;
   Basic.dupvec(1,n,0,z,y);
   for (i=1; i<=n; i++)
     x[i] = (y[i] == 0.0) ? eps : (1.0+eps)*y[i];
   n1=Math.sqrt(Basic.vecvec(1,n,0,x,x))*eps;
   testimpex.deriv(t,x,f1,n);
   for (it=1; it<=5; it++) {
     testimpex.deriv(t,z,f2,n);
     Basic.elmvec(1,n,0,f2,f1,-1.0);
     n2=n1/Math.sqrt(Basic.vecvec(1,n,0,f2,f2));
     Basic.dupvec(1,n,0,z,x);
     Basic.elmvec(1,n,0,z,f2,n2);
   }
   testimpex.deriv(t,z,f2,n);
   Basic.elmvec(1,n,0,f2,f1,-1.0);
   l=Math.sqrt(Basic.vecvec(1,n,0,f2,f2))/n1;
   h2=Math.pow(eps*320.0,1.0/5.0)/(4.0*l);
   System.out.println("EPS = " + fiveDigit.format(eps) +
     "\nInterval of integration = (" + (int)t + "," + (int)tend +
     ")\nMaximally allowed stepsize = " + fiveDigit.format(hmax) +
     "\n\nLipschconst = " + fiveDigit.format(l) + "\nStarting stepsize = " +
     fiveDigit.format(h2) + "\nFunctional eval = " + nfe +
     "\n\n   X     ERROR     Y[1]           Y[2]          NFE     NJE\n");
   Analytic_problems.impex(n,t,tend,y,testimpex,h2,hmax,false,eps,sw,fail);
   System.out.println("\nNumber of functional evaluations = " + nfe +
                   "\nNumber of Jacobian evaluations   = " + nje);
}


public void deriv(double t, double y[], double f1[], int n)
{
  nfe++;
  f1[1]=0.2*(y[2]-y[1]);
  f1[2]=10.0*y[1]-(60.0-0.125*y[3])*y[2]+0.125*y[3];
  f1[3]=1.0;
}
```

```java
public boolean available(double t, double y[], double a[][], int n)
{
  nje++;
  a[1][1] = -0.2;
  a[1][2] = 0.2;
  a[1][3]=a[3][1]=a[3][2]=a[3][3]=0.0;
  a[2][1]=10.0;
  a[2][2]=0.125*y[3]-60.0;
  a[2][3]=0.125*(1.0+y[2]);
  return true;
}


public void update(double sw[], double r1[], int n)
{
  int i;
  double s1,s2;

  for (i=1; i<=n; i++) {
    s1=1.0/sw[i];
    s2=Math.abs(r1[i]);
    if (s1 < s2) sw[i]=1.0/s2;
  }
}


public void control(double tp[], double t, double h, double hnew, double y[][],
                    double err[], int n, double tend)
{
  int i;
  double s,s2,s3,s4;
  double c[] = new double[6];
  double x[] = new double[n+1];

  DecimalFormat fiveDigit = new DecimalFormat("0.00000E0");
  DecimalFormat oneDigit = new DecimalFormat("0.0E0");
  while (true) {
    s=(t-tp[0])/h;
    s2=s*s;
    s3=s2*s;
    s4=s3*s;
    c[3]=(s2-s)/2.0;
    c[4] = -s3/6.0+s2/2.0-s/3.0;
    c[5]=s4/24.0-s3/4.0+11.0*s2/24.0-s/4.0;
    for (i=1; i<=n; i++)
      x[i]=y[1][i]-s*y[2][i]+c[3]*y[3][i]+c[4]*y[4][i]+c[5]*y[5][i];
    System.out.println(" " + tp[0] + "\t" + oneDigit.format(err[3]) + "\t" +
                       fiveDigit.format(x[1]) + "\t" + fiveDigit.format(x[2]) +
                       "\t" + nfe + "\t" + nje);
    if (tp[0] >= tend) break;
```

```
      point++;
      tp[0] = print[point];
      if (tp[0] > t) break;
    }
  }
}
```

Output:

```
The results with IMPEX are:


EPS = 1.00000E-5
Interval of integration = (0,400)
Maximally allowed stepsize = 4.00000E2


Lipschconst = 6.00334E1
Starting stepsize = 1.32001E-3
Functional eval = 7
```

X	ERROR	Y[1]	Y[2]	NFE	NJE
0.0	0.0E0	0.00000E0	0.00000E0	7	0
0.1	6.3E-7	1.49614E-6	1.74014E-4	48	4
1.0	1.5E-6	1.91042E-4	2.08361E-3	87	8
10.0	8.7E-7	1.30148E-2	2.34488E-2	121	9
100.0	1.3E-5	3.06303E-1	3.27552E-1	227	13
400.0	1.3E-5	2.22407E1	2.71091E1	559	30

```
Number of functional evaluations = 559
Number of Jacobian evaluations   = 30
```

Procedure tested:

modifiedtaylor

Example:
Solve the differential equation:
$$Du(t) = -e^t\{u(t) - ln(t)\} + 1/t \tag{1}$$
where $D=d/dt$, with initial condition $u(0.01) = ln(0.01)$ (the analytic solution is $u(t) = ln(t)$) over the ranges $[0.01,e]$ and $[e,e^2]$, using a fourth order modified Taylor series method with third order exact stability polynomial with coefficients $ß_j = 1/j!$ ($j=0,...,3$) $ß_4 = 0.018455702$ (whose range constant is $ß(4) = 6.025$). The required derivatives of $u(t)$ are given by
$$D^2u(t) = e^t\{ln(t) + 1/t - u(t) - Du(t)\} - 1/t^2$$
$$D^3u(t) = e^t\{ln(t) + 2/t - u(t) - 2Du(t) - D^2u(t) - 1/t^2\} + 2/t^3$$
$$D^4u(t) = D^3u(t) - 2(1 + 3/t)/t^3 + e^t[\{1 - (2 - 2/t)/t\}/t - Du(t) - 2D^2u(t) - D^3u(t)].$$
The spectral radius with respect to the eigenvalues lying in the left half-plane of the Jacobian matrix derived from equation (1) is simply e^t.

```java
package numal;

import java.text.DecimalFormat;
import numal.*;

public class Test_modifiedtaylor extends Object implements AP_modifiedtaylor_methods
{
  static double expt,lnt,c0,c1,c2,c3;

  public static void main(String args[]) {

    int j,order,accuracy;
    int k[] = new int[1];
    double te;
    double u[] = new double[1];
    double t[] = new double[1];
    double eta[] = new double[1];
    double rho[] = new double[1];
    double data[] =new double[5];
    double xtmp[] = new double[8];

    Test_modifiedtaylor testmodifiedtaylor = new Test_modifiedtaylor();
    System.out.println("The results with MODIFIEDTAYLOR are:");
    te=0.0;
    order=4;
    accuracy=3;
    data[0]=6.025;   data[1]=1.0; data[2]=0.5;   data[3]=1.0/6.0;
    data[4]=0.018455702;
    t[0]=u[0]=1.0e-2;
    k[0]=0;
    for (j=1; j<=2; j++) {
      te = (j == 1) ? Math.exp(1.0) : te*te;
      Analytic_problems.modifiedtaylor(t,te,0,0,u,testmodifiedtaylor,1.0e-4,k,
                          order,accuracy,data,1.5,1,eta,rho,xtmp);
    }
  }


  public void derivative(double t[], int m0, int m, int i, double a[])
  {
    if (i == 1) {
      expt=Math.exp(t[0]);
      lnt=Math.log(t[0]);
      c0=a[0];
      c1 = a[0] = -expt*c0+1.0/t[0]+expt*lnt;
    }
    if (i == 2) c2=a[0]=expt*(lnt+1.0/t[0]-c0-c1)-1.0/t[0]/t[0];
    if (i == 3)
      c3=a[0]=expt*(lnt+2.0/t[0]-c0-2.0*c1-c2-1.0/t[0]/t[0])+2.0/t[0]/t[0]/t[0];
    if (i == 4)
```

```
    a[0]=c3-2.0*(1.0+3.0/t[0])/t[0]/t[0]/t[0]+
        expt*((1.0-(2.0-2.0/t[0])/t[0])/t[0]-c1-c2*2.0-c3);
}


public double sigma(double t[], int m0, int m)
{
  return Math.exp(t[0]);
}


public double aeta(double t[], int m0, int m)
{
  return 1.0e-5;
}


public double reta(double t[], int m0, int m)
{
  return 1.0e-4;
}


public void out(double t[], double te, int m0, int m, double u[], int k[],
              double eta[], double rho[])
{
  DecimalFormat fiveDigit = new DecimalFormat("0.00000E0");
  if (t[0] == te)
    System.out.println("\nNumber of steps: " + k[0] + "\nSolution:  T = " +
              fiveDigit.format(t[0]) + "    U(T) = " + fiveDigit.format(u[0]));
}
}
```

Output:

```
The results with MODIFIEDTAYLOR are:

Number of steps: 46
Solution:  T = 2.71828E0    U(T) = 1.00003E0

Number of steps: 424
Solution:  T = 7.38906E0    U(T) = 2.00000E0
```

Procedure tested:

eft

Example:
Solve the differential equation:
$$Du(t) = -e^t\{u(t) - ln(t)\} + 1/t \tag{1}$$
where $D=d/dt$, with initial condition $u(0.01) = ln(0.01)$ (the analytic solution is $u(t) = ln(t)$) over the ranges $[0.01, e]$ and $[e, e^2]$, using a third order, first order consistent, exponentially fitted Taylor series method. The equation is solved eight times with different absolute and relative tolerances. The derivatives of (1) are as described in the test program of *modifiedtaylor*.

```
package numal;

import java.text.DecimalFormat;
import numal.*;

public class Test_eft_1 extends Object implements AP_eft_methods {

  static double expt,lnt,u0,u1,u2,accuracy;

  public static void main(String args[]) {

    int j,l;
    int k[] = new int[1];
    double te,te1,te2;
    double u[] = new double[1];
    double t[] = new double[1];
    double eta[] = new double[1];
    double rho[] = new double[1];
    double hs[] = new double[1];

    Test_eft_1 testeft = new Test_eft_1();
    DecimalFormat oneDigit = new DecimalFormat("0.0E0");
    System.out.println("The results with EFT are:\n" +
                " K\tU(TE1)\t\tK\tU(TE2)\t\tRETA");
    te1=Math.exp(1.0);
    te2=Math.exp(2.0);
    accuracy=1.0;
    for (j=1; j<=4; j++) {
      accuracy *= 1.0e-1;
      t[0]=0.01;
      u[0]=Math.log(t[0]);
      k[0]=0;
      hs[0]=0.0;
      for (l=1; l<=2; l++) {
        te = (l == 1) ? te1 : te2;
        Analytic_problems.eft(t,te,0,0,u,testeft,Math.PI,k,1.5,2,eta,rho,1.0e-4,hs);
      }
```

```
        System.out.println(oneDigit.format(accuracy));
    }
  }

  public void derivative(double t[], int m0, int m, int i, double u[])
  {
    if (i == 1) {
      expt=Math.exp(t[0]);
      lnt=Math.log(t[0]);
      u0=u[0];
      u1=u[0]=expt*(lnt-u0)+1.0/t[0];
    } else if (i == 2)
      u2=u[0]=expt*(lnt-u0-u1+1.0/t[0])-1.0/t[0]/t[0];
    else
      u[0]=expt*(lnt-u0-2.0*u1-u2+2.0/t[0]-1.0/t[0]/t[0])+2.0/t[0]/t[0]/t[0];
  }


  public double sigma(double t[], int m0, int m)
  {
    return Math.exp(t[0]);
  }


  public double diameter(double t[], int m0, int m)
  {
    return 2.0*Math.exp(2.0*t[0]/3.0);
  }


  public double aeta(double t[], int m0, int m)
  {
    return accuracy/10.0;
  }


  public double reta(double t[], int m0, int m)
  {
    return accuracy;
  }


  public void out(double t[], double te, int m0, int m, double u[], int k[],
                  double eta[], double rho[])
  {
    DecimalFormat sixDigit = new DecimalFormat("0.000000E0");
    if (t[0] == te)
      System.out.print(k[0] + "\t" + sixDigit.format(u[0]) + "\t");
  }
}
```

Output:

```
The results with EFT are:
  K       U(TE1)       K       U(TE2)       RETA
 15       1.003845E0  42       2.000076E0  1.0E-1
 22       1.001211E0  52       2.000066E0  1.0E-2
 36       1.000109E0  92       2.000020E0  1.0E-3
 56       1.000045E0  171      2.000000E0  1.0E-4
```

```java
package numal;

import java.text.DecimalFormat;
import numal.*;

public class Test_eft_2 extends Object implements AP_eft_methods {

  static double expt,lnt,u0,u1,u2,accuracy;

  public static void main(String args[]) {
    int j,l;
    int k[] = new int[1];
    double te,te1,te2;
    double u[] = new double[1];
    double t[] = new double[1];
    double eta[] = new double[1];
    double rho[] = new double[1];
    double hs[] = new double[1];

    Test_eft_2 testeft = new Test_eft_2();
    DecimalFormat oneDigit = new DecimalFormat("0.0E0");
    System.out.println("Results of EFT " +
                       "with relaxed accuracy conditions for t > 3 :\n" +
                       " K\tU(TE1)\t\tK\tU(TE2)\t\tRETA");
    te1=Math.exp(1.0);
    te2=Math.exp(2.0);
    accuracy=1.0;
    for (j=1; j<=4; j++) {
      accuracy *= 1.0e-1;
      t[0]=0.01;
      u[0]=Math.log(t[0]);
      k[0]=0;
      hs[0]=0.0;
      for (l=1; l<=2; l++) {
        te = (l == 1) ? te1 : te2;
        Analytic_problems.eft(t,te,0,0,u,testeft,Math.PI,k,1.5,2,eta,rho,1.0e-4,hs);
      }
      System.out.println(oneDigit.format(accuracy));
    }
  }
```

```
public void derivative(double t[], int m0, int m, int i, double u[])
{
  if (i == 1) {
    expt=Math.exp(t[0]);
    lnt=Math.log(t[0]);
    u0=u[0];
    u1=u[0]=expt*(lnt-u0)+1.0/t[0];
  } else if (i == 2)
    u2=u[0]=expt*(lnt-u0-u1+1.0/t[0])-1.0/t[0]/t[0];
  else
    u[0]=expt*(lnt-u0-2.0*u1-u2+2.0/t[0]-1.0/t[0]/t[0])+2.0/t[0]/t[0]/t[0];
}


public double sigma(double t[], int m0, int m)
{
  return Math.exp(t[0]);
}


public double diameter(double t[], int m0, int m)
{
  return 2.0*Math.exp(2.0*t[0]/3.0);
}


public double aeta(double t[], int m0, int m)
{
  return accuracy/10.0*((t[0] < 3.0) ? 1.0 : Math.exp(2.0*(t[0]-3.0)));
}


public double reta(double t[], int m0, int m)
{
  return accuracy*((t[0] < 3.0) ? 1.0 : Math.exp(2.0*(t[0]-3.0)));
}


public void out(double t[], double te, int m0, int m, double u[], int k[],
                double eta[], double rho[])
{
  DecimalFormat sixDigit = new DecimalFormat("0.000000E0");
  if (t[0] == te)
    System.out.print(k[0] + "\t" + sixDigit.format(u[0]) + "\t");
}
}
```

Output:

```
Results of EFT with relaxed accuracy conditions for t > 3 :
  K       U(TE1)      K       U(TE2)       RETA
  15      1.003845E0  42      2.000076E0   1.0E-1
  22      1.001211E0  50      2.000050E0   1.0E-2
  36      1.000109E0  68      2.000023E0   1.0E-3
  56      1.000045E0  98      2.000065E0   1.0E-4
```

Procedure tested:

rk2

Example:

The van del Pol equation

$$d^2y/dx^2 = 10(1-y^2)(dy/dx) - y, \quad x \geq 0,$$
$$y = 2, \, dy/dx = 0, \quad x = 0,$$

can be integrated by *rk2*. At the points *x=9.32386578, 18.86305405, 28.40224162,* and *37.94142918* the derivative *dy/dx* vanishes.

```
package numal;

import java.text.DecimalFormat;
import numal.*;

public class Test_rk2 extends Object implements AP_rk2_method {

  public static void main(String args[]) {
    boolean fi;
    int i;
    double x[] = new double[1];
    double y[] = new double[1];
    double z[] = new double[1];
    double e[] = new double[5];
    double d[] = new double[6];
    double b[] = {9.32386578, 18.86305405, 28.40224162, 37.94142918};

    Test_rk2 testrk2 = new Test_rk2();
    DecimalFormat fourDigit = new DecimalFormat("0.0000E0");
    DecimalFormat fiveDigit = new DecimalFormat("0.00000");
    e[1]=e[2]=e[3]=e[4]=1.0e-8;
    System.out.println("RK2 delivers :\n");
    for (i=0; i<=3; i++) {
      fi=(b[i] < 10.0);
      Analytic_problems.rk2(x,0.0,b[i],y,2.0,z,0.0,testrk2,e,d,fi);
      System.out.println("  X = " + fiveDigit.format(x[0]) + "\tY = " +
              fiveDigit.format(y[0]) + "\tDY/DX = " + fourDigit.format(z[0]));
    }
  }
```

```
  public double fxyz(double x[], double y[], double z[])
  {
    return 10.0*(1.0-y[0]*y[0])*z[0]-y[0];
  }
}
```

Output:

```
RK2 delivers :

  X = 9.32387        Y = -2.01429    DY/DX = 7.5501E-8
  X = 18.86305       Y = 2.01429     DY/DX = -7.0980E-6
  X = 28.40224       Y = -2.01429    DY/DX = 1.2710E-5
  X = 37.94143       Y = 2.01429     DY/DX = -1.8301E-5
```

Procedure tested:

rk2n

Example:

The second order differential equation

$$d^2y_1/dx^2 = -5(y_1 + dy_2/dx) + y_2,$$
$$d^2y_2/dx^2 = -5(y_2 + dy_1/dx) + y_1, \quad x \geq 0,$$
$$y_1 = dy_2/dx = 1, \quad y_2 = dy_1/dx = 0, \quad x = 0$$

with analytic solution

$$y_1 = (5/6)e^x + e^{-2x} - (1/2)e^{-3x} - (1/3)e^{-4x},$$
$$y_2 = (5/6)e^x - e^{-2x} + (1/2)e^{-3x} - (1/3)e^{-4x}$$

can be integrated by *rk2n* from 0 to 5 with 1, 2, 3, and 4 as reference points.

```
package numal;

import java.text.DecimalFormat;
import numal.*;

public class Test_rk2n extends Object implements AP_rk2n_method {

  public static void main(String args[]) {
    boolean fi;
    int k;
    double b,expx;
    double y[]  = new double[3];
    double ya[] = new double[3];
    double z[]  = new double[3];
    double za[] = new double[3];
    double e[]  = new double[9];
    double d[]  = new double[8];
    double x[]  = new double[1];
```

```
        Test_rk2n testrk2n = new Test_rk2n();
        DecimalFormat fiveDigit = new DecimalFormat("0.00000E0");
        System.out.println("Results from RK2N :");
        for (k=1; k<=8; k++) e[k]=1.0e-5;
        ya[1]=za[2]=1.0;
        ya[2]=za[1]=0.0;
        b=1.0;
        do {
          fi=(b == 1.0);
          Analytic_problems.rk2n(x,0.0,b,y,ya,z,za,testrk2n,e,d,fi,2);
          expx=Math.exp(-x[0]);
          ya[1] = -expx*(expx*(expx*(expx/3.0+0.5)-1.0)-5.0/6.0);
          ya[2] = -expx*(expx*(expx*(expx/3.0-0.5)+1.0)-5.0/6.0);
          za[1]=expx*(expx*(expx*(expx/0.75+1.5)-2.0)-5.0/6.0);
          za[2]=expx*(expx*(expx*(expx/0.75-1.5)+2.0)-5.0/6.0);
          System.out.println("\n  X = " + fiveDigit.format(x[0]) + "\nY[1]-YEXACT[1] = "
                          + fiveDigit.format(y[1]-ya[1]) + "     Y[2]-YEXACT[2] = " +
                          fiveDigit.format(y[2]-ya[2]) + "\nZ[1]-ZEXACT[1] = " +
                          fiveDigit.format(z[1]-za[1]) + "     Z[2]-ZEXACT[2] = " +
                          fiveDigit.format(z[2]-za[2]));
          b += 1.0;
        } while (b < 5.0);
    }


    public double fxyzj(int n, int j, double x[], double y[], double z[])
    {
      return -5.0*(y[j]+z[j])+((j == 1) ? y[2] : y[1]);
    }
}
```

Output:

```
Results from RK2N :

  X = 1.00000E0
Y[1]-YEXACT[1] = 1.04238E-8     Y[2]-YEXACT[2] = 1.69277E-8
Z[1]-ZEXACT[1] = -4.82341E-8    Z[2]-ZEXACT[2] = -7.26955E-8

  X = 2.00000E0
Y[1]-YEXACT[1] = -7.55890E-8    Y[2]-YEXACT[2] = 1.74176E-8
Z[1]-ZEXACT[1] = 3.53151E-7     Z[2]-ZEXACT[2] = -8.75771E-8

  X = 3.00000E0
Y[1]-YEXACT[1] = -7.60010E-8    Y[2]-YEXACT[2] = -2.16558E-8
Z[1]-ZEXACT[1] = 4.15140E-7     Z[2]-ZEXACT[2] = 7.89331E-8

  X = 4.00000E0
Y[1]-YEXACT[1] = -5.56955E-8    Y[2]-YEXACT[2] = -4.06884E-8
Z[1]-ZEXACT[1] = 3.31358E-7     Z[2]-ZEXACT[2] = 1.94980E-7
```

Procedure tested:

rk3

Example:
Compute the solution of
$$y'' = xy, \quad y(0) = 0, \quad y'(0) = 1$$
at interval [0,1].

```
package numal;

import java.text.DecimalFormat;
import numal.*;

public class Test_rk3 extends Object implements AP_rk3_method {

  public static void main(String args[]) {

    boolean fi;
    int n,i;
    double x3,s,term;
    double b[]={0.25, 0.50, 0.75, 1.00};
    double e[] = new double[6];
    double d[] = new double[6];
    double x[] = new double[1];
    double y[] = new double[1];
    double z[] = new double[1];

    Test_rk3 testrk3 = new Test_rk3();
    DecimalFormat twoDigit = new DecimalFormat("0.00E0");
    DecimalFormat fiveDigit = new DecimalFormat("0.00000");
    e[1]=e[3]=1.0e-6;
    e[2]=e[4]=1.0e-6;
    System.out.println("RK3 delivers :\n");
    for (i=0; i<=3; i++) {
      fi=(b[i] < 0.3);
      Analytic_problems.rk3(x,0.0,b[i],y,0.0,z,1.0,testrk3,e,d,fi);
      x3=x[0]*x[0]*x[0];
      term=x[0];
      s=0.0;
      n=3;
      do {
        s += term;
        term=term*x3/n/(n+1);
        n += 3;
      } while (Math.abs(term) > 1.0e-14);
      System.out.println("Y-YEXACT= " + twoDigit.format(y[0]-s) + "\tX= " + x[0] +
                    "\t\tY= " + fiveDigit.format(y[0]));
    }
  }
}
```

```
  public double fxy(double x[], double y[])
  {
    return x[0]*y[0];
  }
}
```

Output:

```
RK3 delivers :

Y-YEXACT= 1.00E-10      X= 0.25      Y= 0.25033
Y-YEXACT= 1.64E-9       X= 0.5       Y= 0.50522
Y-YEXACT= 3.55E-9       X= 0.75      Y= 0.77663
Y-YEXACT= 5.75E-9       X= 1.0       Y= 1.08534
```

Procedure tested:

rk3n

Example:

Solve the second order differential equation
$$y_1'' = y_2, \quad y_2'' = -y_1$$
with $y_1(0) = y_2(0) = 1$, $y_1'(0) = y_2'(0) = 0$, the exact solutions of which are
$$y_1(x) = \cosh(x/\sqrt{2})*\cos(x/\sqrt{2}) + \sinh(x/\sqrt{2})*\sin(x/\sqrt{2})$$
$$y_2(x) = \cosh(x/\sqrt{2})*\cos(x/\sqrt{2}) - \sinh(x/\sqrt{2})*\sin(x/\sqrt{2}).$$
These equations can be integrated by $rk3n$ because first derivatives are absent from them. Integration is carried out over the intervals $[v, v+1]$, $v=0,1,...,4$.

```
package numal;

import java.text.DecimalFormat;
import numal.*;

public class Test_rk3n extends Object implements AP_rk3n_method {

  public static void main(String args[]) {

    boolean fi;
    int k,i,n,j;
    double b,x2,term;
    double y[]  = new double[3];
    double ya[] = new double[3];
    double z[]  = new double[3];
    double e[]  = new double[9];
    double d[]  = new double[8];
    double x[]  = new double[1];
```

```
      Test_rk3n testrk3n = new Test_rk3n();
      DecimalFormat twoDigit = new DecimalFormat("0.00E0");
      System.out.println("Results from RK3N :\n");
      for (k=1; k<=8; k++) e[k]=1.0e-5;
      fi=true;
      y[1]=y[2]=1.0;
      z[1]=z[2]=0.0;
      b=0.0;
      do {
        b += 1.0;
        Analytic_problems.rk3n(x,0.0,b,y,y,z,z,testrk3n,e,d,fi,2);
        ya[1]=ya[2]=0.0;
        term=1.0;
        x2=x[0]*x[0]*0.5;
        n=1;
        do {
          for (i=1; i<=2; i++) {
            j=(i+n-2)/2;
            ya[i] += term*((j%2 == 0) ? 1 : -1);
          }
          term=term*x2/n/(n*2-1);
          n++;
        } while (Math.abs(term) > 1.0e-14);
        System.out.println(" ABS(YEXACT[1]-Y[1])+" + "ABS(YEXACT[2]-Y[2]) = " +
                        twoDigit.format(Math.abs(y[1]-ya[1])+Math.abs(ya[2]-y[2])));
        fi=false;
      } while (b < 5.0);
  }


  public double fxyj(int n, int j, double x[], double y[])
  {
    return ((j == 1) ? y[2] : -y[1]);
  }
}
```

Output:

```
Results from RK3N :

ABS(YEXACT[1]-Y[1])+ABS(YEXACT[2]-Y[2])  = 1.60E-7
ABS(YEXACT[1]-Y[1])+ABS(YEXACT[2]-Y[2])  = 5.16E-7
ABS(YEXACT[1]-Y[1])+ABS(YEXACT[2]-Y[2])  = 1.34E-6
ABS(YEXACT[1]-Y[1])+ABS(YEXACT[2]-Y[2])  = 3.49E-6
ABS(YEXACT[1]-Y[1])+ABS(YEXACT[2]-Y[2])  = 8.34E-6
```

Procedure tested:

arkmat

Example:
Solve the initial boundary value problem

$$\frac{\partial^2 u(t,x,y)}{\partial t^2} = \frac{\partial^2 u(t,x,y)}{\partial x^2} + \frac{\partial^2 u(t,x,y)}{\partial y^2} \tag{1}$$

on the domain $0 \le t \le 1$, $0 \le x \le \pi$, $0 \le y \le 1$, with initial conditions

$$u(0,x,y) = \sin(x)*\cos(\pi y/2), \quad (\partial/\partial t)u(t,x,y)_{t=0}=0 \quad (0 \le x \le \pi, \; 0 \le y \le 1)$$

and boundary values

$$u(t,x,0) = \sin(x)*\cos\{(1+\pi^2/4)t\}, \quad u(t,x,1) = 0 \quad (0 \le x \le \pi)$$
$$u(t,0,y) = u(t,\pi,y) = 0 \quad (0 \le y \le 1)$$

over the range $0 \le t \le 1$. The analytic solution of the above initial boundary value problem is

$$u(t,x,y) = \cos\{(1+\pi^2/4)t\}*\sin(x)*\cos(\pi y/2).$$

Equation (1) may be reduced to a form of the first order in t by setting

$$v(t,x,y) = Du(t,x,y)$$

where $D = \partial/\partial t$, when

$$D\,v(t,x,y) = \frac{\partial^2 u(t,x,y)}{\partial x^2} + \frac{\partial^2 u(t,x,y)}{\partial y^2}.$$

The above first order form is solved by discretizing the variables x,y, and including the construction of the boundary values in the integration of the resulting set of first order differential equations in t, making use of the conditions

$$D^2u(t,x,1) = 0 \qquad (0 \le x \le \pi)$$
$$D^2u(t,0,y) = D^2u(t,\pi,y) = 0 \quad (0 \le y \le 1)$$

and further use of the symmetry property (easily deduced from the form of the original equation and the initial and boundary conditions) that $u(t,x,-y) = u(t,x,y)$, to assist in the determination of $D^2u(t,x,0)$ $(0 \le x \le \pi)$.

Taking $\_x = \pi/(m-1)$, $\_y = 1/(n-1)$ and for $i=1,...,n$; $j=1,...,m$

$$U^{(i,j)}(t) = u(t,(j-1)\_x,(i-1)\_y)$$
$$U^{(n+i,j)}(t) = v(t,(j-1)\_x,(i-1)\_y)$$

the equations

$$DU^{(i,j)} = U^{(n+i,j)} \quad (i=1,...,n; \; j=1,...,m)$$
$$DU^{(n+i,j)} = \{U^{(i,j+1)} - 2U^{(i,j)} + U^{(i,j-1)}\}/(\_x)^2 + \{U^{(i+1,j)} - 2U^{(i,j)} + U^{(i-1,j)}\}/(\_y)^2$$
$$(i=2,...,n-1; \; j=2,...,m-1)$$

$$DU^{(n+i,1)} = DU^{(n+i,m)} = 0 \quad (i=1,...,n)$$
$$DU^{(2n,j)} = 0 \quad (j=2,...,m-1)$$
$$DU^{(n+1,j)} = \{U^{(i,j+1)} - 2U^{(1,j)} + U^{(1,j-1)}\}/(\_x)^2 + 2\{U^{(2,j)} - U^{(1,j)}\}/(\_y)^2$$
$$(j=2,...,m-1)$$

for the components of the $2n \times m$ matrix U are derived. The initial values for this set of equations are

$$U^{(i,j)} = \sin\{(j-1)\_x\}*\cos\{\pi(i-1)\_y/2\}, \quad U^{(n+i,j)} = 0 \quad (i=1,...,n; \; j=1,...,m).$$

The above set of equations with $m=n=10$ are solved by use of *arkmat* with parameters *type* = 3, *order* = 2 and constant step length in t equal to 0.1.

```
package numal;

import java.text.DecimalFormat;
import numal.*;

public class Test_arkmat extends Object implements AP_arkmat_methods {

  static int tel;
  static double h1k,h2k,hpi,h1,h2;

  public static void main(String args[]) {

    int i,j,n,m,typ;
    int orde[] = new int[1];
    double te,cos1;
    double spr[] = new double[1];
    double t[] = new double[1];
    double u[][] = new double[21][11];

    Test_arkmat testarkmat = new Test_arkmat();
    hpi=0.5*Math.PI;
    h2=1.0/9.0;
    h1=(2.0*hpi)/9.0;
    n=m=10;
    h1k=h1*h1;
    h2k=h2*h2;
    tel=0;
    t[0]=0.0;
    te=1.0;
    for (j=1; j<=m; j++) u[n][j]=Math.sin(h1*(j-1));
    for (i=1; i<=n; i++) {
      cos1=Math.cos(h2*hpi*(i-1));
      for (j=1; j<=m; j++) u[i][j]=u[n][j]*cos1;
    }
    Basic.inimat(n+1,n+n,1,m,u,0.0);
    typ=3;
    orde[0]=2;
    spr[0]=80.0;
    System.out.println("The program delivers:\n\n    X    " +
      "  Y        U(1,X,Y)        U(1,X,Y)\n                          COMPUTED        EXACT\n");
    Analytic_problems.arkmat(t,te,m,n+n,u,testarkmat,typ,orde,spr);
  }


  public void der(int m, int n, double t, double u[][], double du[][])
  {
    int i,j;

    n=n/2;
    for (i=2; i<=n-1; i++)
```

```
      for (j=2; j<=m-1; j++) {
        du[i][j]=u[i+n][j];
        du[i+n][j]=(u[i][j+1]-2.0*u[i][j]+u[i][j-1])/h1k+
                   (u[i+1][j]-2.0*u[i][j]+u[i-1][j])/h2k;
      }
    for (j=1; j<=m; j += m-1) {
      Basic.inimat(n+1,n+n,j,j,du,0.0);
      for (i=1; i<=n; i++) du[i][j]=u[n+1][j];
    }
    for (i=1; i<=n; i += n-1)
      for (j=2; j<=m-1; j++) {
        du[i][j]=u[i+n][j];
        if (i == 1)
          du[n+1][j]=(u[1][j+1]-2.0*u[1][j]+u[1][j-1])/h1k+
                     (2.0*u[2][j]-2.0*u[1][j])/h2k;
        else
          du[2*n][j]=0.0;
      }
  }


  public void out(double t[], double te, int m, int n, double u[][], int type,
                  int order[], double spr[])
  {
    int i;

    DecimalFormat threeDigit = new DecimalFormat("0.000");
    DecimalFormat fiveDigit = new DecimalFormat("0.00000");
    tel++;
    if (t[0] == te) {
      for (i=1; i<=10; i++)
        System.out.println("  " + threeDigit.format((i-1)*h1) + "  " +
          threeDigit.format((i-1)*h2) + "    " + fiveDigit.format(u[i][i]) + "\t" +
          fiveDigit.format(Math.sin(h1*(i-1))*Math.cos(hpi*h2*(i-1))*Math.cos(t[0]*
                           Math.sqrt(1.0+hpi*hpi))));
      System.out.println("\nThe number of integration steps: " + tel +
                         "\n Type is " + type + "  Order is " + order[0]);
    }
  }
}
```

Output:

```
The program delivers:

     X        Y        U(1,X,Y)            U(1,X,Y)
                       COMPUTED            EXACT


  0.000    0.000      0.00000            -0.00000
  0.349    0.111     -0.09520            -0.09673
  0.698    0.222     -0.17072            -0.17347
  1.047    0.333     -0.21198            -0.21540
  1.396    0.444     -0.21323            -0.21666
  1.745    0.556     -0.17892            -0.18180
  2.094    0.667     -0.12239            -0.12436
  2.443    0.778     -0.06214            -0.06314
  2.793    0.889     -0.01679            -0.01706
  3.142    1.000      0.00000            -0.00000


The number of integration steps: 11
  Type is 3    Order is 2
```

Procedure tested:

femlagsym

Example:

Compute approximations to the solution of the boundary value problem
$$-D\{e^x Dy(x)\} + \cos(x)y(x) = e^x\{\sin(x) - \cos\{x\}\} + \sin(2x)/2$$
$$y(0) = y(\pi) = 0$$
where $D = d/dx$, at the points $x_i = i\pi/n$, $i=0,\dots,n$, (approximating the solution on a uniform grid) for the six cases in which $n = 10$, 20, and the order of the method used is 2, 4, and 6. The analytical solution to the above problem is $y(x) = \sin(x)$.

```java
package numal;

import java.text.DecimalFormat;
import numal.*;

public class Test_femlagsym extends Object implements AP_femlagsym_methods {

  public static void main(String args[]) {
    int n,i,order;
    double rho,d;
    double x[] = new double[21];
    double y[] = new double[21];
    double e[] = new double[7];

    Test_femlagsym testfemlagsym = new Test_femlagsym();
    DecimalFormat twoDigit = new DecimalFormat("0.00E0");
```

```
    System.out.println("FEMLAGSYM delivers:\n");
    for (n=10; n<=20; n += 10) {
      e[1]=e[4]=1.0;
      e[2]=e[3]=e[5]=e[6]=0.0;
      for (i=0; i<=n; i++) x[i]=Math.PI*i/n;
      System.out.println("N=" + n);
      for (order=2; order<=6; order += 2) {
        Analytic_problems.femlagsym(x,y,n,testfemlagsym,order,e);
        rho=0.0;
        for (i=0; i<=n; i++) {
          d=Math.abs(y[i]-Math.sin(x[i]));
          if (rho < d) rho=d;
        }
        System.out.println("    ORDER = " + order + "    MAX.ERROR = " +
                            twoDigit.format(rho));
      }
    }
  }


  public double r(double x)
  {
    return Math.cos(x);
  }


  public double p(double x)
  {
    return Math.exp(x);
  }


  public double f(double x)
  {
    return Math.exp(x)*(Math.sin(x)-Math.cos(x))+Math.sin(2.0*x)/2.0;
  }
}
```

Output:

```
FEMLAGSYM delivers:

N=10
      ORDER = 2    MAX.ERROR = 1.36E-2
      ORDER = 4    MAX.ERROR = 7.55E-5
      ORDER = 6    MAX.ERROR = 3.48E-8
N=20
      ORDER = 2    MAX.ERROR = 3.41E-3
      ORDER = 4    MAX.ERROR = 4.79E-6
      ORDER = 6    MAX.ERROR = 5.51E-10
```

Procedure tested:

femlag

Example:

Compute approximations to the solution of the boundary value problem

$$-D^2 y(x) + e^x y(x) = sin(x)\{1+e^x\}$$
$$y(0) = y(\pi) = 0$$

where $D = d/dx$, at the points $x_i = i\pi/n$, $i=0,...,n$, (approximating the solution on a uniform grid) for the six cases in which $n = 10$, 20, and the order of the method used is 2, 4, and 6. The analytical solution to the above problem is $y(x) = sin(x)$.

```java
package numal;

import java.text.DecimalFormat;
import numal.*;

public class Test_femlag extends Object implements AP_femlag_methods {

  public static void main(String args[]) {

    int n,i,order;
    double rho,d;
    double x[]  = new double[21];
    double y[]  = new double[21];
    double e[]  = new double[7];

    Test_femlag testfemlag = new Test_femlag();
    DecimalFormat twoDigit = new DecimalFormat("0.00E0");
    System.out.println("FEMLAG delivers:\n");
    for (n=10; n<=20; n += 10) {
      e[1]=e[4]=1.0;
      e[2]=e[3]=e[5]=e[6]=0.0;
      for (i=0; i<=n; i++) x[i]=Math.PI*i/n;
      System.out.println("N=" + n);
      for (order=2; order<=6; order += 2) {
        Analytic_problems.femlag(x,y,n,testfemlag,order,e);
        rho=0.0;
        for (i=0; i<=n; i++) {
          d=Math.abs(y[i]-Math.sin(x[i]));
          if (rho < d) rho=d;
        }
        System.out.println("     ORDER = " + order + "     MAX.ERROR = " +
                      twoDigit.format(rho));
      }
    }
  }
}
```

```
  public double r(double x)
  {
    return Math.exp(x);
  }


  public double f(double x)
  {
    return Math.sin(x)*(1.0+Math.exp(x));
  }
}
```

Output:

```
FEMLAG delivers:

N=10
     ORDER = 2    MAX.ERROR = 1.60E-3
     ORDER = 4    MAX.ERROR = 1.55E-5
     ORDER = 6    MAX.ERROR = 7.28E-10
N=20
     ORDER = 2    MAX.ERROR = 4.01E-4
     ORDER = 4    MAX.ERROR = 9.80E-7
     ORDER = 6    MAX.ERROR = 9.38E-12
```

Procedure tested:

femlagspher

Example:
Solve the boundary value problem
$$-(y'x^{nc})'/x^{nc} + y = 1 - x^4 + (12 + 4*nc)x^2$$
$$0 < x < 1; \quad y'(0) = y(1) = 0.$$
The analytical solution is $y(x) = 1 - x^4$. We approximate the solution on a uniform grid, i.e., $x_i=i/n$, $i=0,...,n$, with $n = 10, 20$ and the order of the method used is 2 and 4.

```
package numal;

import java.text.DecimalFormat;
import numal.*;
```

```
public class Test_femlagspher extends Object implements AP_femlagspher_methods {

  static int nc;

  public static void main(String args[]) {

    int n,i,order;
    double rho,d;
    double x[] = new double[21];
    double y[] = new double[21];
    double e[] = new double[7];

    Test_femlagspher testfemlagspher = new Test_femlagspher();
    DecimalFormat twoDigit = new DecimalFormat("0.00E0");
    System.out.println("FEMLAGSPHER delivers:\n");
    for (n=10; n<=20; n += 10)
      for (nc=0; nc<=2; nc++) {
        e[2]=e[4]=1.0;
        e[1]=e[3]=e[5]=e[6]=0.0;
        for (i=0; i<=n; i++) x[i]=(float)(i)/(float)(n);
        System.out.println("N= " + n + "    NC=" + nc);
        for (order=2; order<=4; order += 2) {
          Analytic_problems.femlagspher(x,y,n,nc,testfemlagspher,order,e);
          rho=0.0;
          for (i=0; i<=n; i++) {
            d=Math.abs(y[i]-1.0+x[i]*x[i]*x[i]*x[i]);
            if (rho < d) rho=d;
          }
          System.out.println("        ORDER = " + order + "    MAX.ERROR = " +
                            twoDigit.format(rho));
        }
      }
  }


  public double r(double x)
  {
    return 1.0;
  }


  public double f(double x)
  {
    return (12+4*nc)*x*x+1.0-x*x*x*x;
  }
}
```

Output:

```
FEMLAGSPHER delivers:


N= 10      NC=0
           ORDER = 2      MAX.ERROR = 4.37E-3
           ORDER = 4      MAX.ERROR = 2.93E-6
N= 10      NC=1
           ORDER = 2      MAX.ERROR = 1.42E-2
           ORDER = 4      MAX.ERROR = 5.49E-5
N= 10      NC=2
           ORDER = 2      MAX.ERROR = 2.46E-2
           ORDER = 4      MAX.ERROR = 1.27E-4
N= 20      NC=0
           ORDER = 2      MAX.ERROR = 1.09E-3
           ORDER = 4      MAX.ERROR = 1.83E-7
N= 20      NC=1
           ORDER = 2      MAX.ERROR = 3.53E-3
           ORDER = 4      MAX.ERROR = 3.91E-6
N= 20      NC=2
           ORDER = 2      MAX.ERROR = 6.10E-3
           ORDER = 4      MAX.ERROR = 9.26E-6
```

Procedure tested:

femlagskew

Example:

Compute approximations to the solution of the boundary value problem
$$-y'' + cos(x)y' + e^x y = sin(x)\{1 + e^x\} + \{cos(x)\}^2$$
$$0 < x < \pi; \quad y(0) = y(\pi) = 0.$$
The analytical solution is $y(x) = sin(x)$. We approximate the solution on a uniform grid, i.e., $x_i = i\pi/n$, $i=0,...,n$, with $n = 10$, 20, and the order of the method used is 2, 4, and 6.

```
package numal;

import java.text.DecimalFormat;
import numal.*;

public class Test_femlagskew extends Object implements AP_femlagskew_methods {

  public static void main(String args[]) {

    int n,i,order;
    double rho,d;
    double x[]  = new double[21];
    double y[]  = new double[21];
    double e[]  = new double[7];
```

```
    Test_femlagskew testfemlagskew = new Test_femlagskew();
    DecimalFormat twoDigit = new DecimalFormat("0.00E0");
    System.out.println("FEMLAGSKEW delivers:\n");
    for (n=10; n<=20; n += 10) {
      e[1]=e[4]=1.0;
      e[2]=e[3]=e[5]=e[6]=0.0;
      for (i=0; i<=n; i++) x[i]=Math.PI*i/n;
      System.out.println("N=" + n);
      for (order=2; order<=6; order += 2) {
        Analytic_problems.femlagskew(x,y,n,testfemlagskew,order,e);
        rho=0.0;
        for (i=0; i<=n; i++) {
          d=Math.abs(y[i]-Math.sin(x[i]));
          if (rho < d) rho=d;
        }
        System.out.println("    ORDER = " + order + "    MAX.ERROR = " +
                           twoDigit.format(rho));
      }
    }
  }


  public double q(double x)
  {
    return Math.cos(x);
  }


  public double r(double x)
  {
    return Math.exp(x);
  }


  public double f(double x)
  {
    return Math.sin(x)*(1.0+Math.exp(x))+Math.cos(x)*Math.cos(x);
  }
}
```

Output:

```
FEMLAGSKEW delivers:

N=10
     ORDER = 2      MAX.ERROR = 2.95E-3
     ORDER = 4      MAX.ERROR = 2.56E-5
     ORDER = 6      MAX.ERROR = 4.26E-8
N=20
     ORDER = 2      MAX.ERROR = 7.55E-4
     ORDER = 4      MAX.ERROR = 1.68E-6
     ORDER = 6      MAX.ERROR = 6.76E-10
```

Procedure tested:

femhermsym

Example:

Compute approximations to the solution of the boundary value problem
$$y'''' - (cos(x)y')' + e^x y = sin(x)\{1 + e^x + 2cos(x)\}$$
$$0 < x < \pi; \quad y(0) = y(\pi) = 0; \quad y'(0) = 1; \quad y'(\pi) = -1.$$

The analytical solution is $y(x) = sin(x)$. We approximate the solution on a uniform grid, i.e., $x_i = i\pi/n$, $i=0,...,n$, with $n = 5$, 10, and the order of the method used is 4, 6, and 8.

```
package numal;

import java.text.DecimalFormat;
import numal.*;

public class Test_femhermsym extends Object implements AP_femhermsym_methods {

  public static void main(String args[]) {

    int n,i,order;
    double rho1,rho2,d1,d2;
    double x[] = new double[11];
    double y[] = new double[19];
    double e[] = new double[5];

    Test_femhermsym testfemhermsym = new Test_femhermsym();
    DecimalFormat twoDigit = new DecimalFormat("0.00E0");
    System.out.println("FEMHERMSYM delivers:\n");
    for (n=5; n<=10; n += 5) {
      e[1]=e[3]=0.0;  e[2]=1.0;  e[4] = -1.0;
      for (i=0; i<=n; i++) x[i]=Math.PI*i/n;
      System.out.println("N=" + n);
      for (order=4; order<=8; order += 2) {
        Analytic_problems.femhermsym(x,y,n,testfemhermsym,order,e);
```

```
        rho1=rho2=0.0;
        for (i=1; i<=n-1; i++) {
          d1=Math.abs(y[2*i-1]-Math.sin(x[i]));
          if (rho1 < d1) rho1=d1;
          d2=Math.abs(y[2*i]-Math.cos(x[i]));
          if (rho2 < d2) rho2=d2;
        }
        System.out.println("     ORDER=" + order +
          "\n            MAX. ABS(Y[2*I-1]-Y(X[I])) = " + twoDigit.format(rho1) +
          "\n            MAX. ABS(Y[2*I]-Y'(X[I]))  = " + twoDigit.format(rho2));
      }
    }
}


public double p(double x)
{
  return 1.0;
}


public double q(double x)
{
  return Math.cos(x);
}


public double r(double x)
{
  return Math.exp(x);
}


public double f(double x)
{
  return Math.sin(x)*(1.0+Math.exp(x)+2.0*Math.cos(x));
}
}
```

Output:

```
FEMHERMSYM delivers:

N=5
    ORDER=4
            MAX. ABS(Y[2*I-1]-Y(X[I]))  = 4.82E-4
            MAX. ABS(Y[2*I]-Y'(X[I]))   = 4.55E-4
    ORDER=6
            MAX. ABS(Y[2*I-1]-Y(X[I]))  = 5.65E-6
            MAX. ABS(Y[2*I]-Y'(X[I]))   = 2.03E-6
    ORDER=8
            MAX. ABS(Y[2*I-1]-Y(X[I]))  = 2.26E-8
            MAX. ABS(Y[2*I]-Y'(X[I]))   = 1.60E-8
N=10
    ORDER=4
            MAX. ABS(Y[2*I-1]-Y(X[I]))  = 2.66E-5
            MAX. ABS(Y[2*I]-Y'(X[I]))   = 2.87E-5
    ORDER=6
            MAX. ABS(Y[2*I-1]-Y(X[I]))  = 8.40E-8
            MAX. ABS(Y[2*I]-Y'(X[I]))   = 3.57E-8
    ORDER=8
            MAX. ABS(Y[2*I-1]-Y(X[I]))  = 8.08E-11
            MAX. ABS(Y[2*I]-Y'(X[I]))   = 6.64E-11
```

Procedure tested:

nonlinfemlagskew

Example:
Compute approximations to the solution of the boundary value problem
$$(y'x^2)'/x^2 = exp(y) + exp(y') - exp(1-x^2) - exp(2x) - 6$$
$$0 < x < 1; \quad y'(0) = y(1) = 0;$$
The analytical solution is $y(x) = 1 - x^2$. We approximate the solution on a uniform grid, i.e., $x_i = i/n$, $i=0,...,n$.

```
package numal;

import java.text.DecimalFormat;
import numal.*;

public class Test_nonlinfemlagskew extends Object
                                implements AP_nonlinfemlagskew_methods {

    static int nc;
```

```
public static void main(String args[]) {

  int n,i;
  double rho,d;
  double x[] = new double[51];
  double y[] = new double[51];
  double e[] = new double[7];

  Test_nonlinfemlagskew testnonlinfemlagskew = new Test_nonlinfemlagskew();
  DecimalFormat twoDigit = new DecimalFormat("0.00E0");
  System.out.println("NONLINFEMLAGSKEW delivers:\n");
  for (nc=0; nc<=2; nc++)
    for (n=25; n<=50; n += 25) {
      e[2]=e[4]=1.0;
      e[1]=e[3]=e[5]=e[6]=0.0;
      for (i=0; i<=n; i++) {
        x[i]=(double)(i)/(double)(n);
        y[i]=0.0;
      }
      System.out.print(" N = " + n + "    NC = " + nc);
      Analytic_problems.nonlinfemlagskew(x,y,n,testnonlinfemlagskew,nc,e);
      rho=0.0;
      for (i=0; i<=n; i++) {
        d=Math.abs(y[i]-1.0+x[i]*x[i]);
        if (rho < d) rho=d;
      }
      System.out.println("    MAX.ERROR = " + twoDigit.format(rho));
    }
}


public double f(double x, double y, double z)
{
  return Math.exp(y)+Math.exp(z)-Math.exp(1.0-x*x)-Math.exp(-2.0*x)-2.0-2*nc;
}


public double fy(double x, double y, double z)
{
  return Math.exp(y);
}


public double fz(double x, double y, double z)
{
  return Math.exp(z);
}
}
```

Output:

```
NONLINFEMLAGSKEW delivers:

N = 25     NC = 0      MAX.ERROR = 2.47E-4
N = 50     NC = 0      MAX.ERROR = 6.19E-5
N = 25     NC = 1      MAX.ERROR = 1.41E-3
N = 50     NC = 1      MAX.ERROR = 3.99E-4
N = 25     NC = 2      MAX.ERROR = 2.44E-3
N = 50     NC = 2      MAX.ERROR = 7.02E-4
```

Procedure tested:

richardson

Example:

Solve the system of equations in the 144 unknowns $U_{j,l}$ $(j,l=1,...,12)$

$$U_{1,j} = 0, \quad U_{12,j} = \pi^2(j-1)^2h^2 \qquad\qquad (j=1,...,12)$$
$$U_{l,1} = 0, \quad U_{l,12} = \pi^2(l-1)^2h^2 \qquad\qquad (l=1,...,12)$$
$$U_{j,l-1} + U_{j-1,l} + U_{j+1,l} + U_{j,l+1} - 4U_{j,l} = 2h^2(x_l^2+y_j^2) \qquad (j,l=2,...,11)$$

where $x_l = (l-1)h$, $y_j = (j-1)h$ and $h = \pi/11$, by use of *richardson* with parameters $a=0.163$, $b=7.83$, and $n=50$. The $U_{j,l}$ are clearly approximations to the solution of the partial differential equation

$$\{(\partial/\partial x)^2 + (\partial/\partial y)^2\}\hat{u}(x,y) = 2(x^2+y^2)$$

with boundary values $\hat{u}(x,0) = 0$, $\hat{u}(x,\pi) = \pi^2x^2$ $(0 \le x \le \pi)$, $\hat{u}(0,y) = 0$, $\hat{u}(\pi,y) = \pi^2y^2$ $(0 \le y \le \pi)$ which has the analytical solution $\hat{u}(x,y) = x^2y^2$ (approximately $U_{j,l} = \hat{u}\{(l-1)h,(j-1)h\}$).

```java
package numal;

import java.text.DecimalFormat;
import numal.*;

public class Test_richardson extends Object implements AP_richardson_methods {

  static double h,h2;

  public static void main(String args[]) {

    int j,l,lj,uj,ll,ul;
    int n[]  = new int[1];
    int k[]  = new int[1];
    double a,b;
    double discr[]  = new double[3];
    double u[][]  = new double[12][12];
    double rateconv[]  = new double[1];
    double domeigval[]  = new double[1];

    Test_richardson testrichardson = new Test_richardson();
```

```
    System.out.println("RICHARDSON delivers:\n");
    lj=0;  uj=11;  ll=0;  ul=11;  n[0]=50;
    a=0.163;  b=7.83;
    h=Math.PI/(uj-lj);
    h2=h*h;
    for (j=lj; j<=uj; j++)
      for (l=ll; l<=ul; l++)
        u[j][l] = (j==lj || j==uj || l==ll || l==ul) ? (j*h)*(j*h)*(l*h)*(l*h) :
                                                        1.0;
    Analytic_problems.richardson(u,lj,uj,ll,ul,true,testrichardson,a,b,n,discr,k,
                                 rateconv,domeigval);
}


public void residual(int lj, int uj, int ll, int ul, double u[][])
{
  int ujmin1,ulmin1,ljplus1,j,l;
  double u2;
  double u1[] = new double[12];

  ujmin1=uj-1;
  ulmin1=ul-1;
  ljplus1=lj+1;
  for (j=lj; j<=uj; j++) {
    u1[j]=u[j][ll];
    u[j][ll]=0.0;
  }
  for (l=ll+1; l<=ulmin1; l++) {
    u1[lj]=u[lj][l];
    u[lj][l]=0.0;
    for (j=ljplus1; j<=ujmin1; j++) {
      u2=u[j][l];
      u[j][l]=(4.0*u2-u1[j-1]-u1[j]-u[j+1][l]-u[j][l+1])+
              2.0*((j*h)*(j*h)+(l*h)*(l*h))*h2;
      u1[j]=u2;
    }
    u[uj][l]=0.0;
  }
  for (j=lj; j<=uj; j++) u[j][ul]=0.0;
}


public void out(double u[][], int lj, int uj, int ll, int ul, int n[],
                double discr[], int k[], double rateconv[], double domeigval[])
{
  DecimalFormat fiveDigit = new DecimalFormat("0.00000E0");
  if (k[0] == n[0])
    System.out.println("  K    DISCR[1]       DISCR[2]      " + "RATECONV\n  " +
           k[0] + "    " + fiveDigit.format(discr[1]) + "    " +
           fiveDigit.format(discr[2]) + "    " + fiveDigit.format(rateconv[0]));
}
}
```

Output:

```
RICHARDSON delivers:

  K     DISCR[1]    DISCR[2]     RATECONV
  50    1.40183E-4  4.66687E-5   2.92172E-1
```

Procedure tested:

elimination

Example:

Solve the system of equations described in the documentation of the test program for *richardson*. *richardson* is first called precisely as in that documentation but with the value allocated to *a* being 0.326. *elimination* is then called.

```java
package numal;

import java.text.DecimalFormat;
import numal.*;

public class Test_elimination extends Object implements AP_richardson_methods {

  static boolean elim;
  static int nn;
  static double h,h2,d1,d2;

  public static void main(String args[]) {

    int j,l,lj,uj,ll,ul;
    int n[] = new int[1];
    int k[] = new int[1];
    int p[] = new int[1];
    double rateconvr,rateconve,a,b;
    double discr[] = new double[3];
    double u[][] = new double[12][12];
    double rateconv[] = new double[1];
    double domeigval[] = new double[1];

    Test_elimination testelimination = new Test_elimination();
    DecimalFormat fiveDigit = new DecimalFormat("0.00000E0");
    System.out.println("RICHARDSON and ELIMINATION deliver:\n");
    lj=0;  uj=11;  ll=0;  ul=11;  n[0]=50;
    a=0.326;  b=7.83;
    h=Math.PI/(uj-lj);
    h2=h*h;
    for (j=lj; j<=uj; j++)
      for (l=ll; l<=ul; l++)
```

```
         u[j][l] = (j==lj || j==uj || l==ll || l==ul) ? (j*h)*(j*h)*(l*h)*(l*h) :
                                              1.0;
   nn=n[0];
   elim = false;
   Analytic_problems.richardson(u,lj,uj,ll,ul,true,testelimination,a,b,n,discr,k,
                          rateconv,domeigval);
   rateconvr=rateconv[0];
   System.out.println("\n dominant eigenvalue:  " +
                   fiveDigit.format(domeigval[0]) + "\n");
   elim = true;
   Analytic_problems.elimination(u,lj,uj,ll,ul,testelimination,a,b,p,discr,k,
                          rateconv,domeigval);
   rateconve=rateconv[0];
   nn=n[0]+p[0];
   System.out.println("\nTotal number of iterations: " + nn +
                   "\nRate of convergence with respect to\n" +
                   "   the zeroth iterand of RICHARDSON:  " +
                   fiveDigit.format((n[0]*rateconvr+p[0]*rateconve)/nn));
}


public void residual(int lj, int uj, int ll, int ul, double u[][])
{
   int ujmin1,ulmin1,ljplus1,j,l;
   double u2;
   double u1[] = new double[12];

   ujmin1=uj-1;
   ulmin1=ul-1;
   ljplus1=lj+1;
   for (j=lj; j<=uj; j++) {
     u1[j]=u[j][ll];
     u[j][ll]=0.0;
   }
   for (l=ll+1; l<=ulmin1; l++) {
     u1[lj]=u[lj][l];
     u[lj][l]=0.0;
     for (j=ljplus1; j<=ujmin1; j++) {
       u2=u[j][l];
       u[j][l]=(4.0*u2-u1[j-1]-u1[j]-u[j+1][l]-u[j][l+1])+
               2.0*((j*h)*(j*h)+(l*h)*(l*h))*h2;
       u1[j]=u2;
     }
     u[uj][l]=0.0;
   }
   for (j=lj; j<=uj; j++) u[j][ul]=0.0;
}
```

```
public void out(double u[][], int lj, int uj, int ll, int ul, int n[],
                double discr[], int k[], double rateconv[], double domeigval[])
{
   DecimalFormat fiveDigit = new DecimalFormat("0.00000E0");
   if (elim) {
     if (k[0] == n[0])
        System.out.println("  " + k[0] + "     " + fiveDigit.format(discr[1]) + "    "
           + fiveDigit.format(discr[2]) + "    " + fiveDigit.format(rateconv[0]));
   } else {
     if (k[0] == 0)
        d1=d2=1.0;
     else {
        d2=d1;
        d1=domeigval[0];
        n[0] = (Math.abs((d1-d2)/d2) < 1.0e-4) ? k[0] : nn;
        if (k[0] == n[0])
           System.out.println("  K     DISCR[1]      DISCR[2]  " + "   RATECONV\n  " +
              k[0] + "    " + fiveDigit.format(discr[1]) + "   " +
              fiveDigit.format(discr[2]) + "   " + fiveDigit.format(rateconv[0]));
     }
   }
}
}
```

Output:

```
RICHARDSON and ELIMINATION deliver:

  K     DISCR[1]      DISCR[2]       RATECONV
  45    4.99846E-2    8.90386E-3    2.00994E-1

 dominant eigenvalue:   1.62044E-1

  7     3.56387E-6    6.71438E-7    1.36009E0

Total number of iterations: 52
Rate of convergence with respect to
   the zeroth iterand of RICHARDSON:   3.57026E-1
```

Procedure tested:

peide

Example:

The following initial value problem arising from biochemistry

$$y_1(0) = 1.0, \quad y_2(0) = 0.0 \tag{1}$$
$$dy_1/dt = -(1-y_2)y_1 + q_2 y_2 \tag{2}$$
$$dy_2/dt = q_1\{(1-y_2)y_1 - (q_2+q_3)y_2\}$$

is considered. Setting $q_i = exp(p_i)$, $i=1,2,3$, equations (2) become

$$dy_1/dt = -(1-y_2)y_1 + exp(p_2)y_2 = f_1(t,y,p) \tag{3}$$
$$dy_2/dt = exp(p_1)\{(1-y_2)y_1 - (exp(p_2) + exp(p_3))y_2\} = f_2(t,y,p).$$

Data have been obtained from formulae (1,2) with $q_1=1000$, $q_2=0.99$, $q_3=0.01$ (i.e., $p_1=6.907755$, $p_2=-0.01005034$, $p_3=-4.6051702$) by tabulating the values of the function $y_2(t)$ obtained for $t^{(i)}=0.0002i$ ($i=1,...,10$), $t^{(i)}=0.02(i-10)$ ($i=11,...,15$), $t^{(16)}=1.0$, $t^{(17)}=2.0$, $t^{(i)}=5.0(i-17)$ ($i=18,...,23$). The initial guess for the determination of the vector q with the above numerical components $\{q_i\}$ is taken to be $q^{(0)}$, where $q_1^{(0)}=1600$, $q_2^{(0)}=0.8$, $q_3^{(0)}=1.2$ (so that $p^{(0)}$ has the components $p_1^{(0)}=ln(1600)\approx7.377759$, $p_2^{(0)}=ln(0.8)\approx-0.2231436$, $p_3^{(0)}=0.1823216$). The superscripts $i=17$, 19, and 21 specify the selected break-points.

From equation (3), we have

$$\partial f_1/\partial y_1 = y_2-1, \quad \partial f_2/\partial y_2 = y_1+exp(p_2) \tag{4}$$
$$\partial f_2/\partial y_1 = exp(p_1)(1-y_2), \quad \partial f_2/\partial y_2 = -exp(p_1)\{y_1+exp(p_2)+exp(p_3)\}$$
$$\partial f_1/\partial p_1 = \partial f_1/\partial p_3 = 0, \quad \partial f_1/\partial p_2 = exp(p_2)y_2 \tag{5}$$
$$\partial f_2/\partial p_1 = exp(p_1)\{(1-y_2)y_1 - (exp(p_2)+exp(p_3))y_2\}$$
$$\partial f_2/\partial p_2 = -exp(p_1+p_2)y_2, \quad \partial f_2/\partial p_3 = -exp(p_1+p_3)y_2.$$

The initial values (1) are not functions of p.

$$ymax = (1,1) \tag{6}$$

is taken to be an estimate of $(max|y_1(t,p)|, max|y_2(t,p)|)$ over the range $0 \le t \le 30$.

The central terms in formulae (3) are evaluated by use of the procedure *deriv*, the elements of the Jacobian matrix given by formulae (4) are evaluated by use of the procedure *jacdfdy*, those of the Jacobian matrix given by formulae (5) by use of the procedure *jacdfdp*, the initial values (1) and estimates (6) are inserted by use of the procedure *callystart*, the required data for the computations are inserted by use of the procedure *data*, and the procedure *monitor* has been used to monitor the computations.

```
package numal;

import java.text.DecimalFormat;
import numal.*;

public class Test_peide extends Object implements AP_peide_methods {

  public static void main(String args[]) {

    int m,n,nobs;
    int bp[]  = new int[4];
    int nbp[] = new int[1];
    double fa;
    double par[]  = new double[7];
    double res[]  = new double[27];
    double in[]   = new double[7];
    double out[]  = new double[8];
    double jtjinv[][] = new double[4][4];

    Test_peide testpeide = new Test_peide();
    System.out.println("          E S C E P - problem");
    m=3;  n=2;  nobs=23;  nbp[0]=3;
```

```
    par[1]=Math.log(1600.0);  par[2]=Math.log(0.8);
    par[3]=Math.log(1.2);
    in[0]=1.0e-14;  in[3]=in[4]=1.0e-4;  in[5]=50.0;
    in[6]=1.0e-2;   in[1]=1.0e-4;  in[2]=1.0e-5;
    bp[1]=17;  bp[2]=19;  bp[3]=21;
    fa=4.94;
    testpeide.communication(1,fa,n,m,nobs,nbp,par,res,bp,jtjinv,in,out,0,0);
    Analytic_problems.peide(n,m,nobs,nbp,par,res,bp,jtjinv,in,out,testpeide);
    testpeide.communication(2,fa,n,m,nobs,nbp,par,res,bp,jtjinv,in,out,0,0);
}


public void communication(int post, double fa, int n, int m, int nobs, int nbp[],
                          double par[], double res[], int bp[], double jtjinv[][],
                          double in[], double out[], int weight, int nis)
{
    int i,j;
    double c;
    double conf[] = new double[m+1];

    DecimalFormat twoDigit = new DecimalFormat("0.00E0");
    DecimalFormat threeDigit = new DecimalFormat("0.000E0");
    DecimalFormat fiveDigit = new DecimalFormat("0.00000E0");
    if (post == 5) {
      System.out.println("\nThe first residual vector\n  I      RES[I]");
      for (i=1; i<=nobs; i++)
        System.out.println(" " + i + "    " + fiveDigit.format(res[i]));
    } else if (post == 3) {
      System.out.println("\nThe Euclidean norm of the residual " + "vector : " +
                  fiveDigit.format(Math.sqrt(Basic.vecvec(1,nobs,0,res,res))) +
                  "\nCalculated parameters:\n");
      for (i=1; i<=m; i++)
        System.out.println("  " + fiveDigit.format(par[i]));
      System.out.println("Number of integration steps performed: " + nis);
    } else if (post == 4) {
      if (nbp[0] == 0)
        System.out.println("Minimization is started without break-points");
      else {
        System.out.println("Minimization is started with weight =" + weight +
                        "The extra parameters are the observations:");
        for (i=1; i<=nbp[0]; i++)
          System.out.println("  " + bp[i]);
      }
      System.out.println("Starting values of the parameters:");
      for (i=1; i<=m; i++)
        System.out.println("  \n" + fiveDigit.format(par[i]));
      System.out.println("Rel. tol. for the Eucl. norm of the " + "res. vector:" +
        fiveDigit.format(in[3]) +
        "\nAbs. tol. for the Eucl. norm of the res. vector:" +
        fiveDigit.format(in[4]) + "\nRelative starting value of lambda: " +
        fiveDigit.format(in[6]));
```

```
  } else if (post == 1) {
    System.out.println("\nStarting values of the parameters:");
    for (i=1; i<=m; i++)
      System.out.println("   " + fiveDigit.format(par[i]));
    System.out.println("Numer of equations    : " + n +
      "\nNumber of observations: " + nobs +
      "\n\nMachine precision                     : " +
      twoDigit.format(in[0]) +
      "\nRelative local error bound for integration: " + twoDigit.format(in[2]) +
      "\nRelative tolerance for residual           : " + twoDigit.format(in[3]) +
      "\nAbsolute tolerance for residual           : " + twoDigit.format(in[4]) +
      "\nMaximum number of integrations to perform : " + (int)(in[5]) +
      "\nRelative starting value of lambda         : " + twoDigit.format(in[6]) +
      "\nRelative minimal steplength               : " + twoDigit.format(in[1]));
    if (nbp[0] == 0)
      System.out.println("There are no break-points\n");
    else {
      System.out.print("\nBreak-points are the observations:");
      for (i=1; i<=nbp[0]; i++)
        System.out.print("    " + bp[i]);
    }
    System.out.println("\nThe alpha-point of the f-distribution: " + fa);
  } else if (post == 2) {
    if (out[1] == 0.0)
      System.out.println("\nNormal termination of the process\n");
    else if (out[1] == 1.0)
      System.out.println("Number of integrations allowed was exceeded");
    else if (out[1] == 2.0)
      System.out.println("Minimal steplength was decreased four times");
    else if (out[1] == 3.0)
      System.out.println("A call of deriv delivered false");
    else if (out[1] == 4.0)
      System.out.println("A call of jacdfdy delivered false");
    else if (out[1] == 5.0)
      System.out.println("A call of jacdfdp delivered false");
    else if (out[1] == 6.0)
      System.out.println("Precision asked for may not be attained");
    if (nbp[0] == 0)
      System.out.println("Last integration was performed without break-points");
    else {
      System.out.println("The process stopped with Break-points:");
      for (i=1; i<=nbp[0]; i++)
        System.out.println("    " + bp[i]);
    }
    System.out.println("\nEucl. norm of the last residual vector : " +
      fiveDigit.format(out[2]) +
      "\nEucl. norm of the first residual vector: " + fiveDigit.format(out[3]) +
      "\nNumber of integrations performed       : " + (int)(out[4]) +
      "\nLast improvement of the euclidean norm : " + fiveDigit.format(out[6]) +
      "\nCondition number of J'*J               : " + fiveDigit.format(out[7]) +
```

```
            "\nLocal error bound was exceeded (maxim.): " + out[5]);
      System.out.println("\n   Parameters   Confidence Interval");
      for (i=1; i<=m; i++) {
        conf[i]=Math.sqrt(m*fa*jtjinv[i][i]/(nobs-m))*out[2];
        System.out.println("   " + fiveDigit.format(par[i]) + "\t" +
                        fiveDigit.format(conf[i]));
      }
      c = (nobs == m) ? 0.0 : out[2]*out[2]/(nobs-m);
      System.out.println("\nCorrelation matrix        Covariance matrix");
      for (i=1; i<=m; i++) {
        for (j=1; j<=m; j++) {
          if (i == j)
            System.out.print("                          ");
          if (i > j)
            System.out.print("  " + fiveDigit.format(
                            jtjinv[i][j]/Math.sqrt(jtjinv[i][i]*jtjinv[j][j])));
          else
            System.out.print("  " + fiveDigit.format(jtjinv[i][j]*c));
        }
        System.out.println();
      }
      System.out.println("\nThe last residual vector\n\n   I      RES[I]");
      for (i=1; i<=nobs; i++)
        System.out.println("  " + i + "\t" + threeDigit.format(res[i]));
    }
  }

  public boolean jacdfdp(int n, int m, double par[], double y[], double x[],
                      double fp[][])
  {
    double y2;

    y2=y[2];
    fp[1][1] = fp[1][3]=0.0;
    fp[1][2] = y2*Math.exp(par[2]);
    fp[2][1] = Math.exp(par[1])*(y[1]*(1.0-y2)-
              (Math.exp(par[2])+Math.exp(par[3]))*y2);
    fp[2][2] = -Math.exp(par[1]+par[2])*y2;
    fp[2][3] = -Math.exp(par[1]+par[3])*y2;
    return true;
  }

  public void data(int nobs, double tobs[], double obs[], int cobs[])
  {
    int i;
    double a[]={0.0002, 0.0004, 0.0006, 0.0008, 0.001,
          0.0012, 0.0014, 0.0016, 0.0018, 0.002, 0.02, 0.04, 0.06,
          0.08, 0.1, 1.0, 2.0, 5.0, 10.0, 15.0, 20.0, 25.0, 30.0};
    double b[]={0.1648, 0.2753, 0.3493, 0.3990, 0.4322,
          0.4545, 0.4695, 0.4795, 0.4862, 0.4907, 0.4999, 0.4998,
```

```
        0.4998, 0.4998, 0.4998, 0.4986, 0.4973, 0.4936, 0.4872,
        0.4808, 0.4743, 0.4677, 0.4610};

    DecimalFormat fourDigit = new DecimalFormat("0.0000");
    tobs[0]=0.0;
    for (i=1; i<=nobs; i++) {
      tobs[i]=a[i-1];
      obs[i]=b[i-1];
      cobs[i]=2;
    }
    System.out.println("\nThe observations were:\n\n" +
                        "  I      TOBS[I]       COBS[I]   OBS[I]");
    for (i=1; i<=nobs; i++)
      System.out.println("  " + i + "\t" + fourDigit.format(tobs[i]) + "\t\t" +
                         cobs[i] + "\t" + fourDigit.format(obs[i]));
  }

  public void callystart(int n, int m, double par[], double y[], double ymax[])
  {
    y[1]=ymax[1]=ymax[2]=1.0;
    y[2]=0.0;
  }

  public boolean deriv(int n, int m, double par[], double y[], double x[],
                        double df[])
  {
    double y2;

    y2=y[2];
    df[1] = -(1.0-y2)*y[1]+Math.exp(par[2])*y2;
    df[2]=Math.exp(par[1])*((1.0-y2)*y[1]-(Math.exp(par[2])+Math.exp(par[3]))*y2);
    return true;
  }

  public boolean jacdfdy(int n, int m, double par[], double y[], double x[],
                          double fy[][])
  {
    fy[1][1] = -1.0+y[2];
    fy[1][2] = Math.exp(par[2])+y[1];
    fy[2][1] = Math.exp(par[1])*(1.0-y[2]);
    fy[2][2] = -Math.exp(par[1])*(Math.exp(par[2])+Math.exp(par[3])+y[1]);
    return true;
  }


  public void monitor(int post, int ncol, int nrow, double par[], double res[],
                      int weight, int nis[])
  {
  }
}
```

Output:

```
        E S C E P - problem

Starting values of the parameters:
  7.37776E0
  -2.23144E-1
  1.82322E-1
Numer of equations     : 2
Number of observations: 23


Machine precision                           : 1.00E-14
Relative local error bound for integration: 1.00E-5
Relative tolerance for residual             : 1.00E-4
Absolute tolerance for residual             : 1.00E-4
Maximum number of integrations to perform : 50
Relative starting value of lambda           : 1.00E-2
Relative minimal steplength                 : 1.00E-4


Break-points are the observations:    17    19    21
The alpha-point of the f-distribution: 4.94


The observations were:


    I        TOBS[I]      COBS[I]      OBS[I]
    1        0.0002         2         0.1648
    2        0.0004         2         0.2753
    3        0.0006         2         0.3493
    4        0.0008         2         0.3990
    5        0.0010         2         0.4322
    6        0.0012         2         0.4545
    7        0.0014         2         0.4695
    8        0.0016         2         0.4795
    9        0.0018         2         0.4862
   10        0.0020         2         0.4907
   11        0.0200         2         0.4999
   12        0.0400         2         0.4998
   13        0.0600         2         0.4998
   14        0.0800         2         0.4998
   15        0.1000         2         0.4998
   16        1.0000         2         0.4986
   17        2.0000         2         0.4973
   18        5.0000         2         0.4936
   19       10.0000         2         0.4872
   20       15.0000         2         0.4808
   21       20.0000         2         0.4743
   22       25.0000         2         0.4677
   23       30.0000         2         0.4610


Normal termination of the process
```

```
Last integration was performed without break-points

Eucl. norm of the last residual vector : 1.43078E-4
Eucl. norm of the first residual vector: 1.33107E0
Number of integrations performed       : 12
Last improvement of the euclidean norm : 2.22369E-5
Condition number of J'*J               : 2.58288E2
Local error bound was exceeded (maxim.): 37.0

   Parameters    Confidence Interval
   6.90767E0     3.20931E-4
  -1.00394E-2   1.68777E-4
  -4.60529E0    1.94250E-3

Correlation matrix        Covariance matrix
                          6.94986E-9  1.40763E-9  -9.12985E-9
  3.85132E-1                          1.92212E-9  -1.41424E-8
 -2.17039E-1  -6.39289E-1                          2.54609E-7

The last residual vector

   I      RES[I]
   1      1.748E-6
   2     -2.905E-5
   3      2.814E-5
   4     -3.879E-5
   5      3.069E-5
   6      3.101E-5
   7     -2.019E-5
   8     -3.887E-6
   9      1.052E-5
  10      1.391E-5
  11     -5.109E-5
  12      2.384E-5
  13     -1.156E-6
  14     -2.616E-5
  15     -5.116E-5
  16      2.244E-5
  17      6.794E-5
  18     -1.418E-5
  19      2.087E-5
  20     -1.980E-5
  21     -3.476E-5
  22     -2.245E-5
  23      1.886E-5
```

Examples for chapter 6 procedures

Procedure tested:

ei

Example:
Evaluate

$$\int_x^\infty \frac{1}{te^t}\,dt$$

for $x = 0.5$.

```
package numal;

import java.text.DecimalFormat;
import numal.*;

public class Test_ei extends Object {

  public static void main(String args[]) {

    DecimalFormat fiveDigit = new DecimalFormat("0.00000E0");
    System.out.println("EI delivers:  " +
                    (fiveDigit.format(-Special_functions.ei(-0.5))));
  }
}
```

Output:

```
EI delivers:  5.59774E-1
```

Procedure tested:

eialpha

Example:
Evaluate

$$\int_1^\infty \frac{t^i}{e^{xt}}\,dt$$

for $x = 0.25$ and $i=0,1,...,5$.

```
package numal;

import java.text.DecimalFormat;
import numal.*;
```

```
public class Test_eialpha extends Object {

    public static void main(String args[]) {

        int k;
        double a[] = new double[6];

        System.out.println("EIALPHA delivers:");
        Special_functions.eialpha(0.25,5,a);
        DecimalFormat fiveDigit = new DecimalFormat("0.00000E0");
        for (k=0; k<=5; k++) {
            System.out.println(" "+ k + "      " + fiveDigit.format(a[k]));
        }
    }
}
```

Output:

```
EIALPHA delivers:
  0     3.11520E0
  1     1.55760E1
  2     1.27723E2
  3     1.53580E3
  4     2.45758E4
  5     4.91520E5
```

Procedures tested:

enx, nonexpenx

Example:
Evaluate in succession: a)

$$\int_1^\infty \frac{1}{e^{xt}t^n}\,dt$$

for $x = 1.1$ and $n = 40, 41, 42$, by means of a call of *enx*; b)

$$e^x \int_1^\infty \frac{t^n}{e^{xt}}\,dt$$

for $x = 50.1$ and $n = 1$, by means of a call of *nonexpenx*.

```
package numal;

import java.text.DecimalFormat;
import numal.*;
```

```
public class Test_enx extends Object {

    public static void main(String args[]) {

        int i;
        double b[] = new double[2];
        double a[] = new double[43];

        DecimalFormat fiveDigit = new DecimalFormat("0.00000E0");
        System.out.println("ENX and NONEXPENX deliver:");
        Special_functions.enx(1.1,40,42,a);
        for (i=40; i<=42; i++)
            System.out.println(" E(" + i + ",1.1) = " + fiveDigit.format(a[i]));
        Special_functions.nonexpenx(50.1,1,1,b);
        System.out.println("\nEXP(50.1)*E(1,50.1) = " + fiveDigit.format(b[1]));

    }
}
```

Output:

```
ENX and NONEXPENX deliver:
 E(40,1.1) = 8.29521E-3
 E(41,1.1) = 8.09366E-3
 E(42,1.1) = 7.90166E-3

EXP(50.1)*E(1,50.1) = 1.95767E-2
```

Procedures tested:

sincosint, sincosfg

Example:
Evaluate

$$SI(x) = \int_0^x \frac{\sin(t)}{t}\, dt, \quad CI(x) = \gamma + \ln|x| + \int_0^x \frac{\cos(t)-1}{t}\, dt,$$

$$F(x) = CI(x)\sin(x) - \left(SI(x) - \frac{\pi}{2}\right)\cos(x),$$

$$G(x) = -CI(x)\cos(x) - \left(SI(x) - \frac{\pi}{2}\right)\sin(x)$$

for $x = 1$.

```
package numal;

import java.text.DecimalFormat;
import numal.*;

public class Test_sincosint extends Object {
```

```
  public static void main(String args[]) {

    double si[] = new double[1];
    double ci[] = new double[1];
    double f[] = new double[1];
    double g[] = new double[1];

    Special_functions.sincosint(1.0,si,ci);
    Special_functions.sincosfg(1.0,f,g);
    DecimalFormat fiveDigit = new DecimalFormat("0.00000E0");
    System.out.println("SINCOSINT and SINCOSFG deliver:");
    System.out.println(" SI(1) = " + fiveDigit.format(si[0]) + "  CI(1) = " +
      fiveDigit.format(ci[0]) + "\n  F(1) = " + fiveDigit.format(f[0]) +
      "   G(1) = " + fiveDigit.format(g[0]));
  }
}
```

Output:

```
SINCOSINT and SINCOSFG deliver:
 SI(1) = 9.46083E-1  CI(1) = 3.37404E-1
  F(1) = 6.21450E-1    G(1) = 3.43378E-1
```

Procedure tested:

recipgamma

Example:
Evaluate
$$\Gamma'(x) = \{\Gamma(1-x)\}^{-1}, \quad \Gamma_0'(x) = (2+x)^{-1}\{\Gamma'(x) - \Gamma'(-x)\}, \quad \Gamma_e'(x) = \{\Gamma'(x) + \Gamma'(-x)\}/2$$
for $x = 0.4$ and $x = 0$.

```
package numal;

import java.text.DecimalFormat;
import numal.*;

public class Test_recipgamma extends Object {

    public static void main(String args[]) {

        double x;
        double odd[] = new double[1];
        double even[] = new double[1];

        DecimalFormat fiveDigit = new DecimalFormat("0.00000E0");
        System.out.println("RECIPGAMMA delivers:");
        x=Special_functions.recipgamma(0.4,odd,even);
```

```
        System.out.println(" 0.4    " + fiveDigit.format(x) + "   " +
            fiveDigit.format(odd[0]) + "   " + fiveDigit.format(even[0]));
        x=Special_functions.recipgamma(0.0,odd,even);
        System.out.println(" 0.4    " + fiveDigit.format(x) + "    " +
            fiveDigit.format(odd[0]) + "   " + fiveDigit.format(even[0]));
    }
}
```

Output:

```
RECIPGAMMA delivers:
 0.4    6.71505E-1   -5.69444E-1   8.99283E-1
 0.4    1.00000E0    -5.77216E-1   1.00000E0
```

Procedure tested:

gamma

Example:
Evaluate $\Gamma(x)$ for x = -8.5, 0.25, 1.5, and 22.

```
package numal;

import java.text.DecimalFormat;
import numal.*;

public class Test_gamma extends Object {

    public static void main(String args[]) {

        int i;
        double x[]={-8.5, 0.25, 1.5, 22.0};

        DecimalFormat fiveDigit = new DecimalFormat("0.00000E0");
        System.out.println("GAMMA delivers:");
        for (i=0; i<=3; i++)
            System.out.println(" " + x[i] + "\t" +
                            fiveDigit.format(Special_functions.gamma(x[i])));
    }
}
```

Output:

```
GAMMA delivers:
 -8.5       -2.63352E-5
 0.25        3.62561E0
 1.5         8.86227E-1
 22.0        5.10909E19
```

Procedure tested:

loggamma

Example:
Evaluate $ln\{\Gamma(x)\}$ for x = 0.25, 1.5, 12, 15 and 80.

```
package numal;

import java.text.DecimalFormat;
import numal.*;

public class Test_loggamma extends Object {

  public static void main(String args[]) {

    int i;
    double x[]={0.25, 1.5, 12.0, 15.0, 80.0};

    DecimalFormat fiveDigit = new DecimalFormat("0.00000E0");
    System.out.println("LOGGAMMA delivers:\n");
    for (i=0; i<=4; i++)
        System.out.println(" " + x[i] + "\t" +
                        fiveDigit.format(Special_functions.loggamma(x[i])));
  }
}
```

Output:

```
LOGGAMMA delivers:

0.25      1.28802E0
1.5       -1.20782E-1
12.0      1.75023E1
15.0      2.51912E1
80.0      2.69291E2
```

Procedure tested:

incomgam

Example:
Evaluate $\gamma(4,3)$ and $\Gamma(4,3)$.

```
package numal;

import java.text.DecimalFormat;
import numal.*;
```

```
public class Test_incomgam extends Object {

    public static void main(String args[]) {

        double p[] = new double[1];
        double q[] = new double[1];

        DecimalFormat fiveDigit = new DecimalFormat("0.00000E0");
        System.out.println("INCOMGAM delivers:");
        Special_functions.incomgam(3.0,4.0,p,q,1.0*2.0*3.0,1.0e-6);
        System.out.println("  KLGAM and GRGAM are :   " +
                           fiveDigit.format(p[0]) + "   " + fiveDigit.format(q[0]));
    }
}
```

Output:

```
INCOMGAM delivers:
  KLGAM and GRGAM are :   2.11661E0   3.88339E0
```

Procedure tested:

incbeta

Example:
Evaluate $I_x(p,q)$ for $x = 0.3$, $p = 1.4$, and $q = 1.5$.

```
package numal;

import java.text.DecimalFormat;
import numal.*;

public class Test_incbeta extends Object {

  public static void main(String args[]) {

    DecimalFormat fiveDigit = new DecimalFormat("0.00000E0");
    System.out.println("INCBETA delivers:   " +
             fiveDigit.format(Special_functions.incbeta(0.3,1.4,1.5,1.0e-6)));
  }
}
```

Output:

```
INCBETA delivers:  2.79116E-1
```

Procedure tested:

ibpplusn

Example:
Evaluate $I_x(p+n,q)$ for $x = 0.3$, $p = 0.4$, $q = 1.5$, and $n = 0, 1, 2$.

```
package numal;

import java.text.DecimalFormat;
import numal.*;

public class Test_ibpplusn extends Object {

    public static void main(String args[]) {

        double isubx[] = new double[3];

        DecimalFormat fiveDigit = new DecimalFormat("0.00000E0");
        Special_functions.ibpplusn(0.3,0.4,1.5,2,1.0e-6,isubx);
        System.out.println("IBPPLUSN delivers:\n  " + fiveDigit.format(isubx[0]) +
            "   " + fiveDigit.format(isubx[1]) + "   " + fiveDigit.format(isubx[2]));
    }
}
```

Output:

```
IBPPLUSN delivers:
  7.21671E-1   2.79116E-1   9.89329E-2
```

Procedure tested:

ibqplusn

Example:
Evaluate $I_x(p,q+n)$ for $x = 0.3$, $p = 1.4$, $q = 0.5$, and $n = 0, 1, 2$.

```
package numal;

import java.text.DecimalFormat;
import numal.*;

public class Test_ibqplusn extends Object {

    public static void main(String args[]) {

        double isubx[] = new double[3];

        DecimalFormat fiveDigit = new DecimalFormat("0.00000E0");
        Special_functions.ibqplusn(0.3,1.4,0.5,2,1.0e-6,isubx);
```

```
        System.out.println("IBQPLUSN delivers:\n  " + fiveDigit.format(isubx[0]) +
            "   " + fiveDigit.format(isubx[1]) + "    " + fiveDigit.format(isubx[2]));
    }
}
```

Output:

```
IBQPLUSN delivers:
  8.94495E-2    2.79116E-1    4.47287E-1
```

Procedures tested:

errorfunction, nonexperfc

Example:
Evaluate

$$erf(1) = \frac{2}{\sqrt{\pi}} \int_0^1 e^{-t^2} dt$$

$$erfc(1) = \frac{2}{\sqrt{\pi}} \int_1^\infty e^{-t^2} dt$$

$$non\exp erfc(x) = \frac{2e^{x^2}}{\sqrt{\pi}} \int_x^\infty e^{-t^2} dt$$

for $x = 100$.

```
package numal;

import java.text.DecimalFormat;
import numal.*;

public class Test_errorfunction extends Object {

    public static void main(String args[]) {

        double p;
        double erf[]  = new double[1];
        double erfc[] = new double[1];

        Special_functions.errorfunction(1.0,erf,erfc);
        p=Special_functions.nonexperfc(100.0);
        DecimalFormat fiveDigit = new DecimalFormat("0.00000E0");
        System.out.println("ERRORFUNCTION and NONEXPERFC deliver:" +
                    "\n  ERF(1)   =  " + fiveDigit.format(erf[0]) +
                    "\n  ERFC(1)  =  " + fiveDigit.format(erfc[0]) +
                    "\n  NONEXPERFC(100)  =  " + fiveDigit.format(p));
    }
}
```

Output:

```
ERRORFUNCTION and NONEXPERFC deliver:
  ERF(1)  =  8.42701E-1
  ERFC(1) =  1.57299E-1
  NONEXPERFC(100) =  5.64161E-3
```

Procedure tested:

inverseerrorfunction

Example:
Evaluate

$$x = \frac{2}{\sqrt{\pi}} \int_0^{y(x)} e^{-t^2} dt$$

for $x = 0.6$ and $1\text{-}10^{-30}$.

```java
package numal;

import java.text.DecimalFormat;
import numal.*;

public class Test_inverseerrorfunction extends Object {

  public static void main(String args[]) {

    double inverf1[] = new double[1];
    double inverf2[] = new double[1];

    Special_functions.inverseerrorfunction(0.6,0.0,inverf1);
    Special_functions.inverseerrorfunction(1.0,1.0e-30,inverf2);
    DecimalFormat fiveDigit = new DecimalFormat("0.00000E0");
    System.out.println("INVERSEERRORFUNCTION delivers:\n" +
               " X = 0.6,  INVERF = " + fiveDigit.format(inverf1[0]) +
               "\n X = 1.0,  INVERF = " + fiveDigit.format(inverf2[0]));
  }
}
```

Output:

```
INVERSEERRORFUNCTION delivers:
 X = 0.6,  INVERF = 5.95116E-1
 X = 1.0,  INVERF = 8.14862E0
```

Procedures tested:

fresnel, fg

Example:
Evaluate

$$C(x) = \int_0^x \cos\left(\frac{\pi t^2}{2}\right) dt, \quad S(x) = \int_0^x \sin\left(\frac{\pi t^2}{2}\right) dt,$$

$$F(x) = \left(\frac{1}{2} - S(x)\right)\cos\left(\frac{\pi x^2}{2}\right) - \left(\frac{1}{2} - C(x)\right)\sin\left(\frac{\pi x^2}{2}\right),$$

$$G(x) = \left(\frac{1}{2} - C(x)\right)\cos\left(\frac{\pi x^2}{2}\right) + \left(\frac{1}{2} - S(x)\right)\sin\left(\frac{\pi x^2}{2}\right)$$

for $x = 1$.

```
package numal;

import java.text.DecimalFormat;
import numal.*;

public class Test_fresnel extends Object {

    public static void main(String args[]) {

        double c[] = new double[1];
        double s[] = new double[1];
        double f[] = new double[1];
        double g[] = new double[1];

        Special_functions.fresnel(1.0,c,s);
        Special_functions.fg(1.0,f,g);
        DecimalFormat fiveDigit = new DecimalFormat("0.00000E0");
        System.out.println("FRESNEL and FG deliver:");
        System.out.println("  C(1) = " + fiveDigit.format(c[0]) + "   S(1) = " +
            fiveDigit.format(s[0]) + "\n  F(1) = " + fiveDigit.format(f[0]) +
            "   G(1) = " + fiveDigit.format(g[0]));
    }
}
```

Output:

```
FRESNEL and FG deliver:
  C(1) = 7.79893E-1    S(1) = 4.38259E-1
  F(1) = 2.79893E-1    G(1) = 6.17409E-2
```

Procedures tested:

bessj0, bessj1, bessj

Example:

Evaluate $J_k''(x) - J_k'(x)$, $k=0,1$, where $J_k''(x)$ is $J_k(x)$ as evaluated by *bessj* ($k=0,1$) and $J_0'(x)$, $J_1'(x)$ are $J_0(x)$, $J_1(x)$ as evaluated by *bessj0* and *bessj1*, respectively, for $x = 1$, 5, 10, and 25.

```java
package numal;

import java.text.DecimalFormat;
import numal.*;

public class Test_bessj extends Object {

    public static void main(String args[]) {

        int i;
        double j[] = new double[2];
        double x[]={1.0, 5.0, 10.0, 25.0};

        DecimalFormat fiveDigit = new DecimalFormat("0.00000E0");
        for (i=0; i<=3; i++) {
            Special_functions.bessj(x[i],1,j);
            System.out.println("  " + x[i] + "\t" +
                    fiveDigit.format(j[0]-Special_functions.bessj0(x[i])) + "\t" +
                    fiveDigit.format(j[1]-Special_functions.bessj1(x[i])));
        }
    }
}
```

Output:

```
1.0        -3.33067E-16    -5.55112E-16
5.0         3.33067E-16     7.77156E-16
10.0        4.99600E-16     1.24900E-15
25.0       -5.13478E-16    -3.33067E-16
```

Procedure tested:

bessy01

Example:
Evaluate $Y_0(1)$ and $Y_1(1)$.

```java
package numal;

import java.text.DecimalFormat;
import numal.*;
```

```
public class Test_bessy01 extends Object {

  public static void main(String args[]) {

    double x;
    double y0[] = new double[1];
    double y1[] = new double[1];

    x=1.0;
    Special_functions.bessy01(x,y0,y1);
    DecimalFormat fiveDigit = new DecimalFormat("0.00000E0");
    System.out.println("BESSY01 delivers:\n  " + x + "    " +
                    fiveDigit.format(y0[0]) + "    " + fiveDigit.format(y1[0]));
  }
}
```

Output:

```
BESSY01 delivers:
  1.0    8.82570E-2    -7.81213E-1
```

Procedure tested:
bessy

Example:
Evaluate $Y_i(1)$ for $i = 0, 1, 2$.

```
package numal;

import java.text.DecimalFormat;
import numal.*;

public class Test_bessy extends Object {

    public static void main(String args[]) {

        double y[]= new double[3];

        DecimalFormat fiveDigit = new DecimalFormat("0.00000E0");
        Special_functions.bessy(1.0,2,y);
        System.out.print("BESSY delivers:\n  " + fiveDigit.format(y[0]) + "    " +
                    fiveDigit.format(y[1]) + "    " + fiveDigit.format(y[2]));
    }
}
```

Output:

```
BESSY delivers:
  8.82570E-2   -7.81213E-1   -1.65068E0
```

Procedures tested:

besspq0, besspq1

Example:

Verify the relation

$$P_0(x)*P_1(x) + Q_0(x)*Q_1(x) = 1$$

for x = 1, 3, 5, 10.

```
package numal;

import java.text.DecimalFormat;
import numal.*;

public class Test_besspq extends Object {

    public static void main(String args[]) {

        int i;
        double p[] = new double[1];
        double q[] = new double[1];
        double r[] = new double[1];
        double s[] = new double[1];
        double x[]={1.0, 3.0, 5.0, 10.0};

        DecimalFormat twoDigit = new DecimalFormat("0.00E0");
        System.out.println("BESSPQ0 and BESSPQ1 deliver:");
        for (i=0; i<=3; i++) {
            Special_functions.besspq0(x[i],p,q);
            Special_functions.besspq1(x[i],r,s);
            System.out.print("    " +
                        twoDigit.format(Math.abs(p[0]*r[0]+q[0]*s[0]-1.0)));
        }
    }
}
```

Output:

```
BESSPQ0 and BESSPQ1 deliver:
   1.22E-15    2.00E-15    5.55E-16    1.11E-15
```

Procedures tested:
bessi, bessk

Example:
Verify the relation
$$x*\{I_{n-1}(x)*K_{n-1}(x) + I_n(x)*K_{n-1}(x)\} = 1$$
for $x = 1,...,20$, and $n = 1,...,5$.

```
package numal;

import java.text.DecimalFormat;
import numal.*;

public class Test_bessi extends Object {

    public static void main(String args[]) {

        int j,n;
        double x;
        double i[] = new double[6];
        double k[] = new double[6];

        DecimalFormat oneDigit = new DecimalFormat("0.0E0");
        System.out.println("BESSI and BESSK deliver:");
        for (j=1; j<=20; j++) {
            x=j;
            System.out.print("\n  " + x);
            Special_functions.bessi(x,5,i);
            Special_functions.bessk(x,5,k);
            for (n=1; n<=5; n++)
                System.out.print("\t" +
                    oneDigit.format(Math.abs(x*(i[n]*k[n-1]+i[n-1]*k[n])-1.0)));
        }
    }
}
```

Output:

```
BESSI and BESSK deliver:

    1.0      0.0E0      1.1E-16    0.0E0      0.0E0      2.2E-16
    2.0      6.0E-15    6.2E-15    6.2E-15    6.4E-15    6.2E-15
    3.0      5.3E-15    5.1E-15    4.9E-15    5.1E-15    5.1E-15
    4.0      5.1E-15    5.1E-15    5.1E-15    5.1E-15    5.1E-15
    5.0      5.1E-15    4.9E-15    4.9E-15    4.9E-15    4.9E-15
    6.0      1.3E-15    1.6E-15    1.3E-15    1.3E-15    1.3E-15
    7.0      8.9E-16    8.9E-16    8.9E-16    6.7E-16    6.7E-16
    8.0      1.6E-15    1.3E-15    1.3E-15    1.3E-15    1.3E-15
    9.0      1.3E-15    1.3E-15    1.3E-15    1.1E-15    1.3E-15
```

10.0	8.9E-16	1.1E-15	8.9E-16	8.9E-16	8.9E-16
11.0	1.3E-15	1.3E-15	1.3E-15	1.3E-15	1.3E-15
12.0	1.3E-15	1.3E-15	1.3E-15	1.3E-15	1.3E-15
13.0	1.1E-15	1.1E-15	1.1E-15	1.1E-15	1.1E-15
14.0	8.9E-16	8.9E-16	8.9E-16	8.9E-16	1.1E-15
15.0	1.3E-15	1.3E-15	1.3E-15	1.3E-15	1.3E-15
16.0	8.9E-16	8.9E-16	8.9E-16	8.9E-16	8.9E-16
17.0	8.9E-16	1.1E-15	1.1E-15	8.9E-16	8.9E-16
18.0	8.9E-16	8.9E-16	8.9E-16	8.9E-16	8.9E-16
19.0	1.3E-15	1.3E-15	1.3E-15	1.3E-15	1.3E-15
20.0	1.3E-15	1.3E-15	1.3E-15	1.1E-15	1.1E-15

Procedure tested:

bessk01

Example:
Evaluate $K_0(x)$ and $K_1(x)$ for $x = 0.5, 1.5, 2.5$.

```
package numal;

import java.text.DecimalFormat;
import numal.*;

public class Test_bessk01 extends Object {

    public static void main(String args[]) {

        int j;
        double x;
        double k0[] = new double[1];
        double k1[] = new double[1];

        DecimalFormat fiveDigit = new DecimalFormat("0.00000E0");
        System.out.println("BESSK01 delivers:");
        x=0.5;
        for (j=1; j<=3; j++) {
            Special_functions.bessk01(x,k0,k1);
            System.out.println("  " + x + "    " + fiveDigit.format(k0[0]) + "    " +
                            fiveDigit.format(k1[0]));
            x += 1.0;
        }
    }
}
```

Output:

```
BESSK01 delivers:
   0.5    9.24419E-1    1.65644E0
   1.5    2.13806E-1    2.77388E-1
   2.5    6.23476E-2    7.38908E-2
```

Procedure tested:

nonexpbessk01

Example:

Evaluate $e^x K_0(x)$ and $e^x K_1(x)$ for x = 0.5, 1.0, 1.5, 2.0, 2.5.

```
package numal;

import java.text.DecimalFormat;
import numal.*;

public class Test_nonexpbessk01 extends Object {

    public static void main(String args[]) {

        int j;
        double x;
        double k0[] = new double[1];
        double k1[] = new double[1];

        DecimalFormat fiveDigit = new DecimalFormat("0.00000E0");
        System.out.println("NONEXPBESSK01 delivers:");
        x=0.5;
        for (j=1; j<=5; j++) {
            Special_functions.nonexpbessk01(x,k0,k1);
            System.out.println("  " + x + "\t" + fiveDigit.format(k0[0]) + "\t" +
                            fiveDigit.format(k1[0]));
            x += 0.5;
        }
    }
}
```

Output:

```
NONEXPBESSK01 delivers:
   0.5    1.52411E0   2.73101E0
   1.0    1.14446E0   1.63615E0
   1.5    9.58210E-1  1.24317E0
   2.0    8.41568E-1  1.03348E0
   2.5    7.59549E-1  9.00174E-1
```

Procedure tested:

nonexpbessk

Example:
Evaluate $e^x K_i(x)$ for $x = 0.5, 1.0, 1.5, 2.0$, and $i = 0, 1, 2$.

```
package numal;

import java.text.DecimalFormat;
import numal.*;

public class Test_nonexpbessk extends Object {

    public static void main(String args[]) {

        int j;
        double x;
        double k[] = new double[3];

        DecimalFormat fiveDigit = new DecimalFormat("0.00000E0");
        System.out.println("NONEXPBESSK delivers:");
        x=0.5;
        for (j=1; j<=4; j++) {
            Special_functions.nonexpbessk(x,2,k);
            System.out.println("  " + x + "\t" + fiveDigit.format(k[0]) + "\t" +
                        fiveDigit.format(k[1]) + "\t" + fiveDigit.format(k[2]));
            x += 0.5;
        }
    }
}
```

Output:

```
NONEXPBESSK delivers:
   0.5    1.52411E0   2.73101E0   1.24481E1
   1.0    1.14446E0   1.63615E0   4.41677E0
   1.5    9.58210E-1  1.24317E0   2.61576E0
   2.0    8.41568E-1  1.03348E0   1.87505E0
```

Procedure tested:

bessjaplusn

Example:
Evaluate $J_{a+i}(x)$ for $x = 2$, $a = 0.78$, and $i = 0, 1, 2$.

```
package numal;

import java.text.DecimalFormat;
import numal.*;

public class Test_bessjaplusn extends Object {

    public static void main(String args[]) {

        int n;
        double a,x;
        double ja[] = new double[3];

        x=2.0;
        a=0.78;
        n=2;
        Special_functions.bessjaplusn(a,x,n,ja);
        DecimalFormat fiveDigit = new DecimalFormat("0.00000E0");
        System.out.println("BESSJAPLUSN delivers:\n  " + "X = " + x + "    A = " +
                    a + "    N = " + n + "\n  " + fiveDigit.format(ja[0]) + "    " +
                    fiveDigit.format(ja[1]) + "    " + fiveDigit.format(ja[2]));
    }
}
```

Output:

```
BESSJAPLUSN delivers:
  X = 2.0    A = 0.78    N = 2
  5.73061E-1    4.15295E-1    1.66163E-1
```

Procedure tested:

besspqa01

Example:
Verify the relation
$$P_a(x)*P_{a+1}(x) + Q_a(x)*Q_{a+1}(x) = 1$$
for $a = 0$, and $x = 1, 3, 5, 10, 15, 20, 50$.

```
package numal;

import java.text.DecimalFormat;
import numal.*;

public class Test_besspqa01 extends Object {

    public static void main(String args[]) {

        int i;
```

```
        double p[]  = new double[1];
        double q[]  = new double[1];
        double r[]  = new double[1];
        double s[]  = new double[1];
        double x[]={1.0, 3.0, 5.0, 10.0, 15.0, 20.0, 50.0};

        DecimalFormat fiveDigit = new DecimalFormat("0.00000E0");
        System.out.println("BESSPQA01 delivers:\n");
        for (i=0; i<=6; i++) {
            Special_functions.besspqa01(0.0,x[i],p,q,r,s);
            System.out.println(" " + x[i] + "\t" +
                            fiveDigit.format(Math.abs(p[0]*r[0]+q[0]*s[0]-1.0)));
        }
    }
}
```

Output:

```
BESSPQA01 delivers:

 1.0        1.22125E-15
 3.0        1.99840E-15
 5.0        5.55112E-16
 10.0       1.11022E-15
 15.0       1.55431E-15
 20.0       1.33227E-15
 50.0       1.33227E-15
```

Procedure tested:

besszeros

Example:
Compute the first two zeros of $Y_{3.14}(z)$.

```
package numal;

import java.text.DecimalFormat;
import numal.*;

public class Test_besszeros extends Object {

    public static void main(String args[]) {

        int n,d;
        double a;
        double z[]  = new double[3];

        a=3.14;
```

```
        n=d=2;
        Special_functions.besszeros(a,n,z,d);
        DecimalFormat fiveDigit = new DecimalFormat("0.00000E0");
        System.out.println("The first two zeros of " +
                        "the Bessel function Y of the order 3.14:\n  " +
                        fiveDigit.format(z[1]) + "    " + fiveDigit.format(z[2]));
    }
}
```

Output:

```
The first two zeros of the Bessel function Y of the order 3.14:
  4.68478E0    8.27659E0
```

Procedures tested:

spherbessi, nonexpspherbessi

Example:
Compute

$$\sqrt{\frac{\pi}{2x}}\, I_{j+\frac{1}{2}}(x) \quad \text{and} \quad e\!\left(\frac{1}{e^x}\sqrt{\frac{\pi}{2x}}\, I_{j+\frac{1}{2}}(x) \right)$$

for $x = 1$, and $j = 0,...,3$.

```
package numal;

import java.text.DecimalFormat;
import numal.*;

public class Test_spherbessi extends Object {

    public static void main(String args[]) {

        int n;
        double x,expx;
        double i1[] = new double[4];
        double i2[] = new double[4];

        System.out.println("SPHERBESSI and NONEXPSPHERBESSI deliver:");
        x=1.0;
        expx=Math.exp(x);
        n=3;
        Special_functions.spherbessi(x,n,i1);
        Special_functions.nonexpspherbessi(x,n,i2);
        DecimalFormat fiveDigit = new DecimalFormat("0.00000E0");
        for (n=0; n<=3; n++)
            System.out.println("  " + n + "\t" + fiveDigit.format(i1[n]) + "\t" +
                        fiveDigit.format(i2[n]*expx));
```

```
    }
}
```

Output:

```
SPHERBESSI and NONEXPSPHERBESSI deliver:
  0 1.17520E0    1.17520E0
  1 3.67879E-1   3.67879E-1
  2 7.15629E-2   7.15629E-2
  3 1.00651E-2   1.00651E-2
```

Procedures tested:

spherbessk, nonexpspherbessk

Example:
Compute

$$\sqrt{\frac{\pi}{2x}} K_{j+\frac{1}{2}}(x) \quad \text{and} \quad \frac{1}{e^2}\left(e^x \sqrt{\frac{\pi}{2x}} K_{j+\frac{1}{2}}(x)\right)$$

for $x = 2$, and $j = 0,...,3$.

```
package numal;

import java.text.DecimalFormat;
import numal.*;

public class Test_spherbessk extends Object {

    public static void main(String args[]) {

        int n;
        double x,expx;
        double k1[] = new double[4];
        double k2[] = new double[4];

        System.out.println("SPHERBESSK and NONEXPSPHERBESSK deliver:");
        x=2.0;
        expx=Math.exp(-x);
        n=3;
        Special_functions.spherbessk(x,n,k1);
        Special_functions.nonexpspherbessk(x,n,k2);
        DecimalFormat fiveDigit = new DecimalFormat("0.00000E0");
        for (n=0; n<=3; n++)
            System.out.println("  " + n + "\t" + fiveDigit.format(k1[n]) + "\t" +
                            fiveDigit.format(k2[n]*expx));
    }
}
```

Output:

```
SPHERBESSK and NONEXPSPHERBESSK deliver:
   0 1.06292E-1  1.06292E-1
   1 1.59438E-1  1.59438E-1
   2 3.45449E-1  3.45449E-1
   3 1.02306E0   1.02306E0
```

Procedure tested:

airy

Example:

Compute $Ai(x)$, $dAi(x)/dx$, $Bi(x)$, and $dBi(x)/dx$ for $x = 9.654894$.

```java
package numal;

import java.text.DecimalFormat;
import numal.*;

public class Test_airy extends Object {

    public static void main(String args[]) {

        double statictemp[] = new double[26];
        double a[] = new double[1];
        double b[] = new double[1];
        double c[] = new double[1];
        double d[] = new double[1];
        double e[] = new double[1];

        DecimalFormat fiveDigit = new DecimalFormat("0.00000E0");
        Special_functions.airy(9.654894,a,b,c,d,e,true,statictemp);
        System.out.println("AIRY delivers:" + "\n AI (9.654894) = " +
                fiveDigit.format(a[0]*Math.exp(-e[0])) + "\n AID(9.654894) = " +
                fiveDigit.format(b[0]*Math.exp(-e[0])) + "\n BI (9.654894) = " +
                fiveDigit.format(c[0]*Math.exp(e[0])) + "\n BID(9.654894) = " +
                fiveDigit.format(d[0]*Math.exp(e[0])));
    }
}
```

Output:

```
AIRY delivers:
 AI (9.654894) = 3.28735E-10
 AID(9.654894) = -1.02980E-9
 BI (9.654894) = 1.55839E8
 BID(9.654894) = 4.80104E8
```

Procedure tested:
airyzeros

Example:
Compute the i-th zero of $Bi(x)$ and the value of $dBi(x)/dx$ at this point for $i = 3$.

```
package numal;

import java.text.DecimalFormat;
import numal.*;

public class Test_airyzeros extends Object {

    public static void main(String args[]) {

        double b;
        double zbi[] = new double[4];
        double vbid[] = new double[4];

        DecimalFormat fiveDigit = new DecimalFormat("0.00000E0");
        b=Special_functions.airyzeros(3,2,zbi,vbid);
        System.out.println("AIRYZEROS delivers:\n" +
                    " The third zero of BI(X) is  " + fiveDigit.format(b) +
                    "\n" + " The value of (D/DX)BI(X) in this point is  " +
                    fiveDigit.format(vbid[3]));

    }
}
```

Output:

```
AIRYZEROS delivers:
 The third zero of BI(X) is  -4.83074E0
 The value of (D/DX)BI(X) in this point is  8.36991E-1
```

Examples for chapter 7 procedures

Procedure tested:
newton

Example:
Compute the divided differences $\delta_x^j f(x_0)$, $j=0,1,2$, from the function values $f_0=1$, $f_1=f_2=0$, and associated arguments $x_0=0$, $x_1=1/2$. $x_2=1$.

```
package numal;

import java.text.DecimalFormat;
```

```
import numal.*;

public class Test_newton extends Object {

    public static void main(String args[]) {

        double x[]  = new double[3];
        double f[]  = new double[3];

        x[0]=0.0;   x[1]=0.5;   x[2]=1.0;
        f[0]=1.0;   f[1]=f[2]=0.0;
        DecimalFormat fiveDigit = new DecimalFormat("0.00000E0");
        Approximation.newton(2,x,f);
        System.out.println("The Newton coefficients are:\n  " +
                        fiveDigit.format(f[0]) + "    " + fiveDigit.format(f[1]) +
                        "    " + fiveDigit.format(f[2]));
    }
}
```

Output:

```
The Newton coefficients are:
  1.00000E0    -2.00000E0   2.00000E0
```

Procedure tested:

ini

Example:
Determine integer approximations $s(j)$, $j=0,1,2$, to the points in the range $[0,20]$ at which $T_2((x-10)/10)$ assumes its maximal absolute values.

```
package numal;

import numal.*;

public class Test_ini extends Object {

    public static void main(String args[]) {

        int s[] = new int[3];

        Approximation.ini(2,20,s);
        System.out.println("INI selects out of 0,1,...,20 the" + " numbers:\n  " +
                        s[0] + "   " + s[1] + "    " + s[2]);
    }
}
```

Output:

```
INI selects out of 0,1,...,20 the numbers:
  0   10    20
```

Procedure tested:

sndremez

Example:
Given the numbers g_i, $i=0,...,7$, in order: 10, 12, -15, -10, -14, 15, 10, 11 and $s(j)$, $j=0,1,2$, in order: 0, 3, 6 (so that $g_{s(0)}$ = 10, $g_{s(1)}$ = -10, $g_{s(2)}$ = 10) selects an integer sequence $S(j)$, $j=0,1,2$, such that the numbers $g_{S(j)}$ have the properties described in the documentation to *sndremez*.

```
package numal;

import numal.*;

public class Test_sndremez extends Object {

    public static void main(String args[]) {

        int s[] = new int[3];
        double em[] = new double[2];
        double g[] = new double[8];

        System.out.println("SNDREMEZ delivers:\n");
        g[0]=10.0;  g[1]=12.0;  g[2] = -15.0;  g[3] = -10.0;
        g[4] = -14.0;  g[5]=15.0;  g[6]=10.0;  g[7]=11.0;
        em[0]=10.0;  s[0]=0;  s[1]=3;  s[2]=6;
        System.out.println("The numbers:\n S[J]:    " + s[0] + "    " + s[1] + "    " +
                      s[2] + "\n G[S[j]]:" + "    " + g[s[0]] + "    " + g[s[1]] +
                      "    " + g[s[2]]);
        Approximation.sndremez(2,7,s,g,em);
        System.out.println("are exchanged with:\n S[J]:    " + s[0] + "    " + s[1] +
                      "    " + s[2] + "\n G[S[j]]:" + "    " + g[s[0]] + "    " +
                      g[s[1]] + "    " + g[s[2]] + "\n\n" +
                      "The reference set of function values is:\n " + g[0] +
                      "    " + g[1] + "    " + g[2] + "    " + g[3] + "    " +
                      g[4] + "    " + g[5] + "    " + g[6] + "    " + g[7]);
    }
}
```

Output:

```
SNDREMEZ delivers:

The numbers:
 S[J]:    0    3    6
 G[S[j]]:  10.0   -10.0    10.0
are exchanged with:
 S[J]:    0    2    5
 G[S[j]]:  10.0   -15.0    15.0

The reference set of function values is:
 10.0    12.0    -15.0    -10.0    -14.0    15.0    10.0    11.0
```

Procedure tested:

minmaxpol

Example:

Determine c_0 and c_1 such that

$$\max \left| f(y_j) - c_0 - c_1 y_j \right|$$

where $y_j = -1+0.01j$ and $f(y) = 1/(10-y)$, is a minimum as j ranges over the values $0,...,200$.

```java
package numal;

import java.text.DecimalFormat;
import numal.*;

public class Test_minmaxpol extends Object {

    static double f(double x) {
        return 1.0/(x-10.0);
    }

    static void compute(int n, double a, double b) {
        int k,l,m;
        double r,idm;
        double em[] = new double[4];
        double coef[] = new double[n+1];

        em[2]=10*n+5;
        m=100*n+10;
        double y[] = new double[m+1];
        double fy[] = new double[m+1];
        idm=(b-a)/m;
        r=y[0]=a;
        fy[0]=f(r);
        r=y[m]=b;
```

```
        fy[m]=f(r);
        l=m-1;
        for (k=1; k<=l; k++) {
            r=y[k]=a+k*idm;
            fy[k]=f(r);
        }
        DecimalFormat fiveDigit = new DecimalFormat("0.00000E0");
        Approximation.minmaxpol(n,m,y,fy,coef,em);
        System.out.print(" COEF:    ");
        for (k=0; k<=n; k++)
            System.out.print("   " + fiveDigit.format(coef[k]));
        System.out.println("\n EM[0:3]:  " + fiveDigit.format(em[0]) + "   " +
                           fiveDigit.format(em[1]) + "   " + em[2] + "   " + em[3]);
    }

    public static void main(String args[]) {

        int n;

        System.out.println("MINMAXPOL delivers:");
        n=1;
        System.out.println(" Degree = " + n);
        compute(n,-1.0,1.0);
    }
}
```

Output:

```
MINMAXPOL delivers:
 Degree = 1
 COEF:    -1.00504E-1  -1.01010E-2
 EM[0:3]: -5.06310E-4   5.06310E-4   15.0   3.0
```

Appendix A: References

[AbS65]   M. Abramowitz and I.A. Stegun, *Handbook of Mathematical Functions*, Dover Publications, Inc., New York, 1965.

[Ad67]   D.A. Adams, "A Stopping Criterion for Polynomial Root Finding", *CACM* 10, No. 10, October 1967, 655-658.

[AHU74]   A.V. Aho, J.E. Hopcroft, and J.D. Ullman, *The Design and Analysis of Computer Algorithms*, Addison-Wesley, Reading, MA, 1974.

[Bab72]   I. Babuska, "Numerical Stability in Problems of Linear Algebra", *SIAM J. Numer. Anal.* 9, 1972, 53-77.

[Bak77]   M. Bakker, "Galerkin Methods in Spherical Regions", Report NW 50, Mathematical Centre, Amsterdam, 1977.

[BakH76]   M. Bakker, P.W. Hemker, P.J. van der Houwen, S.J. Polak, and M. van Veldhuizen, "Colloquium on Discretization Methods", MC Syllabus NR 27, Mathematisch Centrum, Amsterdam, 1976.

[Be72]   P.A. Beentjes, "An Algol 60 Version of Stabilized Runge Kutta Methods", Report NR 23, Mathematical Centre, Amsterdam, 1972.

[Be74]   P.A. Beentjes, "Some Special Formulas of the England Class of Fifth Order Runge-Kutta Schemes", Report NW 24, Mathematical Centre, Amsterdam, 1974.

[BeDHKW73]   P.A. Beentjes, K. Dekker, H.C. Hemker, S.P.N. van Kampen, and G.M. Willems, "Colloquium Stiff Differential Equations 2", MC Syllabus NR 15.2, Mathematical Centre, Amsterdam, 1973.

[BeDHV74]   P.A. Beentjes, K. Dekker, H.C. Hemker, and M.V. Veldhuizen, "Colloquium Stiff Differential Equations 3", MC Syllabus NR 15.3, Mathematical Centre, Amsterdam, 1974.

[BjG67]   A. Bjorck and G.H. Golub, "Iterative Refinement of Least Squares Solutions by Householder Transformation", *BIT* 7, 1967, 322-337.

[Bla74]   J.M. Blair, "Rational Chebyshev Approximations for the Modified Bessel Functions $I_0(x)$ and $I_1(x)$", *Math. Comp.* 28, 1974, 581-583.

[Bre73]   R.P. Brent, *Algorithms for Minimization without Derivatives*, Prentice-Hall, Englewood Cliffs, NJ, 1973.

[Bro71]   C.G. Broyden, "The Convergence of an Algorithm for Solving Sparse Nonlinear Systems", *Math. Comp.* 25, No.114, 1971, 285-294.

[Bul67]   R. Bulirsch, "Numerical Calculation of the Sine, Cosine and Fresnel Integrals", *Numer. Math.* 9, 1967, 380-385.

[BulS65]   R. Bulirsch and J. Stoer, "Numerical Treatment of Ordinary Differential Equations by Extrapolation Methods", *Numer. Math.* 8, 1965, 1-13.

[BunK77]   J.R. Bunch and L. Kaufman, "Some Stable Methods for Calculating Inertia and Solving Symmetric Linear Systems", *Math. Comp.* 31, 1977, 163-180.

[BunKP76] J.R. Bunch, L. Kaufman, and B.N. Parlett, "Decomposition of a Symmetric Matrix", *Numerische Mathematik* 27, 1976, 95-109.

[Bus72a]   J.C.P. Bus, "Linear Systems with Calculation of Error Bounds and Iterative Refinement", LR 3.4.19, Mathematical Centre, Amsterdam, 1972.

[Bus72b]   J.C.P. Bus, "Minimization of Functions of Several Variables", NR 29, Mathematical Centre, Amsterdam, 1972.

[Bus76]   J.C.P. Bus (Ed.), "Colloquium Numerieke Programmatuur", MC Syllabus NR 29.1a, 29.1b, Mathematical Centre, Amsterdam, 1976.

[BusD75]   J.C.P. Bus and T.J. Dekker, "Two Efficient Algorithms with Guaranteed Convergence for Finding a Zero of a Function", *ACM Transactions on Mathematical Software*, 1975, 330-345.

[BusDK75] J.C.P. Bus, B. van Domselaar, and J. Kok, "Nonlinear Least Squares Estimation", Report NW 17, Mathematical Centre, Amsterdam, 1975.

[Busi71]   P.A. Businger, "Monitoring the Numerical Stability of Gaussian Elimination", *Numer. Math.* 16, 1971, 360-361.

[BusiG65] P. Businger and G.H. Golub, "Linear Least Squares Solution by Householder Transformations", *Numerische Mathematik* 7, 1965, 269-276.

[Cle62]   C.W. Clenshaw, "Chebyshev Series for Mathematical Functions", *Mathematical Tables* Vol. 5, Nat. Physical Lab., Her Majesty's Stationery Office, London, 1962.

[Cod68]   W.J. Cody, "Chebyshev Approximations for the Fresnel Integrals", *Math. Comp.* 22, 1968, 450-453.

[Cod69]   W.J. Cody, "Rational Chebyshev Approximations for the Error Function", *Math. Comp.* 23, 1969, 631-637.

[CodT68] W. J. Cody and H. C. Thacher, Jr., "Rational Chebyshev Approximations for the Exponential Integral ei(x)", *Math. Comp.* 22, 1968, 641-649.

[CodT69] W. J. Cody and H. C. Thacher, Jr., "Chebyshev Approximations for the Exponential Integral ei(x)", *Math. Comp.* 23, 1969, 289-303.

[CooHHS73] T.M.T. Coolen, P.W. Hemker, P.J. van der Houwen, and E. Slagt, "Algol 60 Procedures for Initial and Boundary Value Problems", MC Syllabus NR 20, Mathematical Centre, Amsterdam, 1973.

[Dan69] J.W. Daniel, "Summation of a Series of Positive Terms by Condensation Transformations", *Math. Comp.* 23, 1969, 91-96.

[Dav59] W.C. Davidon, "Variable Metric Methods for Minimization", Argonne National Lab., Report 5990, 1959.

[De72] K. Dekker, "An Algol 60 Version of Exponentially Fitted Runge-Kutta Methods", Report NR 25, Mathematical Centre, Amsterdam, 1972.

[Dek66] T. J. Dekker, "Newton-Laguerre Iteration", Mathematisch Centrum MR 82, Amsterdam, 1966.

[Dek68] T. J. Dekker, "ALGOL 60 Procedures in Numerical Algebra, Part 1", Mathematical Centre Tract 22, Mathematisch Centrum, Amsterdam, 1968.

[Dek71] T. J. Dekker, "Numerical Algebra", MC Syllabus NR 12, Mathematical Centre, Amsterdam, 1971.

[DekHH72] T. J. Dekker, P. W. Hemker, and P. J. van der Houwen, "Colloquium Stiff Differential Equations 1", MC Syllabus NR 15.1, Mathematical Centre, Amsterdam, 1972.

[DekHo68] T. J. Dekker and W. Hoffmann, "ALGOL 60 Procedures in Numerical Algebra, Part 2", Mathematical Centre Tract 23, Mathematisch Centrum, Amsterdam, 1968.

[DekR71] T. J. Dekker and C. J. Roothart, "Introduction to Numerical Analysis", Report CR 24, Mathematical Centre, Amsterdam, 1971.

[Eng69] R. England, "Error Estimates for Runge-Kutta Type Solutions to Systems of Ordinary Differential Equations", *Computer Journal* 12, 1969, 166-169.

[Fle63] R. Fletcher, "A New Approach to Variable Metric Algorithms", *Computer Journal* 6, 1963, 163-168.

[FoP68]   L. Fox and I. B. Parker, *Chebyshev Polynomials in Numerical Analysis*, Oxford University Press, 1968.

[Fox72]   P. Fox, "A Comparative Study of Computer Programs for Integrating Differential Equations", *CACM* 15, 1972, 941-948.

[Fr61]   J.G. Francis, "The QR Transformation, Parts 1 and 2", *Computer Journal* 4, 1961, 265-271, 332-345.

[G61]   W. Gautschi, "Recursive Computation of Certain Integrals", *JACM* 8, 1961, 21-40.

[G64]   W. Gautschi, "Incomplete Beta Function Ratios", *CACM* 7, 1964, 143-144.

[G67]   W. Gautschi, "Computational Aspects of Three-term Recurrence Relations", *SIAM Review* 9, 1967, 24-82.

[G68a]   W. Gautschi, "Construction of Gauss-Christoffel Formulas", *Math. Comp.* 22, 1968, 251-270.

[G68b]   W. Gautschi, "Gaussian Quadrature Formulas", *CACM* 11, No. 6, 1968, 432-436.

[G70]   W. Gautschi, "Generation of Gaussian Quadrature Rules and Orthogonal Polynomials", in "Colloquium Approximatietheorie", MC Syllabus NR 14, Mathematisch Centrum, Amsterdam, 1970.

[G73]   W. Gautschi, "Exponential Integrals", *CACM* 16, 1973, 761-763.

[Ge69]   W.M. Gentleman, "An Error Analysis of Goertzel's (Watt's) Method for Computing Fourier Coefficients", *Comput. J.*, Vol. 12, 1969, 160-165.

[GolP67]   A. A. Goldstein and J. F. Price, "An Effective Algorithm for Minimization", *Numer. Math.* 10, 1967, 184-189.

[GolW69]   G. H. Golub and J. H. Welsch, "Calculation of Gauss Quadrature Rules", *Math. Comp.* 23, 1969, 221-230.

[Gor69]   R.G. Gordon, "Evaluation of Airy Functions", *The Journal of Chemical Physics* 51, 1969, 23-24.

[Gr67]   J.L. Greenstadt, "On the Relative Efficiencies of Gradient Methods", *Math. Comp.* 21, 1967, 360-367.

[GrK69]   R. T. Gregory and D. L. Karney, *A Collection of Matrices for Testing Computational Algorithms*, Wiley-Interscience, New York, 1969.

[H73]   R.W. Hamming, *Numerical Methods for Scientists and Engineers*,
               McGraw-Hill, New York, 1973.

[HaCL68]   J.F. Hart, E.W. Cheney, C.L. Lawson, H.J. Maehly, C.K. Mesztenyi,
               J. R. Rice, H. C. Thacher, Jr., and C. Witzgall, *Computer
               Approximations*, Wiley, New York, 1968.

[Har61]   H.O. Hartley, "The Modified Gauss-Newton Method", *Technometrics*
               3, 1961, 269-280.

[He71]   P.W. Hemker, "An Algol 60 Procedure for the Solution of Stiff
               Differential Equations", Report MR 128, Mathematical Centre,
               Amsterdam, 1971.

[He73]   P.W. Hemker, "A Sequence of Nested Cubature Rules",
               Report NW 3, Mathematical Centre, Amsterdam, 1973.

[He75]   P.W. Hemker, "Galerkin's Method and Lobatto Points", Report 24,
               Mathematisch Centrum, Amsterdam, 1975.

[He80]   P.W. Hemker (Ed.), "NUMAL, Numerical Procedures in ALGOL 60",
               MC Syllabus, 47.1-47.7, Mathematisch Centrum, Amsterdam,
               1980.

[HEFS72]   T. E. Hull, W. H. Enright, B. M. Fellen, and A. E. Sedgwick,
               "Comparing Numerical Methods for Ordinary Differential
               Equations", *SIAM Journal on Numerical Analysis* 9, 1972, 603-637.

[Hu64]   D.B. Hunter, "The Calculation of Certain Bessel Functions",
               Math. Comp. 18, 1964, 123-128.

[K69]   D.E. Knuth, *The Art of Computer Programming, Seminumerical
               Algorithms*, Vol. 2, Addison-Wesley, Reading, MA, 1969.

[K73]   D.E. Knuth, *The Art of Computer Programming, Vol. 3, Sorting and
               Searching*, Addison-Wesley, Reading, MA, 1973.

[L57]   C. Lanczos, *Applied Analysis*, Prentice Hall, Englewood Cliffs, NJ,
               1957.

[Li73]   B. Lindberg, "IMPEX 2, A Procedure for the Solution of Systems of
               Stiff Differential Equations", Report TRITA-NA-7303, Royal Institute
               of Technology, Stockholm, 1973.

[LiW70]   W. Liniger and R. A. Willoughby, "Efficient Integration Methods for
               Stiff Systems of Ordinary Differential Equations", *SIAM J. Numer.
               Anal.* 7, 1970, 47-66.

[Lu69]   Y.L. Luke, *The Special Functions and their Approximations*, 2 Vols.,
               Academic Press, New York, 1969.

[Lu70]   Y.L. Luke, "Evaluation of the Gamma Function by means of Padé Approximations", *SIAM J. Math. Anal.* 1, 1970, 266-281.

[M63]   D.W. Marquardt, "An Algorithm for Least-Squares Estimation of Nonlinear Parameters", *J. SIAM* 11, 1963, 431-441.

[Me64]   G. Meinardus, *Approximation of Function and their Numerical Treatment*, Springer Tracts in Natural Philosophy, Vol. 4, Springer-Verlag, Berlin, 1964.

[MS71]   C. B. Moler and G. W. Stewart, "An algorithm for the Generalized Matrix Eigenvalue Problem Ax=λBx", report STAN-CS-232-71, Stanford University, 1971.

[Mu66]   D.J. Mueller, "Householder's Method for Complex Matrices and Eigensystems of Hermitian Matrices", *Numerische Mathematik* 8, 1966, 72-92.

[N73]   A.C.R. Newbery, "Error Analysis for Fourier Series Evaluation", *Math. Comp.*, Vol. 26, 1973, 923-924.

[Os60]   E.E. Osborne, "On Preconditioning of Matrices", *J. Assoc. Comput. Mach.* 7, 1960, 338-354.

[PaR69]   B.N. Parlett, C. Reinsch, "Balancing a Matrix for Calculation of Eigenvalues and Eigenvectors", *Numerische Mathematik* 13, 1969, 293-304.

[PeW71]   G. Peters and J. H. Wilkinson, "Practical Problems Arising in the Solution of Polynomial Equations", *Journal of the Institute of Mathematics and Its Applications.* 8, No. 1, 1971, 16-35.

[Po70]   M.J.D. Powell, "Rank One Methods for Unconstrained Optimization", in *Integer and Nonlinear Programming*, edited by J. Abadie, North-Holland, 1970.

[Pr92]   W. H. Press, S. A. Teukolsky, W. T. Vetterling, and B. P. Flannery, *Numerical Recipes in C*, 2nd edition, Cambridge University Press, 1992.

[R71]   J.K. Reid, "On the Method of Conjugate Gradients for the Solution of Large Sparse Systems of Linear Equations", in *Large Sparse Sets of Linear Equations*, edited by J.K. Reid, Academic Press, 1971.

[Re67]   C. Reinsch, "A Note on Trigonometric Interpolation", Report No. 6709, Mathematics Department, Technische Universität München, 1967.

[Re71]   C.H. Reinsch, "A Stable, Rational QR Algorithm for the Computation of the Eigenvalues of an Hermitian, Tridiagonal Matrix", *Math. Comp.* 25, 1971, 591-597.

[Rey77]   Th.H.P. Reymer, "Berekening Van Nulpunten Van Reele Polynomen En Foutgrenzen Voor Deze Nulpunten", Doctoraal Scriptie UVA, April 1977.

[Rie77]   H.J.J. te Riele (Ed.), "Colloquium Numerieke Programmatuur", MC Syllabus NR 29.2, Mathematical Centre, Amsterdam, 1977.

[Riv74]   T.J. Rivlin, *The Chebyshev Polynomials*, Wiley, New York, 1974.

[RoF72]   C. J. Roothart and H. Fiolet, "Quadrature Procedures", Report MR 137, Mathematical Centre, Amsterdam, 1972.

[ShT74]   M. Shaw and J. Traub, "On the Number of Multiplications for the Evaluation of a Polynomial and some of its Derivatives", *J. Assoc. Comput. Mach.*, 1974, Vol. 21, No. 1, 161-167.

[Sp67]   H. Spaeth, "The Damped Taylor's Series Method for Minimizing a Sum of Squares and for Solving Systems of Nonlinear Equations", *CACM* 10, 1967, 726-728.

[St72]   J. Stoer, *Einfuehrung in die Numerische Mathematik I*, Heidelberger Taschenbuecher 105, Springer-Verlag, Berlin, 1972.

[StF73]   G. Strang and G. J. Fix, *An Analysis of the Finite Element Method*, Prentice-Hall, Englewood Cliffs, NJ, 1973.

[Str68]   A.J. Strecok, "On the Calculation of the Inverse of the Error Function", *Math. Comp.* 22, 1968, 144-158.

[T75]   N.M. Temme, "On the Numerical Evaluation of the Ordinary Bessel Function of the Third Kind", *J. Comput. Phys.* 19, 1975, 324-337.

[T76a]   N.M. Temme, "On the Numerical Evaluation of the Ordinary Bessel Function of the Second Kind", *J. Comput. Phys.* 21, 1976, 343-350.

[T76b]   N.M. Temme, "Speciale Functies", in "Colloquium Numerieke Programmatuur", J.C.P. Bus (ed.), MC Syllabus NR 29.1b, Mathematical Centre, Amsterdam, 1976.

[T78]   N.M. Temme, "An Algorithm with Algol 60 Implementation for the Calculation of the Zeros of a Bessel Function", Report TW 179, Mathematical Centre, Amsterdam, 1978.

[Vh68]   P.J. van der Houwen, "Finite Difference Methods for Solving Partial Differential Equations", Mathematical Centre Tract 20, Mathematical Centre, Amsterdam, 1968.

[Vh70a]   P.J. van der Houwen, "One-Step Methods for Linear Initial Value
             Problems I, Polynomial Methods", Report TW 119, Mathematical
             Centre, Amsterdam, 1970.

[Vh70b]   P.J. van der Houwen, "One-Step Methods for Linear Initial Value
             Problems II, Polynomial Methods", Report TW 122, Mathematical
             Centre, Amsterdam, 1970.

[Vh71]   P.J. van der Houwen, "Stabilized Runge Kutta Method with Limited
             Storage Requirements", Report TW 124, Mathematical Centre,
             Amsterdam, 1971.

[Vh72]   P.J. van der Houwen, "One-Step Methods with Adaptive Stability
             Functions for the Integration of Differential Equations",
             Lecture Notes of the Conference on Numerische Insbesondere
             Approximationstheoretische Behandlung Von Funktional-
             Gleichungen, Oberwolfach, December 3-12, 1972.

[VhBDS71]   P. J. van der Houwen, P. Beentjes, K. Dekker, and E. Slagt,
             "One Step Methods for Linear Initial Value Problems III, Numerical
             Examples", Report TW 130, Mathematical Centre, Amsterdam,
             1971.

[VhK71]   P. J. van der Houwen and J. Kok, "Numerical Solution of a
             Minimax Problem", Report TW 123, Mathematical Centre,
             Amsterdam, 1971.

[VhV74]   P. J. van der Houwen and J. G. Verwer, "Generalized Linear
             Multistep Methods 1, Development of Algorithms with
             Zero-Parasitic Roots", Report NW 10, Mathematisch Centrum,
             Amsterdam, 1974.

[Vn75]   B. van Domselaar, "Nonlinear Parameter Estimation in Initial Value
             Problems", Report NW 18, Mathematical Centre, Amsterdam, 1975.

[Vnw65]   A. van Wijngaarden, "Course Scientific Computing B, Process
             Analysis", Mathematisch Centrum CR 18, Amsterdam, 1965.

[Vr66]   J.M. Varah, "Eigenvectors of a Real Matrix by Inverse Iteration",
             Technical Report No. CS 34, Stanford University, 1966.

[Vw74]   J.G. Verwer, "Generalized Linear Multistep Methods 2, Numerical
             Applications", Report NW 12, Mathematisch Centrum, Amsterdam,
             1974.

[W45]   G.N. Watson, *A Treatise on the Theory of Bessel Functions*,
             Cambridge University Press, 1945.

[Wi63]   J.H. Wilkinson, *Rounding Errors in Algebraic Processes*, Prentice-Hall, Englewood Cliffs, NJ, 1963.

[Wi65]   J.H. Wilkinson, *The Algebraic Eigenvalue Problem*, Clarendon Press, Oxford, 1965.

[WiR71]   J. H. Wilkinson and C. Reinsch, *Handbook of Automatic Computation, Vol. 2, Linear Algebra*, Springer-Verlag, Berlin, 1971.

[Wo74]   H. Wozniakowski, "Rounding Error Analysis for the Evaluation of a Polynomial and some of its Derivatives", *SIAM J. Numer. Anal.*, Vol. 11, No. 4, 1974, 780-787.

[Wy81]   P. Wynn, "NUMAL in FORTRAN", IIMAS, Universidad Nacional Autónoma de México, Comunicaciones técnicas Nos. 48.0-48.11, 1981.

[Z64]   J.A. Zonneveld, "Automatic Numerical Integration", Mathematical Centre Tract 8, Mathematisch Centrum, Amsterdam, 1964.

Appendix B: Procedures Description

1. Elementary Procedures

INIVEC	initializes a vector with a constant.
INIMAT	initializes a matrix with a constant.
INIMATD	initializes a (co)diagonal of a matrix.
INISYMD	initializes a (co)diagonal of a symmetric matrix, whose upper triangle is stored columnwise in a one-dimensional array.
INISYMROW	initializes a row of a symmetric matrix, whose upper triangle is stored columnwise in a one-dimensional array.
DUPVEC	copies a vector into another vector.
DUPVECROW	copies a row vector into a vector.
DUPROWVEC	copies a vector into a row vector.
DUPVECCOL	copies a column vector into a vector.
DUPCOLVEC	copies a vector into a column vector.
DUPMAT	copies a matrix into another matrix.
MULVEC	stores a constant multiplied by a vector into a vector.
MULROW	stores a constant multiplied by a row vector into a row vector.
MULCOL	stores a constant multiplied by a solumn vector into a column vector.
COLCST	multiplies a column vector by a constant.
ROWCST	multiplies a row vector by a constant.
VECVEC	forms the scalar product of a vector and a vector.
MATVEC	forms the scalar product of a row vector and a vector.
TAMVEC	forms the scalar product of a column vector and a vector.
MATMAT	forms the scalar product of a row vector and a column vector.
TAMMAT	forms the scalar product of a column vector and a column vector.
MATTAM	forms the scalar product of a row vector and a row vector.
SEQVEC	forms the scalar product of two vectors given in one-dimensional arrays, where the mutual spacings between the indices of the first vector change linearly.
SCAPRD1	forms the scalar product of two vectors given in one-dimensional arrays, where the spacings of both vectors are constant.
SYMMATVEC	forms the scalar product of a vector and a row of a symmetric matrix, whose upper triangle is given columnwise in a one-dimensional array.
FULMATVEC	forms the product A*B of a given matrix A and a vector B.
FULTAMVEC	forms the product A^T*B, where A^T is the transpose of a given matrix A and B is a vector.
FULSYMMATVEC	forms the product A*B, where A is a symmetric matrix whose upper triangle is stored columnwise in a one-dimensional array and B is a vector.
RESVEC	calculates the residual vector A*B+X*C, where A is a given matrix, B and C are vectors, and X is a scalar.

SYMRESVEC	calculates the residual vector A*B+X*C, where A is a symmetric matrix whose upper triangle is stored columnwise in a one-dimensional array, B and C are vectors, and X is a scalar.
HSHVECMAT	premultiplies a matrix by a Householder matrix, the vector defining this Householder matrix being given in a one-dimensional array.
HSHCOLMAT	premultiplies a matrix by a Householder matrix, the vector defining this Householder matrix being given as a column in a two-dimensional array.
HSHROWMAT	premultiplies a matrix by a Householder matrix, the vector defining this Householder matrix being given as a row in a two-dimensional array.
HSHVECTAM	postmultiplies a matrix by a Householder matrix, the vector defining this Householder matrix being given in a one-dimensional array.
HSHCOLTAM	postmultiplies a matrix by a Householder matrix, the vector defining this Householder matrix being given as a column in a two-dimensional array.
HSHROWTAM	postmultiplies a matrix by a Householder matrix, the vector defining this Householder matrix being given as a row in a two-dimensional array.
ELMVEC	adds a constant times a vector to a vector.
ELMCOL	adds a constant times a column vector to a column vector.
ELMROW	adds a constant times a row vector to a row vector.
ELMVECCOL	adds a constant times a column vector to a vector.
ELMCOLVEC	adds a constant times a vector to a column vector.
ELMVECROW	adds a constant times a row vector to a vector.
ELMROWVEC	adds a constant times a vector to a row vector.
ELMCOLROW	adds a constant times a row vector to a column vector.
ELMROWCOL	adds a constant times a column vector to a row vector.
MAXELMROW	adds a constant times a row vector to a row vector, it also delivers the subscript of an element of the new row vector which is of maximum absolute value.
ICHVEC	interchanges two vectors given in a one-dimensional array.
ICHCOL	interchanges two columns of a matrix.
ICHROW	interchanges two rows of a matrix.
ICHROWCOL	interchanges a row and a column of a matrix.
ICHSEQVEC	interchanges a row and a column of an upper triangular matrix, which is stored columnwise in a one-dimensional array.
ICHSEQ	interchanges two columns of an upper triangular matrix, which is stored columnwise in a one-dimensional array.
ROTCOL	replaces two column vectors X and Y by two vectors CX+SY and CY-SX.
ROTROW	replaces two row vectors X and Y by two vectors CX+SY and CY-SX.
INFNRMVEC	calculates the infinity norm of a vector.
INFNRMROW	calculates the infinity norm of a row vector.
INFNRMCOL	calculates the infinity norm of a column vector.

INFNRMMAT	calculates the infinity norm of a matrix.
ONENRMVEC	calculates the 1-norm of a vector.
ONENRMROW	calculates the 1-norm of a row vector.
ONENRMCOL	calculates the 1-norm of a column vector.
ONENRMMAT	calculates the 1-norm of a matrix.
ABSMAXMAT	calculates the modulus of the largest element of a matrix and delivers the indices of the maximal element.
REASCL	normalizes the columns of a two-dimensional array.
COMCOLCST	multiplies a complex column vector by a complex number.
COMROWCST	multiplies a complex row vector by a complex number.
COMMATVEC	forms the scalar product of a complex row vector and a complex vector.
HSHCOMCOL	transforms a complex vector into a vector proportional to a unit vector.
HSHCOMPRD	premultiplies a complex matrix with a complex Householder matrix.
ELMCOMVECCOL	adds a complex number times a complex column vector to a complex vector.
ELMCOMCOL	adds a complex number times a complex column vector to a complex column vector.
ELMCOMROWVEC	adds a complex number times a complex vector to a complex row vector.
ROTCOMCOL	replaces two complex column vectors X and Y by two complex vectors CX+SY and CY-SX.
ROTCOMROW	replaces two complex row vectors X and Y by two complex vectors CX+SY and CY-SX.
CHSH2	finds a complex rotation matrix.
COMEUCNRM	calculates the Euclidean norm of a complex matrix with some lower diagonals.
COMSCL	normalizes real and complex eigenvectors.
SCLCOM	normalizes the columns of a complex matrix.
COMABS	calculates the modulus of a complex number.
COMSQRT	calculates the square root of a complex number.
CARPOL	transforms the Cartesian coordinates of a complex number into polar coordinates.
COMMUL	calculates the product of two complex numbers.
COMDIV	calculates the quotient of two complex numbers.
LNGINTADD	computes the sum of long nonnegative integers.
LNGINTSUBTRACT	computes the difference of long nonnegative integers.
LNGINTMULT	computes the product of long nonnegative integers.
LNGINTDIVIDE	computes the quotient with remainder of long nonnegative integers.
LNGINTPOWER	computes U^P, where U is a long nonnegative integer, and P is the positive (single-length) exponent.

2. Algebraic Evaluations

POL	evaluates a polynomial.
TAYPOL	evaluates the first k terms of a Taylor series.
NORDERPOL	evaluates the first k normalized derivatives of a polynomial.

DERPOL	evaluates the first k derivatives of a polynomial.
ORTPOL	evaluates the value of an n-degree orthogonal polynomial, given by a set of recurrence coefficients.
ORTPOLSYM	evaluates the values of an n-degree orthogonal polynomial, given by a set of recurrence coefficients.
ALLORTPOL	evaluates the value of all orthogonal polynomials up to a given degree, given a set of recurrence coefficients.
ALLORTPOLSYM	evaluates the values of all orthogonal polynomials up to a given degree, given a set of recurrence coefficients.
SUMORTPOL	evaluates a finite series expressed in orthogonal polynomials, given by a set of recurrence coefficients.
SUMORTPOLSYM	evaluates a finite series expressed in orthogonal polynomials, given by a set of recurrence coefficients.
CHEPOLSUM	evaluates a finite sum of Chebyshev polynomials.
ODDCHEPOLSUM	evaluates a finite sum of Chebyshev polynomials of odd degree.
CHEPOL	evaluates a Chebyshev polynomial.
ALLCHEPOL	evaluates all Chebyshev polynomials up to a certain degree.
SINSER	evaluates a sine series.
COSSER	evaluates a cosine series.
FOUSER	evaluates a fourier series with equal sine and cosine coefficients.
FOUSER1	evaluates a fourier series.
FOUSER2	evaluates a fourier series.
COMFOUSER	evaluates a complex fourier series with real coefficients.
COMFOUSER1	evaluates a complex fourier series.
COMFOUSER2	evaluates a complex fourier series.
JFRAC	calculates a terminating continued fraction.
POLCHS	transforms a polynomial from power sum into Chebyshev sum form.
CHSPOL	transforms a polynomial from Chebyshev sum into power sum form.
POLSHTCHS	transforms a polynomial from power sum into shifted Chebyshev sum form.
SHTCHSPOL	transforms a polynomial from shifted Chebyshev sum form into power sum form.
GRNNEW	transforms a polynomial from power sum into Newton sum form.
NEWGRN	transforms a polynomial from Newton sum into power sum form.
LINTFMPOL	transforms a polynomial in X into a polynomial in Y (Y=A*X+B).
INTCHS	computes the indefinite integral of a given Chebyshev series.

3.   Linear Algebra

DEC	performs a triangular decomposition with partial pivoting.
GSSELM	performs a triangular decomposition with a combination of partial and complete pivoting.

ONENRMINV   calculates the 1-norm of the inverse of a matrix whose triangularly decomposed form is delivered by GSSELM.

ERBELM   calculates a rough upper bound for the error in the solution of a system of linear equations whose matrix is triangularly decomposed by GSSELM.

GSSERB   performs a triangular decomposition of the matrix of a system of linear equations and calculates an upper bound for the relative error in the solution of that system.

GSSNRI   performs a triangular decomposition, and calculates the 1-norm of the inverse matrix.

DETERM   calculates the determinant of a triangularly decomposed matrix.

SOL   solves a system of linear equations whose matrix has been triangularly decomposed by DEC.

DECSOL   solves a system of linear equations whose order is small relative to the number of binary digits in the number representation.

SOLELM   solves a system of linear equations whose matrix has been triangularly decomposed by GSSELM or GSSERB.

GSSSOL   solves a system of linear equations.

GSSSOLERB   calculates the inverse of a matrix and 1-norm, an upper bound for the error in the inverse matrix is also given.

INV   calculates the inverse of a matrix that has been triangularly decomposed by DEC.

DECINV   calculates the inverse of a matrix whose order is small relative to the number of binary digits in the number representation.

INV1   calculates the inverse of a matrix that has been triangularly decomposed by GSSELM or GSSERB, the 1-norm of the inverse matrix might also be calculated.

GSSINV   calculates the inverse of a matrix.

GSSINVERB   calculates the inverse of a matrix and 1-norm, an upper bound for the error in the inverse matrix is also given.

ITISOL   solves a system of linear equations whose matrix has been triangularly decomposed by GSSELM or GSSERB, this solution is improved iteratively.

GSSITISOL   solves a system of linear equations, and the solution is improved iteratively.

ITISOLERB   solves a system of linear equations whose matrix has been triangularly decomposed by GSSNRI, this solution is improved iteratively, and an upper bound for the error in the solution is calculated.

GSSITISOLERB   solves a system of linear equations, this solution is improved iteratively, and an upper bound for the error in the solution is calculated.

CHLDEC2   calculates the Cholesky decomposition of a positive definite symmetric matrix whose upper triangle is given in a two-dimensional array.

CHLDEC1   calculates the Cholesky decomposition of a positive definite symmetric matrix whose upper triangle is given columnwise in a one-dimensional array.

CHLDETERM2   calculates the determinant of a positive definite symmetric matrix, the Cholesky decomposition being given in a two-dimensional array.

CHLDETERM1   calculates the determinant of a positive definite symmetric matrix, the Cholesky decomposition being given columnwise in a one-dimensional array.

CHLSOL2   solves a system of linear equations if the coefficient matrix has been decomposed by CHLDEC2 or CHLDECSOL2.

CHLSOL1   solves a system of linear equations if the coefficient matrix has been decomposed by CHLDEC1 or CHLDECSOL1.

CHLDECSOL2   solves a positive definite symmetric system of linear equations by Cholesky's square root method, the coefficient matrix should be given in the upper triangle of a two-dimensional array.

CHLDECSOL1   solves a positive definite symmetric system of linear equations by Cholesky's square root method, the coefficient matrix should be given columnwise in a one-dimensional array.

CHLINV2   calculates the inverse of a positive definite symmetric matrix, if the matrix has been decomposed by CHLDEC2 or CHLDECSOL2.

CHLINV1   calculates the inverse of a positive definite symmetric matrix, if the matrix has been decomposed by CHLDEC1 or CHLDECSOL1.

CHLDECINV2   calculates the inverse of a positive definite symmetric matrix by Cholesky's square root method, the coefficient matrix given in the upper triangle of a two-dimensional array.

CHLDECINV1   calculates the inverse of a positive definite symmetric matrix by Cholesky's square root method, the coefficient matrix given columnwise in a one-dimensional array.

DECSYM2   calculates the symmetric decomposition of a symmetric matrix.

DETERMSYM2   calculates the determinant of a symmetric matrix, the symmetric decomposition being given.

SOLSYM2   solves a symmetric system of linear equations if the coefficient matrix has been decomposed by DECSYM2 or DECSOLSYM2.

DECSOLSYM2   solves a symmetric system of linear equations by symmetric decomposition.

LSQORTDEC   delivers the Householder triangularization with column interchanges of a matrix of a linear least squares problem.

LSQDGLINV   calculates the diagonal elements of the inverse of M^TM, where M is the coefficient matrix of a linear least squares problem.

LSQSOL   solves a linear least squares problem if the coefficient matrix has been decomposed by LSQORTDEC.

LSQORTDECSOL	solves a linear least squares problem by Householder triangularization with column interchanges, and calculates the diagonal of the inverse of M^TM, where M is the coefficient matrix.
LSQINV	calculates the inverse of the matrix S^TS, where S is the coefficient matrix of a linear least squares problem.
LSQDECOMP	computes the QR decomposition of a linear least squares problem with linear constraints.
LSQREFSOL	solves a linear least squares problem with linear constraints if the coefficient matrix has been decomposed by LSQDECOMP.
SOLSVDOVR	solves an overdetermined system of linear equations, multiplying the right-hand side by the pseudo-inverse of the given matrix.
SOLOVR	calculates the singular values decomposition, and solves an overdetermined system of linear equations.
SOLSVDUND	solves an underdetermined system of linear equations, multiplying the right-hand side by the pseudo-inverse of the given matrix.
SOLUND	calculates the singular values decomposition, and solves an underdetermined system of linear equations.
HOMSOLSVD	solves the homogeneous system of linear equations AX=0 and $X^TA=0$, where A denotes a matrix and X a vector (the singular value decomposition being given).
HOMSOL	solves the homogeneous system of linear equations AX=0 and $X^TA=0$, where A denotes a matrix and X a vector.
PSDINVSVD	calculates the pseudo-inverse of a matrix (the singular value decomposition being given).
PSDINV	calculates the pseudo-inverse of a matrix.
DECBND	performs a triangular decomposition of a band matrix, using partial pivoting.
DETERMBND	calculates the determinant of a band matrix.
SOLBND	solves a system of linear equations, the matrix being decomposed by DECBND.
DECSOLBND	solves a system of linear equations by Gaussian elimination with partial pivoting if the coefficient matrix is in band form and is stored rowwise in a one-dimensional array.
DECTRI	performs a triangular decomposition of a tridiagonal matrix.
DECTRIPIV	performs a triangular decomposition of a tridiagonal matrix, using partial pivoting.
SOLTRI	solves a tridiagonal system of linear equations, the triangular decomposition being given.
DECSOLTRI	solves a tridiagonal system of linear equations and performs the triangular decomposition without pivoting.
SOLTRIPIV	solves a tridiagonal system of linear equations, the triangular decomposition being given.
DECSOLTRIPIV	solves a tridiagonal system of linear equations, and performs the triangular decomposition with partial pivoting.
CHLDECBND	performs the Cholesky decomposition of a positive definite symmetric band matrix.

CHLDETERMBND   calculates the determinant of a positive definite symmetric band matrix.

CHLSOLBND   solves a positive definite symmetric linear system, the triangular decomposition being given.

CHLDECSOLBND   solves a positive definite symmetric linear system, and performs the triangular decomposition by Cholesky's method.

DECSYMTRI   performs the triangular decomposition of a symmetric tridiagonal matrix.

SOLSYMTRI   solves a symmetric tridiagonal system of linear equations, the triangular decomposition being given.

DECSOLSYMTRI   solves a symmetric tridiagonal system of linear equations, and performs the tridiagonal decomposition.

CONJGRAD   solves a positive definite symmetric system of linear equations by the method of conjugate gradients.

EQILBR   equilibrates a matrix by means of a diagonal similarity transformation.

BAKLBR   performs the back transformation corresponding to EQILBR.

EQILBRCOM   equilibrates a complex matrix.

BAKLBRCOM   transforms the eigenvectors of a complex equilibrated (by EQILBRCOM) matrix into the eigenvectors of the original matrix.

TFMSYMTRI2   transforms a real symmetric matrix into a similar tridiagonal one by means of Householder's transformation.

BAKSYMTRI2   performs the back transformation corresponding to TFMSYMTRI2.

TFMPREVEC   in combination with TFMSYMTRI2 calculates the transforming matrix.

TFMSYMTRI1   transforms a real symmetric matrix into a similar tridiagonal one by means of Householder's transformation.

BAKSYMTRI1   performs the back transformation corresponding to TFMSYMTRI1.

TFMREAHES   transforms a matrix into a similar upper Hessenberg matrix by means of Wilkinson's transformation.

BAKREAHES1   performs the back transformation (on a vector) corresponding to TFMREAHES.

BAKREAHES2   performs the back transformation (on columns) corresponding to TFMREAHES.

HSHHRMTRI   transforms a Hermitian matrix into a similar real symmetric tridiagonal matrix.

HSHHRMTRIVAL   delivers the main diagonal elements and the squares of the codiagonal elements of a Hermitian tridiagonal matrix which is unitary similar with a given Hermitian matrix.

BAKHRMTRI   performs the back transformation corresponding to HSHHRMTRI.

HSHCOMHES   transforms a complex matrix by means of Householder's transformation followed by a complex diagonal transformation into a similar unitary upper Hessenberg matrix with a real nonnegative subdiagonal.

BAKCOMHES   performs the back transformation corresponding to HSHCOMHES.

HSHREABID	transforms a matrix to bidiagonal form, by premultiplying and postmultiplying with orthogonal matrices.
PSTTFMMAT	calculates the postmultiplying matrix from the data generated by HSHREABID.
PRETFMMAT	calculates the premultiplying matrix from the data generated by HSHREABID.
VALSYMTRI	calculates all, or some consecutive, eigenvalues of a symmetric tridiagonal matrix by means of linear interpolation using a Sturm sequence.
VECSYMTRI	calculates eigenvectors of a symmetric tridiagonal matrix by means of inverse iteration.
QRIVALSYMTRI	calculates the eigenvalues of a symmetric tridiagonal matrix by means of QR iteration.
QRISYMTRI	calculates the eigenvalues and eigenvectors of a symmetric tridiagonal matrix by means of QR iteration.
EIGVALSYM2	calculates all (or some) eigenvalues of a symmetric matrix using linear interpolation of a function derived from a Sturm sequence.
EIGSYM2	calculates eigenvalues and eigenvectors by means of inverse iteration.
EIGVALSYM1	calculates all (or some) eigenvalues of a symmetric matrix using linear interpolation of a function derived from a Sturm sequence.
EIGSYM1	calculates eigenvalues and eigenvectors by means of inverse iteration.
QRIVALSYM2	calculates the eigenvalues of a symmetric matrix by means of QR iteration.
QRISYM	calculates all eigenvalues and eigenvectors of a symmetric matrix by means of QR iteration.
QRIVALSYM1	calculates the eigenvalues of a symmetric matrix by means of QR iteration.
MERGESORT	delivers a permutation of indices corresponding to sorting the elements of a given vector into nondecreasing order.
VECPERM	permutes the elements of a given vector according to a given permutation of indices.
ROWPERM	permutes the elements of a given row according to a given permutation of indices.
ORTHOG	orthogonalizes some adjacent matrix columns according to the modified Gram-Schmidt method.
SYMEIGIMP	improves an approximation of a real symmetric eigensystem, and calculates error bounds for the eigenvalues.
REAVALQRI	calculates the eigenvalues of a real upper Hessenberg matrix, provided that all eigenvalues are real, by means of single QR iteration.
REAVECHES	calculates an eigenvector corresponding to a given real eigenvalue of a real upper Hessenberg matrix by means of inverse iteration.
REAQRI	calculates all eigenvalues and eigenvectors of a real upper Hessenberg matrix, provided that all eigenvalues are real, by means of single QR iteration.

COMVALQRI   calculates the real and complex eigenvalues of a real upper Hessenberg matrix by means of double QR iteration.

COMVECHES   calculates the eigenvector corresponding to a given complex eigenvalue of a real upper Hessenberg matrix by means of inverse iteration.

REAEIGVAL   calculates the eigenvalues of a matrix, provided that all eigenvalues are real.

REAEIG1   calculates the eigenvalues and eigenvectors of a matrix, provided that they are all real.

REAEIG3   calculates the eigenvalues and eigenvectors of a matrix, provided that they are all real.

COMEIGVAL   calculates the eigenvalues of a matrix.

COMEIG1   calculates the eigenvalues and eigenvectors of a matrix.

EIGVALHRM   calculates the eigenvalues of a complex Hermitian matrix.

EIGHRM   calculates the eigenvalues and eigenvectors of a complex Hermitian matrix.

QRIVALHRM   calculates the eigenvalues of a complex Hermitian matrix.

QRIHRM   calculates the eigenvalues and eigenvectors of a complex Hermitian matrix.

VALQRICOM   calculates the eigenvalues of a complex upper Hessenberg matrix with a real subdiagonal.

QRICOM   calculates the eigenvalues and eigenvectors of a complex upper Hessenberg matrix.

EIGVALCOM   calculates the eigenvalues of a complex matrix.

EIGCOM   calculates the eigenvalues and eigenvectors of a complex matrix.

QZIVAL   computes generalized eigenvalues by means of QZ iteration.

QZI   computes generalized eigenvalues and eigenvectors by means of QZ iteration.

HSHDECMUL   is an auxiliary procedure for the computation of generalized eigenvalues.

HESTGL3   is an auxiliary procedure for the computation of generalized eigenvalues.

HESTGL2   is an auxiliary procedure for the computation of generalized eigenvalues.

HSH2COL   is an auxiliary procedure for the computation of generalized eigenvalues.

HSH3COL   is an auxiliary procedure for the computation of generalized eigenvalues.

HSH2ROW3   is an auxiliary procedure for the computation of generalized eigenvalues.

HSH2ROW2   is an auxiliary procedure for the computation of generalized eigenvalues.

HSH3ROW3   is an auxiliary procedure for the computation of generalized eigenvalues.

HSH3ROW2   is an auxiliary procedure for the computation of generalized eigenvalues.

QRISNGVALBID   calculates the singular values of a bidiagonal matrix.

QRISNGVALDECBID
                calculates the singular values decomposition of a matrix of which the bidiagonal and the pre- and postmultiplying matrices are given.

QRISNGVAL   calculates the singular values of a given matrix.

QRISNGVALDEC   calculates the singular values decomposition UDV^T, with U and V orthogonal and D positive diagonal.

ZERPOL   calculates all roots of a polynomial with real coefficients by Laguerre's method.

BOUNDS   calculates the error in approximated zeros of a polynomial with real coefficients.

ALLZERORTPOL   calculates all zeros of an orthogonal polynomial.

LUPZERORTPOL   calculates a number of adjacent upper or lower zeros of an orthogonal polynomial.

SELZERORTPOL   calculates a number of adjacent zeros of an orthogonal polynomial.

ALLJACZER   calculates the zeros of a Jacobian polynomial.

ALLLAGZER   calculates the zeros of a Laguerre polynomial.

COMKWD   calculates the roots of a quadratic equation with complex coefficients.

4. Analytic Evaluations

EULER   performs the summation of an alternating infinite series.

SUMPOSSERIES   performs the summation of an infinite series with positive monotonically decreasing terms using the van Wijngaarden transformation.

QADRAT   computes the definite integral of a function of one variable over a finite interval.

INTEGRAL   calculates the definite integral of a function of one variable over a finite or infinite interval or over a number of consecutive intervals.

TRICUB   computes the definite integral of a function of two variables over a triangular domain.

RECCOF   calculates recurrence coefficients of an orthogonal polynomial, a weight function being given.

GSSWTS   calculates the Gaussian weights of a weight function, the recurrence coefficients being given.

GSSWTSSYM   calculates the Gaussian weights of a weight function, the recurrence coefficients being given.

GSSJACWGHTS   computes the abscissae and weights for Gauss-Jacobi quadrature.

GSSLAGWGHTS   computes the abscissae and weights for Gauss-Lagrange quadrature.

JACOBNNF   calculates the Jacobian matrix of an n-dimensional function of n variables using forward differences.

JACOBNMF   calculates the Jacobian matrix of an n-dimensional function of m variables using forward differences.

JACOBNBNDF   calculates the Jacobian matrix of an n-dimensional function of n variables, if the Jacobian is known to be a band matrix.

5. Analytic Problems

ZEROIN	finds (in a given interval) a zero of a function of one variable.
ZEROINRAT	finds (in a given interval) a zero of a function of one variable.
ZEROINDER	finds (in a given interval) a zero of a function of one variable using values of the function and of its derivative.
QUANEWBND	solves a system of nonlinear equations of which the Jacobian (being a band matrix) is given.
QUANEWBND1	solves a system of nonlinear equations of which the Jacobian is a band matrix.
MININ	minimizes a function of one variable in a given interval.
MININDER	minimizes a function of one variable in a given interval, using values of the function and of its derivative.
LINEMIN	minimizes a function of several variables in a given direction.
RNK1UPD	adds a rank-1 matrix to a symmetric matrix.
DAVUPD	adds a rank-2 matrix to a symmetric matrix.
FLEUPD	adds a rank-2 matrix to a symmetric matrix.
PRAXIS	minimizes a function of several variables.
RNK1MIN	minimizes a function of several variables.
FLEMIN	minimizes a function of several variables.
MARQUARDT	calculates the least squares solution of an overdetermined system of nonlinear equations with Marquardt's method.
GSSNEWTON	calculates the least squares solution of an overdetermined system of nonlinear equations with the Gauss-Newton method.
RK1	solves a single first order differential equation by means of a fifth order Runge-Kutta method.
RKE	solves a system of first order differential equations (initial value problem) by means of a fifth order Runge-Kutta method.
RK4A	solves a single first order differential equation by means of a fifth order Runge-Kutta method; the integration is terminated as soon as a condition on X and Y, which is supplied by the user, is satisfied.
RK4NA	solves a system of first order differential equations (initial value problem) by means of a fifth order Runge-Kutta method; the integration is terminated as soon as a condition on X[0], X[1],..., X[n], supplied by the user, is satisfied.
RK5NA	solves a system of first order differential equations (initial value problem) by means of a fifth order Runge-Kutta method; the arc length is introduced as an integration variable; the integration is terminated as soon as a condition on X[0], X[1],..., X[n], supplied by the user, is satisfied.
MULTISTEP	solves a system of first order differential equations (initial value problem) by means of a variable order multistep method Adams-Moulton, Adams-Bashforth, or Gear's method; the order of accuracy is automatic, up to fifth order; this method is suitable for stiff systems.

DIFFSYS	solves a system of first order differential equations (initial value problem) by extrapolation, applied to low order results, a high order of accuracy is obtained; this method is suitable for smooth problems when high accuracy is required.
ARK	solves a system of first order differential equations (initial value problem) by means of a stabilized Runge-Kutta method with limited storage requirements.
EFRK	solves a system of first order differential equations (initial value problem) by means of a first, second, or third order, exponentially fitted Runge-Kutta method; automatic step size control is not provided; this method can be used to solve stiff systems with known eigenvalue spectrum.
EFSIRK	solves an autonomous system of first order differential equations (initial value problem) by means of a third order, exponentially fitted, semi-implicit Runge-Kutta method; this method can be used to solve stiff systems.
EFERK	solves an autonomous system of first order differential equations (initial value problem) by means of an exponentially fitted, third order Runge-Kutta method; this method can be used to solve stiff systems with known eigenvalue spectrum.
LINIGER1VS	solves an autonomous system of first order differential equations (initial value problem) by means of an implicit, exponentially fitted first order one-step method; this method can be used to solve stiff systems.
LINIGER2	solves an autonomous system of first order differential equations (initial value problem) by means of an implicit, exponentially fitted first order one-step method; automatic step size control is not provided; this method can be used to solve stiff systems.
GMS	solves an autonomous system of first order differential equations (initial value problem) by means of a third order multistep method; this method can be used to solve stiff systems.
IMPEX	solves an autonomous system of first order differential equations (initial value problem) by means of the implicit midpoint rule with smoothing and extrapolation; this method is suitable for the integration of stiff differential equations.
MODIFIEDTAYLOR	solves a system of first order differential equations (initial value problem) by means of a first, second, or third order one-step Taylor method; this method can be used to solve large and sparse systems, provided higher order derivatives can easily be obtained.
EFT	solves a system of first order differential equations (initial value problem) by means of a variable order Taylor method; this method can be used to solve stiff systems, with known eigenvalue spectrum, provided higher order derivatives can easily be obtained.

RK2	solves a single second order differential equation (initial value problem) by means of a fifth order Runge-Kutta method.
RK2N	solves a system of second order differential equations (initial value problem) by means of a fifth order Runge-Kutta method.
RK3	solves a single second order differential equation (initial value problem) by means of a fifth order Runge-Kutta method; this method can only be used if the right-hand side of the differential equation does not depend on y'.
RK3N	solves a system of second order differential equations (initial value problem) by means of a fifth order Runge-Kutta method; this method can only be used if the right-hand side of the differential equation does not depend on y'.
ARKMAT	solves a system of first order differential equations (initial boundary-value problem) by means of a stabilized Runge-Kutta method, in particular suitable for systems arising from two-dimensional time-dependent partial differential equations.
FEMLAGSYM	solves a linear two-point boundary-value problem for a second order self-adjoint differential equation by a Ritz-Galerkin method.
FEMLAG	solves a linear two-point boundary-value problem for a second order self-adjoint differential equation by a Ritz-Galerkin method; the coefficient of y' is supposed to be unity.
FEMLAGSPHER	solves a linear two-point boundary-value problem for a second order self-adjoint differential equation with spherical coordinates by a Ritz-Galerkin method.
FEMLAGSKEW	solves a linear two-point boundary-value problem for a second order differential equation by a Ritz-Galerkin method.
FEMHERMSYM	solves a linear two-point boundary-value problem for a fourth order self-adjoint differential equation with Dirichlet boundary conditions by a Ritz-Galerkin method.
NONLINFEMLAGSKEW	
	solves a nonlinear two-point boundary-value problem for a second order differential equation with spherical coordinates by a Ritz-Galerkin method and Newton iteration.
RICHARDSON	solves a system of linear equations with positive real eigenvalues (elliptic boundary value problem) by means of a nonstationary second order iterative method.
ELIMINATION	solves a system of linear equations with positive real eigenvalues (elliptic boundary value problem) by means of a nonstationary second order iterative method, which is an acceleration of Richardson's method.
PEIDE	estimates unknown parameters in a system of first order differential equations; the unknown variables may appear nonlinearly both in the differential equations and its initial values; a set of observed values of some components of the solution of the differential equations must be given.

6.   Special Functions

ARCSINH	computes the inverse hyperbolic sine for a real argument.
ARCCOSH	computes the inverse hyperbolic cosine for a real argument.
ARCTANH	computes the inverse hyperbolic tangent for a real argument.
LOGONEPLUSX	evaluates the logarithmic function $ln(1+x)$.
EI	calculates the exponential integral.
EIALPHA	calculates a sequence of integrals of the form ($e^{-xt}t^n$ dt), from t=1 to t=infinity.
ENX	calculates a sequence of exponential integrals $E(n,x)$ = the integral from 1 to infinity of e^{-xt}/t^n dt.
NONEXPENX	calculates a sequence of integrals $e^x * E(n,x)$, see ENX.
SINCOSINT	calculates the sine integral $SI(x)$ and the cosine integral $CI(x)$.
SINCOSFG	is an auxiliary procedure for the sine and cosine integrals.
RECIPGAMMA	calculates the reciprocal of the gamma function for arguments in the range [0.5,1.5]; moreover odd and even parts are delivered.
GAMMA	calculates the gamma function.
LOGGAMMA	calculates the natural logarithm of the gamma function for positive arguments.
INCOMGAM	calculates the incomplete gamma functions.
INCBETA	calculates the incomplete beta function $I(x,p,q)$ for $0{\leq}x{\leq}1$, $p>0$, $q>0$.
IBPPLUSN	calculates the incomplete beta function ratios $I(x,p+n,q)$ for n=0,1,...,nmax, $0{\leq}x{\leq}1$, $p>0$, $q>0$.
IBQPLUSN	calculates the incomplete beta function ratios $I(x,p,q+n)$ for n=0,1,...,nmax, $0{\leq}x{\leq}1$, $p>0$, $q>0$.
IXQFIX	is an auxiliary procedure for the computation of incomplete bessel functions.
IXPFIX	is an auxiliary procedure for the computation of incomplete bessel functions.
FORWARD	is an auxiliary procedure for the computation of incomplete bessel functions.
BACKWARD	is an auxiliary procedure for the computation of incomplete bessel functions.
ERRORFUNNCTION	computes the error function (erf) and complementary error function (erfc) for a real argument.
NONEXPERFC	computes $erfc(x) * exp(x^2)$; see ERRORFUNCTION.
INVERSEERRORFUNCTION	
	calculates the inverse error function y = inverf(x).
FRESNEL	calculates the fresnel integrals $C(x)$ and $S(x)$.
FG	is an auxiliary procedure for the computation of fresnel integrals.
BESSJ0	calculates the ordinary bessel function of the first kind of order zero.
BESSJ1	calculates the ordinary bessel function of the first kind of order one.
BESSJ	calculates the ordinary bessel functions of the first kind of order l, $l=0,...,n$.

BESSY01	calculates the ordinary bessel functions of the second kind of orders zero and one with argument x, x>0.
BESSY	calculates the ordinary bessel functions of the second kind of order l, l=0,...,n, with argument x, x>0.
BESSPQ0	is an auxiliary procedure for the computation of the ordinary bessel functions of order zero for large values of their argument.
BESSPQ1	is an auxiliary procedure for the computation of the ordinary bessel functions of order one for large values of their argument.
BESSI0	calculates the modified bessel function of the first kind of order zero.
BESSI1	calculates the modified bessel function of the first kind of order one.
BESSI	calculates the modified bessel functions of the first kind of order l, l=0,...,n.
BESSK01	calculates the modified bessel functions of the third kind of orders zero and one with argument x, x>0.
BESSK	calculates the modified bessel functions of the third kind of order l, l=0,...,n, with argument x, x>0.
NONEXPBESSI0	calculates the modified bessel function of the first kind of order zero; the result is multiplied by $e^{-|x|}$.
NONEXPBESSI1	calculates the modified bessel function of the first kind of order one; the result is multiplied by $e^{-|x|}$.
NONEXPBESSI	calculates the modified bessel functions of the first kind of order l, l=0,...,n; the result is multiplied by $e^{-|x|}$.
NONEXPBESSK01	calculates the modified bessel functions of the third kind of orders zero and one with argument x, x>0; the result is multiplied by e^x.
NONEXPBESSK	calculates the modified bessel functions of the third kind of order l, l=0,...,n, with argument x, x>0; the result is multiplied by e^x.
BESSJAPLUSN	calculates the bessel functions of the first kind of order a+k, $0 \leq k \leq n$, $0 \leq a < 1$.
BESSYA01	calculates the bessel functions of the second kind (also called Neumann's functions) of order a and a+1, $a \geq 0$, and argument x>0.
BESSYAPLUSN	calculates the bessel functions of the second kind of order a+n, n=0,...,nmax, $a \geq 0$, and argument x>0.
BESSPQA01	is an auxiliary procedure for the computation of the bessel functions for large values of their argument.
BESSZEROS	calculates zeros of a bessel function (of first or second kind) and of its derivative.
START	is an auxiliary procedure in bessel function procedures.
BESSIAPLUSN	calculates the modified bessel functions of the first kind of order a+n, n=0,...,nmax, $a \geq 0$, and argument $x \geq 0$.
BESSKA01	calculates the modified bessel functions of the third kind of orders a and a+1, $a \geq 0$, and argument x>0.
BESSKAPLUSN	calculates the modified bessel functions of the third kind of order a+n, n=0,...,nmax, $a \geq 0$, and argument x>0.

NONEXPBESSIAPLUSN
    calculates the modified bessel functions of the first kind of order a+n, n=0,...,nmax, a≥0 and argument x≥0, multiplied by e^{-x}.

NONEXPBESSKA01 calculates the modified bessel functions of the third kind of order a and a+1, a≥0 and argument x, x>0, multiplied by the factor e^x.

NONEXPBESSKAPLUSN
    calculates the modified bessel functions of the third kind of order a+n, n=0,...,nmax, a≥0 and argument x>0, multiplied by e^x.

SPHERBESSJ   calculates the spherical bessel functions of the first kind.

SPHERBESSY   calculates the spherical bessel functions of the third kind.

SPHERBESSI   calculates the modified spherical bessel functions of the first kind.

SPHERBESSK   calculates the modified spherical bessel functions of the third kind.

NONEXPSPHERBESSI
    calculates the modified spherical bessel functions of the first kind multiplied by e^{-x}.

NONEXPSPHERBESSK
    calculates the modified spherical bessel functions of the third kind multiplied by e^x.

AIRY   evaluates the Airy functions AI(z) and BI(z) and their derivatives.

AIRYZEROS   computes the zeros and associated values of the Airy functions AI(z) and BI(z) and their derivatives.

7. Interpolation and Approximation

NEWTON   calculates the coefficients of the Newton polynomial through given interpolation points and corresponding function values.

INI   selects a subset of integers out of a given set of integers; it is an auxiliary procedure for MINMAXPOL.

SNDREMEZ   exchanges at most n+1 numbers with numbers out of a reference set; it is an auxiliary procedure for MINMAXPOL.

MINMAXPOL   calculates the coefficients of the polynomial that approximates a function, given for discrete arguments, such that the infinity norm of the error vector is minimized.

Procedures Description (alphabetical order)

ABSMAXMAT	calculates the modulus of the largest element of a matrix and delivers the indices of the maximal element.
AIRY	evaluates the Airy functions AI(z) and BI(z) and their derivatives.
AIRYZEROS	computes the zeros and associated values of the Airy functions AI(z) and BI(z) and their derivatives.
ALLCHEPOL	evaluates all Chebyshev polynomials up to a certain degree.
ALLJACZER	calculates the zeros of a Jacobian polynomial.
ALLLAGZER	calculates the zeros of a Laguerre polynomial.
ALLORTPOL	evaluates the value of all orthogonal polynomials up to a given degree, given a set of recurrence coefficients.
ALLORTPOLSYM	evaluates the values of all orthogonal polynomials up to a given degree, given a set of recurrence coefficients.
ALLZERORTPOL	calculates all zeros of an orthogonal polynomial.
ARCCOSH	computes the inverse hyperbolic cosine for a real argument.
ARCSINH	computes the inverse hyperbolic sine for a real argument.
ARCTANH	computes the inverse hyperbolic tangent for a real argument.
ARK	solves a system of first order differential equations (initial value problem) by means of a stabilized Runge-Kutta method with limited storage requirements.
ARKMAT	solves a system of first order differential equations (initial boundary-value problem) by means of a stabilized Runge-Kutta method, in particular suitable for systems arising from two-dimensional time-dependent partial differential equations.
BACKWARD	is an auxiliary procedure for the computation of incomplete bessel functions.
BAKCOMHES	performs the back transformation corresponding to HSHCOMHES.
BAKHRMTRI	performs the back transformation corresponding to HSHHRMTRI.
BAKLBR	performs the back transformation corresponding to EQILBR.
BAKLBRCOM	transforms the eigenvectors of a complex equilibrated (by EQILBRCOM) matrix into the eigenvectors of the original matrix.
BAKREAHES1	performs the back transformation (on a vector) corresponding to TFMREAHES.
BAKREAHES2	performs the back transformation (on columns) corresponding to TFMREAHES.
BAKSYMTRI1	performs the back transformation corresponding to TFMSYMTRI1.
BAKSYMTRI2	performs the back transformation corresponding to TFMSYMTRI2.
BESSI	calculates the modified bessel functions of the first kind of order l, $l=0,...,n$.
BESSI0	calculates the modified bessel function of the first kind of order zero.

BESSI1	calculates the modified bessel function of the first kind of order one.
BESSIAPLUSN	calculates the modified bessel functions of the first kind of order a+n, n=0,...,nmax, a≥0, and argument x≥0.
BESSJ	calculates the ordinary bessel functions of the first kind of order l, l=0,...,n.
BESSJ0	calculates the ordinary bessel function of the first kind of order zero.
BESSJ1	calculates the ordinary bessel function of the first kind of order one.
BESSJAPLUSN	calculates the bessel functions of the first kind of order a+k, 0≤k≤n, 0≤a<1.
BESSK	calculates the modified bessel functions of the third kind of order l, l=0,...,n, with argument x, x>0.
BESSK01	calculates the modified bessel functions of the third kind of orders zero and one with argument x, x>0.
BESSKA01	calculates the modified bessel functions of the third kind of orders a and a+1, a≥0, and argument x>0.
BESSKAPLUSN	calculates the modified bessel functions of the third kind of order a+n, n=0,...,nmax, a≥0, and argument x>0.
BESSPQ0	is an auxiliary procedure for the computation of the ordinary bessel functions of order zero for large values of their argument.
BESSPQ1	is an auxiliary procedure for the computation of the ordinary bessel functions of order one for large values of their argument.
BESSPQA01	is an auxiliary procedure for the computation of the bessel functions for large values of their argument.
BESSY	calculates the ordinary bessel functions of the second kind of order l, l=0,...,n, with argument x, x>0.
BESSY01	calculates the ordinary bessel functions of the second kind of orders zero and one with argument x, x>0.
BESSYA01	calculates the bessel functions of the second kind (also called Neumann's functions) of order a and a+1, a≥0, and argument x>0.
BESSYAPLUSN	calculates the bessel functions of the second kind of order a+n, n=0,...,nmax, a≥0, and argument x>0.
BESSZEROS	calculates zeros of a bessel function (of first or second kind) and of its derivative.
BOUNDS	calculates the error in approximated zeros of a polynomial with real coefficients.
CARPOL	transforms the Cartesian coordinates of a complex number into polar coordinates.
CHEPOL	evaluates a Chebyshev polynomial.
CHEPOLSUM	evaluates a finite sum of Chebyshev polynomials.
CHLDEC1	calculates the Cholesky decomposition of a positive definite symmetric matrix whose upper triangle is given columnwise in a one-dimensional array.

CHLDEC2	calculates the Cholesky decomposition of a positive definite symmetric matrix whose upper triangle is given in a two-dimensional array.
CHLDECBND	performs the Cholesky decomposition of a positive definite symmetric band matrix.
CHLDECINV1	calculates the inverse of a positive definite symmetric matrix by Cholesky's square root method, the coefficient matrix given columnwise in a one-dimensional array.
CHLDECINV2	calculates the inverse of a positive definite symmetric matrix by Cholesky's square root method, the coefficient matrix given in the upper triangle of a two-dimensional array.
CHLDECSOL1	solves a positive definite symmetric system of linear equations by Cholesky's square root method, the coefficient matrix should be given columnwise in a one-dimensional array.
CHLDECSOL2	solves a positive definite symmetric system of linear equations by Cholesky's square root method, the coefficient matrix should be given in the upper triangle of a two-dimensional array.
CHLDECSOLBND	solves a positive definite symmetric linear system and performs the triangular decomposition by Cholesky's method.
CHLDETERM1	calculates the determinant of a positive definite symmetric matrix, the Cholesky decomposition being given columnwise in a one-dimensional array.
CHLDETERM2	calculates the determinant of a positive definite symmetric matrix, the Cholesky decomposition being given in a two-dimensional array.
CHLDETERMBND	calculates the determinant of a positive definite symmetric band matrix.
CHLINV1	calculates the inverse of a positive definite symmetric matrix, if the matrix has been decomposed by CHLDEC1 or CHLDECSOL1.
CHLINV2	calculates the inverse of a positive definite symmetric matrix, if the matrix has been decomposed by CHLDEC2 or CHLDECSOL2.
CHLSOL1	solves a system of linear equations if the coefficient matrix has been decomposed by CHLDEC1 or CHLDECSOL1.
CHLSOL2	solves a system of linear equations if the coefficient matrix has been decomposed by CHLDEC2 or CHLDECSOL2.
CHLSOLBND	solves a positive definite symmetric linear system, the triangular decomposition being given.
CHSH2	finds a complex rotation matrix.
CHSPOL	transforms a polynomial from Chebyshev sum into power sum form.
COLCST	multiplies a column vector by a constant.
COMABS	calculates the modulus of a complex number.
COMCOLCST	multiplies a complex column vector by a complex number.
COMDIV	calculates the quotient of two complex numbers.
COMEIG1	calculates the eigenvalues and eigenvectors of a matrix.
COMEIGVAL	calculates the eigenvalues of a matrix.

COMEUCNRM	calculates the Euclidean norm of a complex matrix with some lower diagonals.
COMFOUSER	evaluates a complex fourier series with real coefficients.
COMFOUSER1	evaluates a complex fourier series.
COMFOUSER2	evaluates a complex fourier series.
COMKWD	calculates the roots of a quadratic equation with complex coefficients.
COMMATVEC	forms the scalar product of a complex row vector and a complex vector.
COMMUL	calculates the product of two complex numbers.
COMROWCST	multiplies a complex row vector by a complex number.
COMSCL	normalizes real and complex eigenvectors.
COMSQRT	calculates the square root of a complex number.
COMVALQRI	calculates the real and complex eigenvalues of a real upper Hessenberg matrix by means of double QR iteration.
COMVECHES	calculates the eigenvector corresponding to a given complex eigenvalue of a real upper Hessenberg matrix by means of inverse iteration.
CONJGRAD	solves a positive definite symmetric system of linear equations by the method of conjugate gradients.
COSSER	evaluates a cosine series.
DAVUPD	adds a rank-2 matrix to a symmetric matrix.
DEC	performs a triangular decomposition with partial pivoting.
DECBND	performs a triangular decomposition of a band matrix, using partial pivoting.
DECINV	calculates the inverse of a matrix whose order is small relative to the number of binary digits in the number representation.
DECSOL	solves a system of linear equations whose order is small relative to the number of binary digits in the number representation.
DECSOLBND	solves a system of linear equations by Gaussian elimination with partial pivoting if the coefficient matrix is in band form and is stored rowwise in a one-dimensional array.
DECSOLSYM2	solves a symmetric system of linear equations by symmetric decomposition.
DECSOLSYMTRI	solves a symmetric tridiagonal system of linear equations and performs the tridiagonal decomposition.
DECSOLTRI	solves a tridiagonal system of linear equations and performs the triangular decomposition without pivoting.
DECSOLTRIPIV	solves a tridiagonal system of linear equations and performs the triangular decomposition with partial pivoting.
DECSYM2	calculates the symmetric decomposition of a symmetric matrix.
DECSYMTRI	performs the triangular decomposition of a symmetric tridiagonal matrix.
DECTRI	performs a triangular decomposition of a tridiagonal matrix.
DECTRIPIV	performs a triangular decomposition of a tridiagonal matrix, using partial pivoting.
DERPOL	evaluates the first k derivatives of a polynomial.

DETERM	calculates the determinant of a triangularly decomposed matrix.
DETERMBND	calculates the determinant of a band matrix.
DETERMSYM2	calculates the determinant of a symmetric matrix, the symmetric decomposition being given.
DIFFSYS	solves a system of first order differential equations (initial value problem) by extrapolation, applied to low order results, a high order of accuracy is obtained; this method is suitable for smooth problems when high accuracy is required.
DUPCOLVEC	copies a vector into a column vector.
DUPMAT	copies a matrix into another matrix.
DUPROWVEC	copies a vector into a row vector.
DUPVEC	copies a vector into another vector.
DUPVECCOL	copies a column vector into a vector.
DUPVECROW	copies a row vector into a vector.
EFERK	solves an autonomous system of first order differential equations (initial value problem) by means of an exponentially fitted, third order Runge-Kutta method; this method can be used to solve stiff systems with known eigenvalue spectrum.
EFRK	solves a system of first order differential equations (initial value problem) by means of a first, second, or third order, exponentially fitted Runge-Kutta method; automatic step size control is not provided; this method can be used to solve stiff systems with known eigenvalue spectrum.
EFSIRK	solves an autonomous system of first order differential equations (initial value problem) by means of a third order, exponentially fitted, semi-implicit Runge-Kutta method; this method can be used to solve stiff systems.
EFT	solves a system of first order differential equations (initial value problem) by means of a variable order Taylor method; this method can be used to solve stiff systems, with known eigenvalue spectrum, provided higher order derivatives can easily be obtained.
EI	calculates the exponential integral.
EIALPHA	calculates a sequence of integrals of the form ($e^{-xt}t^n$ dt), from t=1 to t=infinity.
EIGCOM	calculates the eigenvalues and eigenvectors of a complex matrix.
EIGHRM	calculates the eigenvalues and eigenvectors of a complex Hermitian matrix.
EIGSYM1	calculates eigenvalues and eigenvectors by means of inverse iteration.
EIGSYM2	calculates eigenvalues and eigenvectors by means of inverse iteration.
EIGVALCOM	calculates the eigenvalues of a complex matrix.
EIGVALHRM	calculates the eigenvalues of a complex Hermitian matrix.
EIGVALSYM1	calculates all (or some) eigenvalues of a symmetric matrix using linear interpolation of a function derived from a Sturm sequence.

EIGVALSYM2	calculates all (or some) eigenvalues of a symmetric matrix using linear interpolation of a function derived from a Sturm sequence.
ELIMINATION	solves a system of linear equations with positive real eigenvalues (elliptic boundary value problem) by means of a nonstationary second order iterative method, which is an acceleration of Richardson's method.
ELMCOL	adds a constant times a column vector to a column vector.
ELMCOLROW	adds a constant times a row vector to a column vector.
ELMCOLVEC	adds a constant times a vector to a column vector.
ELMCOMCOL	adds a complex number times a complex column vector to a complex column vector.
ELMCOMROWVEC	adds a complex number times a complex vector to a complex row vector.
ELMCOMVECCOL	adds a complex number times a complex column vector to a complex vector.
ELMROW	adds a constant times a row vector to a row vector.
ELMROWCOL	adds a constant times a column vector to a row vector.
ELMROWVEC	adds a constant times a vector to a row vector.
ELMVEC	adds a constant times a vector to a vector.
ELMVECCOL	adds a constant times a column vector to a vector.
ELMVECROW	adds a constant times a row vector to a vector.
ENX	calculates a sequence of exponential integrals $E(n,x)$ = the integral from 1 to infinity of e^{-xt}/t^n dt.
EQILBR	equilibrates a matrix by means of a diagonal similarity transformation.
EQILBRCOM	equilibrates a complex matrix.
ERBELM	calculates a rough upper bound for the error in the solution of a system of linear equations whose matrix is triangularly decomposed by GSSELM.
ERRORFUNNCTION	computes the error function (erf) and complementary error function (erfc) for a real argument.
EULER	performs the summation of an alternating infinite series.
FEMHERMSYM	solves a linear two-point boundary-value problem for a fourth order self-adjoint differential equation with Dirichlet boundary conditions by a Ritz-Galerkin method.
FEMLAG	solves a linear two-point boundary-value problem for a second order self-adjoint differential equation by a Ritz-Galerkin method; the coefficient of y' is supposed to be unity.
FEMLAGSKEW	solves a linear two-point boundary-value problem for a second order differential equation by a Ritz-Galerkin method.
FEMLAGSPHER	solves a linear two-point boundary-value problem for a second order self-adjoint differential equation with spherical coordinates by a Ritz-Galerkin method.
FEMLAGSYM	solves a linear two-point boundary-value problem for a second order self-adjoint differential equation by a Ritz-Galerkin method.
FG	is an auxiliary procedure for the computation of fresnel integrals.

FLEMIN   minimizes a function of several variables.
FLEUPD   adds a rank-2 matrix to a symmetric matrix.
FORWARD   is an auxiliary procedure for the computation of incomplete
                        bessel functions.
FOUSER   evaluates a fourier series with equal sine and cosine
                        coefficients.
FOUSER1   evaluates a fourier series.
FOUSER2   evaluates a fourier series.
FRESNEL   calculates the fresnel integrals C(x) and S(x).
FULMATVEC   forms the product A*B of a given matrix A and a vector B.
FULSYMMATVEC   forms the product A*B, where A is a symmetric matrix whose
                        upper triangle is stored columnwise in a one-dimensional
                        array and B is a vector.
FULTAMVEC   forms the product A^T*B, where A^T is the transpose of a given
                        matrix A and B is a vector.
GAMMA   calculates the gamma function.
GMS   solves an autonomous system of first order differential
                        equations (initial value problem) by means of a third order
                        multistep method; this method can be used to solve stiff
                        systems.
GRNNEW   transforms a polynomial from power sum into Newton sum
                        form.
GSSELM   performs a triangular decomposition with a combination of
                        partial and complete pivoting.
GSSERB   performs a triangular decomposition of the matrix of a system
                        of linear equations, and calculates an upper bound for the
                        relative error in the solution of that system.
GSSINV   calculates the inverse of a matrix.
GSSINVERB   calculates the inverse of a matrix and 1-norm, an upper
                        bound for the error in the inverse matrix is also given.
GSSITISOL   solves a system of linear equations, and the solution is
                        improved iteratively.
GSSITISOLERB   solves a system of linear equations, this solution is improved
                        iteratively and an upper bound for the error in the solution is
                        calculated.
GSSJACWGHTS   computes the abscissae and weights for Gauss-Jacobi
                        quadrature.
GSSLAGWGHTS   computes the abscissae and weights for Gauss-Lagrange
                        quadrature.
GSSNEWTON   calculates the least squares solution of an overdetermined
                        system of nonlinear equations with the Gauss-Newton
                        method.
GSSNRI   performs a triangular decomposition and calculates the
                        1-norm of the inverse matrix.
GSSSOL   solves a system of linear equations.
GSSSOLERB   calculates the inverse of a matrix and 1-norm, an upper
                        bound for the error in the inverse matrix is also given.
GSSWTS   calculates the Gaussian weights of a weight function, the
                        recurrence coefficients being given.

GSSWTSSYM	calculates the Gaussian weights of a weight function, the recurrence coefficients being given.
HESTGL2	is an auxiliary procedure for the computation of generalized eigenvalues.
HESTGL3	is an auxiliary procedure for the computation of generalized eigenvalues.
HOMSOL	solves the homogeneous system of linear equations AX=0 and $X^TA=0$, where A denotes a matrix and X a vector.
HOMSOLSVD	solves the homogeneous system of linear equations AX=0 and $X^TA=0$, where A denotes a matrix and X a vector (the singular value decomposition being given).
HSH2COL	is an auxiliary procedure for the computation of generalized eigenvalues.
HSH2ROW2	is an auxiliary procedure for the computation of generalized eigenvalues.
HSH2ROW3	is an auxiliary procedure for the computation of generalized eigenvalues.
HSH3COL	is an auxiliary procedure for the computation of generalized eigenvalues.
HSH3ROW2	is an auxiliary procedure for the computation of generalized eigenvalues.
HSH3ROW3	is an auxiliary procedure for the computation of generalized eigenvalues.
HSHCOLMAT	premultiplies a matrix by a Householder matrix, the vector defining this Householder matrix being given as a column in a two-dimensional array.
HSHCOLTAM	postmultiplies a matrix by a Householder matrix, the vector defining this Householder matrix being given as a column in a two-dimensional array.
HSHCOMCOL	transforms a complex vector into a vector proportional to a unit vector.
HSHCOMHES	transforms a complex matrix by means of Householder's transformation followed by a complex diagonal transformation into a similar unitary upper Hessenberg matrix with a real nonnegative subdiagonal.
HSHCOMPRD	premultiplies a complex matrix with a complex Householder matrix.
HSHDECMUL	is an auxiliary procedure for the computation of generalized eigenvalues.
HSHHRMTRI	transforms a Hermitian matrix into a similar real symmetric tridiagonal matrix.
HSHHRMTRIVAL	delivers the main diagonal elements and the squares of the codiagonal elements of a Hermitian tridiagonal matrix which is unitary similar with a given Hermitian matrix.
HSHREABID	transforms a matrix to bidiagonal form, by premultiplying and postmultiplying with orthogonal matrices.
HSHROWMAT	premultiplies a matrix by a Householder matrix, the vector defining this Householder matrix being given as a row in a two-dimensional array.

HSHROWTAM	postmultiplies a matrix by a Householder matrix, the vector defining this Householder matrix being given as a row in a two-dimensional array.
HSHVECMAT	premultiplies a matrix by a Householder matrix, the vector defining this Householder matrix being given in a one-dimensional array.
HSHVECTAM	postmultiplies a matrix by a Householder matrix, the vector defining this Householder matrix being given in a one-dimensional array.
IBPPLUSN	calculates the incomplete beta function ratios $I(x,p+n,q)$ for $n=0,1,...,nmax$, $0 \leq x \leq 1$, $p>0$, $q>0$.
IBQPLUSN	calculates the incomplete beta function ratios $I(x,p,q+n)$ for $n=0,1,...,nmax$, $0 \leq x \leq 1$, $p>0$, $q>0$.
ICHCOL	interchanges two columns of a matrix.
ICHROW	interchanges two rows of a matrix.
ICHROWCOL	interchanges a row and a column of a matrix.
ICHSEQ	interchanges two columns of an upper triangular matrix, which is stored columnwise in a one-dimensional array.
ICHSEQVEC	interchanges a row and a column of an upper triangular matrix, which is stored columnwise in a one-dimensional array.
ICHVEC	interchanges two vectors given in a one-dimensional array.
IMPEX	solves an autonomous system of first order differential equations (initial value problem) by means of the implicit midpoint rule with smoothing and extrapolation; this method is suitable for the integration of stiff differential equations.
INCBETA	calculates the incomplete beta function $I(x,p,q)$ for $0 \leq x \leq 1$, $p>0$, $q>0$.
INCOMGAM	calculates the incomplete gamma functions.
INFNRMCOL	calculates the infinity norm of a column vector.
INFNRMMAT	calculates the infinity norm of a matrix.
INFNRMROW	calculates the infinity norm of a row vector.
INFNRMVEC	calculates the infinity norm of a vector.
INI	selects a subset of integers out of a given set of integers; it is an auxiliary procedure for MINMAXPOL.
INIMAT	initializes a matrix with a constant.
INIMATD	initializes a (co)diagonal of a matrix.
INISYMD	initializes a (co)diagonal of a symmetric matrix, whose upper triangle is stored columnwise in a one-dimensional array.
INISYMROW	initializes a row of a symmetric matrix, whose upper triangle is stored columnwise in a one-dimensional array.
INIVEC	initializes a vector with a constant.
INTCHS	computes the indefinite integral of a given Chebyshev series.
INTEGRAL	calculates the definite integral of a function of one variable over a finite or infinite interval or over a number of consecutive intervals.
INV	calculates the inverse of a matrix that has been triangularly decomposed by DEC.

INV1   calculates the inverse of a matrix that has been triangularly decomposed by GSSELM or GSSERB, the 1-norm of the inverse matrix might also be calculated.

INVERSEERRORFUNCTION
                      calculates the inverse error function y = inverf(x).

ITISOL   solves a system of linear equations whose matrix has been triangularly decomposed by GSSELM or GSSERB, this solution is improved iteratively.

ITISOLERB   solves a system of linear equations whose matrix has been triangularly decomposed by GSSNRI, this solution is improved iteratively and an upper bound for the error in the solution is calculated.

IXPFIX   is an auxiliary procedure for the computation of incomplete bessel functions.

IXQFIX   is an auxiliary procedure for the computation of incomplete bessel functions.

JACOBNBNDF   calculates the Jacobian matrix of an n-dimensional function of n variables, if the Jacobian is known to be a band matrix.

JACOBNMF   calculates the Jacobian matrix of an n-dimensional function of m variables using forward differences.

JACOBNNF   calculates the Jacobian matrix of an n-dimensional function of n variables using forward differences.

JFRAC   calculates a terminating continued fraction.

LINEMIN   minimizes a function of several variables in a given direction.

LINIGER1VS   solves an autonomous system of first order differential equations (initial value problem) by means of an implicit, exponentially fitted first order one-step method; this method can be used to solve stiff systems.

LINIGER2   solves an autonomous system of first order differential equations (initial value problem) by means of an implicit, exponentially fitted first order one-step method; automatic step size control is not provided; this method can be used to solve stiff systems.

LINTFMPOL   transforms a polynomial in X into a polynomial in Y (Y=A*X+B).

LNGINTADD   computes the sum of long nonnegative integers.

LNGINTDIVIDE   computes the quotient with remainder of long nonnegative integers.

LNGINTMULT   computes the product of long nonnegative integers.

LNGINTPOWER   computes U^P, where U is a long nonnegative integer, and P is the positive (single-length) exponent.

LNGINTSUBTRACT   computes the difference of long nonnegative integers.

LOGGAMMA   calculates the natural logarithm of the gamma function for positive arguments.

LOGONEPLUSX   evaluates the logarithmic function $ln(1+x)$.

LSQDECOMP   computes the QR decomposition of a linear least squares problem with linear constraints.

LSQDGLINV   calculates the diagonal elements of the inverse of M^TM, where M is the coefficient matrix of a linear least squares problem.

LSQINV	calculates the inverse of the matrix S^TS, where S is the coefficient matrix of a linear least squares problem.
LSQORTDEC	delivers the Householder triangularization with column interchanges of a matrix of a linear least squares problem.
LSQORTDECSOL	solves a linear least squares problem by Householder triangularization with column interchanges and calculates the diagonal of the inverse of M^TM, where M is the coefficient matrix.
LSQREFSOL	solves a linear least squares problem with linear constraints if the coefficient matrix has been decomposed by LSQDECOMP.
LSQSOL	solves a linear least squares problem if the coefficient matrix has been decomposed by LSQORTDEC.
LUPZERORTPOL	calculates a number of adjacent upper or lower zeros of an orthogonal polynomial.
MARQUARDT	calculates the least squares solution of an overdetermined system of nonlinear equations with Marquardt's method.
MATMAT	forms the scalar product of a row vector and a column vector.
MATTAM	forms the scalar product of a row vector and a row vector.
MATVEC	forms the scalar product of a row vector and a vector.
MAXELMROW	adds a constant times a row vector to a row vector, it also delivers the subscript of an element of the new row vector which is of maximum absolute value.
MERGESORT	delivers a permutation of indices corresponding to sorting the elements of a given vector into nondecreasing order.
MININ	minimizes a function of one variable in a given interval.
MININDER	minimizes a function of one variable in a given interval, using values of the function and of its derivative.
MINMAXPOL	calculates the coefficients of the polynomial that approximates a function, given for discrete arguments, such that the infinity norm of the error vector is minimized.
MODIFIEDTAYLOR	solves a system of first order differential equations (initial value problem) by means of a first, second, or third order one-step Taylor method; this method can be used to solve large and sparse systems, provided higher order derivatives can easily be obtained.
MULCOL	stores a constant multiplied by a solumn vector into a column vector.
MULROW	stores a constant multiplied by a row vector into a row vector.
MULTISTEP	solves a system of first order differential equations (initial value problem) by means of a variable order multistep method Adams-Moulton, Adams-Bashforth, or Gear's method; the order of accuracy is automatic, up to fifth order; this method is suitable for stiff systems.
MULVEC	stores a constant multiplied by a vector into a vector.
NEWGRN	transforms a polynomial from Newton sum into power sum form.
NEWTON	calculates the coefficients of the Newton polynomial through given interpolation points and corresponding function values.
NONEXPBESSI	calculates the modified bessel functions of the first kind of order l, $l=0,...,n$; the result is multiplied by $e^{-|x|}$.

NONEXPBESSI0   calculates the modified bessel function of the first kind of order zero; the result is multiplied by $e^{-|x|}$.

NONEXPBESSI1   calculates the modified bessel function of the first kind of order one; the result is multiplied by $e\ e^{-|x|}$.

NONEXPBESSIAPLUSN

calculates the modified bessel functions of the first kind of order a+n, n=0,...,nmax, a≥0 and argument x≥0, multiplied by e^{-x}.

NONEXPBESSK   calculates the modified bessel functions of the third kind of order l, l=0,...,n, with argument x, x>0; the result is multiplied by e^x.

NONEXPBESSK01   calculates the modified bessel functions of the third kind of orders zero and one with argument x, x>0; the result is multiplied by e^x.

NONEXPBESSKA01   calculates the modified bessel functions of the third kind of order a and a+1, a≥0 and argument x, x>0, multiplied by the factor e^x.

NONEXPBESSKAPLUSN

calculates the modified bessel functions of the third kind of order a+n, n=0,...,nmax, a≥0 and argument x>0, multiplied by e^x.

NONEXPENX   calculates a sequence of integrals $e^x * E(n,x)$, see ENX.

NONEXPERFC   computes $erfc(x) * exp(x^2)$; see ERRORFUNCTION.

NONEXPSPHERBESSI

calculates the modified spherical bessel functions of the first kind multiplied by e^{-x}.

NONEXPSPHERBESSK

calculates the modified spherical bessel functions of the third kind multiplied by e^x.

NONLINFEMLAGSKEW

solves a nonlinear two-point boundary-value problem for a second order differential equation with spherical coordinates by a Ritz-Galerkin method and Newton iteration.

NORDERPOL   evaluates the first k normalized derivatives of a polynomial.

ODDCHEPOLSUM   evaluates a finite sum of Chebyshev polynomials of odd degree.

ONENRMCOL   calculates the 1-norm of a column vector.

ONENRMINV   calculates the 1-norm of the inverse of a matrix whose triangularly decomposed form is delivered by GSSELM.

ONENRMMAT   calculates the 1-norm of a matrix.

ONENRMROW   calculates the 1-norm of a row vector.

ONENRMVEC   calculates the 1-norm of a vector.

ORTHOG   orthogonalizes some adjacent matrix columns according to the modified Gram-Schmidt method.

ORTPOL   evaluates the value of an n-degree orthogonal polynomial, given by a set of recurrence coefficients.

ORTPOLSYM   evaluates the values of an n-degree orthogonal polynomial, given by a set of recurrence coefficients.

PEIDE
estimates unknown parameters in a system of first order differential equations; the unknown variables may appear nonlinearly both in the differential equations and its initial values; a set of observed values of some components of the solution of the differential equations must be given.

POL
evaluates a polynomial.

POLCHS
transforms a polynomial from power sum into Chebyshev sum form.

POLSHTCHS
transforms a polynomial from power sum into shifted Chebyshev sum form.

PRAXIS
minimizes a function of several variables.

PRETFMMAT
calculates the premultiplying matrix from the data generated by HSHREABID.

PSDINV
calculates the pseudo-inverse of a matrix.

PSDINVSVD
calculates the pseudo-inverse of a matrix (the singular value decomposition being given).

PSTTFMMAT
calculates the postmultiplying matrix from the data generated by HSHREABID.

QADRAT
computes the definite integral of a function of one variable over a finite interval.

QRICOM
calculates the eigenvalues and eigenvectors of a complex upper Hessenberg matrix.

QRIHRM
calculates the eigenvalues and eigenvectors of a complex Hermitian matrix.

QRISNGVAL
calculates the singular values of a given matrix.

QRISNGVALBID
calculates the singular values of a bidiagonal matrix.

QRISNGVALDEC
calculates the singular values decomposition UDV^T, with U and V orthogonal and D positive diagonal.

QRISNGVALDECBID
calculates the singular values decomposition of a matrix of which the bidiagonal and the pre- and postmultiplying matrices are given.

QRISYM
calculates all eigenvalues and eigenvectors of a symmetric matrix by means of QR iteration.

QRISYMTRI
calculates the eigenvalues and eigenvectors of a symmetric tridiagonal matrix by means of QR iteration.

QRIVALHRM
calculates the eigenvalues of a complex Hermitian matrix.

QRIVALSYM1
calculates the eigenvalues of a symmetric matrix by means of QR iteration.

QRIVALSYM2
calculates the eigenvalues of a symmetric matrix by means of QR iteration.

QRIVALSYMTRI
calculates the eigenvalues of a symmetric tridiagonal matrix by means of QR iteration.

QUANEWBND
solves a system of nonlinear equations of which the Jacobian (being a band matrix) is given.

QUANEWBND1
solves a system of nonlinear equations of which the Jacobian is a band matrix.

QZI
computes generalized eigenvalues and eigenvectors by means of QZ iteration.

QZIVAL
computes generalized eigenvalues by means of QZ iteration.

REAEIG1	calculates the eigenvalues and eigenvectors of a matrix, provided that they are all real.
REAEIG3	calculates the eigenvalues and eigenvectors of a matrix, provided that they are all real.
REAEIGVAL	calculates the eigenvalues of a matrix, provided that all eigenvalues are real.
REAQRI	calculates all eigenvalues and eigenvectors of a real upper Hessenberg matrix, provided that all eigenvalues are real, by means of single QR iteration.
REASCL	normalizes the columns of a two-dimensional array.
REAVALQRI	calculates the eigenvalues of a real upper Hessenberg matrix, provided that all eigenvalues are real, by means of single QR iteration.
REAVECHES	calculates an eigenvector corresponding to a given real eigenvalue of a real upper Hessenberg matrix by means of inverse iteration.
RECCOF	calculates recurrence coefficients of an orthogonal polynomial, a weight function being given.
RECIPGAMMA	calculates the reciprocal of the gamma function for arguments in the range [0.5,1.5]; moreover odd and even parts are delivered.
RESVEC	calculates the residual vector A∗B+X∗C, where A is a given matrix, B and C are vectors, and X is a scalar.
RICHARDSON	solves a system of linear equations with positive real eigenvalues (elliptic boundary value problem) by means of a nonstationary second order iterative method.
RK1	solves a single first order differential equation by means of a fifth order Runge-Kutta method.
RK2	solves a single second order differential equation (initial value problem) by means of a fifth order Runge-Kutta method.
RK2N	solves a system of second order differential equations (initial value problem) by means of a fifth order Runge-Kutta method.
RK3	solves a single second order differential equation (initial value problem) by means of a fifth order Runge-Kutta method; this method can only be used if the right-hand side of the differential equation does not depend on y'.
RK3N	solves a system of second order differential equations (initial value problem) by means of a fifth order Runge-Kutta method; this method can only be used if the right-hand side of the differential equation does not depend on y'.
RK4A	solves a single first order differential equation by means of a fifth order Runge-Kutta method; the integration is terminated as soon as a condition on X and Y, which is supplied by the user, is satisfied.
RK4NA	solves a system of first order differential equations (initial value problem) by means of a fifth order Runge-Kutta method; the integration is terminated as soon as a condition on X[0], X[1],..., X[n], supplied by the user, is satisfied.

RK5NA   solves a system of first order differential equations
                        (initial value problem) by means of a fifth order Runge-Kutta
                        method; the arc length is introduced as an integration
                        variable; the integration is terminated as soon as a condition
                        on X[0], X[1],..., X[n], supplied by the user, is satisfied.

RKE   solves a system of first order differential equations
                        (initial value problem) by means of a fifth order Runge-Kutta
                        method.

RNK1MIN   minimizes a function of several variables.

RNK1UPD   adds a rank-1 matrix to a symmetric matrix.

ROTCOL   replaces two column vectors X and Y by two vectors CX+SY
                        and CY-SX.

ROTCOMCOL   replaces two complex column vectors X and Y by two complex
                        vectors CX+SY and CY-SX.

ROTCOMROW   replaces two complex row vectors X and Y by two complex
                        vectors CX+SY and CY-SX.

ROTROW   replaces two row vectors X and Y by two vectors CX+SY and
                        CY-SX.

ROWCST   multiplies a row vector by a constant.

ROWPERM   permutes the elements of a given row according to a given
                        permutation of indices.

SCAPRD1   forms the scalar product of two vectors given in
                        one-dimensional arrays, where the spacings of both vectors
                        are constant.

SCLCOM   normalizes the columns of a complex matrix.

SELZERORTPOL   calculates a number of adjacent zeros of an orthogonal
                        polynomial.

SEQVEC   forms the scalar product of two vectors given in
                        one-dimensional arrays, where the mutual spacings between
                        the indices of the first vector change linearly.

SHTCHSPOL   transforms a polynomial from shifted Chebyshev sum form
                        into power sum form.

SINCOSFG   is an auxiliary procedure for the sine and cosine integrals.

SINCOSINT   calculates the sine integral SI(x) and the cosine integral CI(x).

SINSER   evaluates a sine series.

SNDREMEZ   exchanges at most n+1 numbers with numbers out of a
                        reference set; it is an auxiliary procedure for MINMAXPOL.

SOL   solves a system of linear equations whose matrix has been
                        triangularly decomposed by DEC.

SOLBND   solves a system of linear equations, the matrix being
                        decomposed by DECBND.

SOLELM   solves a system of linear equations whose matrix has been
                        triangularly decomposed by GSSELM or GSSERB.

SOLOVR   calculates the singular values decomposition, and solves an
                        overdetermined system of linear equations.

SOLSVDOVR   solves an overdetermined system of linear equations,
                        multiplying the right-hand side by the pseudo-inverse of the
                        given matrix.

SOLSVDUND   solves an underdetermined system of linear equations, multiplying the right-hand side by the pseudo-inverse of the given matrix.

SOLSYM2   solves a symmetric system of linear equations if the coefficient matrix has been decomposed by DECSYM2 or DECSOLSYM2.

SOLSYMTRI   solves a symmetric tridiagonal system of linear equations, the triangular decomposition being given.

SOLTRI   solves a tridiagonal system of linear equations, the triangular decomposition being given.

SOLTRIPIV   solves a tridiagonal system of linear equations, the triangular decomposition being given.

SOLUND   calculates the singular values decomposition, and solves an underdetermined system of linear equations.

SPHERBESSI   calculates the modified spherical bessel functions of the first kind.

SPHERBESSJ   calculates the spherical bessel functions of the first kind.

SPHERBESSK   calculates the modified spherical bessel functions of the third kind.

SPHERBESSY   calculates the spherical bessel functions of the third kind.

START   is an auxiliary procedure in bessel function procedures.

SUMORTPOL   evaluates a finite series expressed in orthogonal polynomials, given by a set of recurrence coefficients.

SUMORTPOLSYM   evaluates a finite series expressed in orthogonal polynomials, given by a set of recurrence coefficients.

SUMPOSSERIES   performs the summation of an infinite series with positive monotonically decreasing terms using the van Wijngaarden transformation.

SYMEIGIMP   improves an approximation of a real symmetric eigensystem and calculates error bounds for the eigenvalues.

SYMMATVEC   forms the scalar product of a vector and a row of a symmetric matrix, whose upper triangle is given columnwise in a one-dimensional array.

SYMRESVEC   calculates the residual vector $A*B+X*C$, where A is a symmetric matrix whose upper triangle is stored columnwise in a one-dimensional array, B and C are vectors, and X is a scalar.

TAMMAT   forms the scalar product of a column vector and a column vector.

TAMVEC   forms the scalar product of a column vector and a vector.

TAYPOL   evaluates the first k terms of a Taylor series.

TFMPREVEC   in combination with TFMSYMTRI2 calculates the transforming matrix.

TFMREAHES   transforms a matrix into a similar upper Hessenberg matrix by means of Wilkinson's transformation.

TFMSYMTRI1   transforms a real symmetric matrix into a similar tridiagonal one by means of Householder's transformation.

TFMSYMTRI2   transforms a real symmetric matrix into a similar tridiagonal one by means of Householder's transformation.

TRICUB   computes the definite integral of a function of two variables over a triangular domain.

VALQRICOM   calculates the eigenvalues of a complex upper Hessenberg matrix with a real subdiagonal.

VALSYMTRI   calculates all, or some consecutive, eigenvalues of a symmetric tridiagonal matrix by means of linear interpolation using a Sturm sequence.

VECPERM   permutes the elements of a given vector according to a given permutation of indices.

VECSYMTRI   calculates eigenvectors of a symmetric tridiagonal matrix by means of inverse iteration.

VECVEC   forms the scalar product of a vector and a vector.

ZEROIN   finds (in a given interval) a zero of a function of one variable.

ZEROINDER   finds (in a given interval) a zero of a function of one variable using values of the function and of its derivative.

ZEROINRAT   finds (in a given interval) a zero of a function of one variable.

ZERPOL   calculates all roots of a polynomial with real coefficients by Laguerre's method.

Index of Procedures

A

B

C

D

I

J

L

M

N

O

P

Q

R

S

T

V

Z

CD-ROM ACCOMPANIES BOOK
SHELVED AT CIRCULATION DESK

DATE DUE

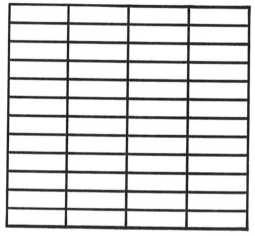

DEMCO